BIG IDEAS MATH®

Advanced 2

A Bridge to Success

TEACHING EDITION

Ron Larson
Laurie Boswell

BIG IDEAS LEARNING®
Erie, Pennsylvania
BigIdeasLearning.com

Big Ideas Learning, LLC
1762 Norcross Road
Erie, PA 16510-3838
USA

For product information and customer support, contact Big Ideas Learning at **1-877-552-7766** or visit us at ***BigIdeasLearning.com***.

Cover Image
Steve Horsley/Shutterstock.com

Printed in the U.S.A.

ISBN 13: 978-1-68033-~~131-8~~
ISBN 10: 1-68033-131-0

2 3 4 5 6 7 8 9 10 WEB 17 16 15

AUTHORS

Ron Larson is a professor of mathematics at Penn State Erie, The Behrend College, where he has taught since receiving his Ph.D. in mathematics from the University of Colorado. Dr. Larson is well known as the lead author of a comprehensive program for mathematics that spans middle school, high school, and college courses. His high school and Advanced Placement books are published by Holt McDougal. Ron's numerous professional activities keep him in constant touch with the needs of students, teachers, and supervisors. Ron and Laurie Boswell began writing together in 1992. Since that time, they have authored over two dozen textbooks. In their collaboration, Ron is primarily responsible for the pupil edition and Laurie is primarily responsible for the teaching edition of the text.

Laurie Boswell is the Head of School and a mathematics teacher at the Riverside School in Lyndonville, Vermont. Dr. Boswell received her Ed.D. from the University of Vermont in 2010. She is a recipient of the Presidential Award for Excellence in Mathematics Teaching. Laurie has taught math to students at all levels, elementary through college. In addition, Laurie was a Tandy Technology Scholar, and served on the NCTM Board of Directors from 2002 to 2005. She currently serves on the board of NCSM, and is a popular national speaker. Along with Ron, Laurie has co-authored numerous math programs.

ABOUT THE BOOK

Big Ideas Math: A Bridge to Success was developed using the same research-based strategy of a balanced approach that has become synonymous with the *Big Ideas Math* series. This approach opens doors to abstract thought, reasoning, and inquiry as students persevere to answer the Essential Questions that introduce each section.

From start to finish, this program was designed with the student in mind. Students are subtly introduced to "Habits of Mind" that help them internalize concepts for a greater depth of understanding. These habits serve students well not only in mathematics, but across all curricula throughout their academic careers. *Big Ideas Math: A Bridge to Success* exposes students to highly motivating and relevant problems. Woven throughout the program are the depth and rigor students need to prepare for career-readiness and other college-level courses.

We consider the Big Ideas Math series to be the crowning jewel of 30 years of achievement in writing educational materials.

Ron Larson   Laurie Boswell

TEACHER REVIEWERS

- Lisa Amspacher
  Milton Hershey School
  Hershey, PA
- Mary Ballerina
  Orange County Public Schools
  Orlando, FL
- Lisa Bubello
  School District of Palm Beach County
  Lake Worth, FL
- Sam Coffman
  North East School District
  North East, PA
- Kristen Karbon
  Troy School District
  Rochester Hills, MI
- Laurie Mallis
  Westglades Middle School
  Coral Springs, FL
- Dave Morris
  Union City Area School District
  Union City, PA
- Bonnie Pendergast
  Tolleson Union High School District
  Tolleson, AZ
- Valerie Sullivan
  Lamoille South Supervisory Union
  Morrisville, VT
- Becky Walker
  Appleton Area School District
  Appleton, WI
- Zena Wiltshire
  Dade County Public Schools
  Miami, FL

STUDENT REVIEWERS

- Mike Carter
- Matthew Cauley
- Amelia Davis
- Wisdom Dowds
- John Flatley
- Nick Ganger
- Hannah Iadeluca
- Paige Lavine
- Emma Louie
- David Nichols
- Mikala Parnell
- Jordan Pashupathi
- Stephen Piglowski
- Robby Quinn
- Michael Rawlings
- Garrett Sample
- Andrew Samuels
- Addie Sedelmyer
- Tyler Steffy
- Erin Taylor
- Reid Wilson

CONSULTANTS

- Patsy Davis
  Educational Consultant
  Knoxville, Tennessee
- Bob Fulenwider
  Mathematics Consultant
  Bakersfield, California
- Linda Hall
  Mathematics Assessment Consultant
  Norman, Oklahoma
- Ryan Keating
  Special Education Advisor
  Gilbert, Arizona
- Michael McDowell
  Project-Based Instruction Specialist
  Fairfax, California
- Sean McKeighan
  Interdisciplinary Advisor
  Norman, Oklahoma
- Bonnie Spence
  Differentiated Instruction Consultant
  Missoula, Montana

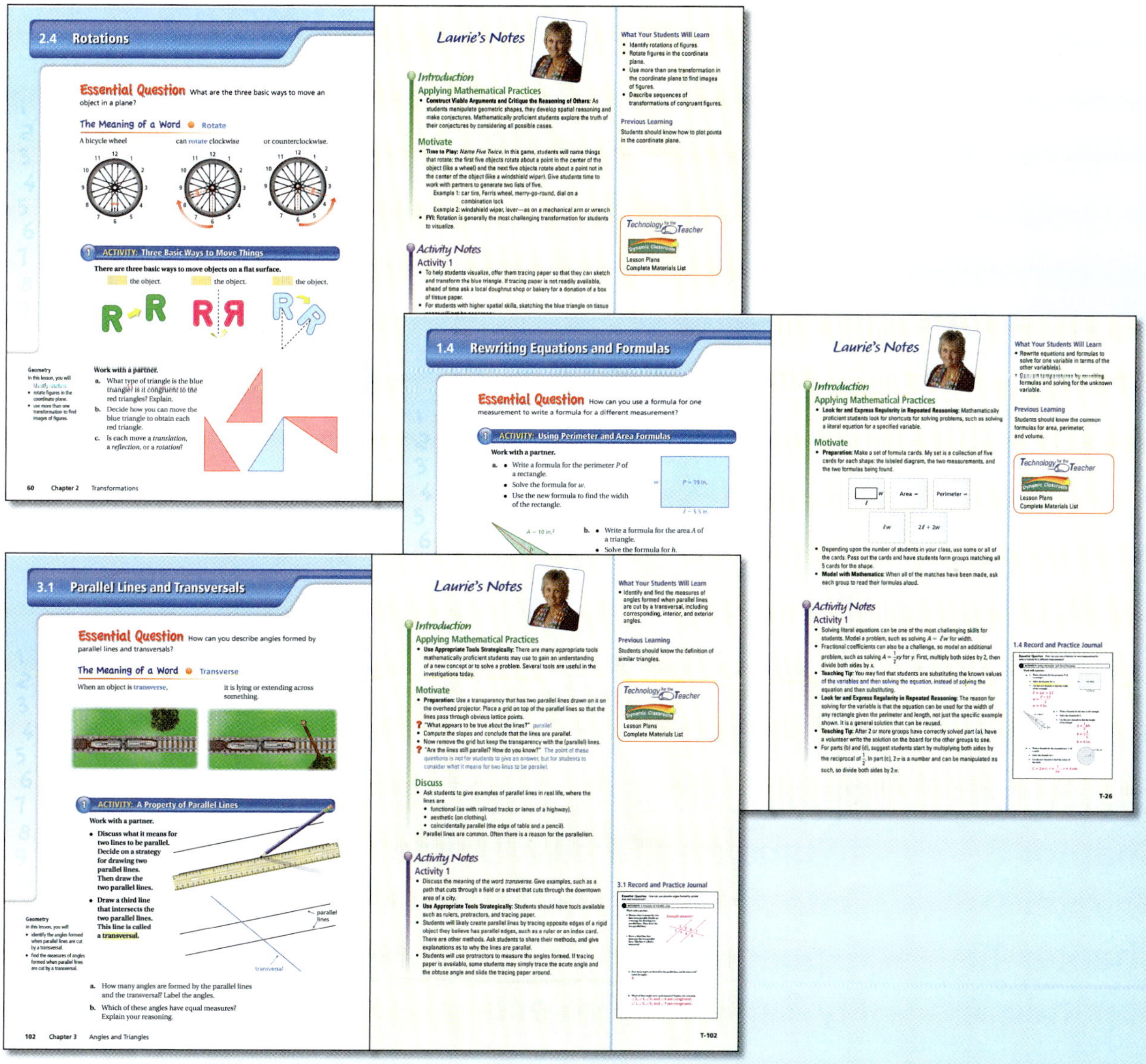

BIG IDEAS MATH

The *Big Ideas Math Advanced* series allows students to complete the material for grades 6, 7, and 8 in two years without skipping any learning objectives.

Chapter 1	Equations
Chapter 2	Transformations
Chapter 3	Angles and Triangles
Chapter 4	Graphing and Writing Linear Equations
Chapter 5	Systems of Linear Equations
Chapter 6	Functions
Chapter 7	Real Numbers and the Pythagorean Theorem
Chapter 8	Volume and Similar Solids
Chapter 9	Data Analysis and Displays
Chapter 10	Exponents and Scientific Notation
Appendix A	My Big Ideas Projects

ADVANCED 2

Using the *Big Ideas Math Advanced* series, students can complete the Advanced Pathway and have the opportunity for conceptual understanding, procedural fluency, and application through the use of focus, coherence, and rigor.

Chapter 11	Inequalities
Chapter 12	Constructions and Scale Drawings
Chapter 13	Circles and Area
Chapter 14	Surface Area and Volume
Chapter 15	Probability and Statistics

11 Inequalities

"Before my school had Big Ideas Math I would always lose test points because I left units off my answers. Now I see why they are so important."

	What You Learned Before	463
Section 11.1	**Writing and Graphing Inequalities**	
	Activity	464
	Lesson	466
Section 11.2	**Solving Inequalities Using Addition or Subtraction**	
	Activity	470
	Lesson	472
	Study Help/Graphic Organizer	476
	11.1–11.2 Quiz	477
Section 11.3	**Solving Inequalities Using Multiplication or Division**	
	Activity	478
	Lesson	480
Section 11.4	**Solving Two-Step Inequalities**	
	Activity	486
	Lesson	488
	11.3–11.4 Quiz	492
	Chapter Review	493
	Chapter Test	496
	Cumulative Assessment	497

Constructions and Scale Drawings

What You Learned Before 501

Section 12.1 Adjacent and Vertical Angles

Activity 502

Lesson 504

Section 12.2 Complementary and Supplementary Angles

Activity 508

Lesson 510

Section 12.3 Triangles

Activity 514

Lesson 516

Extension: Angle Measures of Triangles 520

Study Help/Graphic Organizer 522

12.1–12.3 Quiz 523

Section 12.4 Quadrilaterals

Activity 524

Lesson 526

Section 12.5 Scale Drawings

Activity 530

Lesson 532

12.4–12.5 Quiz 538

Chapter Review 539

Chapter Test 542

Cumulative Assessment 543

"Word problems used to confuse me. Now I understand how to look for patterns and what the question is asking!"

13 Circles and Area

"I like the Big Ideas Math Tutorials because they help explain the math when I am at home."

What You Learned Before 547

Section 13.1 Circles and Circumference
- Activity 548
- Lesson 550

Section 13.2 Perimeters of Composite Figures
- Activity 556
- Lesson 558
- Study Help/Graphic Organizer 562
- 13.1–13.2 Quiz 563

Section 13.3 Areas of Circles
- Activity 564
- Lesson 566

Section 13.4 Areas of Composite Figures
- Activity 570
- Lesson 572
- 13.3–13.4 Quiz 576
- Chapter Review 577
- Chapter Test 580
- Cumulative Assessment 581

14 Surface Area and Volume

	What You Learned Before	585
Section 14.1	**Surface Areas of Prisms**	
	Activity	586
	Lesson	588
Section 14.2	**Surface Areas of Pyramids**	
	Activity	594
	Lesson	596
Section 14.3	**Surface Areas of Cylinders**	
	Activity	600
	Lesson	602
	Study Help/Graphic Organizer	606
	14.1–14.3 Quiz	607
Section 14.4	**Volumes of Prisms**	
	Activity	608
	Lesson	610
Section 14.5	**Volumes of Pyramids**	
	Activity	614
	Lesson	616
	Extension: Cross Sections of Three-Dimensional Figures	620
	14.4–14.5 Quiz	622
	Chapter Review	623
	Chapter Test	626
	Cumulative Assessment	627

“I really like the Graphic Organizers because they show me another way to take notes.”

15

Probability and Statistics

What You Learned Before 631

Section 15.1 Outcomes and Events
Activity 632
Lesson 634

Section 15.2 Probability
Activity 638
Lesson 640

Section 15.3 Experimental and Theoretical Probability
Activity 644
Lesson 646

Section 15.4 Compound Events
Activity 652
Lesson 654

Section 15.5 Independent and Dependent Events
Activity 660
Lesson 662
Extension: Simulations 668
Study Help/Graphic Organizer 670
15.1–15.5 Quiz 671

Section 15.6 Samples and Populations
Activity 672
Lesson 674
Extension: Generating Multiple Samples 678

Section 15.7 Comparing Populations
Activity 680
Lesson 682
15.6–15.7 Quiz 686
Chapter Review 687
Chapter Test 692
Cumulative Assessment 693

"Using the Interactive Manipulatives from the Dynamic Student Edition helps me to see the mathematics that I am learning."

Key Vocabulary Index A1
Student Index A2
Additional Answers A15
Mathematics Reference Sheet B1

LOOKING BACK AT ADVANCED 1

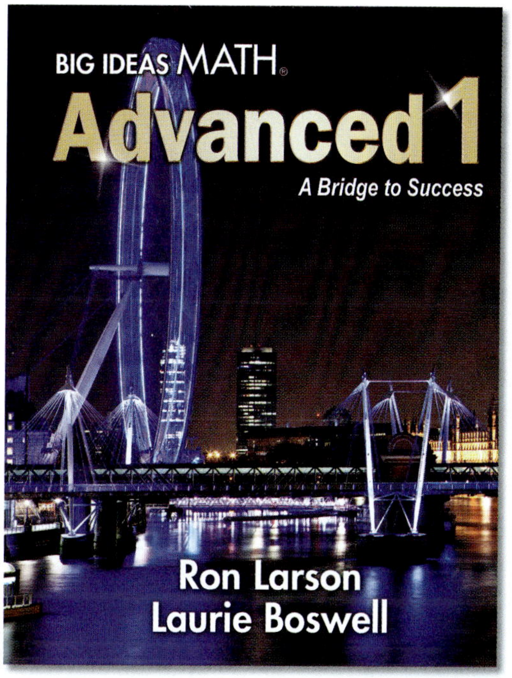

Chapter 1	Numerical Expressions and Factors
Chapter 2	Fractions and Decimals
Chapter 3	Algebraic Expressions and Properties
Chapter 4	Areas of Polygons
Chapter 5	Ratios and Rates
Chapter 6	Integers and the Coordinate Plane
Chapter 7	Equations and Inequalities
Chapter 8	Surface Area and Volume
Chapter 9	Statistical Measures
Chapter 10	Data Displays
Chapter 11	Integers
Chapter 12	Rational Numbers
Chapter 13	Expressions and Equations
Chapter 14	Ratios and Proportions
Chapter 15	Percents
Appendix A	My Big Ideas Projects

PROGRAM OVERVIEW

Print *Also available online*

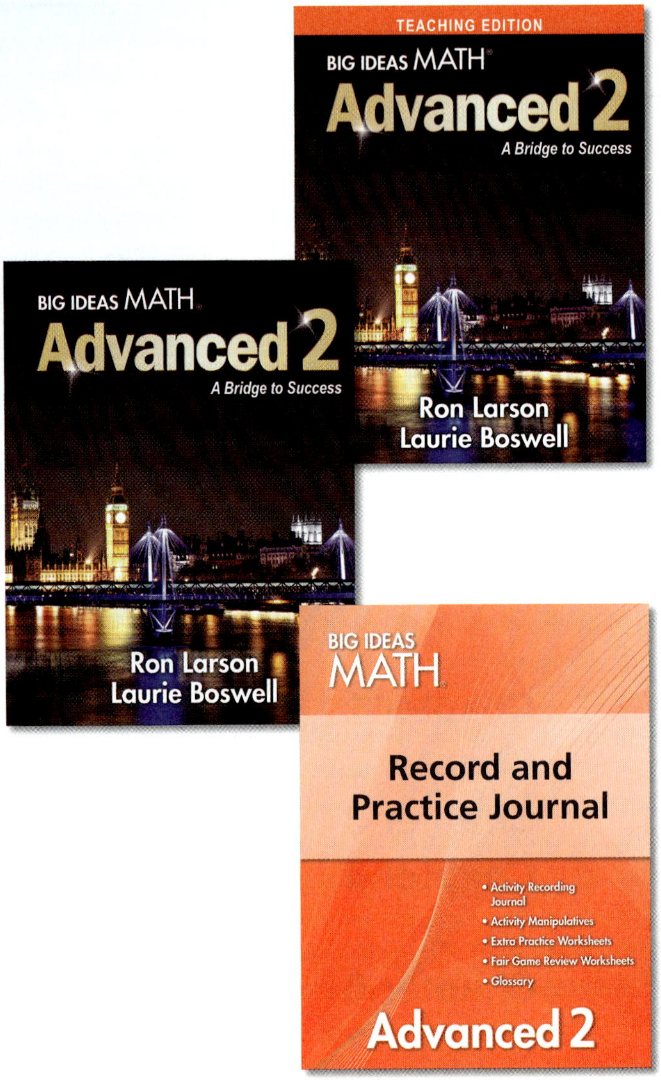

Pupil Edition

Teaching Edition

- Laurie's Notes

Record and Practice Journal

Available in English and Spanish

Resources by Chapter and Assessment Book

- Resources by Chapter
    - Family and Community Involvement: English and Spanish
    - Start Thinking! and Warm Up
    - Extra Practice
    - Enrichment and Extension
    - Puzzle Time
    - Technology Connection
- Assessment Book
    - Quizzes
    - Chapter Tests
    - Cumulative Assessment
    - Alternative Assessment
    - End-of-Course Tests

Technology

Student Edition

With complete English and Spanish Audio

- Dynamic eBook App
  - Lesson Tutorial Videos
  - Dynamic Investigations
- Online Home Edition
  - Multi-Language Glossary
  - Basic Skills Handbook
  - Skills Review Handbook
  - Game Closet

Dynamic Classroom

- Interactive Manipulatives
- Answer Presentation Tool
- Mini-Assessment
- Closure Activity

Dynamic Assessment and Progress Monitoring Tool

- Homework and Assessment Creation
- Progress Monitoring
- Direct Ties to Remediation
- All-In-One Reporting
- Online Chat Tutor

Dynamic Teaching Tools

- Interactive Whiteboard Lesson Library
  - Includes standard and customizable lessons
  - Compatible with SMART®, Promethean®, and Mimeo® technology
- **Exam**View® Assessment Suite
- Real-Life STEM Videos
- Editable Online Resources
  - Lesson Plans
  - Pacing Guides
  - Assessment Book
  - Resources by Chapter

SCOPE AND

Regular Pathway

Grade 6

Ratios and Proportional Relationships	– Understand Ratio Concepts; Use Ratio Reasoning
The Number System	– Perform Fraction and Decimal Operations; Understand Rational Numbers
Expressions and Equations	– Write, Interpret, and Use Expressions, Equations, and Inequalities
Geometry	– Solve Problems Involving Area, Surface Area, and Volume
Statistics and Probability	– Summarize and Describe Distributions; Understand Variability

Grade 7

Ratios and Proportional Relationships	– Analyze Proportional Relationships
The Number System	– Perform Rational Number Operations
Expressions and Equations	– Generate Equivalent Expressions; Solve Problems Using Linear Equations and Inequalities
Geometry	– Understand Geometric Relationships; Solve Problems Involving Angles, Surface Area, and Volume
Statistics and Probability	– Analyze and Compare Populations; Find Probabilities of Events

Grade 8

The Number System	– Approximate Real Numbers; Perform Real Number Operations
Expressions and Equations	– Use Radicals and Integer Exponents; Connect Proportional Relationships and Lines; Solve Systems of Linear Equations
Functions	– Define, Evaluate, and Compare Functions; Model Relationships
Geometry	– Understand Congruence and Similarity; Apply the Pythagorean Theorem; Apply Volume Formulas
Statistics and Probability	– Analyze Bivariate Data

SEQUENCE

Advanced Pathway

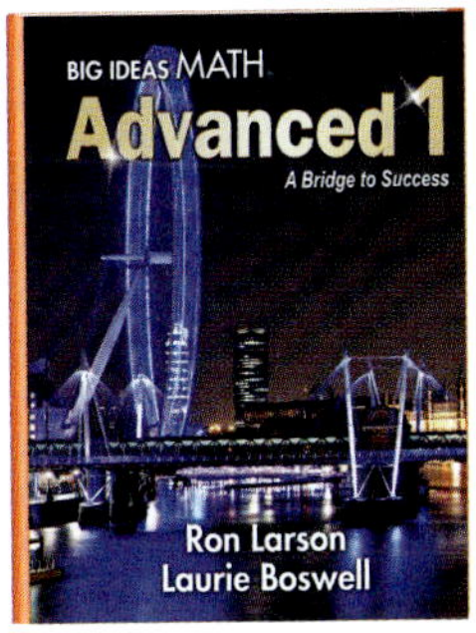

Grade 6 Advanced

Ratios and Proportional Relationships	– Understand Ratio Concepts; Use Ratio Reasoning; Analyze Proportional Relationships
The Number System	– Perform Fraction and Decimal Operations; Understand Rational Numbers; Perform Rational Number Operations
Expressions and Equations	– Write, Interpret, and Use Expressions, Equations, and Inequalities; Generate Equivalent Expressions; Solve Problems Using Linear Equations
Geometry	– Solve Problems Involving Area, Surface Area, and Volume
Statistics and Probability	– Summarize and Describe Distributions; Understand Variability

Grade 7 Advanced

The Number System	– Approximate Real Numbers; Perform Real Number Operations
Expressions and Equations	– Solve Problems Using Linear Inequalities; Use Radicals and Integer Exponents; Connect Proportional Relationships and Lines; Solve Systems of Linear Equations
Functions	– Define, Evaluate, and Compare Functions; Model Relationships
Geometry	– Understand Geometric Relationships; Solve Problems Involving Angles, Surface Area, and Volume; Understand Congruence and Similarity; Apply the Pythagorean Theorem
Statistics and Probability	– Analyze and Compare Populations; Find Probabilities of Events; Analyze Bivariate Data

LEARNING PROGRESSION

Advanced Pathway

Domain	Advanced 1	Advanced 2	Algebra 1
Number and Quantity			
Ratios and Proportional Relationships	Use ratios to solve problems. *Chapter 5* Use proportional relationships to solve problems. *Chapters 14, 15*		
The Number System	Perform operations with multi-digit numbers and find common factors and multiples. *Chapter 1* Divide fractions by fractions. *Chapter 2* Extend understanding of numbers to the rational number system. *Chapter 6* Perform operations with rational numbers. *Chapter 12*	Extend understanding of numbers to the real number system. *Chapter 7*	Use properties of rational exponents, rational numbers, and irrational numbers. *Chapters 6, 9, 10*
Algebra			
Expressions and Equations	Perform arithmetic with algebraic expressions. *Chapter 3* Solve one-variable equations and inequalities. *Chapter 7* Analyze relationships between dependent and independent variables. *Chapter 7* Write equivalent expressions. *Chapter 13* Use numerical and algebraic expressions and equations to solve problems. *Chapter 13*	Understand the connections between proportional relationships, lines, and linear equations. *Chapter 4* Solve linear equations and systems of linear equations. *Chapter 5* Work with radicals and integer exponents. *Chapters 7, 10* Use numerical and algebraic inequalities to solve problems. *Chapter 11*	Perform arithmetic on linear and quadratic polynomials. *Chapters 7–9* Solve linear and absolute value equations and inequalities, and quadratic and exponential equations. *Chapters 1–4, 6–10* Solve linear systems of equations and inequalities, and nonlinear systems of equations. *Chapters 5, 9*

MIDDLE SCHOOL MATH

Domain	Advanced 1	Advanced 2	Algebra 1
Functions			
Functions		Define, evaluate, and compare functions, and use functions to model relationships between quantities. *Chapter 6*	Interpret and analyze linear, quadratic, exponential, absolute value, step, and piecewise functions. *Chapters 3, 4, 6, 8–10* Build linear, quadratic, exponential, and absolute value functions from existing functions. *Chapters 3, 6, 8, 10*
Geometry			
Geometry	Solve real-world and mathematical problems involving area, surface area, and volume. *Chapters 4, 8*	Understand congruence and similarity. *Chapters 2, 3* Use the Pythagorean Theorem. *Chapter 7* Solve problems involving volume of cylinders, cones, and spheres. *Chapter 8* Draw, construct, and describe geometrical figures and describe the relationships between them. *Chapters 12–14* Solve problems involving angle measure, area, surface area, and volume. *Chapters 12–14*	
Statistics and Probability			
Statistics and Probability	Develop understanding of statistical variability and summarize and describe distributions. *Chapters 9, 10*	Investigate patterns of association in bivariate data. *Chapter 9* Make inferences about a population, compare two populations, and use probability models. *Chapter 15*	Summarize, represent, and interpret data using measures of center and spread. *Chapter 11*

PACING GUIDE FOR ADVANCED 2

Chapters 1–15 **142 Days**

Scavenger Hunt (1 Day)

Chapter 1 (7 Days)

Chapter Opener	1 Day
Section 1.1	1 Day
Section 1.2	1 Day
Section 1.3	1 Day
Section 1.4	1 Day
Chapter Review/Chapter Tests	2 Days

Chapter 2 (12 Days)

Chapter Opener	1 Day
Section 2.1	1 Day
Section 2.2	1 Day
Section 2.3	1 Day
Section 2.4	1 Day
Study Help/Quiz	1 Day
Section 2.5	1 Day
Section 2.6	1 Day
Section 2.7	2 Days
Chapter Review/Chapter Tests	2 Days

Chapter 3 (8 Days)

Chapter Opener	1 Day
Section 3.1	1 Day
Section 3.2	1 Day
Section 3.3	1 Day
Section 3.4	2 Days
Chapter Review/Chapter Tests	2 Days

Chapter 4 (12 Days)

Chapter Opener	1 Day
Section 4.1	1 Day
Section 4.2	1 Day
Extension 4.2	1 Day
Section 4.3	1 Day
Study Help/Quiz	1 Day
Section 4.4	1 Day
Section 4.5	1 Day
Section 4.6	1 Day
Section 4.7	1 Day
Chapter Review/Chapter Tests	2 Days

Chapter 5 (8 Days)

Chapter Opener	1 Day
Section 5.1	1 Day
Section 5.2	1 Day
Section 5.3	1 Day
Section 5.4	1 Day
Extension 5.4	1 Day
Chapter Review/Chapter Tests	2 Days

Chapter 6 (9 Days)

Chapter Opener	1 Day
Section 6.1	1 Day
Section 6.2	1 Day
Section 6.3	1 Day
Study Help/Quiz	1 Day
Section 6.4	1 Day
Section 6.5	1 Day
Chapter Review/Chapter Tests	2 Days

Chapter 7 (10 Days)

Chapter Opener	1 Day
Section 7.1	1 Day
Section 7.2	1 Day
Section 7.3	1 Day
Study Help/Quiz	1 Day
Section 7.4	1 Day
Extension 7.4	1 Day
Section 7.5	1 Day
Chapter Review/Chapter Tests	2 Days

Chapter 8 (7 Days)

Chapter Opener	1 Day
Section 8.1	1 Day
Section 8.2	1 Day
Section 8.3	1 Day
Section 8.4	1 Day
Chapter Review/Chapter Tests	2 Days

Chapter 9 (7 Days)

Chapter Opener	1 Day
Section 9.1	1 Day
Section 9.2	1 Day
Section 9.3	1 Day
Section 9.4	1 Day
Chapter Review/Chapter Tests	2 Days

Chapter 10 (11 Days)

Chapter Opener	1 Day
Section 10.1	1 Day
Section 10.2	1 Day
Section 10.3	1 Day
Section 10.4	1 Day
Study Help/Quiz	1 Day
Section 10.5	1 Day
Section 10.6	1 Day
Section 10.7	1 Day
Chapter Review/Chapter Tests	2 Days

Chapter 11 (7 Days)

Chapter Opener	1 Day
Section 11.1	1 Day
Section 11.2	1 Day
Section 11.3	1 Day
Section 11.4	1 Day
Chapter Review/Chapter Tests	2 Days

Chapter 12 (12 Days)

Chapter Opener	1 Day
Section 12.1	1 Day
Section 12.2	1 Day
Section 12.3	2 Days
Extension 12.3	1 Day
Study Help/Quiz	1 Day
Section 12.4	2 Days
Section 12.5	1 Day
Chapter Review/Chapter Tests	2 Days

Chapter 13 (7 Days)

Chapter Opener	1 Day
Section 13.1	1 Day
Section 13.2	1 Day
Section 13.3	1 Day
Section 13.4	1 Day
Chapter Review/Chapter Tests	2 Days

Chapter 14 (11 Days)

Chapter Opener	1 Day
Section 14.1	1 Day
Section 14.2	2 Days
Section 14.3	1 Day
Study Help/Quiz	1 Day
Section 14.4	1 Day
Section 14.5	1 Day
Extension 14.5	1 Day
Chapter Review/Chapter Tests	2 Days

Chapter 15 (13 Days)

Chapter Opener	1 Day
Section 15.1	1 Day
Section 15.2	1 Day
Section 15.3	1 Day
Section 15.4	1 Day
Section 15.5	1 Day
Extension 15.5	1 Day
Study Help/Quiz	1 Day
Section 15.6	1 Day
Extension 15.6	1 Day
Section 15.7	1 Day
Chapter Review/Chapter Tests	2 Days

Mathematical Practices

Make sense of problems and persevere in solving them.

- Multiple representations are presented to help students move from concrete to representative and into abstract thinking
- *Essential Questions* help students focus and analyze
- *In Your Own Words* provide opportunities for students to look for meaning and entry points to a problem

Reason abstractly and quantitatively.

- Visual problem solving models help students create a coherent representation of the problem
- Opportunities for students to decontextualize and contextualize problems are presented in every lesson

Construct viable arguments and critique the reasoning of others.

- *Error Analysis*; *Different Words, Same Question*; and *Which One Doesn't Belong* features provide students the opportunity to construct arguments and critique the reasoning of others
- *Inductive Reasoning* activities help students make conjectures and build a logical progression of statements to explore their conjecture

Model with mathematics.

- Real-life situations are translated into diagrams, tables, equations, and graphs to help students analyze relations and to draw conclusions
- Real-life problems are provided to help students learn to apply the mathematics that they are learning to everyday life

Use appropriate tools strategically.

- *Graphic Organizers* support the thought process of what, when, and how to solve problems
- A variety of tool papers, such as graph paper, number lines, and manipulatives, are available as students consider how to approach a problem
- Opportunities to use the web, graphing calculators, and spreadsheets support student learning

Attend to precision.

- *On Your Own* questions encourage students to formulate consistent and appropriate reasoning
- Cooperative learning opportunities support precise communication

Look for and make use of structure.

- *Inductive Reasoning* activities provide students the opportunity to see patterns and structure in mathematics
- Real-world problems help students use the structure of mathematics to break down and solve more difficult problems

Look for and express regularity in repeated reasoning.

- Opportunities are provided to help students make generalizations
- Students are continually encouraged to check for reasonableness in their solutions

Go to *BigIdeasMath.com* for more information on the Mathematical Practices.

Mathematical Content

Chapter Coverage

1 · 2 · 3 · 4 · 5 · 6 · **7** · 8 · 9 · 10 · 11 · 12 · 13 · 14 · 15

The Number System

- Know that there are numbers that are not rational, and approximate them by rational numbers.

1 · 2 · 3 · **4** · **5** · 6 · **7** · 8 · 9 · **10** · **11** · 12 · 13 · 14 · 15

Expressions and Equations

- Use properties of operations to generate equivalent expressions.
- Solve real-life and mathematical problems using numerical and algebraic expressions and equations.
- Work with radicals and integer exponents.
- Understand the connections between proportional relationships, lines, and linear equations.
- Analyze and solve linear equations and pairs of simultaneous equations.

1 · 2 · 3 · 4 · 5 · **6** · 7 · 8 · 9 · 10 · 11 · 12 · 13 · 14 · 15

Functions

- Define, evaluate, and compare functions.
- Use functions to model relationships between quantities.

1 · **2** · **3** · 4 · 5 · 6 · **7** · **8** · 9 · 10 · 11 · **12** · **13** · **14** · 15

Geometry

- Draw, construct, and describe geometrical figures and describe the relationships between them.
- Solve real-life and mathematical problems involving angle measure, area, surface area, and volume.
- Understand congruence and similarity using physical models, transparencies, or geometry software.
- Understand and apply the Pythagorean Theorem.
- Solve real-world and mathematical problems involving volume of cylinders, cones, and spheres.

1 · 2 · 3 · 4 · 5 · 6 · 7 · 8 · **9** · 10 · 11 · 12 · 13 · 14 · **15**

Statistics and Probability

- Use random sampling to draw inferences about a population.
- Draw informal comparative inferences about two populations.
- Investigate chance processes and develop, use, and evaluate probability models.
- Investigate patterns of association in bivariate data.

11 Inequalities

11.1 Writing and Graphing Inequalities

11.2 Solving Inequalities Using Addition or Subtraction

11.3 Solving Inequalities Using Multiplication or Division

11.4 Solving Two-Step Inequalities

What Your Students Have Learned

- Use and interpret simple equations.
- Determine whether a value is a solution of an inequality.
- Represent constraints with inequalities and recognize that they can have infinitely many solutions.
- Solve one-step equations and inequalities.

What Your Students Will Learn

- Solve one-step inequalities involving integers and rational numbers.
- Solve multi-step inequalities.

Pacing Guide for Chapter 11

Section	Days
Chapter Opener Advanced	1 Day
Section 1 Advanced	1 Day
Section 2 Advanced	1 Day
Section 3 Advanced	1 Day
Section 4 Advanced	1 Day
Chapter Review/ Chapter Tests Advanced	2 Days
Total Chapter 11 Advanced	7 Days
Year-to-Date Advanced	99 Days

Technology for the Teacher

BigIdeasMath.com
Chapter at a Glance
Complete Materials List
Parent Letters: English and Spanish

What Your Students Have Learned

- Graph inequalities on a number line.
- Compare rational numbers and decimals.

Additional Topics for Review

- Writing Expressions
- Evaluating Expressions
- Solving Two-Step Equations
- Converting Between Fractions and Decimals

Try It Yourself

1. 
2. 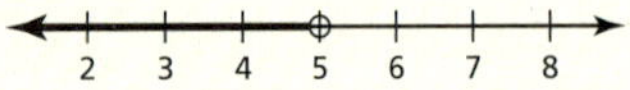
3. 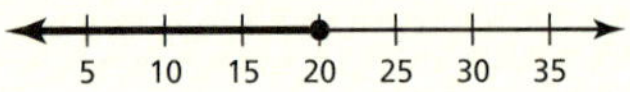
4. 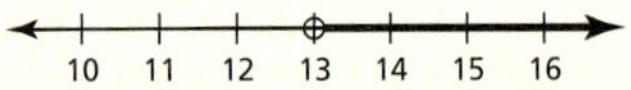
5. <
6. >
7. <
8. <
9. >
10. <

Record and Practice Journal Fair Game Review

1. 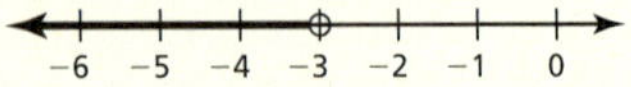
2. 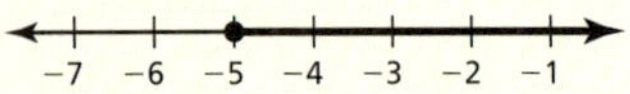
3. 
4. 
5. 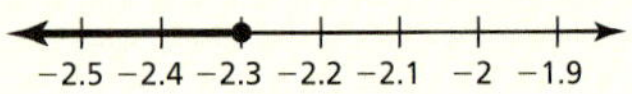
6. 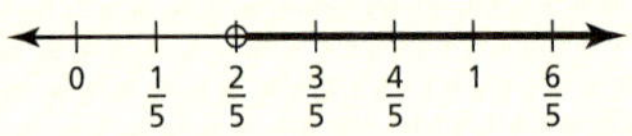
7. 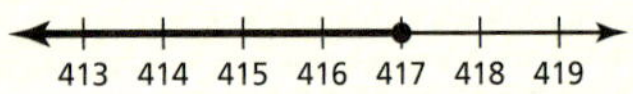
8. >
9. >
10. >
11. <
12. <
13. <
14. your friend; 5.6 ft is about 5 ft and 7 in.

Math Background Notes

Vocabulary Review

- Inequality
- Number Line
- Integers
- Rational Numbers

Graphing Inequalities

- Students should be able to graph inequalities on a number line.
- Remind students that an equation produces a finite number of solutions, but an inequality produces an entire set of solutions. That is why an inequality requires you to shade the number line to describe the solutions.
- Remind students that inequalities containing $\leq$ or $\geq$ will require a closed circle. Inequalities containing $<$ or $>$ will require an open circle.
- **Teaching Tip:** Some students have difficulty deciding which side of the number line to shade. Encourage students to pick a test value on each side of the circle. Substitute each test value for *x*. Only one of the resulting inequalities will be true. Shade the number line on the side of the circle from which the valid test value was selected.

Comparing Numbers

- Students should know how to order integers and rational numbers and work with the number line.
- You may want to discuss the scaling of the number line with students. Each tick mark should count the same amount and be equally spaced on the number line.
- **Common Error:** Students may not use a number line and end up with incorrect comparisons. Encourage students to graph the numbers on a number line, as shown in Example 2, to help them correctly compare the numbers.

Reteaching and Enrichment Strategies

If students need help. . .	If students got it. . .
Record and Practice Journal • Fair Game Review Skills Review Handbook Lesson Tutorials	Game Closet at *BigIdeasMath.com* Start the next section

What You Learned Before

"Move farther back. We still have an inequality."

Graphing Inequalities

Example 1 Graph $x \geq 2$.

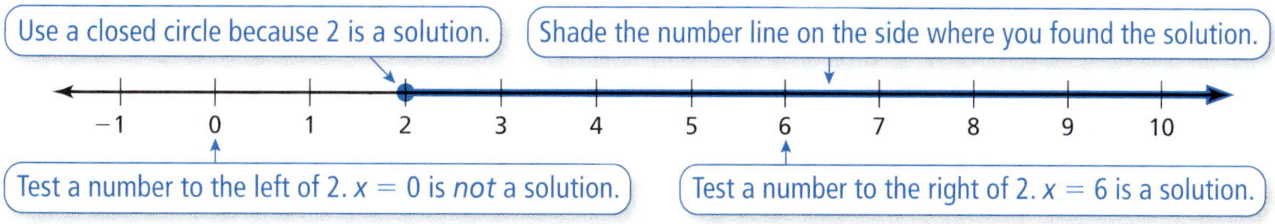

Try It Yourself

Graph the inequality.

1. $x \geq 1$ **2.** $x < 5$ **3.** $x \leq 20$ **4.** $x > 13$

Comparing Numbers

Example 2 Compare $-\frac{1}{3}$ and $-\frac{5}{6}$.

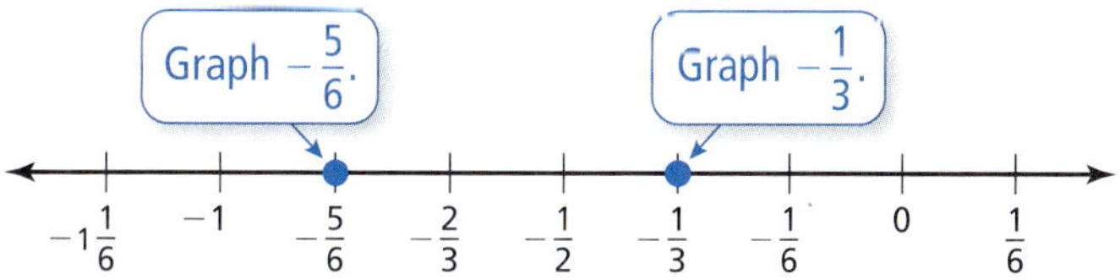

$-\frac{5}{6}$ is to the left of $-\frac{1}{3}$.

So, $-\frac{5}{6} < -\frac{1}{3}$.

Try It Yourself

Copy and complete the statement using < or >.

5. $-\frac{2}{3}$ ☐ $\frac{3}{8}$ **6.** $-\frac{1}{2}$ ☐ $-\frac{7}{8}$

7. $-\frac{1}{5}$ ☐ $\frac{1}{10}$ **8.** -1.4 ☐ 1.2

9. -2.2 ☐ -4.6 **10.** -1.9 ☐ -1.1

11.1 Writing and Graphing Inequalities

Essential Question How can you use a number line to represent solutions of an inequality?

1 ACTIVITY: Understanding Inequality Statements

Work with a partner. Read the statement. Circle each number that makes the statement true, and then answer the questions.

a. "You are in at least 5 of the photos."

−3 −2 −1 0 1 2 3 4 5 6

- What do you notice about the numbers that you circled?
- Is the number 5 included? Why or why not?
- Write four other numbers that make the statement true.

b. "The temperature is less than −4 degrees Fahrenheit."

−7 −6 −5 −4 −3 −2 −1 0 1 2

- What do you notice about the numbers that you circled?
- Can the temperature be exactly −4 degrees Fahrenheit? Explain.
- Write four other numbers that make the statement true.

c. "More than 3 students from our school are in the chess tournament."

−3 −2 −1 0 1 2 3 4 5 6

- What do you notice about the numbers that you circled?
- Is the number 3 included? Why or why not?
- Write four other numbers that make the statement true.

d. "The balance in a yearbook fund is no more than −\$5."

−7 −6 −5 −4 −3 −2 −1 0 1 2

- What do you notice about the numbers that you circled?
- Is the number −5 included? Why or why not?
- Write four other numbers that make the statement true.

Inequalities

In this lesson, you will

- write and graph inequalities.
- use substitution to check whether a number is a solution of an inequality.

Laurie's Notes

Introduction

Applying Mathematical Practices

- **Attend to Precision:** Mathematically proficient students communicate precisely to others. This is done orally, in writing, and in the graphs they construct.

Motivate

- Ask students to write down their heights in two ways.
  - in feet and inches
  - in inches

  For instance, my height is 5 feet 7 inches or 67 inches. Tell students you are going to ask questions about their height written both ways.
- Read each statement below. Have students stand up (or raise their hands) when the statement is true for their height. Discuss each statement before going on to the next one. For instance, for the first statement ask, "Should someone who is 72 inches tall stand?" yes "Should someone who is 64 inches tall stand?" no
  - Your height is greater than 64 inches ($h > 64''$).
  - Your height is at most 5 feet 2 inches ($h \le 5'2''$).
  - Your height is at least 63 inches ($h \ge 63''$).

Activity Notes

Activity 1

- Have students work through the four parts with their partners.
- If you discussed each statement in the Motivate activity, then students should have no difficulty answering the questions posed.
- When students finish, have volunteers share their results.
- **Big Idea:** You want students to notice that there is an infinite number of solutions to each statement (though some solutions may not make sense in the context of the problem).
- Another idea to focus on is that there is a boundary point for the set of solutions and you need to pay attention to the problem wording to understand the boundary point.
- **Attend to Precision:** Although only integers are listed as possible solutions, students should recognize that the solutions in parts (b) and (d) also include non-integer values. For example, a solution in part (b) is -8.3°F and a possible balance in part (d) is $-\$7.50$.

What Your Students Will Learn

- Write and graph inequalities.
- Use substitution to check whether numbers are solutions of inequalities.

Previous Learning

Students should know how to graph numbers on a number line, solve single variable equations, and solve single variable inequalities using whole numbers.

Technology for the Teacher

Dynamic Classroom

Lesson Plans

Complete Materials List

11.1 Record and Practice Journal

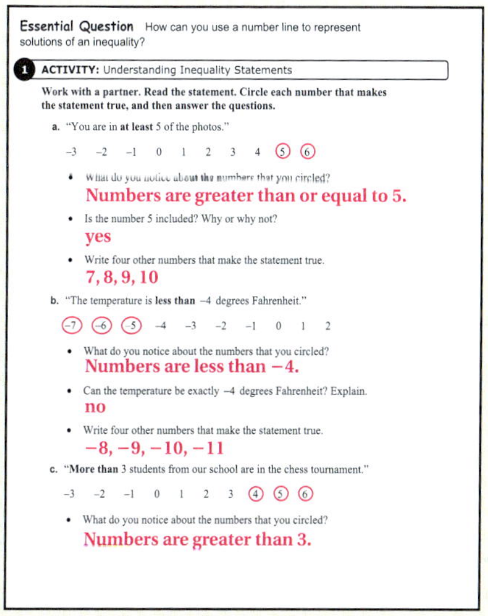

Essential Question How can you use a number line to represent solutions of an inequality?

1 ACTIVITY: Understanding Inequality Statements

Work with a partner. Read the statement. Circle each number that makes the statement true, and then answer the questions.

a. "You are in **at least** 5 of the photos."

–3 –2 –1 0 1 2 3 4 (5) (6)

- What do you notice about the numbers that you circled?

  Numbers are greater than or equal to 5.
- Is the number 5 included? Why or why not?

  yes
- Write four other numbers that make the statement true.

  7, 8, 9, 10

b. "The temperature is **less than** –4 degrees Fahrenheit."

(–7) (–6) (–5) –4 –3 –2 –1 0 1 2

- What do you notice about the numbers that you circled?

  Numbers are less than –4.
- Can the temperature be exactly –4 degrees Fahrenheit? Explain.

  no
- Write four other numbers that make the statement true.

  –8, –9, –10, –11

c. "**More than** 3 students from our school are in the chess tournament."

–3 –2 –1 0 1 2 3 (4) (5) (6)

- What do you notice about the numbers that you circled?

  Numbers are greater than 3.

English Language Learners

Vocabulary and Symbols

Students should review the vocabulary and symbols for inequalities. Have students add to their notebooks a table of symbols and what the symbols mean. Students should add to the table as new phrases are used in the chapter.

Symbol	Phrase
$=$	is equal to
$\neq$	is not equal to
$<$	is less than
$\leq$	is less than or equal to
$>$	is greater than
$\geq$	is greater than or equal to

11.1 Record and Practice Journal

- Is the number 3 included? Why or why not?
  no
- Write four other numbers that make the statement true.
  7, 8, 9, 10

d. "The balance in a yearbook fund is **no more than** –$5."

(–7) (–6) (–5) –4 –3 –2 –1 0 1 2

- What do you notice about the numbers that you circled?
  Numbers are less than or equal to -5.
- Is the number –5 included? Why or why not?
  yes
- Write four other numbers that make the statement true.
  $-8, -9, -10, -11$

2 ACTIVITY: Understanding Inequality Symbols

Work with a partner.

a. Consider the statement "x is a number such that $x > -1.5$."

- Can the number be exactly –1.5? Explain.
  no
- Make a number line. Shade the part of the number line that shows the numbers that make the statement true.
  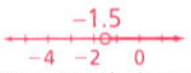
- Write four other numbers that are not integers that make the statement true.
  $-1.4, -1.3, -1.2, -1.1$

b. Consider the statement "x is a number such that $x \leq \frac{5}{2}$."

- Can the number be exactly $\frac{5}{2}$? Explain.
  yes

- Make a number line. Shade the part of the number line that shows the numbers that make the statement true.

- Write four other numbers that are not integers that make the statement true. $\frac{3}{2}, \frac{1}{2}, \frac{1}{4}, \frac{1}{6}$

3 ACTIVITY: Writing and Graphing Inequalities

Work with a partner. Write an inequality for each graph. Then, in words, describe all the values of x that make the inequality true.

a. 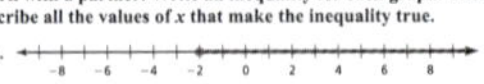

$x \geq -2$

b. 

$x > -2$

c. 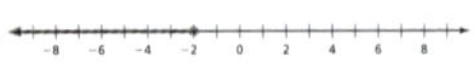

$x \leq -2$

d. 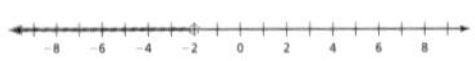

$x < -2$

What Is Your Answer?

4. **IN YOUR OWN WORDS** How can you use a number line to represent solutions of an inequality?
   Shade the numbers that are solutions.

5. **STRUCTURE** Is $x \geq -1.4$ the same as $-1.4 \leq x$? Explain.
   yes

Laurie's Notes

Activity 2

- The questions in this activity have no context.
- Students will consider the boundary point and which direction to look (greater than or less than) for additional solutions. The solutions are not limited to integer values.
- **Attend to Precision:** Students must pay attention to the inequality symbol. In part (a) they need to consider whether the boundary point (-1.5) is a solution.
- When students finish, have volunteers share their results. Pay attention to how students locate the boundary point and how they deal with the strict and non-strict inequality symbols.
- Students may or may not use circles. They may just shade the number line to the left or right of a number or shade through a number.

Activity 3

- The boundary point in each problem is -2. The inequality symbol is what changes for each graph.
- Without giving students any definitions, let them think about the difference between the closed circle and open circle.
- Discuss students' answers for parts (a) and (b), comparing the open and closed circle. Do the same for parts (c) and (d).

What Is Your Answer?

- Some students may not know the language needed to describe the process. They might focus on creating clear examples.
- In Question 5, students may be confused because of the rational number. You could modify the question and ask, is $x \geq 4$ the same as $4 \leq x$?

Closure

- **Writing:** Write an inequality for each graph. Describe all the values of x that make the inequality true.

a. 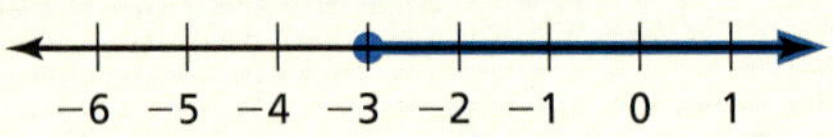

$x \geq -3$; all values of x greater than or equal to -3

b. 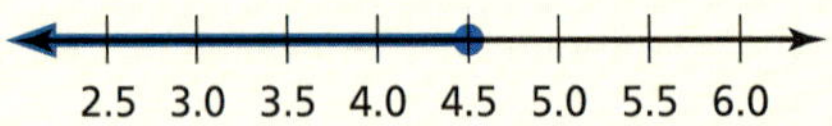

$x \leq 4.5$; all values of x less than or equal to 4.5

2 ACTIVITY: Understanding Inequality Symbols

Work with a partner.

a. Consider the statement "x is a number such that $x > -1.5$."

- Can the number be exactly -1.5? Explain.
- Make a number line. Shade the part of the number line that shows the numbers that make the statement true.
- Write four other numbers that are not integers that make the statement true.

b. Consider the statement "x is a number such that $x \leq \frac{5}{2}$."

- Can the number be exactly $\frac{5}{2}$? Explain.
- Make a number line. Shade the part of the number line that shows the numbers that make the statement true.
- Write four other numbers that are not integers that make the statement true.

3 ACTIVITY: Writing and Graphing Inequalities

Math Practice

Check Progress

All the graphs are similar. So, what can you do to make sure that you have correctly written each inequality?

Work with a partner. Write an inequality for each graph. Then, in words, describe all the values of x that make the inequality true.

a.

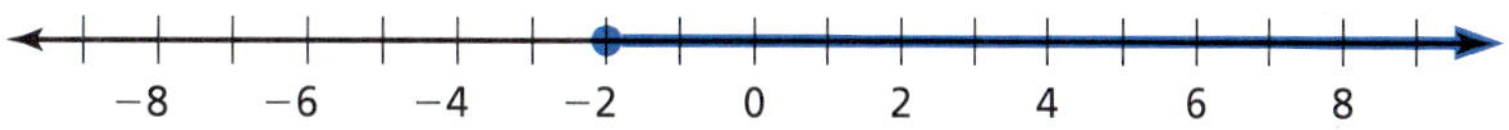

b.

c.

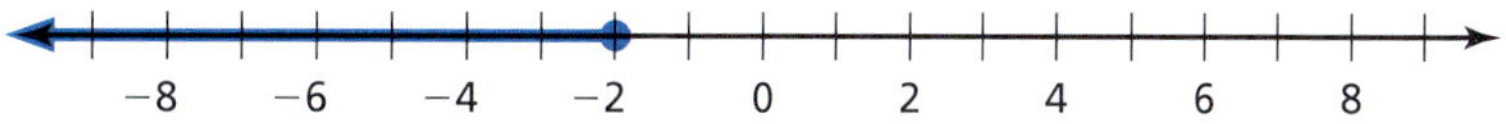

d.

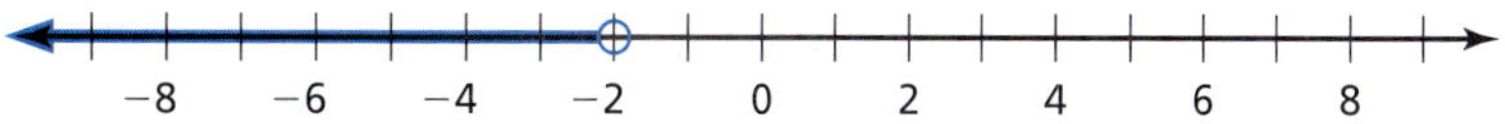

What Is Your Answer?

4. **IN YOUR OWN WORDS** How can you use a number line to represent solutions of an inequality?

5. **STRUCTURE** Is $x \geq -1.5$ the same as $-1.4 \leq x$? Explain.

Practice Use what you learned about writing and graphing inequalities to complete Exercises 4 and 5 on page 468.

11.1 Lesson

Check It Out
Lesson Tutorials
BigIdeasMath.com

Key Vocabulary
inequality, *p. 466*
solution of an inequality, *p. 466*
solution set, *p. 466*
graph of an inequality, *p. 467*

An **inequality** is a mathematical sentence that compares expressions. It contains the symbols $<$, $>$, $\leq$, or $\geq$. To write an inequality, look for the following phrases to determine where to place the inequality symbol.

Inequality Symbols				
Symbol	$<$	$>$	$\leq$	$\geq$
Key Phrases	• is less than • is fewer than	• is greater than • is more than	• is less than or equal to • is at most • is no more than	• is greater than or equal to • is at least • is no less than

EXAMPLE 1 Writing an Inequality

A number q plus 5 is greater than or equal to -7.9. Write this word sentence as an inequality.

A number q plus 5 | is greater than or equal to | -7.9.

$q + 5$ | $\geq$ | -7.9

An inequality is $q + 5 \geq -7.9$.

On Your Own

Now You're Ready
Exercises 6–9

Write the word sentence as an inequality.

1. A number x is at most -10.
2. Twice a number y is more than $-\frac{5}{2}$.

A **solution of an inequality** is a value that makes the inequality true. An inequality can have more than one solution. The set of all solutions of an inequality is called the **solution set**.

Reading
The symbol $\nleq$ means *is not less than or equal to.*

Value of x	$x + 2 \leq -1$	Is the inequality true?
-2	$-2 + 2 \overset{?}{\leq} -1$ $0 \nleq -1$ ✗	no
-3	$-3 + 2 \overset{?}{\leq} -1$ $-1 \leq -1$ ✓	yes
-4	$-4 + 2 \overset{?}{\leq} -1$ $-2 \leq -1$ ✓	yes

Multi-Language Glossary at BigIdeasMath.com

Laurie's Notes

Introduction

Connect

- **Yesterday:** Yesterday students investigated writing and graphing inequalities.
- **Today:** Students will translate inequalities from words to symbols and check to see whether a value is a solution of the inequality.

Motivate

- **Story Time:** You are planning to visit several theme parks and notice in doing your research that some of the rides have height restrictions.

Attraction	Restriction	Inequality
Dinosaur	Minimum is now 40 inches	$h \geq 40$
Primeval Whirl	Must be at least 48 inches	$h \geq 48$
Bay Slide	Must be under 60 inches	$h < 60$

- Ask students to write each as an inequality, where h is the rider's height.
- In today's lesson, they will be translating words to symbols.

Lesson Notes

Discuss

- Write the definition of an inequality.
- Review the four inequality symbols and key phrases or words that suggest each inequality.

Example 1

? Read the problem and ask, "How do you write a number q plus 5 in symbols?" $q + 5$

- For clarity, notice the use of color to help students translate each portion of the inequality.

On Your Own

- **Think-Pair-Share:** Students should read each question independently and then work in pairs to answer the questions. When they have answered the questions, the pair should compare their answers with another group and discuss any discrepancies.

Discuss

- Discuss what is meant by a solution of an inequality. Inequalities can, and generally do, have more than one solution. All of the solutions are collectively referred to as the **solution set**.
- It is helpful to write the inequality and substitute the value you are checking, as shown in the table.
- **Common Error:** Students will often make the mistake of thinking $-2 \geq -1$, forgetting that relationships are reversed on the negative side of 0; $-2 \leq -1$.

Goal Today's lesson is writing and graphing **inequalities**.

Technology for the Teacher

Dynamic Classroom

Lesson Tutorials
Lesson Plans
Answer Presentation Tool

Extra Example 1

A number y minus 3 is less than -15.3. Write this word sentence as an inequality.

$y - 3 < -15.3$

On Your Own

1. $x \leq -10$
2. $2y > -\frac{5}{2}$

Differentiated Instruction

Auditory

Stress to students the importance of reading a statement and translating it into an expression, equation, or inequality. The word "is" plays an important role in the meaning of the statement. For instance, *six less than a number* translates to $x - 6$, while *six is less than a number* translates to $6 < x$.

Extra Example 2

Tell whether -3 is a solution of each inequality.

a. $y + 6 < 4$ solution

b. $\frac{y}{-4} > 4$ not a solution

On Your Own

3. not a solution

4. not a solution

5. solution

Extra Example 3

Graph $p < -3$.

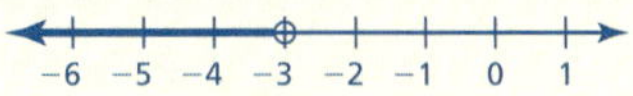

On Your Own

6.

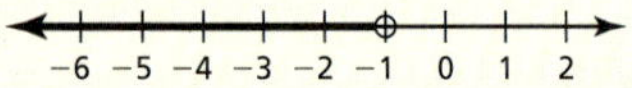

7.

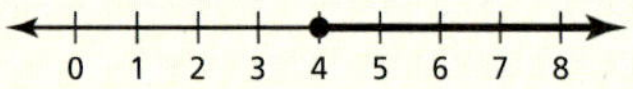

8.

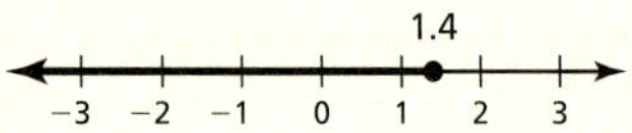

9.

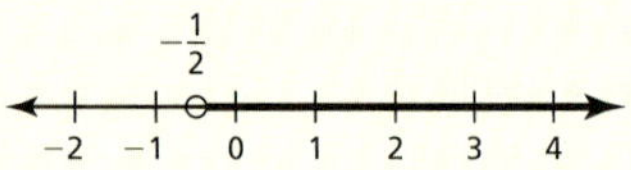

Laurie's Notes

Example 2

? "How do you determine whether -2 is a solution of an inequality?" Substitute -2 for the variable, simplify, and decide whether the resulting inequality is true.

- Work through each example as shown. In part (b), students must recall that the product of two negatives is a positive.

On Your Own

- **Common Error:** In Question 3, when students substitute -5 for x, they may incorrectly see the result $7 > 7$ as a true inequality. Remind students to pay close attention to the inequality symbol.
- Ask volunteers to share their work at the board.

Discuss

- Discuss what is meant by the graph of an inequality. Remind students of the difference between the open and closed circles.

Example 3

- Decide which side of the boundary point to shade by testing one number on each side of the boundary point. It is called a boundary point because all of the values on one side of this value satisfy the inequality while all of the values on the other side of this value do *not* satisfy the inequality.
- Because -8 is not a solution of the inequality, use an open circle for the graph. Use a closed circle only when the boundary point is a solution of the graph (the inequality involves $\leq$ or $\geq$).
- **Use Appropriate Tools Strategically:** The graph of an inequality is a helpful visual tool that allows the solution to be seen. It also shows what values are *not* solutions.

On Your Own

- In Question 9, check to see that students locate $-\frac{1}{2}$ correctly.
- Ask students to share their graphs at the board.

Closure

- **Writing Prompt:** To decide whether a number is a solution of the inequality, you . . .

EXAMPLE 2 Checking Solutions

Tell whether -2 is a solution of each inequality.

a. $y - 5 \geq -6$

$y - 5 \geq -6$	Write the inequality.	$-5.5y < 14$
$-2 - 5 \overset{?}{\geq} -6$	Substitute -2 for y.	$-5.5(-2) \overset{?}{<} 14$
$-7 \ngeq -6$ ✗	Simplify.	$11 < 14$ ✓

b. $-5.5y < 14$

a. -7 is *not* greater than or equal to -6.

So, -2 is *not* a solution of the inequality.

b. 11 is less than 14.

So, -2 is a solution of the inequality.

On Your Own

Now You're Ready Exercises 11–16

Tell whether -5 is a solution of the inequality.

3. $x + 12 > 7$ **4.** $1 - 2p \leq -9$ **5.** $n \div 2.5 \geq -3$

The **graph of an inequality** shows all the solutions of the inequality on a number line. An open circle ○ is used when a number is *not* a solution. A closed circle ● is used when a number is a solution. An arrow to the left or right shows that the graph continues in that direction.

EXAMPLE 3 Graphing an Inequality

Graph $y > -8$.

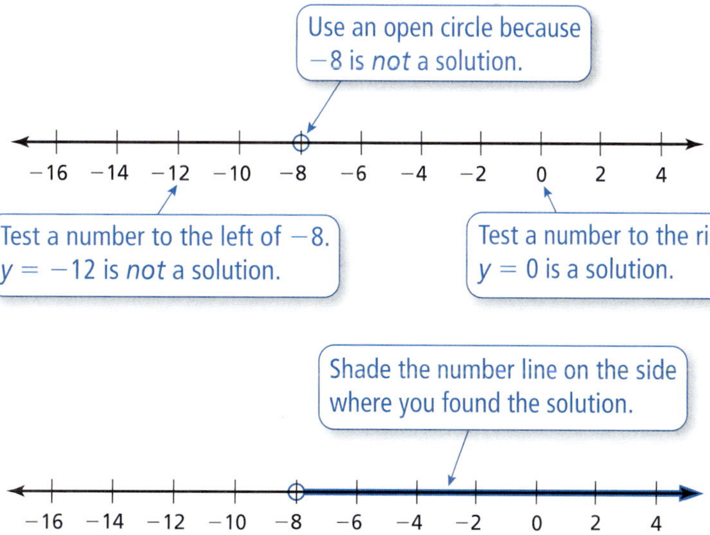

Study Tip

The graph in Example 3 shows that the inequality has *infinitely many* solutions.

On Your Own

Now You're Ready Exercises 17–20

Graph the inequality on a number line.

6. $x < -1$ **7.** $z \geq 4$ **8.** $s \leq 1.4$ **9.** $-\frac{1}{2} < t$

11.1 Exercises

Check It Out
Help with Homework
BigIdeasMath.com

Vocabulary and Concept Check

1. **PRECISION** Should you use an open circle or a closed circle in the graph of the inequality $b \geq -42$? Explain.

2. **DIFFERENT WORDS, SAME QUESTION** Which is different? Write "both" inequalities.

k is less than or equal to -3.	k is no more than -3.
k is at most -3.	k is at least -3.

3. **REASONING** Do $x < 5$ and $5 < x$ represent the same inequality? Explain.

Practice and Problem Solving

Write an inequality for the graph. Then, in words, describe all the values of x that make the inequality true.

4. 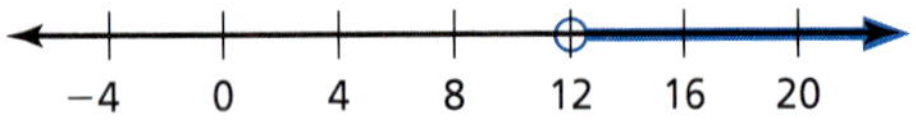

5. 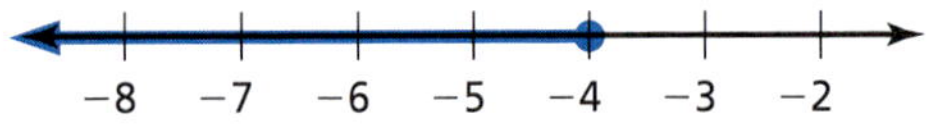

Write the word sentence as an inequality.

1 6. A number y is no more than -8.

7. A number w added to 2.3 is more than 18.

8. A number t multiplied by -4 is at least $-\frac{2}{5}$.

9. A number b minus 4.2 is less than -7.5.

10. **ERROR ANALYSIS** Describe and correct the error in writing the word sentence as an inequality.

> ✗ Twice a number x is at most −24.
>
> $2x \geq -24$

Tell whether the given value is a solution of the inequality.

2 11. $n + 8 \leq 13;\ n = 4$

12. $5h > -15;\ h = -5$

13. $p + 1.4 \leq 0.5;\ p = 0.1$

14. $\frac{a}{6} > -4;\ a = -18$

15. $-\frac{2}{3}s \geq 6;\ s = -9$

16. $\frac{7}{8} - 3k < -\frac{1}{2};\ k = \frac{1}{4}$

Graph the inequality on a number line.

3 17. $r \leq -9$

18. $g > 2.75$

19. $x \geq -3\frac{1}{2}$

20. $z < 1\frac{1}{4}$

21. **FOOD TRUCK** Each day at lunchtime, at least 53 people buy food from a food truck. Write an inequality that represents this situation.

Assignment Guide and Homework Check

Level	Assignment	Homework Check
Advanced	1–5, 6–28 even, 29–32	8, 14, 22, 24, 26

Common Errors

- **Exercises 11–16** Students may not understand when the boundary point is a solution of the inequality. Remind them that the inequality is true for the boundary point only when the inequality symbol is ≤ or ≥.
- **Exercises 17–20** Students may use the wrong type of circle at the boundary point. Remind them to use a closed circle when the boundary point is a solution and an open circle when the boundary point is not a solution.
- **Exercises 17–20** Students may shade on the wrong side of the boundary point. Remind them to test one point on each side of the boundary point.

11.1 Record and Practice Journal

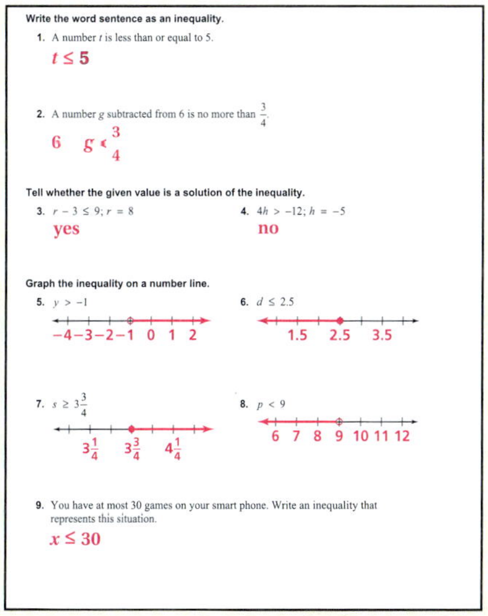
Write the word sentence as an inequality.

1. A number t is less than or equal to 5.
   $t \le 5$
2. A number g subtracted from 6 is no more than $\frac{3}{4}$.
   $6 - g \le \frac{3}{4}$

Tell whether the given value is a solution of the inequality.

3. $r - 3 \le 9;\ r = 8$
   yes
4. $4h > -12;\ h = -5$
   no

Graph the inequality on a number line.

5. $y > -1$
6. $d \le 2.5$
7. $s \ge 3\frac{3}{4}$
8. $p < 9$

9. You have at most 30 games on your smart phone. Write an inequality that represents this situation.
   $x \le 30$

Vocabulary and Concept Check

1. A closed circle would be used because −12 is a solution.
2. k is at least −3; $k \ge -3$; $k \le -3$
3. no; $x < 5$ is all values of x less than 5. $5 < x$ is all values of x greater than 5.

Practice and Problem Solving

4. $x > 12$; all values of x greater than 12
5. $x \le -4$; all values of x less than or equal to −4
6. $y \le -8$
7. $w + 2.3 > 18$
8. $-4t \ge -\frac{2}{5}$
9. $b - 4.2 < -7.5$
10. The inequality symbol is reversed; $2x \le -24$
11. yes
12. no
13. no
14. yes
15. yes
16. no
17. 

18. 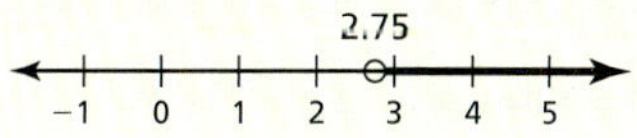

19. 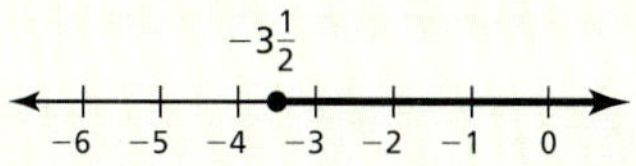

20. 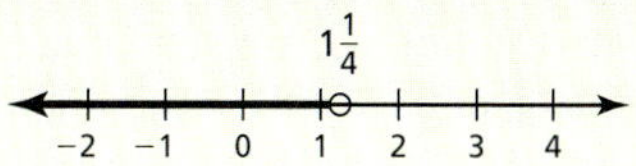

21. $p \ge 53$

Practice and Problem Solving

22. no **23.** yes

24. no **25.** yes

26. **a.** $1.25x > 35$

b. yes; It costs \$56.25 for 45 trips, which is more than the \$35 monthly pass.

27. **a.** any value that is greater than -2

b. any value that is less than or equal to -2; $b \le -2$

c. They represent the entire set of real numbers; yes

28. See *Taking Math Deeper.*

Fair Game Review

29. $p = 11$ **30.** $w = -3.6$

31. $x = -7$ **32.** C

Mini-Assessment

1. A number a is at least 5. Write this word sentence as an inequality. $a \ge 5$
2. Four times a number b is at most -4.73. Write this word sentence as an inequality. $4b \le -4.73$
3. Tell whether -2 is a solution of the inequality $6g - 14 > -21$. not a solution
4. Graph $p \le -1.7$ on a number line.

Taking Math Deeper

Exercise 28

This problem gives students an opportunity to use an inequality to describe a real-life situation.

1 Write an expression that represents the girth of the rectangular package.

$$\text{girth} = w + h + w + h$$
$$= 2w + 2h$$

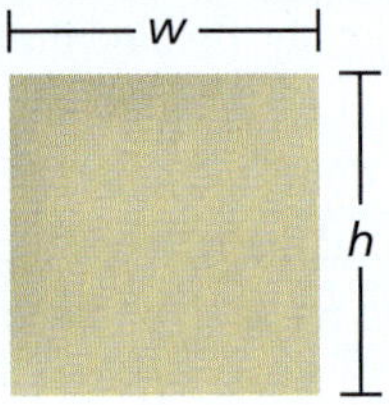

2 The combined length ℓ and girth of the package is represented by the expression $\ell + 2w + 2h$. The combined length and girth can be no more than 108 inches.

a. So, an inequality that represents the allowable dimensions is

$$\ell + 2w + 2h \le 108.$$

3 **b.** You can use the strategy *Guess, Check, and Revise* to find three different sets of allowable dimensions. Organize your results in a table. Below are some examples.

Dimensions (in.)	$\ell + 2w + 2h \overset{?}{\le} 108$	Volume (in.^{3})
$\ell = 12, w = 5, h = 5$	$32 \le 108$ ✓	300
$\ell = 20, w = 8, h = 12$	$60 \le 108$ ✓	1920
$\ell = 30, w = 20, h = 19$	$108 \le 108$ ✓	11,400

Project

Research the pricing offered by the postal service and delivery companies in your area. What are some of their requirements for shipping packages? What is their pricing based on?

Reteaching and Enrichment Strategies

If students need help. . .	If students got it. . .
Resources by Chapter • Practice A and Practice B • Puzzle Time Record and Practice Journal Practice Differentiating the Lesson Lesson Tutorials Skills Review Handbook	Resources by Chapter • Enrichment and Extension • Technology Connection Start the next section

Tell whether the given value is a solution of the inequality.

22. $4k < k + 8;\ k = 3$

23. $\frac{w}{3} \geq w - 12;\ w = 15$

24. $7 - 2y > 3y + 13;\ y = -1$

25. $\frac{3}{4}b - 2 \leq 2b + 8;\ b = -4$

26. **MODELING** A subway ride for a student costs \$1.25. A monthly pass costs \$35.

a. Write an inequality that represents the number of times you must ride the subway for the monthly pass to be a better deal.

b. You ride the subway about 45 times per month. Should you buy the monthly pass? Explain.

27. **LOGIC** Consider the inequality $b > -2$.

a. Describe the values of b that are solutions of the inequality.

b. Describe the values of b that are *not* solutions of the inequality. Write an inequality for these values.

c. What do all the values in parts (a) and (b) represent? Is this true for any inequality?

28. Critical Thinking A postal service says that a rectangular package can have a maximum combined length and *girth* of 108 inches. The girth of a package is the distance around the perimeter of a face that does not include the length.

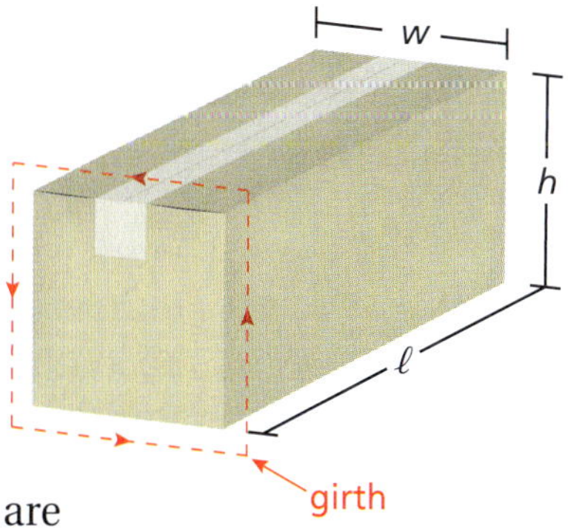

a. Write an inequality that represents the allowable dimensions for the package.

b. Find three different sets of allowable dimensions that are reasonable for the package. Find the volume of each package.

Fair Game Review What you learned in previous grades & lessons

Solve the equation. Check your solution. *(Section 1.1)*

29. $p - 8 = 3$

30. $8.7 + w = 5.1$

31. $x - 2 = -9$

32. **MULTIPLE CHOICE** Which expression has a value less than -5? *(Skills Review Handbook)*

Ⓐ $5 + 8$ Ⓑ $-9 + 5$ Ⓒ $1 + (-8)$ Ⓓ $7 + (-2)$

11.2 Solving Inequalities Using Addition or Subtraction

Essential Question How can you use addition or subtraction to solve an inequality?

1 ACTIVITY: Writing an Inequality

Work with a partner. Members of the Boy Scouts must be less than 18 years old. In 4 years, your friend will still be eligible to be a scout.

a. Which of the following represents your friend's situation? What does x represent? Explain your reasoning.

$x + 4 > 18$ $x + 4 < 18$

$x + 4 \geq 18$ $x + 4 \leq 18$

b. Graph the possible ages of your friend on a number line. Explain how you decided what to graph.

2 ACTIVITY: Writing an Inequality

Work with a partner. Supercooling is the process of lowering the temperature of a liquid or a gas below its freezing point without it becoming a solid. Water can be supercooled to 86°F below its normal freezing point (32°F) and still not freeze.

Inequalities
In this lesson, you will
- solve inequalities using addition or subtraction.
- solve real-life problems.

a. Let x represent the temperature of water. Which inequality represents the temperature at which water can be a liquid or a gas? Explain your reasoning.

$x - 32 > -86$ $x - 32 < -86$

$x - 32 \geq -86$ $x - 32 \leq -86$

b. On a number line, graph the possible temperatures at which water can be a liquid or a gas. Explain how you decided what to graph.

Laurie's Notes

Introduction

Applying Mathematical Practices

- **Model with Mathematics:** Mathematically proficient students are able to make connections between the context of a problem and the graph of the solution. In checking solutions, students verify that solutions make sense (are reasonable) in the context of the problem.

Motivate

- Ask students to name the seven continents. Then have students work with partners to guess a ranked order for the continents based on the coldest temperature recorded on each continent.
- When students finish ranking the continents, have them guess the actual coldest temperature for each continent.
- Tell students that you will return to this data later in the class.

Activity Notes

Activity 1

- Ask students whether any have belonged to organizations such as scouts, 4-H, etc. Students may be aware of age restrictions.
- Partners work together to answer each part. In graphing possible ages, students may intuitively solve the inequality as they would an equation. This process is used in Activity 3.

? "What does x represent?" your friend's age

- **Attend to Precision:** Be sure that students accurately define the variables. The variable x represents your friend's *current* age.

? "What does $x + 4$ represent?" your friend's age in 4 years

- **Model with Mathematics:** In discussing their graphs, students should be able to explain how they chose the domain relative to the limitations stated in the problem and age restrictions in scouting. (The minimum age for a scout is 10.)

Activity 2

- To help students comprehend the problem, discuss the background information for this activity. University of Utah chemists have shown that water does not necessarily have to freeze until its temperature drops to −55°F. This is 87 degrees Fahrenheit colder than the well-known standard of 32°F.
- When students finish, have volunteers share their results to each part.

? "What does the expression $x - 32$ represent?" the number of degrees Fahrenheit below the standard freezing point of 32°F that the water temperature can reach without freezing

- Interpret the graph. It represents the temperatures at which water *can* be a liquid or gas. All numbers not included in the graph are Fahrenheit temperatures at which water must be ice.

What Your Students Will Learn

- Solve inequalities using the Addition Property of Inequality or the Subtraction Property of Inequality.

Previous Learning

Students should know how to graph inequalities, solve single variable equations, and solve single variable inequalities using whole numbers.

Technology for the Teacher

Dynamic Classroom

Lesson Plans
Complete Materials List

11.2 Record and Practice Journal

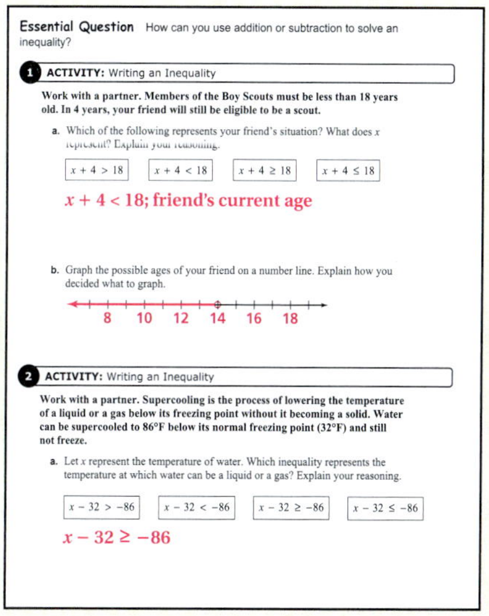

Essential Question How can you use addition or subtraction to solve an inequality?

1 ACTIVITY: Writing an Inequality

Work with a partner. Members of the Boy Scouts must be less than 18 years old. In 4 years, your friend will still be eligible to be a scout.

a. Which of the following represents your friend's situation? What does x represent? Explain your reasoning.

$x + 4 > 18$ $x + 4 < 18$ $x + 4 \geq 18$ $x + 4 \leq 18$

$x + 4 < 18$; friend's current age

b. Graph the possible ages of your friend on a number line. Explain how you decided what to graph.

8 10 12 14 16 18

2 ACTIVITY: Writing an Inequality

Work with a partner. Supercooling is the process of lowering the temperature of a liquid or a gas below its freezing point without it becoming a solid. Water can be supercooled to 86°F below its normal freezing point (32°F) and still not freeze.

a. Let x represent the temperature of water. Which inequality represents the temperature at which water can be a liquid or a gas? Explain your reasoning.

$x - 32 > -86$ $x - 32 < -86$ $x - 32 \geq -86$ $x - 32 \leq -86$

$x - 32 \geq -86$

English Language Learners

Vocabulary

It is important that English learners understand the difference between *is less than* and *is less than or equal to,* as well as *is greater than* and *is greater than or equal to.* Give each student a card with one of the numbers $-10, -9, -8, -7, -6, \dots, 10$. (Include more numbers if your class is larger.) Tell students to stand up if their number *is less than* (say a number), and then *is less than or equal to* that same number. The class should discuss the difference between the two. This can be repeated with *is greater than* and *is greater than or equal to.*

11.2 Record and Practice Journal

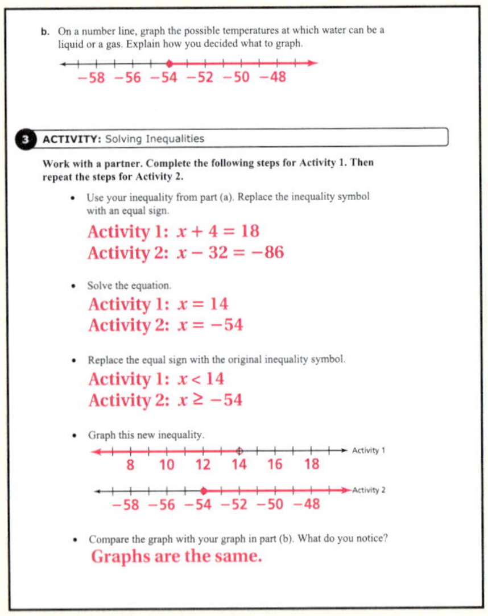

b. On a number line, graph the possible temperatures at which water can be a liquid or a gas. Explain how you decided what to graph.

3 ACTIVITY: Solving Inequalities

Work with a partner. Complete the following steps for Activity 1. Then repeat the steps for Activity 2.

- Use your inequality from part (a). Replace the inequality symbol with an equal sign.
  Activity 1: $x + 4 = 18$
  Activity 2: $x - 32 = -86$
- Solve the equation.
  Activity 1: $x = 14$
  Activity 2: $x = -54$
- Replace the equal sign with the original inequality symbol.
  Activity 1: $x < 14$
  Activity 2: $x \geq -54$
- Graph this new inequality.

- Compare the graph with your graph in part (b). What do you notice?
  Graphs are the same.

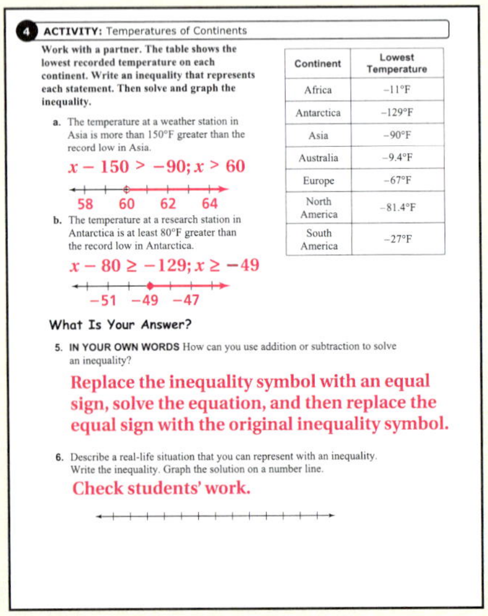

4 ACTIVITY: Temperatures of Continents

Work with a partner. The table shows the lowest recorded temperature on each continent. Write an inequality that represents each statement. Then solve and graph the inequality.

Continent	Lowest Temperature
Africa	−11°F
Antarctica	−129°F
Asia	−90°F
Australia	−9.4°F
Europe	−67°F
North America	−81.4°F
South America	−27°F

a. The temperature at a weather station in Asia is more than 150°F greater than the record low in Asia.

$x - 150 > -90; x > 60$

b. The temperature at a research station in Antarctica is at least 80°F greater than the record low in Antarctica.

$x - 80 \geq -129; x \geq -49$

What Is Your Answer?

5. **IN YOUR OWN WORDS** How can you use addition or subtraction to solve an inequality?
   Replace the inequality symbol with an equal sign, solve the equation, and then replace the equal sign with the original inequality symbol.
6. Describe a real-life situation that you can represent with an inequality. Write the inequality. Graph the solution on a number line.
   Check students' work.

Laurie's Notes

Activity 3

- Students have already graphed the solution, and they have prior experience in solving equations. This activity connects these two skills.
- ? "How do you think addition and subtraction inequalities are solved?" Listen for the same techniques used in solving equations.
- Check that students perform the steps correctly.

Activity 4

- Students can now check their rankings in the Motivate activity and how well they guessed the coldest recorded temperature for each continent.
- Students will use the data in the table to write the two inequalities in this activity.
- If students have trouble writing the inequalities, encourage them to reread the words and circle key words that are helpful in translating the words to symbols.
- When students finish, ask volunteers to share and discuss their work for each part.
- ? "Does the graph help you to consider the reasonableness of your answer?" Answers will vary.

What Is Your Answer?

- In Question 5, some students may say to replace the inequality symbol with an equal sign, solve the equation, and then replace the equation with the original inequality symbol.
- ? "Is it necessary to replace the inequality symbol with an equal sign to solve an inequality?" Answers will vary. This is an opportunity to discuss the Addition and Subtraction Properties of Inequality.

Closure

- Write, solve, and graph an inequality to describe your sister's height now. A ride at the amusement park has a maximum height limit of 48 inches. When your sister grows 3 inches, she will still be able to go on the ride.
  $x + 3 \leq 48; x \leq 45;$

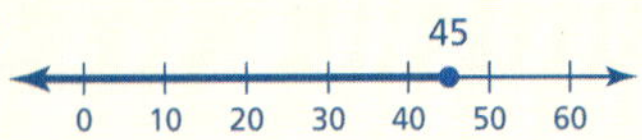

3 ACTIVITY: Solving Inequalities

Math Practice

Interpret Results

What does the solution of the inequality represent?

Work with a partner. Complete the following steps for Activity 1. Then repeat the steps for Activity 2.

- Use your inequality from part (a). Replace the inequality symbol with an equal sign.
- Solve the equation.
- Replace the equal sign with the original inequality symbol.
- Graph this new inequality.
- Compare the graph with your graph in part (b). What do you notice?

4 ACTIVITY: Temperatures of Continents

Work with a partner. The table shows the lowest recorded temperature on each continent. Write an inequality that represents each statement. Then solve and graph the inequality.

a. The temperature at a weather station in Asia is more than 150°F greater than the record low in Asia.

b. The temperature at a research station in Antarctica is at least 80°F greater than the record low in Antarctica.

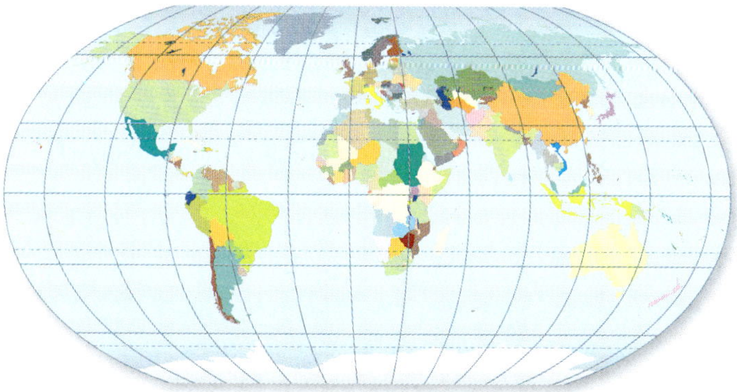

Continent	Lowest Temperature
Africa	−11°F
Antarctica	−129°F
Asia	−90°F
Australia	−9.4°F
Europe	−67°F
North America	−81.4°F
South America	−27°F

What Is Your Answer?

5. **IN YOUR OWN WORDS** How can you use addition or subtraction to solve an inequality?

6. Describe a real-life situation that you can represent with an inequality. Write the inequality. Graph the solution on a number line.

Practice Use what you learned about solving inequalities to complete Exercises 3–5 on page 474.

11.2 Lesson

Check It Out
Lesson Tutorials
BigIdeasMath.com

Key Ideas

Study Tip

You can solve inequalities in the same way you solve equations. Use inverse operations to get the variable by itself.

Addition Property of Inequality

Words When you add the same number to each side of an inequality, the inequality remains true.

Numbers

$$\begin{array}{rcr} -4 & < & 3 \\ \underline{+2} & & \underline{+2} \\ -2 & < & 5 \end{array}$$

Algebra If $a < b$, then $a + c < b + c$.

If $a > b$, then $a + c > b + c$.

Subtraction Property of Inequality

Words When you subtract the same number from each side of an inequality, the inequality remains true.

Numbers

$$\begin{array}{rcr} -2 & < & 2 \\ \underline{-3} & & \underline{-3} \\ -5 & < & -1 \end{array}$$

Algebra If $a < b$, then $a - c < b - c$.

If $a > b$, then $a - c > b - c$.

These properties are also true for $\leq$ and $\geq$.

EXAMPLE 1 Solving an Inequality Using Addition

Solve $x - 5 < -3$. Graph the solution.

$$\begin{array}{rcrl} x - 5 & < & -3 & \text{Write the inequality.} \\ \underline{+5} & & \underline{+5} & \text{Addition Property of Inequality} \\ x & < & 2 & \text{Simplify.} \end{array}$$

Undo the subtraction. → $+5$ $+5$

The solution is $x < 2$.

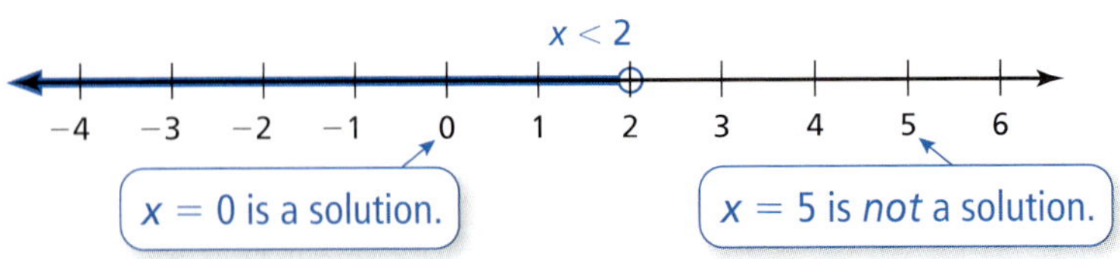

Check:

$x = 0$: $0 - 5 \overset{?}{<} -3$

$-5 < -3$ ✓

$x = 5$: $5 - 5 \overset{?}{<} -3$

$0 \not< -3$ ✗

On Your Own

Solve the inequality. Graph the solution.

1. $y - 6 > -7$

2. $b - 3.8 \leq 1.7$

3. $-\frac{1}{2} > z - \frac{1}{4}$

Laurie's Notes

Introduction

Connect

- **Yesterday:** Yesterday students investigated solving addition and subtraction inequalities.
- **Today:** Students will use Properties of Inequality to solve addition and subtraction inequalities.

Motivate

- It's "Did you know?" time for the students. The application problem involves NASA astronauts. Share some astronaut food facts with the students.
- To reduce launch weight during space shuttle flights, water was removed from certain foods and beverages. To eat these items, astronauts rehydrated them with water produced by the shuttle fuel cells. Rehydratable foods include breakfast cereals with milk, scrambled eggs, casseroles, and appetizers such as shrimp cocktail.

Lesson Notes

Key Ideas

- Write the Key Ideas. These properties should look familiar, as they are similar to the Addition and Subtraction Properties of Equality that students have used in solving equations.
- **Teaching Tip:** Summarize these two properties in the following way: George is older than Martha. In two years, George will still be older than Martha.

George's age > Martha's age	If $a > b$,
George's age + 2 > Martha's age + 2	then $a + c > b + c$.

Two years ago, George was older than Martha.

George's age > Martha's age	If $a > b$,
George's age − 2 > Martha's age − 2	then $a - c > b - c$.

Example 1

? Write the problem. "How do you isolate the variable, meaning get x by itself? Add 5 to each side of the inequality.

- Adding 5 is the inverse operation of subtracting 5.
- Solve, graph, and check.
- Take time to check a solution point. 0 is generally an easy value to work with.
- Also check a point that is not a solution on the graph, and verify that it does not satisfy the inequality.

On Your Own

- These problems integrate review of fraction and decimal operations.

Goal Today's lesson is solving inequalities using addition or subtraction.

Technology for the Teacher

Dynamic Classroom

Lesson Tutorials
Lesson Plans
Answer Presentation Tool

Extra Example 1

Solve $x - 2 \geq -4$. Graph the solution.
$x \geq -2$;

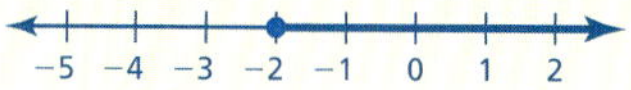

On Your Own

1. $y > -1$;

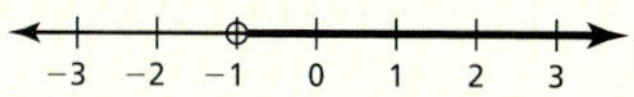

2. $b \leq 5.5$;

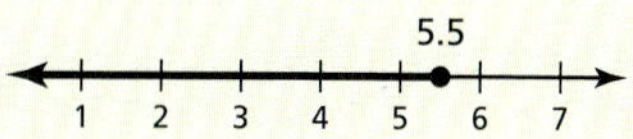

3. $-\frac{1}{4} > z$, or $z < -\frac{1}{4}$;

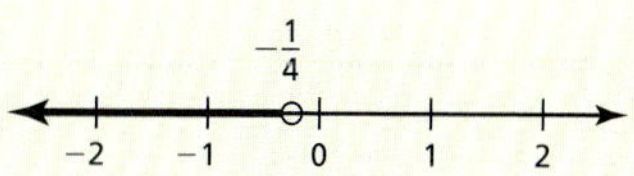

Extra Example 2

Solve $14 > x + 8$. Graph the solution.

$x < 6$;

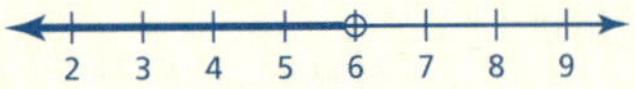

On Your Own

4. $w \le -3$;

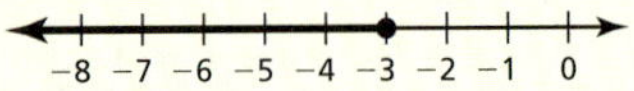

5. $2.5 \ge d$, or $d \le 2.5$;

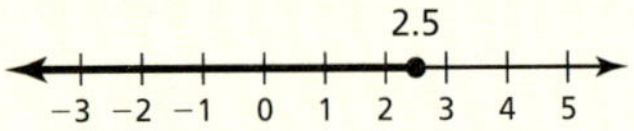

6. $x > \frac{3}{4}$;

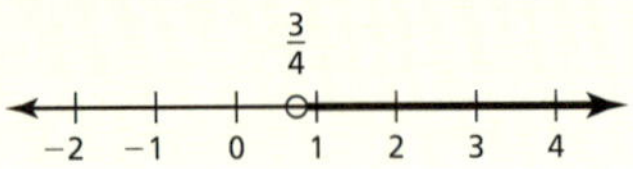

Differentiated Instruction

Auditory

To help auditory learners comprehend the verbal model in Example 3, point at each part of the verbal model as you explain or ask what it represents in the context of the problem.

Extra Example 3

A person can be no taller than $6\frac{5}{12}$ feet to become a fighter pilot for the United States Air Force. Your brother is 6 feet 3 inches tall. Write and solve an inequality that represents how much your brother can grow and still meet the requirement.

$6\frac{1}{4} + h \le 6\frac{5}{12}$; $h \le \frac{1}{6}$; Your brother can grow at most $\frac{1}{6}$ foot, or 2 inches.

On Your Own

7. $5.25 + h \le 6.25$; $h \le 1$; Your cousin can grow no more than 1 foot, or 12 inches.

Laurie's Notes

Example 2

? Write the inequality and ask, "How do you isolate the variable x?" Subtract 14 from each side of the inequality.

- **Common Error:** When students graph the solution they may look at the inequality symbol $\le$ and shade to the left of -1. It helps to rewrite the solution $-1 \le x$ as $x \ge -1$.

On Your Own

- **Think-Pair-Share:** Students should read each question independently and then work in pairs to answer the questions. When they have answered the questions, the pair should compare their answers with another group and discuss any discrepancies.
- Note that the variable is on the right side of the inequality in Question 5.

Example 3

- Ask a volunteer to read the problem.

? "The height limit is written as a decimal number of feet. Your friend's height is written in feet and inches. Can you work with the measurements as they are?" no; One of the forms must be changed to the other form.

? "A height of 6.25 feet is how many inches? Explain."

75 inches; $6.25 \text{ ft} = 6\frac{1}{4} \text{ ft} = 6 \text{ ft} + \frac{1}{4} \text{ ft} = 72 \text{ in.} + 3 \text{ in.} = 75 \text{ in.}$

? "A height of 5 feet 9 inches is how many feet in decimal form? Explain."

5.75 feet; Because $9 \text{ in.} = \frac{3}{4} \text{ ft}$, $5 \text{ ft } 9 \text{ in.} = 5\frac{3}{4} \text{ ft} = 5.75 \text{ ft}$.

- **. Model with Mathematics:** Writing verbal models is an important step in helping students gain confidence in translating and setting up equations and inequalities. Color code the verbal model, if possible.
- **FYI:** The problem is solved using decimal heights, but it could be solved using inches as well.
- Continue to solve the problem as shown. Interpret the answer, $h \le 0.5$, in terms of a decimal number of feet and in terms of inches.

On Your Own

- Remind students to write a verbal model and then define variables before writing the inequality.

Closure

- **Exit ticket:** Solve and graph the solution for $-5.2 + x > 14.8$

$x > 20$;

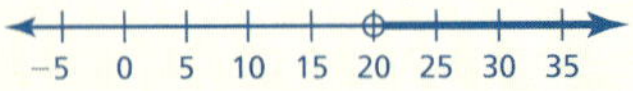

EXAMPLE 2 Solving an Inequality Using Subtraction

Solve $13 \le x + 14$. Graph the solution.

$13 \le x + 14$	Write the inequality.
Undo the addition. → $-14 \quad\quad -14$	Subtraction Property of Inequality
$-1 \le x$	Simplify.

The solution is $x \ge -1$.

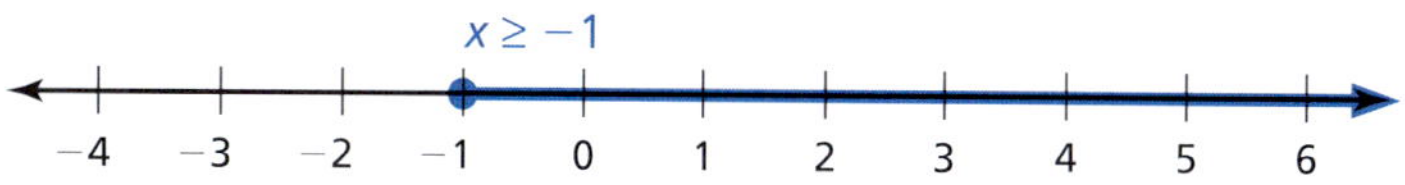

Reading

The inequality $-1 \le x$ is the same as $x \ge -1$.

On Your Own

Now You're Ready
Exercises 3–17

Solve the inequality. Graph the solution.

4. $w + 7 \le 4$ **5.** $12.5 \ge d + 10$ **6.** $x + \frac{3}{4} > 1\frac{1}{2}$

EXAMPLE 3 Real-Life Application

A person can be no taller than 6.25 feet to become an astronaut pilot for NASA. Your friend is 5 feet 9 inches tall. Write and solve an inequality that represents how much your friend can grow and still meet the requirement.

Words	Current height	plus	amount your friend can grow	is no more than	the height limit.
Variable	Let h be the possible amounts your friend can grow.				
Inequality	5.75	$+$	h	$\le$	6.25

5 ft 9 in. = 60 + 9 = 69 in.

$69 \text{ in.} \times \frac{1 \text{ ft}}{12 \text{ in.}} = 5.75 \text{ ft}$

$5.75 + h \le 6.25$	Write the inequality.
$-5.75 \quad\quad -5.75$	Subtraction Property of Inequality
$h \le 0.5$	Simplify.

So, your friend can grow no more than 0.5 foot, or 6 inches.

On Your Own

7. Your cousin is 5 feet 3 inches tall. Write and solve an inequality that represents how much your cousin can grow and still meet the requirement.

11.2 Exercises

Check It Out
Help with Homework
BigIdeasMath.com

Vocabulary and Concept Check

1. **REASONING** Is the inequality $c + 3 > 5$ the same as $c > 5 - 3$? Explain.
2. **WHICH ONE DOESN'T BELONG?** Which inequality does *not* belong with the other three? Explain your reasoning.

$w + \frac{7}{4} > \frac{3}{4}$	$w - \frac{3}{4} > -\frac{7}{4}$	$w + \frac{7}{4} < \frac{3}{4}$	$\frac{3}{4} < w + \frac{7}{4}$

Practice and Problem Solving

Solve the inequality. Graph the solution.

1 2
3. $x + 7 \geq 18$
4. $a - 2 > 4$
5. $3 \leq 7 + g$
6. $8 + k \leq -3$
7. $-12 < y - 6$
8. $n - 4 < 5$
9. $t - 5 \leq -7$
10. $p + \frac{1}{4} \geq 2$
11. $\frac{2}{7} > b + \frac{5}{7}$
12. $z - 4.7 \geq -1.6$
13. $-9.1 < d - 6.3$
14. $\frac{8}{5} > s + \frac{12}{5}$
15. $-\frac{7}{8} \geq m - \frac{13}{8}$
16. $r + 0.2 < -0.7$
17. $h - 6 \leq -8.4$

ERROR ANALYSIS **Describe and correct the error in solving the inequality or graphing the solution of the inequality.**

18. ✗

$$\begin{aligned} x - 7 &> -2 \\ +7 &\quad +7 \\ x &> 9 \end{aligned}$$

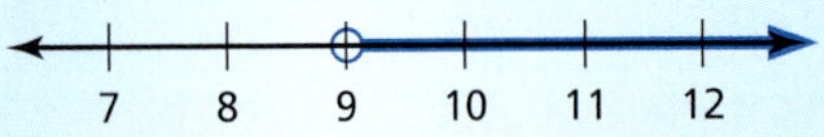

19. ✗

$$\begin{aligned} 8 &\leq x + 3 \\ -3 &\quad -3 \\ 5 &\leq x \end{aligned}$$

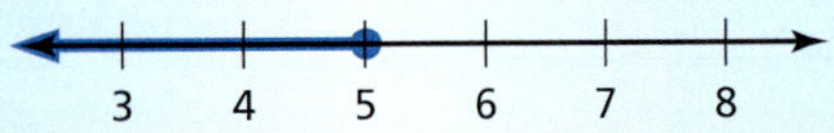

20. **AIRPLANE** A small airplane can hold 44 passengers. Fifteen passengers board the plane.
    a. Write and solve an inequality that represents the additional number of passengers that can board the plane.
    b. Can 30 more passengers board the plane? Explain.

Assignment Guide and Homework Check

Level	Assignment	Homework Check
Advanced	1–5, 6–28 even, 29–33	12, 14, 20, 22, 26

Common Errors

- **Exercises 3–17** Students may use the same operation rather than the inverse operation to isolate the variable. Remind students that when a number is added to the variable, subtract that number from each side. When a number is subtracted from the variable, add that number to each side.
- **Exercises 3–17** Students may shade the number line in the wrong direction when the variable is on the right side of the inequality symbol. Remind them to rewrite the inequality with the variable at the left or to check a value on each side of the boundary point.

11.2 Record and Practice Journal

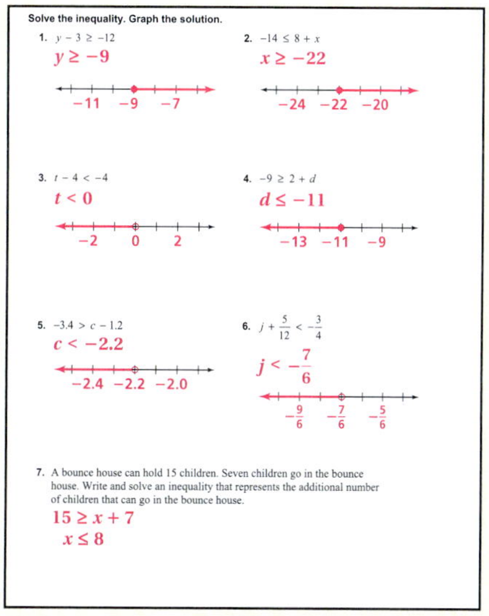
Solve the inequality. Graph the solution.

1. $y - 3 \ge -12$
$y \ge -9$

2. $-14 \le 8 + x$
$x \ge -22$

3. $t - 4 < -4$
$t < 0$

4. $-9 \ge 2 + d$
$d \le -11$

5. $-3.4 > c - 1.2$
$c < -2.2$

6. $j + \frac{5}{12} < -\frac{3}{4}$
$j < -\frac{7}{6}$

7. A bounce house can hold 15 children. Seven children go in the bounce house. Write and solve an inequality that represents the additional number of children that can go in the bounce house.
$15 \ge x + 7$
$x \le 8$

Vocabulary and Concept Check

1. Yes, because of the Subtraction Property of Inequality.
2. $w - \frac{3}{4} > -\frac{7}{4}$ does not belong with the other three because to solve this inequality, you use the Addition Property of Inequality. To solve the other three, you use the Subtraction Property of Inequality.

Practice and Problem Solving

3. $x \ge 11$;

4. $a > 6$;

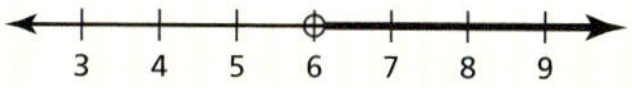

5. $-4 \le g$;

6. $k \le -11$;

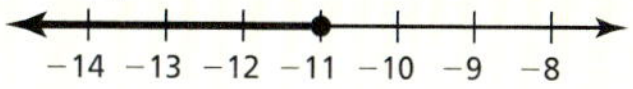

7. $-6 < y$;

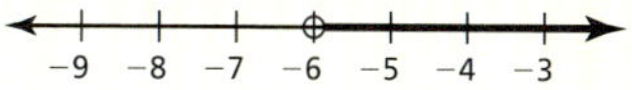

8. $n < 9$;

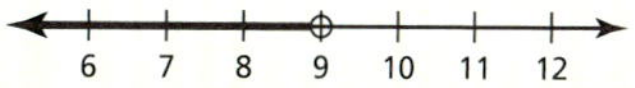

9. $t \le -2$;

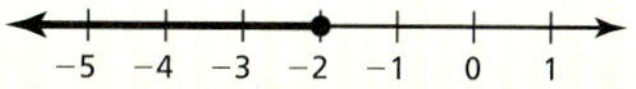

10. $p \ge 1\frac{3}{4}$;

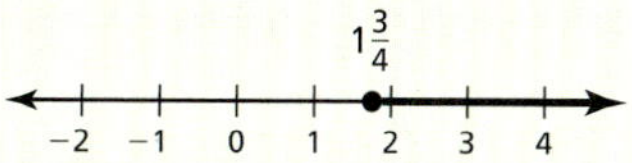

11. $-\frac{3}{7} > b$;

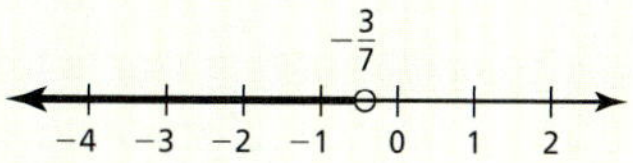

12–20. See Additional Answers.

21. $7 + 7 + x < 28$; $x < 14$ ft

Practice and Problem Solving

22. $x + 3 > 8$; $x > 5$ in.

23. $8 + 8 + 10 + 10 + x \leq 51$; $x \leq 15$ m

24. 4

25. $x - 3 \geq 5$; $x \geq 8$ ft

26. **a.** $x + 14 \leq 35$; $x \leq \$21$

b. It changes the number added to x. So, the inequality becomes $x + 9.8 \leq 35$ and you have more money left.

c. yes; The cost of the shirt and the pants is \$32.80, which is less than \$35.

27. See *Taking Math Deeper*.

28. -14

Fair Game Review

29. $x = 9$ **30.** $w = -27$

31. $b = -22$ **32.** $h = 80$

33. A

Mini-Assessment

1. Solve $x - 4 > 11$. Graph the solution.
$x > 15$;

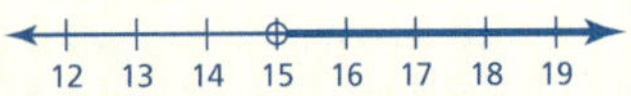

2. Solve $11 \leq w + 3.4$. Graph the solution. $w \geq 7.6$;

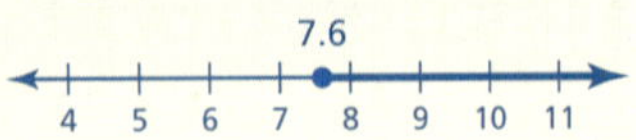

3. Solve $k - \frac{2}{5} < -\frac{4}{5}$. Graph the solution. $k < -\frac{2}{5}$;

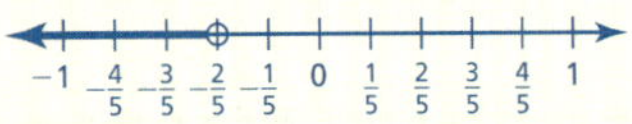

4. The school cafeteria seats 250 students. 203 students are already seated. Write and solve an inequality that represents the additional number of students that can be seated. $203 + s \leq 250$; $s \leq 47$; At most 47 more students can be seated.

Taking Math Deeper

Exercise 27

Some students may not have had experiences with an electrical circuit that overloads and triggers the circuit breaker. This problem is a nice opportunity to familiarize students with the fact that different appliances use different amounts of electricity.

1 Write an inequality.

Let x = amount (in watts) of additional electricity used.

1050 = amount (in watts) of electricity used by the portable heater.

2400 = amount (in watts) of electricity that overloads the circuit.

$x + 1050 < 2400$

2 Solve the inequality.

a. $x + 1050 < 2400$

$x < 1350$ watts

3 Answer the question.

b. Yes, there is more than one possibility.

You can also plug in the following pairs of items without overloading the circuit.

Item	Watts
Aquarium	200
Hair dryer	1200
Television	150
Vacuum cleaner	1100

aquarium and television: $200 + 150 = 350$ watts

aquarium and vacuum cleaner: $200 + 1100 = 1300$ watts

television and vacuum cleaner: $150 + 1100 = 1250$ watts

Project

It might be interesting for students to research other types of items and how much electricity they use. In this example, you can see that the item that generates heat uses a lot of electricity.

Reteaching and Enrichment Strategies

If students need help. . .	If students got it. . .
Resources by Chapter • Practice A and Practice B • Puzzle Time Record and Practice Journal Practice Differentiating the Lesson Lesson Tutorials Skills Review Handbook	Resources by Chapter • Enrichment and Extension • Technology Connection Start the next section

Write and solve an inequality that represents x.

21. The perimeter is less than 28 feet.

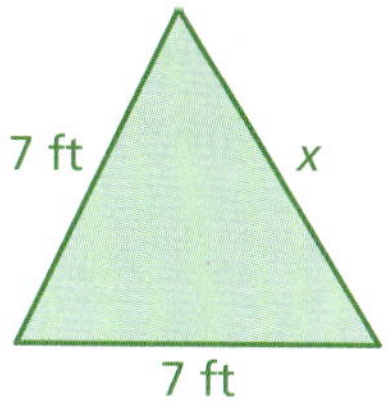

22. The base is greater than the height.

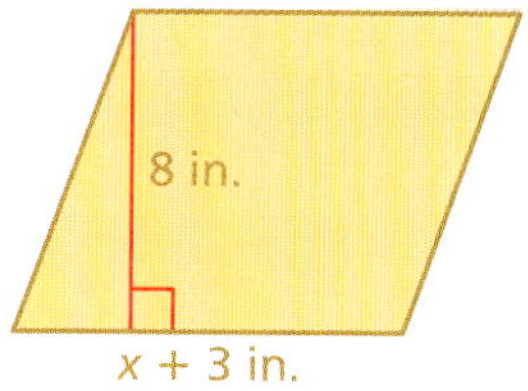

23. The perimeter is less than or equal to 51 meters.

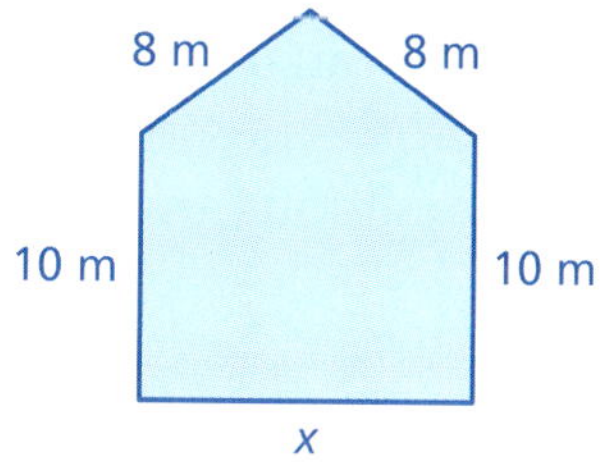

24. **REASONING** The solution of $d + s > -3$ is $d > -7$. What is the value of s?

25. **BIRDFEEDER** The hole for a birdfeeder post is 3 feet deep. The top of the post needs to be at least 5 feet above the ground. Write and solve an inequality that represents the required length of the post.

26. **SHOPPING** You can spend up to \$35 on a shopping trip.

a. You want to buy a shirt that costs \$14. Write and solve an inequality that represents the amount of money you will have left if you buy the shirt.

b. You notice that the shirt is on sale for 30% off. How does this change the inequality?

c. Do you have enough money to buy the shirt that is on sale and a pair of pants that costs \$23? Explain.

27. **POWER** A circuit overloads at 2400 watts of electricity. A portable heater that uses 1050 watts of electricity is plugged into the circuit.

a. Write and solve an inequality that represents the additional number of watts you can plug in without overloading the circuit.

b. In addition to the portable heater, what two other items in the table can you plug in at the same time without overloading the circuit? Is there more than one possibility? Explain.

Item	Watts
Aquarium	200
Hair dryer	1200
Television	150
Vacuum cleaner	1100

28. Number Sense The possible values of x are given by $x + 8 \le 6$. What is the greatest possible value of $7x$?

Fair Game Review What you learned in previous grades & lessons

Solve the equation. Check your solution. *(Section 1.1)*

29. $4x = 36$

30. $\frac{w}{3} = -9$

31. $-2b = 44$

32. $60 = \frac{3}{4}h$

33. **MULTIPLE CHOICE** Which fraction is equivalent to -2.4? *(Skills Review Handbook)*

Ⓐ $-\frac{12}{5}$ Ⓑ $-\frac{51}{25}$ Ⓒ $-\frac{8}{5}$ Ⓓ $-\frac{6}{25}$

11 Study Help

Check It Out
Graphic Organizer
BigIdeasMath.com

You can use a **Y chart** to compare two topics. List differences in the branches and similarities in the base of the Y. Here is an example of a Y chart that compares solving equations and solving inequalities.

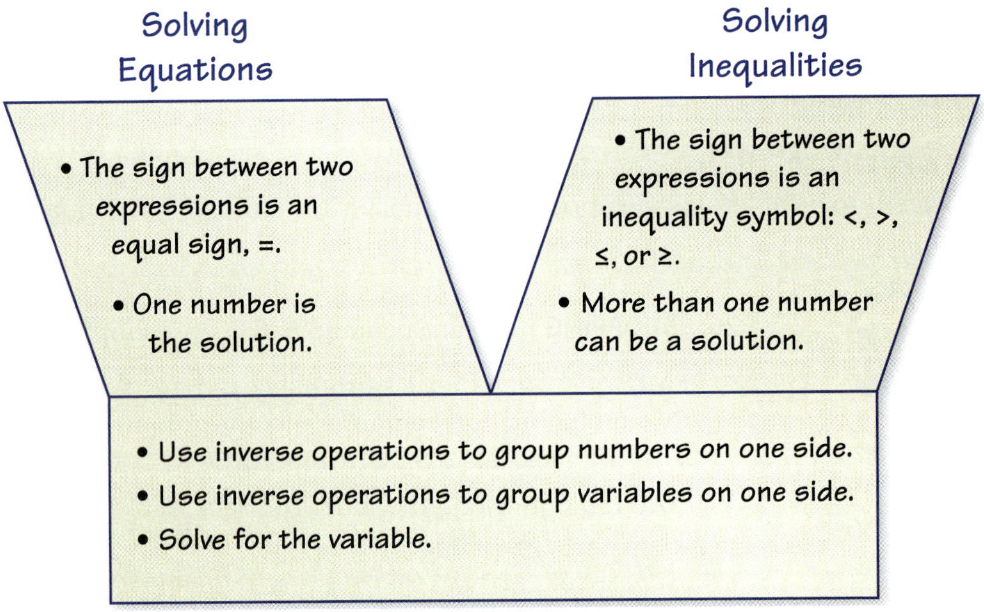

On Your Own

Make Y charts to help you study and compare these topics.

1. writing equations and writing inequalities
2. graphing the solution of an equation and graphing the solution of an inequality
3. graphing inequalities that use > and graphing inequalities that use <
4. graphing inequalities that use > or < and graphing inequalities that use ≥ or ≤
5. solving inequalities using addition and solving inequalities using subtraction

"Hey Descartes, do you have any suggestions for the Y chart I am making?"

After you complete this chapter, make Y charts for the following topics.

6. solving inequalities using multiplication and solving inequalities using division
7. solving two-step equations and solving two-step inequalities
8. Pick two other topics that you studied earlier in this course and make a Y chart to compare them.

Sample Answers

1.

Writing equations
- The sign between two expressions is an equal sign, =.
- One number is the solution.

Writing inequalities
- The sign between two expressions is an inequality symbol: <, >, ≤ or ≥.
- More than one number can be a solution.

Both
- Write one expression on the left and one expression on the right.
- Look for key phrases to determine which operation(s) to use: +, −, ×, or ÷.
- Look for key phrases to determine where to place the equal or inequality sign.

2.

Graphing the solution of an equation
- A solution is represented by a closed circle, ●.
- One number is the solution.

Graphing the solution of an inequality
- The endpoint of the graph can be an open circle, ○, or a closed circle, ●.
- An arrow pointing to the left or the right shows that the graph continues in that direction.
- More than one number can be a solution.

Both
- Solve for the variable.
- Graph the solution on a number line.

3.

Graphing inequalities that use >
- For x greater than a number, use an arrow pointing to the right.

Graphing inequalities that use <
- For x less than a number, use an arrow pointing to the left.

Both
- Use a number line to graph the solution.
- Use an open circle, ○.
- Use an arrow to show that the graph continues.
- Solution usually includes many numbers.

4–5. Available at *BigIdeasMath.com*.

List of Organizers

Available at *BigIdeasMath.com*

Comparison Chart
Concept Circle
Definition (Idea) and Example Chart
Example and Non-Example Chart
Formula Triangle
Four Square
Information Frame
Information Wheel
Notetaking Organizer
Process Diagram
Summary Triangle
Word Magnet
Y Chart

About this Organizer

A **Y Chart** can be used to compare two topics. Students list differences between the two topics in the branches of the Y and similarities in the base of the Y. As with an example and non-example chart, a Y chart serves as a good tool for assessing students' knowledge of a pair of topics that have subtle but important differences. You can include blank Y charts on tests or quizzes for this purpose.

Technology for the Teacher

Editable Graphic Organizer

Answers

1. $y + 2 > -5$
2. $s - 2.4 \geq 8$
3. yes
4. no
5. 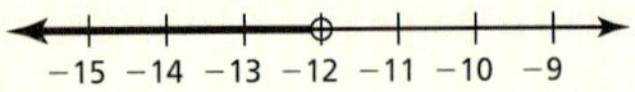
6. 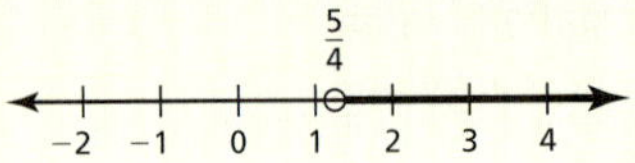
7. 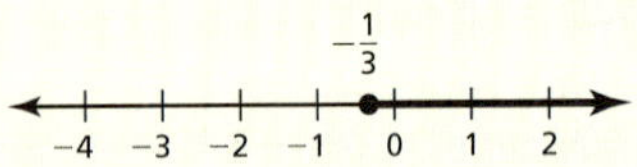
8. 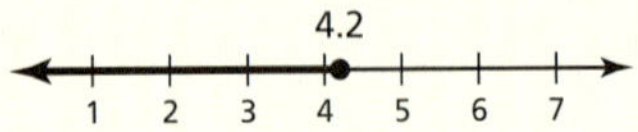
9. $n \leq -8$;

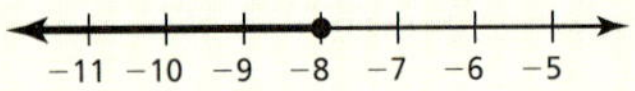

10. $t > \frac{9}{7}$;

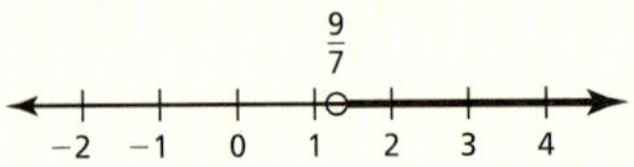

11. $-\frac{7}{4} \geq w$;

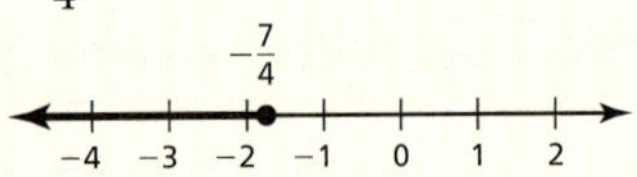

12. $y < -0.8$;

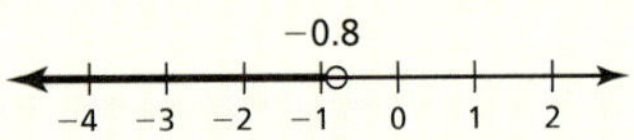

13. $s \geq 1.5$
14. **a.** $j \geq 2$;

$p \geq 25$;

$n \geq 10$;

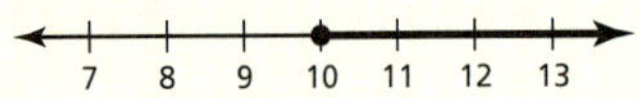

b. yes; $2.5 \geq 2$; $30 \geq 25$; $20 \geq 10$

15. $-\frac{7}{2} + x > -\frac{3}{2}$; $x > 2$

You must move more than 2 feet in the positive direction.

Alternative Quiz Ideas

100% Quiz
Error Notebook
Group Quiz
Homework Quiz
Math Log
Notebook Quiz
Partner Quiz
Pass the Paper

Homework Quiz

A homework notebook provides an opportunity for teachers to check that students are doing their homework regularly. Students keep homework in their notebooks. They should be told to record the page number, problem number, and copy the problem exactly in their homework notebooks. Each day the teacher walks around and visually checks that homework is completed. Periodically, without advance notice, the teacher tells the students to put everything away except their homework notebooks.

Questions are from students' homework.

1. What are the answers to Exercises 1, 5, 10, and 21 on page 128?
2. What are the answers to Exercises 23 and 27 on page 129?
3. What are the answers to Exercises 3–5 on page 134?
4. What is the answer to Exercise 26 on page 135?

Reteaching and Enrichment Strategies

If students need help...	If students got it...
Resources by Chapter • Practice A and Practice B • Puzzle Time Lesson Tutorials *BigIdeasMath.com*	Resources by Chapter • Enrichment and Extension • Technology Connection Game Closet at *BigIdeasMath.com* Start the next section

11.1–11.2 Quiz

Write the word sentence as an inequality. *(Section 11.1)*

1. A number y plus 2 is greater than -5.

2. A number s minus 2.4 is at least 8.

Tell whether the given value is a solution of the inequality. *(Section 11.1)*

3. $8p < -3;\ p = -2$

4. $z + 2 > -4;\ z = -8$

Graph the inequality on a number line. *(Section 11.1)*

5. $x < -12$

6. $v > \frac{5}{4}$

7. $b \geq -\frac{1}{3}$

8. $q \leq 4.2$

Solve the inequality. Graph the solution. *(Section 11.2)*

9. $n + 2 \leq -6$

10. $t - \frac{3}{7} > \frac{6}{7}$

11. $-\frac{3}{4} \geq w + 1$

12. $y - 2.6 < -3.4$

13. **STUDYING** You plan to study at least 1.5 hours for a geography test. Write an inequality that represents this situation. *(Section 11.1)*

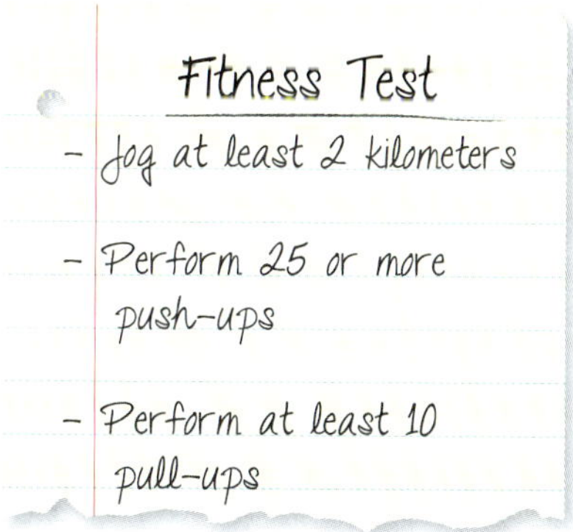

14. **FITNESS TEST** The three requirements to pass a fitness test are shown. *(Section 11.1)*

a. Write and graph three inequalities that represent the requirements.

b. You can jog 2500 meters, perform 30 push-ups, and perform 20 pull-ups. Do you satisfy the requirements of the test? Explain.

15. **NUMBER LINE** Use tape on the floor to make the number line shown. All units are in feet. You are standing at $-\frac{7}{2}$. You want to move to a number greater than $-\frac{3}{2}$. Write and solve an inequality that represents the distance you must move. *(Section 11.2)*

11.3 Solving Inequalities Using Multiplication or Division

Essential Question How can you use multiplication or division to solve an inequality?

1 ACTIVITY: Using a Table to Solve an Inequality

Work with a partner.

- **Copy and complete the table.**
- **Decide which graph represents the solution of the inequality.**
- **Write the solution of the inequality.**

a. $4x > 12$

x	-1	0	1	2	3	4	5
$4x$							
$4x \overset{?}{>} 12$							

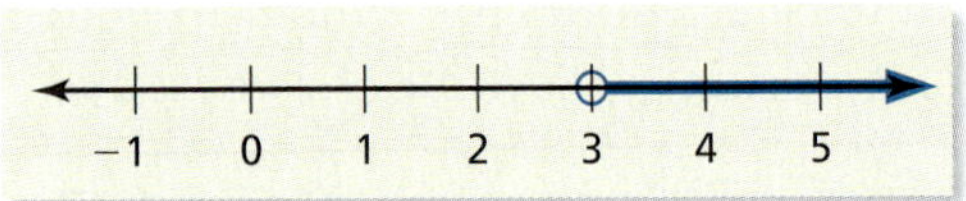

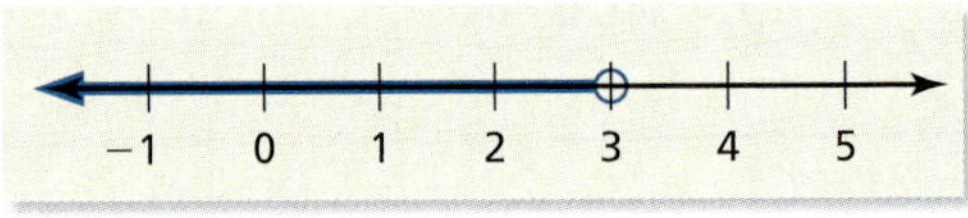

b. $-3x \le 9$

x	-5	-4	-3	-2	-1	0	1
$-3x$							
$-3x \overset{?}{\le} 9$							

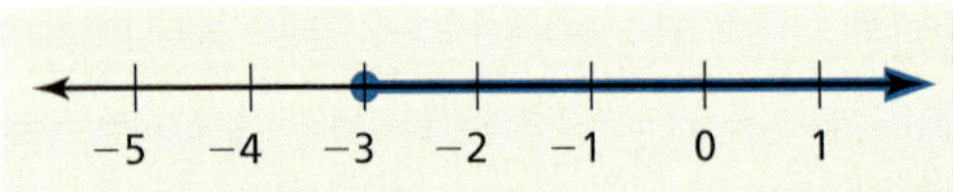

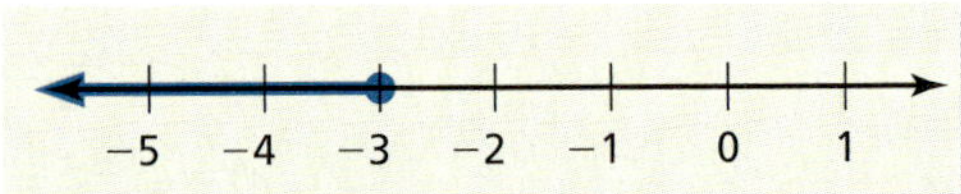

Inequalities

In this lesson, you will

- solve inequalities using multiplication or division.
- solve real-life problems.

2 ACTIVITY: Solving an Inequality

Work with a partner.

a. Solve $-3x \le 9$ by adding $3x$ to each side of the inequality first. Then solve the resulting inequality.

b. Compare the solution in part (a) with the solution in Activity 1(b).

Laurie's Notes

Introduction

Applying Mathematical Practices

- **Construct Viable Arguments and Critique the Reasoning of Others:** Mathematically proficient students use the results of investigations to observe patterns and make conjectures. Although student conjectures are not always correct, encourage them to try. Prior to this activity, students tend to think that inequalities involving multiplication and division will be solved in the same fashion as the related equations. After completing the four activities, students may revise their thoughts.

Motivate

- Ask a series of questions and record the students' solutions.
  - ? "What integers are solutions of $x > 4$?" $5, 6, 7, \ldots$
  - ? "What integers are solutions of $-x > 4$, meaning what numbers have an opposite that is greater than 4?" $-5, -6, -7, \ldots$
  - ? "What integers are solutions of $x < -4$?" $-5, -6, -7, \ldots$
- Leave these 3 problems on the board and refer to them at the end of class.

Activity Notes

Activity 1

- Explain that students are to evaluate one side of the inequality for each value of x and then decide whether the inequality is satisfied. This means, *is the value of x a solution of the inequality?* Students will write *yes* or *no* in the third row of the table to indicate whether the value is a solution or not.
- Using the information in the table, students decide which graph represents the solution. Finally, they write the solution.

? Discuss the results with your students.
  - "What did you find as the solution for part (a)?" $x > 3$
  - "Is this what you would have expected?" Likely, they will say yes.
  - "What did you find as the solution for part (b)?" $x \geq -3$
  - "Is this what you would have expected?" Likely, they will say no.
- Do not tell students a rule at this point. Simply say that perhaps they need to try a few more problems to help figure out what is going on.

Activity 2

- The directions may seem odd to students. It is not obvious to all students that adding $3x$ to each side of the inequality makes the left side 0.
- **Make Sense of Problems and Persevere in Solving Them:** Resist the urge to jump in too quickly to help students. Tell them to reread the directions carefully. Students should be able to combine what they know about solving multi-step equations with what they learned in Section 11.2 to solve the multi-step inequality.
- Ask a volunteer to share his or her work. Discuss how the process and solution compare to the process and solution in Activity 1 part (b).

What Your Students Will Learn

- Solve inequalities using Case 1 or Case 2 of the Multiplication and Division Properties of Inequality.

Previous Learning

Students should know how to solve equations using multiplication and division. Students should be able to evaluate expressions and decide whether a number is a solution of an inequality.

Technology for the Teacher

Dynamic Classroom

Lesson Plans
Complete Materials List

11.3 Record and Practice Journal

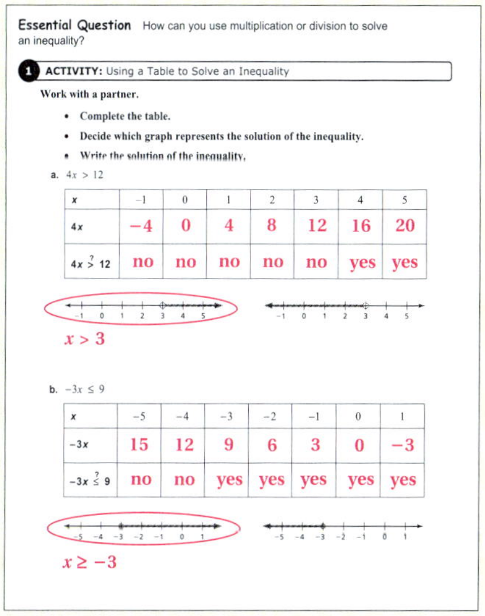

Essential Question How can you use multiplication or division to solve an inequality?

1 ACTIVITY: Using a Table to Solve an Inequality

Work with a partner.

- **Complete the table.**
- **Decide which graph represents the solution of the inequality.**
- **Write the solution of the inequality.**

a. $4x > 12$

x	−1	0	1	2	3	4	5
$4x$	−4	0	4	8	12	16	20
$4x \overset{?}{>} 12$	no	no	no	no	no	yes	yes

$x > 3$

b. $-3x \leq 9$

x	−5	−4	−3	−2	−1	0	1
$-3x$	15	12	9	6	3	0	−3
$-3x \overset{?}{\leq} 9$	no	no	yes	yes	yes	yes	yes

$x \geq -3$

English Language Learners

Pair Activity

Create index cards with problems similar to those in Activities 2 and 4. Pair English learners with English speakers and give 5 cards to each pair. Have students work together to solve the inequalities. When students have completed the problems, check their work and give them another set of cards.

11.3 Record and Practice Journal

2 ACTIVITY: Solving an Inequality

Work with a partner.

a. Solve $-3x \leq 9$ by adding $3x$ to each side of the inequality first. Then solve the resulting inequality.

$-3 \leq x$

b. Compare the solution in part (a) with the solution in Activity 1(b).

They are the same.

3 ACTIVITY: Using a Table to Solve an Inequality

Work with a partner.

- **Complete the table.**
- **Decide which graph represents the solution of the inequality.**
- **Write the solution of the inequality.**

a. $\frac{x}{3} < 1$

x	-1	0	1	2	3	4	5
$\frac{x}{3}$	$-\frac{1}{3}$	0	$\frac{1}{3}$	$\frac{2}{3}$	1	$\frac{4}{3}$	$\frac{5}{3}$
$\frac{x}{3} \overset{?}{<} 1$	yes	yes	yes	yes	no	no	no

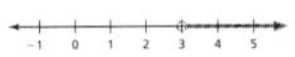

$x < 3$

b. $\frac{x}{-4} \geq \frac{3}{4}$

x	-5	-4	-3	-2	-1	0	1
$\frac{x}{-4}$	$\frac{5}{4}$	1	$\frac{3}{4}$	$\frac{1}{2}$	$\frac{1}{4}$	0	$-\frac{1}{4}$
$\frac{x}{-4} \overset{?}{\geq} \frac{3}{4}$	yes	yes	yes	no	no	no	no

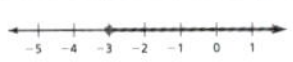
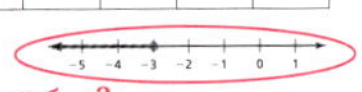

$x \leq -3$

4 ACTIVITY: Writing Rules

Work with a partner. Use a table to solve each inequality.

a. $-2x \leq 10$ $x \geq -5$ b. $-6x > 0$ $x < 0$ c. $\frac{x}{-4} < 1$ $x > -4$ d. $\frac{x}{-8} \geq \frac{1}{8}$ $x \leq -1$

Sample answer:

x	-5	-4	-3	-2	-1	0	1	2	3	4	5
$-2x$	10	8	6	4	2	0	-2	-4	-6	-8	-10
$-6x$	30	24	18	12	6	0	-6	-12	-18	-24	-30
$\frac{x}{-4}$	$\frac{5}{4}$	1	$\frac{3}{4}$	$\frac{1}{2}$	$\frac{1}{4}$	0	$-\frac{1}{4}$	$-\frac{1}{2}$	$-\frac{3}{4}$	-1	$-\frac{5}{4}$
$\frac{x}{-8}$	$\frac{5}{8}$	$\frac{1}{2}$	$\frac{3}{8}$	$\frac{1}{4}$	$\frac{1}{8}$	0	$-\frac{1}{8}$	$-\frac{1}{4}$	$-\frac{3}{8}$	$-\frac{1}{2}$	$-\frac{5}{8}$

Write a set of rules that describes how to solve inequalities like those in Activities 1 and 3. Then use your set of rules to solve each of the four inequalities above.

When multiplying or dividing each side of an inequality by a negative number, switch the direction of the inequality symbol.

What Is Your Answer?

5. **IN YOUR OWN WORDS** How can you use multiplication or division to solve an inequality? See Additional Answers.

Laurie's Notes

Activity 3

- Explain that this activity is similar to Activity 1 except the variables are involved in division instead of multiplication.
- Give time for students to work through the two problems.

? Discuss the results with your students.
  - "What did you find as the solution for part (a)?" $x < 3$
  - "Is this what you would have expected?" Likely, they will say yes.
  - "What did you find as the solution for part (b)?" $x \leq -3$
  - "Is this what you would have expected?" Likely, they will say no.
- Again, resist the temptation to simply tell students a rule. Suggest that trying additional problems might help them.

Activity 4

- Give time for students to work through the four problems with their partners.

? "Did any of the inequalities have solutions that you expected?" Again, listen for the same comments from students as before. The solution of each problem involves the expected number, but the direction of the inequality symbol is reversed.

- After working through all four problems, students should have a sense that solving these inequalities is the same as solving equations except when the coefficient is negative.

What Is Your Answer?

- **Neighbor Check:** Have students work independently and then have their neighbors check their work. Have students discuss any discrepancies.

Closure

- Refer to the three inequalities written at the beginning of class.

$x > 4$ $\quad -x > 4 \quad$ $x < -4$

? "Which inequalities have the same solution?" $-x > 4$ and $x < -4$

? "Is this consistent with what you discovered in the activities? Explain." yes; Listen for comments about the negative coefficient of x and the switching of the inequality symbol.

3 ACTIVITY: Using a Table to Solve an Inequality

Work with a partner.

- **Copy and complete the table.**
- **Decide which graph represents the solution of the inequality.**
- **Write the solution of the inequality.**

a. $\frac{x}{3} < 1$

x	-1	0	1	2	3	4	5
$\frac{x}{3}$							
$\frac{x}{3} \overset{?}{<} 1$							

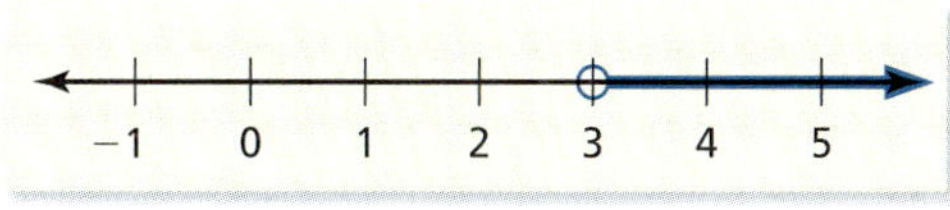

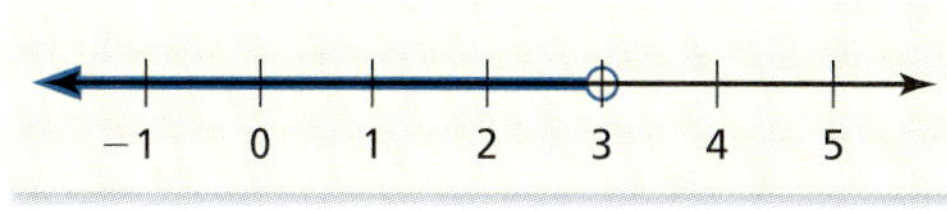

b. $\frac{x}{-4} \geq \frac{3}{4}$

x	-5	-4	-3	-2	-1	0	1
$\frac{x}{-4}$							
$\frac{x}{-4} \overset{?}{\geq} \frac{3}{4}$							

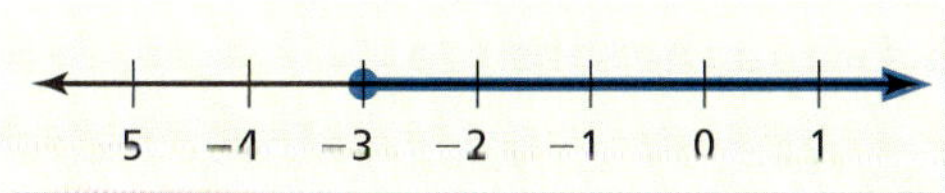

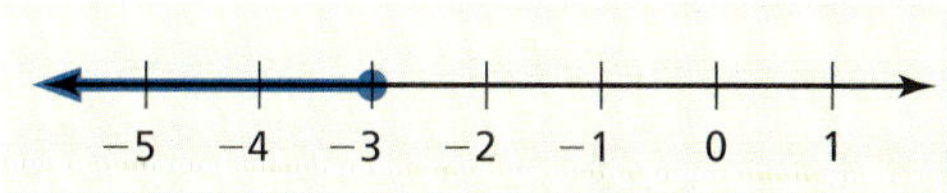

4 ACTIVITY: Writing Rules

Work with a partner. Use a table to solve each inequality.

a. $-2x \leq 10$ **b.** $-6x > 0$ **c.** $\frac{x}{-4} < 1$ **d.** $\frac{x}{-8} \geq \frac{1}{8}$

Write a set of rules that describes how to solve inequalities like those in Activities 1 and 3. Then use your rules to solve each of the four inequalities above.

Math Practice

Analyze Conjectures

When you apply your rules to parts (a)–(d), do you get the same solutions? Explain.

What Is Your Answer?

5. **IN YOUR OWN WORDS** How can you use multiplication or division to solve an inequality?

Practice Use what you learned about solving inequalities using multiplication or division to complete Exercises 4–9 on page 483.

11.3 Lesson

Check It Out
Lesson Tutorials
BigIdeasMath.com

Key Idea

Remember

Multiplication and division are inverse operations.

Multiplication and Division Properties of Inequality (Case 1)

Words When you multiply or divide each side of an inequality by the same *positive* number, the inequality remains true.

Numbers

$-4 < 6$ $\qquad$ $4 > -6$

$2 \cdot (-4) < 2 \cdot 6$ $\qquad$ $\frac{4}{2} > \frac{-6}{2}$

$-8 < 12$ $\qquad$ $2 > -3$

Algebra If $a < b$ and c is positive, then

$$a \cdot c < b \cdot c \quad \text{and} \quad \frac{a}{c} < \frac{b}{c}.$$

If $a > b$ and c is positive, then

$$a \cdot c > b \cdot c \quad \text{and} \quad \frac{a}{c} > \frac{b}{c}.$$

These properties are also true for $\le$ and $\ge$.

EXAMPLE 1 Solving an Inequality Using Multiplication

Solve $\frac{x}{5} \le -3$. Graph the solution.

$\frac{x}{5} \le -3$ $\qquad$ Write the inequality.

Undo the division. → $5 \cdot \frac{x}{5} \le 5 \cdot (-3)$ $\qquad$ Multiplication Property of Inequality

$x \le -15$ $\qquad$ Simplify.

The solution is $x \le -15$.

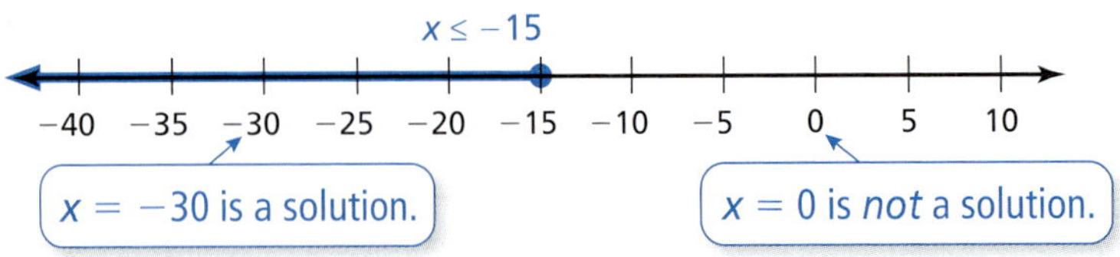

On Your Own

Solve the inequality. Graph the solution.

1. $n \div 3 < 1$
2. $-0.5 \le \frac{m}{10}$
3. $-3 > \frac{2}{3}p$

Laurie's Notes

Introduction

- **Yesterday:** Students gained an intuitive understanding of solving inequalities involving multiplication and division.
- **Today:** Students will use the Multiplication and Division Properties of Inequality to solve inequalities.

Motivate

? "Have you heard of Ultimate?" Answers will vary.

- It is a sport played with a flying disc at colleges, high schools, and some middle schools. There are 10 simple rules, one of which is that there aren't any officials! Pretty cool.
- The popularity of the sport has skyrocketed, but there is *at most* one-fifth the numbers of students playing Ultimate as there are playing lacrosse. If there are 26 students playing Ultimate, what is the minimum number playing lacrosse? $\frac{1}{5}x \geq 26$; $x \geq 130$ players

Lesson Notes

Key Idea

- These properties should look familiar, as they are similar to the Multiplication and Division Properties of Equality used in solving equations.
- Note that the properties are restricted to multiplying and dividing by a *positive* number. This is very important.

Example 1

? "How do you isolate the variable, meaning get x by itself?" Multiply by 5 on each side of the inequality.

- Multiplying by 5 is the inverse operation of dividing by 5.

On Your Own

- **Think-Pair-Share:** Students should read each question independently and then work in pairs to answer the questions. When they have answered the questions, the pair should compare their answers with another group and discuss any discrepancies.
- Division is represented in different ways in Questions 1 and 2. The second representation is more common in algebra (higher mathematics).
- After solving the inequality in Question 3, the result will be $-\frac{9}{2} > p$. Students can also rewrite this as $p < -\frac{9}{2}$. The direction of the inequality symbol is reversed *only* because the solution is being rewritten with the variable on the left side of the inequality statement.

Goal Today's lesson is solving inequalities using multiplication or division.

Technology for the Teacher

Dynamic Classroom

Lesson Tutorials
Lesson Plans
Answer Presentation Tool

Extra Example 1

Solve $\frac{m}{4} > -4$. Graph the solution.

$m > -16$;

On Your Own

1. $n < 3$;

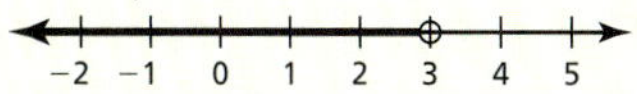

2. $m \geq -5$;

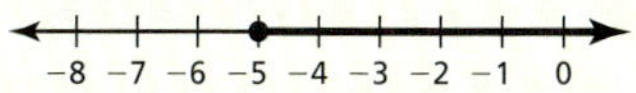

3. $p < -\frac{9}{2}$;

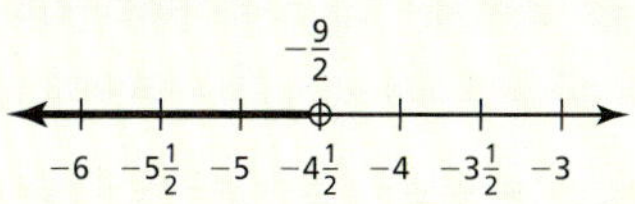

Extra Example 2

Solve $7y \leq -21$. Graph the solution.

$y \leq -3$;

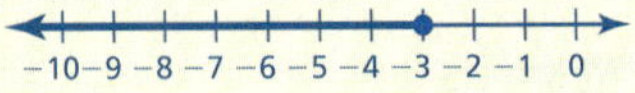

On Your Own

4. $b \geq \frac{1}{2}$;

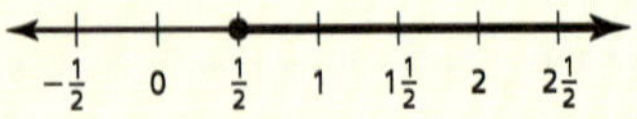

5. $k \leq -2$;

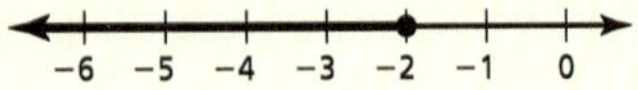

6. $q > -6$;

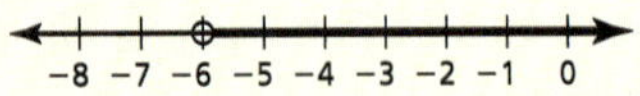

Differentiated Instruction

Visual

Use the inequality $6 < 9$ to show students why it is necessary to reverse the inequality symbol when multiplying or dividing by a negative number.

Add -3 to each side. The result is $3 < 6$, a true statement.

Subtract -3 from each side. The result is $9 < 12$, a true statement.

Multiply each side by -3. If the inequality is *not* reversed, then the statement $-18 < -27$ is false. By reversing the inequality, the statement $-18 > -27$ is true.

Divide each side by -3. If the inequality is *not* reversed, then the statement $-2 < -3$ is false. By reversing the inequality, the statement $-2 > -3$ is true.

Laurie's Notes

Example 2

? "What operation is being performed on x?" multiplication

? "How do you undo a multiplication problem?" divide

- Solve, graph, and check.

On Your Own

- **Think-Pair-Share:** Students should read each question independently and then work in pairs to answer the questions. When they have answered the questions, the pair should compare their answers with another group and discuss any discrepancies.
- **Construct Viable Arguments and Critique the Reasoning of Others:** Notice that although all of the coefficients are positive, sometimes the constant is negative. This is important in helping students understand when the direction of the inequality symbol is going to be reversed. The focus is on the sign of the coefficient, not the sign of the constant.
- For Question 6, remind students that after solving this inequality, the result will be $-6 < q$. Students can rewrite this as $q > -6$. The direction of the inequality symbol is reversed *only* because the two sides of the inequality are being reversed. Reversing the sign has nothing to do with the negative constant (-6).
- These problems integrate review of decimal operations.

Key Idea

- These properties look identical to what they have been using in the lesson, *except* now the direction of the inequality symbol must be reversed for the inequality to remain true because they are multiplying or dividing by a *negative* quantity!
- The short version of the property: When you multiply or divide by a negative quantity, reverse the direction of the inequality symbol.
- **Common Error:** When students solve $2x < -4$, they sometimes reverse the inequality symbol because there's a negative number in the problem. Reverse the inequality symbol when you multiply or divide each side by a negative number to eliminate a negative coefficient. Do not reverse the inequality symbol just because there is a negative constant.

EXAMPLE 2 Solving an Inequality Using Division

Solve $6x > -18$. Graph the solution.

$6x > -18$ Write the inequality.

Undo the multiplication. → $\frac{6x}{6} > \frac{-18}{6}$ Division Property of Inequality

$x > -3$ Simplify.

∴ The solution is $x > -3$.

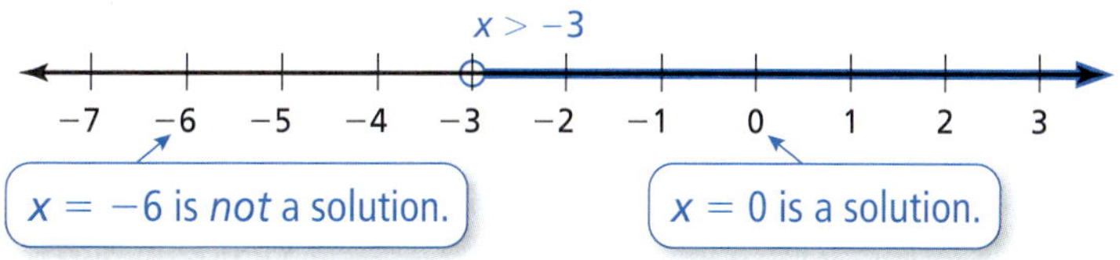

On Your Own

Now You're Ready
Exercises 10–18

Solve the inequality. Graph the solution.

4. $4b \geq 2$ **5.** $12k \leq -24$ **6.** $-15 < 2.5q$

Key Idea

Multiplication and Division Properties of Inequality (Case 2)

Words When you multiply or divide each side of an inequality by the same *negative* number, the direction of the inequality symbol must be reversed for the inequality to remain true.

Numbers

$-4 < 6$ | $4 > -6$

$-2 \cdot (-4) > -2 \cdot 6$ | $\frac{4}{-2} < \frac{-6}{-2}$

$8 > -12$ | $-2 < 3$

Algebra If $a < b$ and c is negative, then

$$a \cdot c > b \cdot c \quad \text{and} \quad \frac{a}{c} > \frac{b}{c}.$$

If $a > b$ and c is negative, then

$$a \cdot c < b \cdot c \quad \text{and} \quad \frac{a}{c} < \frac{b}{c}.$$

These properties are also true for ≤ and ≥.

Common Error

A negative sign in an inequality does not necessarily mean you must reverse the inequality symbol.

Only reverse the inequality symbol when you multiply or divide both sides by a negative number.

EXAMPLE 3 Solving an Inequality Using Multiplication

Solve $-\frac{3}{2}n \le 6$. Graph the solution.

$-\frac{3}{2}n \le 6$	Write the inequality.
$-\frac{2}{3} \cdot \left(-\frac{3}{2}n\right) \ge -\frac{2}{3} \cdot 6$	Use the Multiplication Property of Inequality. Reverse the inequality symbol.
$n \ge -4$	Simplify.

The solution is $n \ge -4$.

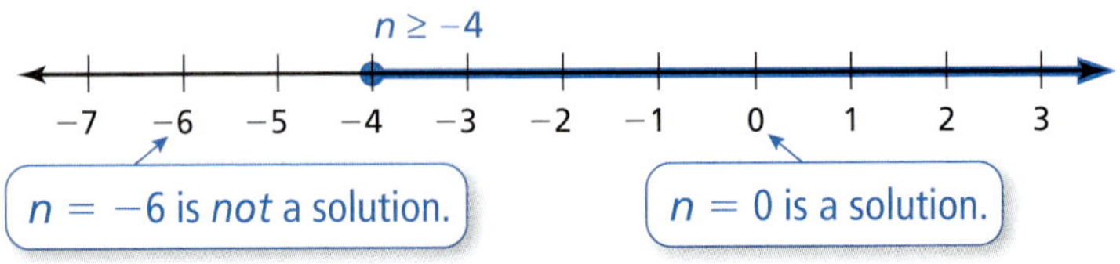

On Your Own

Solve the inequality. Graph the solution.

7. $\frac{x}{-3} > -4$

8. $0.5 \le -\frac{y}{2}$

9. $-12 \ge \frac{6}{5}m$

10. $-\frac{2}{5}h \le -8$

EXAMPLE 4 Solving an Inequality Using Division

Solve $-3z > -4.5$. Graph the solution.

$-3z > -4.5$	Write the inequality.
$\frac{-3z}{-3} < \frac{-4.5}{-3}$	Use the Division Property of Inequality. Reverse the inequality symbol.
$z < 1.5$	Simplify.

Undo the multiplication.

The solution is $z < 1.5$.

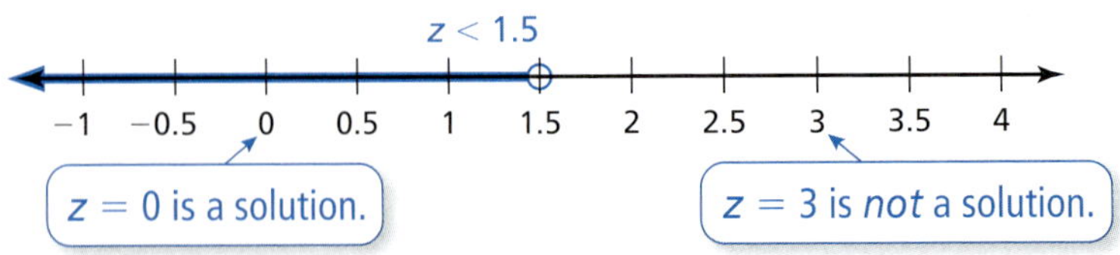

On Your Own

Now You're Ready
Exercises 27–35

Solve the inequality. Graph the solution.

11. $-5z < 35$

12. $-2a > -9$

13. $-1.5 < 3n$

14. $-4.2 \ge -0.7w$

Laurie's Notes

Example 3

- Write the inequality.

? "What operation is performed on n?" n is multiplied by $-\frac{3}{2}$.

? "How do you undo multiplying by $-\frac{3}{2}$?" Divide by $-\frac{3}{2}$.

? "What is equivalent to dividing by $-\frac{3}{2}$?" Multiplying by $-\frac{2}{3}$.

- Solve as usual, but remember to reverse the direction of the inequality symbol when multiplying each side by $-\frac{2}{3}$.
- When graphing, remember to use a closed circle because the inequality is greater than or equal to.

On Your Own

- **Think-Pair-Share:** Students should read each question independently and then work in pairs to answer the questions. When they have answered the questions, the pair should compare their answers with another group and discuss any discrepancies.

Example 4

- Write the inequality.

? "What operation is performed on z?" z is multiplied by -3.

? "How do you undo multiplying by -3?" Divide by -3.

- Solve as usual, but remember to reverse the direction of the inequality symbol when dividing each side by -3. Note that the quotient of two negatives is positive.

? "Should you use an open or closed circle?" Use an open circle because the inequality is strictly less than.

On Your Own

- **Neighbor Check:** Have students work independently and then have their neighbors check their work. Have students discuss any discrepancies.
- Have students share their work at the board.

Closure

- **Exit Ticket:** Solve and graph.

$\frac{x}{-3} \leq -9$ $x \geq 27$;

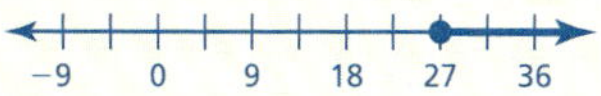

$-8 > 4x$ $x < -2$;

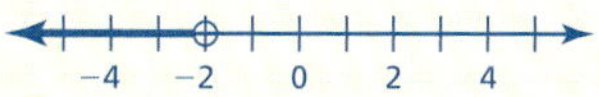

Extra Example 3

Solve $-\frac{4}{3}b > 8$. Graph the solution.

$b < -6$;

On Your Own

7. $x < 12$;

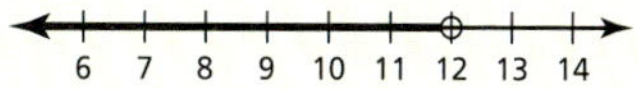

8. $y \leq -1$;

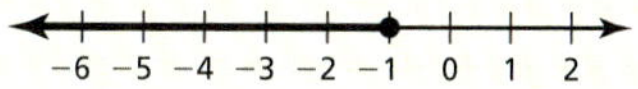

9. $m \leq -10$;

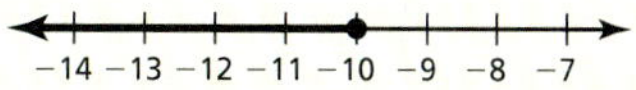

10. $h \geq 20$;

Extra Example 4

Solve $-2w \leq -5.2$. Graph the solution.

$w \geq 2.6$;

On Your Own

11. $z > -7$;

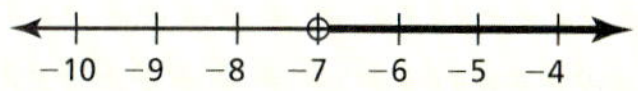

12. $a < \frac{9}{2}$;

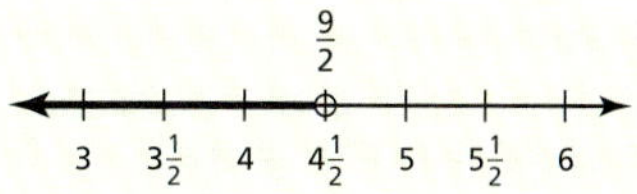

13. $n > -0.5$;

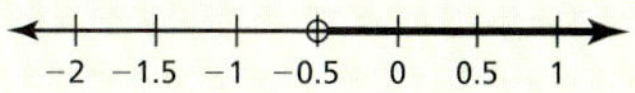

14. $w \geq 6$;

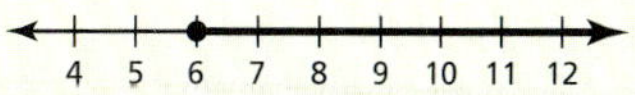

Vocabulary and Concept Check

1. Multiply each side by 3.
2. The first inequality will be divided by a positive number. The second inequality will be divided by a negative number. Because this inequality is divided by a negative number, the direction of the inequality symbol will be reversed.
3. *Sample answer:* $-4x < 16$

Practice and Problem Solving

4. $x < 1$
5. $x \geq -1$
6. $x < -3$
7. $x \leq -35$
8. $x < -\frac{2}{5}$
9. $x \leq \frac{3}{2}$
10. $n > 10$;

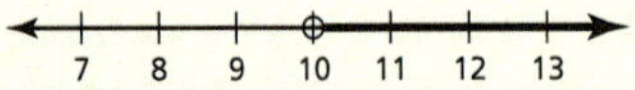

11. $c \leq -36$;

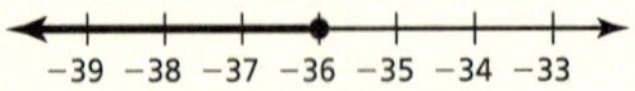

12. $m < 5$;

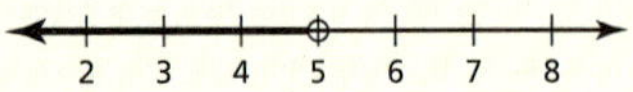

13. $x < -32$;

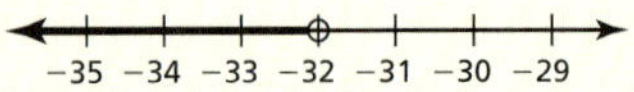

14. $w \geq 15$;

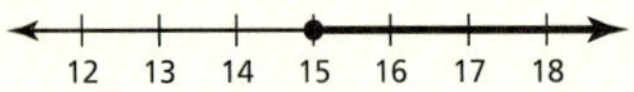

15. $k > 2$;

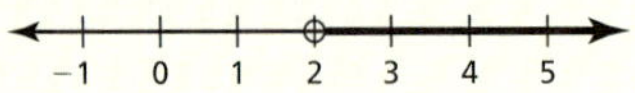

16. $x \leq -\frac{5}{12}$;

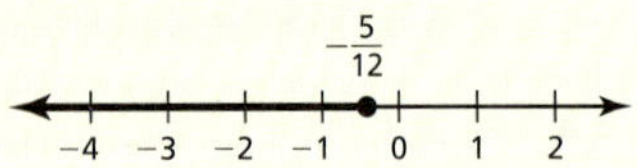

17–23. See Additional Answers.

24. $9.2x \geq 299$; $x \geq 32.5$ h

Assignment Guide and Homework Check

Level	Assignment	Homework Check
Advanced	1–9, 14–18 even, 19, 20–46 even, 48–52	14, 18, 26, 30, 34

Common Errors

- **Exercises 10–18** Students may perform the same operation on both sides instead of the opposite operation when solving the inequality. Remind them that solving inequalities is similar to solving equations.
- **Exercises 10–18** When there is a negative in the inequality, students may reverse the direction of the inequality symbol. Remind them that they only reverse the direction when they are multiplying or dividing by a negative number. All of these exercises keep the same inequality symbol.
- **Exercises 10–18** Students may shade the wrong direction when the variable is on the right side of the inequality instead of the left. Remind them to rewrite the inequality by reversing the right and left sides and reversing the inequality symbol.

11.3 Record and Practice Journal

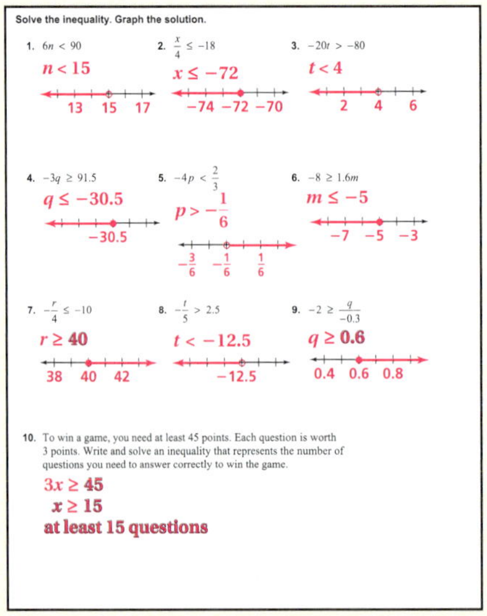

Solve the inequality. Graph the solution.

1. $6n < 90$ — $n < 15$ (13 15 17)
2. $\frac{x}{4} \leq -18$ — $x \leq -72$ (−74 −72 −70)
3. $-20t > -80$ — $t < 4$ (2 4 6)
4. $-3q \geq 91.5$ — $q \leq -30.5$ (−30.5)
5. $-4p < \frac{2}{3}$ — $p > -\frac{1}{6}$ ($-\frac{3}{6}$ $-\frac{1}{6}$ $\frac{1}{6}$)
6. $-8 \geq 1.6m$ — $m \leq -5$ (−7 −5 −3)
7. $-\frac{r}{4} \leq -10$ — $r \geq 40$ (38 40 42)
8. $-\frac{t}{5} > 2.5$ — $t < -12.5$ (−12.5)
9. $-2 \geq \frac{q}{-0.3}$ — $q \geq 0.6$ (0.4 0.6 0.8)
10. To win a game, you need at least 45 points. Each question is worth 3 points. Write and solve an inequality that represents the number of questions you need to answer correctly to win the game.

$3x \geq 45$
$x \geq 15$
at least 15 questions

11.3 Exercises

Check It Out
Help with Homework
BigIdeasMath.com

Vocabulary and Concept Check

1. **WRITING** Explain how to solve $\frac{x}{3} < -2$.
2. **PRECISION** Explain how solving $4x < -16$ is different from solving $-4x < 16$.
3. **OPEN-ENDED** Write an inequality that you can solve using the Division Property of Inequality where the direction of the inequality symbol must be reversed.

Practice and Problem Solving

Use a table to solve the inequality.

4. $2x < 2$
5. $-3x \leq 3$
6. $-6x > 18$
7. $\frac{x}{-5} \geq 7$
8. $\frac{x}{-1} > \frac{2}{5}$
9. $\frac{x}{3} \leq \frac{1}{2}$

Solve the inequality. Graph the solution.

1 2 10. $2n > 20$
11. $\frac{c}{9} \leq -4$
12. $2.2m < 11$
13. $-16 > x \div 2$
14. $\frac{1}{6}w \geq 2.5$
15. $7 < 3.5k$
16. $3x \leq -\frac{5}{4}$
17. $4.2y \leq -12.6$
18. $11.3 > \frac{b}{4.3}$
19. **ERROR ANALYSIS** Describe and correct the error in solving the inequality.

$$\frac{x}{3} < -9$$
$$3 \cdot \frac{x}{3} > 3 \cdot (-9)$$
$$x > -27$$

Write the word sentence as an inequality. Then solve the inequality.

20. The quotient of a number and 4 is at most 5.
21. A number divided by 7 is less than -3.
22. Six times a number is at least -24.
23. The product of -2 and a number is greater than 30.
24. **SMART PHONE** You earn \$9.20 per hour at your summer job. Write and solve an inequality that represents the number of hours you need to work in order to buy a smart phone that costs \$299.

25. **AVOCADOS** You have \$9.60 to buy avocados for a guacamole recipe. Avocados cost \$2.40 each.

   a. Write and solve an inequality that represents the number of avocados you can buy.

   b. Are there infinitely many solutions in this context? Explain.

26. **SCIENCE PROJECT** Students in a science class are divided into 6 equal groups with at least 4 students in each group for a project. Write and solve an inequality that represents the number of students in the class.

Solve the inequality. Graph the solution.

③ ④ 27. $-5n \le 15$

28. $-7w > 49$

29. $-\frac{1}{3}h \ge 8$

30. $-9 < -\frac{1}{5}x$

31. $-3y < -14$

32. $-2d \ge 26$

33. $4.5 > -\frac{m}{6}$

34. $\frac{k}{-0.25} \le 36$

35. $-2.4 > \frac{b}{-2.5}$

36. **ERROR ANALYSIS** Describe and correct the error in solving the inequality.

✗

$$-3m \ge 9$$

$$\frac{-3m}{-3} \ge \frac{9}{-3}$$

$$m \ge -3$$

37. **TEMPERATURE** It is currently 0°C outside. The temperature is dropping 2.5°C every hour. Write and solve an inequality that represents the number of hours that must pass for the temperature to drop below −20°C.

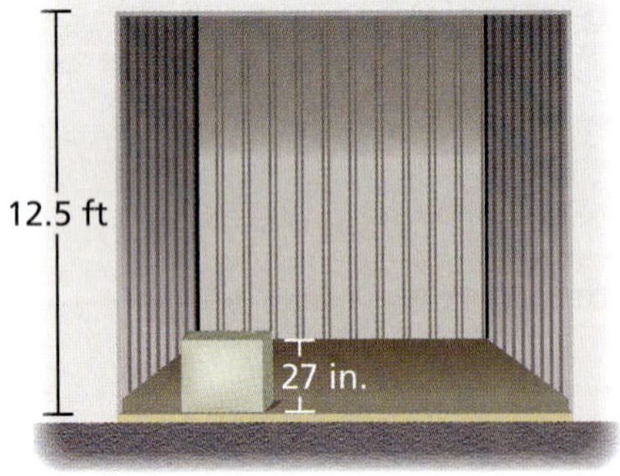

38. **STORAGE** You are moving some of your belongings into a storage facility.

   a. Write and solve an inequality that represents the number of boxes that you can stack vertically in the storage unit.

   b. Can you stack 6 boxes vertically in the storage unit? Explain.

Write and solve an inequality that represents x.

39. Area $\ge$ 120 cm^2

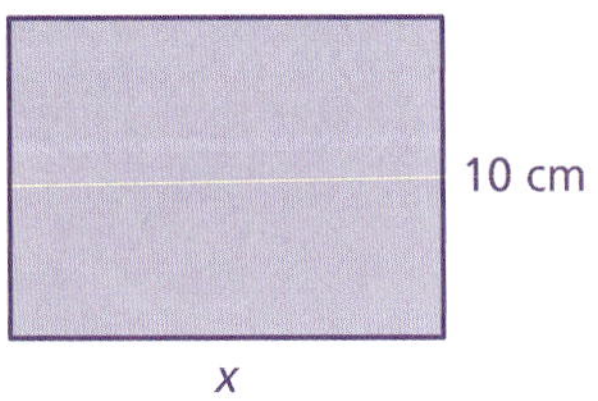

40. Area < 20 ft^2

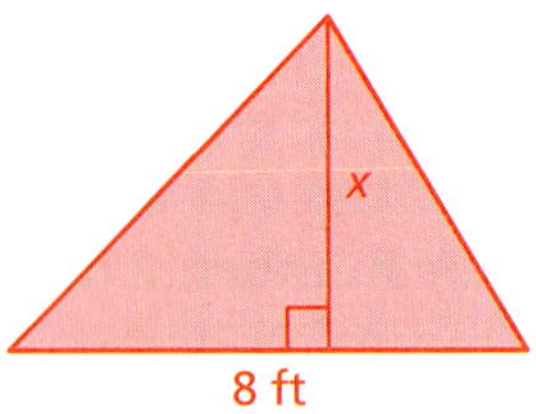

Common Errors

- **Exercises 27–35** Students may forget to reverse the inequality symbol when multiplying or dividing by a negative number. Remind them of this rule. Encourage students to substitute values into the original inequality to check that the solution is correct.
- **Exercise 41** Students may write an incorrect inequality before solving. They may write $\frac{x}{4} < 100$ because there are four friends. However, the student is included in the trip as well, so there are 5 people going on the trip. The inequality should be $\frac{x}{5} < 100$.

Differentiated Instruction

Visual

Students may question why they are asked to graph the solution of an inequality, but not asked to graph the solution of an equation. The graph of an equation is just a point on the number line. The graph of an inequality provides more information, because of the infinite number of possible solutions.

Practice and Problem Solving

25. **a.** $2.40x \le 9.60$; $x \le 4$ avocados

b. no; You must buy a whole number of avocados.

26. $\frac{x}{6} \ge 4$; $x \ge 24$; There are at least 24 students.

27. $n \ge -3$;

28. $w < -7$;

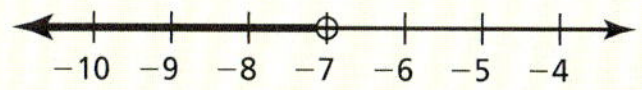

29. $h \le -24$;

30. $x < 45$;

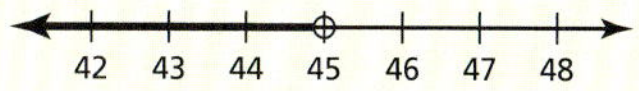

31. $y > \frac{14}{3}$;

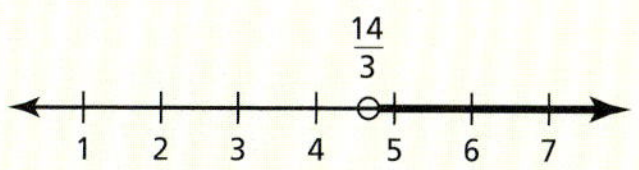

32. $d \le -13$;

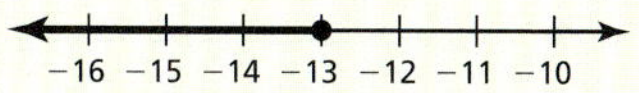

33. $m > -27$;

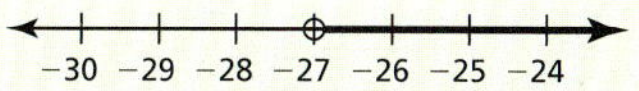

34. $k \ge -9$;

35–39. See Additional Answers.

40. $4x < 20$; $x < 5$ ft

Practice and Problem Solving

41. $\frac{x}{5} < 100$; $x < \$500$

42. *Sample answer:* Consider the inequality $5 > 3$. If you multiply or divide each side by -1 without reversing the direction of the inequality symbol, you obtain $-5 > -3$, which is not true. So, whenever you multiply or divide an inequality by a negative number, you must reverse the direction of the inequality symbol to obtain a true statement.

43. *Answer should include, but is not limited to:* Use the correct number of months that the novel has been out.

44. See *Taking Math Deeper.*

45. $n \geq -12$ and $n \leq -5$;

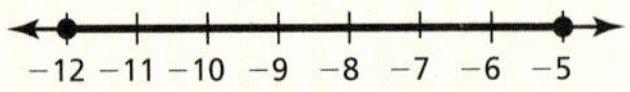

46–47. See Additional Answers.

Fair Game Review

48. $w = 8$ **49.** $v = 45$

50. $x = 7$ **51.** $m = 4$

52. B

Mini-Assessment

Solve and graph the inequality.

1. $\frac{b}{6} > -11$ $b > -66$;

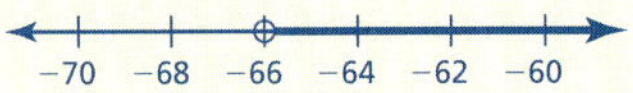

2. $4c \leq -28$ $c \leq -7$;

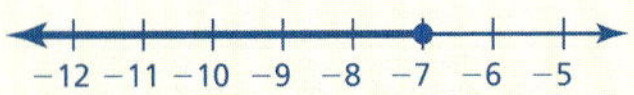

3. $-\frac{6}{7}k > -12$ $k < 14$;

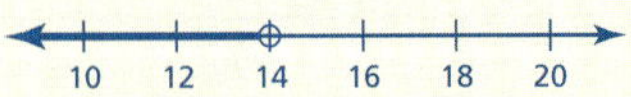

4. $-4b \leq 9.6$ $b \geq -2.4$;

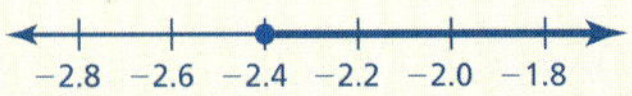

Taking Math Deeper

Exercise 44

Double (or compound) inequalities, like those in Exercises 44–47, can often be written using a single inequality statement.

1 Begin by graphing each inequality.

$$4m > -4$$
$$m > -1$$

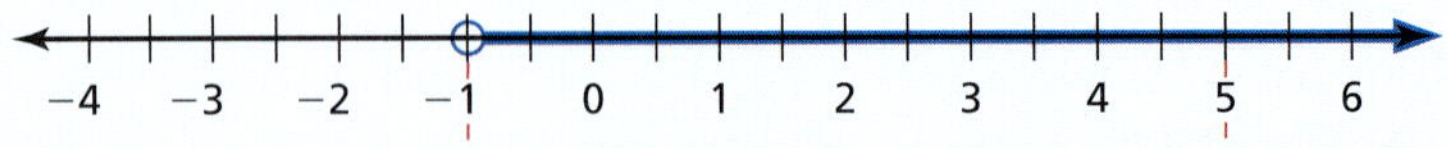

$$3m < 15$$
$$m < 5$$

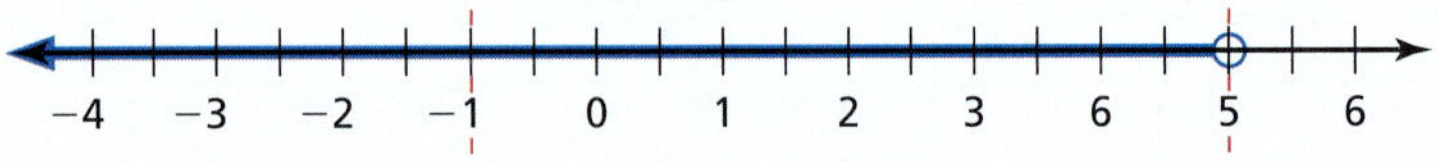

2 Combine the two graphs.

3 The numbers that satisfy both inequalities are all numbers greater than -1 and less than 5. If you rewrite $m > -1$ as $-1 < m$, then you can write the statement as a single inequality, $-1 < m < 5$.

Project

Use the newspaper, Internet, TV, radio, or any other source to record and graph at least 10 different uses of inequalities.

Reteaching and Enrichment Strategies

If students need help. . .	If students got it. . .
Resources by Chapter • Practice A and Practice B • Puzzle Time Record and Practice Journal Practice Differentiating the Lesson Lesson Tutorials Skills Review Handbook	Resources by Chapter • Enrichment and Extension • Technology Connection Start the next section

41. AMUSEMENT PARK You and four friends are planning a visit to an amusement park. You want to keep the cost below \$100 per person. Write and solve an inequality that represents the total cost of visiting the amusement park.

42. LOGIC When you multiply or divide each side of an inequality by the same negative number, you must reverse the direction of the inequality symbol. Explain why.

43. PROJECT Choose two novels to research.

a. Use the Internet or a magazine to complete the table.

b. Use the table to find and compare the average number of copies sold per month for each novel. Which novel do you consider to be the most successful? Explain.

c. Assume each novel continues to sell at the average rate. Write and solve an inequality that represents the number of months it will take for the total number of copies sold to exceed twice the current number sold.

Author	Name of Novel	Release Date	Current Number of Copies Sold
1.			
2.			

Number Sense **Describe all numbers that satisfy *both* inequalities. Include a graph with your description.**

44. $4m > -4$ and $3m < 15$

45. $\frac{n}{3} \geq -4$ and $\frac{n}{-5} \geq 1$

46. $2x \geq -6$ and $2x \geq 6$

47. $-\frac{1}{2}s > -7$ and $\frac{1}{3}s < 12$

Fair Game Review What you learned in previous grades & lessons

Solve the equation. Check your solution. *(Section 1.2)*

48. $-2w + 4 = -12$

49. $\frac{v}{5} - 6 = 3$

50. $3(x - 1) = 18$

51. $\frac{m + 200}{4} = 51$

52. MULTIPLE CHOICE What is the value of $\frac{2}{3} + \left(-\frac{5}{7}\right)$? *(Skills Review Handbook)*

Ⓐ $-\frac{3}{4}$ Ⓑ $-\frac{1}{21}$ Ⓒ $\frac{7}{10}$ Ⓓ $1\frac{8}{21}$

11.4 Solving Two-Step Inequalities

Essential Question How can you use an inequality to describe the dimensions of a figure?

1 ACTIVITY: Areas and Perimeters of Figures

Work with a partner.

- **Use the given condition to choose the inequality that you can use to find the possible values of the variable. Justify your answer.**
- **Write four values of the variable that satisfy the inequality you chose.**

a. You want to find the values of x so that the area of the rectangle is more than 22 square units.

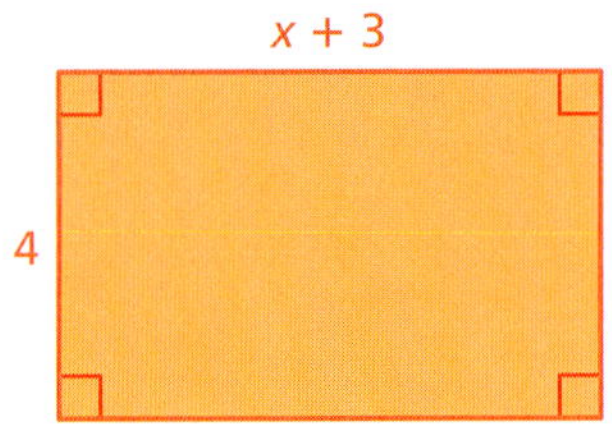

$4x + 12 > 22$ $\qquad$ $4x + 3 > 22$

$4x + 12 \geq 22$ $\qquad$ $2x + 14 > 22$

b. You want to find the values of x so that the perimeter of the rectangle is greater than or equal to 28 units.

$x + 7 \geq 28$ $\qquad$ $4x + 12 \geq 28$ $\qquad$ $2x + 14 \geq 28$ $\qquad$ $2x + 14 \leq 28$

c. You want to find the values of y so that the area of the parallelogram is fewer than 41 square units.

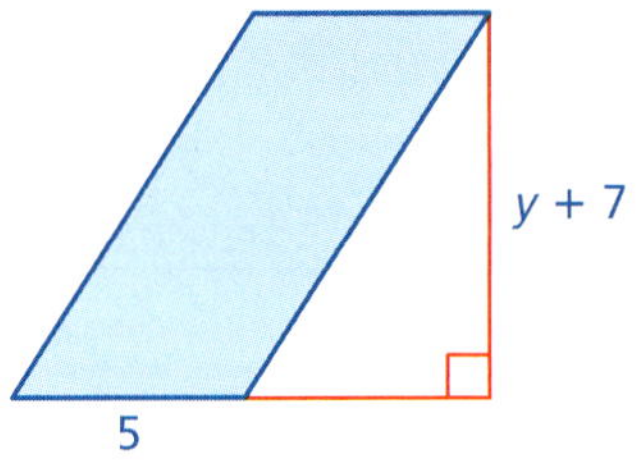

$5y + 7 < 41$ $\qquad$ $5y + 35 < 41$

$5y + 7 \leq 41$ $\qquad$ $5y + 35 \leq 41$

d. You want to find the values of z so that the area of the trapezoid is at most 100 square units.

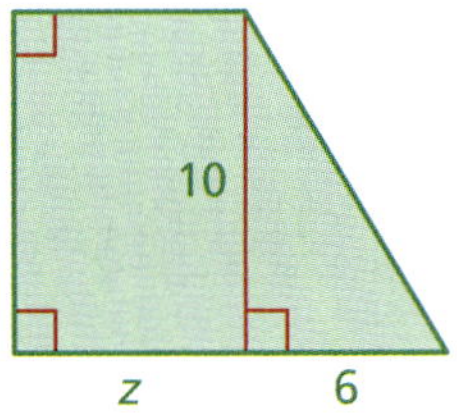

$5z + 30 \leq 100$ $\qquad$ $10z + 30 \leq 100$

$5z + 30 < 100$ $\qquad$ $10z + 30 < 100$

Inequalities
In this lesson, you will
- solve multi-step inequalities.
- solve real-life problems.

Laurie's Notes

Introduction

Applying Mathematical Practices

- **Construct Viable Arguments and Critique the Reasoning of Others:** In these activities, students must choose the inequality that solves the stated problem and then justify that choice.

Motivate

- Display a set of tangram pieces arranged as shown. Tell students that the area of the entire square formed by the tangrams is 16 square units.
- Ask students to find the area of each tangram piece. gold triangle = 4, green triangle = 2, red triangle = 1, turquoise square = 2, purple parallelogram = 2

? "Did you use any area formulas?" no; Students likely used fractional relationships.

- Explain that this activity involves area formulas. Review formulas if necessary.

Activity Notes

Activity 1

- This activity reviews perimeters of rectangles and areas of parallelograms and triangles.
- Students know how to use the Distributive Property and combine like terms. Encourage them to use these skills to first write algebraic expressions for the perimeter or area.
- Students are not asked to solve the inequalities, though some may.
- When students finish, ask volunteers to discuss parts (a) and (b). In part (a), some students may have initially written $4 + (x + 3) + 4 + (x + 3)$ while others may have written $2(4) + 2(x + 3)$.
- **Construct Viable Arguments and Critique the Reasoning of Others:** In each part, listen carefully to the student's justification for choosing the inequality. There should be a clear reason for the choice.
- **Extension:** Ask what happens to the shape of the rectangle as x varies.
- In part (c), students need to interpret the figure correctly. The parallelogram has a base of 5 units and a height of $y + 7$ units.
- **Common Error:** Students may forget to use the Distributive Property when they find the expression for area and incorrectly write $5(y + 7) = 5y + 7$.
- **Extension:** Ask what happens to the shape of the parallelogram as y varies.
- In part (d), students could find the sum of the areas of the rectangle and right triangle.
- **Extension:** Ask what happens to the shape of the trapezoid as z varies.

What Your Students Will Learn

- Use the properties of inequalities to solve and graph solutions of multi-step inequalities.

Previous Learning

Students should know how to solve two-step equations.

Technology for the Teacher

Dynamic Classroom

Lesson Plans
Complete Materials List

11.4 Record and Practice Journal

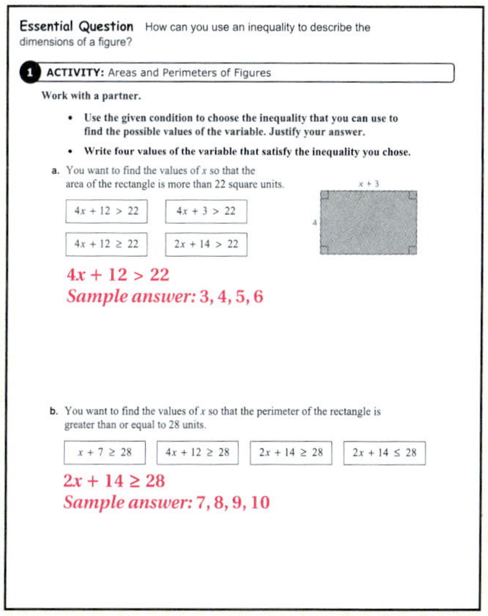

Essential Question How can you use an inequality to describe the dimensions of a figure?

1 ACTIVITY: Areas and Perimeters of Figures

Work with a partner.

- **Use the given condition to choose the inequality that you can use to find the possible values of the variable. Justify your answer.**
- **Write four values of the variable that satisfy the inequality you chose.**

a. You want to find the values of x so that the area of the rectangle is more than 22 square units.

$4x + 12 > 22$ | $4x + 3 > 22$ | $4x + 12 \geq 22$ | $2x + 14 > 22$

$4x + 12 > 22$
Sample answer: 3, 4, 5, 6

b. You want to find the values of x so that the perimeter of the rectangle is greater than or equal to 28 units.

$x + 7 \geq 28$ | $4x + 12 \geq 28$ | $2x + 14 \geq 28$ | $2x + 14 \leq 28$

$2x + 14 \geq 28$
Sample answer: 7, 8, 9, 10

Differentiated Instruction

Auditory

In each part of the Activities, ask a volunteer to say out loud the key phrase ("is fewer than," "is no more than," etc.) of the given condition that determines the correct inequality symbol to use. Discuss which inequality symbol represents the phrase, and why.

11.4 Record and Practice Journal

c. You want to find the values of y so that the area of the parallelogram is fewer than 41 square units.

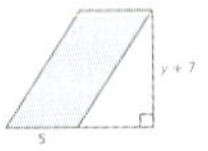

$5y + 7 < 41$	$5y + 35 < 41$
$5y + 7 \le 41$	$5y + 35 \le 41$

$5y + 35 < 41$

Sample answer: 0.5, 0.8, 1, 1.1

d. You want to find the values of z so that the area of the trapezoid is at most 100 square units.

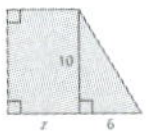

$5z + 30 \le 100$	$10z + 30 \le 100$
$5z + 30 < 100$	$10z + 30 < 100$

$10z + 30 \le 100$

Sample answer: 4, 5, 6, 7

2 ACTIVITY: Volumes of Rectangular Prisms

Work with a partner.

- **Use the given condition to choose the inequality that you can use to find the possible values of the variable. Justify your answer.**
- **Write four values of the variable that satisfy the inequality you chose.**

a. You want to find the values of x so that the volume of the rectangular prism is at least 50 cubic units.

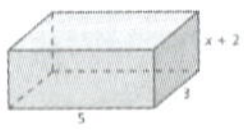

$15x + 30 > 50$	$x + 10 \ge 50$
$15x + 30 \ge 50$	$15x + 2 \ge 50$

$15x + 30 \ge 50$

Sample answer: 2, 3, 4, 5

b. You want to find the values of x so that the volume of the rectangular prism is no more than 36 cubic units.

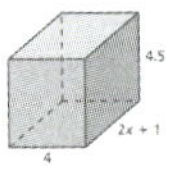

$8x + 4 < 36$	$36x + 18 < 36$
$2x + 9.5 \le 36$	$36x + 18 \le 36$

$36x + 18 \le 36$

Sample answer: $\frac{1}{8}, \frac{1}{6}, \frac{1}{4}, \frac{1}{3}$

What Is Your Answer?

3. **IN YOUR OWN WORDS** How can you use an inequality to describe the dimensions of a figure?

   Bound the area, perimeter, or volume by a number and find possibilities for the missing side.

4. Use what you know about solving equations and inequalities to describe how you can solve a two-step inequality. Give an example to support your explanation.

   Solve a two-step inequality in the same way you solve a two-step equation. Use inverse operations to isolate the variable.

Laurie's Notes

Activity 2

- This activity reviews volumes of rectangular prisms.
- Students will need to use the Distributive Property to find the volume of each prism.
- **Construct Viable Arguments and Critique the Reasoning of Others:** In each part, listen carefully to the student's justification for the inequality he or she chose. Make sure the student mentions using the Distributive Property.
- Check that students translate "at least" and "no more than" correctly.
- Have students share and discuss values of the variable that satisfy each inequality. This will remind students that each problem has infinitely many solutions.

? "Does x need to be a whole number in each problem?" no

- **Extension:** Ask what happens to the shape of the rectangular prism as x varies in each problem.

What Is Your Answer?

- In Question 4, students are likely to say that solving two-step inequalities is just like solving two-step equations. Make sure they mention how they are different: When the coefficient is negative, you need to multiply or divide each side by a negative quantity and reverse the direction of the inequality symbol.

Closure

- **Exit Ticket:** Solve each inequality.

  $4x + 5 \ge 21$ $x \ge 4$ and $\frac{x}{4} - 3 \ge 13$ $x \ge 64$

2 ACTIVITY: Volumes of Rectangular Prisms

Work with a partner.

- **Use the given condition to choose the inequality that you can use to find the possible values of the variable. Justify your answer.**
- **Write four values of the variable that satisfy the inequality you chose.**

> **Math Practice**
> **State the Meaning of Symbols**
> What inequality symbols do the phrases *at least* and *no more than* represent? Explain.

a. You want to find the values of x so that the volume of the rectangular prism is at least 50 cubic units.

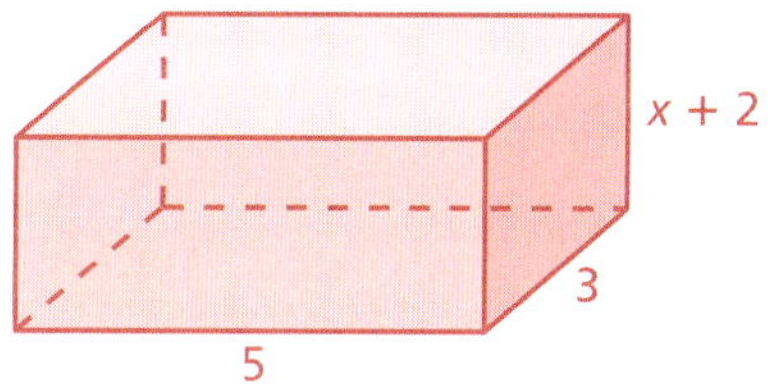

$15x + 30 > 50$	$x + 10 \geq 50$	$15x + 30 \geq 50$	$15x + 2 \geq 50$

b. You want to find the values of x so that the volume of the rectangular prism is no more than 36 cubic units.

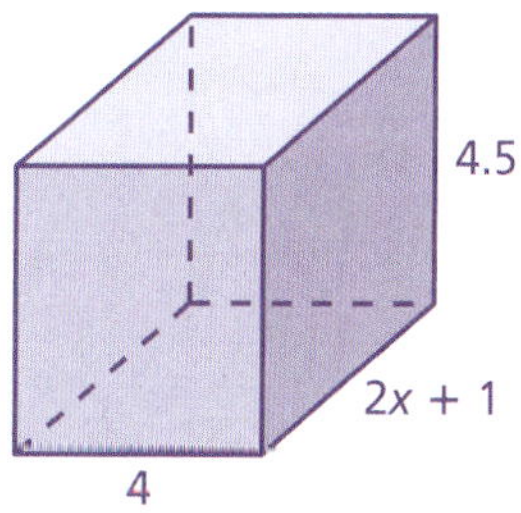

$8x + 4 < 36$	$36x + 18 < 36$	$2x + 9.5 \leq 36$	$36x + 18 \leq 36$

What Is Your Answer?

3. **IN YOUR OWN WORDS** How can you use an inequality to describe the dimensions of a figure?
4. Use what you know about solving equations and inequalities to describe how you can solve a two-step inequality. Give an example to support your explanation.

Practice Use what you learned about solving two-step inequalities to complete Exercises 3 and 4 on page 490.

11.4 Lesson

Check It Out
Lesson Tutorials
BigIdeasMath.com

You can solve two-step inequalities in the same way you solve two-step equations.

EXAMPLE 1 Solving Two-Step Inequalities

a. Solve $5x - 4 \geq 11$. Graph the solution.

$5x - 4 \geq 11$ Write the inequality.

Step 1: Undo the subtraction. → $+4 \quad +4$ Addition Property of Inequality

$5x \geq 15$ Simplify.

Step 2: Undo the multiplication. → $\frac{5x}{5} \geq \frac{15}{5}$ Division Property of Inequality

$x \geq 3$ Simplify.

The solution is $x \geq 3$.

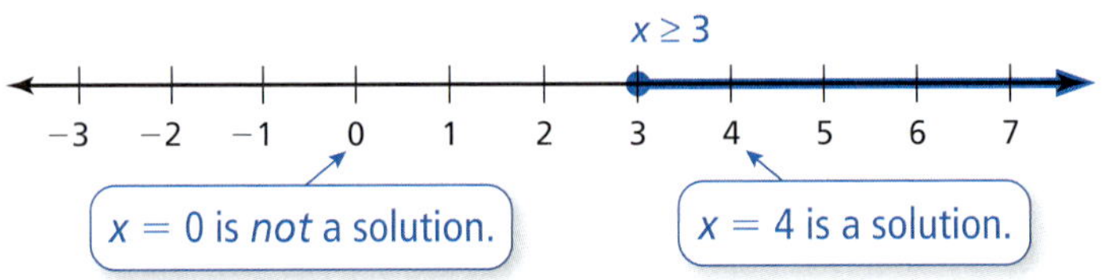

b. Solve $\frac{b}{-3} + 4 < 13$. Graph the solution.

$\frac{b}{-3} + 4 < 13$ Write the inequality.

Step 1: Undo the addition. → $-4 \quad -4$ Subtraction Property of Inequality

$\frac{b}{-3} < 9$ Simplify.

Step 2: Undo the division. → $-3 \cdot \frac{b}{-3} > -3 \cdot 9$ Use the Multiplication Property of Inequality. Reverse the inequality symbol.

$b > -27$ Simplify.

The solution is $b > -27$.

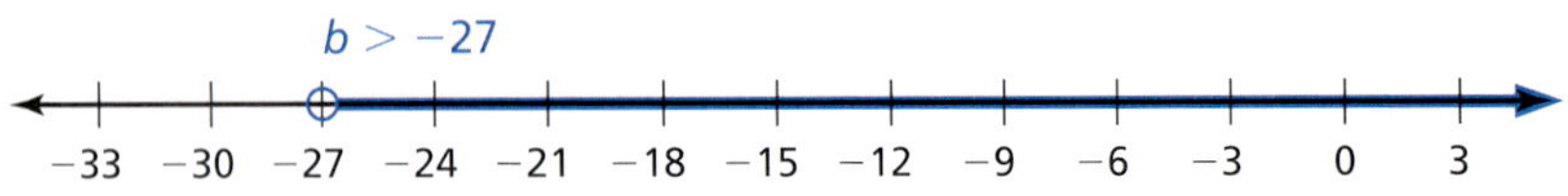

On Your Own

Now You're Ready
Exercises 5–10

Solve the inequality. Graph the solution.

1. $6y - 7 > 5$
2. $4 - 3d \geq 19$
3. $\frac{w}{-4} + 8 > 9$

Laurie's Notes

Introduction

Connect

- **Yesterday:** Students developed an intuitive understanding of solving two-step inequalities.
- **Today:** Students will solve and graph two-step inequalities.

Motivate

- Share the following scenario with students.
  - Student A has test scores of 82, 94, 86, and 81.
  - Student B has test scores of 92, 98, 88 and 94.
- Student A wants to achieve an average of at least 85, and Student B wants to achieve an average of at least 90. What must each student score on the 5th and final test to meet their goals?

? "Have any of you ever wondered what score you needed on a particular test to have a certain average?" Answers will vary.

? "Do you think it is mathematically possible for Student A and Student B to achieve their goals?" Answers will vary.

- Tell students you will return to this problem at the end of class.

Lesson Notes

Discuss

- You solve two-step inequalities in much the same way you solve two-step equations. You only need to remember to change the direction of the inequality symbol if you multiply or divide by a negative quantity.
- Recall that solving an equation undoes the evaluating in reverse order. The goal is to isolate the variable.

Example 1

? "What operations are being performed on the left side of the inequality?" multiplication and subtraction

? "What is the first step in isolating the variable, meaning getting the x-term by itself?" Add 4 to each side of the inequality.

- Notice that subtracting 4 would have been the last step if evaluating the left side, so its inverse operation is the first step in solving the inequality.

? "To solve for x, what is the last step?" Divide both sides by 5.

- Because you are dividing by a positive quantity, the inequality symbol does not change. The solution is $x \geq 3$. Graph and check.
- Part (b) is solved in a similar fashion.

? "To solve $\frac{b}{-3} < 9$, what do you need to do?" Multiply both sides by -3 and change the direction of the inequality symbol.

- Graph and check. Remember to use an open circle because the variable cannot equal -27.
- Point out to students that the number line has increments of 3 units.

On Your Own

- These are all straightforward problems. Students should not be confused by them.

Goal Today's lesson is solving two-step inequalities.

Technology for the Teacher

Dynamic Classroom

Lesson Tutorials
Lesson Plans
Answer Presentation Tool

Extra Example 1

a. Solve $17 \leq 3y - 4$. Graph the solution. $y \geq 7$;

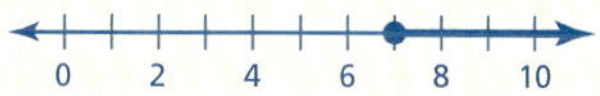

b. Solve $\frac{x}{-2} - 8 \geq -6$. Graph the solution. $x \leq -4$;

On Your Own

1. $y > 2$;

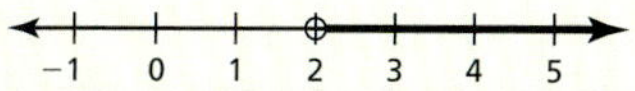

2. $d \leq -5$;

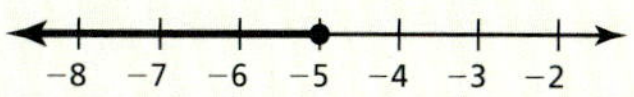

3. $w < -4$;

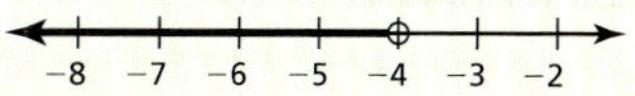

English Language Learners

Vocabulary

Give English learners the opportunity to use precise language to solve an inequality. Write the inequality $-2x + 4 > 8$ on the board. Have one student come to the board. For each step of the solution, call on another student to give the instruction for solving. The instructions should be given in complete sentences. The instructions for the inequality are:

(1) Subtract 4 from each side.
(2) Simplify.
(3) Divide each side by -2.
Reverse the inequality symbol.
(4) Simplify.

Extra Example 2

Solve $12 > -2(y - 4)$. Graph the solution. $y > -2$;

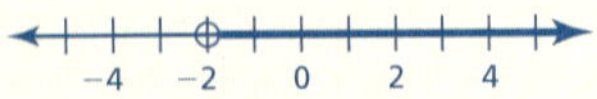

Extra Example 3

In Example 3, the contestant wants to lose an average of at least 7 pounds per month. How many pounds must the contestant lose in the fifth month to meet the goal? at least 1 pound

On Your Own

4. $k < 8$;

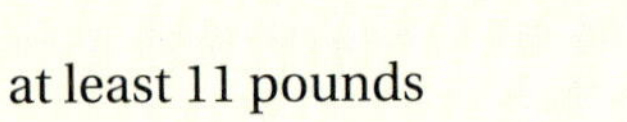

5. $n > 2$;

6. $y \geq -14$;

7. at least 11 pounds

Laurie's Notes

Example 2

- **Look for and Make Use of Structure:** There is another way to solve Example 2. The inequality has two factors on the left side: -7 and $(x + 3)$. Instead of distributing, divide both sides by -7. Dividing by a negative number changes the direction of the inequality symbol.

$-7(x + 3) \leq 28$	Write the inequality.
$\dfrac{-7(x+3)}{-7} \geq \dfrac{28}{-7}$	Divide each side by -7. Reverse the inequality symbol.
$x + 3 \geq -4$	Simplify.
$\underline{-3} \quad \underline{-3}$	Subtract 3 from each side.
$x \geq -7$	Simplify.

- Discuss each method with students.

Example 3

- Before beginning to solve the example, talk about different total numbers of pounds that average at least 8 pounds per month for 5 months.
- ? "Would a total of 35 pounds be enough?" no
- ? "Would a total of 45 pounds be enough?" yes
- Ask students to interpret the information in the Progress Report. They should recognize that 34 pounds were lost in the first four months.
- ? "What are you trying to solve for in this problem?" the number of pounds the contestant needs to lose in the 5th month to meet the goal
- ? "What inequality symbol is needed if the goal is "at least" 8 pounds per month?" $\geq$
- You may want to show students a verbal model for the inequality.
- **Common Error:** Students may want to subtract 34 from each side before dealing with the division by 5. The division must be undone first.
- **Connection:** This problem has a different context but is of the same type as the question in the Motivate activity.

On Your Own

- **Common Error:** If students solve Question 5 by distributing the -4, it is very possible they will write $-4n - 40$ instead of $-4n + 40$. For the factor $n - 10$, they need to remember to *add the opposite* so that the initial equation could be written as $-4[n + (-10)] < 32$. Then distribute the -4.
- Students may need guidance on Question 6. Distributing 0.5 results in $-3 \leq 4 + 0.5y$.

Closure

- Have students write and solve the inequalities for each of the students in the Motivate activity.

Student A: $(82 + 94 + 86 + 81 + a) \div 5 \geq 85$; $a \geq 82$
Student B: $(92 + 98 + 88 + 94 + b) \div 5 \geq 90$; $b \geq 78$

It is mathematically possible for both students to reach their goals.

EXAMPLE 2 Graphing an Inequality

Which graph represents the solution of $-7(x+3) \leq 28$?

Ⓐ 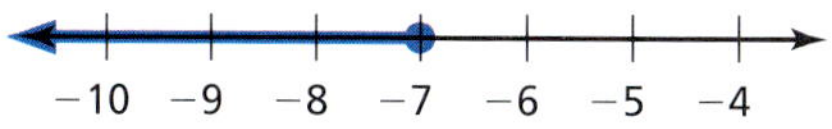

Ⓑ 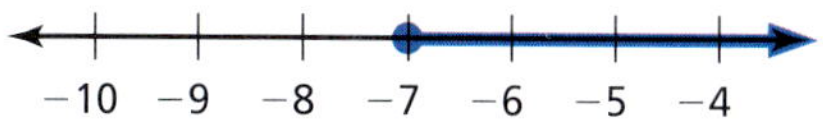

Ⓒ 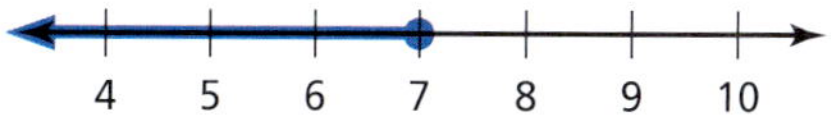

Ⓓ 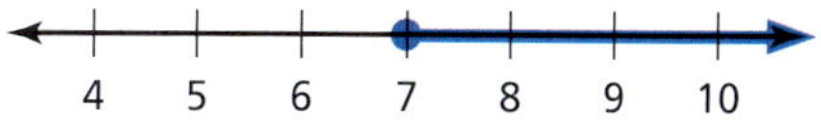

	$-7(x+3) \leq 28$	Write the inequality.
	$-7x - 21 \leq 28$	Distributive Property
Step 1: Undo the subtraction. →	$+21 \quad +21$	Addition Property of Inequality
	$-7x \leq 49$	Simplify.
Step 2: Undo the multiplication. →	$\dfrac{-7x}{-7} \geq \dfrac{49}{-7}$	Use the Division Property of Inequality. Reverse the inequality symbol.
	$x \geq -7$	Simplify.

The correct answer is Ⓑ.

EXAMPLE 3 Real-Life Application

Progress Report	
Month	**Pounds Lost**
1	12
2	9
3	5
4	8

Remember

In Example 3, the average is equal to the sum of the pounds lost divided by the number of months.

A contestant in a weight-loss competition wants to lose an average of at least 8 pounds per month during a 5-month period. How many pounds must the contestant lose in the fifth month to meet the goal?

Write and solve an inequality. Let x be the number of pounds lost in the fifth month.

The phrase *at least* means *greater than or equal to.*

$\dfrac{12+9+5+8+x}{5} \geq 8$	
$\dfrac{34+x}{5} \geq 8$	Simplify.
$5 \cdot \dfrac{34+x}{5} \geq 5 \cdot 8$	Multiplication Property of Inequality
$34 + x \geq 40$	Simplify.
$x \geq 6$	Subtract 34 from each side.

So, the contestant must lose at least 6 pounds to meet the goal.

On Your Own

Now You're Ready
Exercises 12–17

Solve the inequality. Graph the solution.

4. $2(k-5) < 6$
5. $-4(n-10) < 32$
6. $-3 \leq 0.5(8+y)$
7. **WHAT IF?** In Example 3, the contestant wants to lose an average of at least 9 pounds per month. How many pounds must the contestant lose in the fifth month to meet the goal?

11.4 Exercises

Check It Out
Help with Homework
BigIdeasMath.com

Vocabulary and Concept Check

1. **WRITING** Compare and contrast solving two-step inequalities and solving two-step equations.

2. **OPEN-ENDED** Describe how to solve the inequality $3(a + 5) < 9$.

Practice and Problem Solving

Match the inequality with its graph.

3. $\frac{t}{3} - 1 \geq -3$

A. 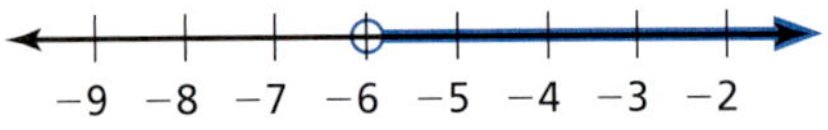

B. 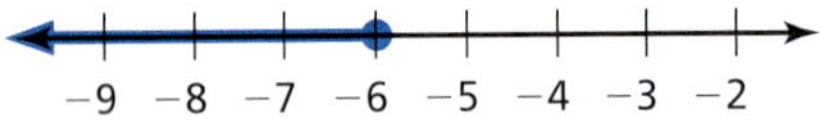

C. 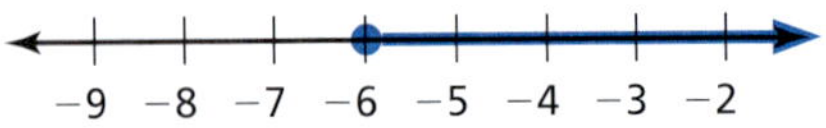

4. $5x + 7 \leq 32$

A. 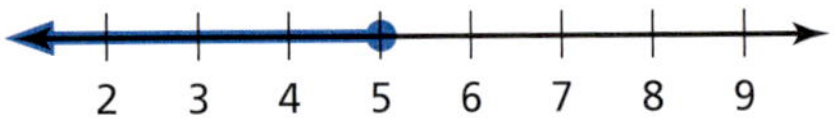

B. 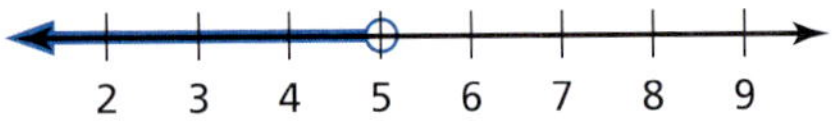

C. 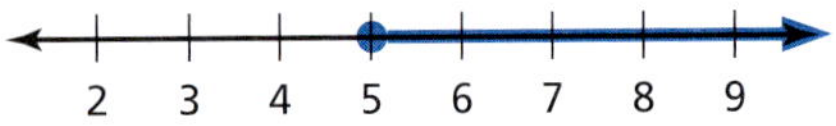

Solve the inequality. Graph the solution.

1 5. $8y - 5 < 3$

6. $3p + 2 \geq -10$

7. $2 > 8 - \frac{4}{3}h$

8. $-2 > \frac{m}{6} - 7$

9. $-1.2b - 5.3 \geq 1.9$

10. $-1.3 \geq 2.9 - 0.6r$

11. **ERROR ANALYSIS** Describe and correct the error in solving the inequality.

✗
$\frac{x}{3} + 4 < 6$
$x + 4 < 18$
$x < 14$

Solve the inequality. Graph the solution.

2 12. $5(g + 4) > 15$

13. $4(w - 6) \leq -12$

14. $-8 \leq \frac{2}{5}(k - 2)$

15. $-\frac{1}{4}(d + 1) < 2$

16. $7.2 > 0.9(n + 8.6)$

17. $20 \geq -3.2(c - 4.3)$

18. **UNICYCLE** The first jump in a unicycle high-jump contest is shown. The bar is raised 2 centimeters after each jump. Solve the inequality $2n + 10 \geq 26$ to find the number of additional jumps needed to meet or exceed the goal of clearing a height of 26 centimeters.

Assignment Guide and Homework Check

Level	Assignment	Homework Check
Advanced	1–4, 6–10 even, 11, 12–24 even, 25–27	10, 14, 20, 22

Common Errors

- **Exercises 5–10** Students may incorrectly multiply or divide before adding to or subtracting from both sides. Remind them that they should work backward through the order of operations, or that they should start away from the variable and move toward it.
- **Exercises 5–10, 12–17** Students may forget to reverse the inequality symbol when multiplying or dividing by a negative number. Encourage them to write the inequality symbol that they should have in the solution before solving.
- **Exercises 12–17** If students distribute before solving, they may forget to distribute the number to the second term. Remind them that they need to distribute to everything within the parentheses. Encourage students to draw arrows to represent the multiplication.

11.4 Record and Practice Journal

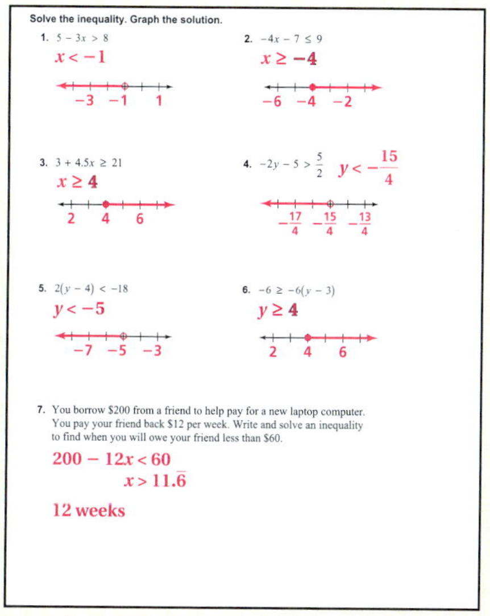
Solve the inequality. Graph the solution.

1. $5 - 3x > 8$ $x < -1$
2. $-4x - 7 \le 9$ $x \ge -4$
3. $3 + 4.5x \ge 21$ $x \ge 4$
4. $-2y - 5 > \frac{5}{2}$ $y < -\frac{15}{4}$
5. $2(y - 4) < -18$ $y < -5$
6. $-6 \ge -6(y - 3)$ $y \ge 4$
7. You borrow $200 from a friend to help pay for a new laptop computer. You pay your friend back $12 per week. Write and solve an inequality to find when you will owe your friend less than $60.

$200 - 12x < 60$
$x > 11.\overline{6}$
12 weeks

Vocabulary and Concept Check

1. *Sample answer:* They use the same techniques, but when solving an inequality, you must be careful to reverse the inequality symbol when you multiply or divide by a negative number.
2. *Sample answer:* Divide both sides by 3 and then subtract 5 from both sides.

Practice and Problem Solving

3. C 4. A

5. $y < 1$;

6. $p \ge -4$;

7. $h > \frac{9}{2}$;

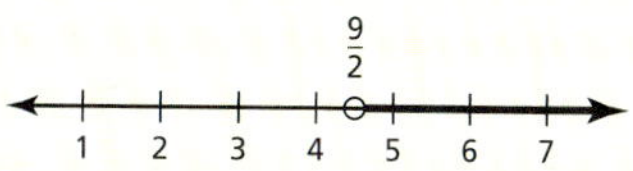

8. $m < 30$;

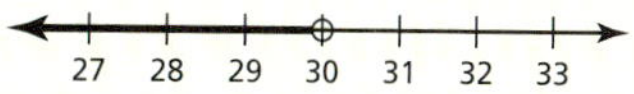

9. $b \le -6$;

10. $r \ge 7$;

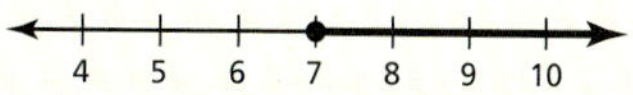

11. They did not perform the operations in the proper order.

$$\frac{x}{3} + 4 < 6$$
$$\frac{x}{3} < 2$$
$$x < 6$$

12–17. See Additional Answers.

18. $n \ge 8$ additional jumps

Practice and Problem Solving

19. $x \geq 4$;

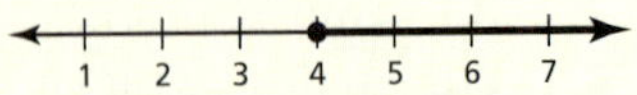

20. $d > 6$;

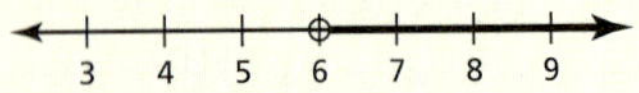

21. $-12x - 38 < -200$;
$x > 13.5$ min

22. See *Taking Math Deeper.*

23. **a.** $9.5(70 + x) \geq 1000$;
$x \geq 35\frac{5}{19}$, which means that at least 36 more tickets must be sold.

b. Because each ticket costs \$1 more, fewer tickets will be needed for the theater to earn \$1000.

24. $r \geq 8$ units

Fair Game Review

25.

Flutes	7	21	28
Clarinets	4	12	16

7 : 4, 21 : 12, and 28 : 16

26.

Boys	6	3	30
Girls	10	5	50

6 : 10, 3 : 5, and 30 : 50

27. A

Mini-Assessment

Solve the inequality. Graph the solution.

1. $2x + 4 < 10$ $x < 3$;

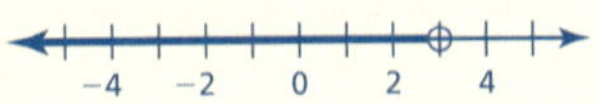

2. $3 \leq \frac{y}{-5} + 7$ $y \leq 20$;

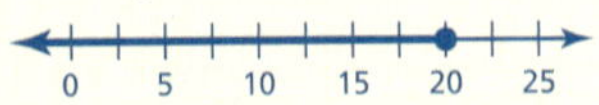

3. $-4.2 - 1.1b \leq 2.4$ $b \geq -6$;

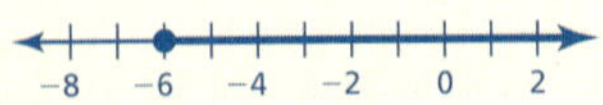

4. $\frac{2}{3}m + \frac{2}{3} \geq -\frac{1}{3}$ $m \geq -\frac{3}{2}$;

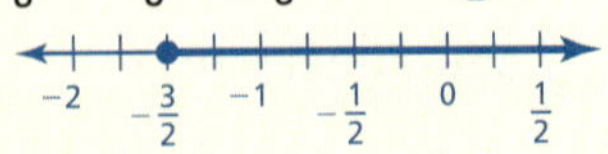

Taking Math Deeper

Exercise 22

Inequality problems can throw students off, simply because of the "inequality." A good way to approach the problem is to imagine that it is an "equality" problem. After the "equation" is written, decide which way the inequality symbol should point.

1 Write an equation.

$$75 + 15x = 140$$

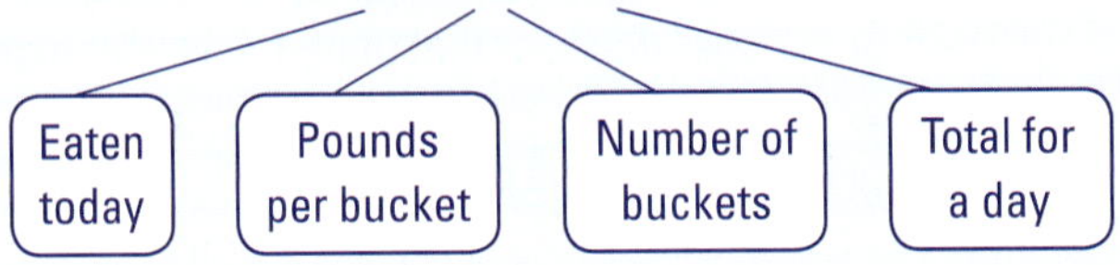

2 **a.** Write an inequality and solve it. Recall the whale needs to eat at least 140 pounds of fish each day.

$75 + 15x \geq 140$	Write inequality.
$15x > 65$	Subtract 75 from each side.
$x \geq 4\frac{1}{3}$	Divide each side by 15.

The whale needs to eat at least $4\frac{1}{3}$ more buckets of fish.

3 **b.** If you only have a choice of whole buckets, then the whale should be given 5 more buckets of food. With 4 buckets, the whale would only get $75 + 60 = 135$ pounds of fish.

Project

Select a marine park. Research the amount of fish needed to feed all of the animals at the park. Determine the cost of the fish, not including storage and personnel needed to feed the animals. Using the entrance fee to the park, how many visitors are needed every day to meet the cost? What are some ways that park officials might help offset the daily costs?

Reteaching and Enrichment Strategies

If students need help. . .	If students got it. . .
Resources by Chapter • Practice A and Practice B • Puzzle Time Record and Practice Journal Practice Differentiating the Lesson Lesson Tutorials Skills Review Handbook	Resources by Chapter • Enrichment and Extension • Technology Connection Start the next section

Solve the inequality. Graph the solution.

19. $9x - 4x + 4 \geq 36 - 12$

20. $3d - 7d + 2.8 < 5.8 - 27$

21. SCUBA DIVER A scuba diver is at an elevation of -38 feet. The diver starts moving at a rate of -12 feet per minute. Write and solve an inequality that represents how long it will take the diver to reach an elevation deeper than -200 feet.

22. KILLER WHALES A killer whale has eaten 75 pounds of fish today. It needs to eat at least 140 pounds of fish each day.

a. A bucket holds 15 pounds of fish. Write and solve an inequality that represents how many more buckets of fish the whale needs to eat.

b. Should the whale eat *four* or *five* more buckets of fish? Explain.

23. REASONING A student theater charges \$9.50 per ticket.

a. The theater has already sold 70 tickets. Write and solve an inequality that represents how many more tickets the theater needs to sell to earn at least \$1000.

b. The theater increases the ticket price by \$1. Without solving an inequality, describe how this affects the total number of tickets needed to earn at least \$1000.

24. Problem Solving For what values of r will the area of the shaded region be greater than or equal to 12 square units?

Fair Game Review What you learned in previous grades & lessons

Find the missing values in the ratio table. Then write the equivalent ratios. *(Skills Review Handbook)*

25.

Flutes	7		28
Clarinets	4	12	

26.

Boys	6	3	
Girls	10		50

27. MULTIPLE CHOICE What is the volume of the cube? *(Skills Review Handbook)*

Ⓐ 8 ft^3 Ⓑ 16 ft^3

Ⓒ 24 ft^3 Ⓓ 32 ft^3

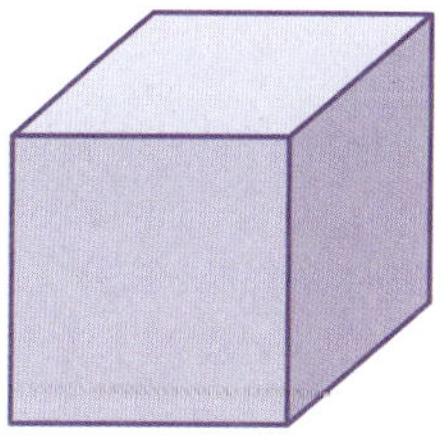

11.3–11.4 Quiz

Check It Out
Progress Check
BigIdeasMath.com

Solve the inequality. Graph the solution. *(Section 11.3 and Section 11.4)*

1. $3p \le 18$

2. $2x > -\frac{3}{5}$

3. $\frac{r}{3} \ge -5$

4. $-\frac{z}{8} < 1.5$

5. $3n + 2 \le 11$

6. $-2 < 5 - \frac{k}{2}$

7. $1.3m - 3.8 < -1.2$

8. $4.8 \ge 0.3(12 - y)$

Write the word sentence as an inequality. Then solve the inequality. *(Section 11.3)*

9. The quotient of a number and 5 is less than 4.

10. Six times a number is at least -14.

11. PEPPERS You have \$18 to buy peppers. Peppers cost \$1.50 each. Write and solve an inequality that represents the number of peppers you can buy. *(Section 11.3)*

12. MOVIES You have a gift card worth \$90. You want to buy several movies that cost \$12 each. Write and solve an inequality that represents the number of movies you can buy and still have at least \$30 on the gift card. *(Section 11.4)*

13. CHOCOLATES Your class sells boxes of chocolates to raise \$500 for a field trip. You earn \$6.25 for each box of chocolates sold. Write and solve an inequality that represents the number of boxes your class must sell to meet or exceed the fundraising goal. *(Section 11.3)*

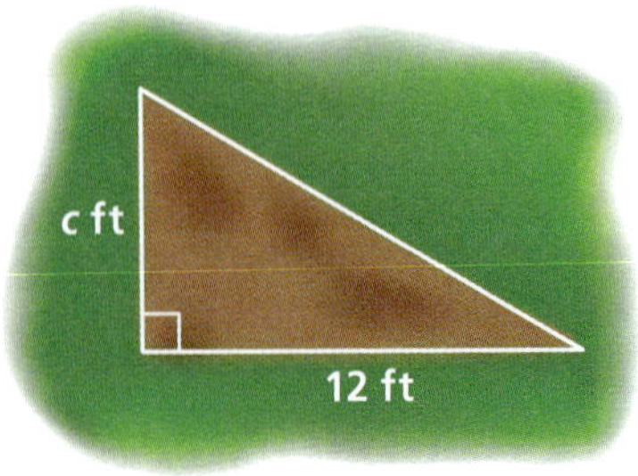

14. FENCE You want to put up a fence that encloses a triangular region with an area greater than or equal to 60 square feet. What is the least possible value of c? Explain. *(Section 11.3)*

Alternative Assessment Options

Math Chat	Student Reflective Focus Question
Structured Interview	Writing Prompt

Math Chat

- Have students work in pairs. One student describes how to write and graph inequalities, giving examples. The other student probes for more information. Students then switch roles and repeat the process for how to solve one- and two-step inequalities.
- The teacher should walk around the classroom listening to the pairs and asking questions to ensure understanding.

Study Help Sample Answers

Remind students to complete Graphic Organizers for the rest of the chapter.

6.

Solving inequalities using multiplication

- Use the Multiplication Property of Inequality (Case 1): When you multiply each side of an inequality by the same *positive* number, the inequality remains true.

  Example: $\frac{x}{5} > 10$

  $\frac{x}{5} \cdot 5 > 10 \cdot 5$

  $x > 50$

- Use the Multiplication Property of Inequality (Case 2): When you multiply each side of an inequality by the same *negative* number, the direction of the inequality symbol must be reversed for the inequality to remain true.

  Example: $-\frac{1}{2}x \le 10$

  $-\frac{2}{1} \cdot \left(-\frac{1}{2}x\right) \ge -\frac{2}{1} \cdot 10$

  $x \ge -20$

Solving inequalities using division

- Use the Division Property of Inequality (Case 1): When you divide each side of an inequality by the same *positive* number, the inequality remains true.

  Example: $5x > 10$

  $\frac{5x}{5} > \frac{10}{5}$

  $x > 2$

- Use the Division Property of Inequality (Case 2): When you divide each side of an inequality by the same *negative* number, the direction of the inequality symbol must be reversed for the inequality to remain true.

  Example: $-3x \ge 9$

  $\frac{-3x}{-3} \le \frac{9}{-3}$

  $x \le -3$

- Use inverse operations to group numbers on one side.
- Use inverse operations to group variables on one side.
- Solve for the variable.

7–8. Available at *BigIdeasMath.com*.

Reteaching and Enrichment Strategies

If students need help...	If students got it...
Resources by Chapter • Practice A and Practice B • Puzzle Time Lesson Tutorials *BigIdeasMath.com*	Resources by Chapter • Enrichment and Extension • Technology Connection Game Closet at *BigIdeasMath.com* Start the Chapter Review

Answers

1. $p \le 6$;

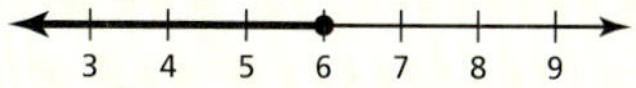

2. $x > -\frac{3}{10}$;

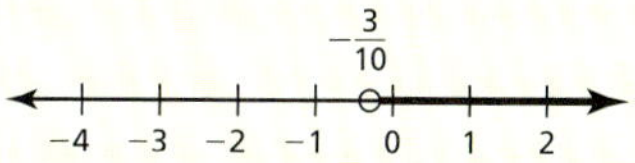

3. $r \ge -15$;

4. $z > -12$;

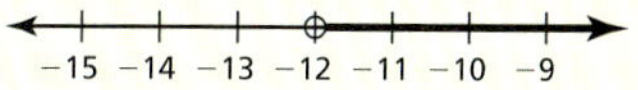

5. $n \le 3$;

6. $k < 14$;

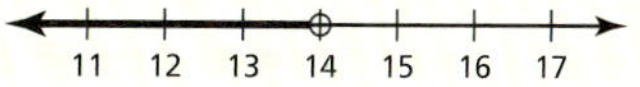

7. $m < 2$;

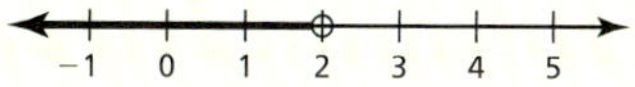

8. $-4 \le y$;

9. $q \div 5 < 4; q < 20$

10. $6t \ge -14; t \ge -\frac{7}{3}$

11. $1.5p \le 18; p \le 12$ peppers

12. $90 - 12c \ge 30; c \le 5$ movies

13. $6.25b \ge 500; b \ge 80$ boxes

14. 10 feet; $\frac{1}{2}(12)c \ge 60 \longrightarrow c \ge 10$; The least value for which $c \ge 10$ is 10.

Technology for the Teacher

Online Assessment
Assessment Book
ExamView® Assessment Suite

For the Teacher

Additional Review Options

- *BigIdeasMath.com*
- Online Assessment
- Game Closet at *BigIdeasMath.com*
- Vocabulary Help
- Resources by Chapter

Answers

1. $w > -3$
2. $y - \frac{1}{2} \leq -\frac{3}{2}$
3. yes
4. no
5. 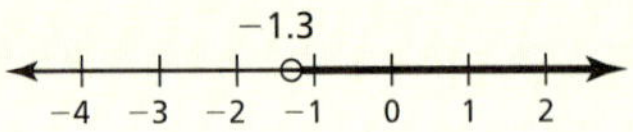

6. 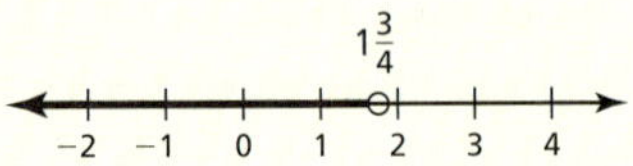

7. $h \geq 42$

Review of Common Errors

Exercises 5 and 6

- Students may use the wrong type of circle at the boundary point. Remind them to use a closed circle when the boundary point is a solution and an open circle when the boundary point is not a solution.

Exercises 5 and 6

- Students may shade on the wrong side of the boundary point. Remind them to test one point on each side of the boundary point.

Exercises 8–10

- Students may use the same operation rather than the inverse operation to isolate the variable. Remind students that when a number is added to the variable, subtract that number from each side. When a number is subtracted from the variable, add that number to each side.

Exercises 10 and 13

- Students may shade the number line in the wrong direction when the variable is on the right side of the inequality symbol. Remind them to rewrite the inequality with the variable at the left or to check a value on each side of the boundary point.

Exercises 11–13

- Students may perform the same operation on both sides instead of the opposite operation when solving the inequality. Remind them that solving inequalities is similar to solving equations.

Exercises 11–13

- When there is a negative in the inequality, students may reverse the direction of the inequality symbol. Remind them that they only reverse the direction when they are multiplying or dividing by a negative number.

Exercises 14–16

- Students may incorrectly multiply or divide before adding to or subtracting from both sides. Remind them that they should work backward through the order of operations, or that they should start away from the variable and move toward it.

Exercises 14–19

- Students may forget to reverse the inequality symbol when multiplying or dividing by a negative number. Encourage them to write the inequality symbol that they should have in the solution before solving.

Exercises 17–19

- If students distribute before solving, they may forget to distribute the number to the second term. Remind them that they need to distribute to everything within the parentheses. Encourage students to draw arrows to represent the multiplication.

11 Chapter Review

Check It Out
Vocabulary Help
BigIdeasMath.com

Review Key Vocabulary

inequality, *p. 466*
solution of an inequality, *p. 466*
solution set, *p. 466*
graph of an inequality, *p. 467*

Review Examples and Exercises

11.1 Writing and Graphing Inequalities *(pp. 464–469)*

a. Six plus a number x is at most $-\frac{1}{4}$. Write this word sentence as an inequality.

Six plus a number x is at most $-\frac{1}{4}$.

$$6 + x \quad \leq \quad -\frac{1}{4}$$

An inequality is $6 + x \leq -\frac{1}{4}$.

b. Graph $m > 3$.

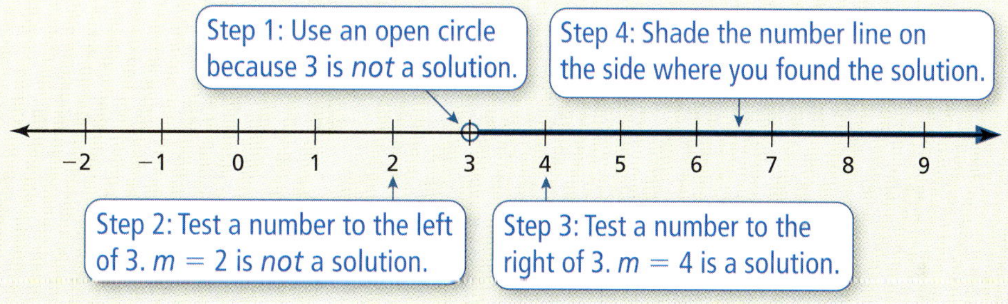

Exercises

Write the word sentence as an inequality.

1. A number w is greater than -3.
2. A number y minus $\frac{1}{2}$ is no more than $-\frac{3}{2}$.

Tell whether the given value is a solution of the inequality.

3. $5 + j > 8; j = 7$
4. $6 \div n \leq -5; n = -3$

Graph the inequality on a number line.

5. $q > -1.3$
6. $s < 1\frac{3}{4}$
7. **BUMPER CARS** You must be at least 42 inches tall to ride the bumper cars at an amusement park. Write an inequality that represents this situation.

11.2 Solving Inequalities Using Addition or Subtraction (pp. 470–475)

Solve $-5 < m - 3$. Graph the solution.

	$-5 < m - 3$	Write the inequality.
Undo the subtraction. →	$\underline{+3} \quad \underline{+3}$	Addition Property of Inequality
	$-2 < m$	Simplify.

∴ The solution is $m > -2$.

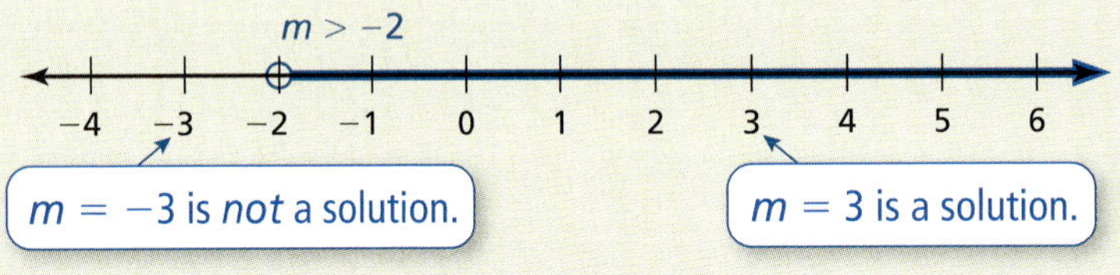

Exercises

Solve the inequality. Graph the solution.

8. $d + 12 < 19$
9. $t - 4 \le -14$
10. $-8 \le z + 6.4$

11.3 Solving Inequalities Using Multiplication or Division (pp. 478–485)

Solve $\frac{c}{-3} \ge -2$. Graph the solution.

	$\frac{c}{-3} \ge -2$	Write the inequality.
Undo the division. →	$-3 \cdot \frac{c}{-3} \le -3 \cdot (-2)$	Use the Multiplication Property of Inequality. Reverse the inequality symbol.
	$c \le 6$	Simplify.

∴ The solution is $c \le 6$.

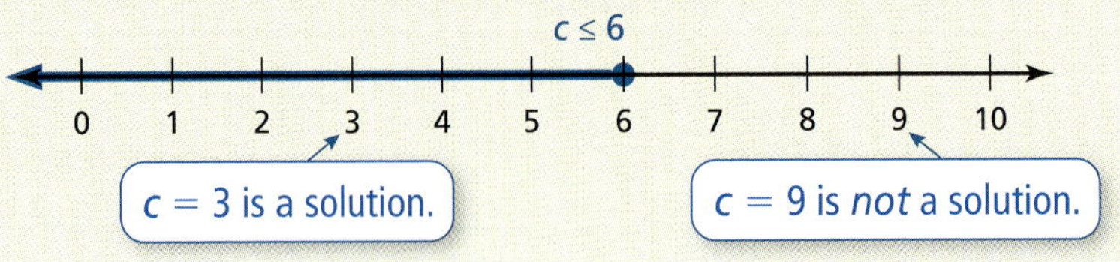

Exercises

Solve the inequality. Graph the solution.

11. $6q < -18$
12. $-\frac{r}{3} \le 6$
13. $-4 > -\frac{4}{3}s$

Review Game

Inequalities

Materials

- Questions (solving inequalities) written on index cards (one question per card) from the chapter homework, quizzes, examples, or tests—at least as many cards as students
- pencil
- paper
- 10 pre-made number lines per pair of students

Directions

Play in pairs. Each pair is a team. Each player is given a question card and solves the inequality on the card. Both players graph their solutions on the same number line. If the shadings on the graph intersect, then the team earns 1 point for the round. If not, then the team does not earn a point.

After each round, each student on the team passes their card in opposite directions to another team along a set rotation to ensure unique pairs of cards. This continues for ten rounds.

Who wins?

The team with the most points at the end of ten rounds wins.

Alternate Point Scoring: In each round, after all the teams complete their graphs, a coin toss can be used to randomly determine whether a point is earned for intersecting graphs or for non-intersecting graphs.

For the Student Additional Practice

- Lesson Tutorials
- Multi-Language Glossary
- Self-Grading Progress Check
- *BigIdeasMath.com*
  Dynamic Student Edition
  Student Resources

Answers

8. $d < 7$;

9. $t \leq -10$;

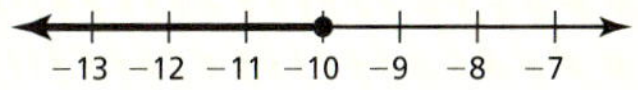

10. $z \geq -14.4$;

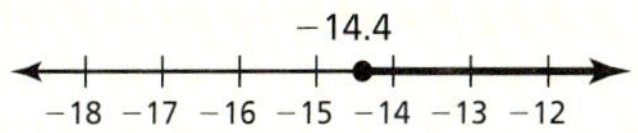

11. $q < -3$;

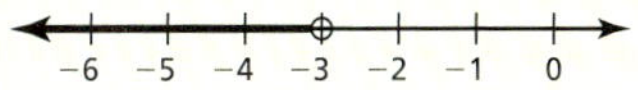

12. $r \geq -18$;

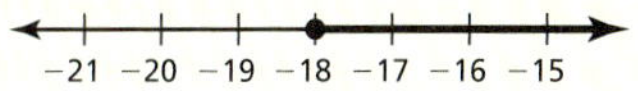

13. $3 < s$;

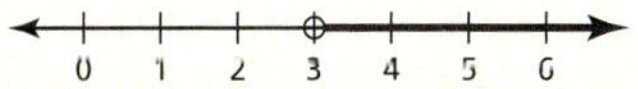

14. $x > 4$;

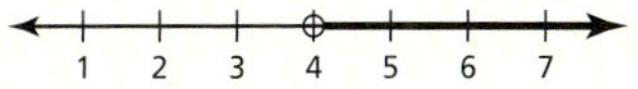

15. $z \geq -8$;

16. $t > -7$;

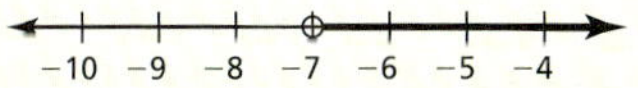

17. $q < -13$;

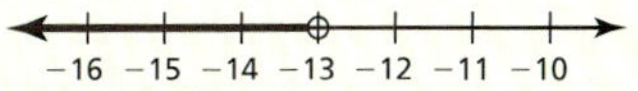

18. $p \geq -21$;

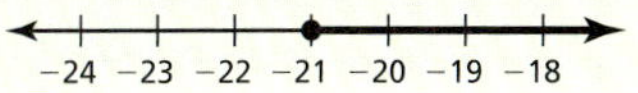

19. $j \geq 0.5$;

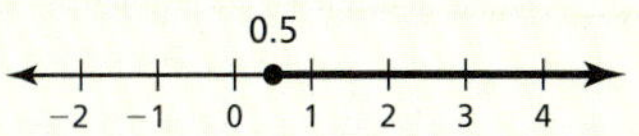

My Thoughts on the Chapter

What worked. . .

Teacher Tip

Not allowed to write in your teaching edition? Use sticky notes to record your thoughts.

What did not work. . .

What I would do differently. . .

11.4 Solving Two-Step Inequalities *(pp. 486–491)*

a. Solve $6x - 8 \le 10$. Graph the solution.

	$6x - 8 \le 10$	Write the inequality.
Step 1: Undo the subtraction. →	$\underline{+8} \quad \underline{+8}$	Addition Property of Inequality
	$6x \le 18$	Simplify.
Step 2: Undo the multiplication. →	$\dfrac{6x}{6} \le \dfrac{18}{6}$	Division Property of Inequality
	$x \le 3$	Simplify.

∴ The solution is $x \le 3$.

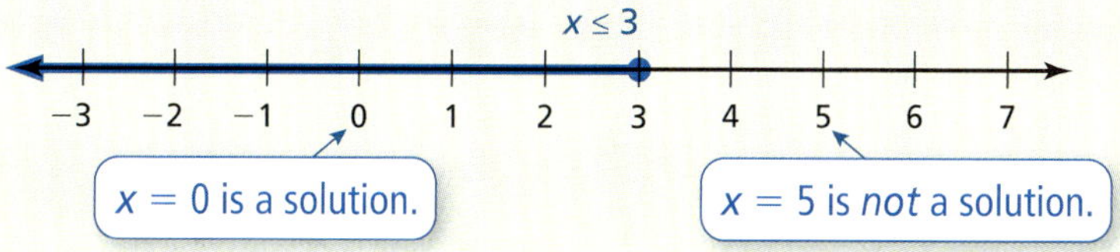

b. Solve $\dfrac{q}{-4} + 7 < 11$. Graph the solution.

	$\dfrac{q}{-4} + 7 < 11$	Write the inequality.
Step 1: Undo the addition. →	$\underline{-7} \quad \underline{-7}$	Subtraction Property of Inequality
	$\dfrac{q}{-4} < 4$	Simplify.
Step 2: Undo the division. →	$-4 \cdot \dfrac{q}{-4} > -4 \cdot 4$	Use the Multiplication Property of Inequality. Reverse the inequality symbol.
	$q > -16$	Simplify.

∴ The solution is $q > -16$.

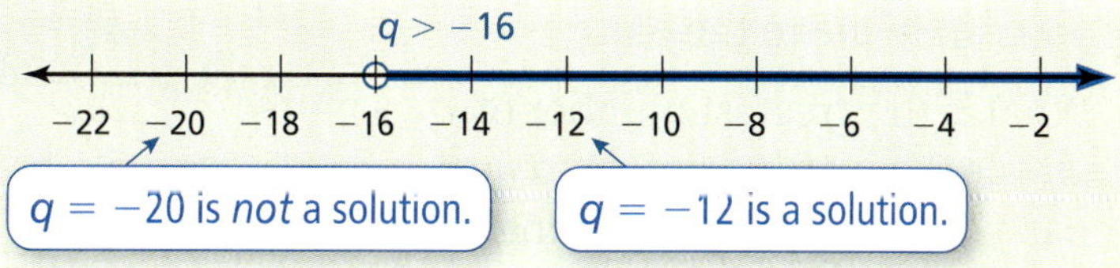

Exercises

Solve the inequality. Graph the solution.

14. $3x + 4 > 16$

15. $\dfrac{z}{-2} - 6 \le -2$

16. $-2t - 5 < 9$

17. $7(q + 2) < -77$

18. $-\dfrac{1}{3}(p + 9) \le 4$

19. $1.2(j + 3.5) \ge 4.8$

11 Chapter Test

Check It Out
Test Practice
BigIdeasMath.com

Write the word sentence as an inequality.

1. A number k plus 19.5 is less than or equal to 40.

2. A number q multiplied by $\frac{1}{4}$ is greater than -16.

Tell whether the given value is a solution of the inequality.

3. $n - 3 \leq 4;\ n = 7$

4. $-\frac{3}{7}m < 1;\ m = -7$

5. $-4c \geq 7;\ c = -2$

6. $-2.4m > -6.8;\ m = -3$

Solve the inequality. Graph the solution.

7. $w + 4 \leq 3$

8. $x - 4 > -6$

9. $-\frac{2}{9} + y \leq \frac{5}{9}$

10. $-6z \geq 36$

11. $-5.2 \geq \frac{p}{4}$

12. $4k - 8 \geq 20$

13. $\frac{4}{7} - b \geq -\frac{1}{7}$

14. $-0.6 > -0.3(d + 6)$

15. **GUMBALLS** You have \$2.50. Each gumball in a gumball machine costs \$0.25. Write and solve an inequality that represents the number of gumballs you can buy.

16. **PARTY** You can spend no more than \$100 on a party you are hosting. The cost per guest is \$8.

    a. Write and solve an inequality that represents the number of guests you can invite to the party.

    b. What is the greatest number of guests that you can invite to the party? Explain your reasoning.

17. **BASEBALL CARDS** You have \$30 to buy baseball cards. Each pack of cards costs \$5. Write and solve an inequality that represents the number of packs of baseball cards you can buy and still have at least \$10 left.

Test Item References

Chapter Test Questions	Section to Review
1–6	11.1
7–9	11.2
10, 11, 15, 16	11.3
12–14, 17	11.4

Test-Taking Strategies

Remind students to quickly look over the entire test before they start so that they can budget their time. When writing word phrases as inequalities, students can get confused by the subtle differences in wording. Encourage students to think carefully about which inequality symbol is implied by the wording. Have students use the **Stop** and **Think** strategy.

Common Errors

- **Exercises 3–6** Students may not understand when the boundary point is a solution of the inequality. Remind them that the inequality is true for the boundary point only when the inequality symbol is $\leq$ or $\geq$.
- **Exercises 7–14** Students may perform the same operation on both sides instead of the opposite operation when solving the inequality. Remind them that solving inequalities is similar to solving equations.
- **Exercises 7–14** Students may reverse the direction of the inequality symbol just because there is a negative number in the inequality. Remind them that you only reverse the direction of the inequality symbol when you multiply each side by a negative number, or when you reverse the right and left side of the inequality.
- **Exercises 7–14** Students may shade in the wrong direction when the variable is on the right side of the inequality instead of the left. Remind them to rewrite the inequality by reversing the right and left sides and reversing the inequality symbol.
- **Exercise 14** If students distribute before solving, they may forget to distribute the number to the second term. Remind them that they need to distribute to everything within the parentheses. Encourage students to draw arrows to represent the multiplication.

Reteaching and Enrichment Strategies

If students need help. . .	If students got it. . .
Resources by Chapter • Practice A and Practice B • Puzzle Time Record and Practice Journal Practice Differentiating the Lesson Lesson Tutorials *BigIdeasMath.com* Skills Review Handbook	Resources by Chapter • Enrichment and Extension • Technology Connection Game Closet at *BigIdeasMath.com* Start Cumulative Assessment

Answers

1. $k + 19.5 \leq 40$
2. $\frac{1}{4}q > -16$
3. yes
4. no
5. yes
6. yes
7. $w \leq -1$;

8. $x > -2$;

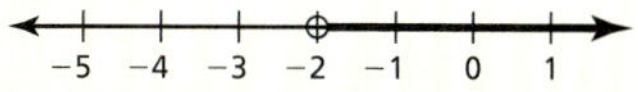

9. $y \leq \frac{7}{9}$;

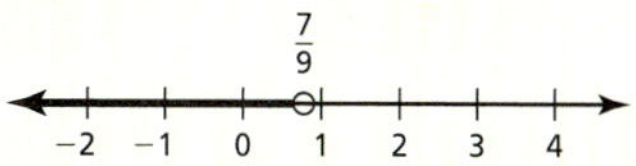

10. $z \leq -6$;

11. $p \leq -20.8$;

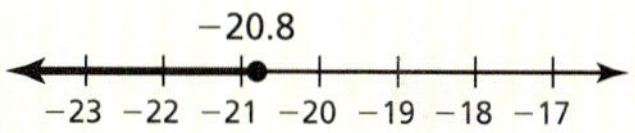

12. $k \geq 7$;

13. $b \leq \frac{5}{7}$;

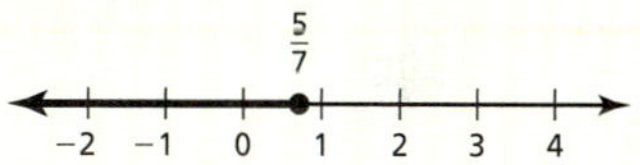

14. $d > -4$;

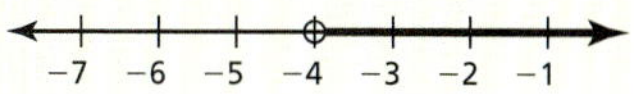

15. $0.25g \leq 2.50$; $g \leq 10$ gumballs
16. a. $8g \leq 100$; $g \leq 12.5$

    b. Twelve guests because 12 is the largest whole number that satisfies the inequality.
17. $30 - 5c \geq 10$; $c \leq 4$ packs of cards

Technology for the Teacher

Online Assessment
Assessment Book
ExamView® Assessment Suite

Test-Taking Strategies

Available at *BigIdeasMath.com*

After Answering Easy Questions, Relax
Answer Easy Questions First
Estimate the Answer
Read All Choices before Answering
Read Question before Answering
Solve Directly or Eliminate Choices
Solve Problem before Looking at Choices
Use Intelligent Guessing
Work Backwards

About this Strategy

When taking a multiple choice test, be sure to read each question carefully and thoroughly. When taking a timed test, it is often best to skim the test and answer the easy questions first. Be careful that you record your answer in the correct position on the answer sheet.

Answers

1. B
2. I
3. A
4. F

Technology for the Teacher

Performance Tasks
Online Assessment
Assessment Book
ExamView® Assessment Suite

Item Analysis

1. **A.** The student makes at least two sign errors while performing the operations.
   B. Correct answer
   C. The student makes a sign error while performing the operations.
   D. The student makes one or more sign errors while performing the operations.

2. **F.** The student inverts the product.
   G. The student inverts the first factor before multiplying.
   H. The student inverts the second factor before multiplying.
   I. Correct answer

3. **A.** Correct answer
   B. The student makes sign errors and fails to reverse the inequality symbol while solving the inequality.
   C. The student makes a sign error while solving the inequality.
   D. The student fails to reverse the inequality symbol while solving the inequality.

4. **F.** Correct answer
   G. The student subtracts 6 from each side, and then divides each side by 5.
   H. The student divides each side by 5, and then adds 6 to each side instead of subtracting.
   I. The student subtracts 5 from each side, and then subtracts 6 from each side.

11 Cumulative Assessment

Test-Taking Strategy
Answer Easy Questions First

"Scan the test and **answer the easy questions first**. Because 2 < 3, A is correct."

1. What is the value of the expression below when $x = -5$, $y = 3$, and $z = -1$?

$$\frac{x^2 - 3y}{z}$$

A. -34 **C.** 16

B. -16 **D.** 34

2. What is the value of the expression below?

$$-\frac{3}{8} \cdot \frac{2}{5}$$

F. $-\frac{20}{3}$ **H.** $-\frac{15}{16}$

G. $-\frac{16}{15}$ **I.** $-\frac{3}{20}$

3. Which graph represents the inequality below?

$$\frac{x}{-4} - 8 \geq -9$$

A.

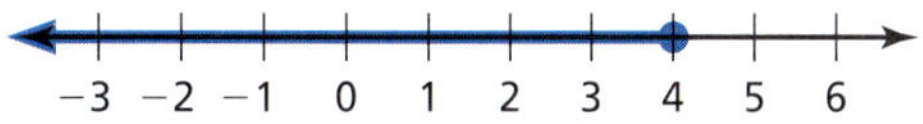

C.

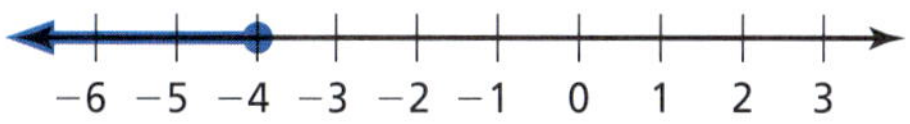

B.

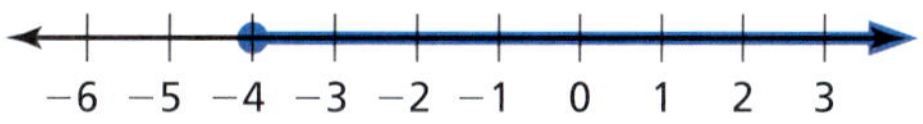

D.

4. Which value of p makes the equation below true?

$$5(p + 6) = 25$$

F. -1 **H.** 11

G. $3\frac{4}{5}$ **I.** 14

5. You set up the lemonade stand. Your profit is equal to your revenue from lemonade sales minus your cost to operate the stand. Your cost is \$8. How many cups of lemonade must you sell to earn a profit of \$30?

A. 4

B. 44

C. 60

D. 76

6. Which value is a solution of the inequality below?

$$3 - 2y < 7$$

F. -6

G. -3

H. -2

I. -1

7. What value of y makes the equation below true?

$$12 - 3y = -6$$

8. What is the mean distance of the four points from -3?

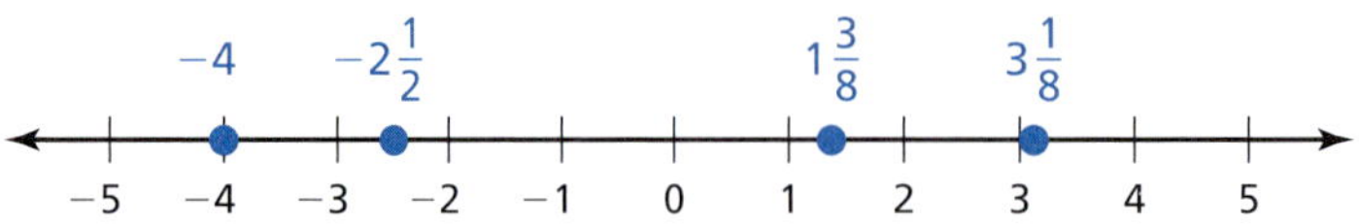

A. $-\frac{1}{2}$

B. $2\frac{1}{2}$

C. 3

D. $7\frac{1}{8}$

Item Analysis (continued)

5. **A.** The student writes and solves the equation $(8 - 0.50)x = 30$.

   B. The student writes and solves the equation $0.50x = 30 - 8$.

   C. The student finds the number of cups to earn a revenue of $30.

   D. Correct answer

6. **F.** Because the inequality symbol is $<$, the student chooses the least possible answer.

   G. The student fails to reverse the inequality symbol and gets $y < -2$.

   H. The student solves the problem as an equation instead of as an inequality.

   I. Correct answer

7. **Gridded response:** Correct answer: 6

   Common error: The student makes a sign error.

8. **A.** The student finds the mean of the four points.

   B. The student finds the difference of each number and -3 and then finds the mean of the results.

   C. Correct answer

   D. The student finds the distance (the range) between the greatest and least points.

Answers

5. D
6. I
7. 6
8. C

Answers

9. F

10. $-\frac{11}{24}$

11. *Part A* at least 48 more T-shirts; The inequality is $20 + 10t \geq 500$.

Part B at least 63 T-shirts; The inequality is $8t \geq 500$.

Part C Your friend; at least 13 more T-shirts; You must sell at least 50 total T-shirts and your friend must sell at least 63 total T-shirts.

12. A

Item Analysis (continued)

9. **F.** Correct answer

G. The student fails to reverse the inequality symbol when dividing each side by a negative number.

H. The student fails to reverse the inequality symbol when dividing each side by a negative number and misinterprets the non-strict inequality symbol.

I. The student makes a sign error in solving the inequality.

10. **Gridded response:** Correct answer: $-\frac{11}{24}$

Common error: The student makes an arithmetic error in finding a common denominator.

11. **4 points** The student demonstrates a thorough understanding of writing and solving inequalities. The student identifies the correct quantities, operations, and inequality symbols, and the solutions are clear, neat, and correct. In Part A, you must sell at least 48 more T-shirts. In Part B, your friend must sell at least 63 T-shirts. In Part C, your friend must sell 13 more T-shirts; He must sell 63, you must sell a total of 50.

3 points The student demonstrates an essential but less than thorough understanding of writing and solving inequalities. There may be one error made, but subsequent work is consistent with the error.

2 points The student demonstrates a partial understanding of writing and solving inequalities. The student's work and explanations demonstrate a lack of essential understanding. The student sets up one or more of the three problem situations incorrectly.

1 point The student demonstrates limited understanding. The student's response is incomplete and exhibits many errors.

0 points The student provides no response, or a response that is completely incorrect, incomprehensible, or fails to demonstrate sufficient understanding of writing and solving inequalities.

12. **A.** Correct answer

B. The student makes a sign error in multiplying.

C. The student fails to handle the signs correctly when subtracting the negative fraction in the expression in the problem statement.

D. The student inverts the dividend instead of the divisor before multiplying.

9. Martin graphed the solution of the inequality $-4x + 18 > 6$ in the box below.

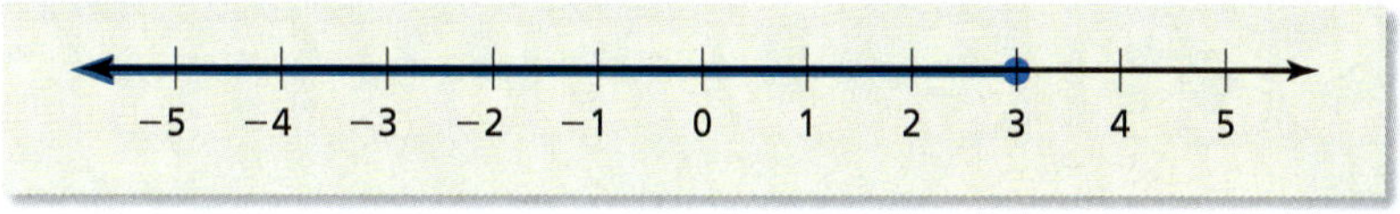

What should Martin do to correct the error that he made?

F. Use an open circle at 3 and shade to the left of 3.

G. Use an open circle at 3 and shade to the right of 3.

H. Use a closed circle and shade to the right of 3.

I. Use an open circle and shade to the left of -3.

10. What is the value of the expression below?

$$\frac{5}{12} - \frac{7}{8}$$

11. You are selling T-shirts to raise money for a charity. You sell the T-shirts for \$10 each.

Think Solve Explain

Part A You have already sold 2 T-shirts. How many more T-shirts must you sell to raise at least \$500? Explain.

Part B Your friend is raising money for the same charity and has not sold any T-shirts previously. He sells the T-shirts for \$8 each. What is the total number of T-shirts he must sell to raise at least \$500? Explain.

Part C Who has to sell more T-shirts in total? How many more? Explain.

12. Which expression is equivalent to the expression below?

$$-\frac{2}{3} - \left(-\frac{4}{9}\right)$$

A. $-\frac{1}{3} + \frac{1}{9}$

B. $-\frac{2}{3} \times \left(-\frac{1}{3}\right)$

C. $-\frac{1}{3} - \frac{7}{9}$

D. $\frac{3}{2} \div \left(-\frac{1}{3}\right)$

12 Constructions and Scale Drawings

12.1 Adjacent and Vertical Angles

12.2 Complementary and Supplementary Angles

12.3 Triangles

12.4 Quadrilaterals

12.5 Scale Drawings

"Move 4 of the lines to make 3 equilateral triangles."

"Well done, Descartes!"

"I'm at 3rd base. You are running to 1st base, and Fluffy is running to 2nd base."

"Should I throw the ball to 2nd to get Fluffy out or throw it to 1st to get you out?"

What Your Students Have Learned

- Classify two dimensional figures into categories based on properties.
- Interpret multiplication as scaling.
- Convert standard measurement units within a measurement system.
- Draw polygons in the coordinate plane given vertices and find lengths of sides.
- Understand ratios and describe ratio relationships.
- Use ratio reasoning to convert measurement units.

What Your Students Will Learn

- Use supplementary, complementary, vertical, and adjacent angles.
- Draw geometric shapes with given conditions, focusing on triangles and quadrilaterals.
- Reproduce a scale drawing at a different scale.
- Represent proportional relationships with equations.
- Use proportionality to solve ratio problems.
- Use scale drawings to compute actual lengths and areas.
- Find scale factors using scale drawings and models.

Pacing Guide for Chapter 12

Chapter Opener Advanced	1 Day
Section 1 Advanced	1 Day
Section 2 Advanced	1 Day
Section 3 Advanced	3 Days
Study Help / Quiz Advanced	1 Day
Section 4 Advanced	2 Days
Section 5 Advanced	1 Day
Chapter Review/ Chapter Tests Advanced	2 Days
Total Chapter 12 Advanced	12 Days
Year-to-Date Advanced	111 Days

Technology for the Teacher

BigIdeasMath.com
Chapter at a Glance
Complete Materials List
Parent Letters: English and Spanish

What Your Students Have Learned

- Measure angles and draw angles using a protractor.

Additional Topics for Review

- Lines
- Intersection
- Coordinate Plane
- Graphing Ordered Pairs
- Polygons
- Triangles
- Ratios
- Proportions
- Perimeter
- Area

Try It Yourself

1. 70°; acute
2. 90°; right
3. 115°; obtuse
4. 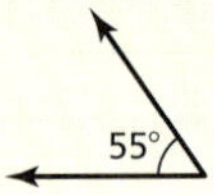

5. 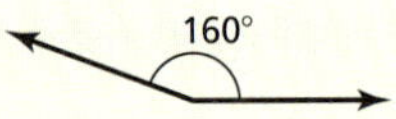

6. 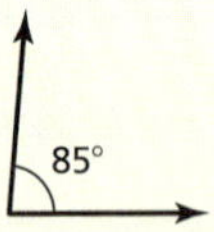

7. 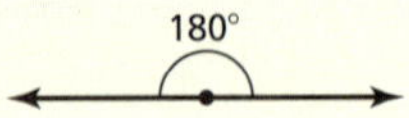

Record and Practice Journal Fair Game Review

1. 60°; acute
2. 90°; right
3. 120°; obtuse
4. 65°; acute
5. 180°; straight
6. 165°; obtuse

7–10. See Additional Answers.

Math Background Notes

Vocabulary Review

- Angle
- Degrees
- Protractor
- Acute
- Obtuse
- Right
- Straight
- Vertex
- Endpoint
- Ray

Measuring Angles

- Students should know how to measure angles using a protractor.
- Remind students that an angle can be classified by its measure. A right angle is 90°, an acute angle is less than 90°, an obtuse angle is between 90° and 180°, and a straight angle is 180°.
- **Common Error:** Students may use the wrong set of angles on a protractor. Encourage them to decide which set to use by comparing the angle measure to 90°.
- **Teaching Tip:** Ask students to find real-life examples of angles, such as the hands of a clock.
- **Representation:** Show students how to measure an angle when one of the rays does not pass through the 0° mark on a protractor. Suppose in Example 1(a), the center of the protractor is aligned at the angle's vertex but the rays pass through the 40° mark and the 60° mark. Students can find the angle measure by subtracting 40 from 60.

Drawing Angles

- Students should know how to draw angles of specified measure.
- **Common Error:** Again, students may use the wrong set of angles on a protractor. Encourage them to decide which set to use by comparing the angle measure to 90°.

Reteaching and Enrichment Strategies

If students need help. . .	If students got it. . .
Record and Practice Journal • Fair Game Review Skills Review Handbook Lesson Tutorials	Game Closet at *BigIdeasMath.com* Start the next section

What You Learned Before

"Look at this baby crocodile! Isn't it cute?"

Measuring Angles

Example 1 Use a protractor to find the measure of each angle. Then classify the angle as *acute*, *obtuse*, *right*, or *straight*.

a.

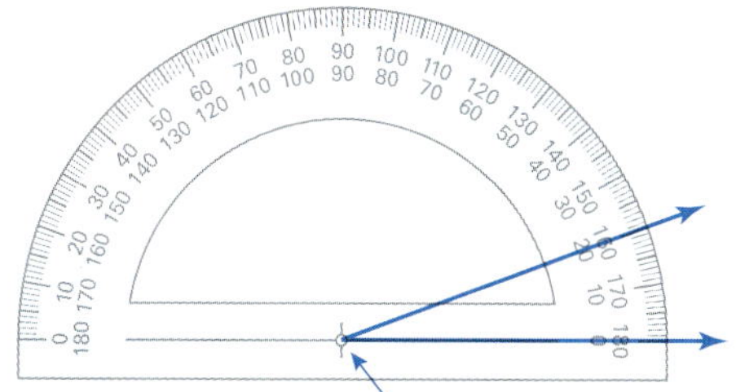

b.

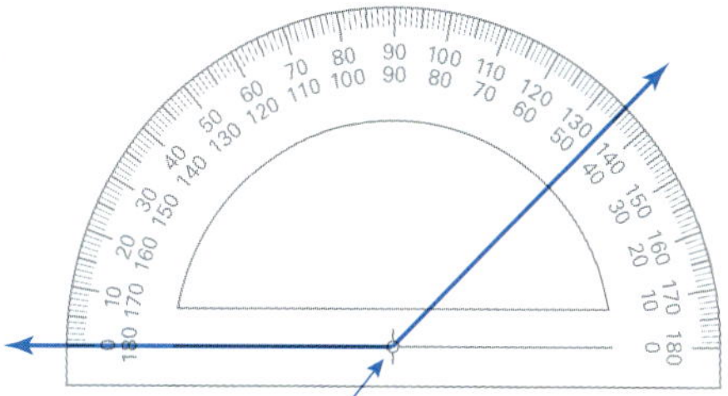

Align the center of the protractor with the angle's vertex.

∴ The angle measure is 20°. So, the angle is acute.

∴ The angle measure is 135°. So, the angle is obtuse.

Drawing Angles

Example 2 Use a protractor to draw a 45° angle.

Draw a ray. Place the center of the protractor on the endpoint of the ray and align the protractor so the ray passes through the 0° mark. Make a mark at 45°. Then draw a ray from the endpoint at the center of the protractor through the mark at 45°.

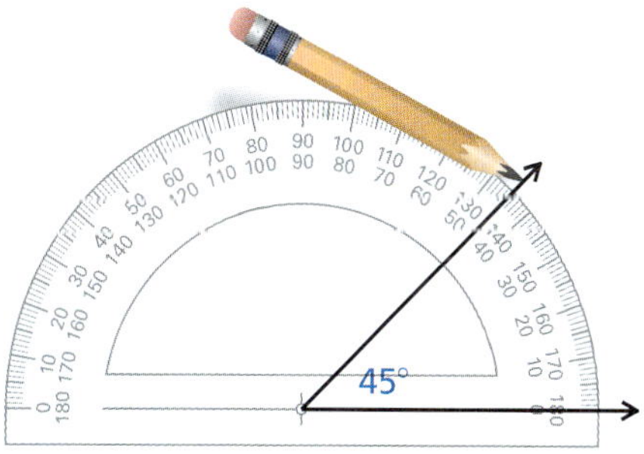

Try It Yourself

Use a protractor to find the measure of the angle. Then classify the angle as *acute*, *obtuse*, *right*, or *straight*.

1.

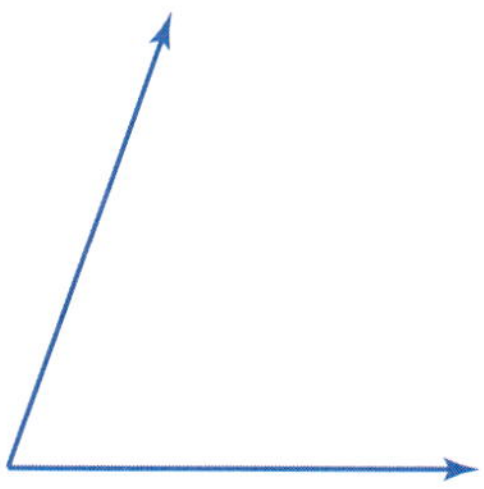

2.

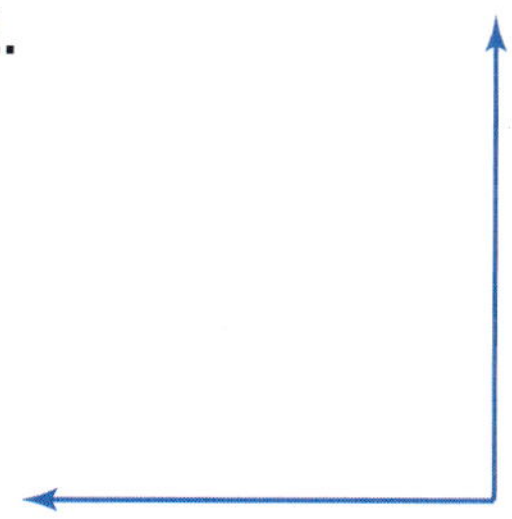

3.

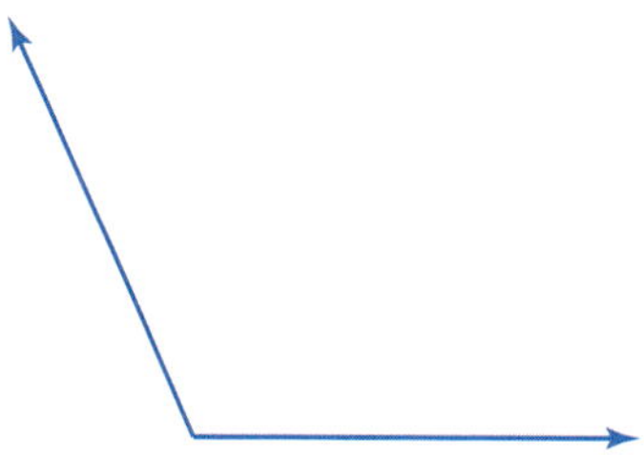

Use a protractor to draw an angle with the given measure.

4. 55° **5.** 160° **6.** 85° **7.** 180°

12.1 Adjacent and Vertical Angles

Essential Question What can you conclude about the angles formed by two intersecting lines?

Classification of Angles

Acute:	***Right:***	***Obtuse:***	***Straight:***
Less than 90°	Equal to 90°	Greater than 90° and less than 180°	Equal to 180°

1 ACTIVITY: Drawing Angles

Work with a partner.

a. Draw the hands of the clock to represent the given type of angle.

Acute Straight Right Obtuse

b. What is the measure of the angle formed by the hands of the clock at the given time?

9:00 6:00 12:00

Geometry

In this lesson, you will

- identify adjacent and vertical angles.
- find angle measures using adjacent and vertical angles.

The Meaning of a Word ● Adjacent

When two states are **adjacent**,

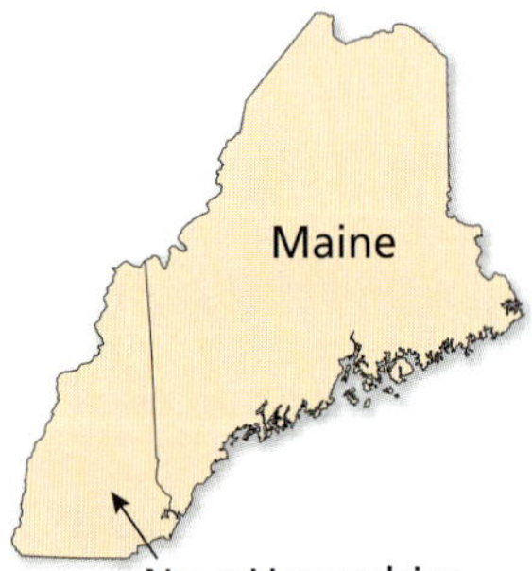

they are next to each other and they share a common border.

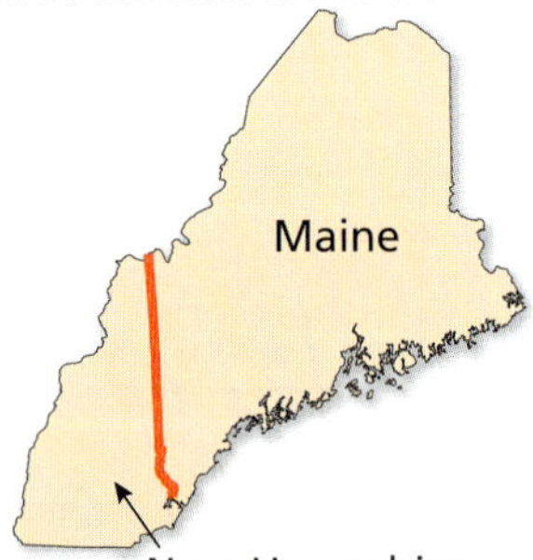

Laurie's Notes

Introduction

Applying Mathematical Practices

- **Construct Viable Arguments and Critique the Reasoning of Others:** Throughout this chapter, students will investigate and make conjectures about geometric properties. Students need practice giving supporting evidence for their conjectures.

Motivate

- **Preparation:** Make a model to practice estimation skills with angle measures. Cut two circles (6-inch diameter) out of file folders. Cut a slit in each. On one circle, label every 10°. The second circle is shaded. Insert one circle into the other so the angle measure faces you and the shaded angle faces the students.
- Ask students to estimate the measure of the shaded angle. You can read the answer from your side of the model. Repeat several times.

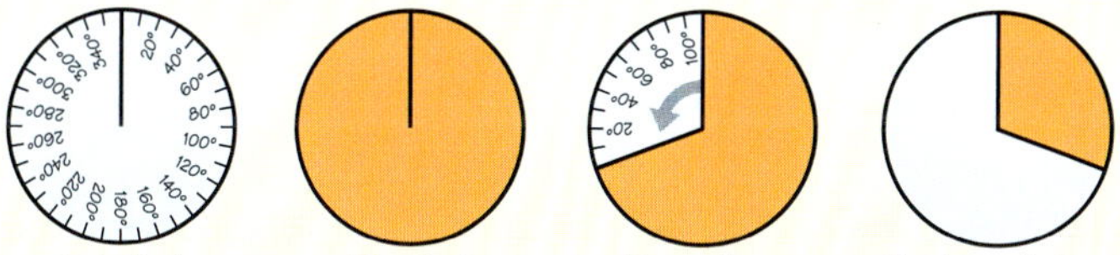

Labeled　Shaded　Your view　Students' view

Activity Notes

Discuss

? "What names do you use to classify angles, and what does each mean?" acute: less than 90°, right: 90°, obtuse: greater than 90° and less than 180°, straight: 180°

- **Caution:** Do not draw every angle in this chapter with the initial ray horizontal and extending rightward. Use varied orientation to gauge students' understanding of reading angle measures.

Activity 1

- In part (a), students could share what time they drew on each clock face.
- **FYI:** Students may ask about angles greater than 180°. If so, tell them that they will study such angles in future math classes. Snowboarders and skateboarders may be familiar with such angles as 360°, 540°, and 720°.
- **Extension:** A *reflex angle* is greater than 180° and less than 360°. For instance, the clockwise angle formed by 2 and 12 is a reflex angle.

The Meaning of the Word

- Discuss the meaning of *adjacent*, using the given map to illustrate.

? "Can you give other examples of objects that are adjacent?" Answers will vary.

What Your Students Will Learn

- Identify adjacent and vertical angles.
- Find angle measures using adjacent and vertical angles.
- Construct angles using a protractor.

Previous Learning

Students should know basic vocabulary associated with angles.

Technology for the Teacher

Dynamic Classroom

Lesson Plans
Complete Materials List

12.1 Record and Practice Journal

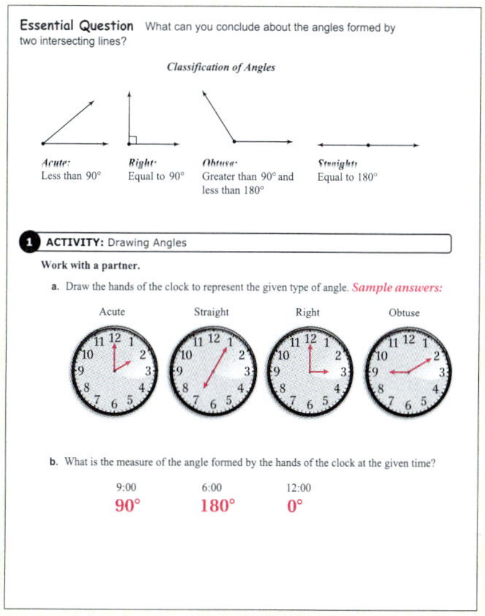

Essential Question What can you conclude about the angles formed by two intersecting lines?

Classification of Angles

Acute: Less than 90°　*Right:* Equal to 90°　*Obtuse:* Greater than 90° and less than 180°　*Straight:* Equal to 180°

1 ACTIVITY: Drawing Angles

Work with a partner.

a. Draw the hands of the clock to represent the given type of angle. *Sample answers:*

Acute　Straight　Right　Obtuse

b. What is the measure of the angle formed by the hands of the clock at the given time?

9:00	6:00	12:00
90°	180°	0°

English Language Learners

Illustrate

Explain to English learners that the name *right angle* does not come from the orientation of the angle opening to the right, as shown in the activity. Students might think that if the angle opens to the left, it is called a *left angle*. Point out that any angle that measures 90° is a *right angle*.

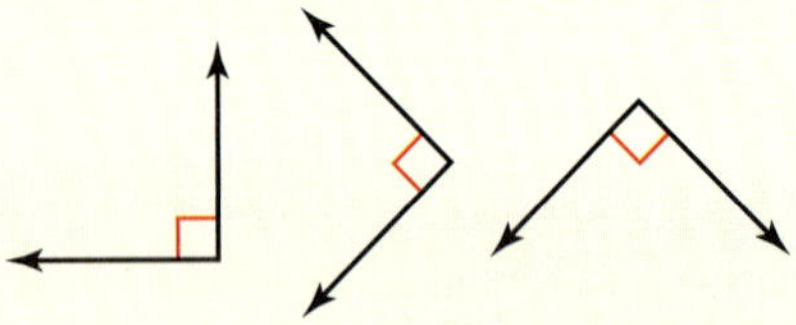

12.1 Record and Practice Journal

2 ACTIVITY: Naming Angles

Work with a partner. Some angles, such as $\angle A$, can be named by a single letter. When this does not clearly identify an angle, you should use three letters, as shown.

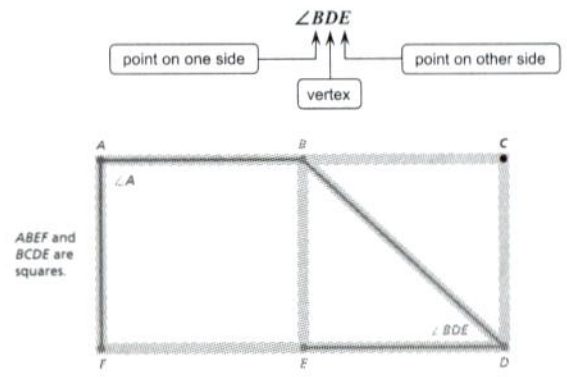

a. Name all of the right angles, acute angles, and obtuse angles.

right: $\angle AFE$, $\angle ABE$, $\angle BEF$, $\angle BAF$, $\angle BED$, $\angle BCD$, $\angle CDE$, $\angle CBE$
acute: $\angle CBD$, $\angle EBD$, $\angle BDE$, $\angle BDC$
obtuse: $\angle ABD$

b. Which pairs of angles do you think are *adjacent*? Explain.

Sample answer: **$\angle ABE$ and $\angle CBE$, $\angle FEB$ and $\angle DEB$, $\angle EBD$ and $\angle CBD$, $\angle EDB$ and $\angle CDB$, $\angle ABD$ and $\angle CBD$**

They share a common side and a common vertex.

3 ACTIVITY: Measuring Angles

Work with a partner.

a. How many angles are formed by the intersecting roads? Number the angles.

4

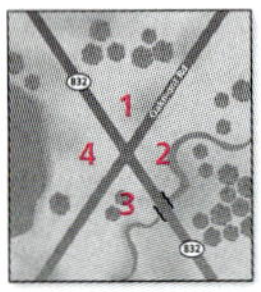

b. **CHOOSE TOOLS** Measure each angle formed by the intersecting roads. What do you notice?

$\angle 1 = 70°$
$\angle 2 = 110°$
$\angle 3 = 70°$
$\angle 4 = 110°$

What Is Your Answer?

4. **IN YOUR OWN WORDS** What can you conclude about the angles formed by two intersecting lines?

Sample answer: **Four angles are formed with four pairs of adjacent angles.**

5. Draw two acute angles that are adjacent.

Check students' work.

Laurie's Notes

Activity 2

- This activity reviews how angles are named. Discuss when there is a need for three letters instead of one. Also discuss that $\angle EBD$ and $\angle DBE$ name the same angle (because the vertex position is the same).

? "What other angles could be named using just one letter?" $\angle C$, $\angle F$

- Although the right angles are not labeled in this diagram, it is assumed that those which appear to be right angles are right angles.

Activity 3

- Do not assume that all students will recall how to measure an angle. It might be helpful to review how to use a protractor before students begin part (b).
- **Attend to Precision:** Discuss the need for precise measurement. It is important for students to take their time and carefully align the center of the protractor with the vertex of the angle.
- Walk around and observe to make sure students are measuring the angles correctly.
- If students ask, tell them that it is not necessary to measure the two straight angles.

? "What do you notice about the angles you measured?" Depending on how carefully students measured, they should have two pairs of corresponding angles that have the same measure.

What Is Your Answer?

- **Think-Pair-Share:** Students should read each question independently and then work in pairs to answer the questions. When they have answered the questions, the pair should compare their answers with another group and discuss any discrepancies.

Closure

- Look around the room. Name angles that appear to be acute, right, obtuse, or straight. (You could also include reflex angles, if you discussed these earlier.)

2 ACTIVITY: Naming Angles

Work with a partner. Some angles, such as $\angle A$, can be named by a single letter. When this does not clearly identify an angle, you should use three letters, as shown.

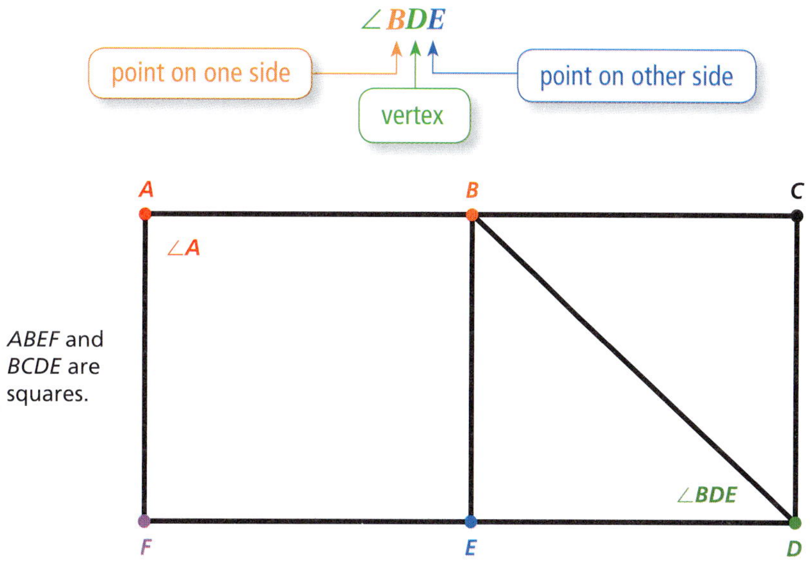

a. Name all the right angles, acute angles, and obtuse angles.

b. Which pairs of angles do you think are *adjacent?* Explain.

> **Math Practice**
>
> **Justify Conclusions**
>
> When you name an angle, does the order in which you write the letters matter? Explain.

3 ACTIVITY: Measuring Angles

Work with a partner.

a. How many angles are formed by the intersecting roads? Number the angles.

b. **CHOOSE TOOLS** Measure each angle formed by the intersecting roads. What do you notice?

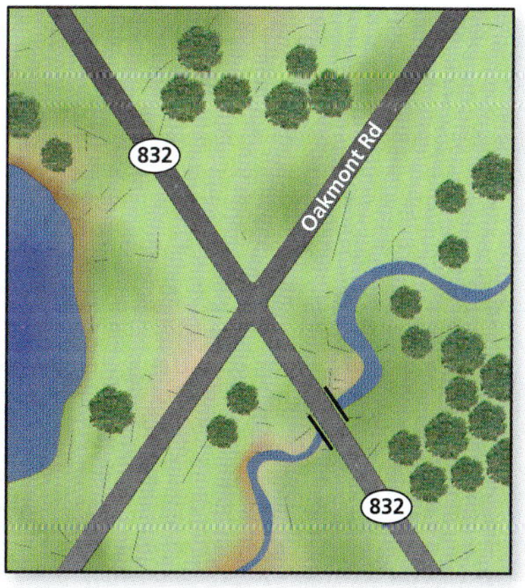

What Is Your Answer?

4. **IN YOUR OWN WORDS** What can you conclude about the angles formed by two intersecting lines?

5. Draw two acute angles that are adjacent.

Practice Use what you learned about angles and intersecting lines to complete Exercises 3 and 4 on page 506.

12.1 Lesson

Check It Out
Lesson Tutorials
BigIdeasMath.com

Key Vocabulary

adjacent angles, *p. 504*
vertical angles, *p. 504*
congruent angles, *p. 504*

Key Ideas

Adjacent Angles

Words Two angles are **adjacent angles** when they share a common side and have the same vertex.

Examples

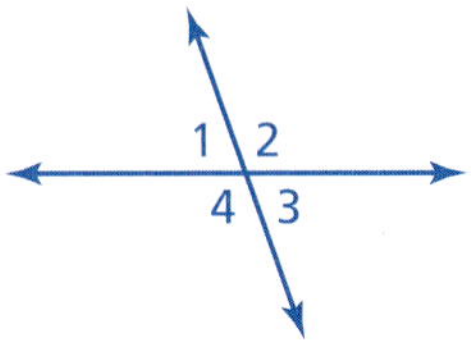

$\angle 1$ and $\angle 2$ are adjacent.

$\angle 2$ and $\angle 4$ are not adjacent.

Vertical Angles

Words Two angles are **vertical angles** when they are opposite angles formed by the intersection of two lines. Vertical angles are **congruent angles**, meaning they have the same measure.

Examples

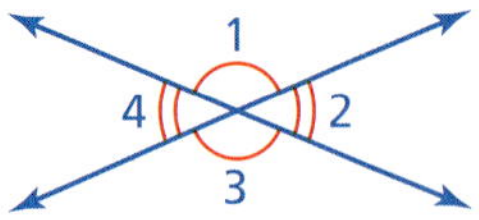

$\angle 1$ and $\angle 3$ are vertical angles.

$\angle 2$ and $\angle 4$ are vertical angles.

EXAMPLE 1 Naming Angles

Use the figure shown.

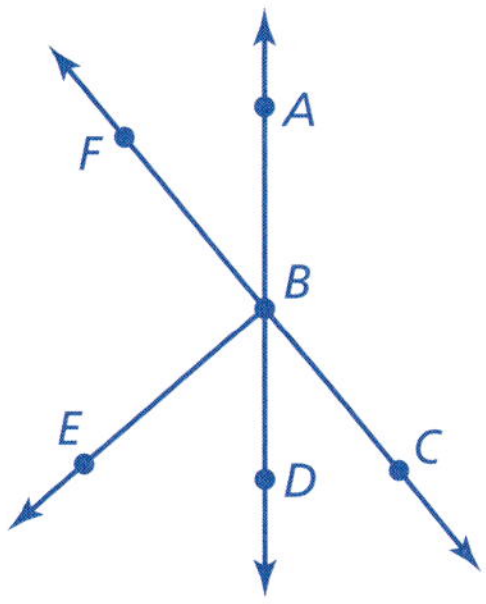

a. Name a pair of adjacent angles.

$\angle ABC$ and $\angle ABF$ share a common side and have the same vertex B.

So, $\angle ABC$ and $\angle ABF$ are adjacent angles.

b. Name a pair of vertical angles.

$\angle ABF$ and $\angle CBD$ are opposite angles formed by the intersection of two lines.

So, $\angle ABF$ and $\angle CBD$ are vertical angles.

On Your Own

Now You're Ready
Exercises 5 and 6

Name two pairs of adjacent angles and two pairs of vertical angles in the figure.

1.

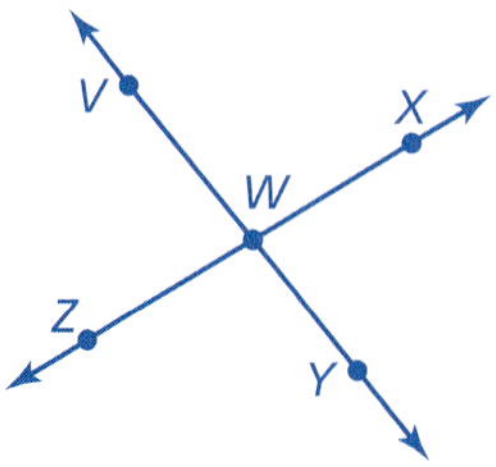

2.

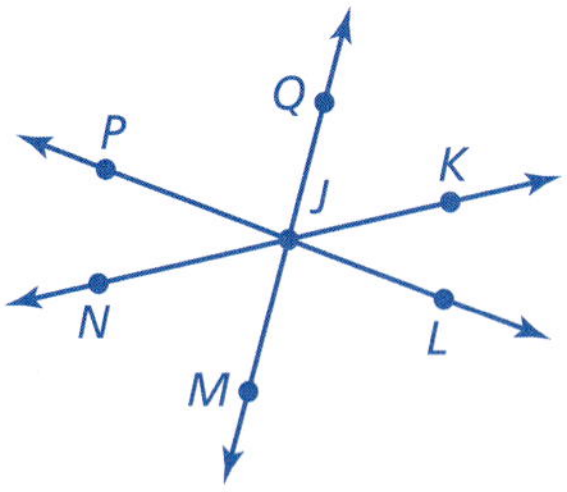

Multi-Language Glossary at BigIdeasMath.com

Laurie's Notes

Introduction

Connect

- **Yesterday:** Students explored drawing, naming, and measuring angles, and investigated relationships between angles.
- **Today:** Students will identify adjacent or vertical angles.

Motivate

- Because this chapter will be focusing on geometry, students should know we credit Euclid for the study of geometry. He is often called the Father of Geometry. Euclid was a Greek mathematician best known for his 13 books on geometry known as *The Elements*. This work influenced the development of Western mathematics for more than 2000 years.

Lesson Notes

Key Ideas

? "Does anyone know what the term *congruent angles* means?" two angles that have the same measure

- Draw two pairs of congruent angles and show students the marks used to indicate that the angles are congruent.

? "What does the word *adjacent* mean?" side-by-side

- Write the Key Ideas.
- **Model:** When two lines intersect, two pairs of vertical angles are formed. Vertical angles are congruent. Demonstrate this with a pair of scissors that have straight blades.

Example 1

- Draw the figure and ask students to identify the lines and rays. There are 5 rays, and two pairs of rays are also collinear (on the same line). Only $\overrightarrow{BE}$ is not part of a line.
- **FYI:** $\angle ABE$ and $\angle CBE$ are adjacent angles. Some students may give this response to part (a).

? "Is $\angle ABE$ adjacent to $\angle FBE$? Explain." Students may say yes because the definition is satisfied; however, adjacent angles do not overlap. In a geometry class, this will be included in the definition.

? "Name a pair of vertical angles." Listen for the two pairs of vertical angles. One pair is acute and the other pair is obtuse.

? **Construct Viable Arguments and Critique the Reasoning of Others:** "When two lines intersect, will there always be one pair of obtuse angles and one pair of acute angles? Explain." no; The two lines could form four right angles.

On Your Own

- **Construct Viable Arguments and Critique the Reasoning of Others:** Question 2 has many correct answers. Students should be able to explain why their answers are correct.

Goal Today's lesson is identifying **adjacent** or **vertical** angles.

Technology for the Teacher

Dynamic Classroom

Lesson Tutorials
Lesson Plans
Answer Presentation Tool

Extra Example 1

Use the figure shown.

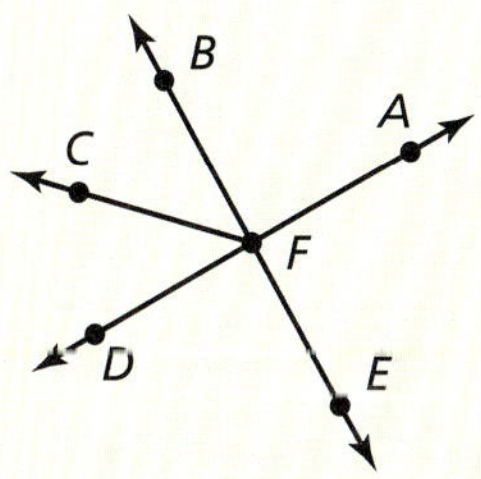

a. Name a pair of adjacent angles.
Sample answer: $\angle AFB$ and $\angle CFB$

b. Name a pair of vertical angles.
Sample answer: $\angle AFB$ and $\angle EFD$

On Your Own

1. *Sample answers:* adjacent: $\angle XWY$ and $\angle ZWY$, $\angle XWY$ and $\angle XWV$; vertical: $\angle VWX$ and $\angle ZWY$, $\angle YWX$ and $\angle ZWV$

2. *Sample answers:* adjacent: $\angle LJM$ and $\angle LJK$, $\angle LJM$ and $\angle NJM$; vertical: $\angle KJL$ and $\angle PJN$, $\angle PJQ$ and $\angle MJL$

Extra Example 2

Tell whether the angles are *adjacent* or *vertical*. Then find the value of *x*.

a.

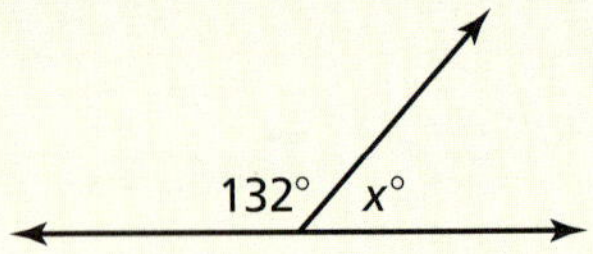

adjacent; 48

b.

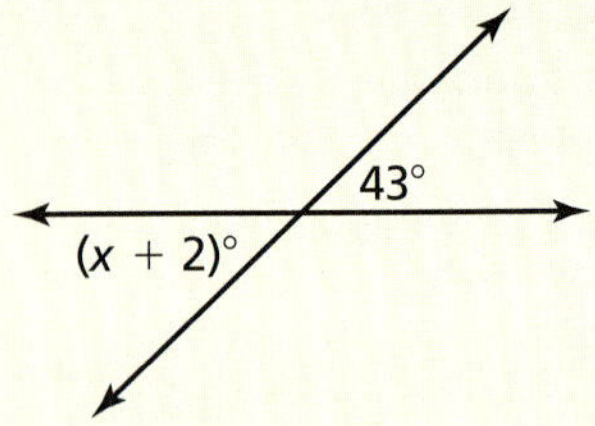

vertical; 41

Extra Example 3

Draw a pair of vertical angles with a measure of 55°.

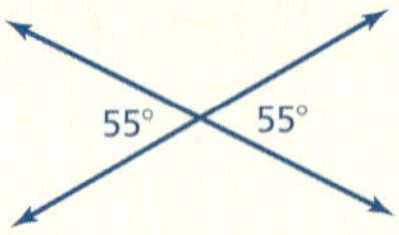

On Your Own

3. adjacent; 95

4. vertical; 90

5. adjacent; 11

6.

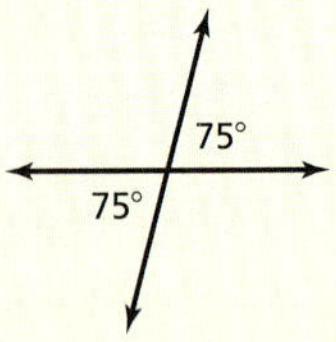

Differentiated Instruction

Visual

Help students visualize vertical angles. Draw vertical angles on the board or overhead.

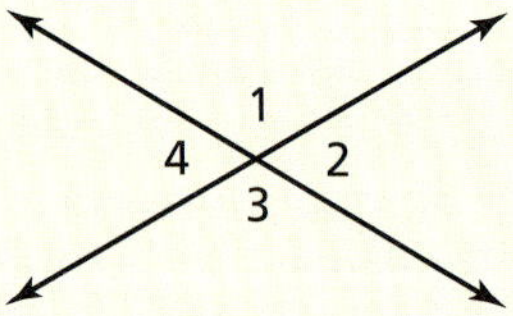

Point out that the lines creating vertical angles form an "X" and that vertical angles do *not* share sides.

Laurie's Notes

Example 2

- Work through each part as shown.
- In part (b), remind students of the corner mark used to designate a right angle.
- Also in part (b), point out the information in the *remember* box. If the angles overlap, then their measures would not sum to the measure of the larger angle.

? Construct Viable Arguments and Critique the Reasoning of Others: Extension: "In part (a), what are the measures of the two remaining angles? How do you know?" 110°; Any two adjacent angles in the figure form a straight angle, which measures 180°, and 180° − 70° = 110°.

Example 3

- Ask students to use their protractors to draw a 40° angle. The first step is not obvious to many students! Draw a ray. Its endpoint will be the vertex of the angle. Then use the protractor to place a mark at 40°, and draw another ray from the vertex through the mark.
- A second method is to draw a line and place a point on the line to represent the vertex. Then place the protractor on the point and place a mark at 40°, and draw a line through the mark and the vertex.
- **Common Error:** When the protractor has two scales (clockwise and counterclockwise), students may draw an angle of 140°. If this happens, then ask whether a 40° angle is acute or obtuse.

On Your Own

- **Think-Pair-Share:** Students should read each question independently and then work in pairs to answer the questions. When they have answered the questions, the pair should compare their answers with another group and discuss any discrepancies.

Closure

- True or False?
  1. Vertical angles are always acute. false
  2. Adjacent angles could be acute. true
  3. Adjacent angles could be obtuse. true
  4. Vertical angles are congruent. true
  5. Adjacent angles could be congruent. true

EXAMPLE 2 Using Adjacent and Vertical Angles

Tell whether the angles are *adjacent* or *vertical*. Then find the value of x.

> **Remember**
>
> You can add angle measures. When two or more adjacent angles form a larger angle, the sum of the measures of the smaller angles is equal to the measure of the larger angle.

a.

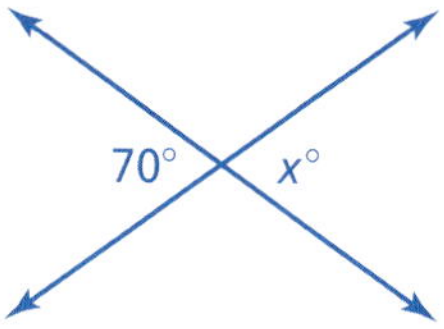

The angles are vertical angles. Because vertical angles are congruent, the angles have the same measure.

∴ So, the value of x is 70.

b.

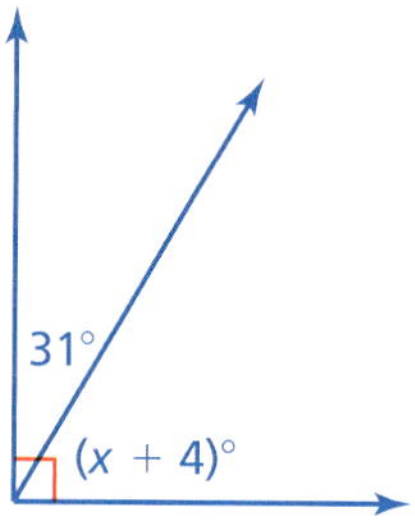

The angles are adjacent angles. Because the angles make up a right angle, the sum of their measures is 90°.

$(x + 4) + 31 = 90$ Write equation.

$x + 35 = 90$ Combine like terms.

$x = 55$ Subtract 35 from each side.

∴ So, the value of x is 55.

EXAMPLE 3 Constructing Angles

Draw a pair of vertical angles with a measure of 40°.

Step 1: Use a protractor to draw a 40° angle.

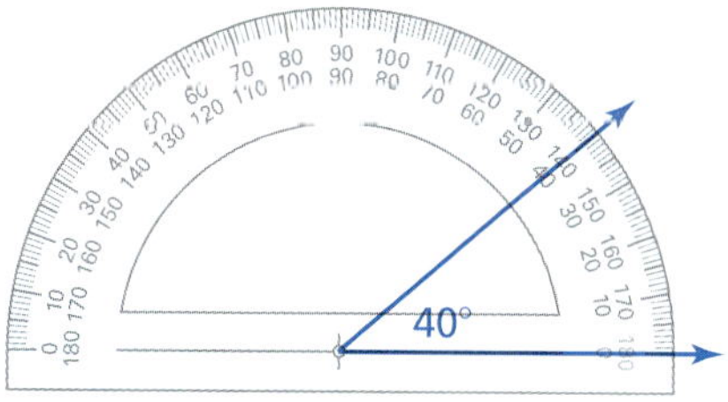

Step 2: Use a straightedge to extend the sides to form two intersecting lines.

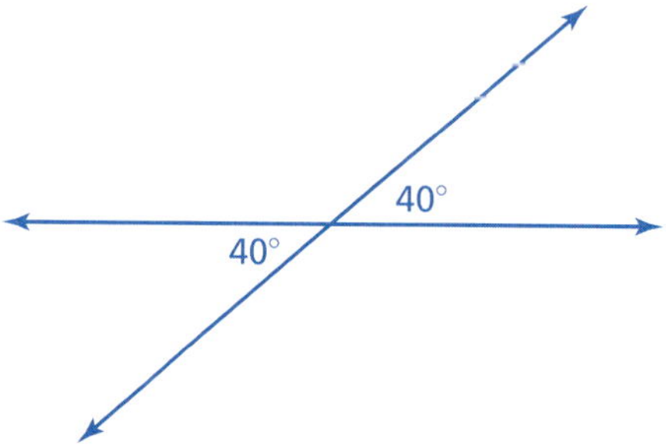

On Your Own

Now You're Ready
Exercises 8–17

Tell whether the angles are *adjacent* or *vertical*. Then find the value of x.

3.

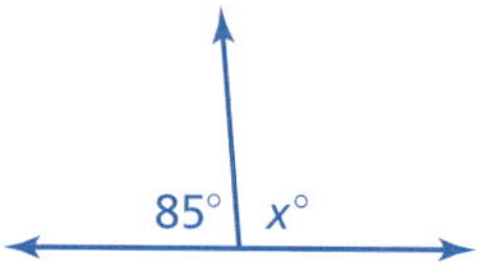

4.

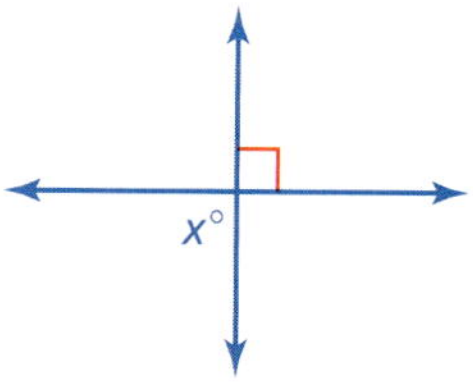

5.

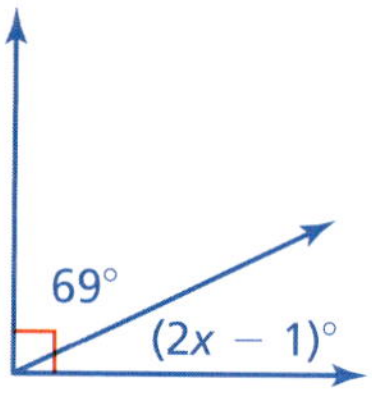

6. Draw a pair of vertical angles with a measure of 75°.

12.1 Exercises

Check It Out
Help with Homework
BigIdeasMath.com

Vocabulary and Concept Check

1. **VOCABULARY** When two lines intersect, how many pairs of vertical angles are formed? How many pairs of adjacent angles are formed?

2. **REASONING** Identify the congruent angles in the figure. Explain your reasoning.

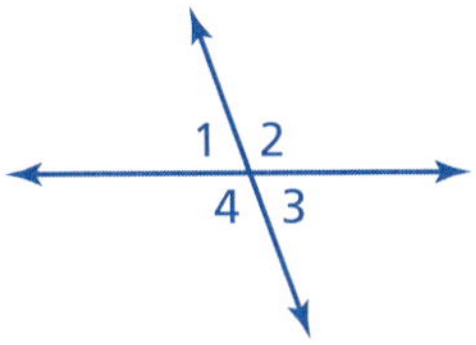

Practice and Problem Solving

Use the figure at the right.

3. Measure each angle formed by the intersecting lines.
4. Name two angles that are adjacent to $\angle ABC$.

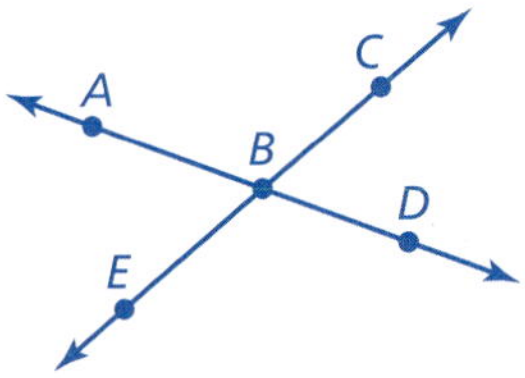

Name two pairs of adjacent angles and two pairs of vertical angles in the figure.

1 5.

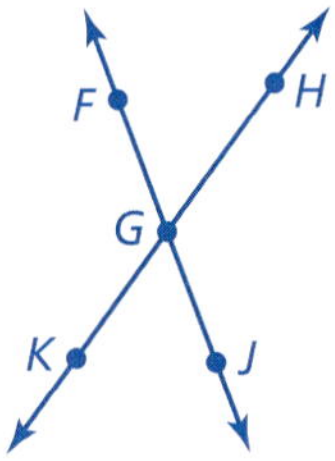

6.

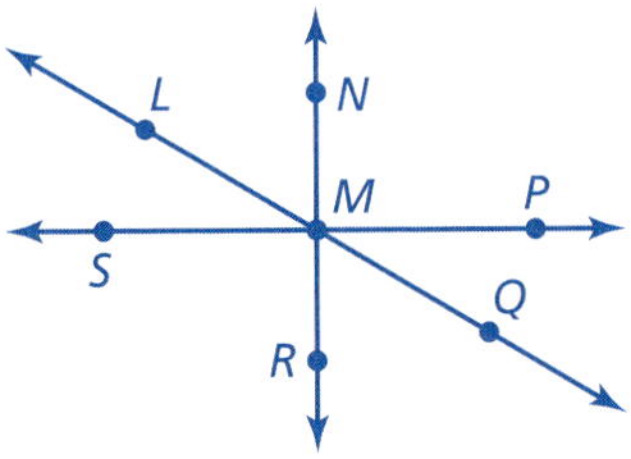

7. **ERROR ANALYSIS** Describe and correct the error in naming a pair of vertical angles.

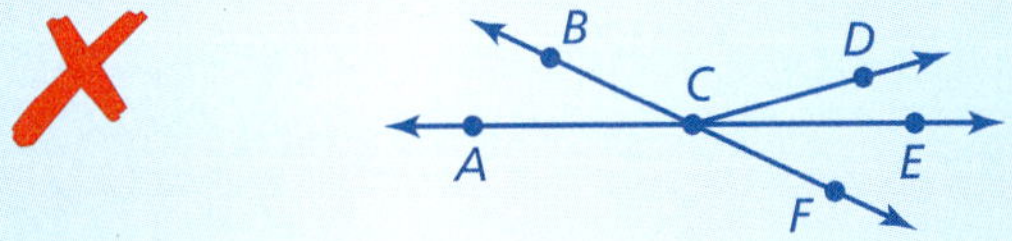

Tell whether the angles are *adjacent* or *vertical*. Then find the value of *x*.

2 8.

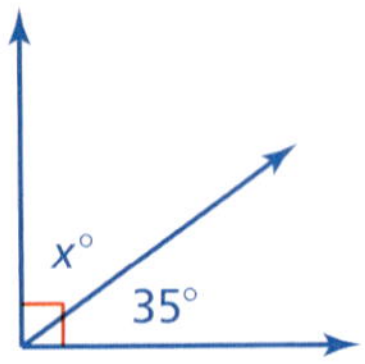

9.

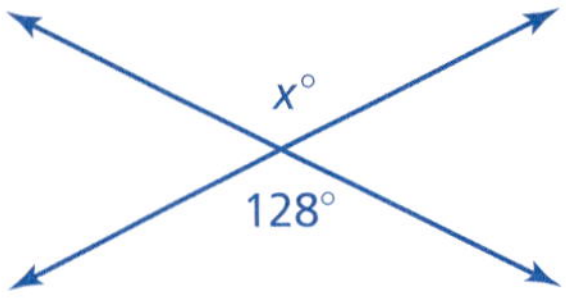

10.

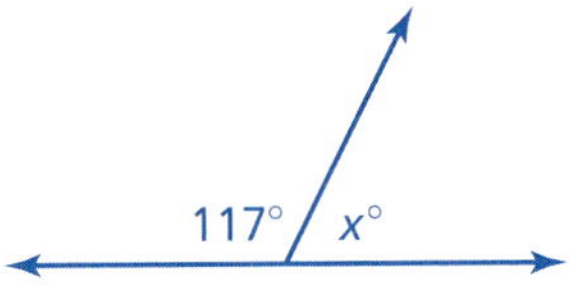

11.

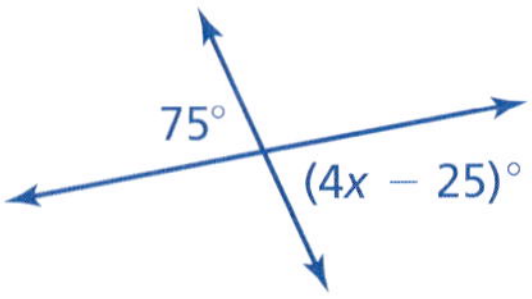

12.

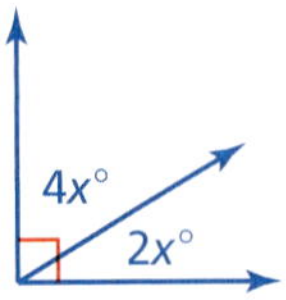

13.

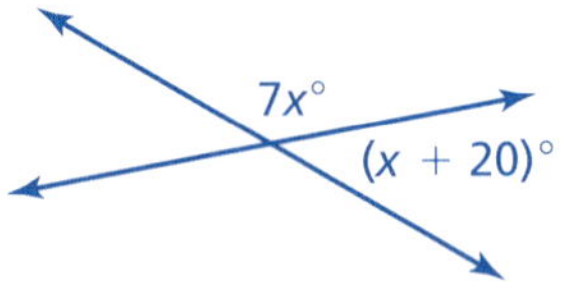

Assignment Guide and Homework Check

Level	Assignment	Homework Check
Advanced	1–4, 6, 7, 8–24 even, 25–30	10, 12, 18, 20, 24

Common Errors

- **Exercises 8–13** Students may think that there is not enough information to determine the value of *x*. Ask them to think about the information given in each figure. For instance, Exercise 8 shows two angles making up a right angle. So, the sum of the two angle measures must be 90°. Students can use this information to set up and solve a simple equation to find the value of *x*.
- **Exercise 26** Students may guess at an answer because they are unsure of how to solve this problem. To get them on the right track, you may need to give them the hint that the sum of the angle measures of a triangle is 180°. Students have not formally learned this yet, but they have seen it in earlier material, such as activities.

12.1 Record and Practice Journal

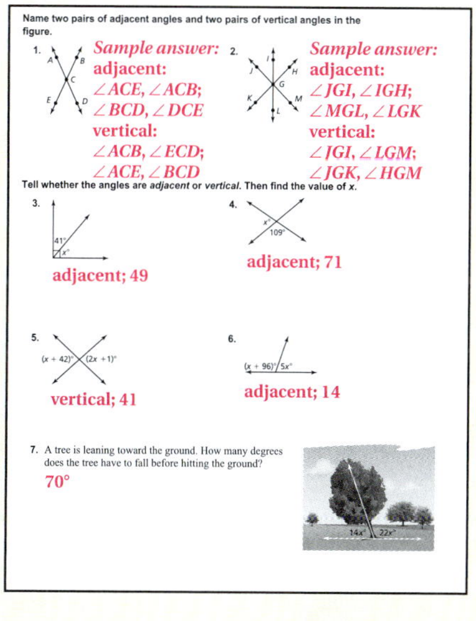
Name two pairs of adjacent angles and two pairs of vertical angles in the figure.

1. *Sample answer:* adjacent: $\angle ACE, \angle ACB$; $\angle BCD, \angle DCE$ vertical: $\angle ACB, \angle ECD$; $\angle ACE, \angle BCD$

2. *Sample answer:* adjacent: $\angle JGI, \angle IGH$; $\angle MGL, \angle LGK$ vertical: $\angle JGI, \angle LGM$; $\angle JGK, \angle HGM$

Tell whether the angles are *adjacent* or *vertical*. Then find the value of *x*.

3. adjacent; 49

4. adjacent; 71

5. vertical; 41

6. adjacent; 14

7. A tree is leaning toward the ground. How many degrees does the tree have to fall before hitting the ground?
70°

Vocabulary and Concept Check

1. two pairs; four pairs
2. $\angle 1$ is congruent to $\angle 3$; $\angle 2$ is congruent to $\angle 4$; They are congruent because they are vertical angles.

Practice and Problem Solving

3. $\angle ABC = 120°$, $\angle CBD = 60°$, $\angle DBE = 120°$, $\angle ABE = 60°$
4. $\angle ABE$, $\angle CBD$
5. *Sample answer:* adjacent: $\angle FGH$ and $\angle HGJ$, $\angle FGK$ and $\angle KGJ$; vertical: $\angle FGH$ and $\angle JGK$, $\angle FGK$ and $\angle JGH$
6. *Sample answer:* adjacent: $\angle SML$ and $\angle LMN$, $\angle SMR$ and $\angle RMQ$; vertical: $\angle NMP$ and $\angle SMR$, $\angle LMN$ and $\angle RMQ$
7. $\angle ACB$ and $\angle BCD$ are adjacent angles, not vertical angles.
8. adjacent; 55
9. vertical; 128
10. adjacent; 63
11. vertical; 25
12. adjacent; 15
13. adjacent; 20

Practice and Problem Solving

14. 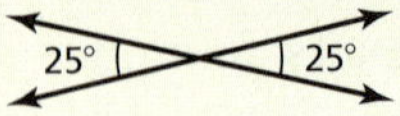

15–25. See Additional Answers.

26. See *Taking Math Deeper.*

Fair Game Review

27. $n < -9$;

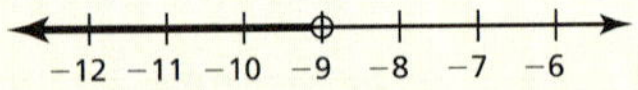

28. $x \geq -34$;

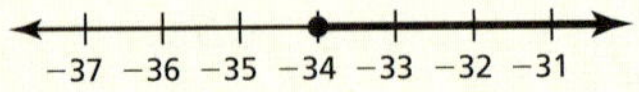

29. $m < 4$;

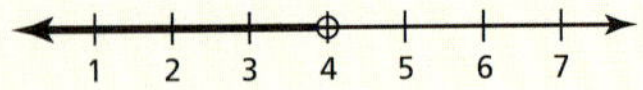

30. B

Mini-Assessment

Use the figure shown.

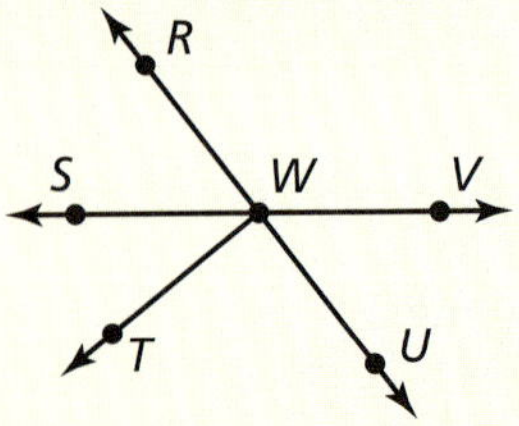

1. Name a pair of adjacent angles.
   Sample answer: $\angle RWS$ and $\angle TWS$
2. Name a pair of vertical angles.
   Sample answer: $\angle RWS$ and $\angle VWU$
3. Tell whether the angles are *adjacent* or *vertical.* Then find the value of *x*.

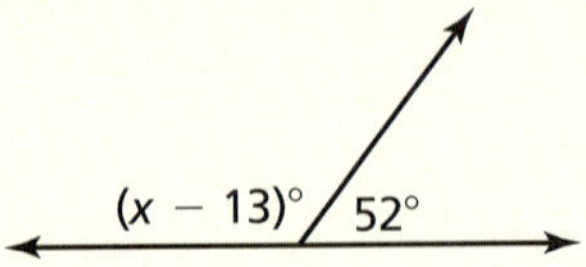

adjacent; 141

4. Draw a pair of vertical angles with a measure of 62°.

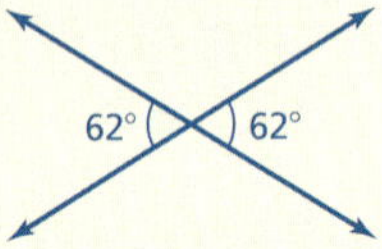

Taking Math Deeper

Exercise 26

This exercise may be difficult for students. They may not be aware that the sum of the interior angle measures of a triangle is 180°, which is formally presented later in this chapter. Students can solve the problem by carefully constructing the triangle that is created by the ground, wall, and ladder.

1 Use the diagram to find as many angle measures as possible.

- You can assume that the angle created by the ground and the wall is 90°.
- Because a straight angle is 180°, the acute angle created by the ladder and the ground is 180° − 120° = 60°.

2 Construct the triangle.

Draw a "backwards L" to represent the right angle. Then use a protractor to draw the 60° angle.

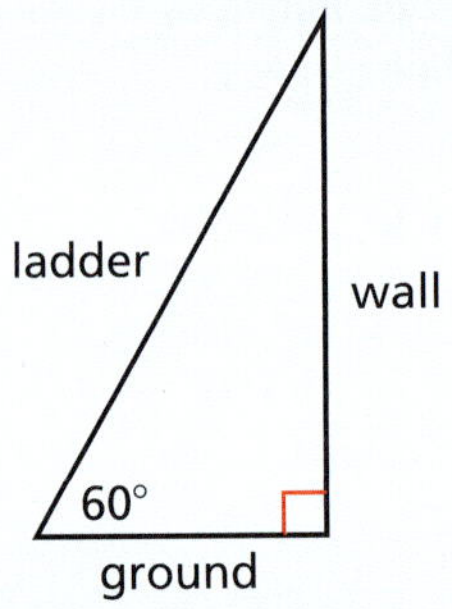

3 No matter how large or small you draw the "backwards L," when you measure the remaining angle, you see that it is 30°.

So, the ladder shown is not leaning at a safe angle.

Project

Have students lean straight objects, such as a yardstick, up against a wall and observe the angles formed. Tell them to move the base of the object closer to and farther from the wall, noting its affect on the two acute angles.

Reteaching and Enrichment Strategies

If students need help. . .	If students got it. . .
Resources by Chapter • Practice A and Practice B • Puzzle Time Record and Practice Journal Practice Differentiating the Lesson Lesson Tutorials Skills Review Handbook	Resources by Chapter • Enrichment and Extension • Technology Connection Start the next section

Draw a pair of vertical angles with the given measure.

3 **14.** 25° **15.** 85° **16.** 110° **17.** 135°

18. IRON CROSS The iron cross is a skiing trick in which the tips of the skis are crossed while the skier is airborne. Find the value of x in the iron cross shown.

19. OPEN-ENDED Draw a pair of adjacent angles with the given description.

a. Both angles are acute.

b. One angle is acute, and one is obtuse.

c. The sum of the angle measures is 135°.

20. PRECISION Explain two procedures that you can use to draw adjacent angles with given measures.

Determine whether the statement is *always, sometimes,* or *never* true.

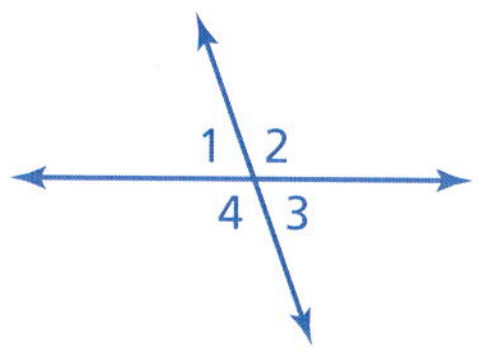

21. When the measure of $\angle 1$ is 70°, the measure of $\angle 3$ is 110°.

22. When the measure of $\angle 4$ is 120°, the measure of $\angle 1$ is 60°.

23. $\angle 2$ and $\angle 3$ are congruent.

24. The measure of $\angle 1$ plus the measure of $\angle 2$ equals the measure of $\angle 3$ plus the measure of $\angle 4$.

25. REASONING Draw a figure in which $\angle 1$ and $\angle 2$ are acute vertical angles, $\angle 3$ is a right angle adjacent to $\angle 2$, and the sum of the measure of $\angle 1$ and the measure of $\angle 4$ is 180°.

26. Structure For safety reasons, a ladder should make a 15° angle with a wall. Is the ladder shown leaning at a safe angle? Explain.

Fair Game Review What you learned in previous grades & lessons

Solve the inequality. Graph the solution. *(Section 11.3)*

27. $-6n > 54$ **28.** $-\frac{1}{2}x \le 17$ **29.** $-1.6 < \frac{m}{-2.5}$

30. MULTIPLE CHOICE What is the slope of the line that passes through the points (2, 3) and (6, 8)? *(Skills Review Handbook)*

Ⓐ $\frac{4}{5}$ Ⓑ $\frac{5}{4}$ Ⓒ $\frac{4}{3}$ Ⓓ $\frac{3}{2}$

12.2 Complementary and Supplementary Angles

Essential Question How can you classify two angles as complementary or supplementary?

1 ACTIVITY: Complementary and Supplementary Angles

Work with a partner.

a. The graph represents the measures of *complementary angles*. Use the graph to complete the table.

x		20°		30°	45°		75°
y	80°		65°	60°		40°	

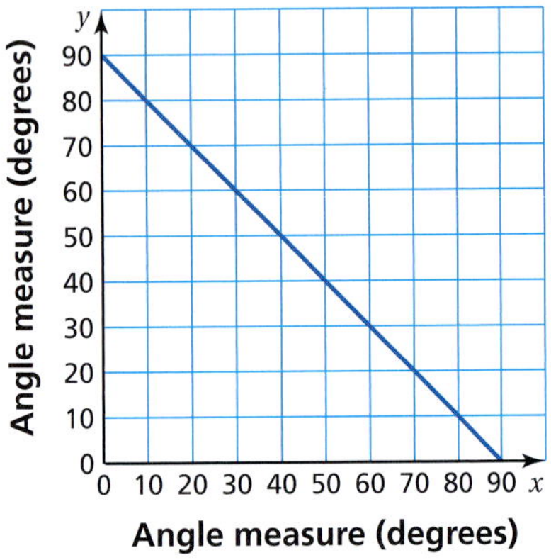

b. How do you know when two angles are complementary? Explain.

c. The graph represents the measures of *supplementary angles*. Use the graph to complete the table.

x	20°		60°	90°		140°	
y		150°		90°	50°		30°

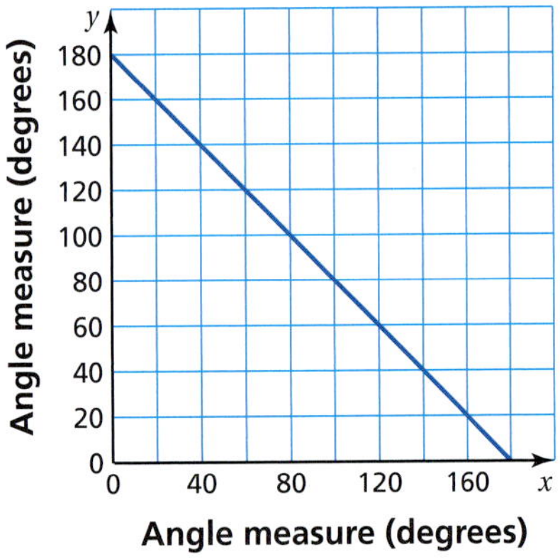

d. How do you know when two angles are supplementary? Explain.

2 ACTIVITY: Exploring Rules About Angles

Geometry
In this lesson, you will
- classify pairs of angles as complementary, supplementary, or neither.
- find angle measures using complementary and supplementary angles.

Work with a partner. Copy and complete each sentence with *always*, *sometimes*, or *never*.

a. If x and y are complementary angles, then both x and y are ________ acute.

b. If x and y are supplementary angles, then x is ________ acute.

c. If x is a right angle, then x is ________ acute.

d. If x and y are complementary angles, then x and y are ________ adjacent.

e. If x and y are supplementary angles, then x and y are ________ vertical.

Laurie's Notes

Introduction

Applying Mathematical Practices

- **Construct Viable Arguments and Critique the Reasoning of Others:** Students will make conjectures about the relationships between two angles that are complementary and two angles that are supplementary.

Motivate

- When students arrive, start making complimentary remarks such as, "Your sweater is nice. I like your glasses. Your shoelaces are cool." Make enough remarks so students catch on that you are giving compliments.
- Next, make a few remarks such as, "Your necklace complements your outfit. We have a full complement of faculty members. The great dessert complements a nice meal."
- You want students to recognize the difference between the homonyms *compliment* and *complement*. Today they will investigate complementary angles.

Activity Notes

Activity 1

- Complementary and supplementary angles are not defined in this activity. Instead, students will read ordered pairs from the graphs to complete the tables. From the tables, students should be able to recognize the relationship between two angles that are complementary and the relationship between two angles that are supplementary.
- This activity is a good review of reading ordered pairs from a graph.
- **Construct Viable Arguments and Critique the Reasoning of Others:** In each part, students need to explain the relationship between the two angles.

? "Are x and y proportional in part (a) or part (b)? Explain." no; In both parts, neither graph passes through the origin.

Activity 2

- Give time for partners to discuss the problems. The graphs and tables of values from Activity 1 should help students think through their answers.
- **Construct Viable Arguments and Critique the Reasoning of Others: Teaching Tip:** When answers are *sometimes* true, it is important to give students a sample of when the statement is true and when the statement is false. For example, in part (b), $x = 75°$ and $y = 105°$ which makes x acute, or $x = 105°$ and $y = 75°$ which makes x obtuse.

What Your Students Will Learn

- Classify pairs of angles as complementary, supplementary, or neither.
- Find angle measures using complementary and supplementary angles.
- Construct complementary and supplementary angles using a protractor.

Previous Learning

Students should know basic vocabulary associated with angles.

Technology for the Teacher

Dynamic Classroom

Lesson Plans
Complete Materials List

12.2 Record and Practice Journal

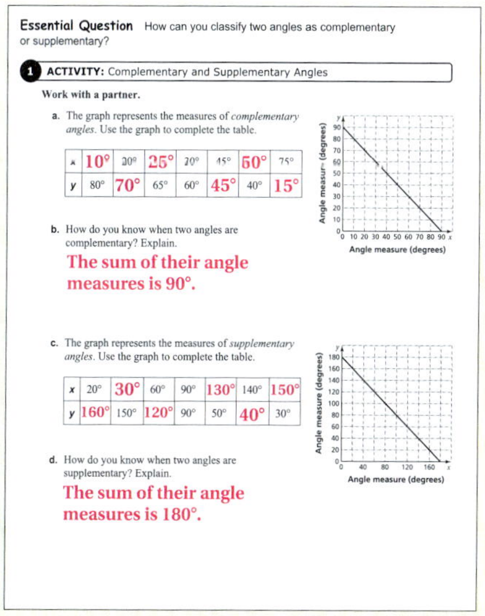

Essential Question How can you classify two angles as complementary or supplementary?

1 ACTIVITY: Complementary and Supplementary Angles

Work with a partner.

a. The graph represents the measures of *complementary angles*. Use the graph to complete the table.

x	10°	20°	25°	30°	45°	50°	75°
y	80°	70°	65°	60°	45°	40°	15°

b. How do you know when two angles are complementary? Explain.

The sum of their angle measures is 90°.

c. The graph represents the measures of *supplementary angles*. Use the graph to complete the table.

x	20°	30°	60°	90°	130°	140°	150°
y	160°	150°	120°	90°	50°	40°	30°

d. How do you know when two angles are supplementary? Explain.

The sum of their angle measures is 180°.

English Language Learners

Vocabulary

For English language learners, discuss the meanings of the words *complement* and *compliment.* As mentioned in the Motivate section, the words are homonyms. They sound the same when they are pronounced, but they have different meanings.

Complement means either of two parts needed to complete the whole.

Compliment is an expression of praise, commendation, or admiration.

12.2 Record and Practice Journal

2 ACTIVITY: Exploring Rules About Angles

Work with a partner. Complete each sentence with *always*, *sometimes*, or *never*.

a. If x and y are complementary angles, then both x and y are **always** acute.

b. If x and y are supplementary angles, then x is **sometimes** acute.

c. If x is a right angle, then x is **never** acute.

d. If x and y are complementary angles, then x and y are **sometimes** adjacent.

e. If x and y are supplementary angles, then x and y are **sometimes** vertical.

3 ACTIVITY: Classifying Pairs of Angles

Work with a partner. Tell whether the two angles shown on the clocks are *complementary*, *supplementary*, or *neither*. Explain your reasoning.

a.

supplementary

b.

complementary

c.

neither

d.

supplementary

4 ACTIVITY: Identifying Angles

Work with a partner. Use a protractor and the figure shown.

a. Name four pairs of complementary angles and four pairs of supplementary angles.

complementary: $\angle 1$ and $\angle 2$, $\angle 3$ and $\angle 4$, $\angle 5$ and $\angle 6$, $\angle 7$ and $\angle 8$

supplementary: $\angle 9$ and $\angle 10$, $\angle 10$ and $\angle 11$, $\angle 11$ and $\angle 12$, $\angle 9$ and $\angle 12$

b. Name two pairs of vertical angles.

$\angle 9$ and $\angle 11$, $\angle 10$ and $\angle 12$

What Is Your Answer?

5. **IN YOUR OWN WORDS** How can you classify two angles as complementary or supplementary? Give examples of each type.

complementary: sum of angle measures is 90°

supplementary: sum of angle measures is 180°

Laurie's Notes

Activity 3

- Many students think that complementary or supplementary angles must always be adjacent because teachers often draw them only that way! There is nothing in either definition (given in the lesson) that requires the angles to be adjacent.
- This activity focuses on the relationships between the angles and not on how they are drawn.
- After students have finished, discuss the answers.

Activity 4

- **Use Appropriate Tools Strategically:** Students will use protractors to justify their answers for this activity. Also, you should discuss the use of numbers instead of three letters to name the angles. When a figure contains many angles, it is much easier to number them.
- Have students measure the angles to the nearest whole degree.

What Is Your Answer?

- **Think-Pair-Share:** Students should read each question independently and then work in pairs to answer the question. When they have answered the question, the pair should compare their answer with another group and discuss any discrepancies.

Closure

- Look around the room. Name angles that appear to be complementary or supplementary.

3 ACTIVITY: Classifying Pairs of Angles

Work with a partner. Tell whether the two angles shown on the clocks are *complementary*, *supplementary*, or *neither*. Explain your reasoning.

a.

b.

c.

d.

4 ACTIVITY: Identifying Angles

Work with a partner. Use a protractor and the figure shown.

a. Name four pairs of complementary angles and four pairs of supplementary angles.

b. Name two pairs of vertical angles.

Math Practice

Use Definitions

How can you use the definitions of *complementary*, *supplementary*, and *vertical angles* to answer the questions?

What Is Your Answer?

5. **IN YOUR OWN WORDS** How can you classify two angles as complementary or supplementary? Give examples of each type.

Practice Use what you learned about complementary and supplementary angles to complete Exercises 3–5 on page 512.

12.2 Lesson

Check It Out
Lesson Tutorials
BigIdeasMath.com

Key Vocabulary

complementary angles, *p. 510*

supplementary angles, *p. 510*

Key Ideas

Complementary Angles

Words Two angles are **complementary angles** when the sum of their measures is 90°.

Examples

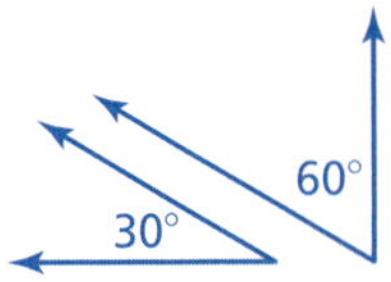

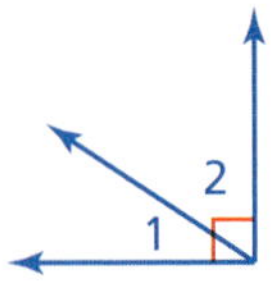

$\angle 1$ and $\angle 2$ are complementary angles.

Supplementary Angles

Words Two angles are **supplementary angles** when the sum of their measures is 180°.

Examples

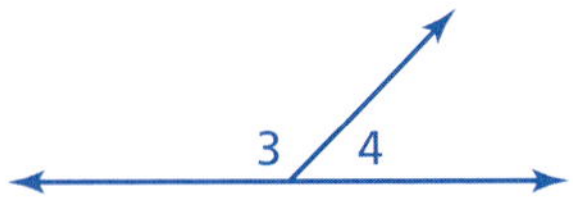

$\angle 3$ and $\angle 4$ are supplementary angles.

EXAMPLE 1 Classifying Pairs of Angles

Tell whether the angles are *complementary*, *supplementary*, or *neither*.

a.

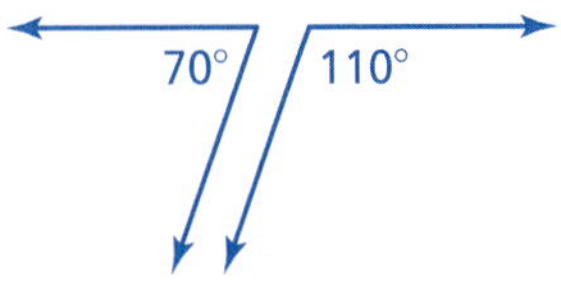

$70° + 110° = 180°$

∴ So, the angles are supplementary.

b.

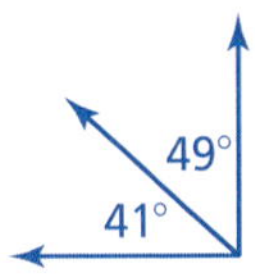

$41° + 49° = 90°$

∴ So, the angles are complementary.

c.

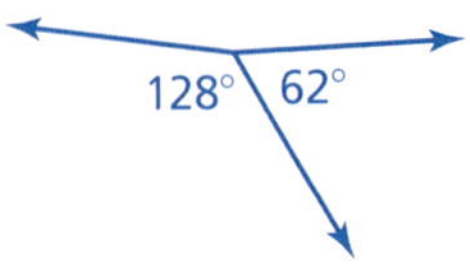

$128° + 62° = 190°$

∴ So, the angles are *neither* complementary nor supplementary.

On Your Own

Now You're Ready
Exercises 6–11

Tell whether the angles are *complementary*, *supplementary*, or *neither*.

1.

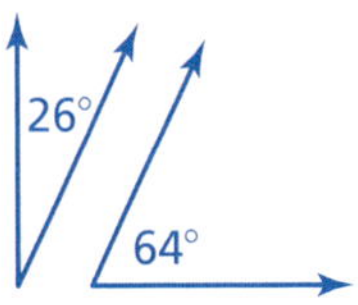

2.

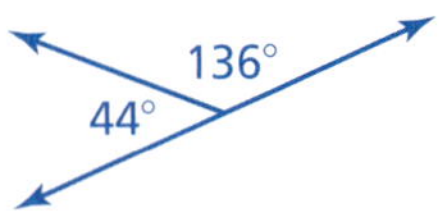

3.

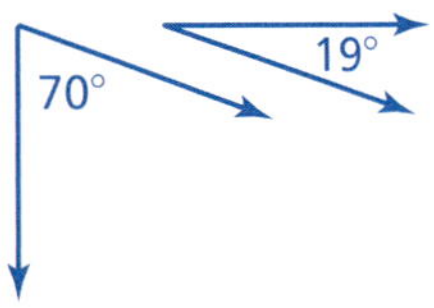

Multi-Language Glossary at BigIdeasMath.com

Laurie's Notes

Introduction

Connect

- **Yesterday:** Students explored complementary and supplementary angles.
- **Today:** Students will classify several pairs of angles as complementary or supplementary.

Motivate

- Yesterday, the homonyms *compliment* and *complement* were used.
- Ask students to give different meanings for the word *supplement.* Student answers may include some or all of the following.
  - a part added to a book or document
  - a part added to a newspaper that might be a special feature
  - something added to complete a deficiency (such as a dietary supplement)
  - something added to support (such as a learning supplement or tutor)

Lesson Notes

Key Ideas

- Write the Key Ideas. Define and sketch complementary angles and supplementary angles.
- In the figures, the angles are drawn with an orientation to suggest that the sum is 90° (complementary) or 180° (supplementary), but they do not need to have this orientation. For example, $\angle A$ and $\angle B$ below are complementary, however, it is not immediately obvious because of their orientation.

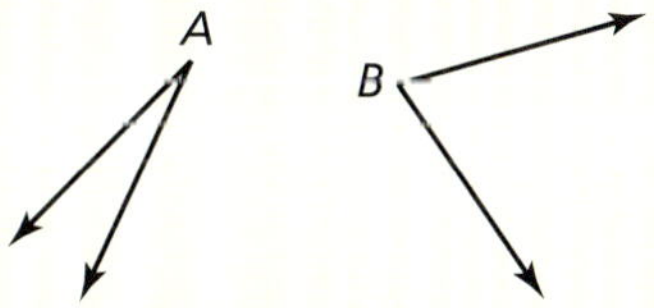

Example 1

- **Construct Viable Arguments and Critique the Reasoning of Others:** In this example, students sum the angle measures and determine if they add to 90°, 180°, or neither. Make sure students do not rely on their eyesight. They should actually add the angle measures.

On Your Own

- **Think-Pair-Share:** Students should read each question independently and then work in pairs to answer the questions. When they have answered the questions, the pair should compare their answers with another group and discuss any discrepancies.

Goal Today's lesson is classifying angles as **complementary** or **supplementary**.

Technology for the Teacher

Dynamic Classroom

Lesson Tutorials
Lesson Plans
Answer Presentation Tool

Extra Example 1

Tell whether the angles are *complementary, supplementary,* or *neither.*

a.

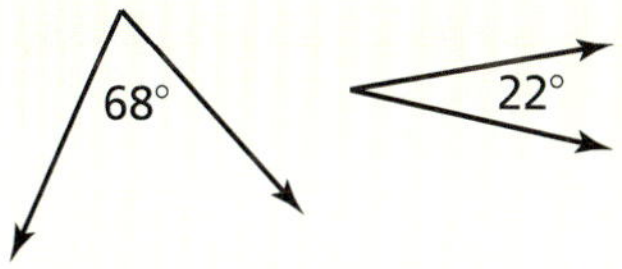

complementary

b.

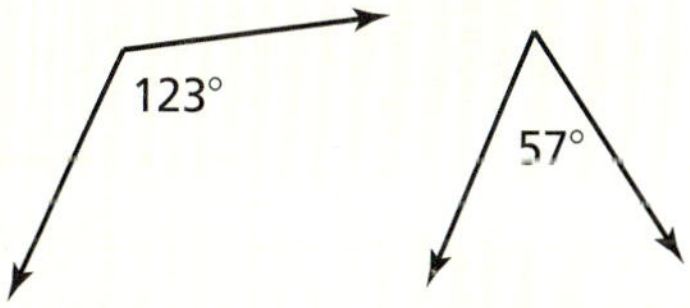

supplementary

c.

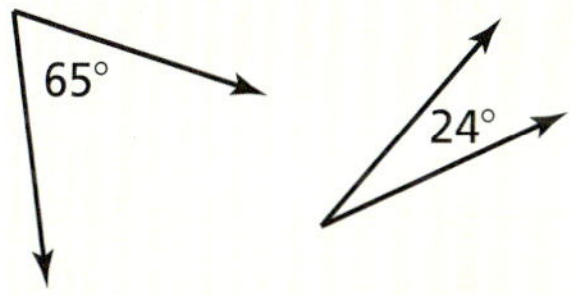

neither

On Your Own

1. complementary
2. supplementary
3. neither

Extra Example 2

Tell whether the angles are *complementary* or *supplementary*. Then find the value of x.

a.

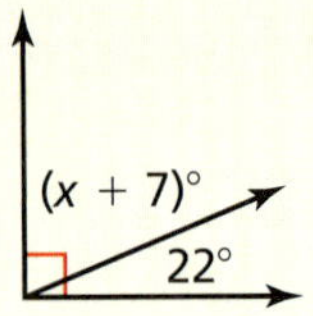

complementary; 61

b.

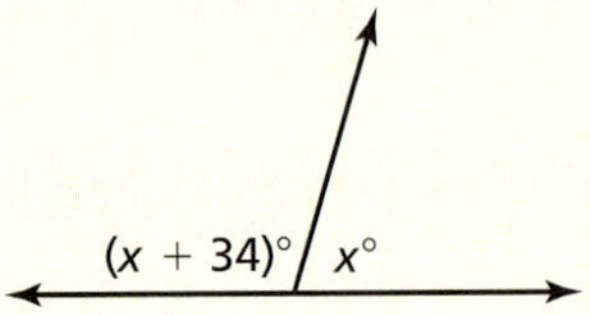

supplementary; 73

Extra Example 3

Draw a pair of adjacent supplementary angles so that one angle has a measure of 70°.

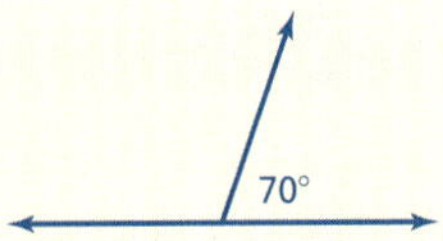

On Your Own

4. supplementary; 33.5°
5. complementary; 31°
6. 

Differentiated Instruction

Inclusion

Students may have problems remembering the measures of *complementary* and *supplementary* angles. Point out that c comes before s in the alphabet and 90 comes before 180 numerically.

Laurie's Notes

Example 2

- In this example, students practice writing and solving equations.
- ? "What do you know about the two angles in part (a)? Explain." The sum of their measures is 90° because they make up a right angle.
- Work through part (a) as shown.
- ? "What do you know about the two angles in part (b)? Explain." The sum of their measures is 180° because they make up a straight angle.
- Work through part (b) as shown.
- Have students check their answers by substituting each value of x in the corresponding figure.

Example 3

- **Use Appropriate Tools Strategically:** Ask students to use their protractors to draw the supplementary angles. Draw a line and place a point on the line to represent the vertex. Then place the protractor on the point and place a mark at 60°, and draw a ray from the vertex through the mark.

On Your Own

- **Think-Pair-Share:** Students should read each question independently and then work in pairs to answer the questions. When they have answered the questions, the pair should compare their answers with another group and discuss any discrepancies.

Closure

- True or False?
  1. Supplementary angles could both be acute. false
  2. Supplementary angles could be congruent. true
  3. Complementary angles sum to 180°. false
  4. Complementary angles could be obtuse. false
  5. Every angle has a complement and a supplement. false

EXAMPLE 2 Using Complementary and Supplementary Angles

Tell whether the angles are *complementary* or *supplementary*. Then find the value of x.

a.

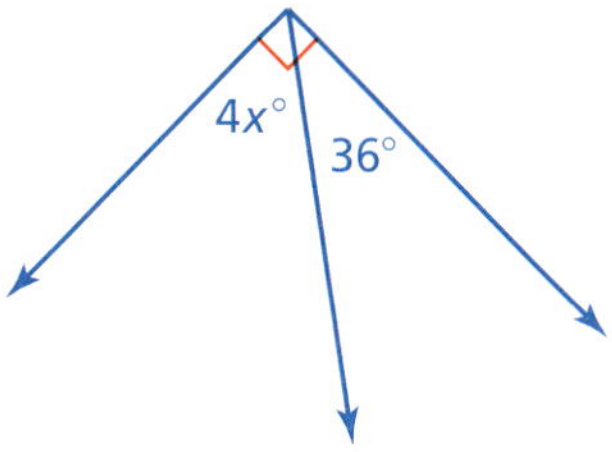

The two angles make up a right angle. So, the angles are complementary angles, and the sum of their measures is 90°.

$$4x + 36 = 90 \quad \text{Write equation.}$$
$$4x = 54 \quad \text{Subtract 36 from each side.}$$
$$x = 13.5 \quad \text{Divide each side by 4.}$$

b.

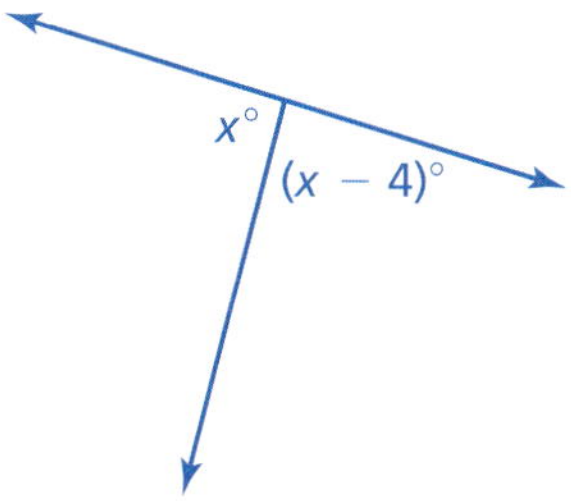

The two angles make up a straight angle. So, the angles are supplementary angles, and the sum of their measures is 180°.

$$x + (x - 4) = 180 \quad \text{Write equation.}$$
$$2x - 4 = 180 \quad \text{Combine like terms.}$$
$$2x = 184 \quad \text{Add 4 to each side.}$$
$$x = 92 \quad \text{Divide each side by 2.}$$

EXAMPLE 3 Constructing Angles

Draw a pair of adjacent supplementary angles so that one angle has a measure of 60°.

Step 1: Use a protractor to draw a 60° angle.

Step 2: Extend one of the sides to form a line.

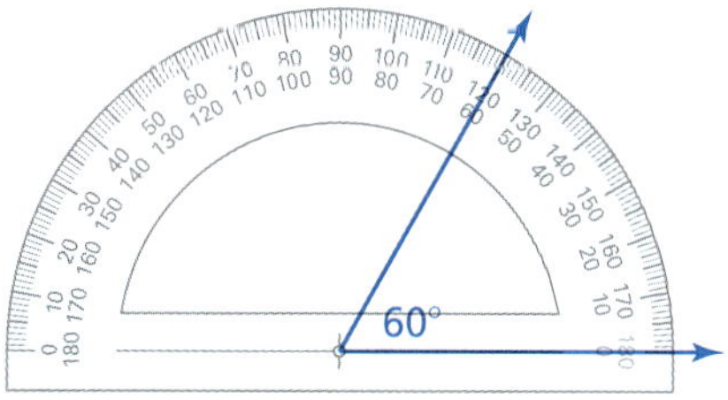

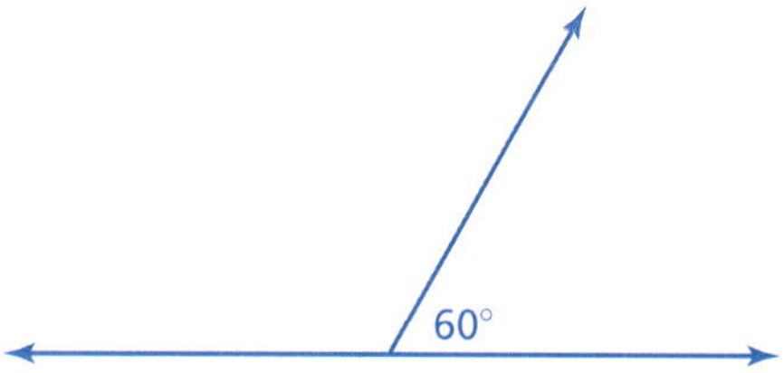

On Your Own

Now You're Ready
Exercises 12–14 and 17–20

Tell whether the angles are *complementary* or *supplementary*. Then find the value of x.

4.

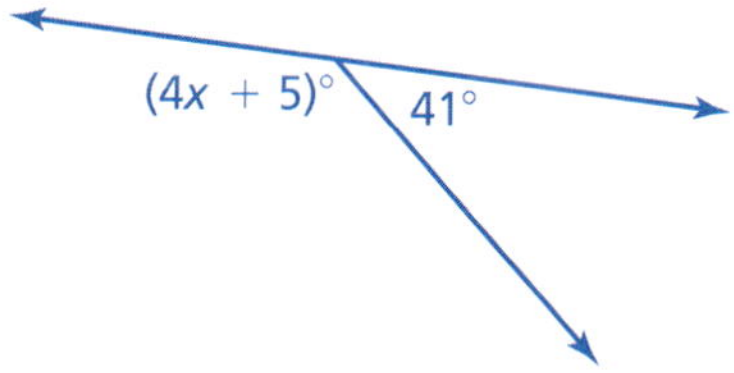

5.

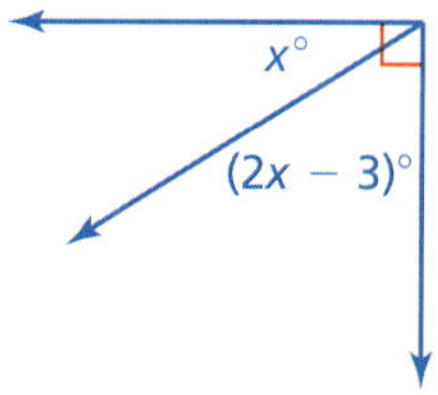

6. Draw a pair of adjacent supplementary angles so that one angle has a measure of 15°.

12.2 Exercises

Check It Out
Help with Homework
BigIdeasMath.com

Vocabulary and Concept Check

1. **VOCABULARY** Explain how complementary angles and supplementary angles are different.

2. **REASONING** Can adjacent angles be supplementary? complementary? neither? Explain.

Practice and Problem Solving

Tell whether the statement is *always*, *sometimes*, or *never* true. Explain.

3. If x and y are supplementary angles, then x is obtuse.

4. If x and y are right angles, then x and y are supplementary angles.

5. If x and y are complementary angles, then y is a right angle.

Tell whether the angles are *complementary*, *supplementary*, or *neither*.

1 6.

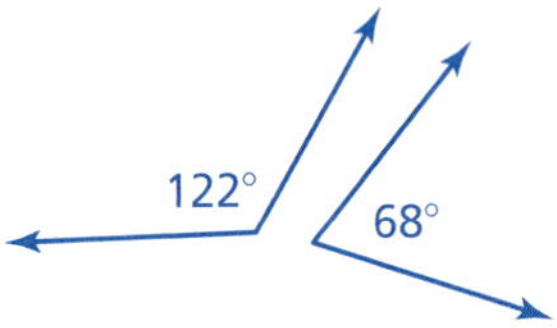

7.

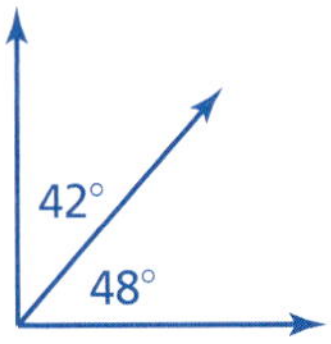

8.

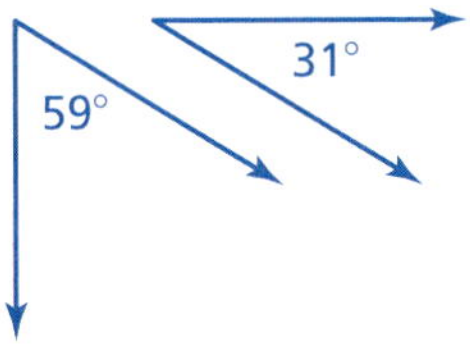

9.

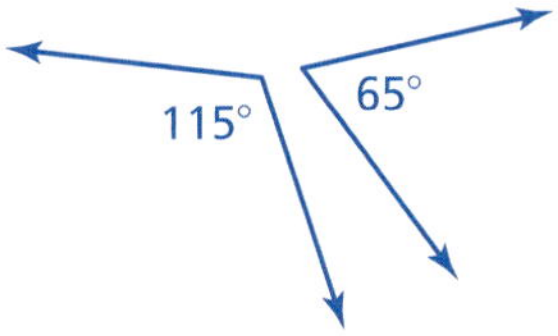

10.

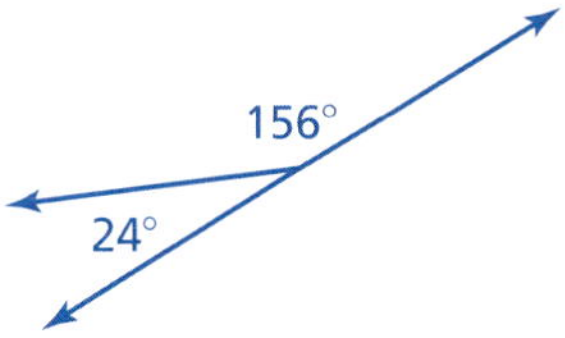

11.

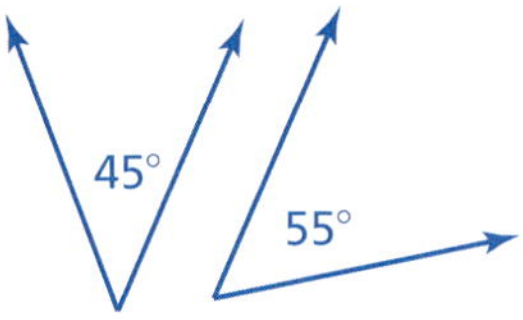

Tell whether the angles are *complementary* or *supplementary*. Then find the value of x.

2 12.

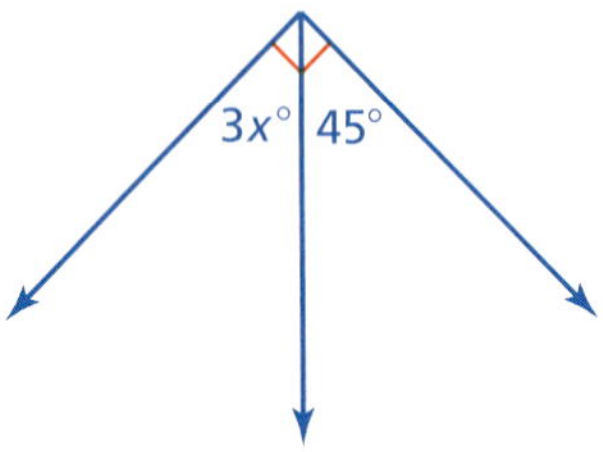

13.

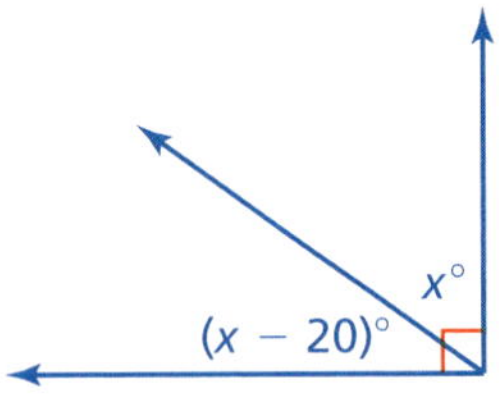

14.

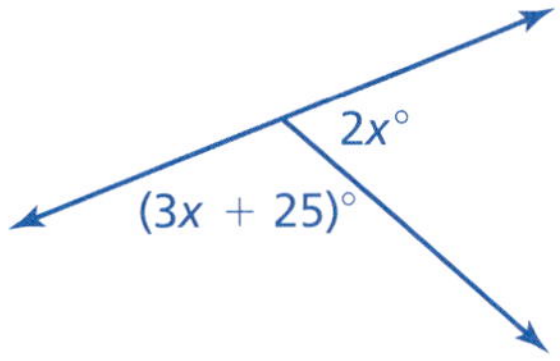

15. **INTERSECTION** What are the measures of the other three angles formed by the intersection?

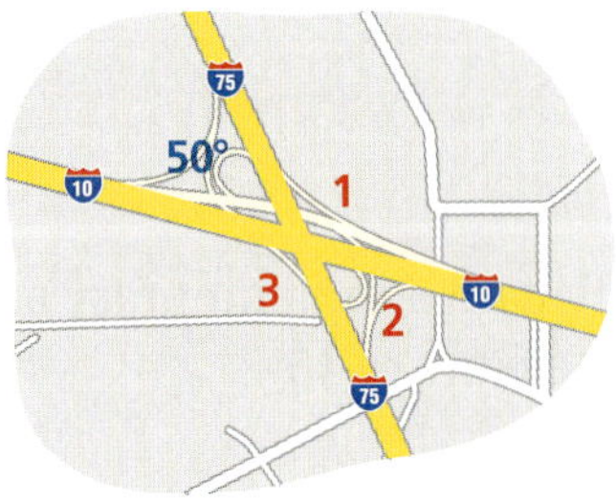

16. **TRIBUTARY** A tributary joins a river at an angle. Find the value of x.

Assignment Guide and Homework Check

Level	Assignment	Homework Check
Advanced	1–5, 6–26 even, 27–31	6, 14, 22, 24

For Your Information

- **Exercise 23** Students may not understand what a *vanishing point* is. A vanishing point is a point in a perspective drawing to which parallel lines appear to converge.

Common Errors

- **Exercises 6–11** Students may mix up the terms *supplementary* and *complementary*. Remind them of the definitions and use the alliteration that complementary angles are corners and supplementary angles are straight.
- **Exercises 12–14** Students may think that there is not enough information to determine the value of x. Remind them of the definitions they have learned in the lesson and ask whether either could apply. For instance, Exercise 12 shows two angles making up a right angle, so a student can use the definition of complementary angles to find x.

12.2 Record and Practice Journal

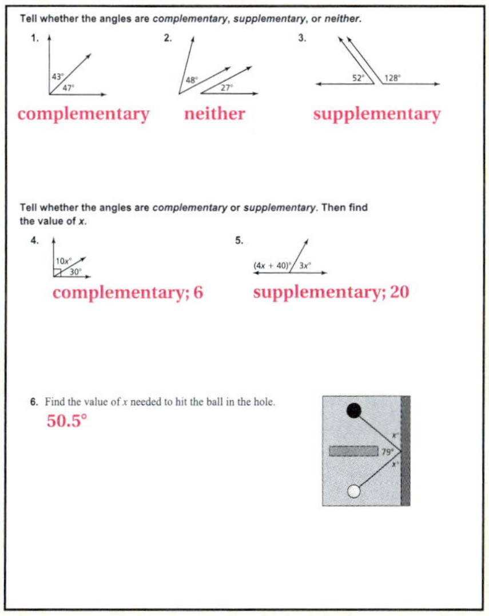

Tell whether the angles are *complementary*, *supplementary*, or *neither*.

1. complementary
2. neither
3. supplementary

Tell whether the angles are *complementary* or *supplementary*. Then find the value of x.

4. complementary; 6
5. supplementary; 20

6. Find the value of x needed to hit the ball in the hole.
50.5°

Vocabulary and Concept Check

1. The sum of the measures of two complementary angles is 90°. The sum of the measures of two supplementary angles is 180°.
2. Adjacent angles are not defined by their measure, so they can be complementary, supplementary, or neither.

Practice and Problem Solving

3. sometimes; Either x or y may be obtuse.
4. always; $90° + 90° = 180°$
5. never; Because x and y must both be less than 90° and greater than 0°.
6. neither
7. complementary
8. complementary
9. supplementary
10. supplementary
11. neither
12. complementary; 15
13. complementary; 55
14. supplementary; 31
15. $\angle 1 = 130°$, $\angle 2 = 50°$, $\angle 3 = 130°$
16. 53

Practice and Problem Solving

17–23. See Additional Answers.

24. yes; *Sample answer:* $\angle LMQ$ is a straight angle. By removing $\angle NMP$, the remaining two angles ($\angle LMN$ and $\angle PMQ$) have a sum of 90°.

25. 54°

26. See *Taking Math Deeper.*

27. $x = 10; y = 20$

Fair Game Review

28. $x = -15$ **29.** $n = -\frac{5}{12}$

30. $y = -9.3$ **31.** B

Mini-Assessment

Tell whether the angles are *complementary, supplementary,* or *neither.*

1.

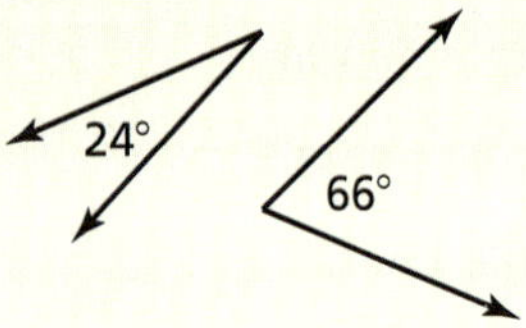

complementary

2.

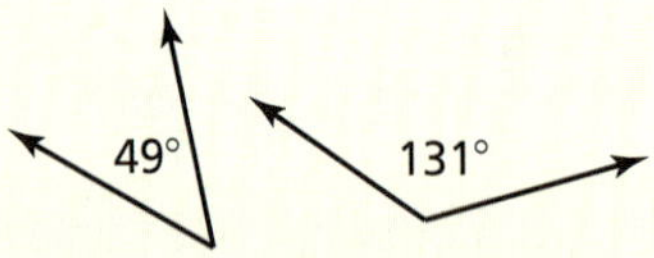

supplementary

3.

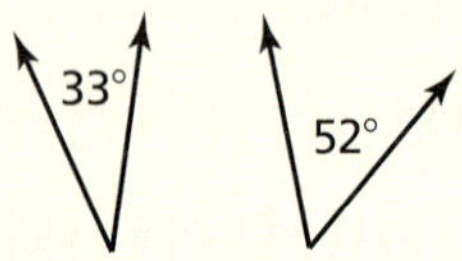

neither

4. Tell whether the angles are *complementary* or *supplementary.* Then find the value of *x*.

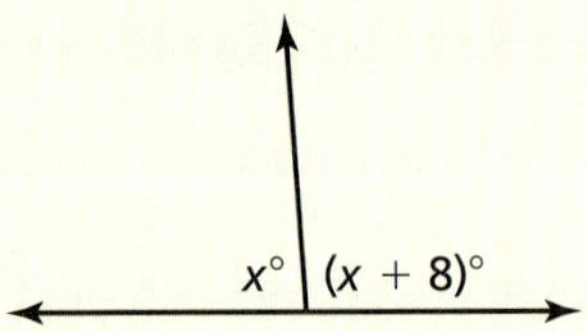

supplementary; 86

Taking Math Deeper

Exercise 26

This exercise is a good lesson for students. The definition of vertical angles is related to the *position* of the angles. However, the definitions of complementary angles and supplementary angles are only based on the *measures* of the angles and not on the position of the angles.

1 Draw the angles.

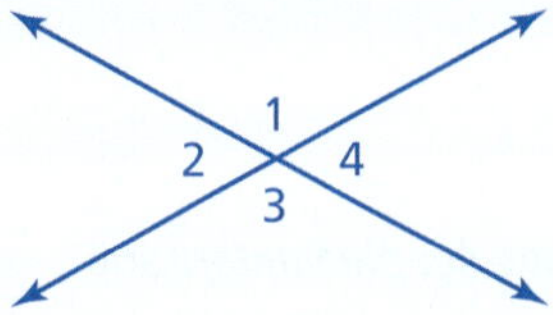

2 $\angle 2$ and $\angle 4$ are complementary.

$\angle 2 = \angle 4$ Vertical angles

$\angle 2 + \angle 4 = 90$ Complementary angles

Solving this implies that $\angle 2 = 45°$ and $\angle 4 = 45°$.

3 $\angle 2$ and $\angle 4$ are supplementary.

$\angle 2 = \angle 4$ Vertical angles

$\angle 2 + \angle 4 = 180$ Supplementary angles

Solving this implies that $\angle 2 = 90°$ and $\angle 4 = 90°$.

Project

Look around your classroom, school, home, or anywhere you go. Find examples of complementary and supplementary angles. How do you know they are complementary or supplementary? What is the most common angle you find?

Reteaching and Enrichment Strategies

If students need help. . .	If students got it. . .
Resources by Chapter • Practice A and Practice B • Puzzle Time Record and Practice Journal Practice Differentiating the Lesson Lesson Tutorials Skills Review Handbook	Resources by Chapter • Enrichment and Extension • Technology Connection Start the next section

Draw a pair of adjacent supplementary angles so that one angle has the given measure.

3 **17.** 20° **18.** 35° **19.** 80° **20.** 130°

21. PRECISION Explain two procedures that you can use to draw two adjacent complementary angles. Then draw a pair of adjacent complementary angles so that one angle has a measure of 30°.

22. OPEN-ENDED Give an example of an angle that can be a supplementary angle but cannot be a complementary angle. Explain.

23. VANISHING POINT The vanishing point of the picture is represented by point B.

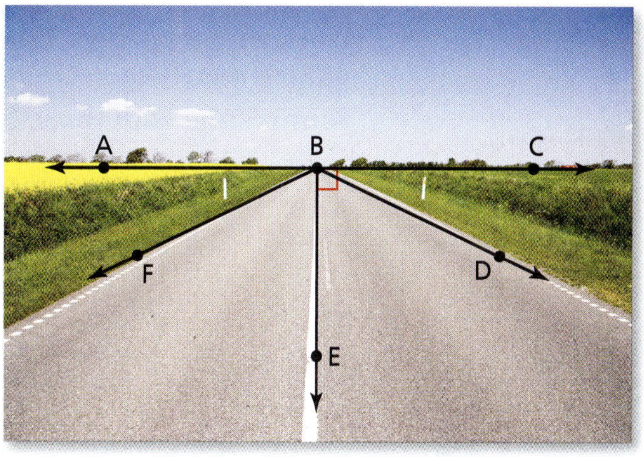

a. The measure of $\angle ABD$ is 6.2 times greater than the measure of $\angle CBD$. Find the measure of $\angle CBD$.

b. $\angle FBE$ and $\angle EBD$ are congruent. Find the measure of $\angle FBE$.

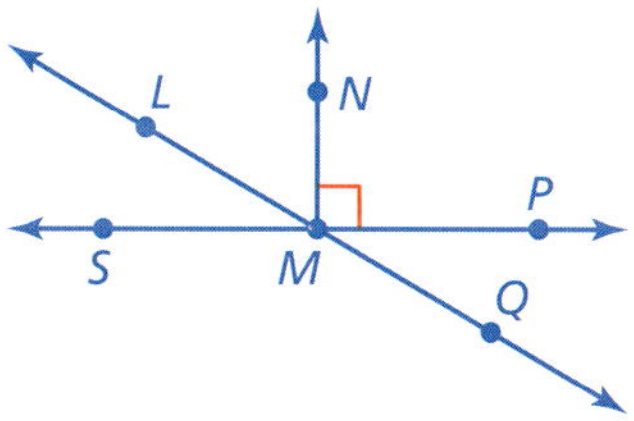

24. LOGIC Your friend says that $\angle LMN$ and $\angle PMQ$ are complementary angles. Is she correct? Explain.

25. RATIO The measures of two complementary angles have a ratio of 3 : 2. What is the measure of the larger angle?

26. REASONING Two angles are vertical angles. What are their measures if they are also complementary angles? supplementary angles?

27. Problem Solving Find the values of x and y.

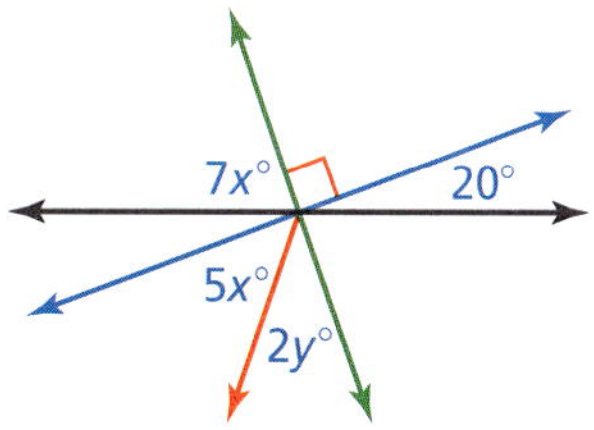

Fair Game Review What you learned in previous grades & lessons

Solve the equation. Check your solution. *(Section 1.1)*

28. $x + 7 = -8$ **29.** $\frac{1}{3} = n + \frac{3}{4}$ **30.** $-12.7 = y - 3.4$

31. MULTIPLE CHOICE Which decimal is equal to 3.7%? *(Skills Review Handbook)*

Ⓐ 0.0037 Ⓑ 0.037 Ⓒ 0.37 Ⓓ 3.7

12.3 Triangles

Essential Question How can you construct triangles?

1 ACTIVITY: Constructing Triangles Using Side Lengths

Work with a partner. Cut different-colored straws to the lengths shown. Then construct a triangle with the specified straws if possible. Compare your results with those of others in your class.

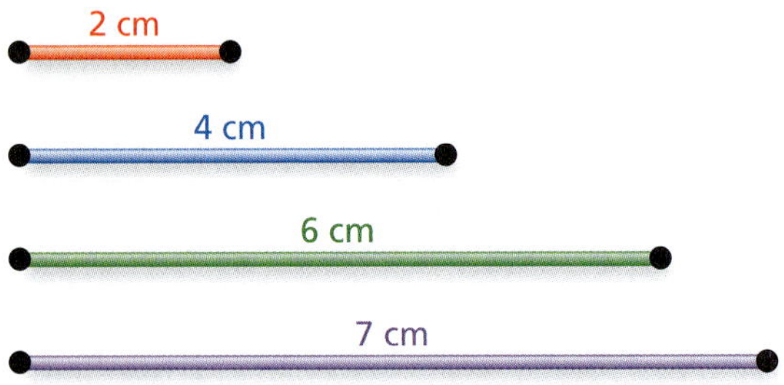

a. blue, green, purple

b. red, green, purple

c. red, blue, purple

d. red, blue, green

2 ACTIVITY: Using Technology to Draw Triangles (Side Lengths)

Work with a partner. Use geometry software to draw a triangle with the two given side lengths. What is the length of the third side of your triangle? Compare your results with those of others in your class.

a. 4 units, 7 units

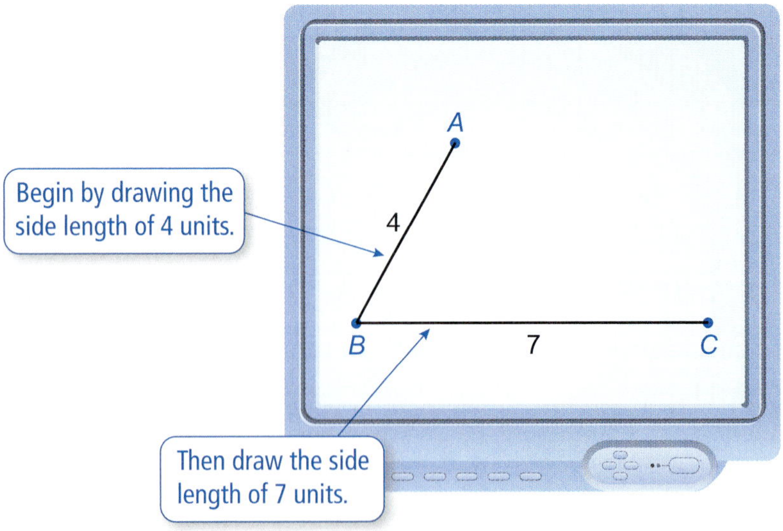

b. 3 units, 5 units

c. 2 units, 8 units

d. 1 unit, 1 unit

Geometry

In this lesson, you will

- construct triangles with given angle measures.
- construct triangles with given side lengths.

Laurie's Notes

Introduction

Applying Mathematical Practices

- **Use Appropriate Tools Strategically:** Students will investigate four activities, including two with technology and two without technology. It is important for students to select tools strategically as they develop understanding of mathematical concepts. Discussion of different approaches is essential.

Motivate

- Play a quick game that will help students remember vocabulary relating to triangles. Divide the class into two groups. Give a vocabulary word and each group must write the definition on a piece of paper and hand it to you. Definitions must be written in complete sentences. The first team with a correct definition gets a point. The team with the most points at the end wins.
- Some examples: obtuse angle, acute angle, right angle, scalene triangle, isosceles triangle, right triangle, equilateral triangle, equiangular triangle

Activity Notes

Words of Wisdom

- If geometry software is available, then let students experience exploring with it. The activities may take more than a day to complete, especially when students are not familiar with the software.

Activity 1

- **Teacher Tip:** To save time, you could pre-cut the straws and prepare reclosable bags with the pieces necessary for each pair of students ahead of time. In place of straws, you could use colored pipe cleaners or uncooked linguine.
- In this activity, students investigate whether different combinations of three side lengths form a triangle.
- Tell students to place the straws end-to-end and pivot the outer two straws to form a triangle, if possible.

? "What conclusions can you make?" In parts (a) and (b), it is possible to form a triangle, but in parts (c) and (d), it is not.

Activity 2

- **Construct Viable Arguments and Critique the Reasoning of Others:** Each part of the activity has a range of correct answers. Encourage students to find these ranges and explain their reasoning.
- **FYI:** With some geometry software, you can round the side lengths from 0 to 15 decimal places. When rounding to 0 decimal places, it is possible to obtain erroneous results, such as a triangle with side lengths of 3, 4, and 7 units. To be safe, use a setting of 1 or 2 decimal places.
- **Use Appropriate Tools Strategically:** Discuss any discoveries students make using the software.

What Your Students Will Learn

- Classify triangles when given angle measures or side lengths.
- Construct triangles with given angle measures or side lengths.

Previous Learning

Students should know how to classify two-dimensional figures based on properties, draw polygons, and draw angles.

Technology for the Teacher

Dynamic Classroom

Lesson Plans
Complete Materials List

12.3 Record and Practice Journal

Essential Question How can you construct triangles?

1 ACTIVITY: Constructing Triangles Using Side Lengths

Work with a partner. Cut different-colored straws to the lengths shown. Then construct a triangle with the specified straws, if possible. Compare your results with those of others in your class.

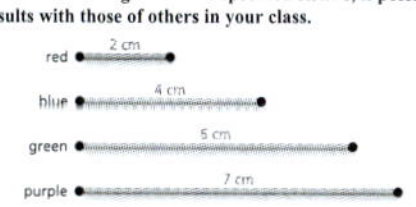

a. blue, green, purple
possible

b. red, green, purple
possible

c. red, blue, purple
not possible

d. red, blue, green
not possible

2 ACTIVITY: Using Technology to Draw Triangles (Side Lengths)

Work with a partner. Use geometry software to draw a triangle with the two given side lengths. What is the length of the third side of your triangle? Compare your results with those of others in your class.

a. 4 units, 7 units

Check students' work.

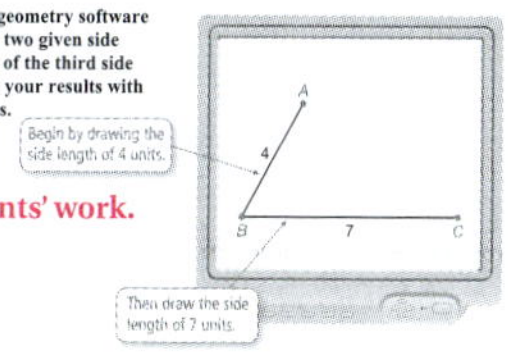

Differentiated Instruction

Kinesthetic

When talking about right, acute, and obtuse angles of a triangle, ask students if it is possible to draw a triangle with 2 right angles. Students should see by drawing the two right angles with a common side that the remaining two sides of the right angles will never meet. So, no triangle can be formed with 2 right angles. Ask students if it is possible for a triangle to have 2 obtuse angles. Students should reach the same conclusion. No triangle can be formed with 2 obtuse angles.

12.3 Record and Practice Journal

b. 3 units, 5 units c. 2 units, 8 units d. 1 unit, 1 unit

Check students' work.

3 ACTIVITY: Constructing Triangles Using Angle Measures

Work with a partner. Two angle measures of a triangle are given. Draw the triangle. What is the measure of the third angle? Compare your results with those of others in your class.

a. 40°, 70°

70°

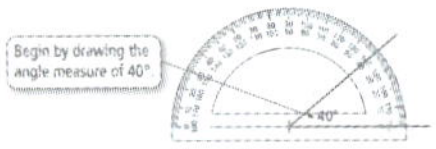

b. 60°, 75°
45°

c. 90°, 30°
60°

d. 100°, 40°
40°

4 ACTIVITY: Using Technology to Draw Triangles (Angle Measures)

Work with a partner. Use geometry software to draw a triangle with the two given angle measures. What is the measure of the third angle? Compare your results with those of others in your class.

a. 45°, 55° 80°

b. 50°, 40° 90°

c. 110°, 35° 35°

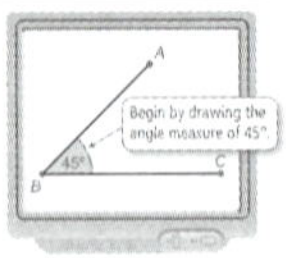

What Is Your Answer?

5. **IN YOUR OWN WORDS** How can you construct triangles?

Check students' work.

6. **REASONING** Complete the table below for each set of side lengths in Activity 2. Write a rule that compares the sum of any two side lengths to the third side length. The sum of any two side lengths must be greater than the remaining side.

Side Length			
Sum of Other Two Side Lengths			

7. **REASONING** Use a table to organize the angle measures of each triangle you formed in Activity 3. Include the sum of the angle measures. Then describe the pattern in the table and write a conclusion based on the pattern.

	$\angle 1$	$\angle 2$	$\angle 3$	$\angle 1 + \angle 2 + \angle 3$
a.	40°	70°	70°	180°
b.	60°	75°	45°	180°
c.	90°	30°	60°	180°
d.	100°	40°	40°	180°

Laurie's Notes

Activity 3

- In this activity, students construct triangles using two given angles, and then find the measure of the third angle. No constraints are given for the side lengths. One way to accomplish this is to draw the first given angle at one end of a segment, draw the second given angle at the other end, and extend the rays for the two angles until they intersect. Then, use a protractor to measure the third angle formed by the intersection.
- After students finish this activity, discuss the results.

? "In part (a), what is the measure of the third angle?" 70°, if students construct the triangle and measure the angle accurately

- **Extension:** Show students two different triangles that satisfy the requirements of part (a), where one triangle is clearly larger than the other. Probe students about the different corresponding side lengths but the same corresponding angle measures.

Activity 4

- If necessary, show students how to use the software to construct a triangle with two given angle measures.
- Set the software to round angle measures to the nearest whole degree.
- After students finish this activity, discuss the results.

? "In part (a), what is the measure of the third angle?" 80°, if students construct the triangle accurately

- **Use Appropriate Tools Strategically:** Discuss any further discoveries students make using the software.

What Is Your Answer?

- Have students work in pairs.

Closure

- **Exit Ticket:**
  - If the lengths of two sides of a triangle are 3 cm and 5 cm, then what are some possible lengths for the third side? *Sample answers:* 4 cm, 2.3 cm, 7 cm
  - If the measures of two angles of a triangle are 45° and 65°, then what is the measure of the third angle? 70°

3 ACTIVITY: Constructing Triangles Using Angle Measures

Work with a partner. Two angle measures of a triangle are given. Draw the triangle. What is the measure of the third angle? Compare your results with those of others in your class.

a. 40°, 70°

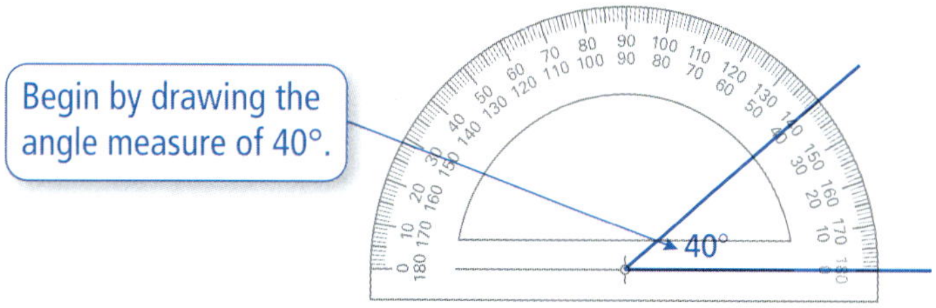

b. 60°, 75° **c.** 90°, 30° **d.** 100°, 40°

4 ACTIVITY: Using Technology to Draw Triangles (Angle Measures)

Math Practice

Recognize Usefulness of Tools

What are some advantages and disadvantages of using geometry software to draw a triangle?

Work with a partner. Use geometry software to draw a triangle with the two given angle measures. What is the measure of the third angle? Compare your results with those of others in your class.

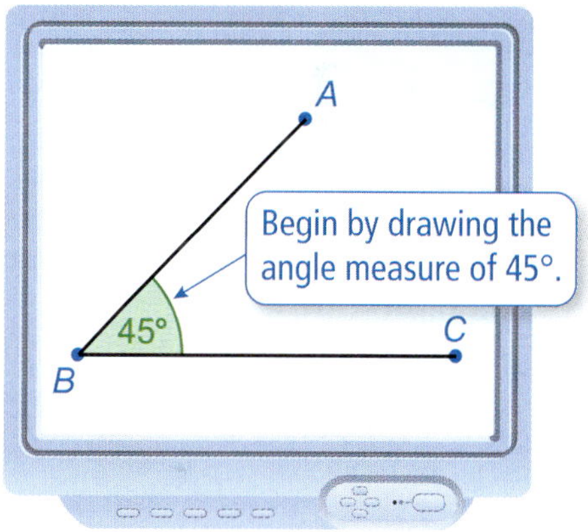

a. 45°, 55°

b. 50°, 40°

c. 110°, 35°

What Is Your Answer?

5. **IN YOUR OWN WORDS** How can you construct triangles?

6. **REASONING** Complete the table below for each set of side lengths in Activity 2. Write a rule that compares the sum of any two side lengths to the third side length.

Side Length			
Sum of Other Two Side Lengths			

7. **REASONING** Use a table to organize the angle measures of each triangle you formed in Activity 3. Include the sum of the angle measures. Then describe the pattern in the table and write a conclusion based on the pattern.

Practice Use what you learned about constructing triangles to complete Exercises 3–5 on page 518.

12.3 Lesson

Check It Out
Lesson Tutorials
BigIdeasMath.com

Key Vocabulary
congruent sides, *p. 516*

You can use side lengths and angle measures to classify triangles.

Key Ideas

Classifying Triangles Using Angles

acute triangle

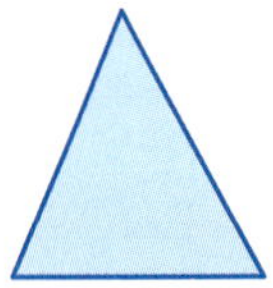

all acute angles

obtuse triangle

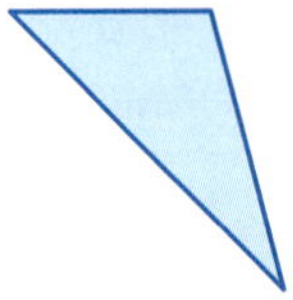

1 obtuse angle

right triangle

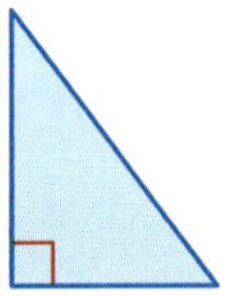

1 right angle

equiangular triangle

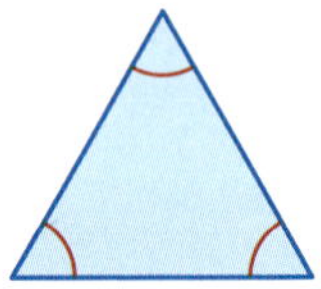

3 congruent angles

Classifying Triangles Using Sides

Congruent sides have the same length.

scalene triangle

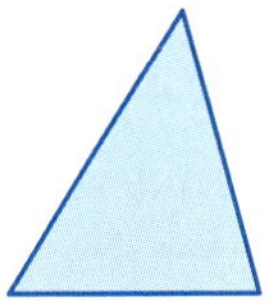

no congruent sides

isosceles triangle

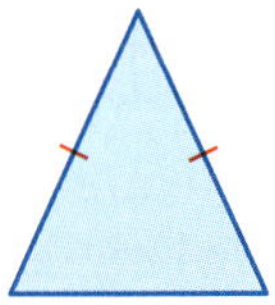

at least 2 congruent sides

equilateral triangle

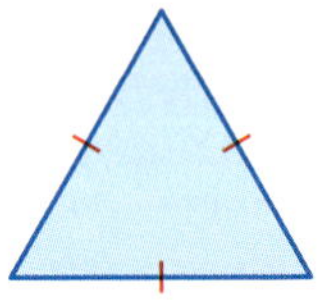

3 congruent sides

Reading

Red arcs indicate congruent angles.
Red tick marks indicate congruent sides.

EXAMPLE 1 Classifying Triangles

Classify each triangle.

a.

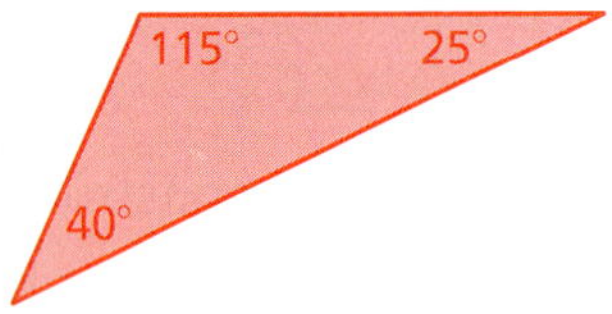

The triangle has one obtuse angle and no congruent sides.

So, the triangle is an obtuse scalene triangle.

b.

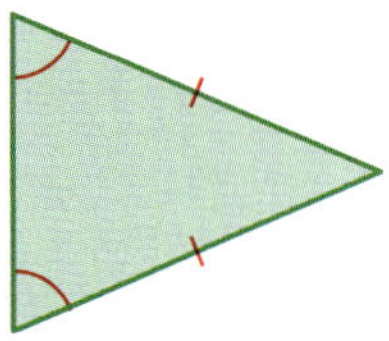

The triangle has all acute angles and two congruent sides.

So, the triangle is an acute isosceles triangle.

On Your Own

Now You're Ready
Exercises 6–11

Classify the triangle.

1.

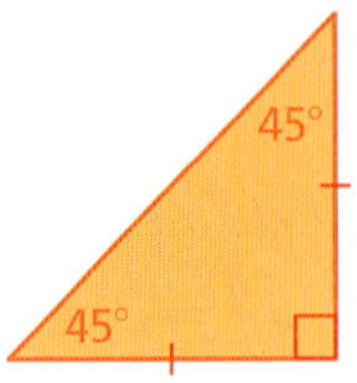

2.

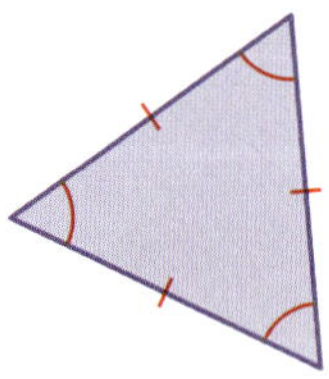

Multi-Language Glossary at BigIdeasMath.com

Laurie's Notes

Introduction

Connect

- **Yesterday:** Students investigated constructing triangles using side lengths and angle measures.
- **Today:** Students will classify and further investigate constructing triangles.

Motivate

- On the overhead or interactive board, display a collection of polygons such as the following.

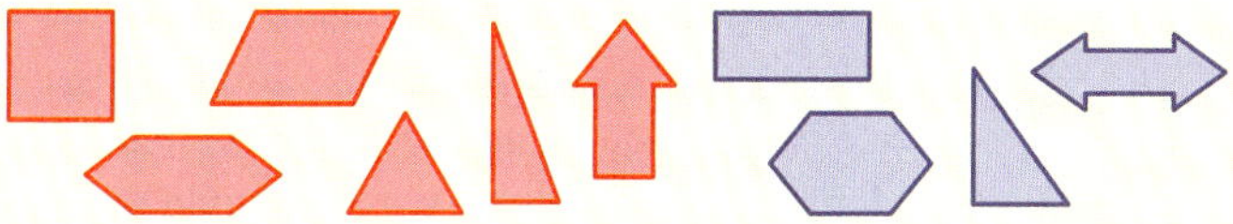

- Ask students how they could sort or classify these polygons. They may suggest sorting by color, number of sides, whether the polygon has a right angle, and so on.

Lesson Notes

Key Ideas

- Terminology used in the Key Ideas should be familiar to students. Some of this terminology is used in a previous course.
- ? "What does *congruent* mean?" the same length or measure
- Recall that sides as well as angles can be congruent.
- ? "How can triangles be classified using angles, that is, what names for triangles refer to the angles?" Listen for the four different names for triangles classified using angles.
- ? "How can triangles be classified using sides, that is, what names for triangles refer to the sides?" Listen for the three different names for triangles classified using sides.
- Sketch and identify each type of triangle shown in the Key Ideas, being sure to mark the right angle and the congruent angles and sides.
- ? "Is it possible to classify a triangle using angles *and* sides?" yes; Listen for examples such as an acute isosceles triangle.

Example 1

- You could have students explore the figures with protractors and rulers.
- **Attend to Precision:** Caution students against making assumptions. When angles or sides are congruent, they will be marked as such. When an angle is a right angle, it will be marked as such.

On Your Own

- **Neighbor Check:** Have students work independently and then have their neighbors check their work. Have students discuss any discrepancies.

Goal Today's lesson is classifying and constructing triangles.

Technology for the Teacher

Dynamic Classroom

Lesson Tutorials
Lesson Plans
Answer Presentation Tool

Extra Example 1

Classify each triangle.

a.

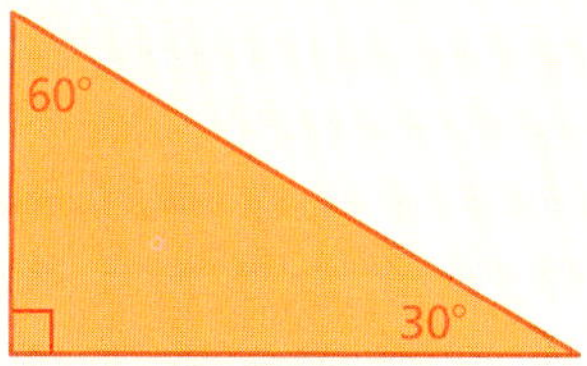

right scalene triangle

b.

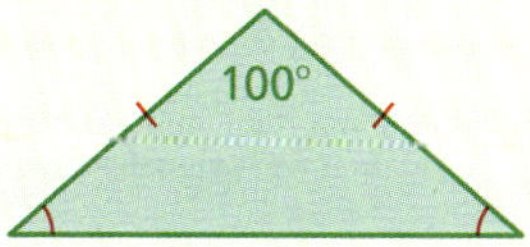

obtuse isosceles triangle

On Your Own

1. right isosceles triangle
2. equilateral equiangular triangle

Extra Example 2

Draw a triangle with angle measures of 35°, 45°, and 100°. Then classify the triangle.

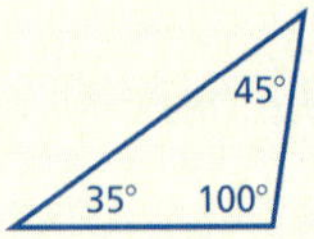

obtuse scalene triangle

Extra Example 3

Draw a triangle with a 3-centimeter side and a 5-centimeter side that meet at a 75° angle. Then classify the triangle.

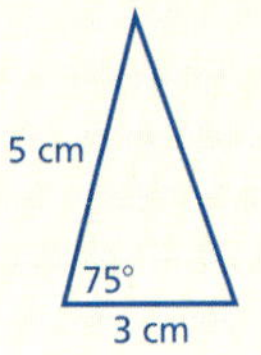

Not actual size

acute scalene triangle

On Your Own

3. 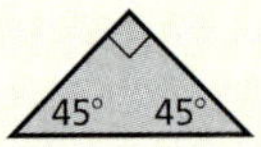

right isosceles triangle

4. See Additional Answers.

English Language Learners

Illustrate

Have students copy the empty table into their notebooks and then complete it with triangles that represent both attributes.

	Acute	Right	Obtuse
Scalene	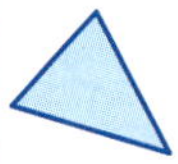		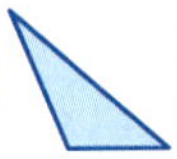
Isosceles	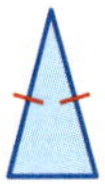	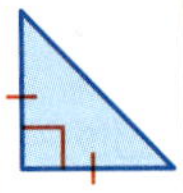	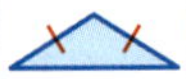
Equilateral	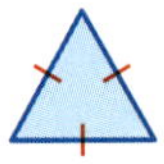	not possible	not possible

Laurie's Notes

Example 2

- Work through the example as shown.
- **Construct Viable Arguments and Critique the Reasoning of Others:** "Does the order in which you draw the angles matter? Explain." no; After you draw two angles, the third angle should have the measure of the remaining angle.
- **Attend to Precision:** Students should measure to verify that the third angle has the desired measure.
- **Teaching Tip:** Encourage students not to make tiny drawings. It can be difficult to measure angles when the side lengths are shorter than the radius of the protractor.
- The triangle is classified as right scalene. Students should compare their triangles to those of their neighbors. They should realize that many different sized triangles can have angles of 30°, 60°, and 90°. They should understand that every neighbor's triangle should be right scalene. Point this out to students if they do not make this conclusion on their own.

Example 3

- Explain that students are going to construct and classify a triangle using three given pieces of information: two sides and the angle between the sides.
- Follow the steps shown. If you wish, have students draw the sides in inches instead of centimeters in order to make the triangle slightly larger and easier to draw.
- Measure the third side. It should be approximately 1.56 centimeters (or inches).
- The triangle is classified as obtuse scalene. Students should compare their triangles to those of their neighbors. This time, each student should have the same sized triangle.
- **Extension:** Repeat the construction but with the 4 centimeter leg horizontal (instead of the 3 centimeter leg). Ask students whether they think this triangle is different from the one in the example.

On Your Own

- **Think-Pair-Share:** Students should read each question independently and then work in pairs to answer the questions. When they have answered the questions, the pair should compare their answers with another group and discuss any discrepancies.

Closure

- Sketch a triangle that is (a) right isosceles and (b) obtuse scalene. Label the sides and angles.

EXAMPLE 2 Constructing a Triangle Using Angle Measures

Draw a triangle with angle measures of 30°, 60°, and 90°. Then classify the triangle.

Step 1: Use a protractor to draw the 30° angle.

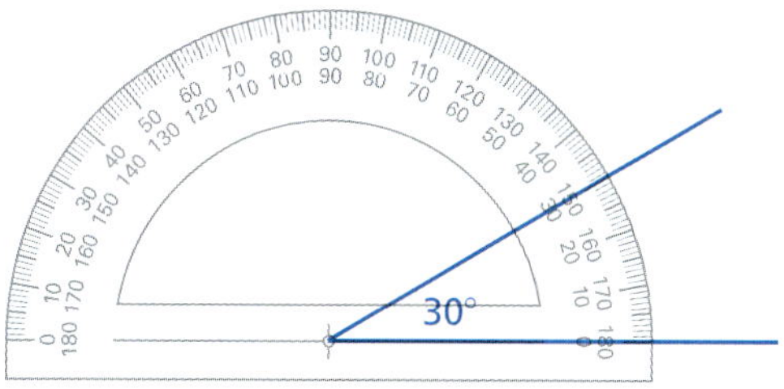

Step 2: Use a protractor to draw the 60° angle.

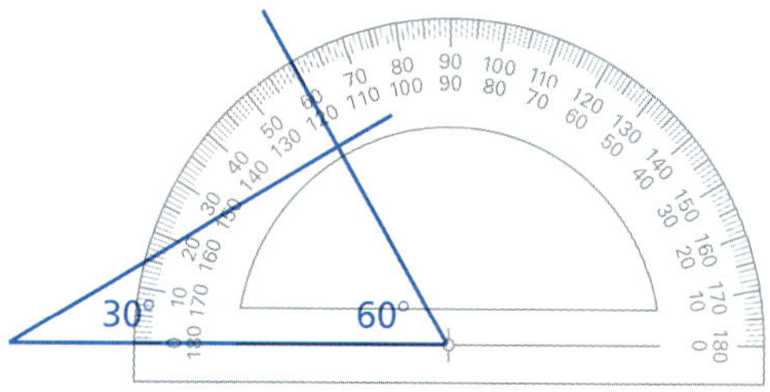

Step 3: The protractor shows that the measure of the remaining angle is 90°.

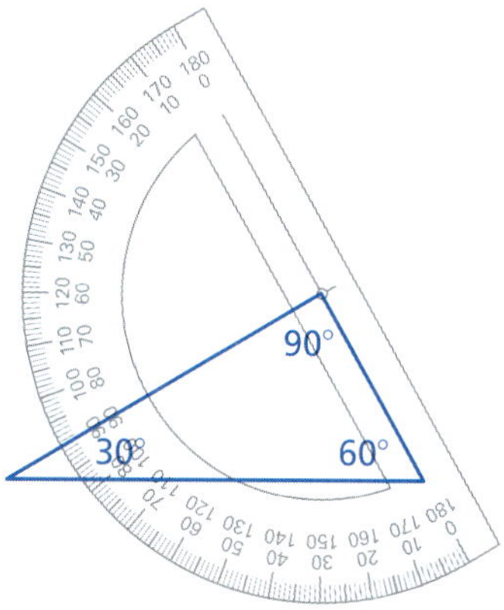

Study Tip

After drawing the first two angles, make sure you check the remaining angle.

∴ The triangle is a right scalene triangle.

EXAMPLE 3 Constructing a Triangle Using Side Lengths

Draw a triangle with a 3-centimeter side and a 4-centimeter side that meet at a 20° angle. Then classify the triangle.

Step 1: Use a protractor to draw a 20° angle.

Step 2: Use a ruler to mark 3 centimeters on one ray and 4 centimeters on the other ray.

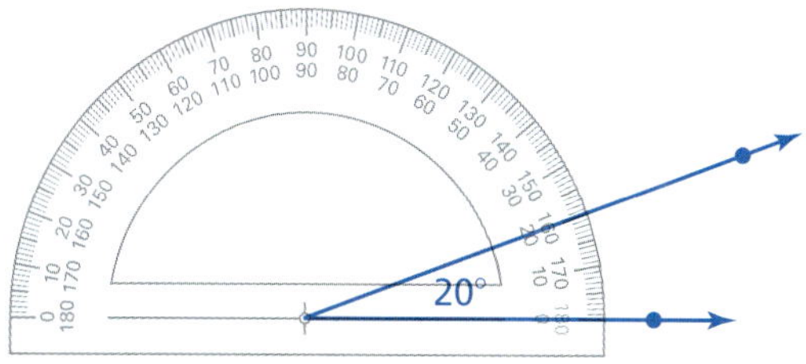

Step 3: Draw the third side to form the triangle.

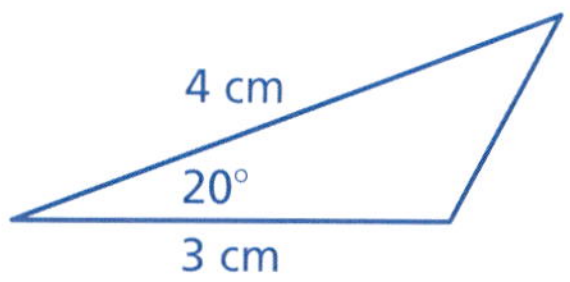

∴ The triangle is an obtuse scalene triangle.

On Your Own

Now You're Ready
Exercises 14–19

3. Draw a triangle with angle measures of 45°, 45°, and 90°. Then classify the triangle.

4. Draw a triangle with a 1-inch side and a 2-inch side that meet at a 60° angle. Then classify the triangle.

12.3 Exercises

Check It Out
Help with Homework
BigIdeasMath.com

Vocabulary and Concept Check

1. **WRITING** How can you classify triangles using angles? using sides?
2. **DIFFERENT WORDS, SAME QUESTION** Which is different? Find "both" answers.

Construct an equilateral triangle.	Construct a triangle with 3 congruent sides.
Construct an equiangular triangle.	Construct a triangle with no congruent sides.

Practice and Problem Solving

Construct a triangle with the given description.

3. side lengths: 4 cm, 6 cm
4. side lengths: 5 cm, 12 cm
5. angles: 65°, 55°

Classify the triangle.

1 6.

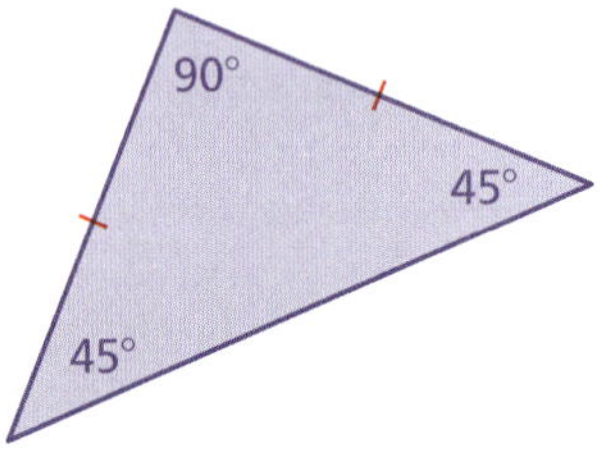

7.

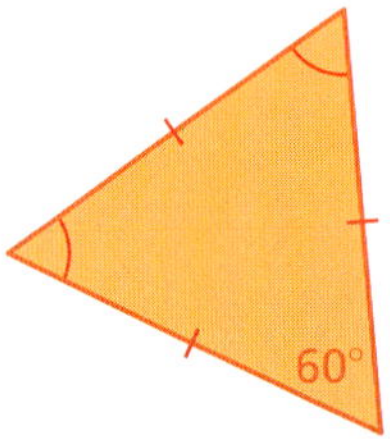

8.

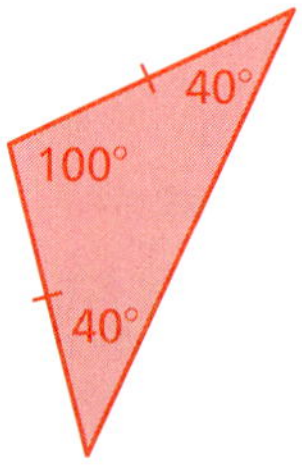

9.

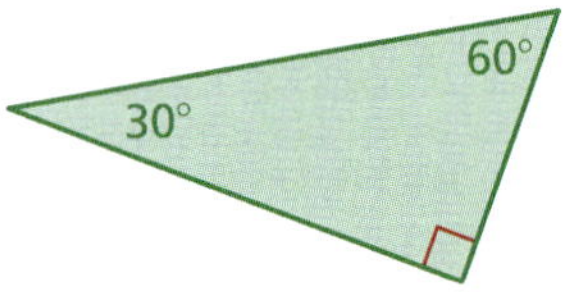

10.

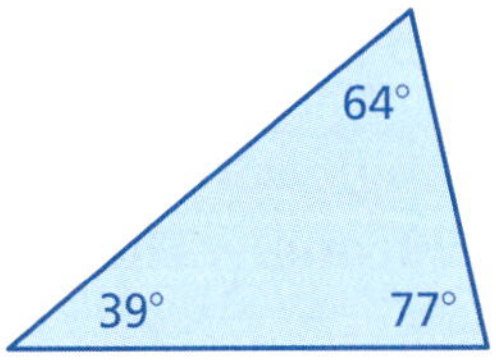

11.

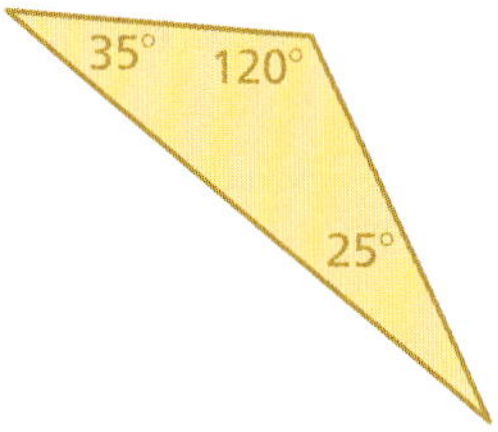

12. **ERROR ANALYSIS** Describe and correct the error in classifying the triangle.

✗

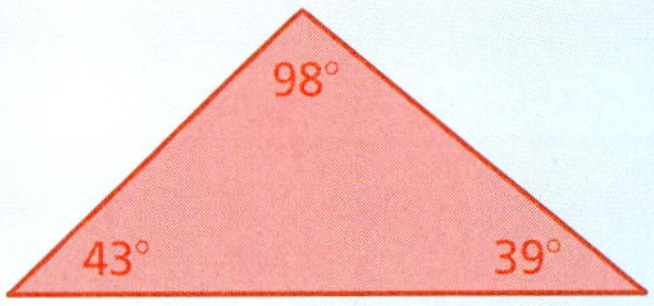

The triangle is acute and scalene because it has two acute angles and no congruent sides.

13. **MOSAIC TILE** A mosaic is a pattern or picture made of small pieces of colored material. Classify the yellow triangle used in the mosaic.

Assignment Guide and Homework Check

Level	Assignment	Homework Check
Advanced	1–5, 6–26 even, 27–31	10, 16, 20, 24, 27

Common Errors

- **Exercises 6–11** Students may classify the triangle using sides only and forget to classify it using angles (or vice versa). Remind students to classify the triangles using sides *and* angles.

12.3 Record and Practice Journal

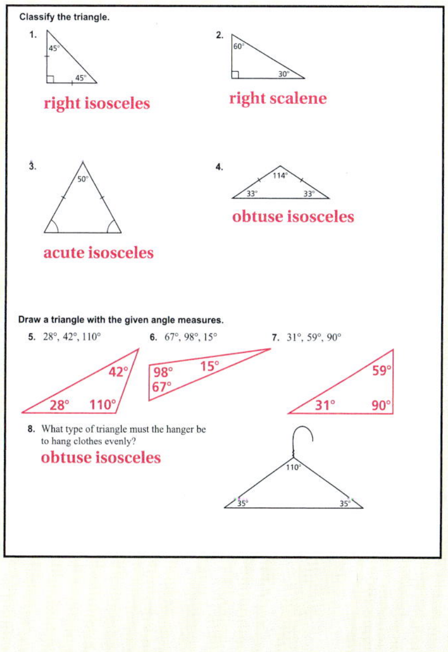
Classify the triangle.

1. right isosceles
2. right scalene
3. acute isosceles
4. obtuse isosceles

Draw a triangle with the given angle measures.

5. 28°, 42°, 110°
6. 67°, 98°, 15°
7. 31°, 59°, 90°

8. What type of triangle must the hanger be to hang clothes evenly? obtuse isosceles

Vocabulary and Concept Check

1. *Angles:* When a triangle has 3 acute angles, it is an acute triangle. When a triangle has 1 obtuse angle, it is an obtuse triangle. When a triangle has 1 right angle, it is a right triangle. When a triangle has 3 congruent angles, it is an equiangular triangle.
Sides: When a triangle has no congruent sides, it is a scalene triangle. When a triangle has 2 congruent sides, it is an isosceles triangle. When a triangle has 3 congruent sides, it is an equilateral triangle.

2. Construct a triangle with no congruent sides;

(no congruent sides)

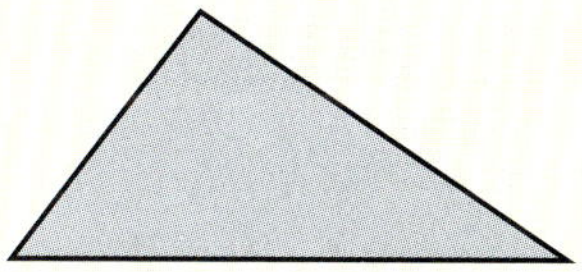

(equilateral, equiangular, 3 congruent sides)

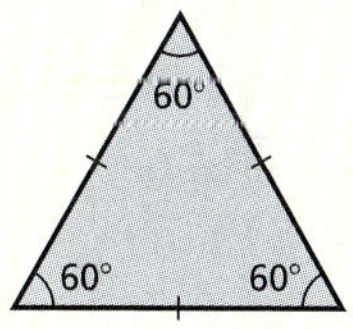

Practice and Problem Solving

3–5. See Additional Answers.

6. right isosceles

7. equilateral equiangular

8. obtuse isosceles

9. right scalene

10. acute scalene

11. obtuse scalene

12–13. See Additional Answers.

Practice and Problem Solving

14–20. See Additional Answers.

21. no; The sum of the angle measures must be 180°.

22. See *Taking Math Deeper.*

23–27. See Additional Answers.

Fair Game Review

28. yes; The equation can be written as $y = kx$ where $k = \frac{1}{2}$.

29. no; The equation cannot be written as $y = kx$.

30. no; The equation cannot be written as $y = kx$.

31. B

Mini-Assessment

Classify the triangle.

1.

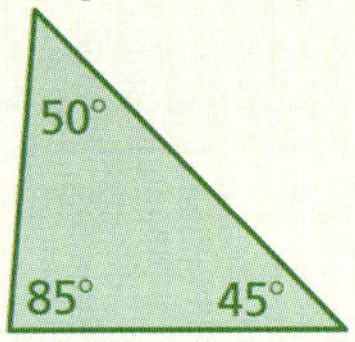

acute scalene triangle

2.

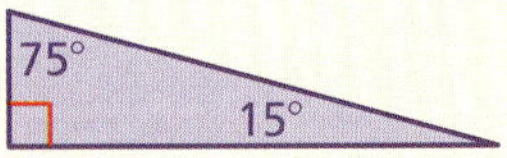

right scalene triangle

3. Draw a triangle with angle measures of 20°, 70°, and 90°. Then classify the triangle.

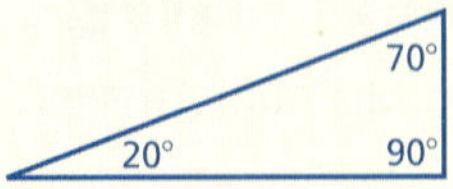

right scalene triangle

4. Draw a triangle with a 3-centimeter side and a 4-centimeter side that meet at a 15° angle. Then classify the triangle.

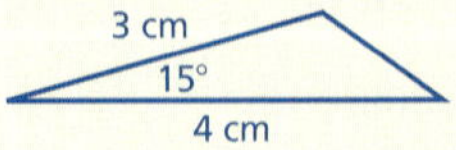

Not actual size

obtuse scalene triangle

Taking Math Deeper

Exercise 22

First, see whether it is possible to construct a triangle with one angle measure of 60° and one 4-centimeter side. If it is, see whether you can construct another triangle that is different from the first one.

1. Use a protractor to draw a 60° angle. Make one of the rays 4 centimeters long. Choose a length for the second ray and draw the third side.

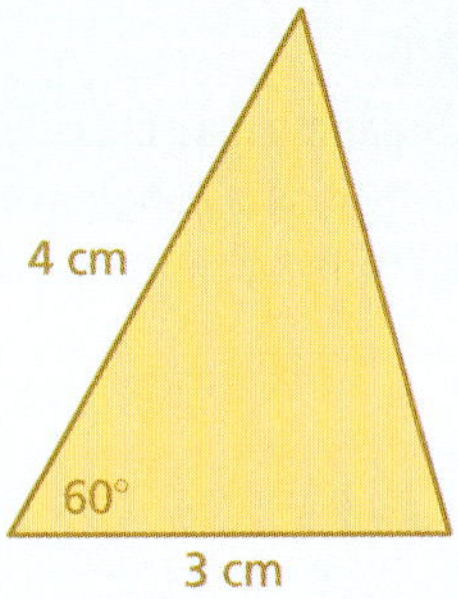

2. Now that you have drawn one triangle, stop and think about what you might be able to change to create a different triangle that still satisfies the given description.

In the triangle above, you chose the length of the second ray to be 3 centimeters. Notice that you can change this length without changing the 60° angle or the 4-centimeter side. However, it changes the other side length and angles.

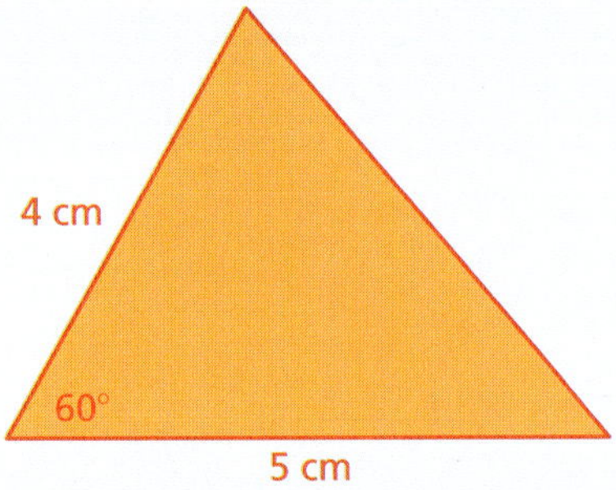

3. Answer the question.

So, you can construct *many* triangles with one angle measure of 60° and one 4-centimeter side.

Project

Tell students to create three problems that give descriptions of triangles. One description should lead to *no* possible triangle, another should lead to *one* possible triangle, and the other should lead to *many* possible triangles. Students can exchange problems and determine how many triangles are possible.

Reteaching and Enrichment Strategies

If students need help...	If students got it...
Resources by Chapter • Practice A and Practice B • Puzzle Time Record and Practice Journal Practice Differentiating the Lesson Lesson Tutorials Skills Review Handbook	Resources by Chapter • Enrichment and Extension • Technology Connection Start the next section

Draw a triangle with the given angle measures. Then classify the triangle.

2 **14.** 15°, 75°, 90° **15.** 20°, 60°, 100° **16.** 30°, 30°, 120°

Draw a triangle with the given description.

3 **17.** a triangle with a 2-inch side and a 3-inch side that meet at a 40° angle

18. a triangle with a 45° angle connected to a 60° angle by an 8-centimeter side

19. an acute scalene triangle

20. **LOGIC** You are constructing a triangle. You draw the first angle, as shown. Your friend says that you must be constructing an acute triangle. Is your friend correct? Explain your reasoning.

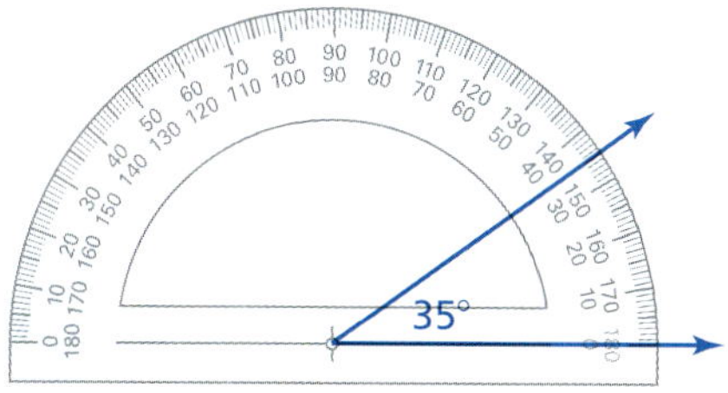

Determine whether you can construct *many, one,* or *no* triangle(s) with the given description. Explain your reasoning.

21. a triangle with angle measures of 50°, 70°, and 100°

22. a triangle with one angle measure of 60° and one 4-centimeter side

23. a scalene triangle with a 3-centimeter side and a 7-centimeter side

24. an isosceles triangle with two 4-inch sides that meet at an 80° angle

25. an isosceles triangle with two 2-inch sides and one 5-inch side

26. a right triangle with three congruent sides

27. Critical Thinking Consider the three isosceles triangles.

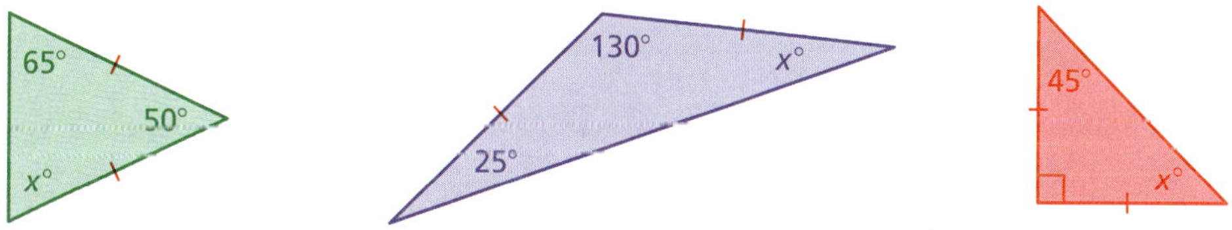

a. Find the value of x for each triangle.

b. What do you notice about the angle measures of each triangle?

c. Write a rule about the angle measures of an isosceles triangle.

Fair Game Review What you learned in previous grades & lessons

Tell whether x and y show direct variation. Explain your reasoning. If so, find the constant of proportionality. *(Skills Review Handbook)*

28. $x = 2y$ **29.** $y - x = 6$ **30.** $xy = 5$

31. **MULTIPLE CHOICE** A savings account earns 6% simple interest per year. The principal is \$800. What is the balance after 18 months? *(Skills Review Handbook)*

Ⓐ \$864 Ⓑ \$872 Ⓒ \$1664 Ⓓ \$7200

Extension 12.3 Angle Measures of Triangles

Check It Out
Lesson Tutorials
BigIdeasMath.com

Key Idea

Sum of the Angle Measures of a Triangle

Words The sum of the angle measures of a triangle is 180°.

Algebra $x + y + z = 180$

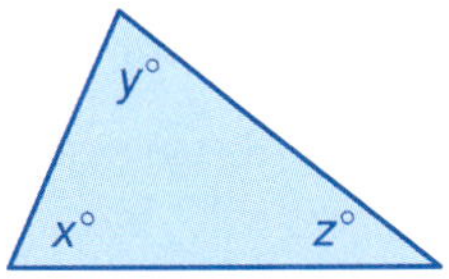

EXAMPLE 1 Finding Angle Measures

Geometry

In this extension, you will

- understand that the sum of the angle measures of any triangle is 180°.
- find missing angle measures in triangles.

Find each value of x. Then classify each triangle.

a.

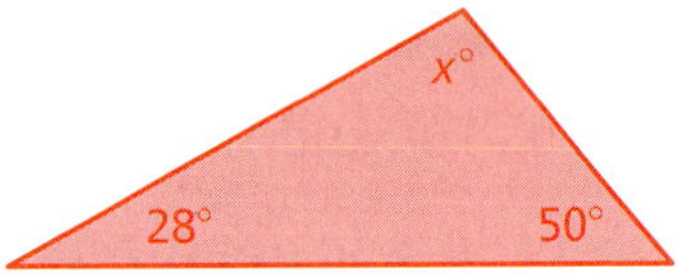

$$x + 28 + 50 = 180$$
$$x + 78 = 180$$
$$x = 102$$

The value of x is 102. The triangle has one obtuse angle and no congruent sides. So, it is an obtuse scalene triangle.

b.

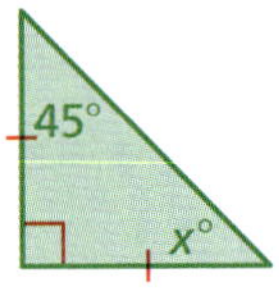

$$x + 45 + 90 = 180$$
$$x + 135 = 180$$
$$x = 45$$

The value of x is 45. The triangle has a right angle and two congruent sides. So, it is a right isosceles triangle.

Practice

Find the value of x. Then classify the triangle.

1.

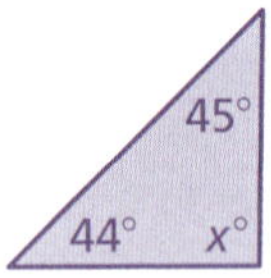

2.

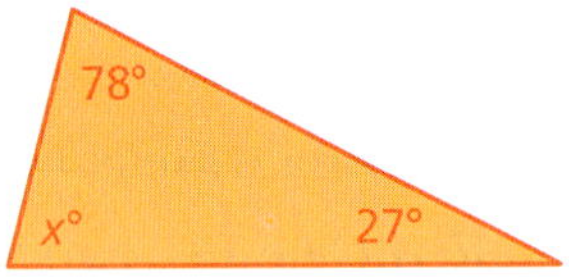

3.

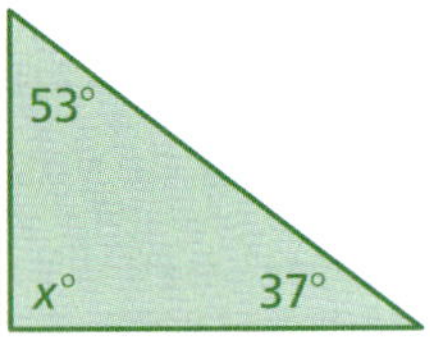

4.

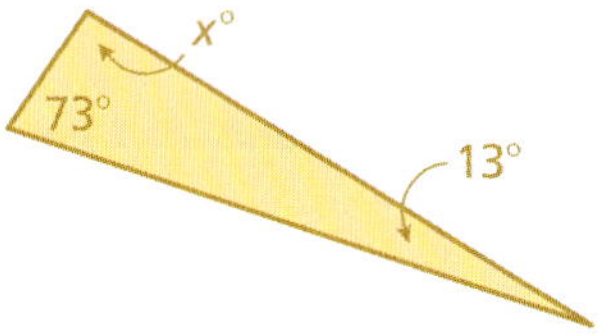

5.

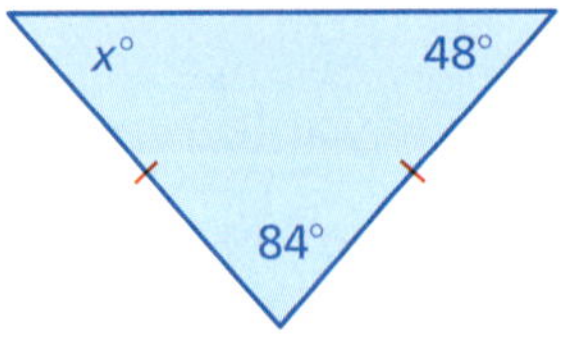

6.

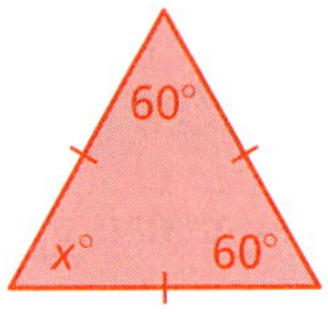

Tell whether a triangle can have the given angle measures. If not, change the first angle measure so that the angle measures form a triangle.

7. $76.2°, 81.7°, 22.1°$

8. $115.1°, 47.5°, 93°$

9. $5\frac{2}{3}°, 64\frac{1}{3}°, 87°$

10. $31\frac{3}{4}°, 53\frac{1}{2}°, 94\frac{3}{4}°$

Laurie's Notes

Introduction

Connect

- **Yesterday:** Students classified and constructed triangles.
- **Today:** Students will find the missing angle measure of a triangle and classify the triangle.

Motivate

? Discuss the Ohio State flag.
The blue triangle represents hills and valleys. The red and white stripes represent roads and waterways. The 13 leftmost stars represent the 13 original colonies. The 4 stars on the right bring the total to 17, representing that Ohio was the 17th state admitted to the Union.

Lesson Notes

Key Idea

- The property is written with variables to suggest that you can solve an equation to find the third angle when you know the other two angles. This is also called the *Triangle Sum Theorem.*
- ? "What type of angles are the remaining angles of a right triangle? a triangle with an obtuse angle?" Both are acute.
- ? "Do you think an obtuse triangle could have a right angle? Explain." no; The sum of the angle measures would be greater than 180°.

Example 1

- Some students may argue that all they need to do is add the angle measures and subtract from 180. Remind them that they are practicing a *process*, one that works when the three angle measures are given as algebraic expressions, such as $(x + 10)°$, $(x + 20)°$, and $(x + 30)°$.

Practice

- **Construct Viable Arguments and Critique the Reasoning of Others:** Students may observe that the isosceles triangles in Exercises 5 and 6 have a pair of congruent angles. This is revisited in Exercises 11–13.
- Exercises 7–10 provide practice with fractions and decimals.

What Your Students Will Learn

- Understand that the sum of the angle measures of any triangle is 180°.
- Find missing angle measures in triangles using the Triangle Sum Theorem.

Goal Today's lesson is finding the missing angle measure of a triangle and classifying the triangle.

Technology for the Teacher

Dynamic Classroom

Lesson Tutorials
Lesson Plans
Answer Presentation Tool

Extra Example 1

Find each value of *x*. Then classify each triangle.

a.

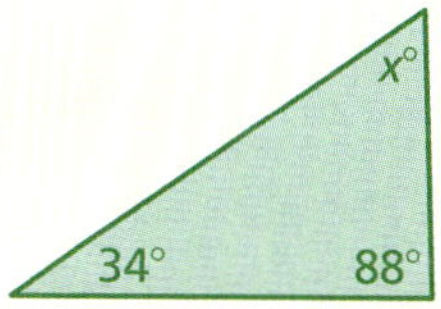

58; acute scalene triangle

b.

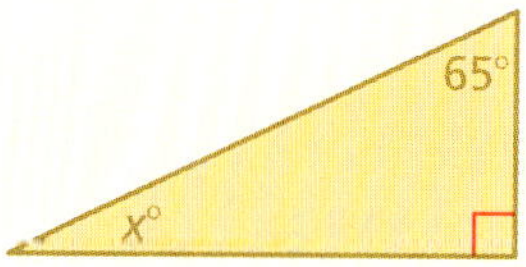

25; right scalene triangle

Practice

1. 91; obtuse scalene triangle
2. 75; acute scalene triangle
3. 90; right scalene triangle
4. 94; obtuse scalene triangle
5. 48; acute isosceles triangle
6. 60; equiangular equilateral triangle
7. yes
8. no; 39.5°
9. no; $28\frac{2}{3}$
10. yes

Record and Practice Journal Extension 12.3 Practice

1–11. See Additional Answers.

Extra Example 2

Find each value of x. Then classify each triangle.

a.

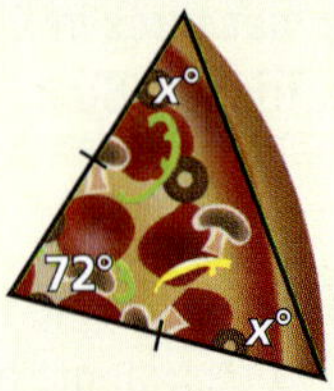

54; acute isosceles triangle

b.

60; equiangular equilateral triangle

Practice

11. 67.5; acute isosceles triangle

12. 60; equiangular equilateral triangle

13. 24; obtuse isosceles triangle

14. 25; right scalene triangle

15. 35; obtuse scalene triangle

16–17. See Additional Answers.

Mini-Assessment

Find the value of x. Then classify the triangle.

1.

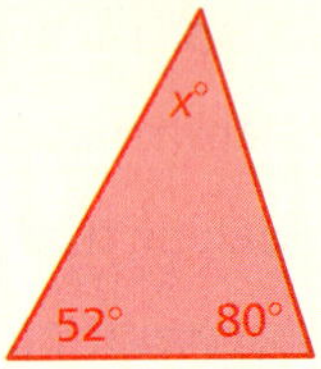

48; acute scalene triangle

2.

18; obtuse isosceles triangle

3.

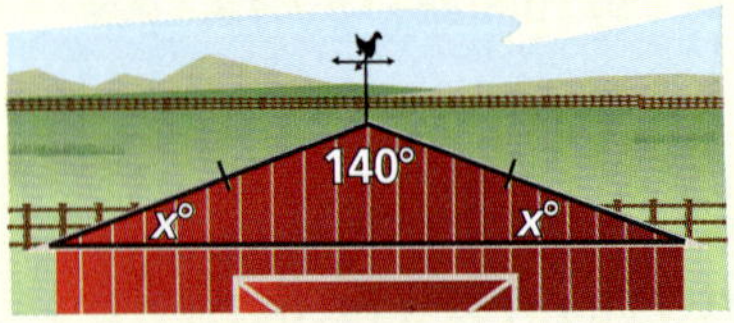

20; obtuse isosceles triangle

Laurie's Notes

Example 2

- Share the symbolism of each flag.
  - **Jamaica:** The yellow divides the flag into four triangles and represents sunshine and natural resources. Black represents the burdens overcome by the people and the hardships in the future. Green represents the land and hope for the future.
  - **Cuba:** The blue stripes refer to the three old divisions of the island and the two white stripes represent the strength of the independent ideal. The red triangle symbolizes equality, fraternity and freedom, and the blood shed in the struggle for independence. The white star symbolizes the absolute freedom among the Cuban people.
- Set up and solve the equations as shown.

? "How would you classify the green triangle on the Jamaican flag?" obtuse isosceles

Practice

- In both examples, students set up and solved simple equations to find the value of x. This experience should help them with Exercises 11–15 and 17.
- **Construct Viable Arguments and Critique the Reasoning of Others:** For Exercise 16, ask a volunteer to explain his or her reasoning.

Closure

- **Exit Ticket:** Find the value of x. Then classify the triangle.

a.

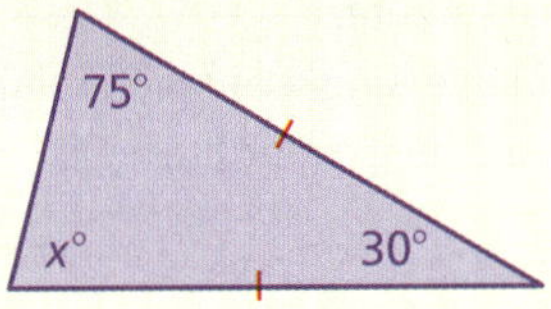

75; acute isosceles triangle

b.

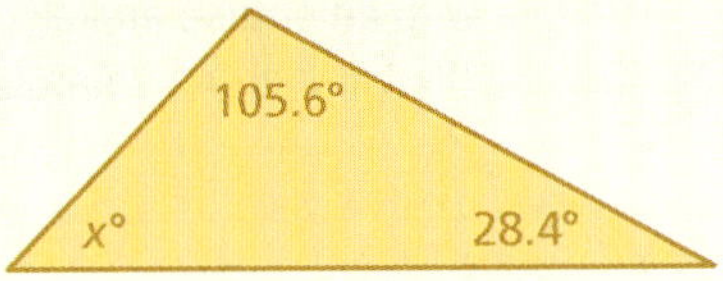

46; obtuse scalene triangle

EXAMPLE 2 Finding Angle Measures

Math Practice

Analyze Givens

What information is given in the problem? How can you use this information to answer the question?

Find each value of x. Then classify each triangle.

a. Flag of Jamaica

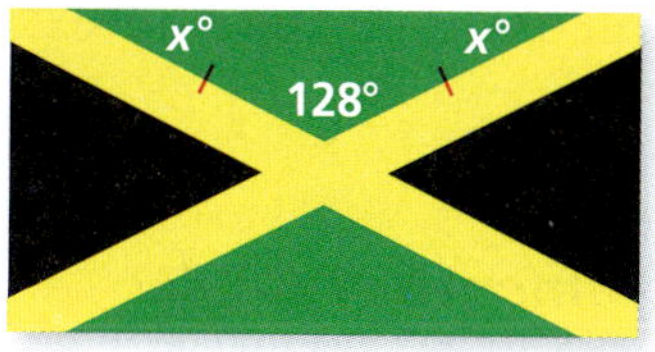

$$x + x + 128 = 180$$
$$2x + 128 = 180$$
$$2x = 52$$
$$x = 26$$

The value of x is 26. The triangle has one obtuse angle and two congruent sides. So, it is an obtuse isosceles triangle.

b. Flag of Cuba

$$x + x + 60 = 180$$
$$2x + 60 = 180$$
$$2x = 120$$
$$x = 60$$

The value of x is 60. All three angles are congruent. So, it is an equilateral and equiangular triangle.

Practice

Find the value of x. Then classify the triangle.

11.

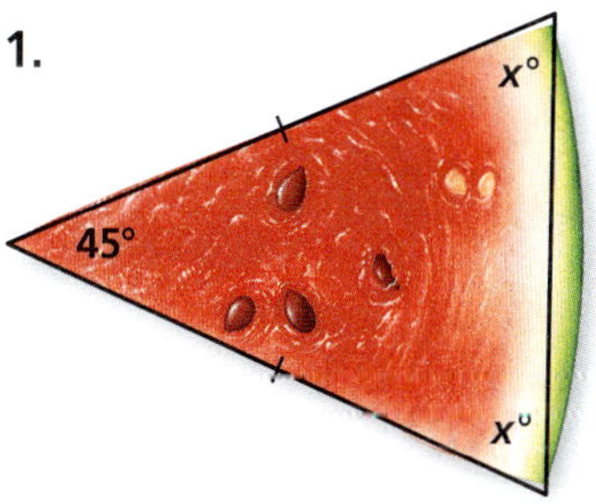

12.

13.

14.

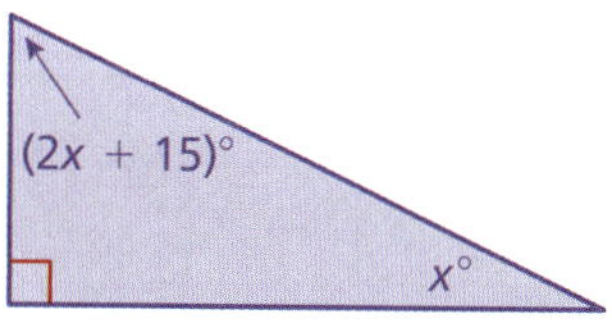

15.

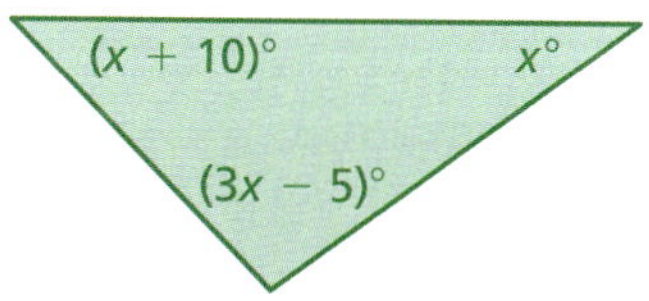

16. REASONING Explain why all triangles have at least two acute angles.

17. CARDS One method of stacking cards is shown.

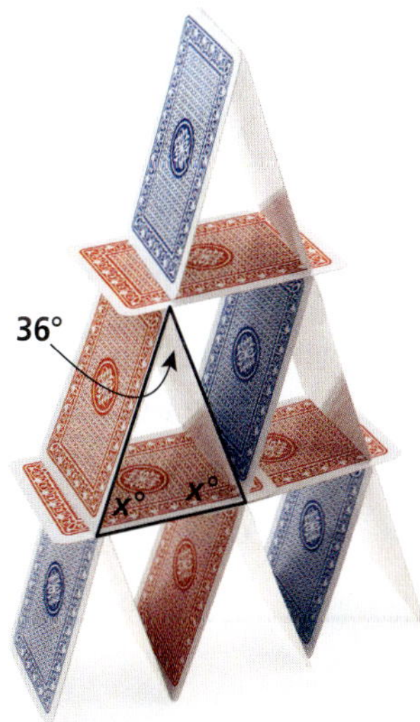

a. Find the value of x.

b. Describe how to stack the cards with different angles. Is the value of x limited? If so, what are the limitations? Explain your reasoning.

12 Study Help

Check It Out
Graphic Organizer
BigIdeasMath.com

You can use an **example and non-example chart** to list examples and non-examples of a vocabulary word or item. Here is an example and non-example chart for complementary angles.

Complementary Angles

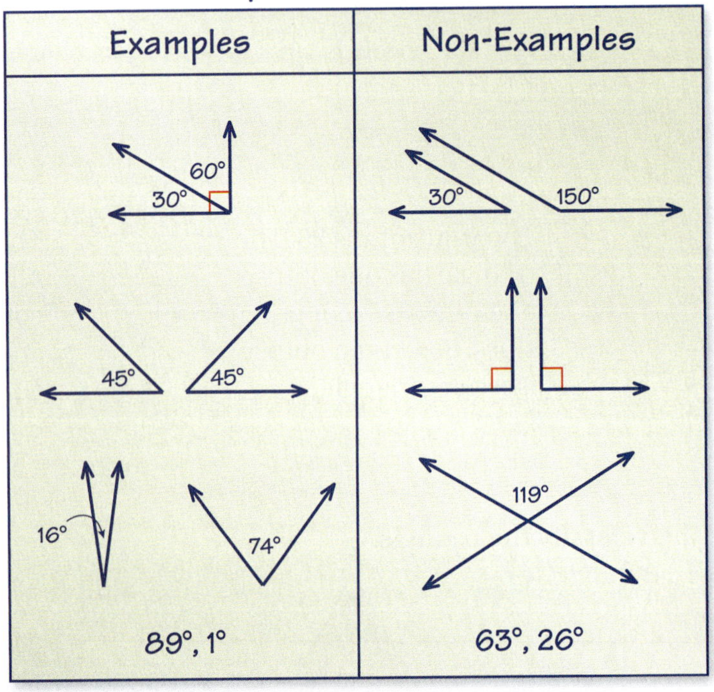

On Your Own

Make example and non-example charts to help you study these topics.

1. adjacent angles
2. vertical angles
3. supplementary angles

After you complete this chapter, make example and non-example charts for the following topics.

4. quadrilaterals
5. scale factor

"What do you think of my example & non-example chart for popular cat toys?"

Sample Answers

1. Adjacent angles

Examples

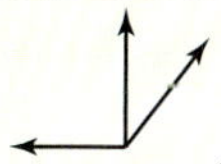

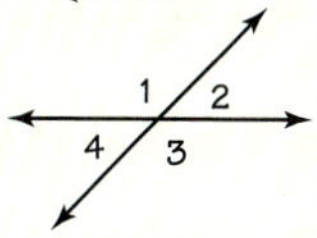

$\angle 1$ and $\angle 2$
$\angle 2$ and $\angle 3$
$\angle 3$ and $\angle 4$
$\angle 4$ and $\angle 1$

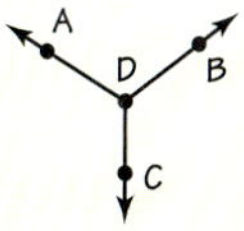

$\angle ADB$ and $\angle BDC$
$\angle BDC$ and $\angle CDA$
$\angle CDA$ and $\angle ADB$

Non-Examples

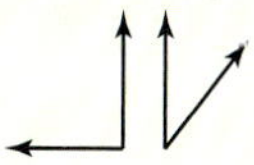

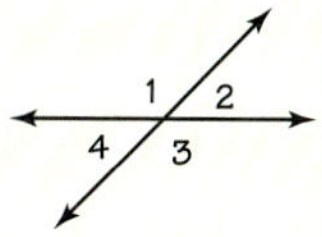

$\angle 1$ and $\angle 3$
$\angle 2$ and $\angle 4$

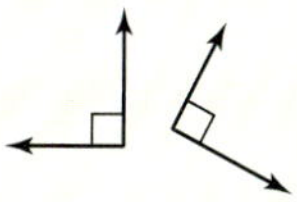

2. Vertical angles

Examples

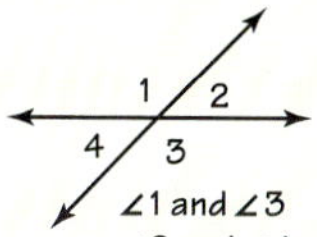

$\angle 1$ and $\angle 3$
$\angle 2$ and $\angle 4$

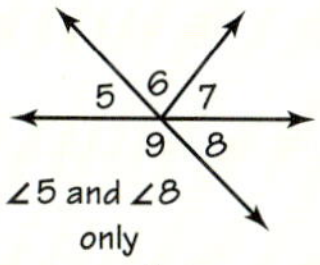

$\angle 5$ and $\angle 8$ only

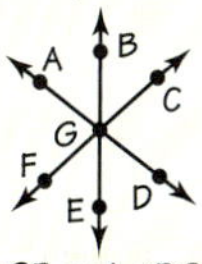

$\angle AGB$ and $\angle DGE$
$\angle AGF$ and $\angle DGC$
$\angle BGC$ and $\angle EGF$

Non-Examples

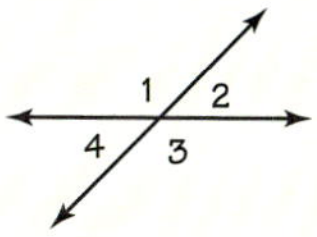

$\angle 1$ and $\angle 2$
$\angle 2$ and $\angle 3$
$\angle 3$ and $\angle 4$
$\angle 4$ and $\angle 1$

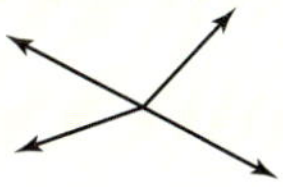

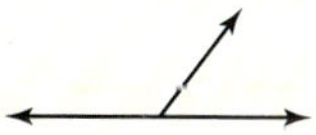

3. Supplementary angles

Examples

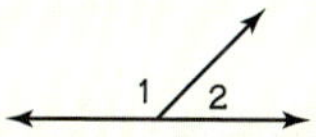

$\angle 1$ and $\angle 2$

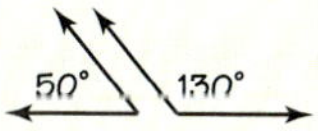

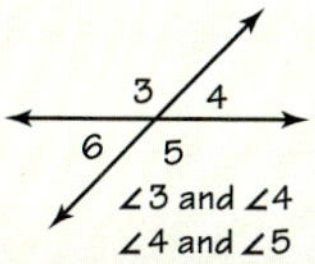

$\angle 3$ and $\angle 4$
$\angle 4$ and $\angle 5$
$\angle 5$ and $\angle 6$
$\angle 6$ and $\angle 3$

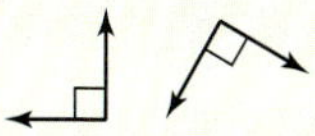

117°, 63°

Non-Examples

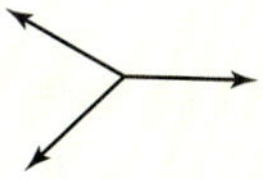

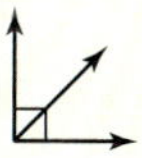

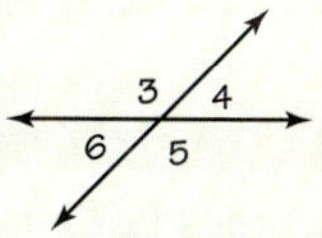

$\angle 3$ and $\angle 5$
$\angle 4$ and $\angle 6$

104°, 62°

List of Organizers

Available at *BigIdeasMath.com*

Comparison Chart
Concept Circle
Definition (Idea) and Example Chart
Example and Non-Example Chart
Formula Triangle
Four Square
Information Frame
Information Wheel
Notetaking Organizer
Process Diagram
Summary Triangle
Word Magnet
Y Chart

About this Organizer

An **Example and Non-Example Chart** can be used to list examples and non-examples of a vocabulary word or term. Students write examples of the word or term in the left column and non-examples in the right column. This type of organizer serves as a good tool for assessing students' knowledge of pairs of topics that have subtle but important differences, such as complementary and supplementary angles. Blank example and non-example charts can be included on tests or quizzes for this purpose.

Technology for the Teacher

Editable Graphic Organizer

Answers

1. *Sample answer:* adjacent: $\angle PQR$ and $\angle RQS$, $\angle PQT$ and $\angle TQS$; vertical: $\angle PQR$ and $\angle TQS$, $\angle PQT$ and $\angle RQS$
2. *Sample answer:* adjacent: $\angle YUZ$ and $\angle ZUV$, $\angle ZUV$ and $\angle VUW$; vertical: $\angle YUX$ and $\angle VUW$, $\angle YUV$ and $\angle XUW$
3. adjacent; 146
4. adjacent; 16
5. vertical; 49
6. supplementary; 50
7. complementary; 24
8. 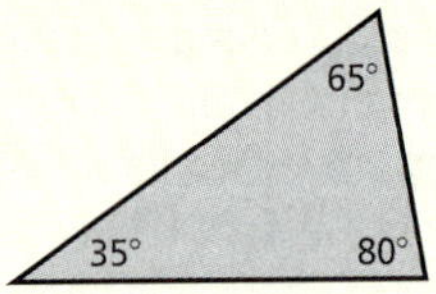

9. See Additional Answers.
10. 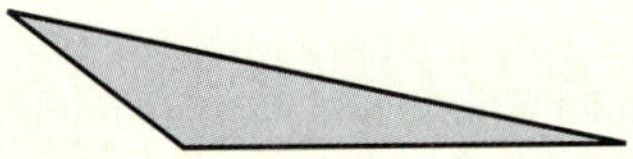
11. 115; obtuse scalene
12. 45; right isosceles
13. 80; equilateral equiangular
14. Use vertical angles to find that the measure of $\angle 2$ is $115°$. Use supplementary angles to find that the measure of $\angle 3$ is $65°$. Then use supplementary angles to find that the measure of $\angle 2$ is $115°$.

Technology for the Teacher

Online Assessment
Assessment Book
ExamView® Assessment Suite

Alternative Quiz Ideas

100% Quiz	Math Log
Error Notebook	**Notebook Quiz**
Group Quiz	Partner Quiz
Homework Quiz	Pass the Paper

Notebook Quiz

A notebook quiz is used to check students' notebooks. Students should be told at the beginning of the course what the expectations are for their notebooks: notes, class work, homework, date, problem number, goals, definitions, or anything else that you feel is important for your class. They also need to know that it is their responsibility to obtain the notes when they miss class.

1. On a certain day, how was this vocabulary term defined?
2. For Section 12.1, what is the answer to On Your Own Question 5?
3. For Section 12.3 Extension, what is the answer to Practice Question 1?
4. For Section 12.4, what is the answer to the Essential Question?
5. On a certain day, what was the homework assignment?

Give the students 5 minutes to answer these questions.

Reteaching and Enrichment Strategies

If students need help. . .	If students got it. . .
Resources by Chapter • Practice A and Practice B • Puzzle Time Lesson Tutorials *BigIdeasMath.com*	Resources by Chapter • Enrichment and Extension • Technology Connection Game Closet at *BigIdeasMath.com* Start the next section

12.1–12.3 Quiz

Check It Out
Progress Check
BigIdeasMath.com

Name two pairs of adjacent angles and two pairs of vertical angles in the figure. *(Section 12.1)*

1. 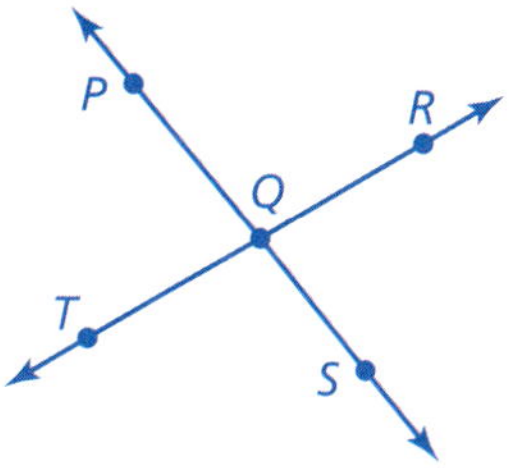

2. 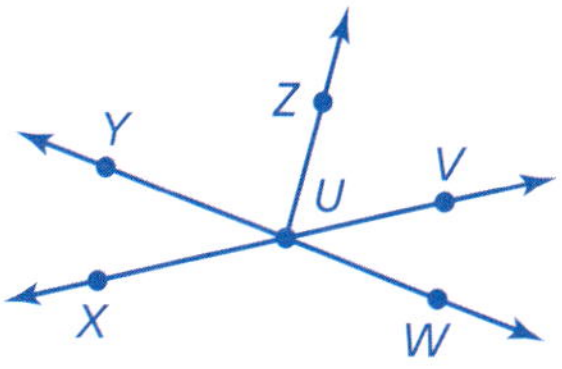

Tell whether the angles are *adjacent* or *vertical*. Then find the value of x. *(Section 12.1)*

3. 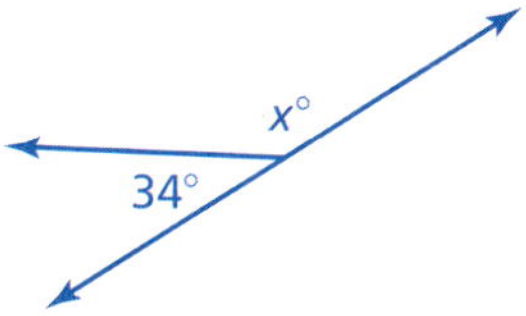

4. 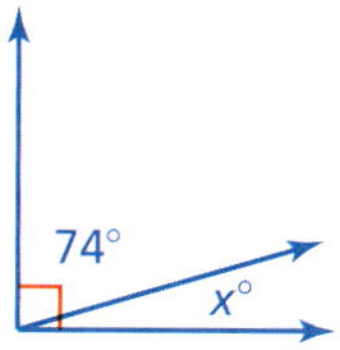

5. 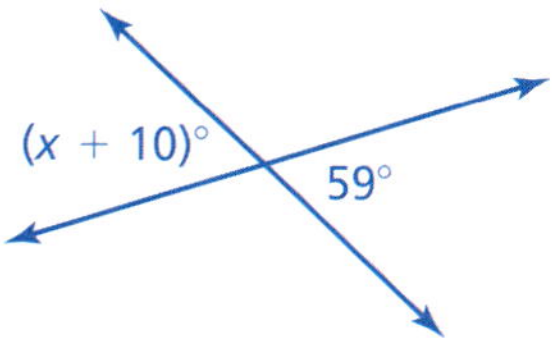

Tell whether the angles are *complementary* or *supplementary*. Then find the value of x. *(Section 12.2)*

6. 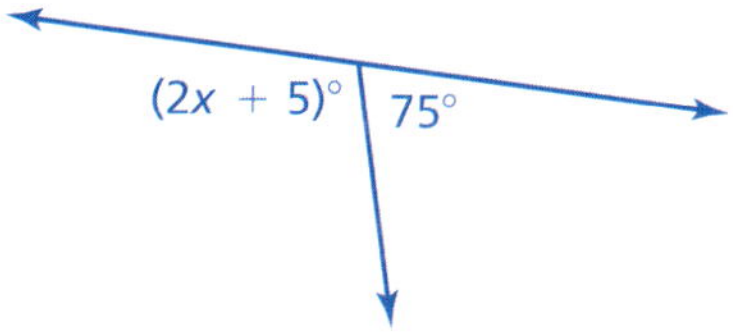

7. 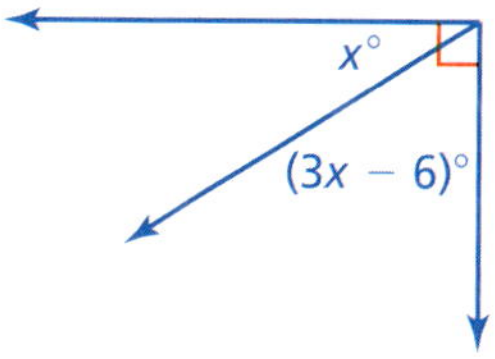

Draw a triangle with the given description. *(Section 12.3)*

8. a triangle with angle measures of 35°, 65°, and 80°

9. a triangle with a 5-centimeter side and a 7-centimeter side that meet at a 70° angle

10. an obtuse scalene triangle

Find the value of x. Then classify the triangle. *(Section 12.3)*

11. 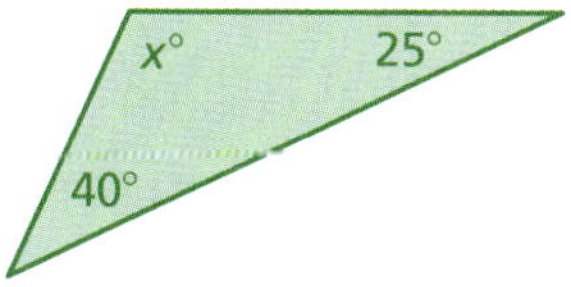

12. 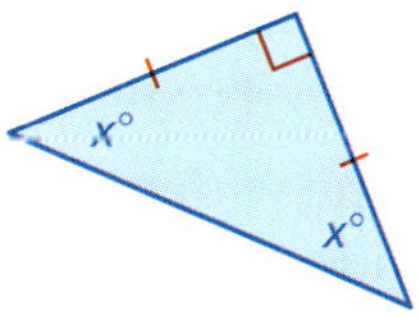

13. 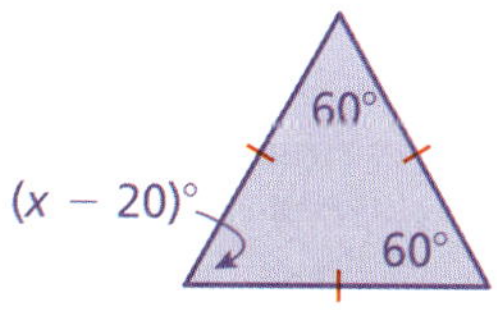

14. RAILROAD CROSSING Describe two ways to find the measure of $\angle 2$. *(Section 12.1 and Section 12.2)*

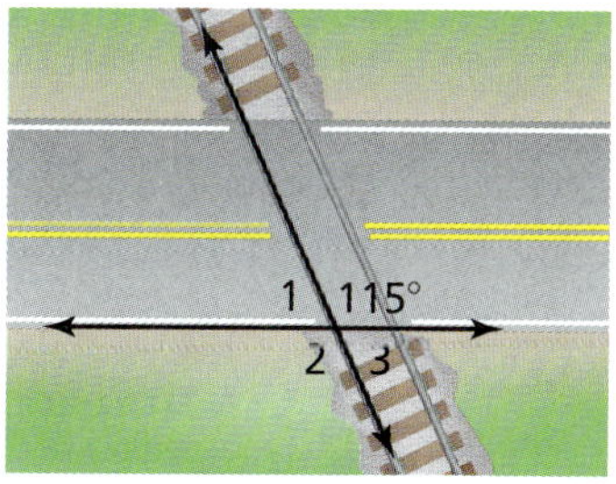

12.4 Quadrilaterals

Essential Question How can you classify quadrilaterals?

Quad means *four* and *lateral* means *side*. So, *quadrilateral* means a polygon with *four sides*.

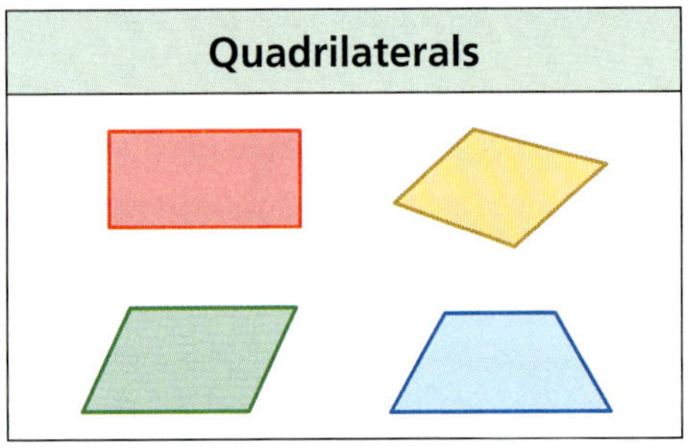

1 ACTIVITY: Using Descriptions to Form Quadrilaterals

Work with a partner. Use a geoboard to form a quadrilateral that fits the given description. Record your results on geoboard dot paper.

a. Form a quadrilateral with exactly one pair of parallel sides.

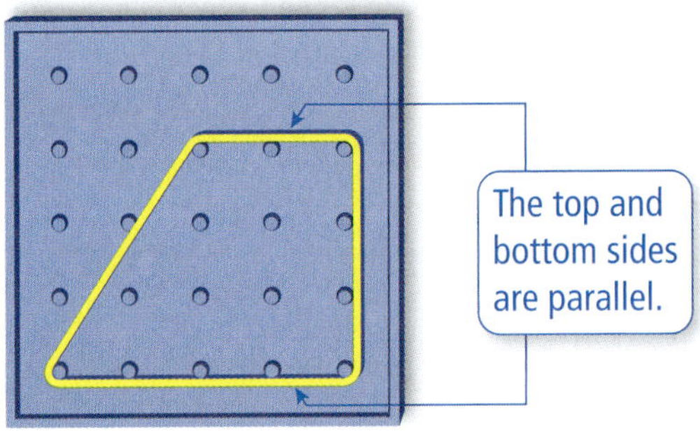

b. Form a quadrilateral with four congruent sides and four right angles.

c. Form a quadrilateral with four right angles that is *not* a square.

d. Form a quadrilateral with four congruent sides that is *not* a square.

e. Form a quadrilateral with two pairs of congruent adjacent sides and whose opposite sides are *not* congruent.

f. Form a quadrilateral with congruent and parallel opposite sides that is *not* a rectangle.

Geometry

In this lesson, you will

- understand that the sum of the angle measures of any quadrilateral is 360°.
- find missing angle measures in quadrilaterals.
- construct quadrilaterals.

2 ACTIVITY: Naming Quadrilaterals

Work with a partner. Match the names *square, rectangle, rhombus, parallelogram, trapezoid,* and *kite* with your 6 drawings in Activity 1.

Laurie's Notes

Introduction

Applying Mathematical Practices

- **Construct Viable Arguments and Critique the Reasoning of Others:** Students will be making observations and statements about the attributes of quadrilaterals. Listen carefully to their reasoning.

Motivate

- ? "Have any of you created designs on grid or dot paper?"
- Quilters often sketch the design they are going to use before they begin to make a quilt. Share an example of a quilt design that contains different geometric shapes. The one shown is called *Symmetry in Motion.*
- ? "What geometric shapes do you see in this design?" triangles, a square, trapezoids, and parallelograms
- You want to gain a sense of what pre-knowledge students have. Do not dwell on the attributes of the different quadrilaterals at this time.

Activity Notes

Write

- Write "quad" on the board, and ask students to think about different words that begin with this prefix. Some examples include *quadrilateral, quadrant, quadruple, quadruplets, quadruped,* and *quadrillion.*

Activity 1

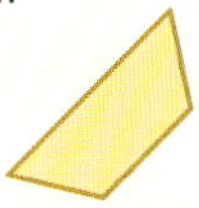

- Observe the orientations of students' quadrilaterals as they form them. For part (a), ask if the parallel sides need to be horizontal or vertical. Do some students form quadrilaterals with a diagonal orientation such as the one shown?
- Students are sometimes challenged to make quadrilaterals that have a particular attribute. For instance, part (e) may not be obvious to all students. Without giving away the answer, assure students that it is possible and have them explore different quadrilaterals on their geoboards.
- For part (e), explain to students that *adjacent sides* mean that the sides share a common vertex.

Activity 2

- **Construct Viable Arguments and Critique the Reasoning of Others:** After students have completed this activity, ask them if it is possible for a quadrilateral to have more than one name. For instance, a rectangle might also be called a parallelogram. However, if students are not yet ready to recognize that all rectangles are parallelograms, but not all parallelograms are rectangles, do not dwell on this.
- Look around the room and identify different types of quadrilaterals.

What Your Students Will Learn

- Classify quadrilaterals using the most specific names possible.
- Understand that the sum of the angle measures of any quadrilateral is 360°.
- Find missing angle measures in quadrilaterals.
- Construct quadrilaterals using a protractor.

Previous Learning

Students should be familiar with plane figures such as squares, rectangles, parallelograms, trapezoids, and rhombi.

Technology for the Teacher

Dynamic Classroom

Lesson Plans
Complete Materials List

12.4 Record and Practice Journal

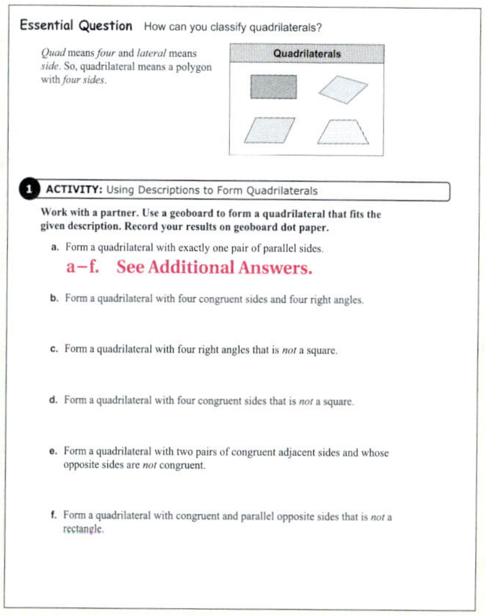

Essential Question How can you classify quadrilaterals?

Quad means *four* and *lateral* means *side*. So, quadrilateral means a polygon with *four sides*.

Quadrilaterals

1 ACTIVITY: Using Descriptions to Form Quadrilaterals

Work with a partner. Use a geoboard to form a quadrilateral that fits the given description. Record your results on geoboard dot paper.

a. Form a quadrilateral with exactly one pair of parallel sides.

a–f. See Additional Answers.

b. Form a quadrilateral with four congruent sides and four right angles.

c. Form a quadrilateral with four right angles that is *not* a square.

d. Form a quadrilateral with four congruent sides that is *not* a square.

e. Form a quadrilateral with two pairs of congruent adjacent sides and whose opposite sides are *not* congruent.

f. Form a quadrilateral with congruent and parallel opposite sides that is *not* a rectangle.

English Language Learners

Pronunciation

English learners may have difficulty pronouncing words such as *parallelogram* and *quadrilateral*. Pronounce these words slowly and have students repeat after you. You may find it helpful to write these words on the board broken down phonetically:

pa-ruh-**lel**-uh-gram
kwa-druh-**la**-tuh-rul

12.4 Record and Practice Journal

2 ACTIVITY: Naming Quadrilaterals

Work with a partner. Match the names *square*, *rectangle*, *rhombus*, *parallelogram*, *trapezoid*, and *kite* with your 6 drawings in Activity 1.

a. trapezoid
b. square
c. rectangle
d. rhombus
e. kite
f. parallelogram

3 ACTIVITY: Forming Quadrilaterals

Work with a partner. Form each quadrilateral on your geoboard. Then move *only one* vertex to create the new type of quadrilateral. Record your results below.

a. Trapezoid ⇨ Kite

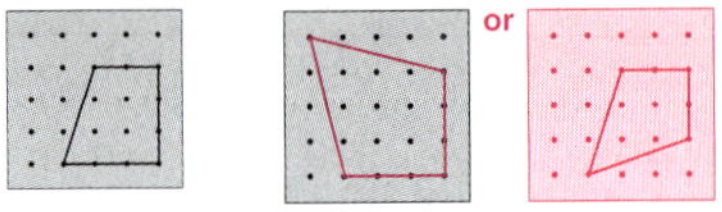

b. Kite ⇨ Rhombus (*not* a square)

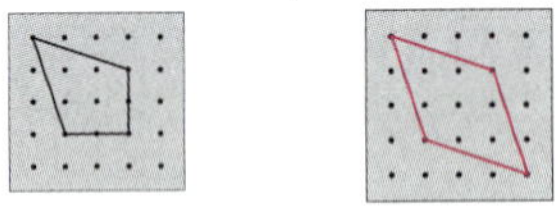

4 ACTIVITY: Using Technology to Draw Quadrilaterals

Work with a partner. Use geometry software to draw a quadrilateral that fits the given description.

a. a square with a side length of 3 units

b. a rectangle with a width of 2 units and a length of 5 units

c. a parallelogram with side lengths of 6 units and 1 unit

d. a rhombus with a side length of 4 units

a–d. Check students' work.

What Is Your Answer?

5. **REASONING** Measure the angles of each quadrilateral you formed in Activity 1. Record your results in a table. Include the sum of the angle measures. Then describe the pattern in the table and write a conclusion based on the pattern.

The sum of the angle measures of any quadrilateral is 360°.

	∠1	∠2	∠3	∠4	∠1 + ∠2 + ∠3 + ∠4
a.	55°	125°	90°	90°	360°
b.	90°	90°	90°	90°	360°
c.	90°	90°	90°	90°	360°
d.	140°	40°	140°	40°	360°
e.	115°	40°	115°	90°	360°
f.	45°	135°	45°	135°	360°

6. **IN YOUR OWN WORDS** How can you classify quadrilaterals? Explain using properties of sides and angles.

You can classify quadrilaterals by the properties of the sides and angles.

Laurie's Notes

Activity 3

- In this activity, students are asked to keep three vertices fixed and change one vertex by moving the rubber band. Two side lengths will change and two side lengths will remain the same.
- If the desired quadrilateral is not formed on their first attempt at moving the rubber band, students should start from the beginning.

? "Do you think there is more than one way to do each part of the activity?" Part (a) has two answers. Part (b) has one answer. Have students share how they answered each part.

Activity 4

- Depending upon the software used, there may or may not be enough time to complete this activity. If the software does not have a menu for selecting a particular type of quadrilateral, then it will take longer.
- **Use Appropriate Tools Strategically** and **Construct Viable Arguments and Critique the Reasoning of Others:** There is more than one way to draw these quadrilaterals. Ask students to share their methods.

What Is Your Answer?

- For Question 6, ask students to include sketches with their explanations.

Closure

- Distribute to each student an index card on which a quadrilateral has been drawn. There should be quadrilaterals from this investigation as well as scalene quadrilaterals. Give a series of quadrilateral attributes and as you do so, students stand up if the quadrilateral on their card has the attribute.

3 ACTIVITY: Forming Quadrilaterals

Work with a partner. Form each quadrilateral on your geoboard. Then move *only one* vertex to create the new type of quadrilateral. Record your results on geoboard dot paper.

a. Trapezoid ⇨ Kite

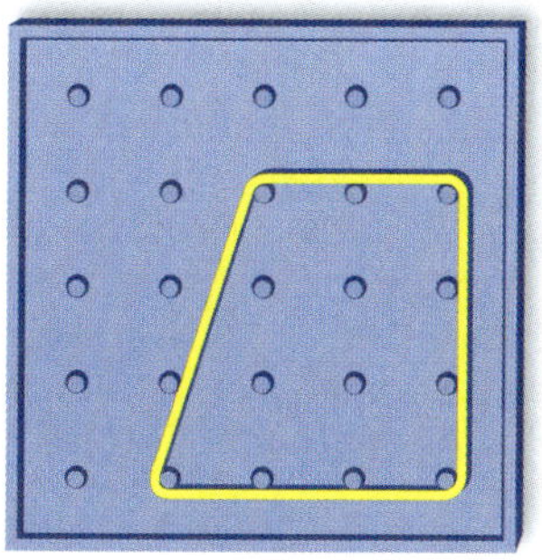

b. Kite ⇨ Rhombus (*not* a square)

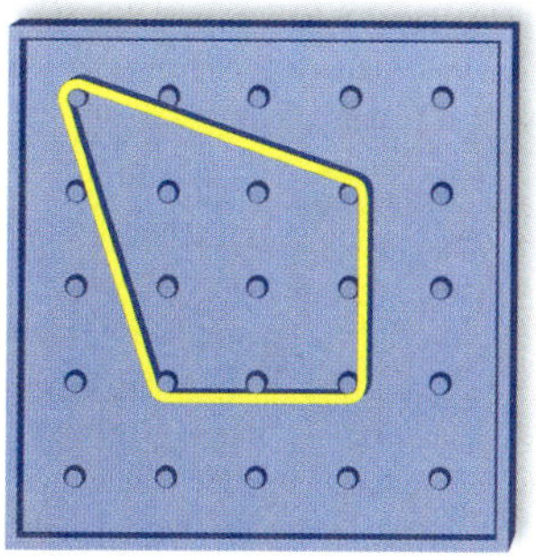

4 ACTIVITY: Using Technology to Draw Quadrilaterals

Math Practice

Use Technology to Explore

How does geometry software help you learn about the characteristics of a quadrilateral?

Work with a partner. Use geometry software to draw a quadrilateral that fits the given description.

a. a square with a side length of 3 units

b. a rectangle with a width of 2 units and a length of 5 units

c. a parallelogram with side lengths of 6 units and 1 unit

d. a rhombus with a side length of 4 units

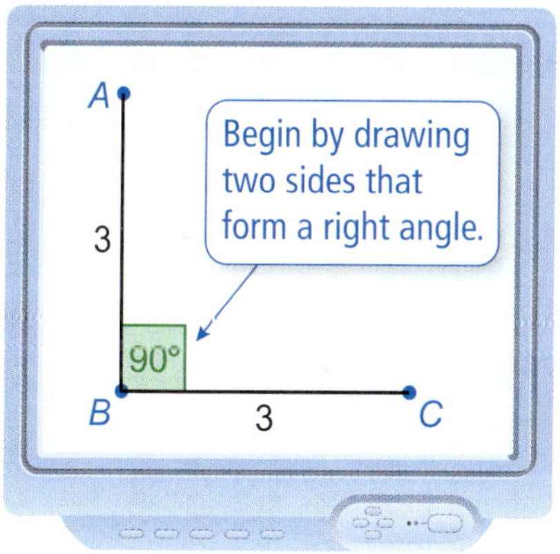

What Is Your Answer?

5. **REASONING** Measure the angles of each quadrilateral you formed in Activity 1. Record your results in a table. Include the sum of the angle measures. Then describe the pattern in the table and write a conclusion based on the pattern.

6. **IN YOUR OWN WORDS** How can you classify quadrilaterals? Explain using properties of sides and angles.

Practice Use what you learned about quadrilaterals to complete Exercises 4–6 on page 528.

12.4 Lesson

Check It Out
Lesson Tutorials
BigIdeasMath.com

Key Vocabulary
kite, *p. 526*

A quadrilateral is a polygon with four sides. The diagram shows properties of different types of quadrilaterals and how they are related. When identifying a quadrilateral, use the name that is most specific.

Reading
Red arrows indicate parallel sides.

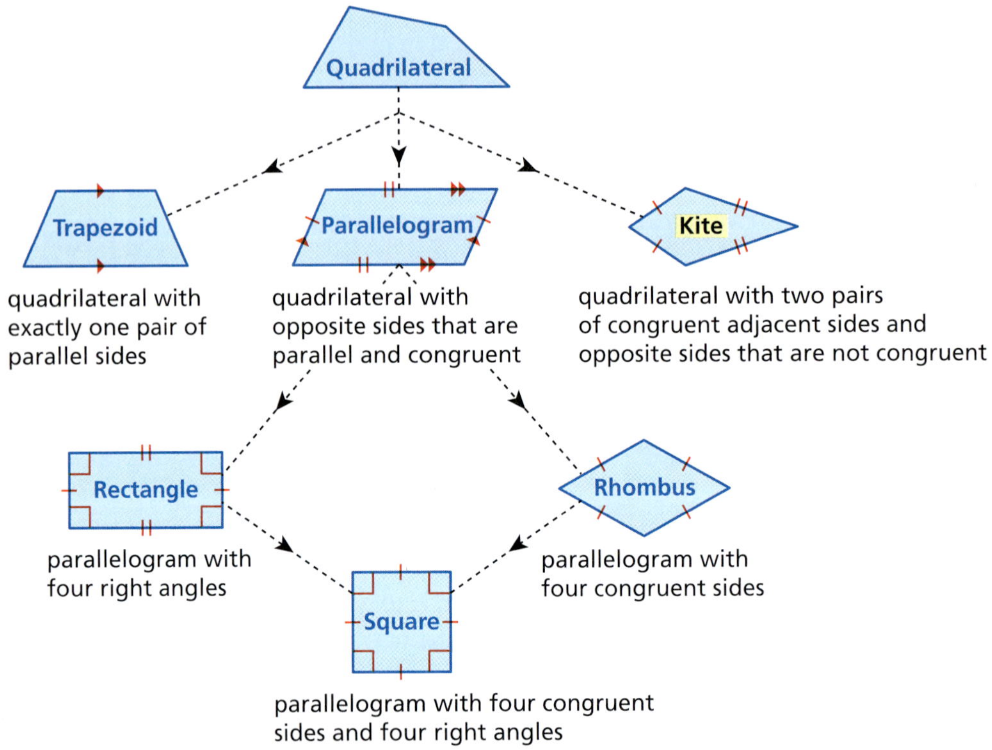

EXAMPLE 1 Classifying Quadrilaterals

Classify the quadrilateral.

a.

The quadrilateral has four congruent sides and four right angles.

So, the quadrilateral is a square.

b.

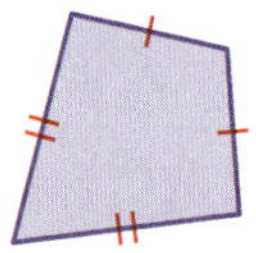

The quadrilateral has two pairs of congruent adjacent sides and opposite sides that are not congruent.

So, the quadrilateral is a kite.

Study Tip
In Example 1(a), the square is also a parallelogram, a rectangle, and a rhombus. Square is the most specific name.

On Your Own

Now You're Ready
Exercises 4–9

Classify the quadrilateral.

1.

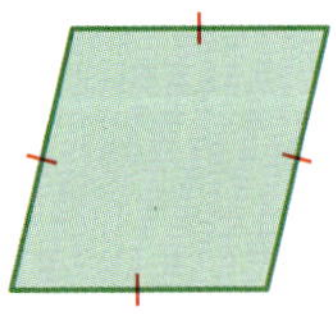

2.

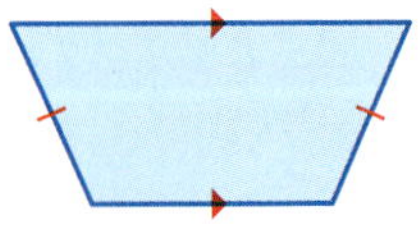

3.

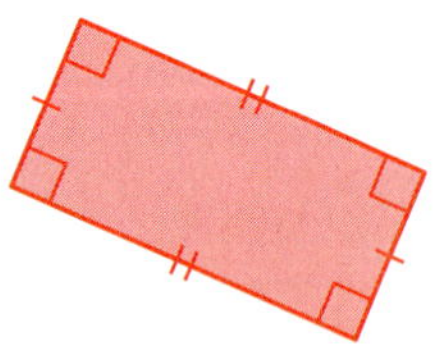

Multi-Language Glossary at BigIdeasMath.com

Laurie's Notes

Introduction

Connect

- **Yesterday:** Students explored properties of quadrilaterals.
- **Today:** Students will classify quadrilaterals by the attributes they possess.

Motivate

- **Preparation:** Make several quadrilaterals out of bendable drinking straws. You can make a slit in one end of a straw and insert it into another straw.
- To make a square, use four full-length straws. To make a rectangle or a kite, cut off 1 to 2 inches from two straws and use two full-length straws.

- The square can be flexed to make a rhombus. The rectangle can be flexed to make a parallelogram.
- Use the models to help students develop a sense of the relationship between the various quadrilaterals.

Lesson Notes

Discuss

- Discuss the quadrilateral classifications of trapezoids, parallelograms, and kites. These groups are disjoint, that is, a quadrilateral cannot belong to more than one of these classifications. Discuss the marks on the quadrilaterals.

? "What do the arrows mean?" parallel sides "What do the tick marks mean?" congruent sides

- Discuss the difference between *opposite* sides and *adjacent* sides.
- The dashed line from the parallelogram means that rectangles and rhombuses (or rhombi) are parallelograms with additional attributes. A rectangle is a parallelogram with four right angles, and a rhombus is a parallelogram with four congruent sides. A square is a parallelogram with *both* of these attributes.

Example 1

? "What properties does the quadrilateral in Example 1(a) have?" four congruent sides and four right angles "What is it?" square

- Although students could say that the quadrilateral in Example 1(a) is a parallelogram, which is correct, they should use the most specific name when identifying a quadrilateral.

On Your Own

- **Neighbor Check:** Have students work independently and then have their neighbors check their work. Have students discuss any discrepancies.

Goal Today's lesson is classifying quadrilaterals.

Technology for the Teacher

Dynamic Classroom

Lesson Tutorials
Lesson Plans
Answer Presentation Tool

Extra Example 1

Classify the quadrilateral.

a.

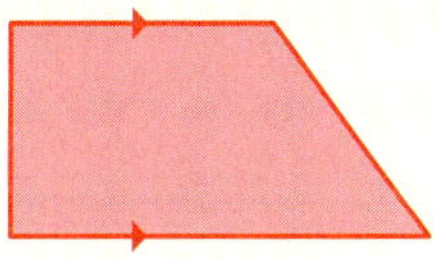

trapezoid

b.

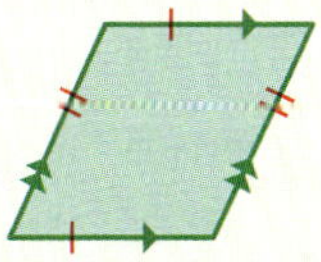

parallelogram

On Your Own

1. rhombus
2. trapezoid
3. rectangle

Differentiated Instruction

Visual

Bring in examples of figures that have the same size and shape (congruent) and figures that have the same shape but not necessarily the same size (similar). Ask students to identify the figures that are congruent. Then ask students to identify figures that are not congruent.

Extra Example 2

Find the value of x. $x = 80$

Extra Example 3

Draw a parallelogram with a 50° angle and a 130° angle.

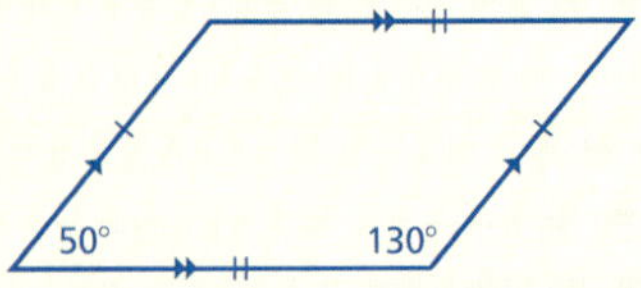

On Your Own

4. $x = 80$
5. $x = 65$
6. *Sample answer:*

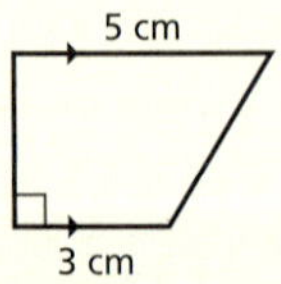

Not actual size

Laurie's Notes

Key Idea

- ? "What observation did you have yesterday regarding the sum of the angle measures of a quadrilateral?" The sum is 360°.
- **Model with Mathematics:** Without measuring, the same conclusion can be reached in the following manner. Cut out a paper quadrilateral. On the overhead projector, tear off the four angles and arrange them about a point. Students should make the observation that the angles complete one revolution or 360°.
- ? **Extension:** "Does this work for any quadrilateral?" yes
- ? "If you know three of the four angle measures of a quadrilateral, could you find the missing angle measure?" yes "How?" Let x represent the missing angle measure. Then write an expression for the sum of the four angle measures, set it equal to 360, and solve.

Example 2

- Although students may want to solve this problem in their heads, work through the example to review equation-solving skills.

Example 3

- You could begin the construction by drawing a pair of parallel lines. Trace lines along both edges of a ruler, and then draw the angles from one of the lines through the other. Ask students whether they prefer this method or the method shown in the example.
- Students may realize that they do not need to "force" corresponding sides to be parallel or congruent. Ask them why they think this is so and whether this would work for any pair of angles. Try to lead them to see that the angles need to be supplementary to construct a parallelogram.
- **Extension:** Investigate the relationships between opposite angles.
- If time allows, repeat the construction with 120° on the bottom left and 60° on the bottom right. Ask students whether they think this parallelogram is different from the one in the example.

On Your Own

- After students finish Question 4, ask if they have an observation about the angles of a parallelogram. Opposite angles are congruent and adjacent angle measures sum to 180°.

Closure

- **Exit Ticket:** Find the value of x.

a.

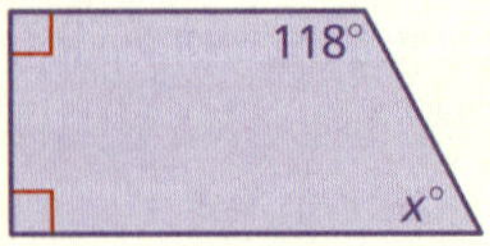

$x = 62$

b.

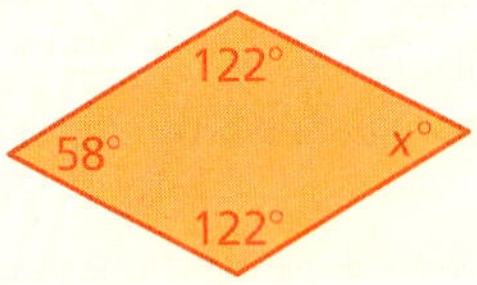

$x = 58$

Key Idea

Sum of the Angle Measures of a Quadrilateral

Words The sum of the angle measures of a quadrilateral is 360°.

Algebra $w + x + y + z = 360$

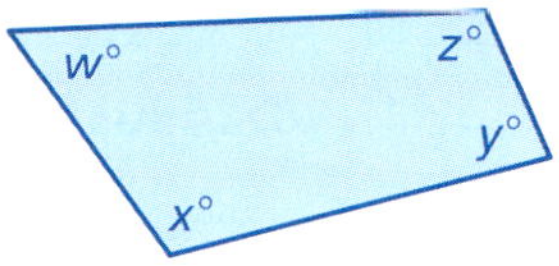

EXAMPLE 2 Finding an Angle Measure of a Quadrilateral

Find the value of *x*.

$$
\begin{aligned}
70 + 75 + 115 + x &= 360 && \text{Write an equation.} \\
260 + x &= 360 && \text{Combine like terms.} \\
-260 \quad &\quad -260 && \text{Subtraction Property of Equality} \\
x &= 100 && \text{Simplify.}
\end{aligned}
$$

The value of x is 100.

EXAMPLE 3 Constructing a Quadrilateral

Draw a parallelogram with a 60° angle and a 120° angle.

Step 1: Draw a line.

Step 2: Draw a 60° angle and a 120° angle that each have one side on the line.

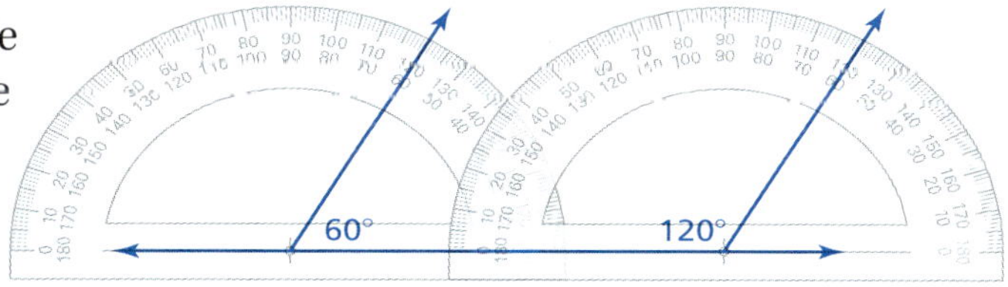

Step 3: Draw the remaining side. Make sure that both pairs of opposite sides are parallel and congruent.

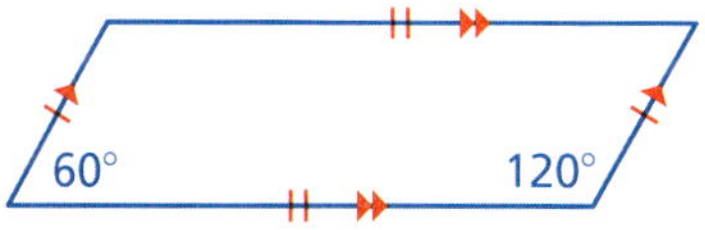

On Your Own

Now You're Ready
Exercises 10–12 and 14–17

Find the value of *x*.

4.

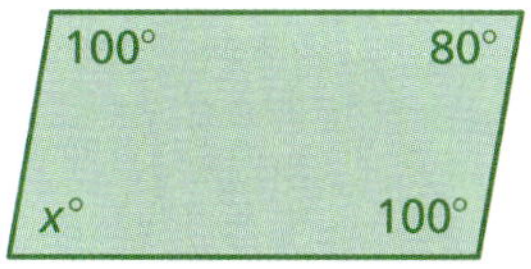

5.

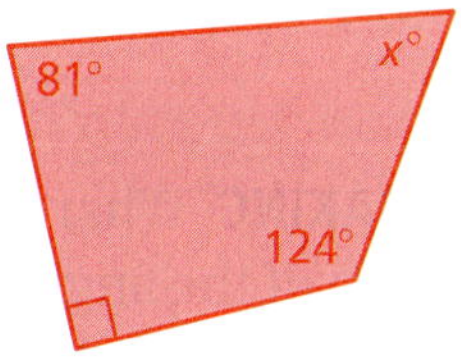

6. Draw a right trapezoid whose parallel sides have lengths of 3 centimeters and 5 centimeters.

12.4 Exercises

Check It Out
Help with Homework
BigIdeasMath.com

Vocabulary and Concept Check

1. **VOCABULARY** Which statements are true?

   a. All squares are rectangles.
   b. All squares are parallelograms.
   c. All rectangles are parallelograms.
   d. All squares are rhombuses.
   e. All rhombuses are parallelograms.

2. **REASONING** Name two types of quadrilaterals with four right angles.

3. **WHICH ONE DOESN'T BELONG?** Which type of quadrilateral does *not* belong with the other three? Explain your reasoning.

   rectangle | parallelogram | square | kite

Practice and Problem Solving

Classify the quadrilateral.

1 4. 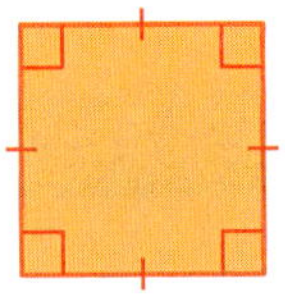

5. 

6. 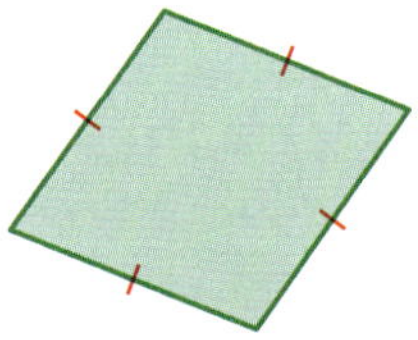

7. 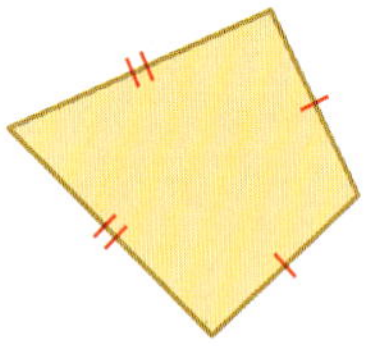

8. 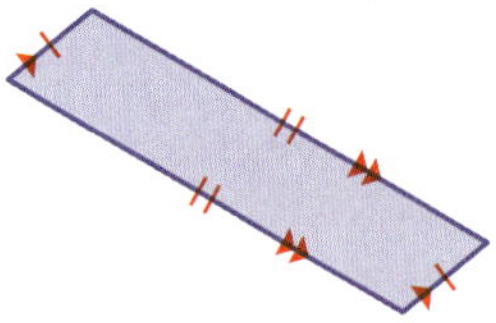

9. 

Find the value of *x*.

2 10. 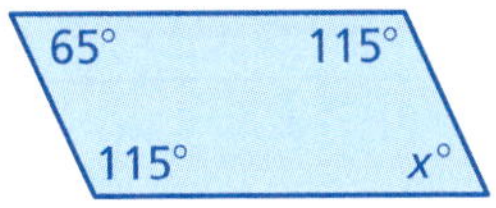

11. 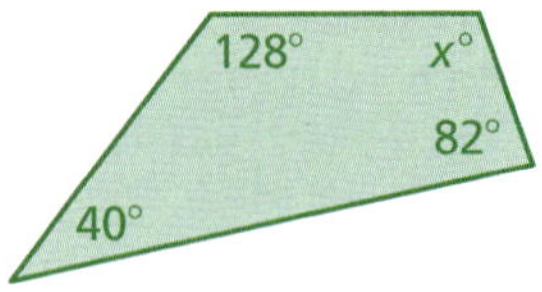

12. 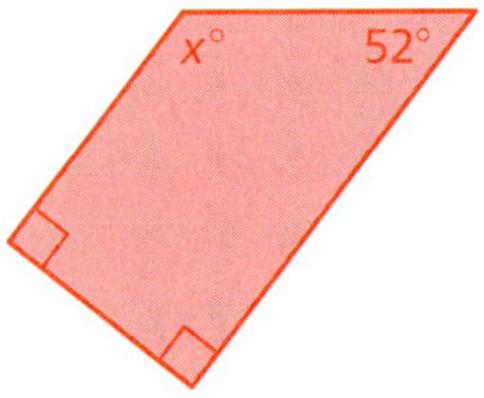

13. **KITE MAKING** What is the measure of the angle at the tail end of the kite?

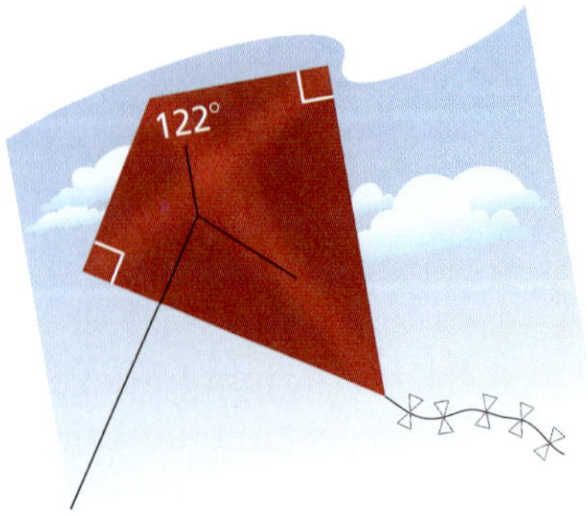

Assignment Guide and Homework Check

Level	Assignment	Homework Check
Advanced	1–6, 8–24 even, 25–30	1, 8, 12, 24

Common Errors

- **Exercises 1 and 18–23** Students may have difficulty answering these questions correctly. Encourage them to learn the diagram at the top of page 294 to help them answer these questions correctly.
- **Exercises 4–9** Students may not use the most specific name when identifying the quadrilaterals. For instance, they may identify the quadrilateral in Exercise 9 as a parallelogram instead of a rectangle. Remind students to identify the quadrilaterals with the most specific name.
- **Exercises 10–13** Students may forget that the sum of the angle measures of a quadrilateral is 360° and that a right angle measures 90°. Remind students of these facts.

12.4 Record and Practice Journal

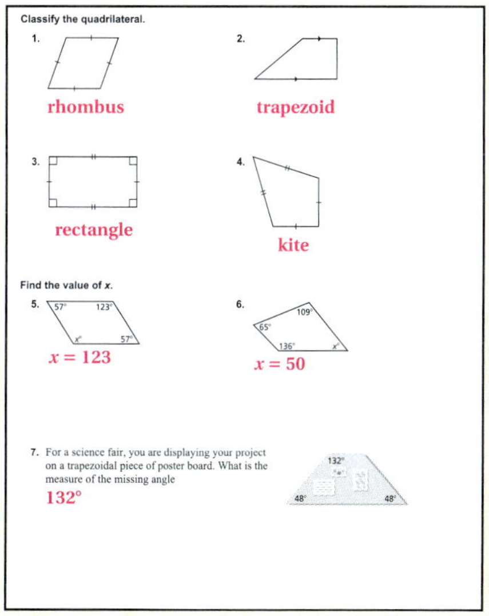
Classify the quadrilateral.

1. rhombus
2. trapezoid
3. rectangle
4. kite

Find the value of x.

5. $x = 123$
6. $x = 50$
7. For a science fair, you are displaying your project on a trapezoidal piece of poster board. What is the measure of the missing angle
132°

Vocabulary and Concept Check

1. all of them
2. rectangle, square
3. kite; It is the only type of quadrilateral listed that does not have opposite sides that are parallel and congruent.

Practice and Problem Solving

4. square
5. trapezoid
6. rhombus
7. kite
8. parallelogram
9. rectangle
10. 65
11. 110
12. 128
13. 58°
14. 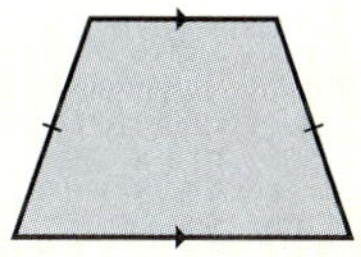
15. 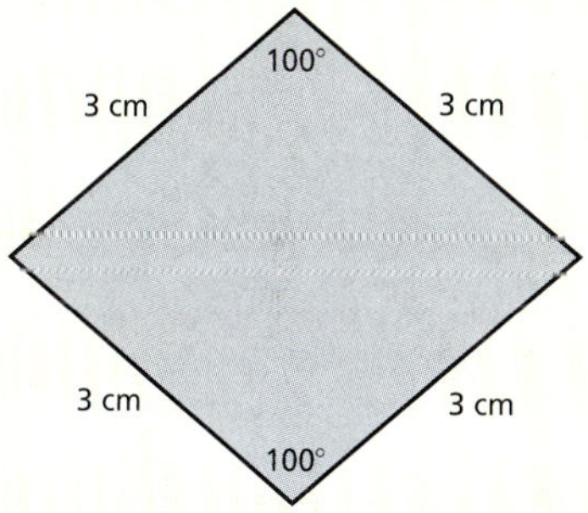

16. 

17. 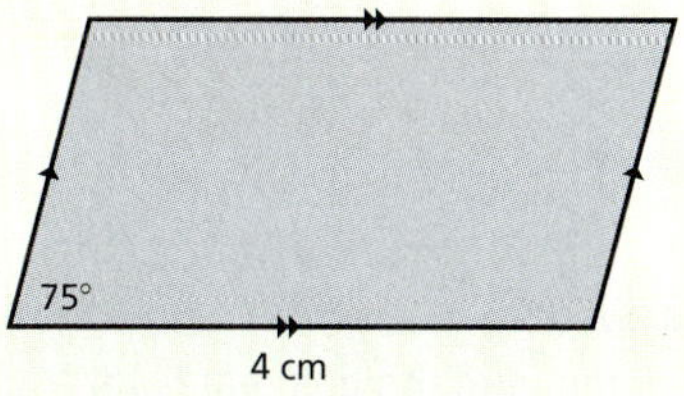

18. always
19. always
20. sometimes
21. never
22. never
23. sometimes

Practice and Problem Solving

24. See *Taking Math Deeper.*

25–26. See Additional Answers.

Fair Game Review

27. $\frac{1}{4}$ **28.** $\frac{2}{3}$

29. $\frac{6}{5}$ **30.** B

Mini-Assessment

1. Classify the quadrilateral.

a.

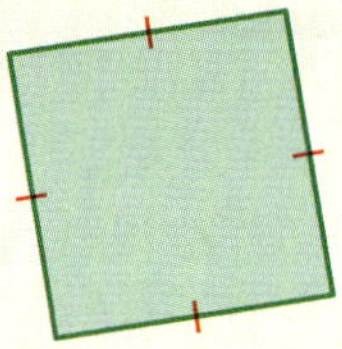

rhombus

b.

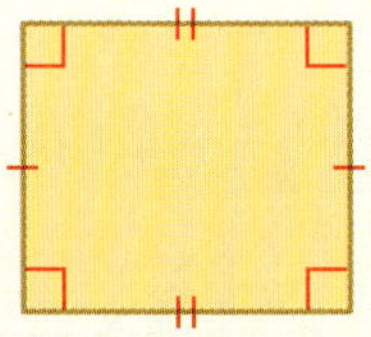

rectangle

2. Find the value of x.

a.

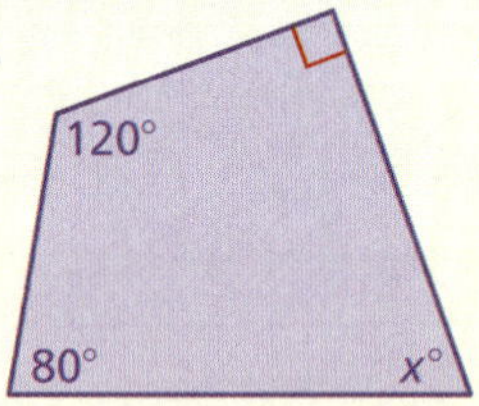

$x = 70$

b.

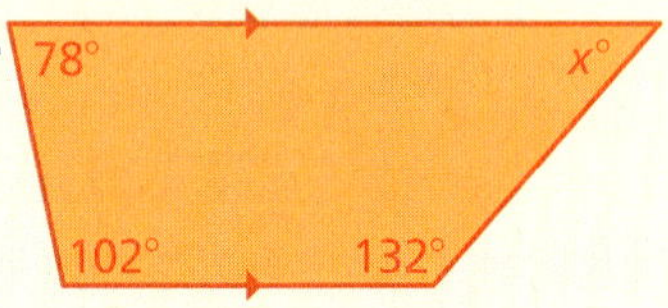

$x = 48$

3. What is the measure of the top right angle of the gold bar? 102°

Taking Math Deeper

Exercise 24

One way to begin the problem is to make a diagram of the old door, the new door, and the piece you remove.

1 Draw a diagram.

a. The new door is a quadrilateral with exactly one pair of parallel sides. So, it is a trapezoid.

2 Write and solve an equation to find the value of x.

There are 360° in a trapezoid. However, rather than setting the sum of the four angle measures equal to 360° and solving, you can write and solve a simpler equation.

Notice that the two angles at the top of the door are supplementary. So, the angles at the bottom of the door must also be supplementary in order for the four angle measures to add up to 360°.

$$x + 91.5 = 180$$

$$x = 88.5$$

3 Answer the question.

b. So, the new angle at the bottom left side of the door is 88.5°.

Reteaching and Enrichment Strategies

If students need help. . .	If students got it. . .
Resources by Chapter • Practice A and Practice B • Puzzle Time Record and Practice Journal Practice Differentiating the Lesson Lesson Tutorials Skills Review Handbook	Resources by Chapter • Enrichment and Extension • Technology Connection Start the next section

Draw a quadrilateral with the given description.

3 **14.** a trapezoid with a pair of congruent, nonparallel sides

15. a rhombus with 3-centimeter sides and two 100° angles

16. a parallelogram with a 45° angle and a 135° angle

17. a parallelogram with a 75° angle and a 4-centimeter side

Copy and complete using *always*, *sometimes*, or *never*.

18. A square is __?__ a rectangle.

19. A square is __?__ a rhombus.

20. A rhombus is __?__ a square.

21. A parallelogram is __?__ a trapezoid.

22. A trapezoid is __?__ a kite.

23. A rhombus is __?__ a rectangle.

24. **DOOR** The dashed line shows how you cut the bottom of a rectangular door so it opens more easily.

a. Identify the new shape of the door. Explain.

b. What is the new angle at the bottom left side of the door? Explain.

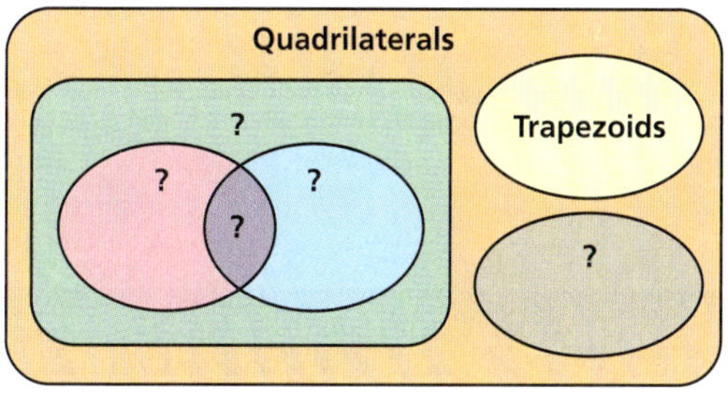

25. **VENN DIAGRAM** The diagram shows that some quadrilaterals are trapezoids, and all trapezoids are quadrilaterals. Copy the diagram. Fill in the names of the types of quadrilaterals to show their relationships.

26. **Structure** Consider the parallelogram.

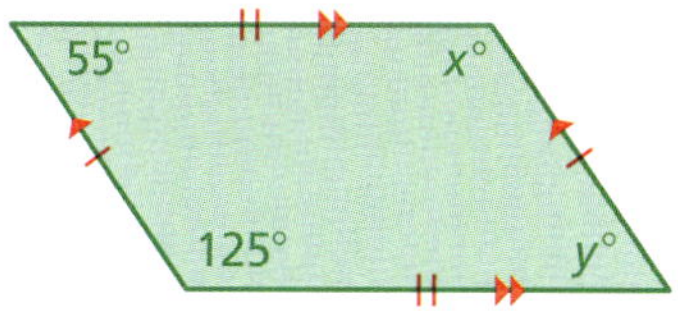

a. Find the values of x and y.

b. Make a conjecture about opposite angles in a parallelogram.

c. In polygons, consecutive interior angles share a common side. Make a conjecture about consecutive interior angles in a parallelogram.

Fair Game Review What you learned in previous grades & lessons

Write the ratio as a fraction in simplest form. *(Skills Review Handbook)*

27. 3 turnovers : 12 assists

28. 18 girls to 27 boys

29. 42 pens : 35 pencils

30. **MULTIPLE CHOICE** Computer sales decreased from 40 to 32. What is the percent of decrease? *(Skills Review Handbook)*

Ⓐ 8% Ⓑ 20% Ⓒ 25% Ⓓ 80%

12.5 Scale Drawings

Essential Question How can you enlarge or reduce a drawing proportionally?

1 ACTIVITY: Comparing Measurements

Work with a partner. The diagram shows a food court at a shopping mall. Each centimeter in the diagram represents 40 meters.

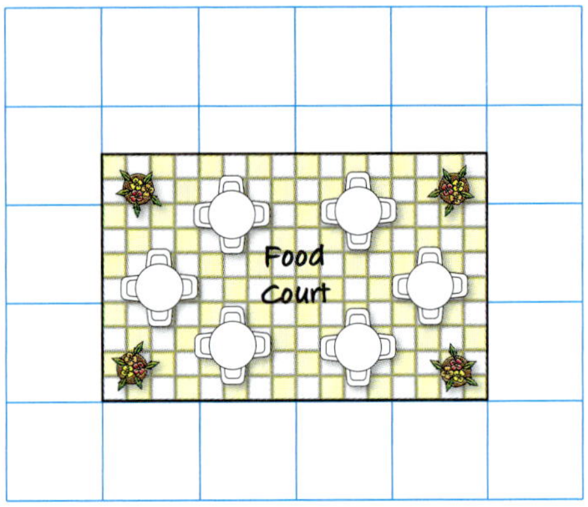

a. Find the length and the width of the drawing of the food court.

length: ____ cm width: ____ cm

b. Find the actual length and width of the food court. Explain how you found your answers.

length: ____ m width: ____ m

c. Find the ratios $\frac{\text{drawing length}}{\text{actual length}}$ and $\frac{\text{drawing width}}{\text{actual width}}$. What do you notice?

2 ACTIVITY: Recreating a Drawing

Work with a partner. Draw the food court in Activity 1 on the grid paper so that each centimeter represents 20 meters.

a. What happens to the size of the drawing?

b. Find the length and the width of your drawing. Compare these dimensions to the dimensions of the original drawing in Activity 1.

Geometry

In this lesson, you will
- use scale drawings to find actual distances.
- find scale factors.
- use scale drawings to find actual perimeters and areas.
- recreate scale drawings at a different scale.

Laurie's Notes

Introduction

Applying Mathematical Practices

- **Attend to Precision:** Mathematically proficient students understand that precision in measurement and in labeling units is important.

Motivate

- For each activity in this section, students should work with partners.
- Give each pair of students a map of the United States. Be sure the map has a scale on it.
- Tell students you are ready to start planning your summer vacation and you would like their input.
- Have students mark the map with a dot to represent your current location. Then tell them you would like to take a road trip this summer and not travel more than 1500 miles from home. As a benchmark, tell them that the distance between New York to Los Angeles is about 2800 miles. Give them a few minutes to decide where you should take your road trip.
- Make a list of the various locations students selected. Return to this list at the end of the period.

Activity Notes

Activity 1

- Distribute centimeter rulers to students. Check to see that all students know how to measure and read lengths correctly.
- **FYI:** Not all rulers start with 0 flush at the end of the ruler. It could be 0.1 to 0.2 centimeters from the edge.
- **Attend to Precision:** Have students measure accurately to the nearest 0.1 centimeter.
- After they have finished measuring, compare results.

? "To answer part (b), did anyone set up a ratio table or write a proportion?" Answers will vary.

? "What do you notice about the ratios you found in part (c)?" They are both equal to 1 cm : 40 m.

- **Connection:** The units are important when writing the ratios, so they should be included. It is given that each centimeter in the diagram represents 40 meters.

Activity 2

- Point out to students that each centimeter on the grid represents 20 meters.
- After students have finished the new drawing, discuss part (a).

? **Construct Viable Arguments and Critique the Reasoning of Others**: "Why did the dimensions double?" 1 centimeter represents only half of what it did in Activity 1.

What Your Students Will Learn

- Use scale drawings to find actual distances, perimeters, and areas.
- Use actual measurements to find distances in models.
- Find scale factors.
- Recreate scale drawings at a different scale.

Previous Learning

Students should know common units of length, both U.S. customary and metric systems. Students should also know how to solve proportions.

Technology for the Teacher

Dynamic Classroom

Lesson Plans

Complete Materials List

12.5 Record and Practice Journal

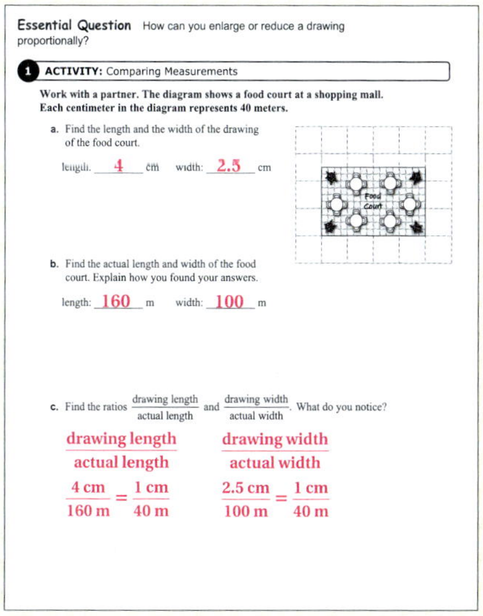

Essential Question How can you enlarge or reduce a drawing proportionally?

1 ACTIVITY: Comparing Measurements

Work with a partner. The diagram shows a food court at a shopping mall. Each centimeter in the diagram represents 40 meters.

a. Find the length and the width of the drawing of the food court.

length: 4 cm width: 2.5 cm

b. Find the actual length and width of the food court. Explain how you found your answers.

length: 160 m width: 100 m

c. Find the ratios $\frac{\text{drawing length}}{\text{actual length}}$ and $\frac{\text{drawing width}}{\text{actual width}}$. What do you notice?

$\frac{\text{drawing length}}{\text{actual length}}$ $\frac{\text{drawing width}}{\text{actual width}}$

$\frac{4 \text{ cm}}{160 \text{ m}} = \frac{1 \text{ cm}}{40 \text{ m}}$ $\frac{2.5 \text{ cm}}{100 \text{ m}} = \frac{1 \text{ cm}}{40 \text{ m}}$

Differentiated Instruction

Kinesthetic

Materials needed: poster board, ruler, colored markers or pencils, copies of the design below.

Students are to recreate the design in a larger size. They will need to choose a ratio to enlarge the design, measure the design, and set up proportions to find the new dimensions.

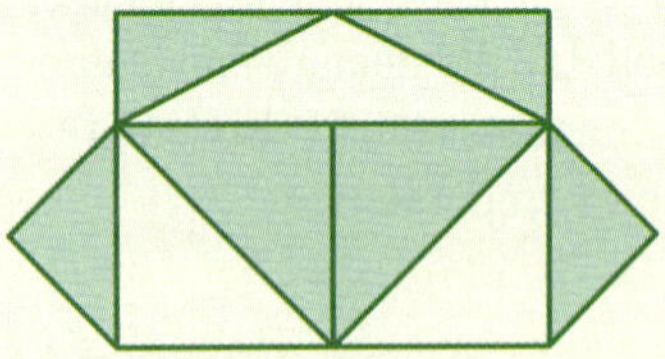

12.5 Record and Practice Journal

2 ACTIVITY: Recreating a Drawing

Work with a partner. Draw the food court in Activity 1 on the grid paper so that each centimeter represents 20 meters.

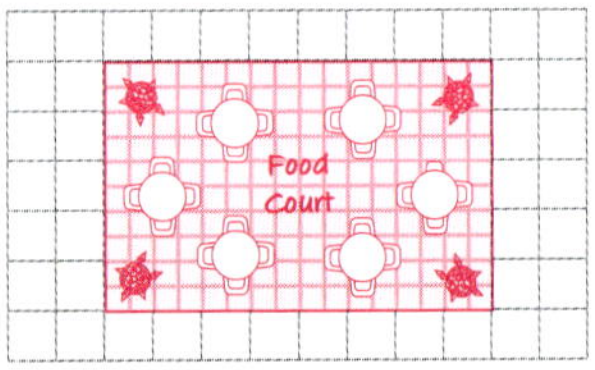

a. What happens to the size of the drawing?
The size of the drawing increases.

b. Find the length and the width of your drawing. Compare these dimensions to the dimensions of the original drawing in Activity 1.
8 cm; 5 cm; The dimensions are doubled.

3 ACTIVITY: Comparing Measurements

Work with a partner. The diagram shows a sketch of a painting. Each unit in the sketch represents 8 inches.

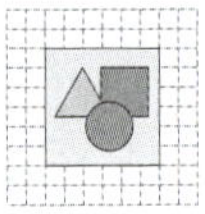

a. Find the length and the width of the sketch.

length: 6 units width: 6 units

b. Find the actual length and width of the painting. Explain how you found your answers.

length: 48 in. width: 48 in.

c. Find the ratios $\frac{\text{sketch length}}{\text{actual length}}$ and $\frac{\text{sketch width}}{\text{actual width}}$. What do you notice?
$\frac{6 \text{ units}}{48 \text{ in.}} = \frac{1 \text{ unit}}{8 \text{ in.}}$; $\frac{6 \text{ units}}{48 \text{ in.}} = \frac{1 \text{ unit}}{8 \text{ in.}}$

4 ACTIVITY: Recreating a Drawing

Work with a partner. Let each unit in the grid paper represent 2 feet. Now sketch the painting in Activity 3 onto the grid paper.

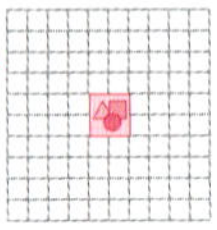

a. What happens to the size of the sketch?
The size of the drawing decreases.

b. Find the length and the width of your sketch. Compare these dimensions to the dimensions of the original sketch in Activity 3.
2 units; 2 units; The dimensions are reduced by $\frac{1}{3}$.

What Is Your Answer?

5. **IN YOUR OWN WORDS** How can you enlarge or reduce a drawing proportionally?
Increase or decrease each dimension of the drawing by the same factor.

6. Complete the table for both the food court and the painting. Food court

	Actual Object	Original Drawing	Your Drawing
Perimeter	520 m	13 cm	26 cm
Area	16,000 m^2	10 cm^2	40 cm^2

See Additional Answers for the painting.
Compare the measurements in each table. What conclusions can you make?
See Additional Answers.

7. **RESEARCH** Look at some maps in your school library or on the Internet. Make a list of the different scales used on the maps.
Check students' work.

8. When you view a map on the Internet, how does the scale change when you zoom out? How does the scale change when you zoom in?
zoom out: scale decreases
zoom in: scale increases

Laurie's Notes

Activity 3

- This activity is similar to Activity 1. Point out that each unit in the sketch represents 8 inches.
- ? "To answer part (b), did anyone set up a ratio table or write a proportion?" Answers will vary.
- ? "What do you notice about the ratios you found in part (c)?" They are both equal to 1 unit: 8 in.

Activity 4

- Point out to students that each unit on the grid represents 2 feet.
- After students have finished the new sketch, discuss part (a).
- ? **Construct Viable Arguments and Critique the Reasoning of Others:** "Why did the dimensions decrease by a factor of $\frac{1}{3}$?" Because 2 ft = 24 in. = 3×8 in., 1 unit represents 3 times what it did in Activity 3.

What Is Your Answer?

- Question 6 explores a big idea in mathematics. When the dimensions change by a factor of k, the perimeter also changes by a factor of k, but the area changes by a factor of k^2. This idea is presented at the end of the lesson.

Closure

- Refer to the map of the United States given to students at the beginning of class. Point out the scale on the map. Have students determine the actual distance from your home to the location they selected.

3 ACTIVITY: Comparing Measurements

Work with a partner. The diagram shows a sketch of a painting. Each unit in the sketch represents 8 inches.

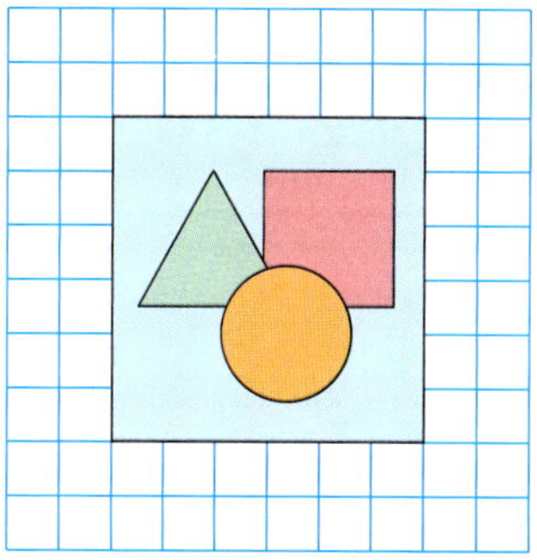

a. Find the length and the width of the sketch.

length: ___ units width: ___ units

b. Find the actual length and width of the painting. Explain how you found your answers.

length: ___ in. width: ___ in.

c. Find the ratios $\frac{\text{sketch length}}{\text{actual length}}$ and $\frac{\text{sketch width}}{\text{actual width}}$.

What do you notice?

4 ACTIVITY: Recreating a Drawing

Math Practice

Specify Units

How do you know whether to use feet or units for each measurement?

Work with a partner. Let each unit in the grid paper represent 2 feet. Now sketch the painting in Activity 3 onto the grid paper.

a. What happens to the size of the sketch?

b. Find the length and the width of your sketch. Compare these dimensions to the dimensions of the original sketch in Activity 3.

What Is Your Answer?

5. **IN YOUR OWN WORDS** How can you enlarge or reduce a drawing proportionally?

6. Complete the table for both the food court and the painting.

	Actual Object	Original Drawing	Your Drawing
Perimeter			
Area			

Compare the measurements in each table. What conclusions can you make?

7. **RESEARCH** Look at some maps in your school library or on the Internet. Make a list of the different scales used on the maps.

8. When you view a map on the Internet, how does the scale change when you zoom out? How does the scale change when you zoom in?

Practice Use what you learned about enlarging or reducing drawings to complete Exercises 4–7 on page 535.

12.5 Lesson

Check It Out
Lesson Tutorials
BigIdeasMath.com

Key Vocabulary
scale drawing, *p. 532*
scale model, *p. 532*
scale, *p. 532*
scale factor, *p. 533*

Key Ideas

Scale Drawings and Models

A **scale drawing** is a proportional, two-dimensional drawing of an object. A **scale model** is a proportional, three-dimensional model of an object.

Scale

The measurements in scale drawings and models are proportional to the measurements of the actual object. The **scale** gives the ratio that compares the measurements of the drawing or model with the actual measurements.

$\frac{1 \text{ in.}}{10 \text{ mi}}$ ← drawing distance (1 in.), actual distance (10 mi)

1 in. : 10 mi ← drawing (1 in.), actual (10 mi)

Study Tip
Scales are written so that the drawing distance comes first in the ratio.

EXAMPLE 1 Finding an Actual Distance

What is the actual distance *d* between Cadillac and Detroit?

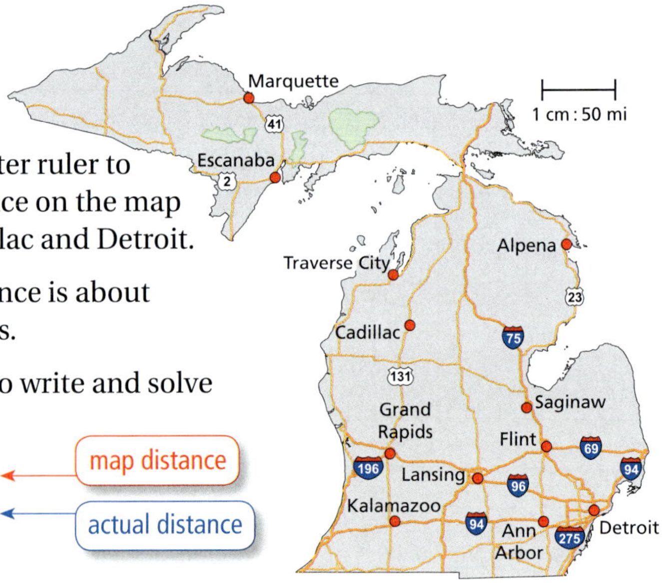

Step 1: Use a centimeter ruler to find the distance on the map between Cadillac and Detroit.

The map distance is about 3.5 centimeters.

Step 2: Use the scale to write and solve a proportion.

$$\frac{1 \text{ cm}}{50 \text{ mi}} = \frac{3.5 \text{ cm}}{d \text{ mi}}$$ ← map distance (3.5 cm), actual distance (d mi)

$d = 50 \cdot 3.5$ Cross Products Property

$d = 175$ Multiply.

So, the distance between Cadillac and Detroit is about 175 miles.

On Your Own

Now You're Ready
Exercises 8–11

1. What is the actual distance between Traverse City and Marquette?

Multi-Language Glossary at BigIdeasMath.com

Laurie's Notes

Introduction

Connect

- **Yesterday:** Students compared measurements and recreated drawings.
- **Today:** Students will use a scale drawing to find a missing measure.

Motivate

- Show the class some items that have scales written on them: map (print one from the Internet), matchbook car, blueprint, or floor plan. Ask about the meaning of the scale for each item.
- **Trivia:** The world's biggest baseball bat is 120 feet and leans against the Louisville Slugger Museum & Factory in Kentucky. It is an exact-scale replica of Babe Ruth's 34-inch Louisville Slugger bat.

Lesson Notes

Key Ideas

- Be careful and consistent with your language today. Continually refer to the ratio of the scale drawing (or scale model) to the actual object.
- Review the *Study Tip*.

Example 1

- Have students explore the map. Ask if they see the scale. You can use the width of your baby finger to approximate 1 centimeter. Ask students to use their baby fingers to approximate the distance across the bottom of Michigan. about 160 mi
- Use centimeter rulers to measure the distance from Cadillac to Detroit.
- **Common Error:** Students might measure in inches instead of in centimeters, or they may confuse centimeters with millimeters.

? "What is the map distance from Cadillac to Detroit?" 3.5 cm

- Set up the proportion using the language, "1 centimeter is to 50 miles as 3.5 centimeters is to what?"
- **Common Question:** Students will often ask how you can use the Cross Products Property because you are multiplying two different units together (50 mi and 3.5 cm). Because both numerators are the same unit (cm) and both denominators are the same unit (mi), it is okay to multiply. If the units were not the same in the numerators and were the same in the denominators, this could not be done. A quick way to explain this to students is to use a simple problem: If the scale is 1 cm : 5 ft, then 2 cm would be what actual distance? 10 ft
- **Attend to Precision:** Encourage students to write the units when they write the initial proportion. This ensures that the proportion has been set up correctly. When the numeric answer has been found, label with the correct units of measure.
- **Check for Reasonableness:** If the map distance is 3.5 centimeters, the actual distance should be 3.5 times the scale distance of 50 miles. Three times 50 miles is 150 miles, so an answer of 175 miles seems reasonable.

Goal Today's lesson is using **scale drawings** to find missing measurements.

Technology for the Teacher

Dynamic Classroom

Lesson Tutorials
Lesson Plans
Answer Presentation Tool

Extra Example 1

Using the map from Example 1, what is the distance *d* between Detroit and Marquette? about 350 mi

On Your Own

1. about 150 mi

Laurie's Notes

Extra Example 2

The Earth's crust has a thickness of 80 kilometers on some of the continents. Using the scale model from Example 2, how thick is the crust of the model?
0.16 in.

On Your Own

2. 5.8 in.

Extra Example 3

A sketch of a fashion designer's shirt is 9 centimeters long. The actual shirt is 1 meter long.

a. What is the scale of the drawing? 9 cm : 1 m

b. What is the scale factor of the drawing? 9 : 100

On Your Own

3. 1 : 200

English Language Learners

Illustrate

Students have seen maps in classrooms and perhaps on road trips with their families. Hand out road maps to students in small groups. Have students find distances between cities on the map. Ask students how they can find the distances between cities using proportions.

Example 2

- This question is looking for a dimension of the scale model, not for an actual distance as in Example 1. Be careful with the language: scale to actual.
- Explain that this is an example of a scale model (3-Dimensional), not a scale drawing (2-Dimensional).
- Note that the units are written in the original proportion.
- **Check for Reasonableness:** Because the scale is 1 in. : 500 km, 2 inches would represent 1000 kilometers and 4 inches would represent 2000 kilometers. An answer of 4.6 inches is reasonable.

On Your Own

- **Think-Pair-Share:** Students should read each question independently and then work in pairs to answer the questions. When they have answered the questions, the pair should compare their answers with another group and discuss any discrepancies.

Write

- Write and discuss the definition of a scale factor.
- Connect the definition to yesterday's activities. For instance, in Activity 1, the scale used was 1 cm : 40 m. Because 40 meters is equal to 4000 centimeters, the scale factor would be 1 cm : 4000 cm = 1 : 4000.

Example 3

- **FYI:** The Sergeant Floyd Monument is located in Sioux City, Iowa. It memorializes Sgt. Charles Floyd, the only man who died on the Lewis and Clark Expedition. The monument is the nation's first nationally registered historical landmark.
- Read the problem and ask students to write the ratio of the model height to the actual height and then find the scale.

? "How can you find the scale factor when you know the scale?" Write the scale with the same units and simplify.

? "What conversion factor can you use?" 1 foot = 12 inches

- Work through the rest of the problem as shown.
- **Extension:** Use $\frac{1}{12}$: 10 to show that the answer is the same when converting inches to feet.

On Your Own

? "Is the scale model larger or smaller than the actual item? Explain." The scale model is smaller because 1 millimeter is shorter than 20 centimeters.

EXAMPLE 2 Finding a Distance in a Model

The liquid outer core of Earth is 2300 kilometers thick. A scale model of the layers of Earth has a scale of 1 in. : 500 km. How thick is the liquid outer core of the model?

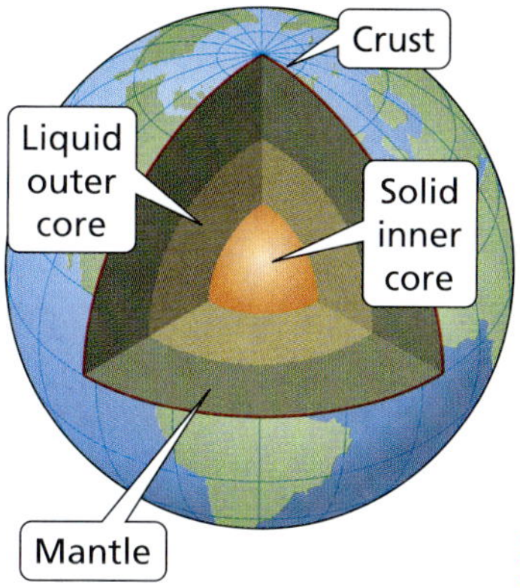

Ⓐ 0.2 in. Ⓑ 4.6 in. Ⓒ 0.2 km Ⓓ 4.6 km

$$\frac{1 \text{ in.}}{500 \text{ km}} = \frac{x \text{ in.}}{2300 \text{ km}}$$ ← model thickness / actual thickness

$$\frac{1 \text{ in.}}{500 \text{ km}} \cdot 2300 \text{ km} = \frac{x \text{ in.}}{2300 \text{ km}} \cdot 2300 \text{ km}$$ Multiplication Property of Equality

$$4.6 = x$$ Simplify.

So, the liquid outer core of the model is 4.6 inches thick. The correct answer is Ⓑ.

On Your Own

2. The mantle of Earth is 2900 kilometers thick. How thick is the mantle of the model?

A scale can be written without units when the units are the same. A scale without units is called a **scale factor**.

EXAMPLE 3 Finding a Scale Factor

A scale model of the Sergeant Floyd Monument is 10 inches tall. The actual monument is 100 feet tall.

a. What is the scale of the model?

$$\frac{\text{model height}}{\text{actual height}} = \frac{10 \text{ in.}}{100 \text{ ft}} = \frac{1 \text{ in.}}{10 \text{ ft}}$$

The scale is 1 in. : 10 ft.

b. What is the scale factor of the model?

Write the scale with the same units. Use the fact that 1 ft = 12 in.

$$\text{scale factor} = \frac{1 \text{ in.}}{10 \text{ ft}} = \frac{1 \text{ in.}}{120 \text{ in.}} = \frac{1}{120}$$

The scale factor is 1 : 120.

On Your Own

Now You're Ready Exercises 12–16

3. A drawing has a scale of 1 mm : 20 cm. What is the scale factor of the drawing?

EXAMPLE 4 Finding an Actual Perimeter and Area

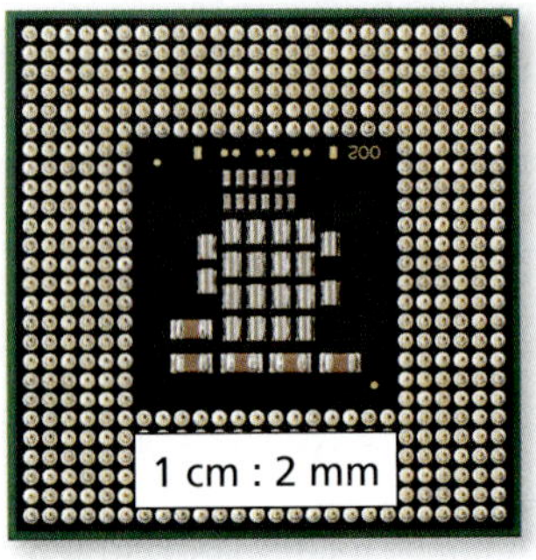

The scale drawing of a computer chip helps you see the individual components on the chip.

a. Find the perimeter and the area of the computer chip in the scale drawing.

When measured using a centimeter ruler, the scale drawing of the computer chip has a side length of 4 centimeters.

∴ So, the perimeter of the computer chip in the scale drawing is $4(4) = 16$ centimeters, and the area is $4^2 = 16$ square centimeters.

b. Find the actual perimeter and area of the computer chip.

$$\frac{1 \text{ cm}}{2 \text{ mm}} = \frac{4 \text{ cm}}{s \text{ mm}}$$

drawing distance (numerators); actual distance (denominators)

$s = 2 \cdot 4$ — Cross Products Property

$s = 8$ — Multiply.

The side length of the actual computer chip is 8 millimeters.

∴ So, the actual perimeter of the computer chip is $4(8) = 32$ millimeters, and the actual area is $8^2 = 64$ square millimeters.

c. Compare the ratios $\frac{\text{drawing perimeter}}{\text{actual perimeter}}$ and $\frac{\text{drawing area}}{\text{actual area}}$ to the scale factor.

Use the fact that 1 cm = 10 mm.

$$\text{scale factor} = \frac{1 \text{ cm}}{2 \text{ mm}} = \frac{10 \cancel{\text{mm}}}{2 \cancel{\text{mm}}} = \frac{5}{1}$$

$$\frac{\text{drawing perimeter}}{\text{actual perimeter}} = \frac{16 \text{ cm}}{32 \text{ mm}} = \frac{1 \text{ cm}}{2 \text{ mm}} = \frac{5}{1}$$

$$\frac{\text{drawing area}}{\text{actual area}} = \frac{16 \text{ cm}^2}{64 \text{ mm}^2} = \frac{1 \text{ cm}^2}{4 \text{ mm}^2} = \left(\frac{1 \text{ cm}}{2 \text{ mm}}\right)^2 = \left(\frac{5}{1}\right)^2$$

∴ So, the ratio of the perimeters is equal to the scale factor, and the ratio of the areas is equal to the square of the scale factor.

Study Tip

The ratios tell you that the perimeter of the drawing is 5 times the actual perimeter, and the area of the drawing is $5^2 = 25$ times the actual area.

On Your Own

Now You're Ready
Exercises 22 and 23

4. **WHAT IF?** The scale of the drawing of the computer chip is 1 cm : 3 mm. How do the answers in parts (a)–(c) change? Justify your answer.

Laurie's Notes

Example 4

- Read the problem and point out the scale drawing.
- **Construct Viable Arguments and Critique the Reasoning of Others:** "Is the actual chip larger or smaller than the chip in the scale drawing? Explain." smaller; According to the scale, 1 centimeter represents only 2 millimeters.
- Measure to find the side length of the chip in the scale drawing.
- "What is the perimeter of the chip in the scale drawing?" $4(4) = 16$ cm
- "What is the area of the chip in the scale drawing?" $4^2 = 16$ cm^2
- To find the actual perimeter and area of the chip, you can begin by setting up and solving a proportion to find the side length of the actual chip. Be careful to label units and use precise language. As shown in the solution, the numerator of each ratio is a drawing distance and the denominator of each ratio is an actual distance.
- Some students will want to bypass the step of writing the proportion and use mental math. Remind them that they are practicing a *process* that will enable them to solve more difficult problems that they may not be able to solve using mental math.
- Knowing the side length of the actual chip enables you to find the actual perimeter and area of the chip.
- **Attend to Precision:** Work slowly through part (c). Remind students that when finding the scale factor, it is necessary to have the same units in the numerator and in the denominator.
- Make sure students understand the manipulation of the square units in the step $\frac{1 \text{ cm}^2}{4 \text{ mm}^2} = \left(\frac{1 \text{ cm}}{2 \text{ mm}}\right)^2$.

Closure

- **Exit Ticket:** A common model train scale is called the HO Scale, where the scale factor is 1 : 87. If the diameter of a wheel on a model train is 0.3 inch, what is the diameter of the actual wheel? 26.1 in.

Extra Example 4

The scale drawing of a miniature glass mosaic tile helps you see the detail on the tile.

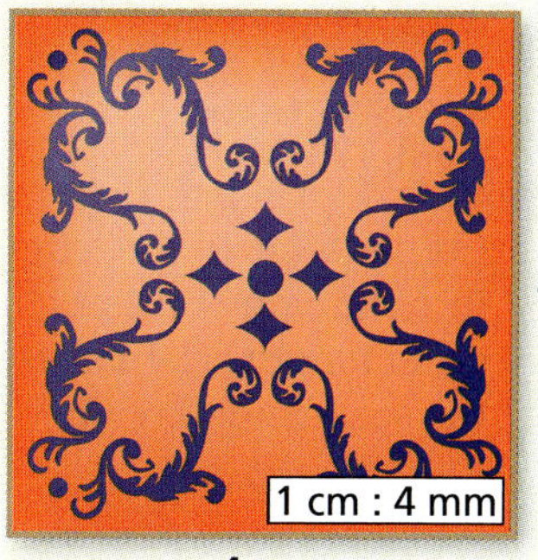

a. Find the perimeter and area of the tile in the scale drawing. 16 cm, 16 cm^2

b. Find the actual perimeter and area of the tile. 64 mm, 256 mm^2

c. Compare the ratios $\frac{\text{drawing perimeter}}{\text{actual perimeter}}$ and $\frac{\text{drawing area}}{\text{actual area}}$ to the scale factor. The ratio of the perimeters is equal to the scale factor, $\frac{5}{2}$, and the ratio of the areas is equal to the square of the scale factor, $\left(\frac{5}{2}\right)^2$.

On Your Own

4. **a.** does not change; The size of the drawing does not change, just the scale.

b. actual perimeter = 48 mm, actual area = 144 mm^2; The side length of the actual computer chip increases to 12 millimeters, so the actual perimeter and area increase accordingly.

c. See Additional Answers.

Vocabulary and Concept Check

1. A scale is the ratio that compares the measurements of the drawing or model with the actual measurements. A scale factor is a scale without any units.
2. larger; because 2 cm > 1 mm
3. Convert one of the lengths into the same units as the other length. Then, form the scale and simplify.

Practice and Problem Solving

4. 25 ft
5. 10 ft by 10 ft
6. 50 ft; 35 ft
7. 112.5%
8. 100 mi
9. 50 mi
10. 200 mi
11. 110 mi
12. 75 in.
13. 15 in.
14. 3.84 m
15. 21.6 yd
16. 17.5 mm
17. The 5 cm should be in the numerator.

$$\frac{1 \text{ cm}}{20 \text{ m}} = \frac{5 \text{ cm}}{x \text{ m}}$$

$$x = 100 \text{ m}$$

Assignment Guide and Homework Check

Level	Assignment	Homework Check
Advanced	1–7, 10–16 even, 17, 18–30 even, 31–36	10, 14, 18, 22

Common Errors

- **Exercises 8–11** When measuring with a centimeter ruler, students may not start at zero on the ruler. As a result, they will get an incorrect distance. Ask students to estimate the distance before measuring so they can check the reasonableness of their measurement.
- **Exercises 12–16** Students may mix up the proportion values when solving for the missing dimension. Remind them that the model dimension is in the numerator and the actual dimension is in the denominator in both ratios.

12.5 Record and Practice Journal

Find the missing dimension. Use the scale factor 1 : 8.

Item	Model	Actual
1. Statue	Height: 168 in.	Height: 112 ft
2. Painting	Width: 2500 cm	Width: 200 m
3. Alligator	Height: 9.6 in.	Height: 6.4 ft
4. Train	Length: 36.5 in.	Length: ______ ft $24\frac{1}{3}$

5. The diameter of the moon is 2160 miles. A model has a scale of 1 in. : 150 mi. What is the diameter of the model?
14.4 in.

6. A map has a scale of 1 in. : 4 mi.
   a. You measure 3 inches between your house and the movie theater. How many miles is it from your house to the movie theater?
   12 mi
   b. It is 17 miles to the mall. How many inches is that on the map?
   4.25 in.

12.5 Exercises

Check It Out
Help with Homework
BigIdeasMath.com

Vocabulary and Concept Check

1. **VOCABULARY** Compare and contrast the terms *scale* and *scale factor*.

2. **CRITICAL THINKING** The scale of a drawing is 2 cm : 1 mm. Is the scale drawing *larger* or *smaller* than the actual object? Explain.

3. **REASONING** How would you find the scale factor of a drawing that shows a length of 4 inches when the actual object is 8 feet long?

Practice and Problem Solving

Use the drawing and a centimeter ruler. Each centimeter in the drawing represents 5 feet.

4. What is the actual length of the flower garden?

5. What are the actual dimensions of the rose bed?

6. What are the actual perimeters of the perennial beds?

7. The area of the tulip bed is what percent of the area of the rose bed?

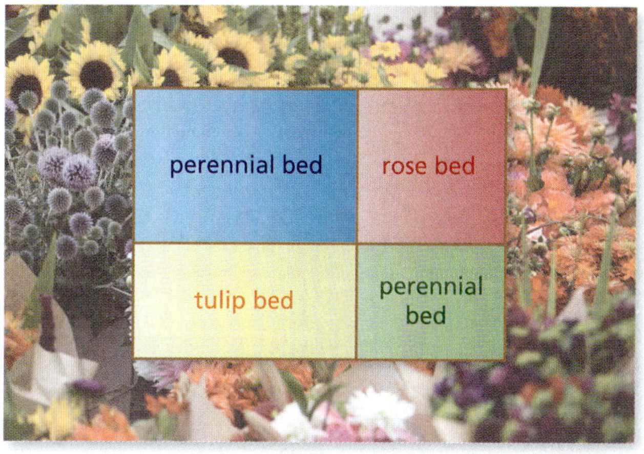

Use the map in Example 1 to find the actual distance between the cities.

1 8. Kalamazoo and Ann Arbor

9. Lansing and Flint

10. Grand Rapids and Escanaba

11. Saginaw and Alpena

Find the missing dimension. Use the scale factor 1 : 12.

	Item	Model	Actual
2 3 12.	Mattress	Length: 6.25 in.	Length: ____ in.
13.	Corvette	Length: ____ in.	Length: 15 ft
14.	Water tower	Depth: 32 cm	Depth: ____ m
15.	Wingspan	Width: 5.4 ft	Width: ____ yd
16.	Football helmet	Diameter: ____ mm	Diameter: 21 cm

17. **ERROR ANALYSIS** A scale is 1 cm : 20 m. Describe and correct the error in finding the actual distance that corresponds to 5 centimeters.

✗ $$\frac{1 \text{ cm}}{20 \text{ m}} = \frac{x \text{ m}}{5 \text{ cm}}$$

$$x = 0.25 \text{ m}$$

Use a centimeter ruler to measure the segment shown. Find the scale of the drawing.

18.

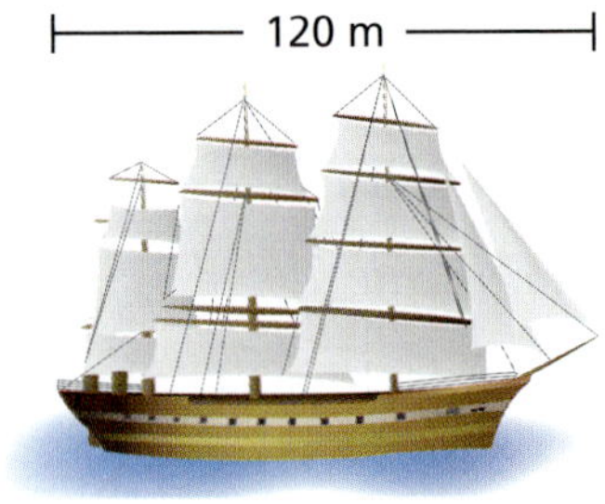

19.

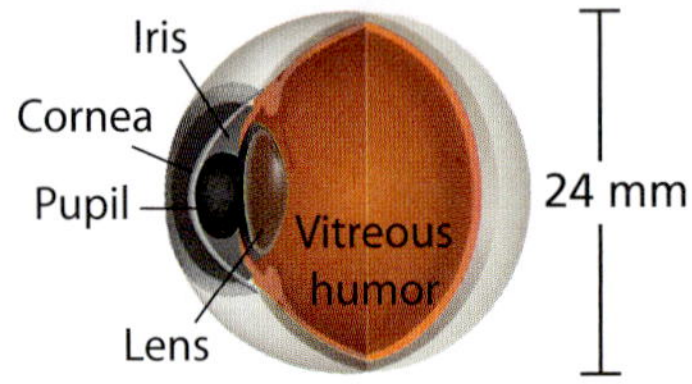

20. **REASONING** You know the length and the width of a scale model. What additional information do you need to know to find the scale of the model?

21. **OPEN-ENDED** You are in charge of creating a billboard advertisement with the dimensions shown.

a. Choose a product. Then design the billboard using words and a picture.

b. What is the scale factor of your design?

④ **22.** **CENTRAL PARK** Central Park is a rectangular park in New York City.

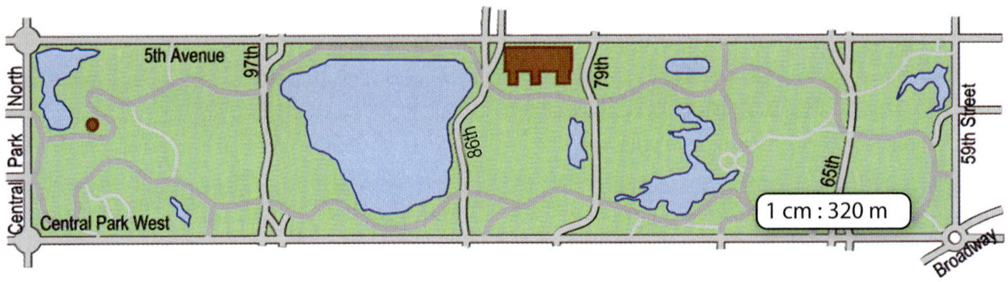

a. Find the perimeter and the area of Central Park in the scale drawing.

b. Find the actual perimeter and area of Central Park.

23. **ICON** You are designing an icon for a mobile app.

a. Find the perimeter and the area of the icon in the scale drawing.

b. Find the actual perimeter and area of the icon.

24. **CRITICAL THINKING** Use the results of Exercises 22 and 23 to make a conjecture about the relationship between the scale factor of a drawing and the ratios $\frac{\text{drawing perimeter}}{\text{actual perimeter}}$ and $\frac{\text{drawing area}}{\text{actual area}}$.

Common Errors

- **Exercise 30** Students may count the squares in the blueprint of the bathroom and use that as the area of the bathroom. Remind them that they need to find the actual length and width of each room and then find the area.

Practice and Problem Solving

18. 4 cm; 1 cm : 30 m

19. 2.4 cm; 1 cm : 10 mm

20. The length or width of the actual item

21. **a.** *Answer should include, but is not limited to:* Make sure words and picture match the product.

b. Answers will vary.

22. **a.** 30 cm; 31.25 cm^2

b. 9600 m; 3,200,000 m^2

23. **a.** 16 cm; 16 cm^2

b. 40 mm; 100 mm^2

24. The ratio of the perimeters is the scale factor and the ratio of the areas is the square of the scale factor.

Differentiated Instruction

Kinesthetic

Materials needed: large piece of construction paper, ruler, and protractor. Have students work in pairs to draw a large right triangle on the paper. Record the lengths of the sides and the measures of the angles in a table. Connect the midpoints of each side of the large triangle and record the side lengths and angle measures of the second triangle in the table. Connect the midpoints of the sides of the second triangle to form a third triangle. Record the side lengths and angle measures of the third triangle in the table. Have students determine if the triangles are scale drawings of each other. If they are, then have students find the ratios of the perimeters and areas of each pair of triangles.

Practice and Problem Solving

25. See Additional Answers.

26. 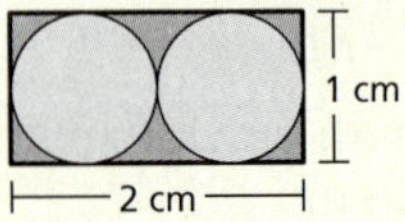

27. 15 ft^2

28. 4.5 ft^2

29. 3 ft^2

30. **a.** \$480

b. \$1536

c. tile; Because \$5 per square foot is greater than \$2 per square foot, the tile has a higher unit cost.

31. See *Taking Math Deeper*.

Fair Game Review

32–35. 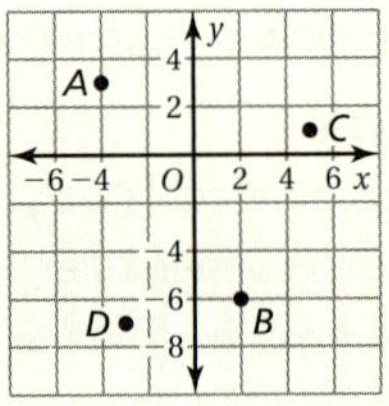

36. D

Mini-Assessment

Find the missing dimension. Use the scale factor 1 : 6.

	Model	Actual
1.	12 in.	72 in.
2.	3 ft	18 ft
3.	20 cm	120 cm
4.	2 yd	12 yd

5. A fish in an aquarium is 4 feet long. A scale model of the fish is 2 inches long. What is the scale factor?
1 : 24

Taking Math Deeper

Exercise 31

This is a short, but nice problem. It requires that students make a reasonable estimate for the radius or diameter of a baseball. Then it requires that students use this information to determine the reasonableness of making a scale model.

1 Draw and label a diagram.

Model diameter (baseball): 3 in.

Actual radius (Earth): 6378 km

Model diameter: x in.

Actual radius (Sun): 695,500 km

2 Write and solve a proportion.

$$\frac{x}{695{,}500} = \frac{3}{6378}$$

$$695{,}500 \cdot \frac{x}{695{,}500} = 695{,}500 \cdot \frac{3}{6378}$$

$$x \approx 327 \text{ in.}$$

3 Answer the question.

327 inches is equal to 27.25 feet. The model for the Sun would have to be about the width of a classroom. So, it is not reasonable to use a baseball for Earth. A reasonable model for Earth would have a diameter of one-quarter inch and the diameter for the model of the Sun would be 27.26 inches.

Project

Make a scale drawing of the solar system. Make sure students include the scale they used.

Reteaching and Enrichment Strategies

If students need help. . .	If students got it. . .
Resources by Chapter • Practice A and Practice B • Puzzle Time Record and Practice Journal Practice Differentiating the Lesson Lesson Tutorials Skills Review Handbook	Resources by Chapter • Enrichment and Extension • Technology Connection Start the next section

Recreate the scale drawing so that it has a scale of 1 cm : 4 m.

25. 

26. 

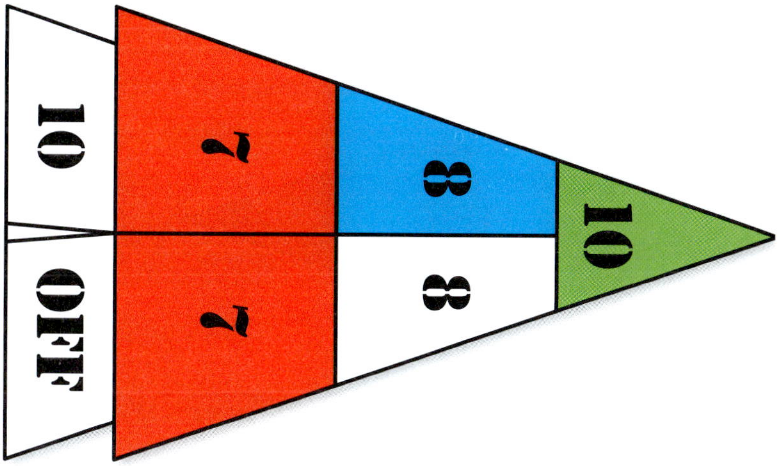

The shuffleboard diagram has a scale of 1 cm : 1 ft. Find the actual area of the region.

27. red region

28. blue region

29. green region

30. **BLUEPRINT** In a blueprint, each square has a side length of $\frac{1}{4}$ inch.

   a. Ceramic tile costs \$5 per square foot. How much would it cost to tile the bathroom?

   b. Carpet costs \$18 per square yard. How much would it cost to carpet the bedroom and living room?

   c. Which has a greater unit cost, the tile or the carpet? Explain.

Reduced Drawing of Blueprint

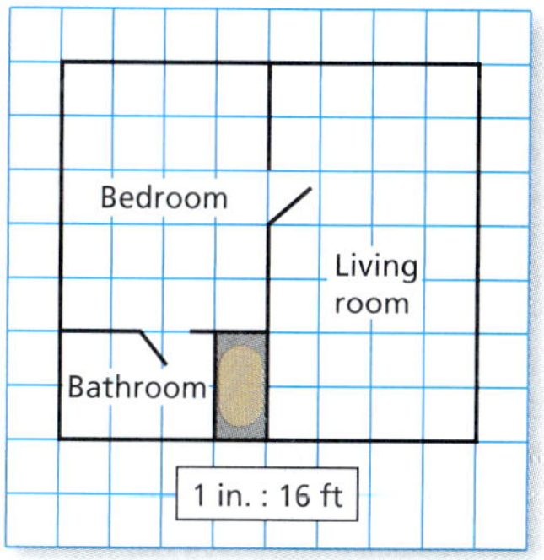

31. **Modeling** You are making a scale model of the solar system. The radius of Earth is 6378 kilometers. The radius of the Sun is 695,500 kilometers. Is it reasonable to choose a baseball as a model of Earth? Explain your reasoning.

Fair Game Review What you learned in previous grades & lessons

Plot and label the ordered pair in a coordinate plane. *(Skills Review Handbook)*

32. $A(-4, 3)$ 33. $B(2, -6)$ 34. $C(5, 1)$ 35. $D(-3, -7)$

36. **MULTIPLE CHOICE** Which set of numbers is ordered from least to greatest? *(Skills Review Handbook)*

   Ⓐ $\frac{7}{20}$, 32%, 0.45 Ⓑ 17%, 0.21, $\frac{3}{25}$ Ⓒ 0.88, $\frac{7}{8}$, 93% Ⓓ 57%, $\frac{11}{16}$, 5.7

12.4–12.5 Quiz

Check It Out
Progress Check
BigIdeasMath.com

Classify the quadrilateral. *(Section 12.4)*

1.

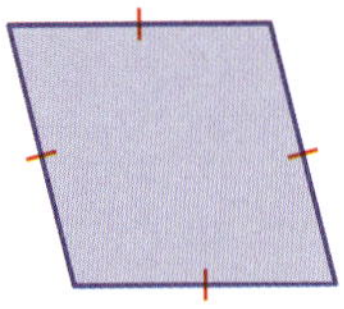

2.

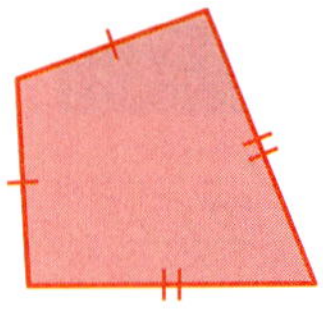

Find the value of *x*. *(Section 12.4)*

3.

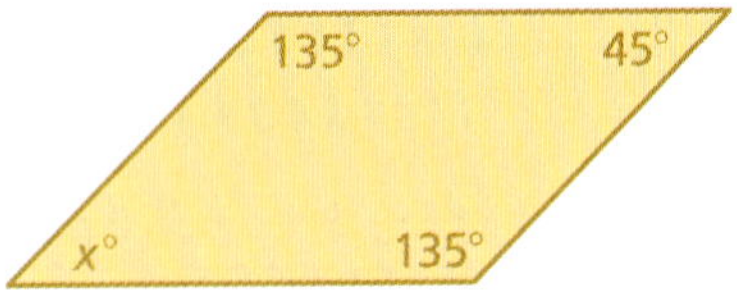

4.

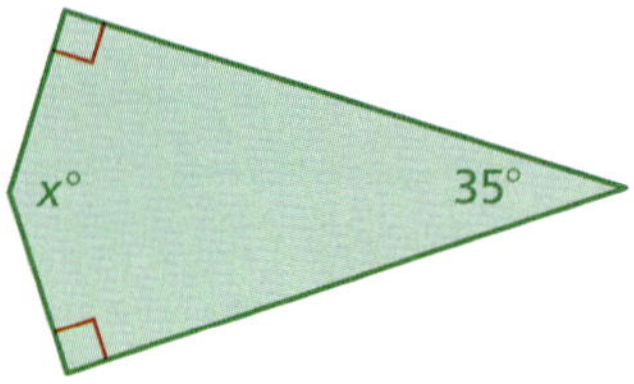

Draw a quadrilateral with the given description. *(Section 12.4)*

5. a rhombus with 2-centimeter sides and two 50° angles

6. a parallelogram with a 65° angle and a 5-centimeter side

Find the missing dimension. Use the scale factor 1 : 20. *(Section 12.5)*

	Item	Model	Actual
7.	Basketball player	Height: ____ in.	Height: 90 in.
8.	Dinosaur	Length: 3.75 ft	Length: ____ ft

9. **SHED** The side of the storage shed is in the shape of a trapezoid. Find the value of *x*. *(Section 12.4)*

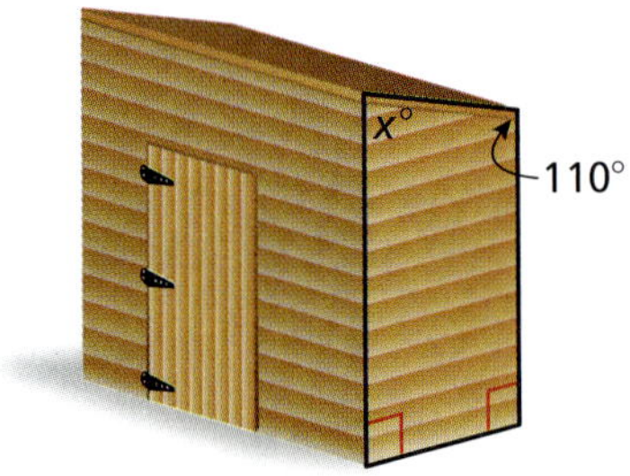

10. **DOLPHIN** A dolphin in an aquarium is 12 feet long. A scale model of the dolphin is $3\frac{1}{2}$ inches long. What is the scale factor of the model? *(Section 12.5)*

11. **SOCCER** A scale drawing of a soccer field is shown. The actual soccer field is 300 feet long. *(Section 12.5)*

a. What is the scale of the drawing?

b. What is the scale factor of the drawing?

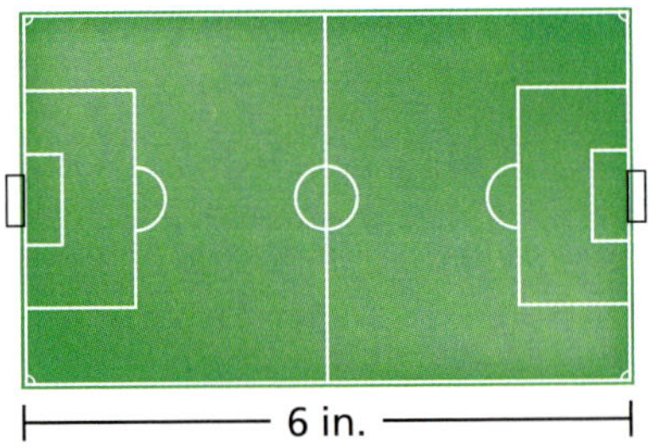

Alternative Assessment Options

Math Chat	Student Reflective Focus Question
Structured Interview	Writing Prompt

Math Chat

- Put students in pairs to complete and discuss the exercises from the quiz. The discussion should include classifying quadrilaterals and using scale drawings.
- The teacher should walk around the classroom listening to the pairs and ask questions to ensure understanding.

Study Help Sample Answers

Remind students to complete Graphic Organizers for the rest of the chapter.

4. Quadrilaterals

Examples	Non-Examples
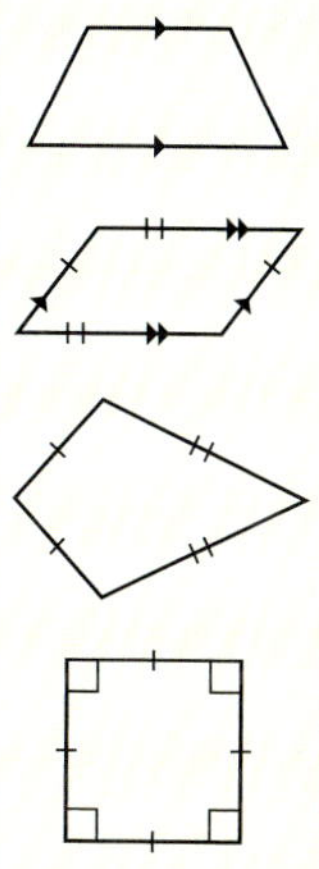	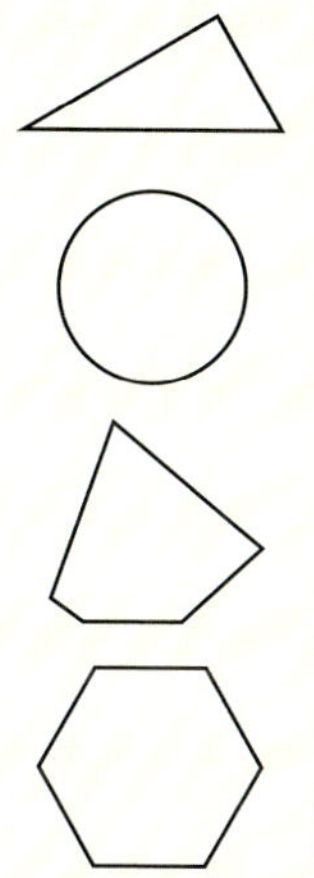

5. Scale factor

Examples	Non-Examples
5 : 1 1 : 200 1 : 1 3 : 2	1 cm : 2 mm 1 mm : 20 cm 12 in. : 1 ft 3 mi : 2 in.

Answers

1. rhombus
2. kite
3. 45
4. 145
5. 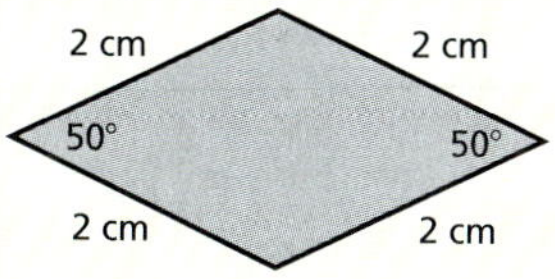

6. See Additional Answers.
7. 4.5 in.
8. 75 ft
9. 70
10. 7 : 288
11. **a.** 1 in. : 50 ft
    b. 1 : 600

Reteaching and Enrichment Strategies

If students need help. . .	If students got it. . .
Resources by Chapter • Practice A and Practice B • Puzzle Time Lesson Tutorials *BigIdeasMath.com*	Resources by Chapter • Enrichment and Extension • Technology Connection Game Closet at *BigIdeasMath.com* Start the Chapter Review

Technology for the Teacher

Online Assessment
Assessment Book
ExamView® Assessment Suite

For the Teacher
Additional Review Options

- *BigIdeasMath.com*
- Online Assessment
- Game Closet at *BigIdeasMath.com*
- Vocabulary Help
- Resources by Chapter

Answers

1. adjacent; 21
2. vertical; 81

Review of Common Errors

Exercises 1 and 2

- Students may think that there is not enough information to determine the value of x. Ask them to think about the information given in each figure. For instance, Exercise 1 shows two angles making up a right angle. So, the sum of the two angle measures must be 90°. Students can use this information to set up and solve a simple equation to find the value of x.

Exercises 3 and 4

- Students may think that there is not enough information to determine the value of x. Remind them of the definitions they have learned in the lesson and ask whether either could apply. For instance, Exercise 3 shows two angles making up a right angle, so a student can use the definition of complementary angles to find x.

Exercises 5 and 6

- Students may have difficulty using a protractor to construct the triangles. A quick review of the methods presented in Section 12.3 may be helpful.

Exercises 7 and 8

- Students may think that there is not enough information to find the value of x. Remind them that the sum of the angle measures of a triangle is 180°, and (in the case of Exercise 7), a box in a vertex signifies a right angle, which has a measure of 90°.
- Students may classify the triangle using sides only and forget to classify it using angles (or vice versa). Remind students to classify the triangles using sides *and* angles.

Exercises 9 and 10

- Students may forget that the sum of the angle measures of a quadrilateral is 360°. Remind them of this fact.

Exercise 11

- Students may forget what a rhombus is. If so, refer them to the diagram at the top of page 294.

Exercises 12 and 13

- When measuring with a centimeter ruler, students may not start at zero on the ruler. As a result, they will get an incorrect measurement and scale. Remind students to check their answers for reasonableness.

12 Chapter Review

Check It Out
Vocabulary Help
BigIdeasMath.com

Review Key Vocabulary

adjacent angles, *p. 504*
vertical angles, *p. 504*
congruent angles, *p. 504*
complementary angles, *p. 510*
supplementary angles, *p. 510*
congruent sides, *p. 516*
kite, *p. 526*
scale drawing, *p. 532*
scale model, *p. 532*
scale, *p. 532*
scale factor, *p. 533*

Review Examples and Exercises

12.1 Adjacent and Vertical Angles *(pp. 502–507)*

Tell whether the angles are *adjacent* or *vertical*. Then find the value of *x*.

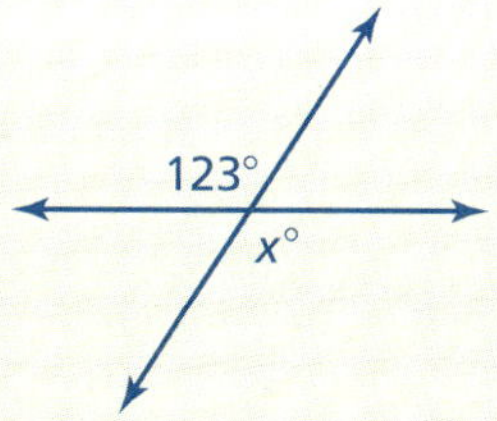

The angles are vertical angles. Because vertical angles are congruent, the angles have the same measure.

So, the value of x is 123.

Exercises

Tell whether the angles are *adjacent* or *vertical*. Then find the value of *x*.

1.

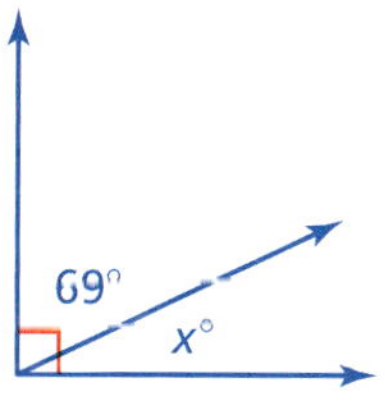

2.

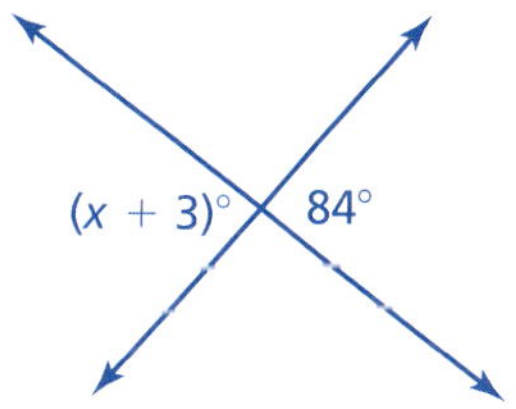

12.2 Complementary and Supplementary Angles *(pp. 508–513)*

Tell whether the angles are *complementary* or *supplementary*. Then find the value of *x*.

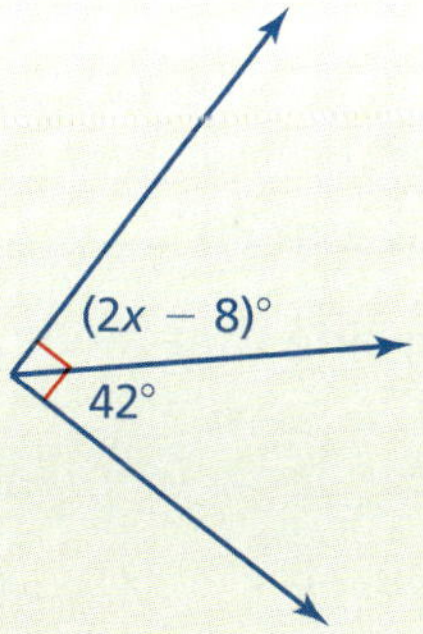

The two angles make up a right angle. So, the angles are complementary angles, and the sum of their measures is 90°.

$(2x - 8) + 42 = 90$	Write equation.
$2x + 34 = 90$	Combine like terms.
$2x = 56$	Subtract 34 from each side.
$x = 28$	Divide each side by 2.

So, the value of x is 28.

Exercises

Tell whether the angles are *complementary* or *supplementary*. Then find the value of x.

3.

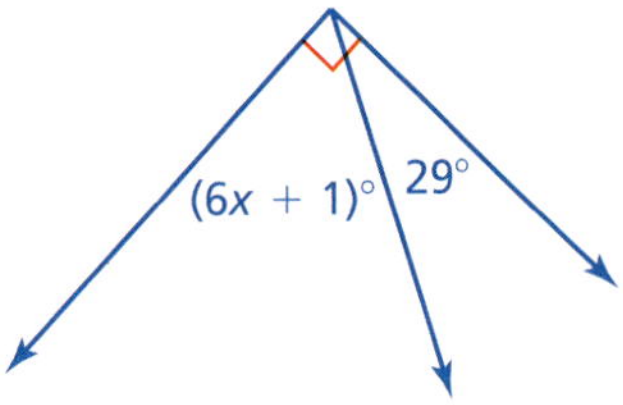

4.

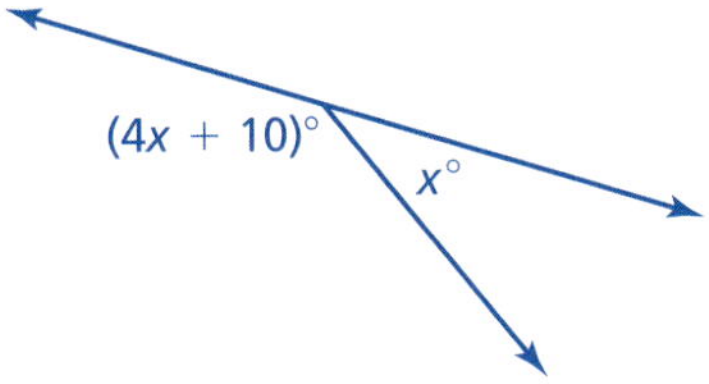

12.3 Triangles *(pp. 514–521)*

Draw a triangle with a 3.5-centimeter side and a 4-centimeter side that meet at a 25° angle. Then classify the triangle.

Step 1: Use a protractor to draw a 25° angle.

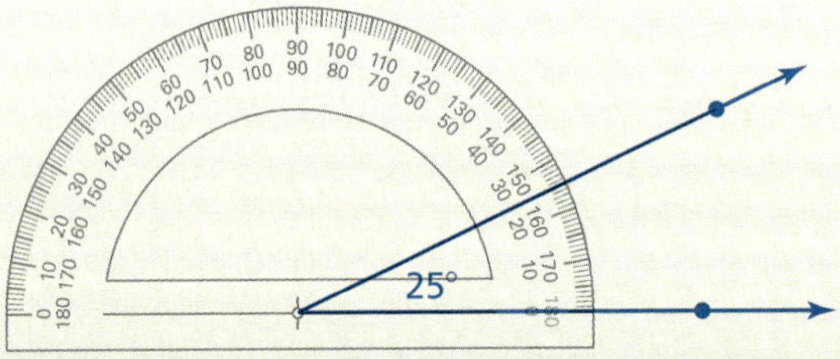

Step 2: Use a ruler to mark 3.5 centimeters on one ray and 4 centimeters on the other ray.

Step 3: Draw the third side to form the triangle.

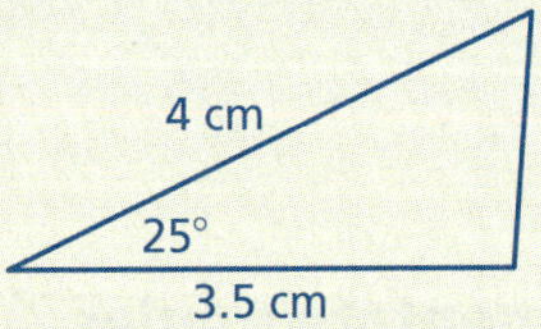

∴ The triangle is an obtuse scalene triangle.

Exercises

Draw a triangle with the given description.

5. a triangle with angle measures of 40°, 50°, and 90°
6. a triangle with a 3-inch side and a 4-inch side that meet at a 30° angle

Find the value of x. Then classify the triangle.

7.

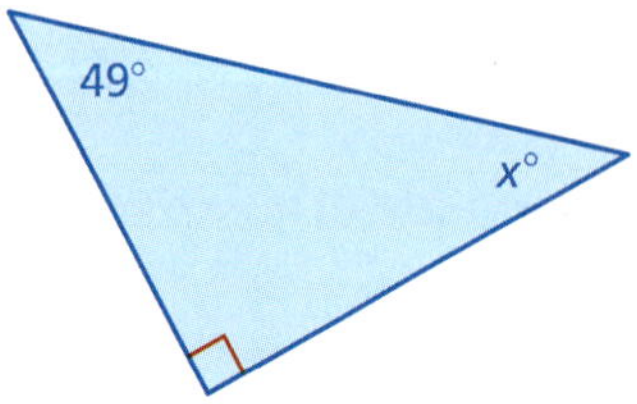

8.

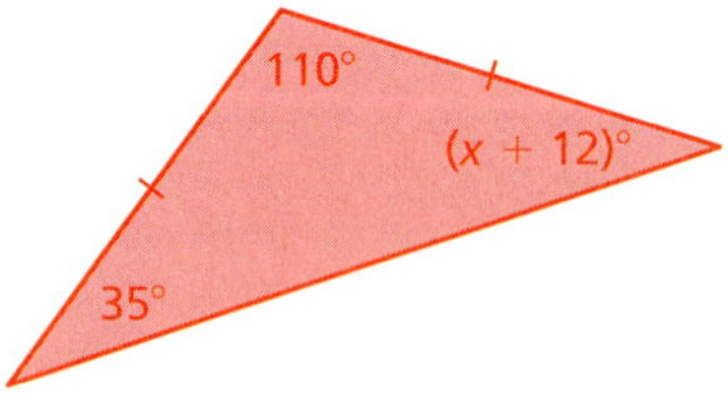

Review Game

Constructions and Figures

Materials per group:

- game cards
- pencil
- paper

Directions

Play in groups of two. One student shuffles the cards and places them face down. The other student turns over a card. Both students write as many details as possible about the figure (right angle, equal sides, etc.).

Who wins?

The student with the most correct details wins the round. Play continues until all cards are turned.

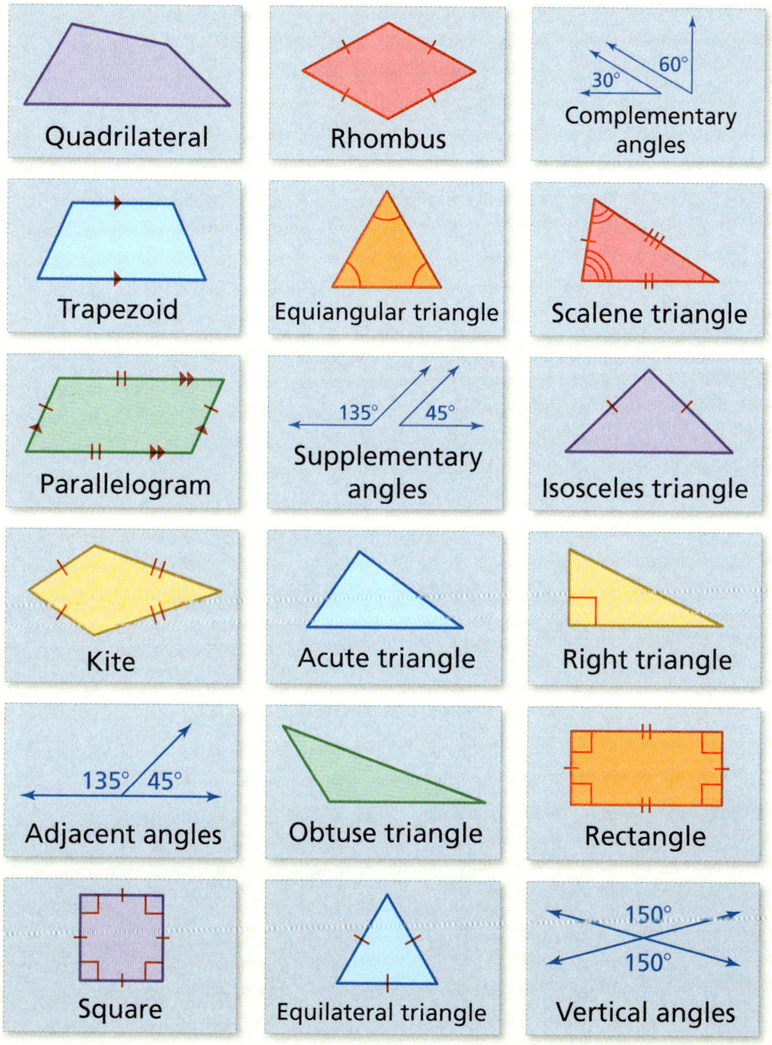

For the Student Additional Practice

- Lesson Tutorials
- Multi-Language Glossary
- Self-Grading Progress Check
- *BigIdeasMath.com*
  Dynamic Student Edition
  Student Resources

Answers

3. complementary; 10
4. supplementary; 34
5. 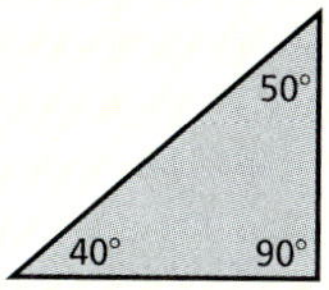

6. See Additional Answers.
7. 41; right scalene
8. 23; isosceles obtuse
9. 52
10. 147
11. 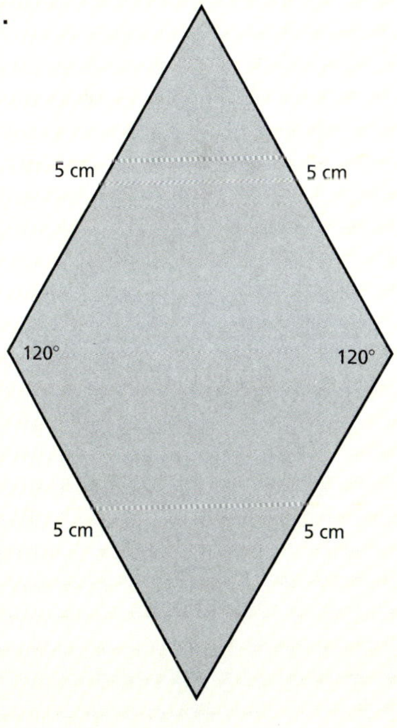

12. 6 cm; 1 cm : 5 in.
13. 2.5 cm; 1 cm : 3 in.

My Thoughts on the Chapter

Teacher Tip

Not allowed to write in your teaching edition? Use sticky notes to record your thoughts.

What worked. . .

What did not work. . .

What I would do differently. . .

12.4 Quadrilaterals (pp. 524–529)

Draw a parallelogram with a 50° angle and a 130° angle.

Step 1: Draw a line.

Step 2: Draw a 50° angle and a 130° angle that each have one side on the line.

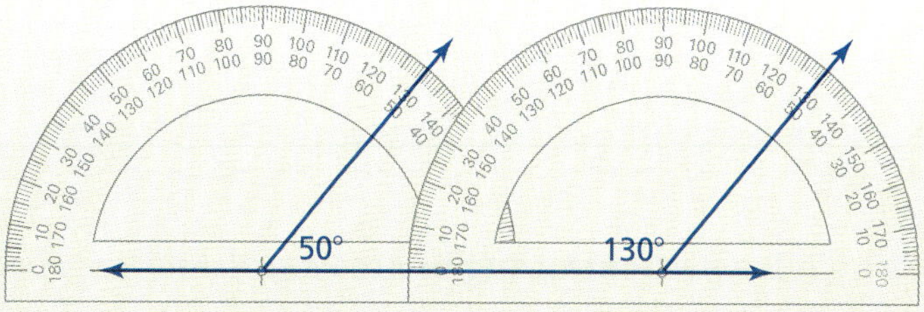

Step 3: Draw the remaining side. Make sure that both pairs of opposite sides are parallel and congruent.

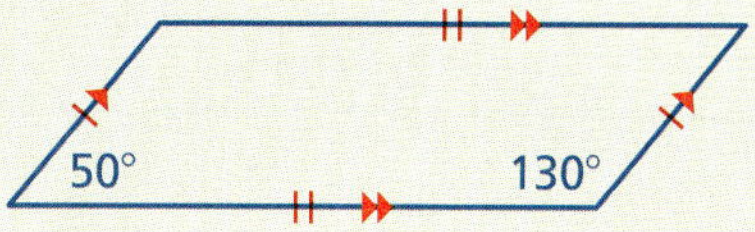

Exercises

Find the value of x.

9.

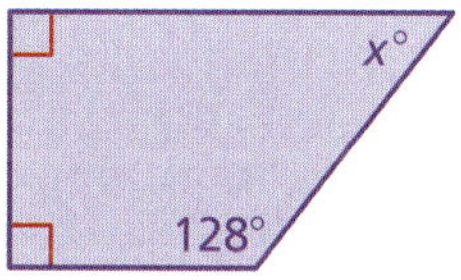

10.

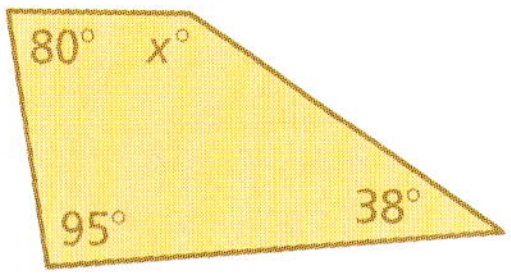

11. Draw a rhombus with 5-centimeter sides and two 120° angles.

12.5 Scale Drawings (pp. 530–537)

A lighthouse is 160 feet tall. A scale model of the lighthouse has a scale of 1 in. : 8 ft. How tall is the model of the lighthouse?

$$\frac{1 \text{ in.}}{8 \text{ ft}} = \frac{x \text{ in.}}{160 \text{ ft}}$$

model height (x in.) — actual height (160 ft)

$$\frac{1 \text{ in.}}{8 \cancel{\text{ft}}} \cdot 160 \cancel{\text{ft}} = \frac{x \text{ in.}}{160 \text{ ft}} \cdot 160 \text{ ft}$$ Multiplication Property of Equality

$$20 = x$$ Simplify.

So, the model of the lighthouse is 20 inches tall.

Exercises

Use a centimeter ruler to measure the segment shown. Find the scale of the drawing.

12.

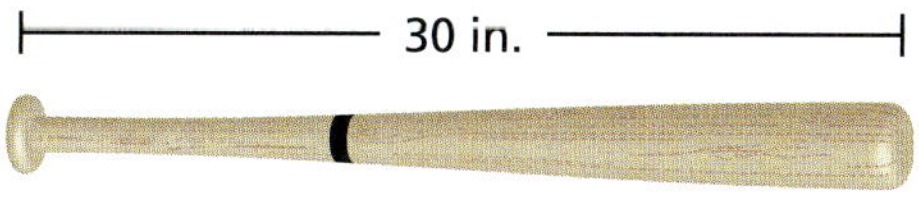

13.

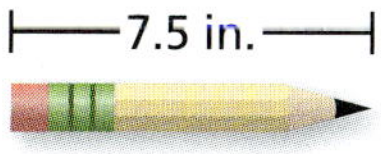

12 Chapter Test

Check It Out
Test Practice
BigIdeasMath.com

Tell whether the angles are *adjacent* or *vertical*. Then find the value of *x*.

1.

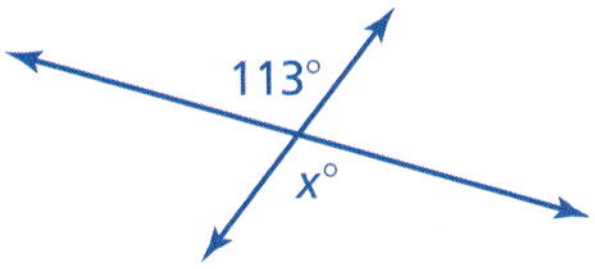

2.

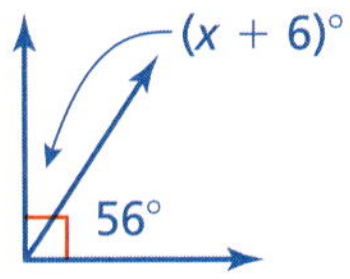

Tell whether the angles are *complementary* or *supplementary*. Then find the value of *x*.

3.

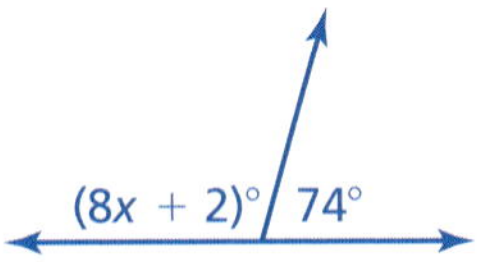

4.

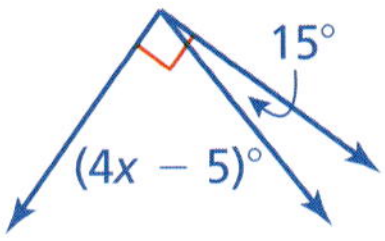

Draw a triangle with the given angle measures. Then classify the triangle.

5. 10°, 80°, 90°

6. 30°, 40°, 110°

Draw a triangle with the given description.

7. a triangle with a 5-inch side and a 6-inch side that meet at a 50° angle

8. a right isosceles triangle

Find the value of *x*. Then classify the triangle.

9.

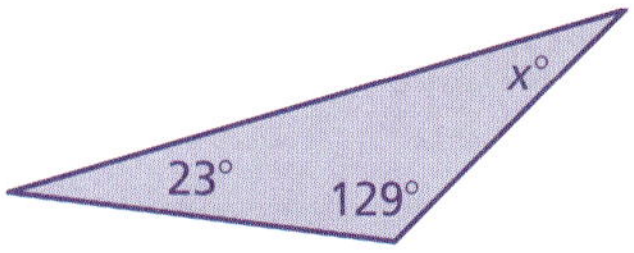

10.

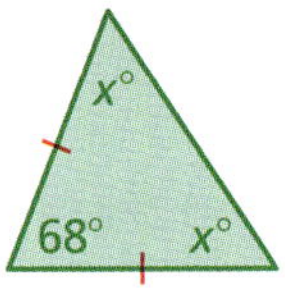

11.

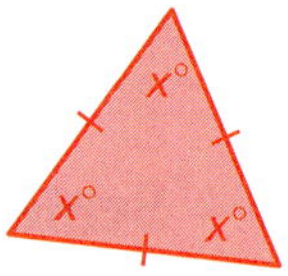

Find the value of *x*.

12.

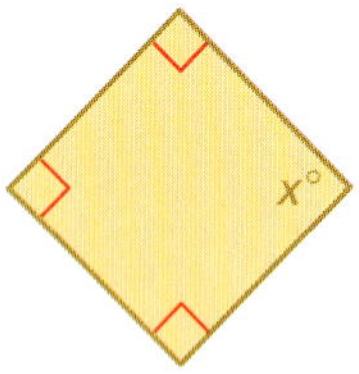

13.

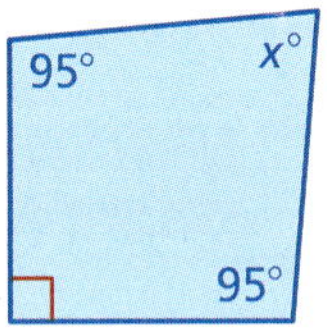

14.

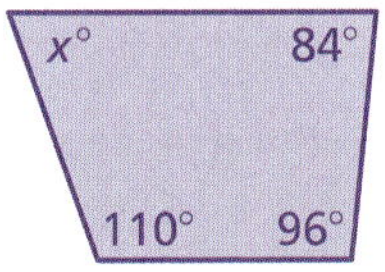

Draw a quadrilateral with the given description.

15. a rhombus with 6-centimeter sides and two 80° angles

16. a parallelogram with a 20° angle and a 160° angle

17. **FISH** Use a centimeter ruler to measure the fish. Find the scale factor of the drawing.

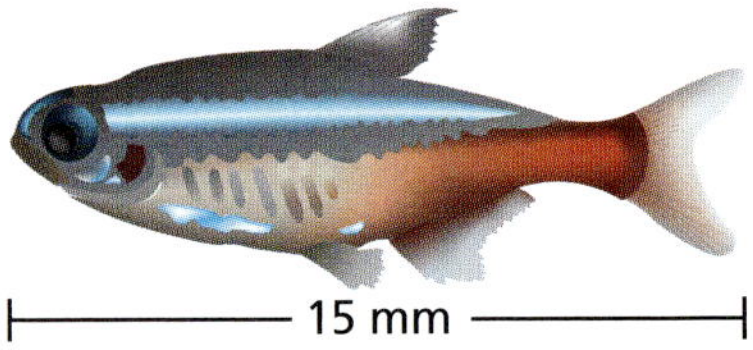

18. **CAD** An engineer is using computer-aided design (CAD) software to design a component for a space shuttle. The scale of the drawing is 1 cm : 60 in. The actual length of the component is 12.5 feet. What is the length of the component in the drawing?

Test Item References

Chapter Test Questions	Section to Review
1, 2	12.1
3, 4	12.2
5–11	12.3
12–16	12.4
17, 18	12.5

Test-Taking Strategies

Remind students to quickly look over the entire test before they start so that they can budget their time. This chapter contains many definitions that some students may find difficult to keep straight. Encourage them to jot down definitions on the back of the test before they start. Students need to **Stop** and **Think** as they work through the test.

Common Errors

- **Exercises 1–4** Students may think that there is not enough information to determine the value of *x*. Remind them of the definitions they have learned and ask whether either could apply. For instance, Exercise 4 shows two angles making up a right angle, so a student can use the definition of complementary angles to find *x*.
- **Exercises 5–8** Students may have difficulty using a protractor to construct the triangles. A quick review of the methods presented in Section 12.3 may be helpful.
- **Exercises 9–14** Students may forget that the sum of the angle measures of a triangle is 180° or that the sum of the angle measures of a quadrilateral is 360°. Remind them of these facts.
- **Exercises 9–11** Students may classify the triangle using sides only and forget to classify it using angles (or vice versa). Remind students to classify the triangles using sides *and* angles.
- **Exercises 15 and 16** Students may forget what a rhombus or a parallelogram is. If so, refer them to the diagram at the top of page 294.

Reteaching and Enrichment Strategies

If students need help. . .	If students got it. . .
Resources by Chapter • Practice A and Practice B • Puzzle Time Record and Practice Journal Practice Differentiating the Lesson Lesson Tutorials *BigIdeasMath.com* Skills Review Handbook	Resources by Chapter • Enrichment and Extension • Technology Connection Game Closet at *BigIdeasMath.com* Start Cumulative Assessment

Answers

1. vertical; 113
2. adjacent; 28
3. supplementary; 13
4. complementary; 20
5. 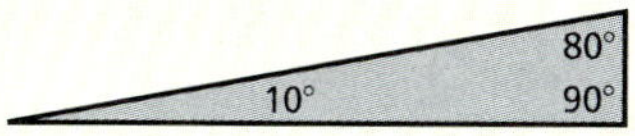

right scalene triangle
6. 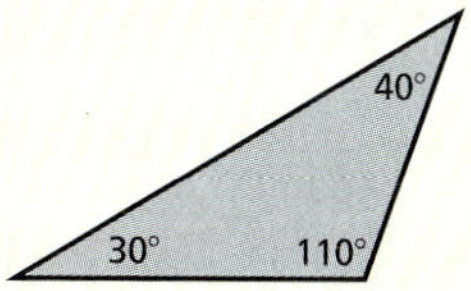

obtuse scalene triangle
7. See Additional Answers.
8. 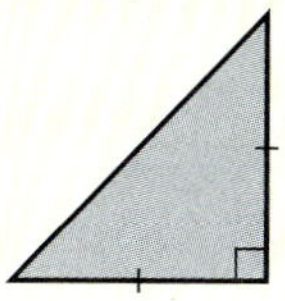
9. 28; obtuse scalene triangle
10. 56; acute isosceles triangle
11. 60; equilateral equiangular triangle
12. 90
13. 80
14. 70
15. See Additional Answers.
16. 

17. 5 cm; 10 : 3
18. 2.5 cm

Technology for the Teacher

Online Assessment
Assessment Book
ExamView® Assessment Suite

Test-Taking Strategies

Available at *BigIdeasMath.com*

After Answering Easy Questions, Relax
Answer Easy Questions First
Estimate the Answer
Read All Choices before Answering
Read Question before Answering
Solve Directly or Eliminate Choices
Solve Problem before Looking at Choices
Use Intelligent Guessing
Work Backwards

About this Strategy

When taking a multiple choice test, be sure to read each question carefully and thoroughly. Sometimes it is easier to solve the problem and then look for the answer among the choices.

Answers

1. D
2. 75%
3. H

Technology for the Teacher

Performance Tasks
Online Assessment
Assessment Book
ExamView® Assessment Suite

Item Analysis

1. **A.** The student divides 1 by 9 instead of dividing 9 by 1.

   B. The student confuses the meanings of x and y.

   C. The student looks from (0, 0) to (1, 9), not recognizing that the line segment between the points represents calories being burned between $x = 0$ and $x = 1$.

   D. Correct answer

2. **Gridded Response:** Correct answer: 75%

   Common Error: The student finds what percent 10 is of 40, and gets an answer of 25%.

3. **F.** The student finds the value of $2 - 6 - 9$.

   G. The student finds the value of $-2 + 6 - 9$.

   H. Correct answer

   I. The student finds the value of $-2 + 6 - (-9)$.

12 Cumulative Assessment

Test-Taking Strategy
Solve Problem Before Looking at Choices

"**Solve the problem before looking at the choices.** You know area increases as the square of the scale. So, it's 8 in.²"

1. The number of calories you burn by playing basketball is proportional to the number of minutes you play. Which of the following is a valid interpretation of the graph below?

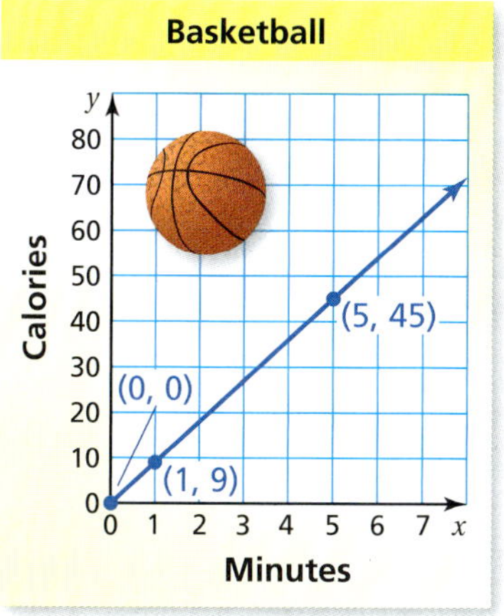

A. The unit rate is $\frac{1}{9}$ calorie per minute.

B. You burn 5 calories by playing basketball for 45 minutes.

C. You do not burn any calories if you do not play basketball for at least 1 minute.

D. You burn an additional 9 calories for each minute of basketball you play.

2. A lighting store is holding a clearance sale. The store is offering discounts on all the lamps it sells. As the sale progresses, the store will increase the percent of discount it is offering.

You want to buy a lamp that has an original price of \$40. You will buy the lamp when its price is marked down to \$10. What percent discount will you have received?

3. What is the value of the expression below?

$$2 - 6 - (-9)$$

F. -13 **H.** 5

G. -5 **I.** 13

4. What is the solution to the proportion below?

$$\frac{8}{12} = \frac{x}{18}$$

5. Which graph represents the inequality below?

$$-5 - 6x \leq -23$$

A. 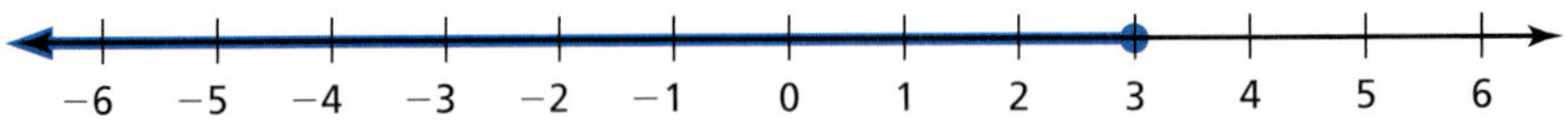

B. 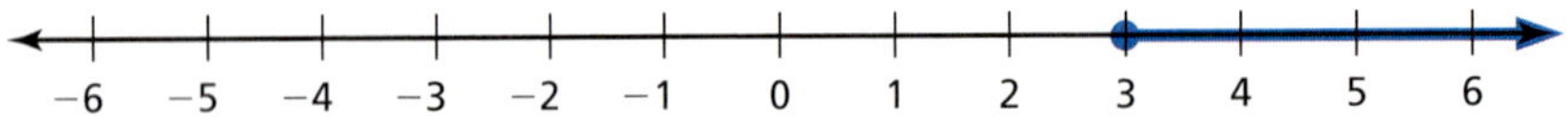

C. 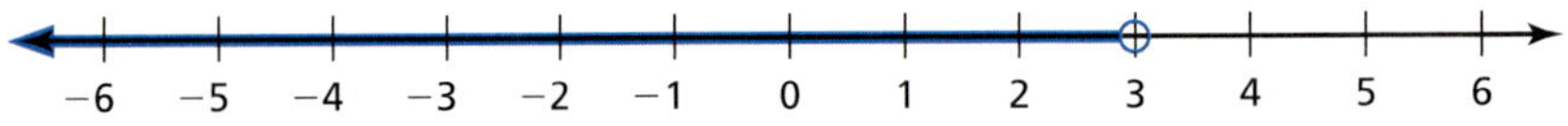

D. 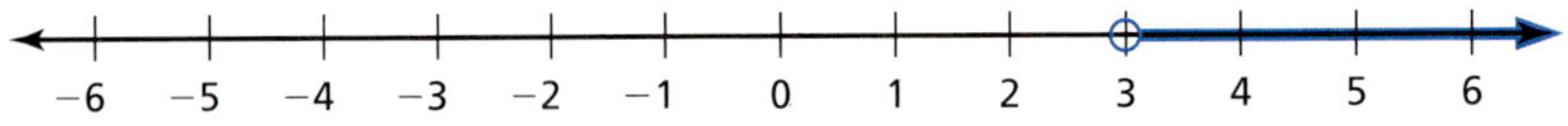

6. You are building a scale model of a park that is planned for a city. The model uses the scale below.

1 centimeter = 2 meters

The park will have a rectangular reflecting pool with a length of 20 meters and a width of 12 meters. In your scale model, what will be the area of the reflecting pool?

F. 60 cm^2

G. 120 cm^2

H. 480 cm^2

I. 960 cm^2

7. The quantities x and y are proportional. What is the missing value in the table?

x	y
$\frac{5}{7}$	10
$\frac{9}{7}$	18
$\frac{15}{7}$	30
4	

A. 38

B. 42

C. 46

D. 56

Item Analysis (continued)

4. **Gridded Response:** Correct answer: 12

   Common Error: The student incorrectly uses the Cross Products Property, getting an answer of 27.

5. **A.** The student does not reverse the inequality symbol.

   B. Correct answer

   C. The student does not reverse the inequality symbol and excludes $x = 3$.

   D. The student excludes $x = 3$.

6. **F.** Correct answer

   G. The student computes the area in square meters but then uses the given scale factor, which is for length, not area.

   H. The student reverses the relationship between the actual park and the scale model. The student also computes the area in square meters but then uses the given scale factor, which is for length, not area.

   I. The student reverses the relationship between the actual park and the scale model.

7. **A.** The student adds 8 to 30 because $18 - 10 = 8$.

   B. The student adds 12 to 30 because $30 - 18 = 12$.

   C. The student thinks the amount being added to y is being increased by 4 each time.

   D. Correct answer

Answers

4. 12
5. B
6. F
7. D

Answers

8. H

9. C

10. *Part A* 90 miles

Part B $3\frac{1}{4}$ inches

Item Analysis (continued)

8. **F.** The student confuses a straight angle with a right angle.

G. The student confuses a straight angle with a right angle and adds instead of subtracts to find the measure of $\angle 2$.

H. Correct answer

I. The student adds instead of subtracts to find the measure of $\angle 2$.

9. **A.** The student subtracts incorrectly. The original difference of -50 is correct.

B. The student subtracts incorrectly. The original expression, $\frac{c}{5} + 15$, is correct.

C. Correct answer

D. The student identifies the operation error but then thinks that division by a number is undone by multiplication by the opposite of the number.

10. **2 points** The student demonstrates a thorough understanding of working with scale drawings. In Part A, the student correctly determines that the actual distance is 90 miles. In Part B, the student correctly determines that the distance on the map should be $3\frac{1}{4}$ inches. The student provides clear and complete work and explanations.

1 point The student demonstrates a partial understanding of working with scale drawings. The student provides some correct work and explanation.

0 points The student demonstrates insufficient understanding of working with scale drawings. The student is unable to make any meaningful progress toward a correct answer.

8. $\angle 1$ and $\angle 2$ form a straight angle. $\angle 1$ has a measure of 28°. What is the measure of $\angle 2$?

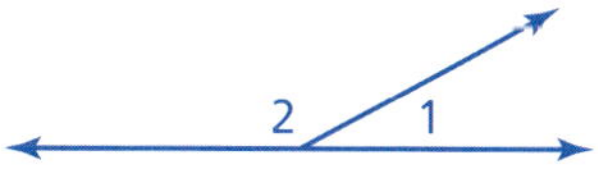

F. 62°

G. 118°

H. 152°

I. 208°

9. Brett solved the equation in the box below.

$$\frac{c}{5} - (-15) = -35$$
$$\frac{c}{5} + 15 = -35$$
$$\frac{c}{5} + 15 - 15 = -35 - 15$$
$$\frac{c}{5} = -50$$
$$\frac{c}{5} = \frac{-50}{5}$$
$$c = -10$$

What should Brett do to correct the error that he made?

A. Subtract 15 from -35 to get -20.

B. Rewrite $\frac{c}{5} - (-15)$ as $\frac{c}{5} - 15$.

C. Multiply each side of the equation by 5 to get $c = -250$.

D. Multiply each side of the equation by -5 to get $c = 250$.

10. A map of the state where Donna lives has the scale shown below.

Think
Solve
Explain

$$\frac{1}{2} \text{ inch} = 10 \text{ miles}$$

Part A Donna measured the distance between her town and the state capital on the map. Her measurement was $4\frac{1}{2}$ inches. Based on Donna's measurement, what is the actual distance, in miles, between her town and the state capital? Show your work and explain your reasoning.

Part B Donna wants to mark her favorite campsite on the map. She knows that the campsite is 65 miles north of her town. What distance on the map, in inches, represents an actual distance of 65 miles? Show your work and explain your reasoning.

13 Circles and Area

13.1 Circles and Circumference

13.2 Perimeters of Composite Figures

13.3 Areas of Circles

13.4 Areas of Composite Figures

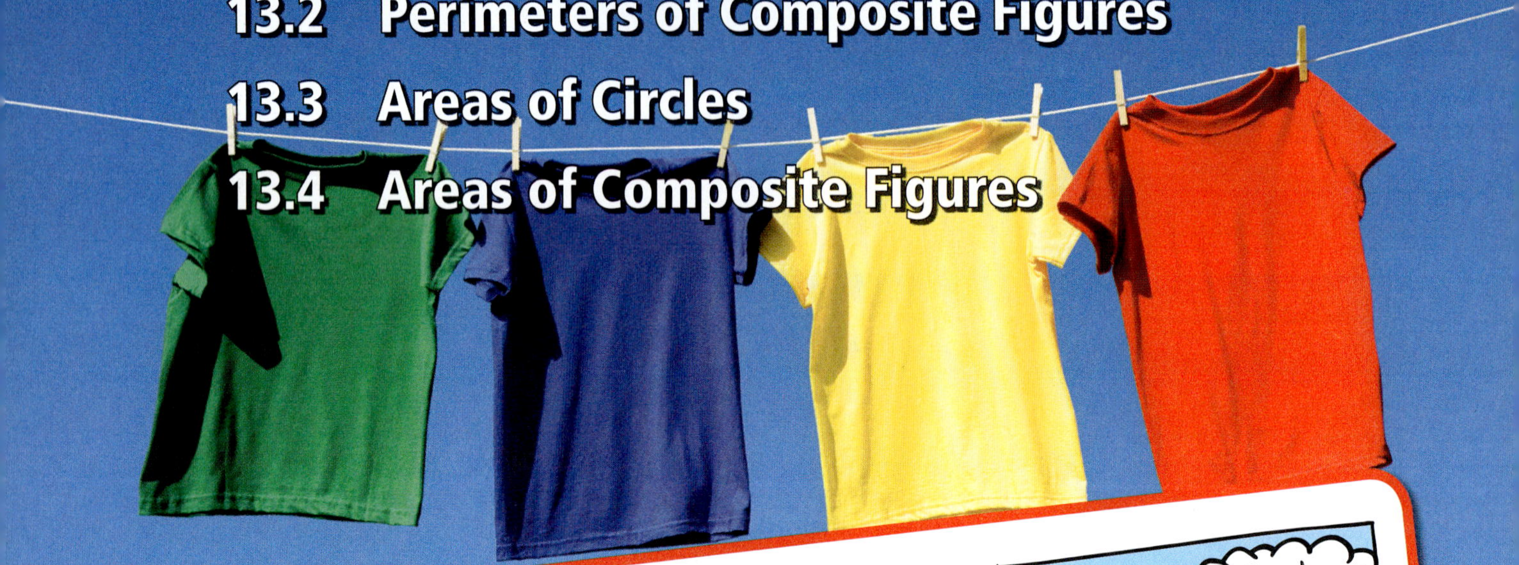

What Your Students Have Learned

- Find areas of rectangles with fractional side lengths.
- Classify two-dimensional figures into categories based on properties.
- Use formulas to find the areas of parallelograms, triangles, and trapezoids.
- Write and evaluate numerical expressions involving whole-number exponents.

What Your Students Will Learn

- Understand pi and its estimates.
- Use values of pi to estimate and calculate the circumference and area of circles.
- Find perimeters and areas of composite two-dimensional figures, including semi-circles.

Pacing Guide for Chapter 13

Chapter Opener Advanced	1 Day
Section 1 Advanced	1 Day
Section 2 Advanced	1 Day
Section 3 Advanced	1 Day
Section 4 Advanced	1 Day
Chapter Review/ Chapter Tests Advanced	2 Days
Total Chapter 13 Advanced	7 Days
Year-to-Date Advanced	118 Days

Technology for the Teacher

BigIdeasMath.com
Chapter at a Glance
Complete Materials List
Parent Letters: English and Spanish

What Your Students Have Learned

- Classify shapes and polygons used in figures.
- Square numbers and use the order of operations to evaluate expressions.

Additional Topics for Review

- Exponents
- Perimeter
- Area

Try It Yourself

1. trapezoids
2. circles
3. trapezoid, triangle
4. triangles
5. rectangle, triangle
6. rectangle, triangles
7. 25
8. 144
9. 12
10. 196
11. 243
12. 188

Record and Practice Journal Fair Game Review

1. trapezoid and triangle
2. rectangles
3. rectangle and trapezoids
4. trapezoids, rectangles, circles
5. trapezoids, parallelogram
6. 49
7. 121
8. 100
9. 700
10. 144
11. 405
12. 744
13. 566
14. 5000 meters

Math Background Notes

Vocabulary Review

- Evaluate
- Simplify
- Exponent

Classifying Figures

- Students should know how to classify figures.
- **Teaching Tip:** Encourage students to redraw each figure on a sheet of paper. By constructing it themselves, students are more likely to identify the polygons used in the figure.
- **Teaching Tip:** Tactile learners might benefit from doing these examples using polygon templates. Allow them to build the shapes they see using the polygon templates. This could help students to distinguish which polygons they are seeing.

Squaring Numbers and Using Order of Operations

- Students should know the order of operations.
- You may want to review the correct order of operations with students. Many students probably learned the pneumonic device *Please Excuse My Dear Aunt Sally*. Ask a volunteer to explain why this phrase is helpful.
- **Common Error:** Remind students that the exponent describes how many times the base acts as a factor. The exponent itself will not appear as a factor.

Reteaching and Enrichment Strategies

If students need help. . .	If students got it. . .
Record and Practice Journal • Fair Game Review Skills Review Handbook Lesson Tutorials	Game Closet at *BigIdeasMath.com* Start the next section

What You Learned Before

Classifying Figures

Identify the basic shapes in the figure.

Example 1

Rectangle, right triangle

Example 2

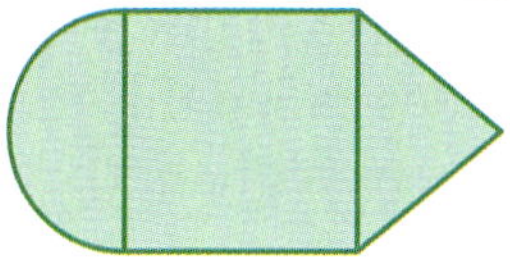

Semicircle, square, and triangle

Try It Yourself

Identify the basic shapes in the figure.

1.

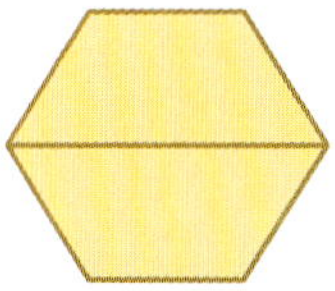

2.

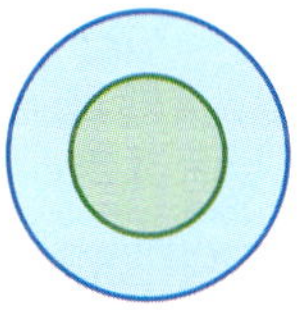

3.

4.

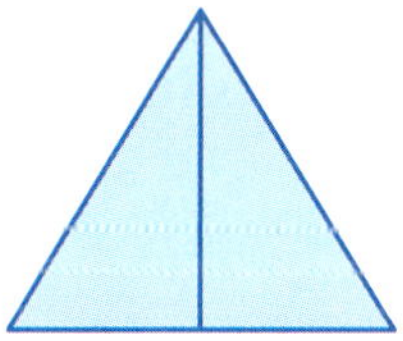

5.

6.

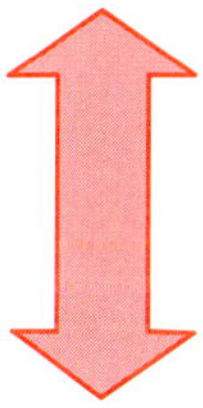

Squaring Numbers and Using Order of Operations

Example 3 **Evaluate 4^2.**

$$4^2 = 4 \cdot 4 = 16$$

4^2 means to multiply 4 by itself.

Example 4 **Evaluate $3 \cdot 6^2$.**

$$3 \cdot 6^2 = 3 \cdot (6 \cdot 6) = 3 \cdot 36 = 108$$

Use order of operations. Evaluate the exponent, and then multiply.

Try It Yourself

Evaluate the expression.

7. 5^2

8. 12^2

9. $3 \cdot 2^2$

10. $4 \cdot 7^2$

11. $3(1 + 8)^2$

12. $2(3 + 7)^2 - 3 \cdot 4$

13.1 Circles and Circumference

Essential Question How can you find the circumference of a circle?

Archimedes was a Greek mathematician, physicist, engineer, and astronomer.

Archimedes discovered that in any circle the ratio of circumference to diameter is always the same. Archimedes called this ratio pi, or π (a letter from the Greek alphabet).

$$\pi = \frac{\text{circumference}}{\text{diameter}}$$

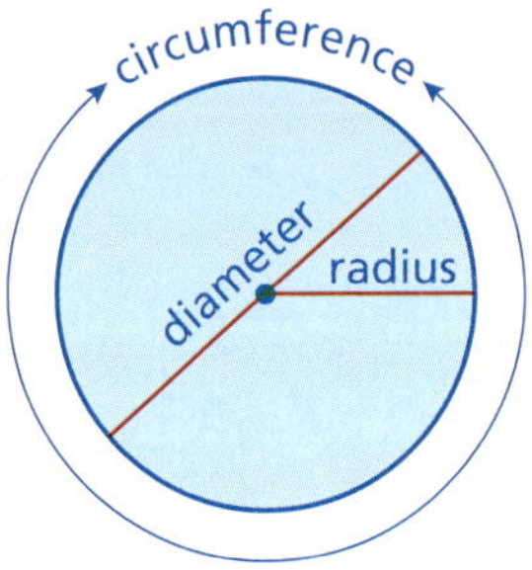

In Activities 1 and 2, you will use the same strategy Archimedes used to approximate π.

1 ACTIVITY: Approximating Pi

Work with a partner. Copy the table. Record your results in the table.

- **Measure the perimeter of the large square in millimeters.**
- **Measure the diameter of the circle in millimeters.**
- **Measure the perimeter of the small square in millimeters.**
- **Calculate the ratios of the two perimeters to the diameter.**
- **The average of these two ratios is an approximation of π.**

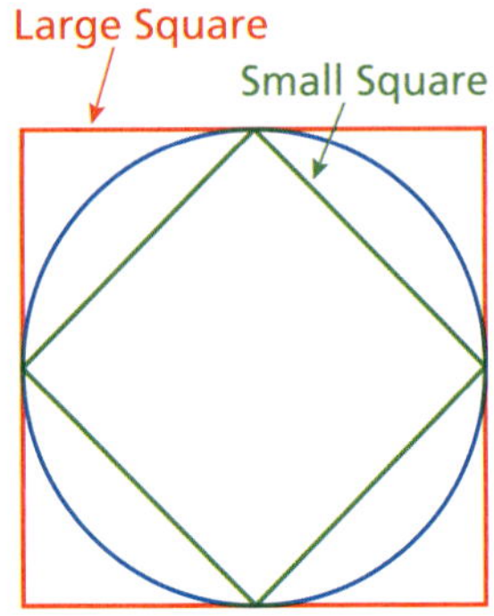

Sides	Large Perimeter	Diameter of Circle	Small Perimeter	$\frac{\text{Large Perimeter}}{\text{Diameter}}$	$\frac{\text{Small Perimeter}}{\text{Diameter}}$	Average of Ratios
4						
6						
8						
10						

Geometry

In this lesson, you will

- describe a circle in terms of radius and diameter.
- understand the concept of pi.
- find circumferences of circles and perimeters of semicircles.

A page from *Sir Cumference and the First Round Table* by Cindy Neuschwander

2 ACTIVITY: Approximating Pi

Math Practice

Make Conjectures

How can you use the results of the activity to find an approximation of pi?

Continue your approximation of pi. Complete the table from Activity 1 using a hexagon (6 sides), an octagon (8 sides), and a decagon (10 sides).

a.

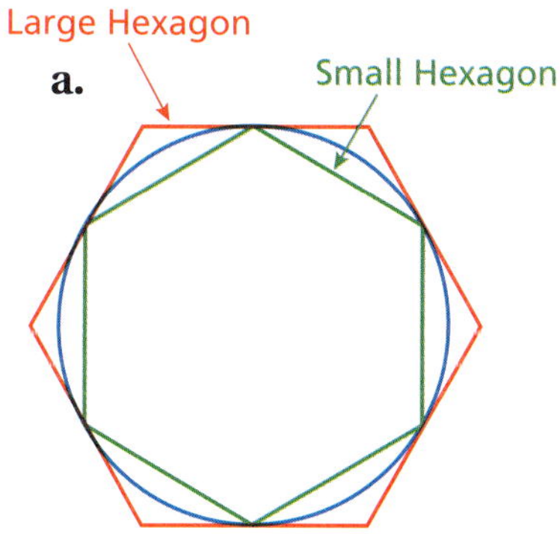

b.

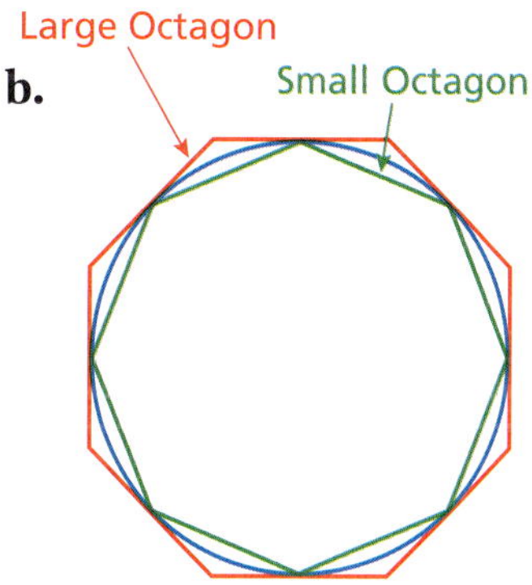

c.

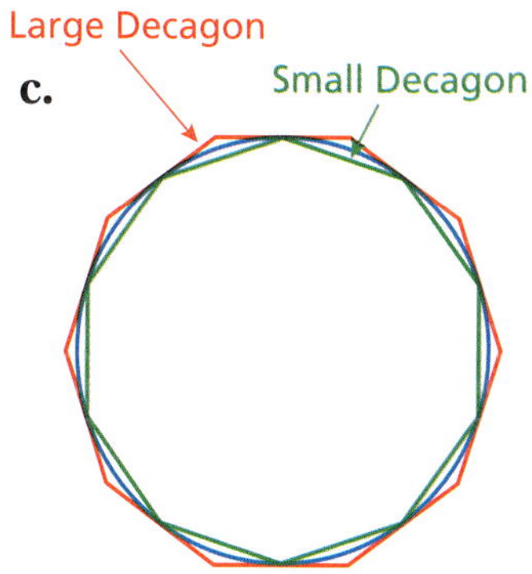

d. From the table, what can you conclude about the value of π? Explain your reasoning.

e. Archimedes calculated the value of π using polygons with 96 sides. Do you think his calculations were more or less accurate than yours?

What Is Your Answer?

3. IN YOUR OWN WORDS Now that you know an approximation for pi, explain how you can use it to find the circumference of a circle. Write a formula for the circumference C of a circle whose diameter is d.

4. CONSTRUCTION Use a compass to draw three circles. Use your formula from Question 3 to find the circumference of each circle.

Practice Use what you learned about circles and circumference to complete Exercises 9–11 on page 553.

13.1 Lesson

Check It Out
Lesson Tutorials
BigIdeasMath.com

Key Vocabulary
circle, *p. 550*
center, *p. 550*
radius, *p. 550*
diameter, *p. 550*
circumference, *p. 551*
pi, *p. 551*
semicircle, *p. 552*

A **circle** is the set of all points in a plane that are the same distance from a point called the **center**.

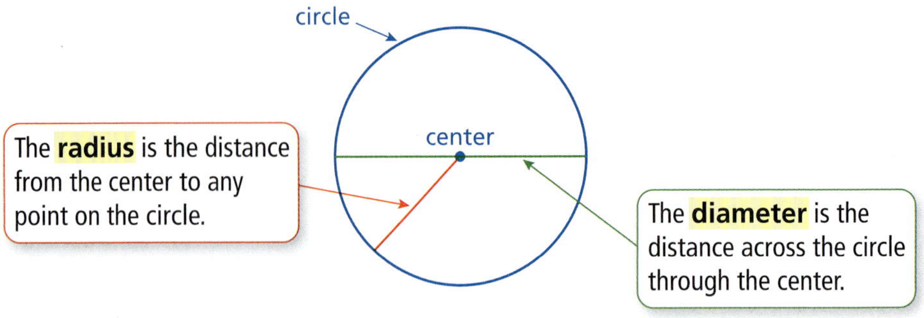

Key Idea

Radius and Diameter

Words The diameter d of a circle is twice the radius r. The radius r of a circle is one-half the diameter d.

Algebra **Diameter:** $d = 2r$ **Radius:** $r = \dfrac{d}{2}$

EXAMPLE 1 Finding a Radius and a Diameter

a. The diameter of a circle is 20 feet. Find the radius.

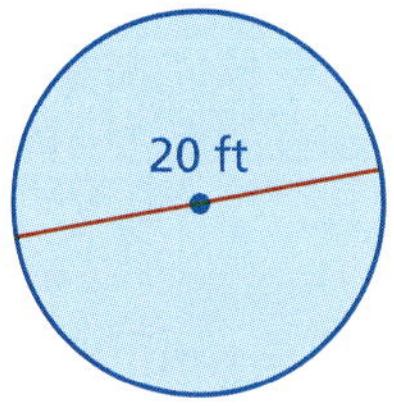

$r = \dfrac{d}{2}$ Radius of a circle

$= \dfrac{20}{2}$ Substitute 20 for *d*.

$= 10$ Divide.

The radius is 10 feet.

b. The radius of a circle is 7 meters. Find the diameter.

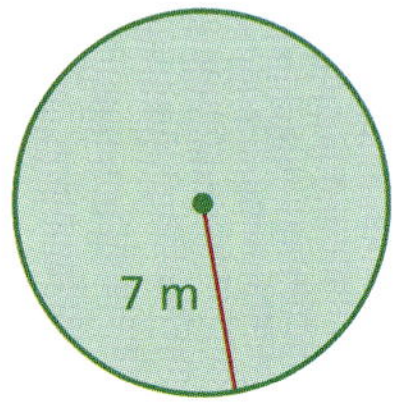

$d = 2r$ Diameter of a circle

$= 2(7)$ Substitute 7 for *r*.

$= 14$ Multiply.

The diameter is 14 meters.

On Your Own

Now You're Ready
Exercises 3–8

1. The diameter of a circle is 16 centimeters. Find the radius.
2. The radius of a circle is 9 yards. Find the diameter.

Multi-Language Glossary at BigIdeasMath.com

Laurie's Notes

Introduction

Connect

- **Yesterday:** Students investigated a technique for calculating the value of pi.
- **Today:** Students will use the formula for circumference to solve real-life problems.

Motivate

- **Whole Class Activity:** How observant are your students? Tell them that you are going to give them 1 minute to write a list of objects in the room that have a special characteristic. Give them time to get scrap paper and a pencil.
- Announce that they need to list objects that are circular or have a circle on them. Go!
- Items will vary but expected items include: clock and/or watch face, bottom of coffee cup, pencil's eraser, pupils of your eyes, Person X's glasses, metal feet on the chair, etc.

Lesson Notes

Key Idea

- Draw a circle on the board and label the center, a radius, and a diameter. Discuss each.
- Write the Key Idea.
- Discuss with students that if you know either the diameter or radius of a circle, you can find the other. There is a 2 : 1 relationship between **diameter** and **radius**.

Example 1

- Work through each example.
- Remind students to label answers with the appropriate units.

On Your Own

- It helps some students to draw a sketch of a circle and label the given information. These students need to see the relationship instead of only reading the given information.

Goal Today's lesson is estimating and calculating the **circumference** of **circles** using common estimates for **pi**.

Technology for the Teacher

Dynamic Classroom

Lesson Tutorials
Lesson Plans
Answer Presentation Tool

Extra Example 1

a. The diameter of a circle is 4 yards. Find the radius. 2 yd

b. The radius of a circle is 15 millimeters. Find the diameter. 30 mm

On Your Own

1. 8 cm
2. 18 yd

English Language Learners

Vocabulary

Use everyday objects, such as a clock, to discuss the meaning of *circle, center, radius, diameter,* and *circumference.* Have students add these terms to their notebooks along with a sketch of a circle with the parts labeled. Students should also include pi with its symbol, π, and its approximations, 3.14 and $\frac{22}{7}$, in their notebooks.

Extra Example 2

a. Find the circumference of a flying disc with a radius of 8 centimeters. Use 3.14 for π. about 50.24 cm

b. Find the circumference of a clock with a diameter of 21 inches. Use $\frac{22}{7}$ for π. about 66 in.

On Your Own

3. about 12.56 cm
4. about 44 ft
5. about 28.26 in.

Laurie's Notes

Key Idea

- **Make Sense of Problems and Persevere in Solving Them:** Read the information at the top of the page. Note the two approximations of pi and how each is used in the problems that follow. Note the *Study Tip.*
- Write the Key Idea.
- **Common Misconception:** Students may know that pi is a number; however, when they see it in a formula, they can become confused. Students may ask if π is a variable. Pi is a constant whose value is approximately 3.14 or $\frac{22}{7}$. In each formula, remind students that πd means π times the diameter and $2\pi r$ means 2 times π times the radius.
- **Big Idea:** The ratio of the circumference to the diameter is pi. To get the formula for circumference, multiply both sides of the equation by the diameter.

$$\frac{\text{circumference}}{\text{diameter}} = \text{pi}$$

$$\frac{C}{d} = \pi$$

$$C = \pi d$$

Example 2

- Work through each example.
- **Make Sense of Problems and Persevere in Solving Them:** Point out to students when to use the different forms of the circumference formula given a radius or a diameter.
- Remind students that the symbol $\approx$ means *approximately equal to.*

? "Why is the equal sign replaced by the approximately equal ($\approx$) sign?" 3.14 is an approximation for π.

? "Why is $2 \times 3.14 \times 5$ equal to $2 \times 5 \times 3.14$?" Commutative Property of Multiplication

- Note that in part (b), the diameter is a multiple of 7, so $\frac{22}{7}$ is used as the approximation for π.

On Your Own

- **Think-Pair-Share:** Students should read each question independently and then work in pairs to answer the questions. When they have answered the questions, the pair should compare their answers with another group and discuss any discrepancies.
- Check which estimate for π students used and why.

The distance around a circle is called the **circumference**. The ratio $\frac{\text{circumference}}{\text{diameter}}$ is the same for *every* circle and is represented by the Greek letter π, called **pi**. The value of π can be approximated as 3.14 or $\frac{22}{7}$.

Study Tip

When the radius or diameter is a multiple of 7, it is easier to use $\frac{22}{7}$ as the estimate of π.

Key Idea

Circumference of a Circle

Words The circumference C of a circle is equal to π times the diameter d or π times twice the radius r.

Algebra $C = \pi d$ or $C = 2\pi r$

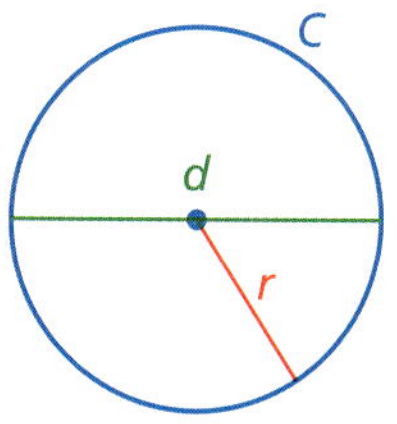

EXAMPLE 2 Finding Circumferences of Circles

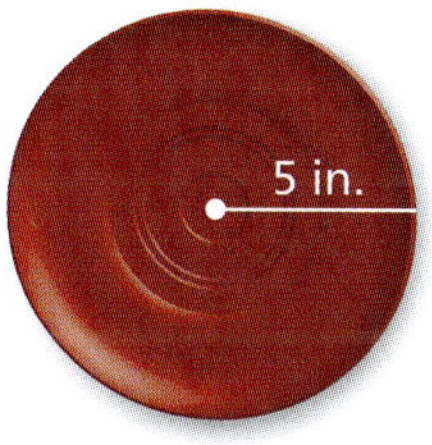

a. Find the circumference of the flying disc. Use 3.14 for π.

$C = 2\pi r$	Write formula for circumference.
$\approx 2 \cdot 3.14 \cdot 5$	Substitute 3.14 for π and 5 for r.
$= 31.4$	Multiply.

The circumference is about 31.4 inches.

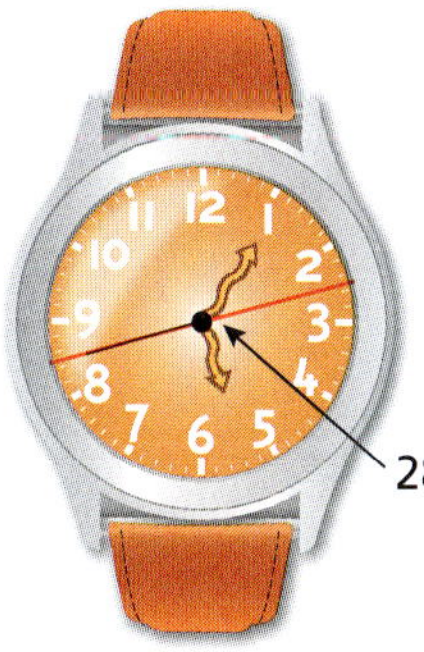

b. Find the circumference of the watch face. Use $\frac{22}{7}$ for π.

$C = \pi d$	Write formula for circumference.
$\approx \frac{22}{7} \cdot 28$	Substitute $\frac{22}{7}$ for π and 28 for d.
$= 88$	Multiply.

The circumference is about 88 millimeters.

On Your Own

Now You're Ready
Exercises 9–11

Find the circumference of the object. Use 3.14 or $\frac{22}{7}$ for π.

3. 

4. 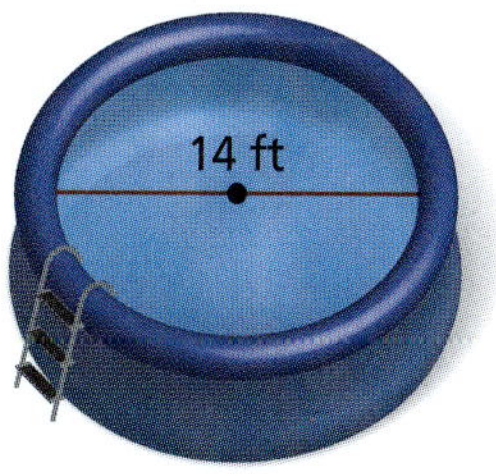

5. 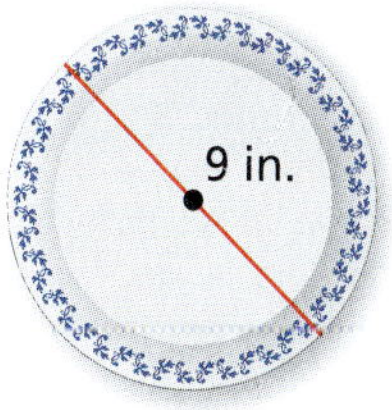

EXAMPLE 3 Estimating a Diameter

C = 31.4 in.

The circumference of the roll of caution tape decreases 10.5 inches after a construction worker uses some of the tape. Which is the best estimate of the diameter of the roll after the decrease?

Ⓐ 5 inches Ⓑ 7 inches Ⓒ 10 inches Ⓓ 12 inches

After the decrease, the circumference of the roll is $31.4 - 10.5 = 20.9$ inches.

$C = \pi d$	Write formula for circumference.
$20.9 \approx 3.14 \cdot d$	Substitute 20.9 for C and 3.14 for π.
$21 \approx 3d$	Round 20.9 up to 21. Round 3.14 down to 3.
$7 = d$	Divide each side by 3.

∴ The correct answer is Ⓑ.

On Your Own

6. **WHAT IF?** The circumference of the roll of tape decreases 5.25 inches. Estimate the diameter of the roll after the decrease.

EXAMPLE 4 Finding the Perimeter of a Semicircular Region

A semicircle is one-half of a circle. Find the perimeter of the semicircular region.

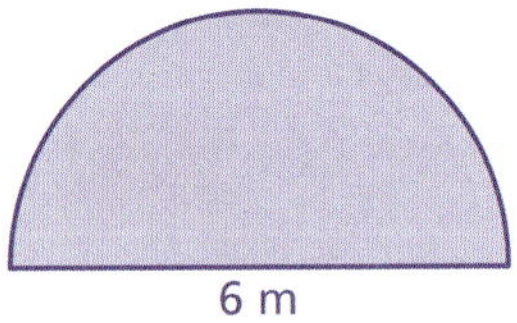

The straight side is 6 meters long. The distance around the curved part is one-half the circumference of a circle with a diameter of 6 meters.

$\frac{C}{2} = \frac{\pi d}{2}$	Divide the circumference by 2.
$\approx \frac{3.14 \cdot 6}{2}$	Substitute 3.14 for π and 6 for d.
$= 9.42$	Simplify.

∴ So, the perimeter is about $6 + 9.42 = 15.42$ meters.

On Your Own

Now You're Ready
Exercises 15 and 16

Find the perimeter of the semicircular region.

7. 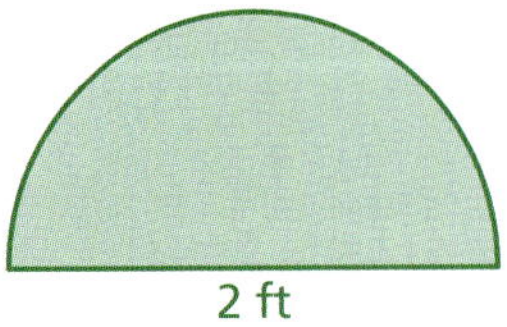

8. 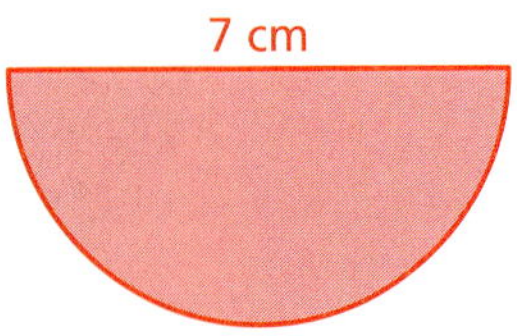

9. 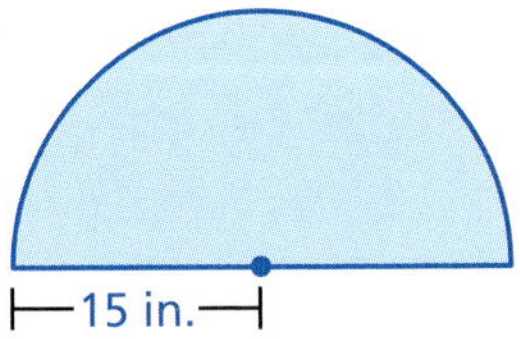

Laurie's Notes

Example 3

- Ask a volunteer to read the given information.
- Note that this multiple choice question asks for the best *estimate*. Rounding the circumference and π is sufficient to select the correct answer.

Example 4

- Draw the diagram. It may be helpful to draw the other half of the semicircle as a dotted line.
- Another way to visualize the distance around the curved part is to overlap a circle with a diameter of 6 meters with a semicircle with a straight side of 6 meters.
- **Common Error:** Students may find half the circumference and forget to add the distance across the diameter.

Closure

- **Exit Ticket:** A peso has a diameter of 21 millimeters.
  a. What is the radius of the peso? 10.5 mm
  b. What is the circumference of the peso? about 66 mm

Extra Example 3

In Example 3, the circumference of the roll of tape decreases 7.5 inches. Estimate the diameter of the roll after the decrease. about 8 in.

On Your Own

6. about 9 in.

Extra Example 4

Find the perimeter of a semicircular region with a radius of 5 yards. about 25.7 yd

On Your Own

7. about 5.14 ft

8. about 18 cm

9. about 77.1 in.

Vocabulary and Concept Check

1. The radius is one-half the diameter.
2. the distance from the center to any point on the circle; This phrase describes the radius of a circle, whereas the other phrases describe the circumference of a circle.

Practice and Problem Solving

3. 2.5 cm
4. 14 mm
5. $1\frac{3}{4}$ in.
6. 12 cm
7. 4 in.
8. 1.6 ft
9. about 31.4 in.
10. about 44 in.
11. about 56.52 in.
12. *Sample answer:* A lawn game has two circular targets with 28-inch diameters. You lost one. You want to use a length of wire to make a replacement.

    $C = \pi d \approx \frac{22}{7} \cdot 28 = 88$

    You need a piece of wire 88 inches long.

Assignment Guide and Homework Check

Level	Assignment	Homework Check
Advanced	1, 2, 4–8 even, 9–11, 12–24 even, 26–29	16, 18, 20, 24

Common Errors

- **Exercises 3–8** Students may confuse what they are finding and double the diameter or halve the radius. Remind them that the radius is half the diameter. Encourage students to draw a line representing the radius or diameter for each problem so that they have a visual reference.
- **Exercises 9–11** Students may use the wrong formula for circumference when given a radius or diameter. Remind them of the different equations. If students are struggling with the two equations, tell them to use only one equation and to convert the dimension given to the one in the chosen formula.
- **Exercises 9–11** Students may use $\frac{22}{7}$ when it would be easier to use 3.14 and get frustrated. Remind them to use $\frac{22}{7}$ when the radius or diameter is a multiple of 7.

13.1 Record and Practice Journal

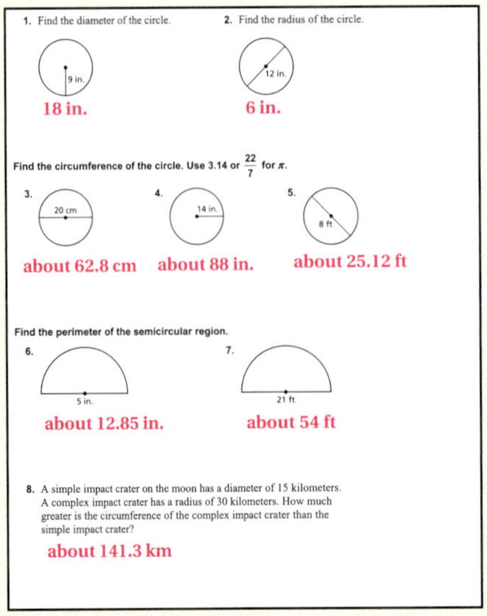
1. Find the diameter of the circle.

9 in.

18 in.

2. Find the radius of the circle.

12 in.

6 in.

Find the circumference of the circle. Use 3.14 or $\frac{22}{7}$ for π.

3. 20 cm — about 62.8 cm

4. 14 in. — about 88 in.

5. 8 ft — about 25.12 ft

Find the perimeter of the semicircular region.

6. 5 in. — about 12.85 in.

7. 21 ft — about 54 ft

8. A simple impact crater on the moon has a diameter of 15 kilometers. A complex impact crater has a radius of 30 kilometers. How much greater is the circumference of the complex impact crater than the simple impact crater?

about 141.3 km

13.1 Exercises

Check It Out
Help with Homework
BigIdeasMath.com

Vocabulary and Concept Check

1. **VOCABULARY** What is the relationship between the radius and the diameter of a circle?

2. **WHICH ONE DOESN'T BELONG?** Which phrase does *not* belong with the other three? Explain your reasoning.

the distance around a circle	π times twice the radius
π times the diameter	the distance from the center to any point on the circle

Practice and Problem Solving

Find the radius of the button.

1 3.

4.

5.

Find the diameter of the object.

6.

7.

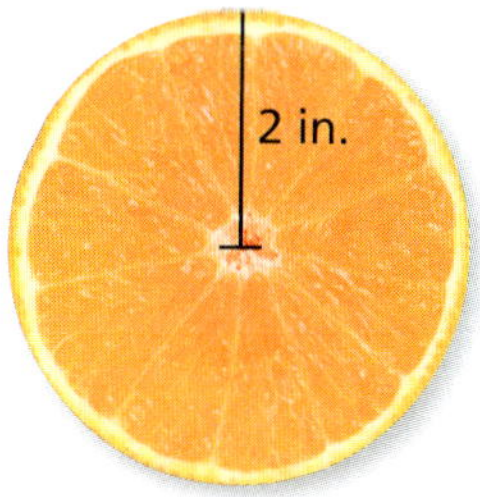

8.

Find the circumference of the pizza. Use 3.14 or $\frac{22}{7}$ for π.

2 9.

10.

11.

12. **CHOOSE TOOLS** Choose a real-life circular object. Explain why you might need to know its circumference. Then find the circumference.

13. **SINKHOLE** A circular sinkhole has a circumference of 75.36 meters. A week later, it has a circumference of 150.42 meters.

   a. Estimate the diameter of the sinkhole each week.

   b. How many times greater is the diameter of the sinkhole now compared to the previous week?

14. **REASONING** Consider the circles *A*, *B*, *C*, and *D*.

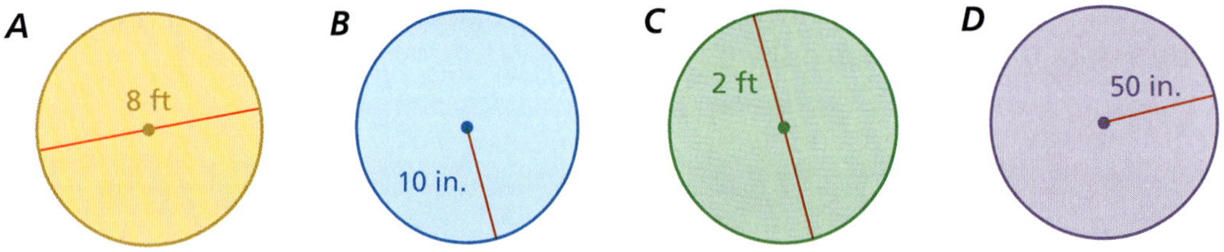

   a. Without calculating, which circle has the greatest circumference?

   b. Without calculating, which circle has the least circumference?

Find the perimeter of the window.

4 15.

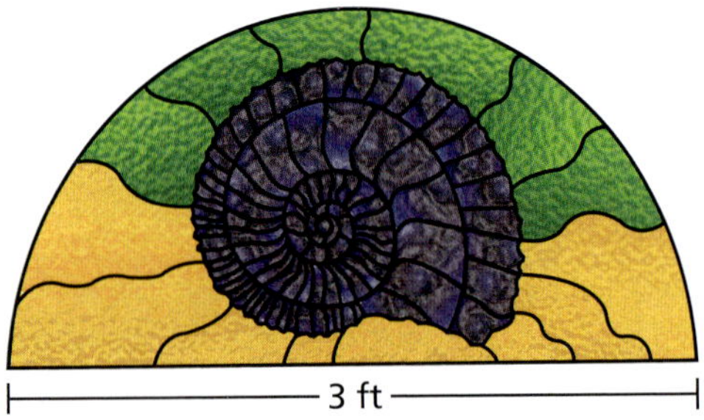

16.

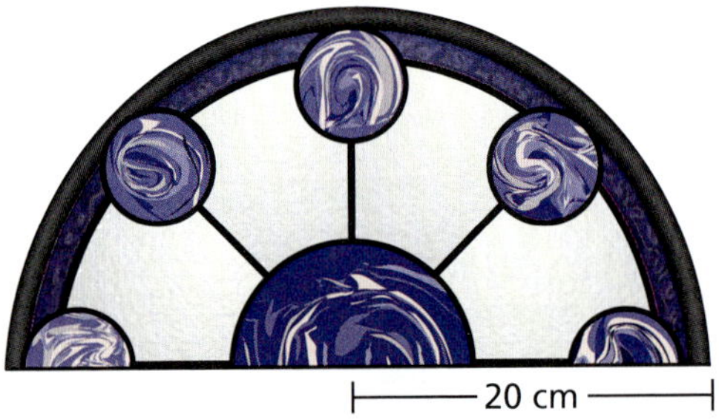

Find the circumferences of both circles.

17.

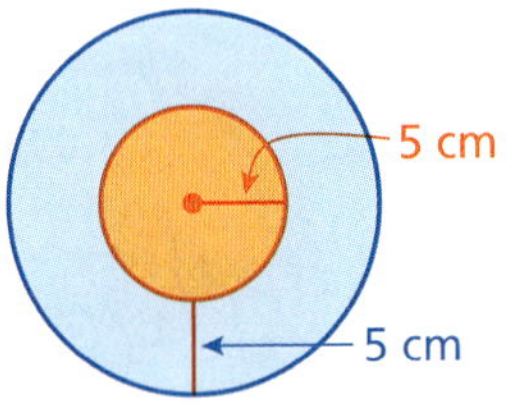

18.

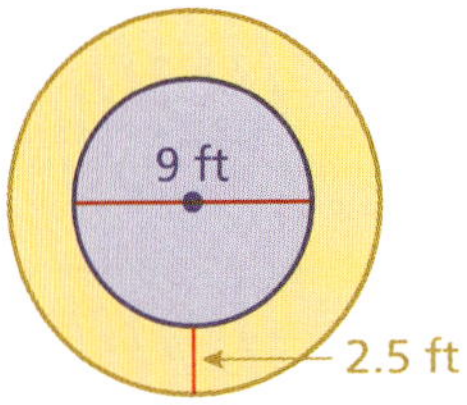

19.

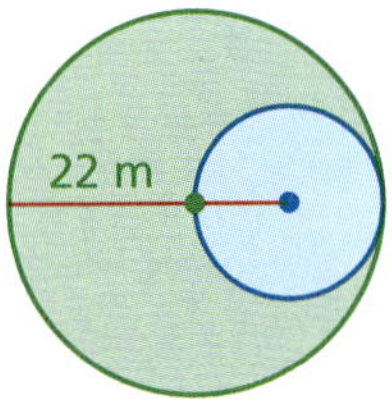

20. **STRUCTURE** Because the ratio $\frac{\text{circumference}}{\text{diameter}}$ is the same for every circle, is the ratio $\frac{\text{circumference}}{\text{radius}}$ the same for every circle? Explain.

21. **WIRE** A wire is bent to form four semicircles. How long is the wire?

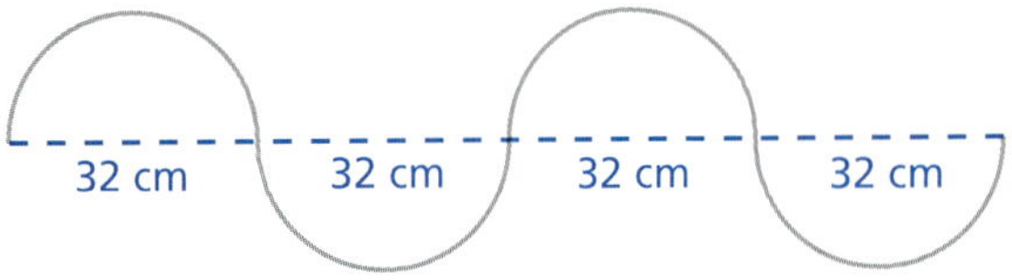

22. **CRITICAL THINKING** Explain how to draw a circle with a circumference of π^2 inches. Then draw the circle.

Common Errors

- **Exercise 14** Students may try to compare the radii or diameters without converting to the same units. Remind them that they need to have all the same units before comparing.
- **Exercises 15 and 16** Students may find the circumference and forget to divide it in half. Remind them that because it is not a whole circle, they must divide it in half.
- **Exercises 15 and 16** Students may forget to add the diameter onto the perimeter after they have found the circumference part. Remind them that the perimeter includes all the sides, not just the circular part.
- **Exercises 17–19** Students may use the incorrect radius or diameter for the larger or smaller circle. Remind them that they will need to figure out the radius or diameter of the other circle.

Practice and Problem Solving

13. **a.** about 25 m; about 50 m

b. about 2 times greater

14. **a.** D

b. B

15. about 7.71 ft

16. about 102.8 cm

17. about 31.4 cm; about 62.8 cm

18. about 28.26 ft; about 44 ft

19. about 69.08 m; about 138.16 m

20. yes; Because

$$\frac{\text{circumference}}{\text{radius}} = \frac{2\pi r}{r} = \frac{2\pi \not{r}}{\not{r}} = 2\pi,$$

the ratio is the same for every circle.

21. about 200.96 cm

22. The circle has a diameter of π inches, so use a diameter of about 3.1 inches.

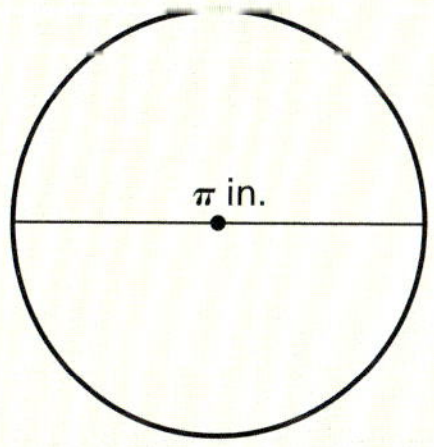

Differentiated Instruction

Auditory

For students that confuse radius and diameter, emphasize that a **di**ameter **di**vides a circle in half. In an exercise when the radius is needed in a formula and the diameter is given, emphasize that you **di**vide the **di**ameter to find the radius.

Practice and Problem Solving

23. See *Taking Math Deeper.*

24. **a.** small tire: about 127 rotations; large tire: about 38 rotations

b. *Sample answer:* A bicycle with large wheels would allow you to travel farther with each rotation of the pedal.

25. **a.** about 254.34 mm; First find the length of the minute hand. Then find $\frac{3}{4}$ of the circumference of a circle whose radius is the length of the minute hand.

b. See Additional Answers.

Fair Game Review

26. 22 ft **27.** 20 m

28. 65 in. **29.** D

Mini-Assessment

Find the circumference of the circle. Use 3.14 or $\frac{22}{7}$ for π.

1.

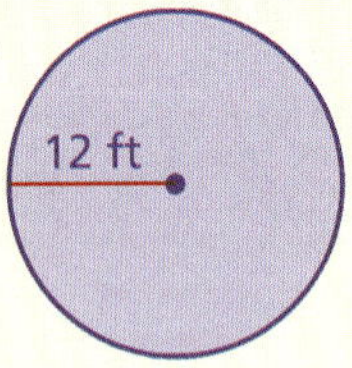

about 75.36 ft

2.

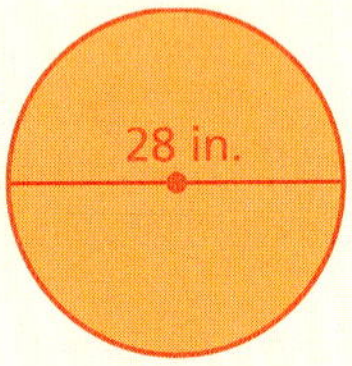

about 88 in.

3.

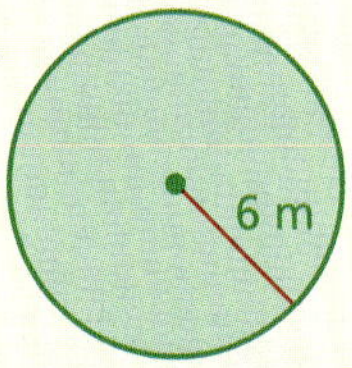

about 37.68 m

4.

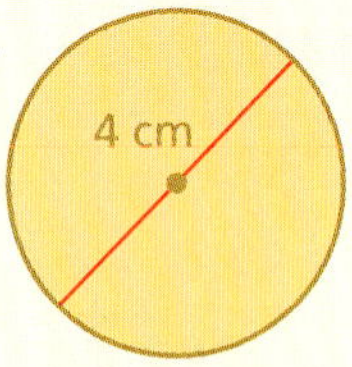

about 12.56 cm

5. Find the perimeter of the window.
about 20.56 ft

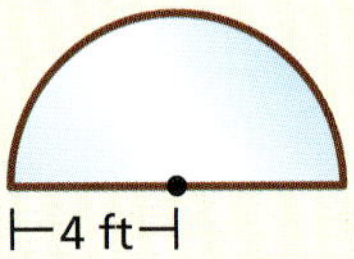

Taking Math Deeper

Exercise 23

Speed records for flights around the world are kept by Fédération Aéronautique Internationale. See *fai.org.* There are many different categories, depending on the type of plane and other factors.

1 **a.** Find the circumference of the Tropic of Cancer.

$$C = 2\pi r$$
$$= 11{,}708\pi$$
$$\approx 36{,}763 \text{ km}$$

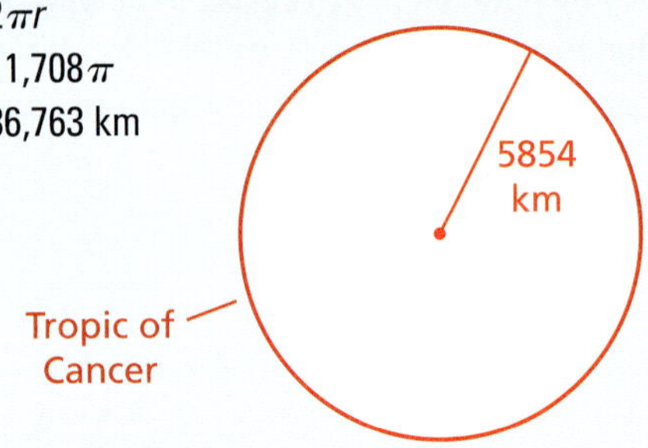

2 Here is a route for a recent attempt to break the speed record for a flight around the world.

3 **b.** One record was set by Claude Delorme of France. He flew around the world at an average speed of 1231 kilometers per hour (westbound).

At this rate, it took Claude about 30 hours to fly around the world. If your students estimate a time that is less than this, you might point out that they should try for the next world record!

Project

Draw a picture of the path you would take to fly around the world in an attempt to set a new world record.

Reteaching and Enrichment Strategies

If students need help. . .	If students got it. . .
Resources by Chapter • Practice A and Practice B • Puzzle Time Record and Practice Journal Practice Differentiating the Lesson Lesson Tutorials Skills Review Handbook	Resources by Chapter • Enrichment and Extension • Technology Connection Start the next section

23. **AROUND THE WORLD** "Lines" of latitude on Earth are actually circles. The Tropic of Cancer is the northernmost line of latitude at which the Sun appears directly overhead at noon. The Tropic of Cancer has a radius of 5854 kilometers.

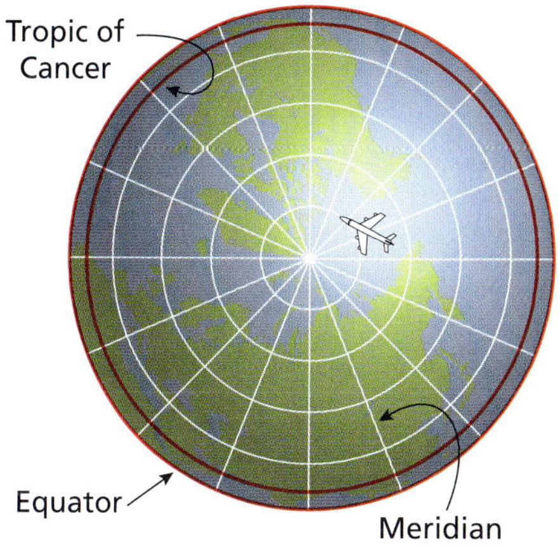

To qualify for an around-the-world speed record, a pilot must cover a distance no less than the circumference of the Tropic of Cancer, cross all meridians, and land on the same airfield where he started.

a. What is the minimum distance that a pilot must fly to qualify for an around-the-world speed record?

b. **RESEARCH** Estimate the time it would take for a pilot to qualify for the speed record.

24. **PROBLEM SOLVING** Bicycles in the late 1800s looked very different than they do today.

a. How many rotations does each tire make after traveling 600 feet? Round your answers to the nearest whole number.

b. Would you rather ride a bicycle made with two large wheels or two small wheels? Explain.

25. **Logic** The length of the minute hand is 150% of the length of the hour hand.

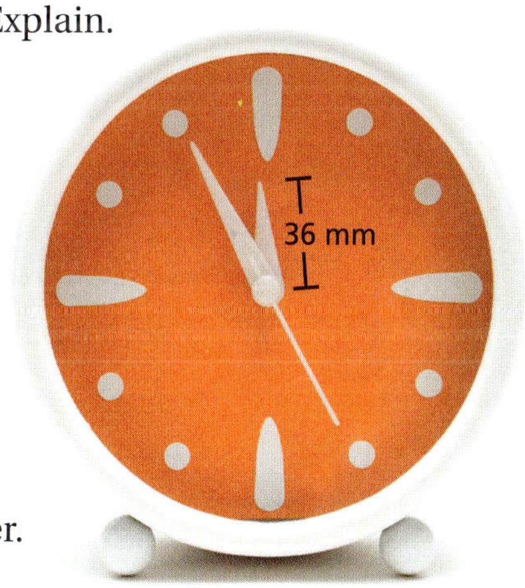

a. What distance will the tip of the minute hand move in 45 minutes? Explain how you found your answer.

b. In 1 hour, how much farther does the tip of the minute hand move than the tip of the hour hand? Explain how you found your answer.

Fair Game Review What you learned in previous grades & lessons

Find the perimeter of the polygon. *(Skills Review Handbook)*

26.

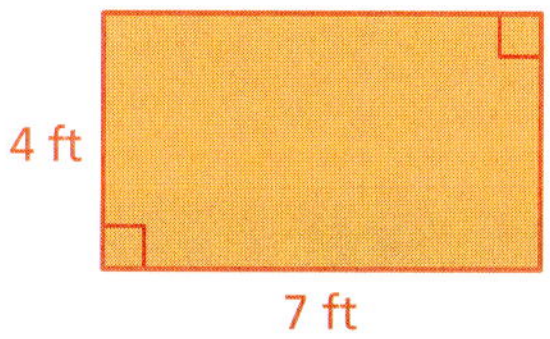

27.

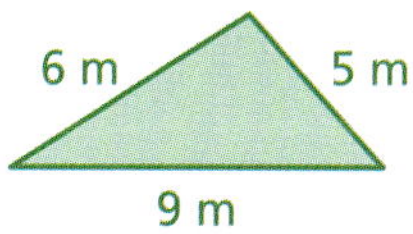

28.

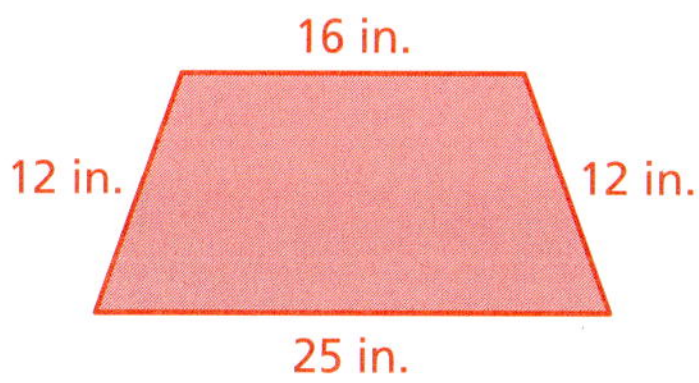

29. **MULTIPLE CHOICE** What is the median of the data set? *(Skills Review Handbook)*

12, 25, 16, 9, 5, 22, 27, 20

Ⓐ 7 Ⓑ 16 Ⓒ 17 Ⓓ 18

13.2 Perimeters of Composite Figures

Essential Question How can you find the perimeter of a composite figure?

1 ACTIVITY: Finding a Pattern

Work with a partner. Describe the pattern of the perimeters. Use your pattern to find the perimeter of the tenth figure in the sequence. (Each small square has a perimeter of 4.)

a. 

b. 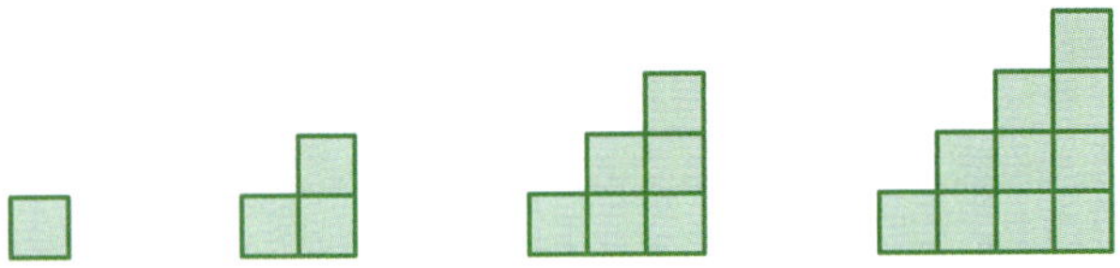

c. 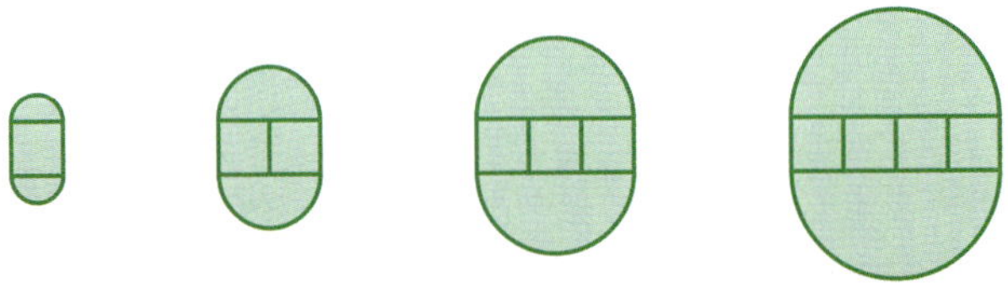

2 ACTIVITY: Combining Figures

Work with a partner.

a. A rancher is constructing a rectangular corral and a trapezoidal corral, as shown. How much fencing does the rancher need to construct both corrals?

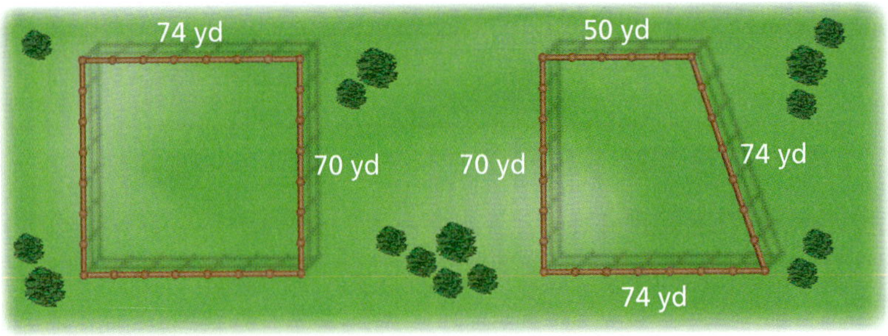

b. Another rancher is constructing one corral by combining the two corrals above, as shown. Does this rancher need more or less fencing? Explain your reasoning.

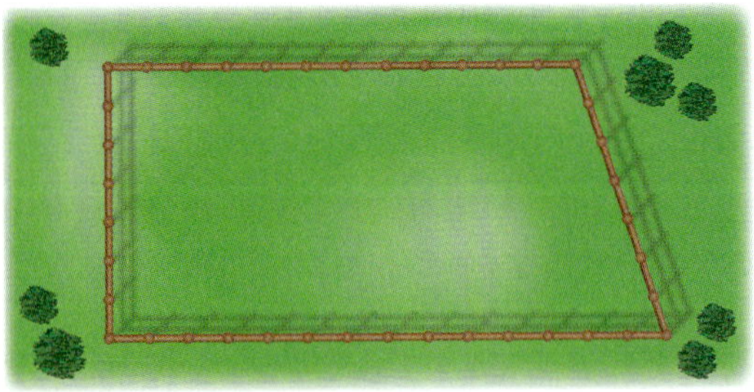

c. How can the rancher in part (b) combine the two corrals to use even less fencing?

Geometry
In this lesson, you will
- find perimeters of composite figures.

Laurie's Notes

Introduction

Applying Mathematical Practices

- **Construct Viable Arguments and Critique the Reasoning of Others:** In the first activity, students describe and generalize a pattern. The ability to offer evidence for their conjectures is good practice for students.

Motivate

- Draw the following "equations." Ask students how they would find the perimeter of the last figure knowing the dimensions of the figures on the left side of the equation.

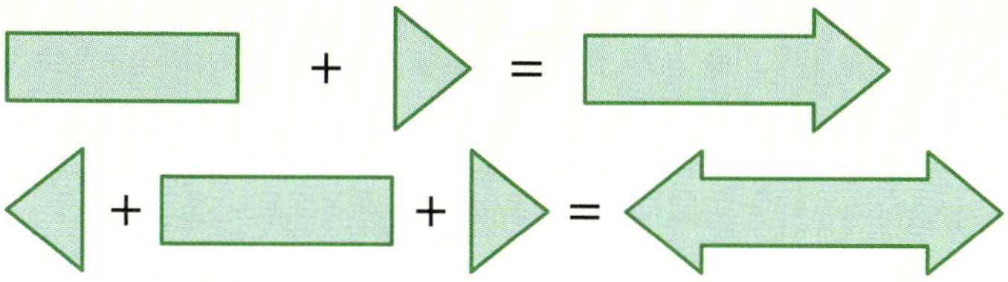

- These puzzles should help focus students' thinking on how they will find the perimeter of a composite figure, without defining composite figures!

Activity Notes

Activity 1

- Students should describe the pattern of the perimeter in words. For instance, the figures in part (a) have perimeters which are *increasing by 2* each time.
- **Common Error:** Remind students that perimeter is the distance *around* the figure, and that interior segments are not included.
- **Representation:** For part (c), if students leave their answers in terms of π, they are more likely to see the pattern.

? "How is the perimeter changing in part (a)? Can you explain why?"
increasing by 2; Explanations will vary.

? "How is the perimeter changing in part (b)? Can you explain why?"
increasing by 4; Explanations will vary.

? "How is the perimeter changing in part (c)? Can you explain why?"
increasing by π; Explanations will vary.

Activity 2

- It may not be obvious that the second rancher uses less fencing.
- Only two sides of the rectangular corral are labeled. Be sure that students have computed the perimeter correctly.
- **Model with Mathematics:** For some students it would be helpful to have cut-out pieces of the two shapes that they can manipulate. This helps them to see the perimeter that is lost when the pieces are moved together.
- **Construct Viable Arguments and Critique the Reasoning of Others:** Listen for student reasoning as they describe their solutions for part (c).
- **Extension:** Ask the students about the area enclosed by ranchers. Are they the same or different? Explain. The area is the same for both ranchers. The area does not change when the two polygons are combined.

What Your Students Will Learn

- Estimate perimeters of figures using grid paper.
- Find perimeters of composite figures using formulas.

Previous Learning

Students should know how to find the perimeter and circumference for common shapes.

Technology for the Teacher

Dynamic Classroom

Lesson Plans
Complete Materials List

13.2 Record and Practice Journal

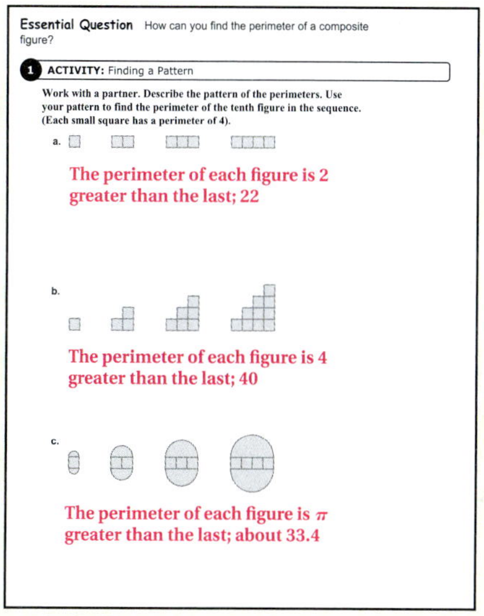

Essential Question How can you find the perimeter of a composite figure?

1 **ACTIVITY:** Finding a Pattern

Work with a partner. Describe the pattern of the perimeters. Use your pattern to find the perimeter of the tenth figure in the sequence. (Each small square has a perimeter of 4).

a. The perimeter of each figure is 2 greater than the last; 22

b. The perimeter of each figure is 4 greater than the last; 40

c. The perimeter of each figure is π greater than the last; about 33.4

English Language Learners

Visual

Have students find rectangular, trapezoidal, triangular, and circular regions in the school and on the school grounds. Then ask students to calculate the perimeter of these regions.

13.2 Record and Practice Journal

2 ACTIVITY: Combining Figures

Work with a partner.

a. A rancher is constructing a rectangular corral and a trapezoidal corral, as shown. How much fencing does the rancher need to construct both corrals?

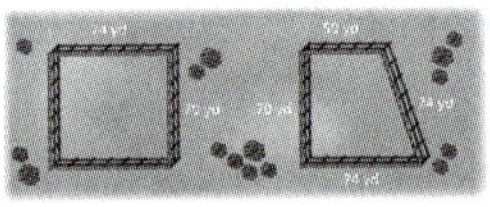

556 yards

b. Another rancher is constructing one corral by combining the two corrals above, as shown. Does this rancher need more or less fencing? Explain your reasoning.

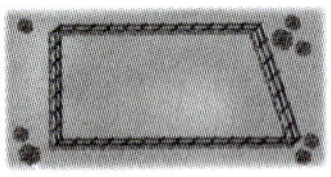

less

c. How can the rancher in part (b) combine the two corrals to use even less fencing?

Sample answer: Combine by stacking corrals so the 74-yd side is shared.

3 ACTIVITY: Submitting a Bid

Work with a partner. You want to bid on a tiling contract. You will be supplying and installing the tile that borders the swimming pool shown on the next page. In the figure, each grid square represents 1 square foot.

- **Your cost for the tile is $4 per linear foot.**
- **It takes about 15 minutes to prepare, install, and clean each foot of tile.**

a. How many tiles do you need for the border?

133 tiles

b. Write a bid for how much you will charge to supply and install the tile. Include what you want to charge as an hourly wage. Estimate what you think your profit will be. Check students' work.

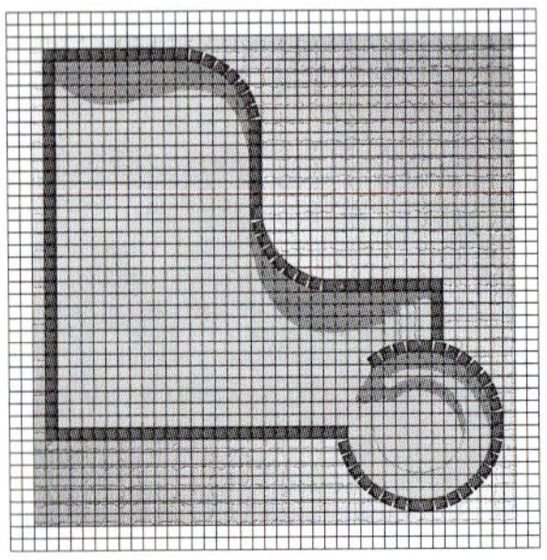

What Is Your Answer?

4. IN YOUR OWN WORDS How can you find the perimeter of a composite figure? Use a semicircle, a triangle, and a parallelogram to draw a composite figure. Label the dimensions. Find the perimeter of the figure.

Check students' work.

Laurie's Notes

Activity 3

- **Model with Mathematics:** Share with your students that builders and contractors submit bids for work that they want to do. If more than one bid is received, the consumer selects the builder or contractor based upon a number of factors, one of which is the cost that is quoted.
- In this problem, assume that each pair of students is bidding on the job. They could even have a name for their two-person company. Their bid sheet (work done) should be neat and organized and easily understood by the pool's owner—you!
- Explain the term *$4 per linear foot* so that all students understand. In construction, when the width of material (in this case, the tile) is predetermined, the cost is given in terms of the length, not in terms of area (square feet).
- Students should count the number of tiles surrounding the pool and use the scale given on the diagram to determine how much tile is needed.
- Make sure students include a labor charge based on the information given. Students will set their own hourly wage. Is it realistic?
- Discuss general results with the whole class. Compare the quotes, separating out the material and labor.
- **Extension:** List the hourly wages and the number of hours of labor that each company charges. Use this data and find the mean, median, and range of the hourly wages and the total labor charge.

What Is Your Answer?

- **Neighbor Check:** Have students work independently and then have their neighbors check their work. Have students discuss any discrepancies.

Closure

- Find the perimeter of the arrow. 60 cm

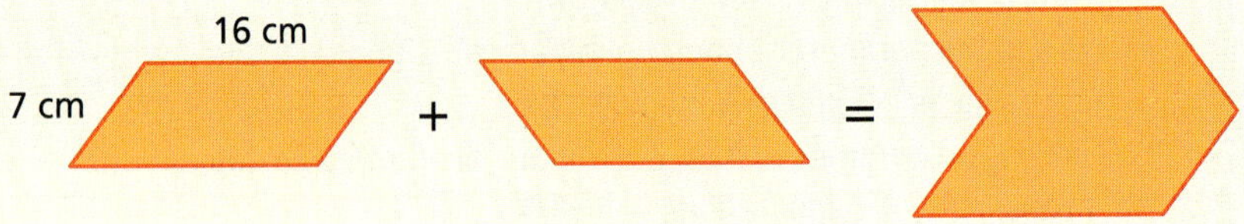

3 ACTIVITY: Submitting a Bid

Work with a partner. You want to bid on a tiling contract. You will be supplying and installing the brown tile that borders the swimming pool. In the figure, each grid square represents 1 square foot.

- **Your cost for the tile is $4 per linear foot.**
- **It takes about 15 minutes to prepare, install, and clean each foot of tile.**

a. How many brown tiles do you need for the border?

b. Write a bid for how much you will charge to supply and install the tile. Include what you want to charge as an hourly wage. Estimate what you think your profit will be.

Math Practice

Communicate Precisely

What do you need to include to create an accurate bid? Explain.

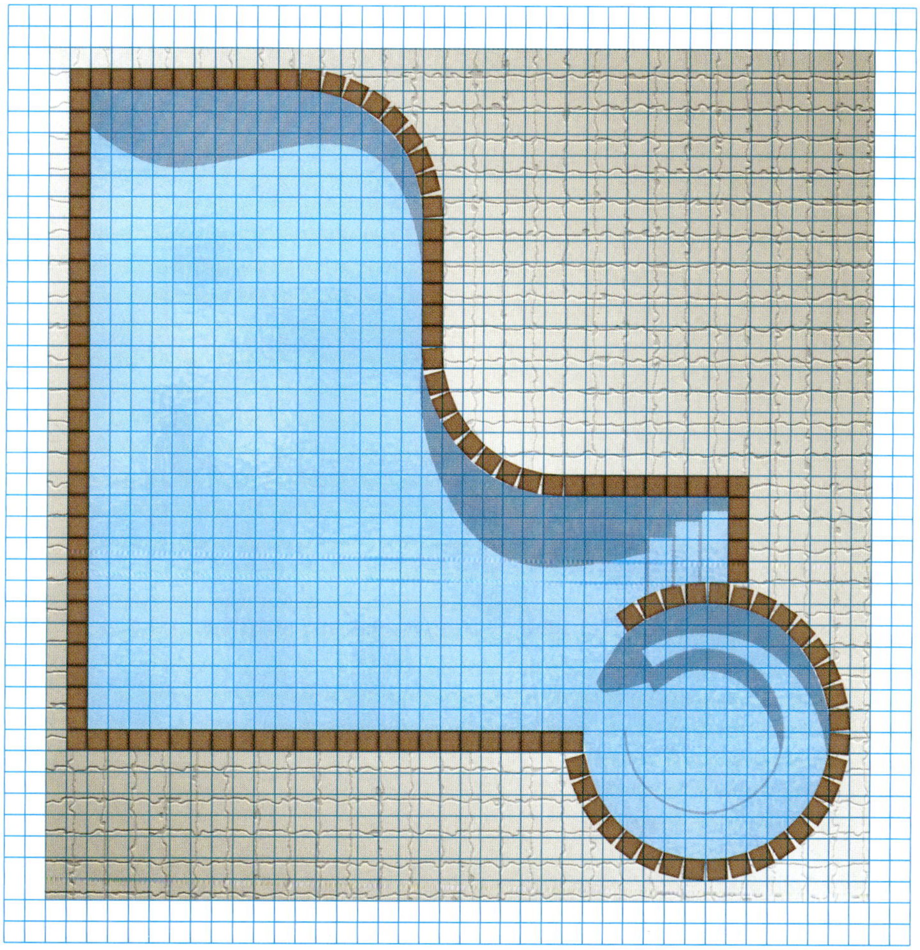

What Is Your Answer?

4. **IN YOUR OWN WORDS** How can you find the perimeter of a composite figure? Use a semicircle, a triangle, and a parallelogram to draw a composite figure. Label the dimensions. Find the perimeter of the figure.

Practice Use what you learned about perimeters of composite figures to complete Exercises 3–5 on page 560.

13.2 Lesson

Check It Out
Lesson Tutorials
BigIdeasMath.com

Key Vocabulary
composite figure, *p. 558*

A **composite figure** is made up of triangles, squares, rectangles, semicircles, and other two-dimensional figures. Here are two examples.

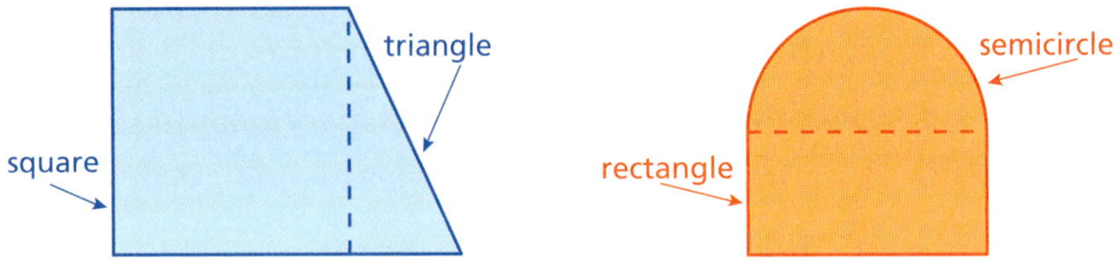

To find the perimeter of a composite figure, find the distance around the figure.

EXAMPLE 1 Estimating a Perimeter Using Grid Paper

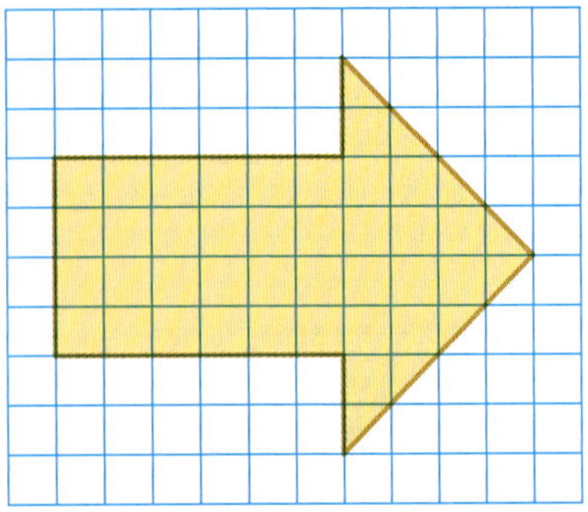

Estimate the perimeter of the arrow.

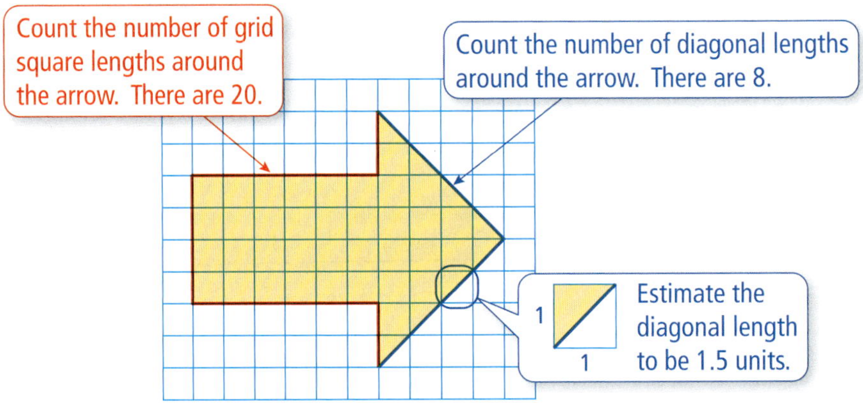

Length of 20 grid square lengths: $20 \times 1 = 20$ units

Length of 8 diagonal lengths: $8 \times 1.5 = 12$ units

So, the perimeter is about $20 + 12 = 32$ units.

On Your Own

Now You're Ready
Exercises 3–8

Estimate the perimeter of the figure.

1.

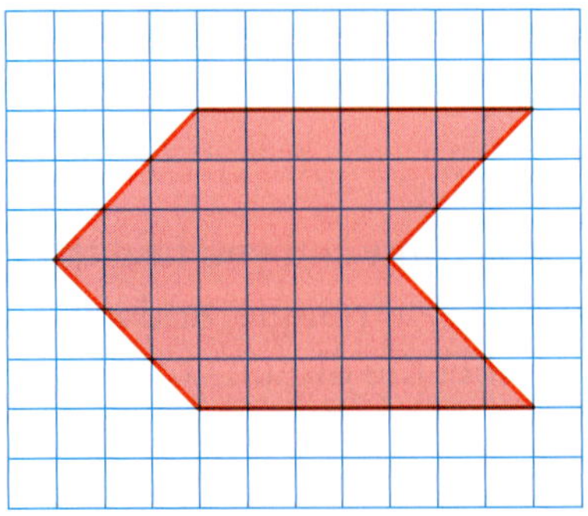

2.

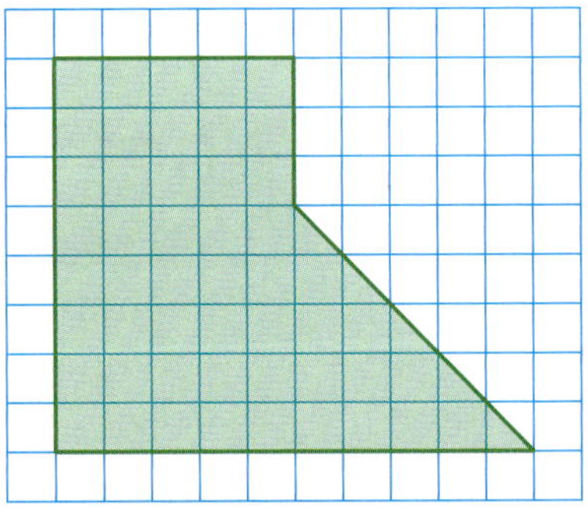

Multi-Language Glossary at BigIdeasMath.com

Laurie's Notes

Introduction

Connect

- **Yesterday:** Students used a variety of problem-solving skills to find the perimeter of several composite figures.
- **Today:** Students will use formulas to find the perimeter of composite figures.

Motivate

- If you have a set of tangrams, use them to share a puzzle with students. Place the 7 pieces on the overhead and ask a volunteer to rearrange the pieces into a square. The solution is shown. Then rearrange the pieces to form the bird shown. Ask how the perimeter changes from the square to the bird.

- The goal is not to find the perimeter of the bird, only to recognize that as more segments are exposed (on the perimeter), the perimeter will increase.
- The figures that students work with today are composed of common geometric shapes. Students should be familiar with how to find the perimeter of each.
- Introduce the vocabulary word *composite figure.*

Lesson Notes

Example 1

- The arrow is a shape that students discussed at the beginning of yesterday's lesson.
- Students are asked to estimate the perimeter, because the slanted lengths are irrational numbers.
- **Common Error:** Students may think the diagonal of the square is 1 unit because the side lengths are 1 unit. Have them think about the *shortcut* of walking across the diagonal of a square versus walking the two side lengths.

On Your Own

- Question 1 is very similar to the first example.
- Ask a volunteer to share his or her thinking and solution to the problem.

Goal Today's lesson is finding the perimeters of **composite figures**.

Technology for the Teacher

Dynamic Classroom

Lesson Tutorials
Lesson Plans
Answer Presentation Tool

Extra Example 1

Estimate the perimeter of the arrow.

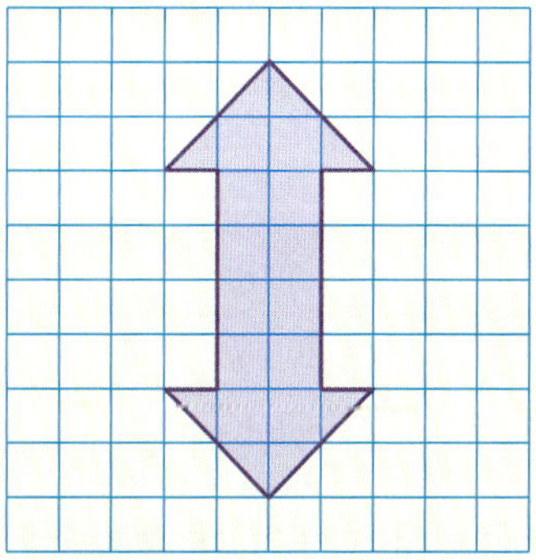

about 24 units

On Your Own

1. about 32 units
2. about 33.5 units

Extra Example 2

The figure is made of a semicircle and a square. Find the perimeter.

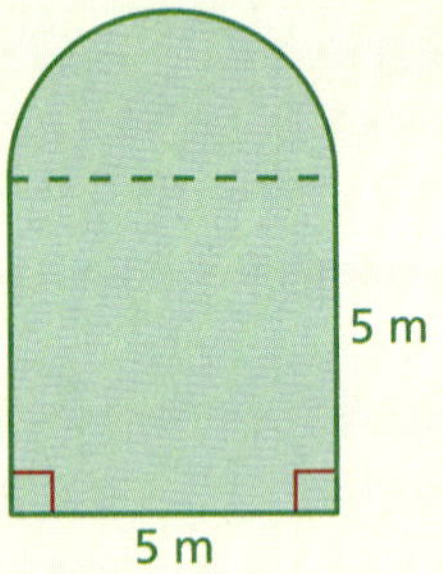

about 22.85 m

Extra Example 3

Find the perimeter of the running track in Example 3 when the radius of the semicircle is 20 yards and the length of the rectangle is 50 yards. about 225.6 yd

On Your Own

3. about 74.82 cm
4. about 41.12 m

Differentiated Instruction

Auditory

Review the two formulas for circumference, $C = \pi d$ and $C = 2\pi r$. On the board or overhead, derive the formulas for diameter, $d = \frac{C}{\pi}$, and radius, $r = \frac{C}{2\pi}$. Have students copy the four formulas in their notebooks to use as a quick reference when doing word problems. Remind students that the circumference of a semicircle is half of a full circle.

Laurie's Notes

Example 2

- Draw the diagram for the problem.
- ? "What is the diameter of the semicircle?" 10 ft
- Explain to students that although the third side of the triangle is shown (which is also the diameter of the circle), it is not part of the perimeter.
- ? **Construct Viable Arguments and Critique the Reasoning of Others** and **Attend to Precision:** "Could $\frac{22}{7}$ be used for π?" yes "Why do you think 3.14 was used?" The diameter is 10, which is not a multiple of 7.

Example 3

- Draw the diagram for the problem.
- ? "What are the dimensions of the rectangle?" 100 m by 64 m
- ? "What is the diameter of the semicircles?" 64 m
- Make sure students understand that they have to add both straightaways of the running track. This is where the 100 + 100 comes from in the answer statement.
- Point out to students that in Examples 2 and 3 the answers are stated to be *about* a certain number. The answers are approximations because 3.14 is only an approximation for π.

On Your Own

- Students may think there is not enough information to solve Question 4. The quadrilateral is a square, so the diameter of the semicircles is 8 meters.
- **Neighbor Check:** Have students work independently and then have their neighbors check their work. Have students discuss any discrepancies.
- When students have finished, ask volunteers to show their work at the board.

Closure

- Find the perimeter of the room shown. 64 ft

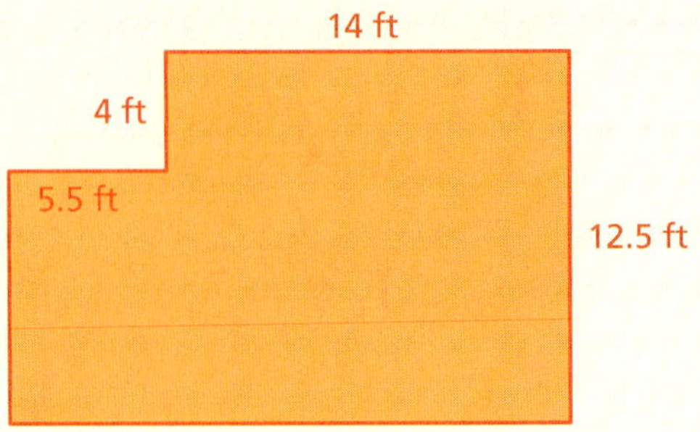

EXAMPLE 2 Finding a Perimeter

The figure is made up of a semicircle and a triangle. Find the perimeter.

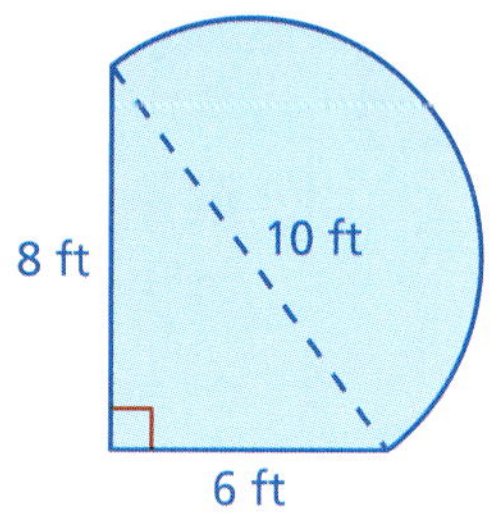

The distance around the triangular part of the figure is $6 + 8 = 14$ feet.

The distance around the semicircle is one-half the circumference of a circle with a diameter of 10 feet.

$\frac{C}{2} = \frac{\pi d}{2}$ Divide the circumference by 2.

$\approx \frac{3.14 \cdot 10}{2}$ Substitute 3.14 for π and 10 for d.

$= 15.7$ Simplify.

So, the perimeter is about $14 + 15.7 = 29.7$ feet.

EXAMPLE 3 Finding a Perimeter

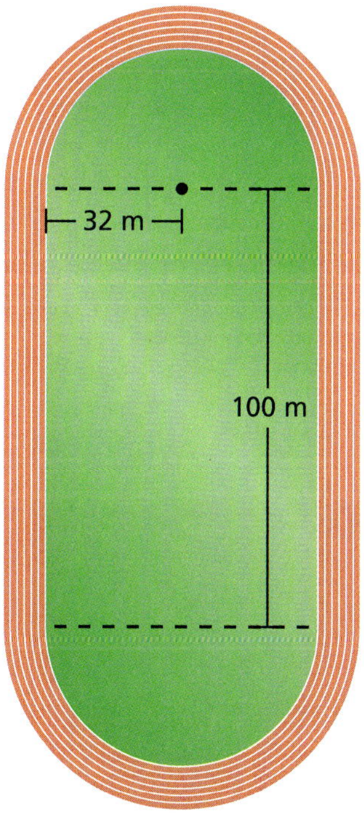

The running track is made up of a rectangle and two semicircles. Find the perimeter.

The semicircular ends of the track form a circle with a radius of 32 meters. Find its circumference.

$C = 2\pi r$ Write formula for circumference.

$\approx 2 \cdot 3.14 \cdot 32$ Substitute 3.14 for π and 32 for r.

$= 200.96$ Multiply.

So, the perimeter is about $100 + 100 + 200.96 = 400.96$ meters.

Now You're Ready
Exercises 9–11

On Your Own

3. The figure is made up of a semicircle and a triangle. Find the perimeter.

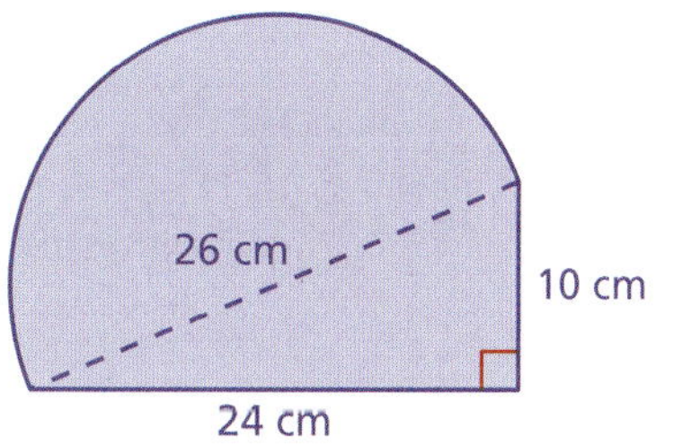

4. The figure is made up of a square and two semicircles. Find the perimeter.

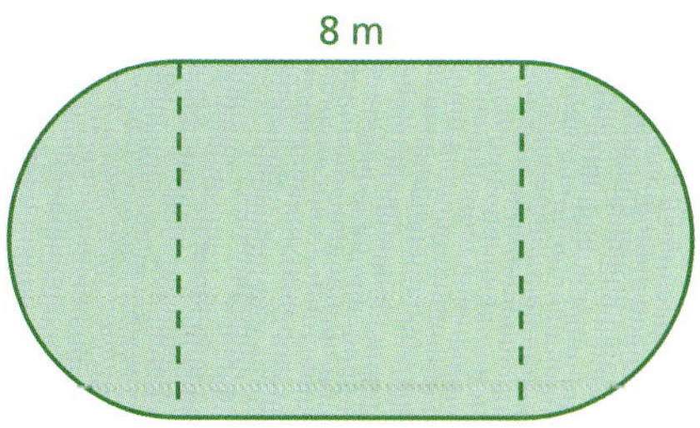

13.2 Exercises

Check It Out
Help with Homework
BigIdeasMath.com

Vocabulary and Concept Check

1. **REASONING** Is the perimeter of the composite figure equal to the sum of the perimeters of the individual figures? Explain.

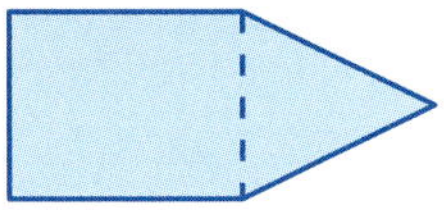

2. **OPEN-ENDED** Draw a composite figure formed by a parallelogram and a trapezoid.

Practice and Problem Solving

Estimate the perimeter of the figure.

1 3. 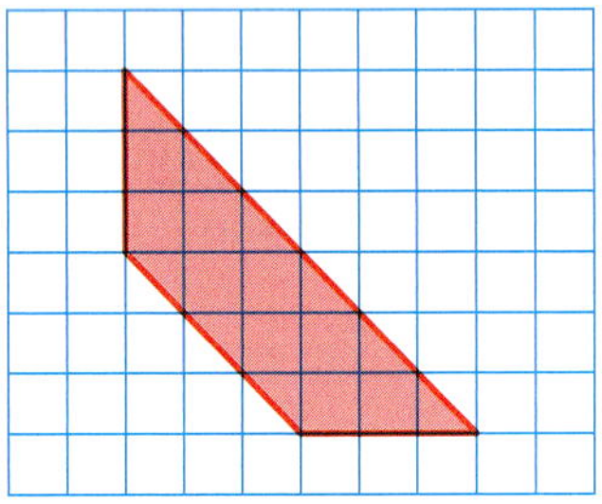

4. 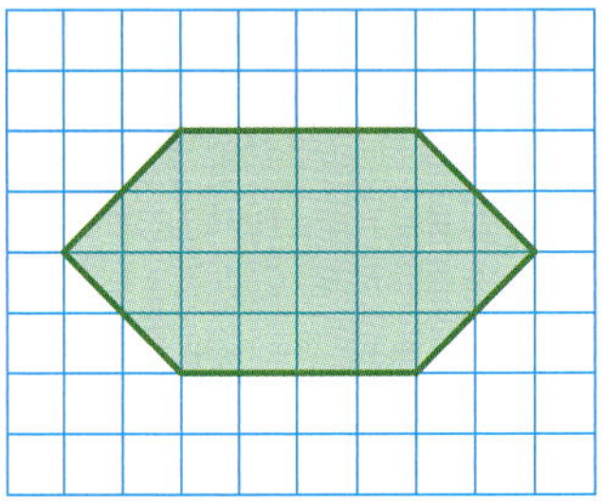

5. 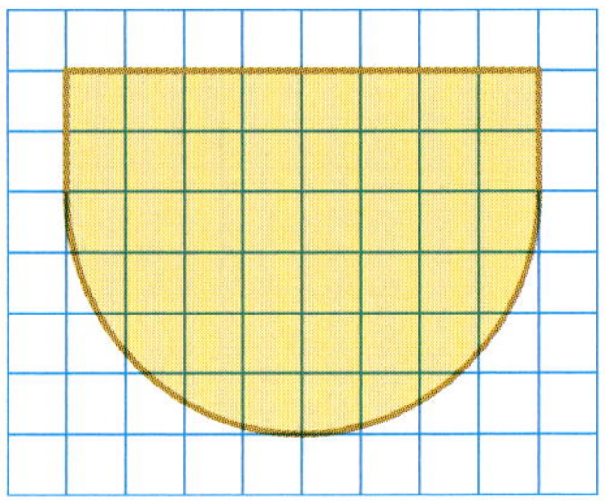

6. 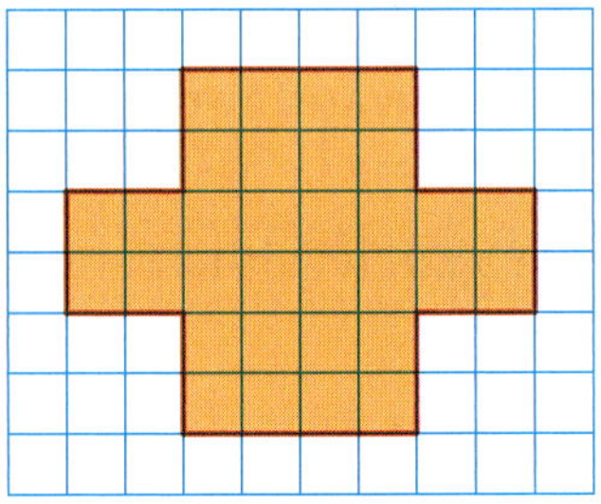

7. 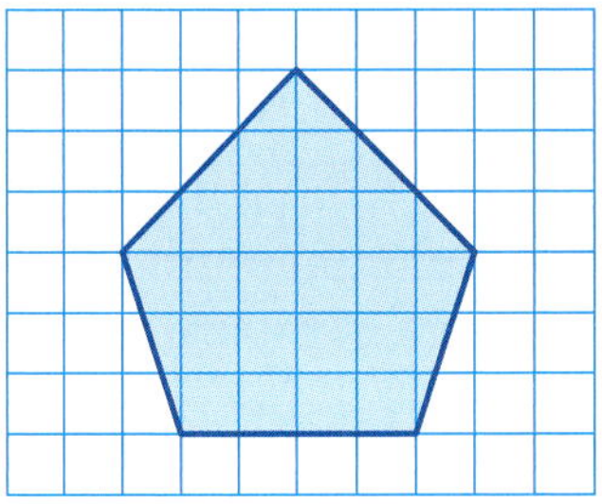

8. 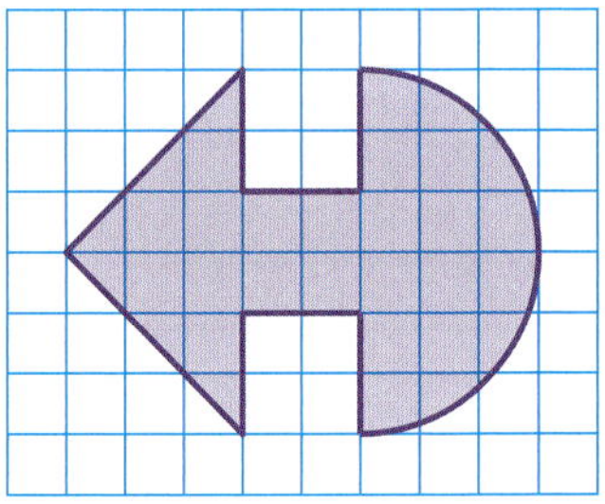

Find the perimeter of the figure.

2 9. 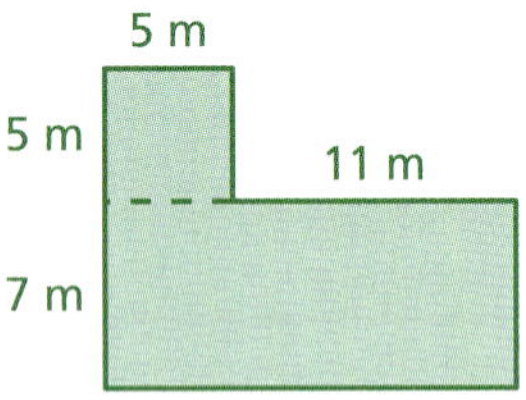

10. 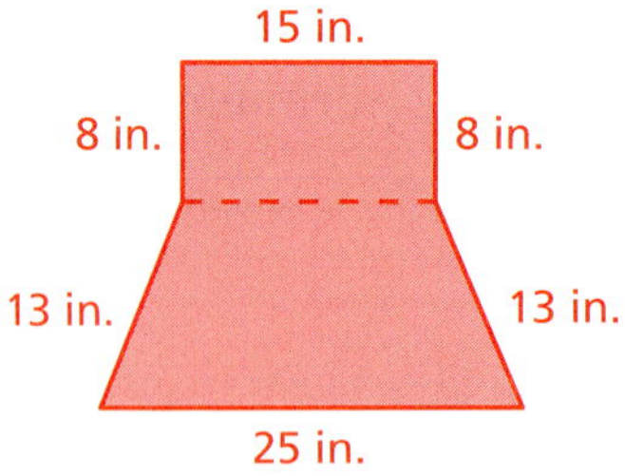

11. 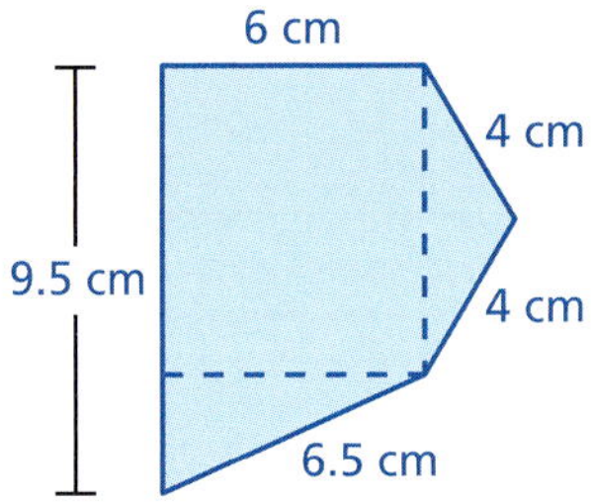

12. **ERROR ANALYSIS** Describe and correct the error in finding the perimeter of the figure.

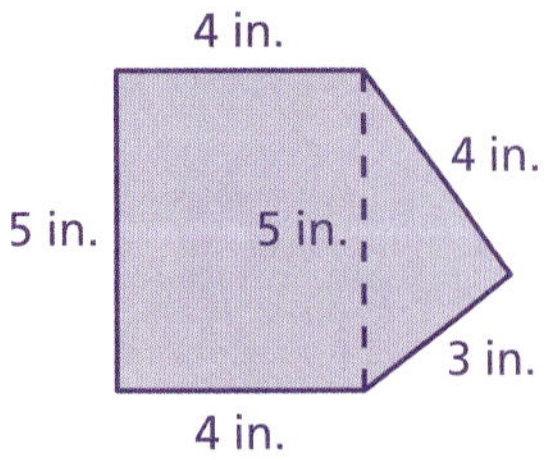

✗ Perimeter $= 4 + 3 + 4 + 5 + 4 + 5$

$= 25$ in.

Assignment Guide and Homework Check

Level	Assignment	Homework Check
Advanced	1–5, 8–14 even, 17–24	8, 10, 12, 14

Common Errors

- **Exercises 1, 9–11, 13–15** Students may include the dotted lines in their calculation of the perimeter. Remind them that the dotted lines are for reference and sometimes give information to find another length. Only the outside lengths are counted.
- **Exercises 3–8** Students may count the sides of the squares inside the figure. Remind them that they are finding the perimeter, so they only need to count the outside lengths.
- **Exercises 5 and 8** Students may have difficulty estimating the curved portions of the figure. Give students tracing paper and tell them to trace the line as straight instead of curved and compare it with the length of the side of a square to help them estimate the length.

13.2 Record and Practice Journal

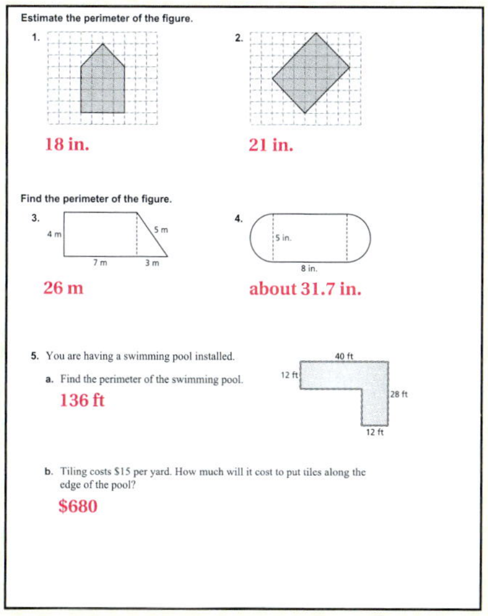

Estimate the perimeter of the figure.

1. 18 in.

2. 21 in.

Find the perimeter of the figure.

3. 26 m

4. about 31.7 in.

5. You are having a swimming pool installed.

   a. Find the perimeter of the swimming pool.
   136 ft

   b. Tiling costs $15 per yard. How much will it cost to put tiles along the edge of the pool?
   $680

Vocabulary and Concept Check

1. no; The perimeter of the composite figure does not include the measure of the shared side.
2. *Sample answer:*

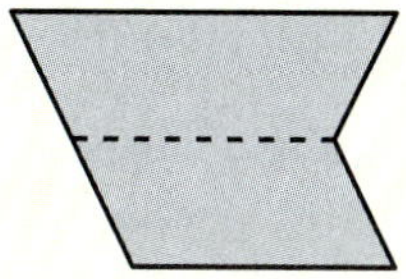

Practice and Problem Solving

3. 19.5 units
4. 20 units
5. 25.5 units
6. 28 units
7. 19 units
8. 30 units
9. 56 m
10. 82 in.
11. 30 cm
12. The length of the rectangle was counted twice.
    Perimeter $= 4 + 3 + 4 + 5 + 4$
    $= 20$ in.

Practice and Problem Solving

13. about 26.85 in.

14. about 50.26 in.

15. about 36.84 ft

16. $16,875

17. See *Taking Math Deeper.*

18. See Additional Answers.

19. *Sample answer:* By adding the triangle shown by the dashed line to the L-shaped figure, you *reduce* the perimeter.

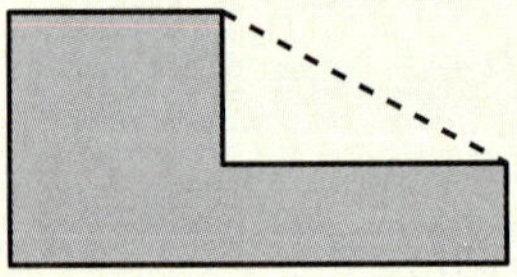

Fair Game Review

20. 19.35 **21.** 279.68

22. 153.86 **23.** 205

24. D

Mini-Assessment

Find the perimeter of the figure.

1.

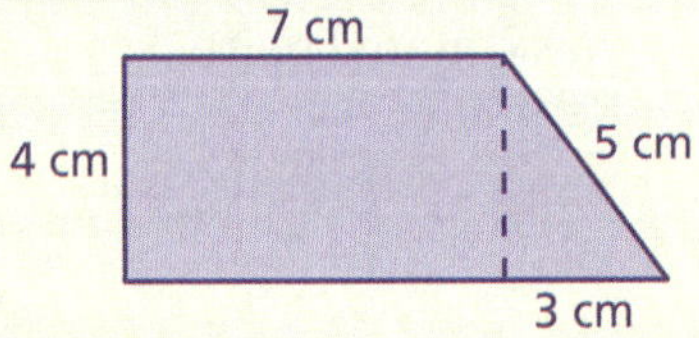

26 cm

2.

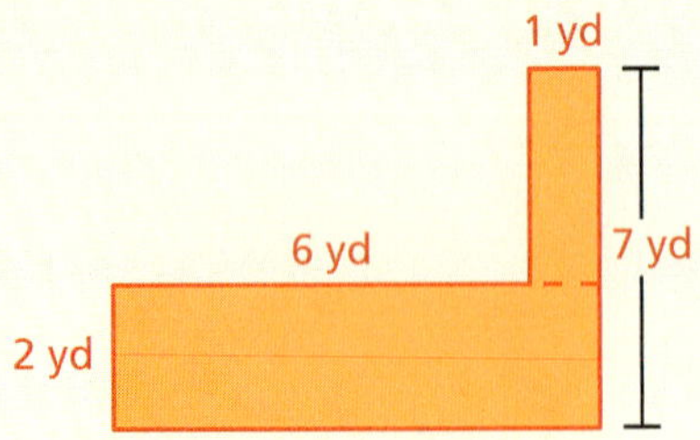

28 yd

3. Find the perimeter of the garden.

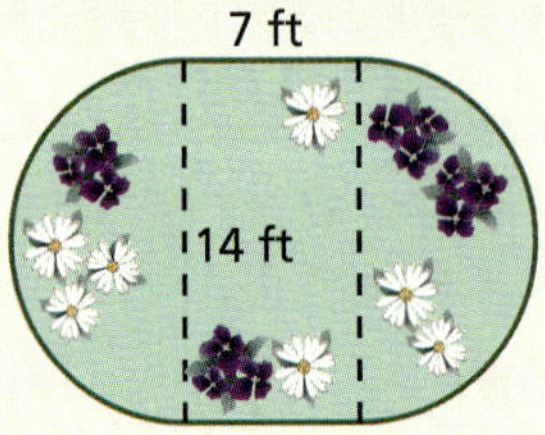

about 58 ft

Taking Math Deeper

Exercise 17

Notice that the dimensions of the field and the distance in the unit rate are multiples of 3. So, you can convert feet to yards and use smaller numbers in the calculations.

1 Convert the baseball field dimensions to yards and your running rate to yards per second.

Radius: $225\ \cancel{\text{ft}} \times \dfrac{1\text{ yd}}{3\ \cancel{\text{ft}}} = 75\text{ yd}$

Sides: $300\ \cancel{\text{ft}} \times \dfrac{1\text{ yd}}{3\ \cancel{\text{ft}}} = 100\text{ yd}$

Rate: $\dfrac{9\ \cancel{\text{ft}}}{1\text{ sec}} \times \dfrac{1\text{ yd}}{3\ \cancel{\text{ft}}} = \dfrac{3\text{ yd}}{1\text{ sec}}$

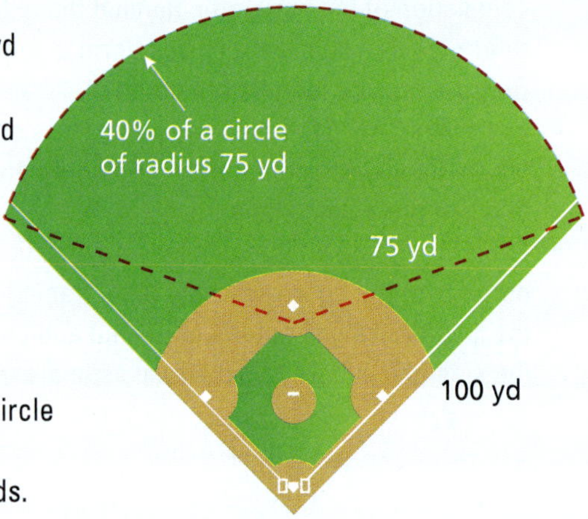

2 Find the distance around the field.

The circumference C of a circle with a radius of 75 yards is $C \approx 2 \cdot 3.14 \cdot 75 = 471$ yards.

The distance around the curved part of the field is 40% of this circumference. The other two sides of the field are 100 yards each. So, the perimeter of the field is about $100 + 100 + 0.4(471) = 388.4$ yards.

3 Divide the distance by the rate to find the time.

$$388.4\text{ yd} \div \frac{3\text{ yd}}{1\text{ sec}} = 388.4\ \cancel{\text{yd}} \times \frac{1\text{ sec}}{3\ \cancel{\text{yd}}} \approx 129.5\text{ sec}$$

It takes you about 129.5 seconds, or about 2 minutes and 10 seconds to run around the baseball field.

Project

Research the dimensions of several different baseball fields. Estimate how long it would take you to run around the perimeter of each field.

Reteaching and Enrichment Strategies

If students need help. . .	If students got it. . .
Resources by Chapter • Practice A and Practice B • Puzzle Time Record and Practice Journal Practice Differentiating the Lesson Lesson Tutorials Skills Review Handbook	Resources by Chapter • Enrichment and Extension • Technology Connection Start the next section

Find the perimeter of the figure.

13.

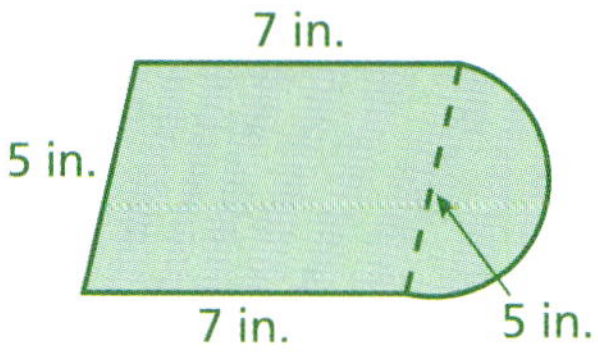

14.

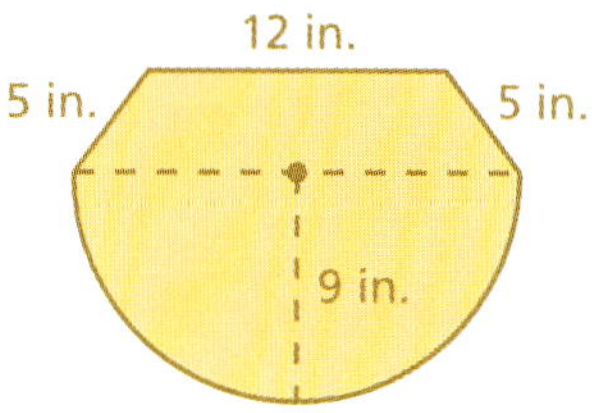

15.

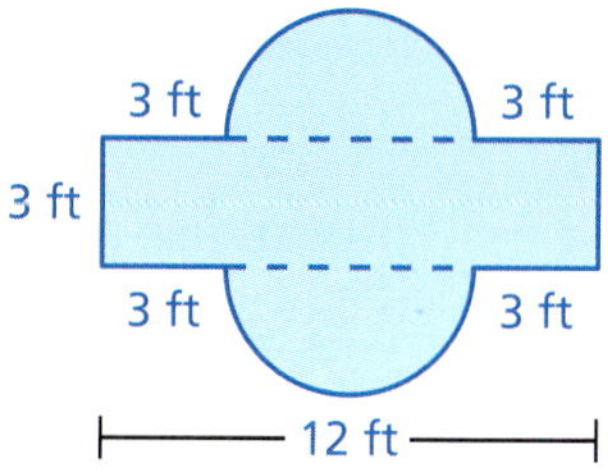

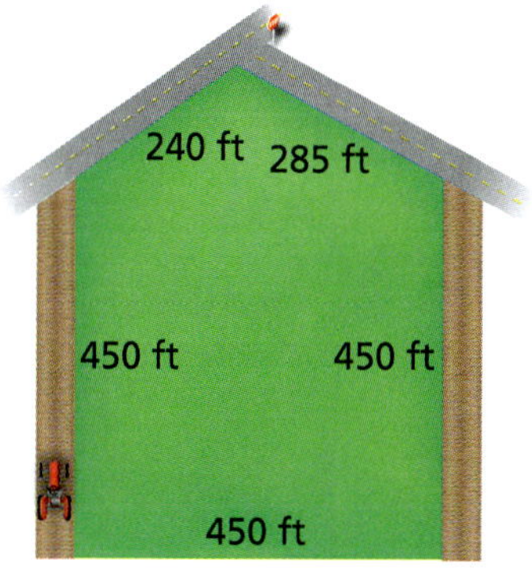

16. PASTURE A farmer wants to fence a section of land for a horse pasture. Fencing costs \$27 per yard. How much will it cost to fence the pasture?

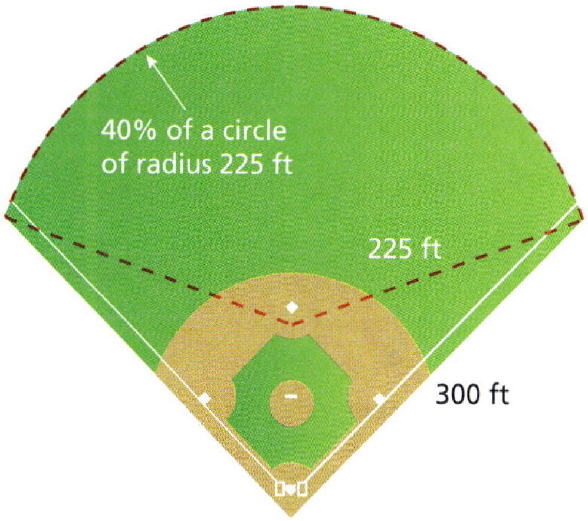

17. BASEBALL You run around the perimeter of the baseball field at a rate of 9 feet per second. How long does it take you to run around the baseball field?

18. TRACK In Example 3, the running track has six lanes. Explain why the starting points for the six runners are staggered. Draw a diagram as part of your explanation.

19. Critical Thinking How can you add a figure to a composite figure without increasing its perimeter? Draw a diagram to support your answer.

Fair Game Review What you learned in previous grades & lessons

Evaluate the expression. *(Skills Review Handbook)*

20. $2.15(3)^2$ **21.** $4.37(8)^2$ **22.** $3.14(7)^2$ **23.** $8.2(5)^2$

24. MULTIPLE CHOICE Which expression is equivalent to $(5y + 4) - 2(7 - 2y)$? *(Skills Review Handbook)*

Ⓐ $y - 10$ Ⓑ $9y + 18$ Ⓒ $3y - 10$ Ⓓ $9y - 10$

13 Study Help

Check It Out
Graphic Organizer
BigIdeasMath.com

You can use a **word magnet** to organize formulas or phrases that are associated with a vocabulary word or term. Here is an example of a word magnet for circle.

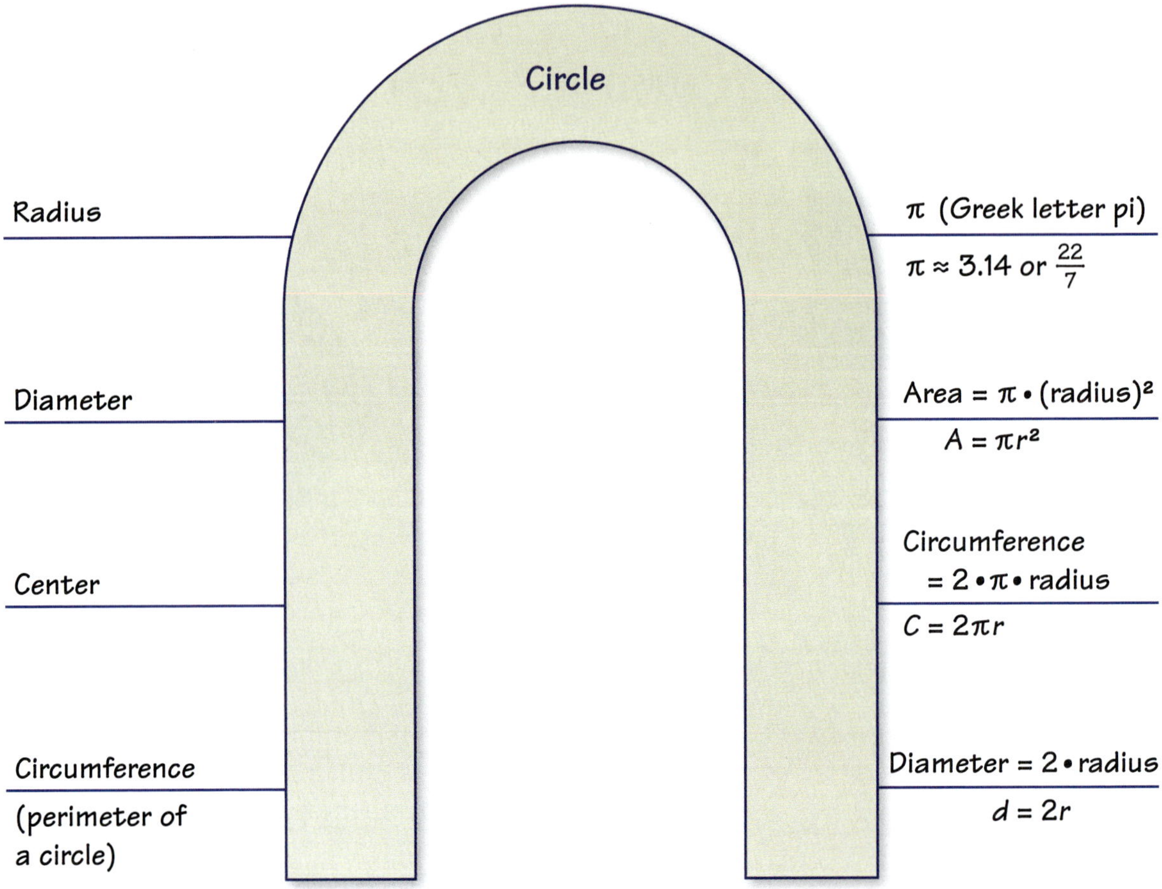

On Your Own

Make word magnets to help you study these topics.

1. semicircle
2. composite figure
3. perimeter

After you complete this chapter, make word magnets for the following topics.

4. area of a circle
5. area of a composite figure

"I'm trying to make a word magnet for happiness, but I can only think of two words."

Sample Answers

1.

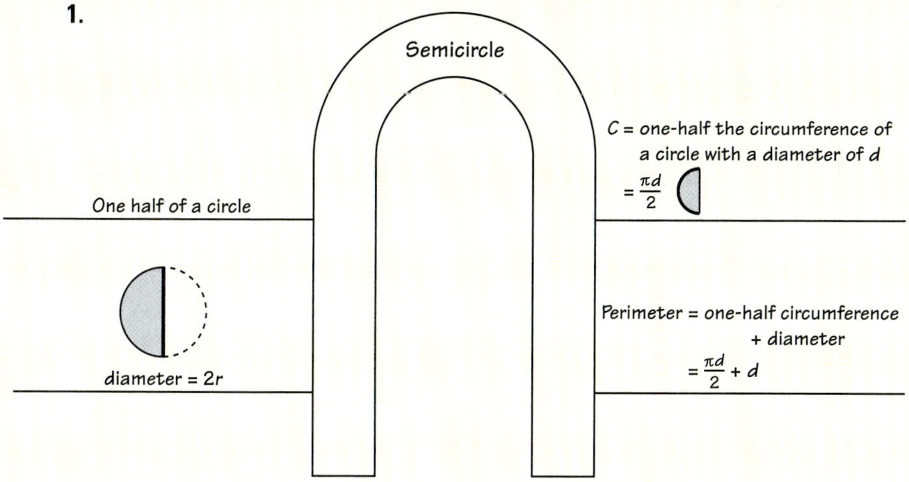

2.

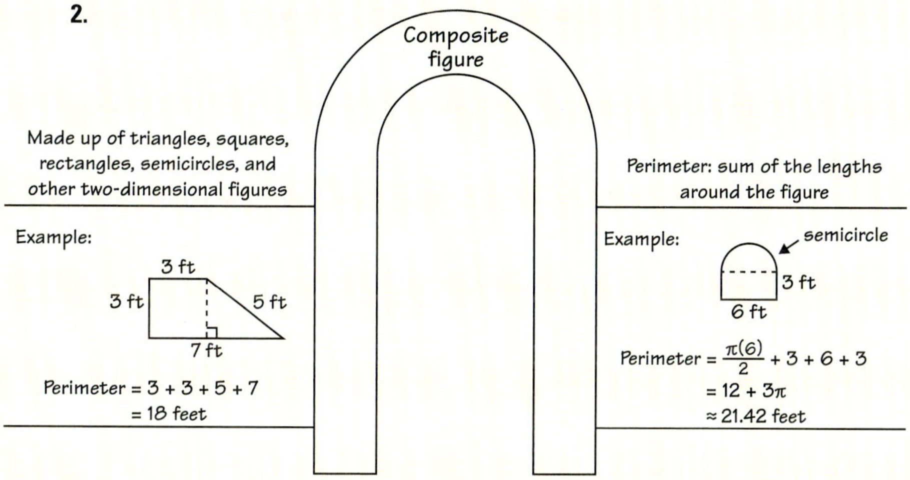

3.

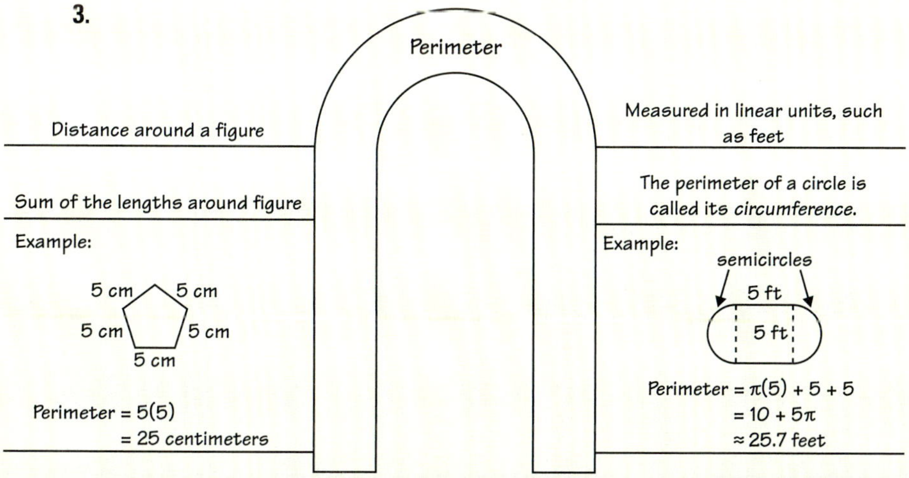

List of Organizers

Available at *BigIdeasMath.com*

Comparison Chart
Concept Circle
Definition (Idea) and Example Chart
Example and Non-Example Chart
Formula Triangle
Four Square
Information Frame
Information Wheel
Notetaking Organizer
Process Diagram
Summary Triangle
Word Magnet
Y Chart

About this Organizer

A **Word Magnet** can be used to organize information associated with a vocabulary word or term. As shown, students write the word or term inside the magnet. Students write associated information on the blank lines that "radiate" from the magnet. Associated information can include, but is not limited to: other vocabulary words or terms, definitions, formulas, procedures, examples, and visuals. This type of organizer serves as a good summary tool because any information related to a topic can be included.

Technology for the Teacher

Editable Graphic Organizer

Answers

1. 18 cm
2. 22 in.
3. 18 units
4. 21.4 units
5. 21 units
6. about 37.68 mm
7. about 4.71 ft
8. about 22 cm
9. 88 in.
10. about 49.12 ft
11. about 7.71 ft
12. about 25.12 mm
13. 60 ft
14. about 15.68 in.

Technology for the Teacher

Online Assessment
Assessment Book
ExamView® Assessment Suite

Alternative Quiz Ideas

100% Quiz	**Math Log**
Error Notebook	Notebook Quiz
Group Quiz	Partner Quiz
Homework Quiz	Pass the Paper

Math Log

Ask students to keep a math log for the chapter. Have them include diagrams, definitions, and examples. Everything should be clearly labeled. It might be helpful if they put the information in a chart. Students can add to the log as new topics are introduced.

Reteaching and Enrichment Strategies

If students need help. . .	If students got it. . .
Resources by Chapter • Practice A and Practice B • Puzzle Time Lesson Tutorials *BigIdeasMath.com*	Resources by Chapter • Enrichment and Extension • Technology Connection Game Closet at *BigIdeasMath.com* Start the next section

13.1–13.2 Quiz

Check It Out
Progress Check
BigIdeasMath.com

1. The diameter of a circle is 36 centimeters. Find the radius. *(Section 13.1)*

2. The radius of a circle is 11 inches. Find the diameter. *(Section 13.1)*

Estimate the perimeter of the figure. *(Section 13.2)*

3. 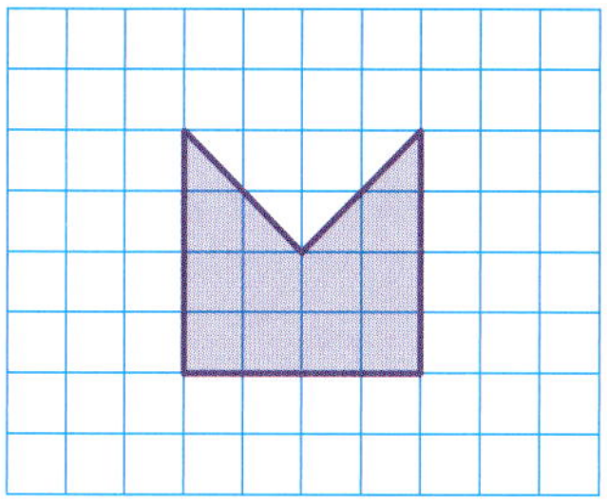

4. 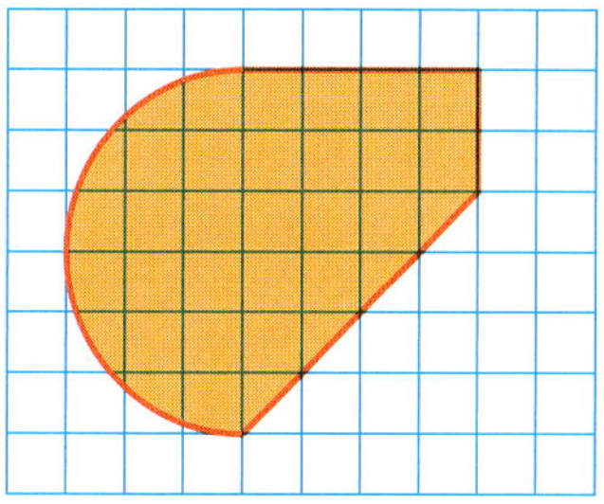

5. 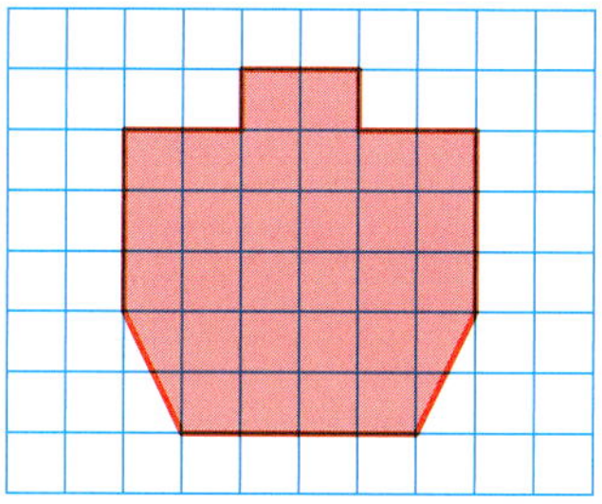

Find the circumference of the circle. Use 3.14 or $\frac{22}{7}$ for π. *(Section 13.1)*

6. 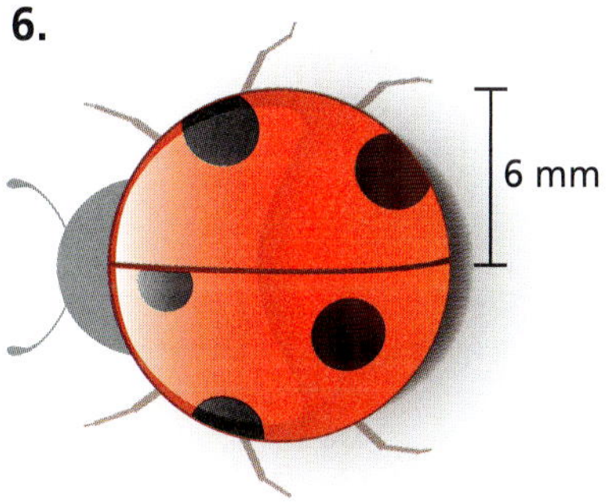

7. 

8. 

Find the perimeter of the figure. *(Section 13.1 and Section 13.2)*

9. 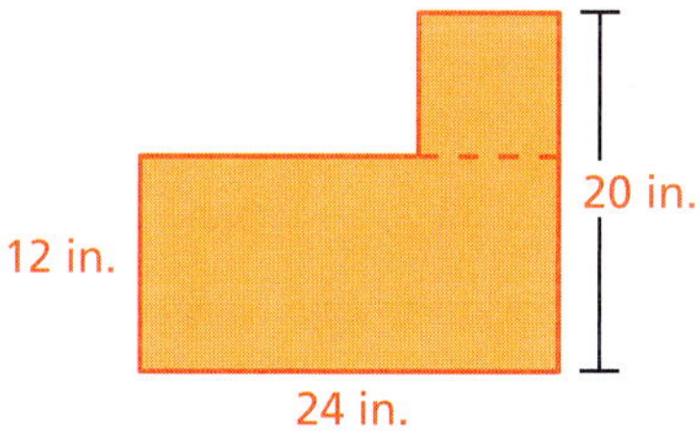

10. 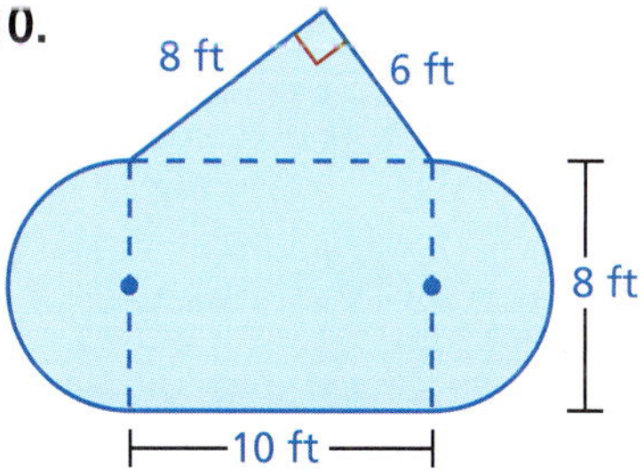

11. 

12. **BUTTON** What is the circumference of a circular button with a diameter of 8 millimeters? *(Section 13.1)*

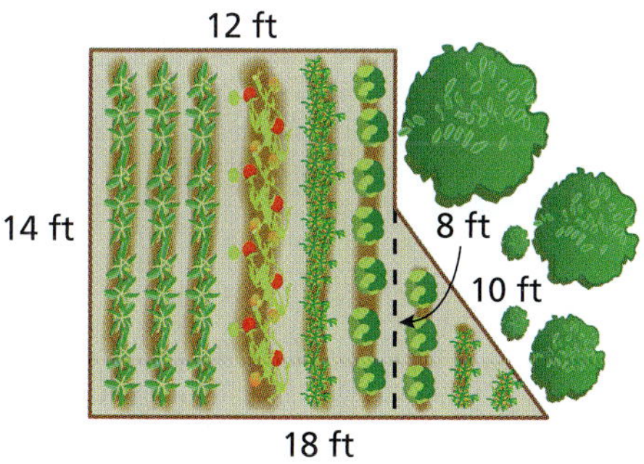

13. **GARDEN** You want to fence part of a yard to make a vegetable garden. How many feet of fencing do you need to surround the garden? *(Section 13.2)*

14. **BAKING** A baker is using two circular pans. The larger pan has a diameter of 12 inches. The smaller pan has a diameter of 7 inches. How much greater is the circumference of the larger pan than that of the smaller pan? *(Section 13.1)*

13.3 Areas of Circles

Essential Question How can you find the area of a circle?

1 ACTIVITY: Estimating the Area of a Circle

Work with a partner. Each square in the grid is 1 unit by 1 unit.

a. Find the area of the large 10-by-10 square.

b. Copy and complete the table.

Region			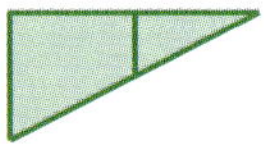
Area (square units)			

c. Use your results to estimate the area of the circle. Explain your reasoning.

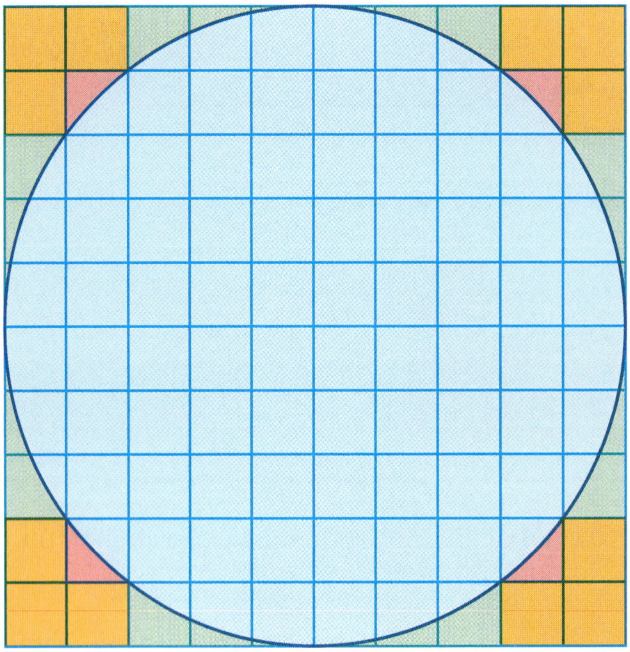

d. Fill in the blanks. Explain your reasoning.

$$\text{Area of large square} = \underline{\quad} \cdot 5^2 \text{ square units}$$

$$\text{Area of circle} \approx \underline{\quad} \cdot 5^2 \text{ square units}$$

e. What dimension of the circle does 5 represent? What can you conclude?

Geometry
In this lesson, you will
- find areas of circles and semicircles.

Laurie's Notes

Introduction

Applying Mathematical Practices

- **Model with Mathematics:** Students are able to develop the formula for the area of a circle from working through two different activities.

Motivate

- Sketch this diagram.

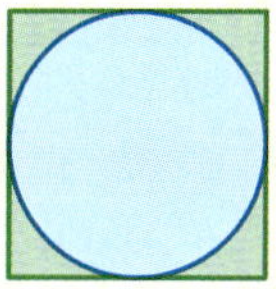

? "The lawn at my house is a square. A rotating sprinkler reaches each side but misses the four corners. What percent of my lawn do you think is getting watered?"

- Students should visually estimate at this stage. They should guess close to 75%. Record their guesses and return to them at the end of the class.

Activity Notes

Activity 1

- **Model with Mathematics:** Using square units to measure the area of a circle is a difficult concept for some students. I have had students ask how the square pieces fit into the curve. This activity helps students visualize how to count the area of a circle and then compute it.
- Students should observe that the circle is inscribed in a large square that has an area of 100 square units.
- As a class, discuss the area of each colored region. The orange square is a unit square. The green triangle is close to half of a 1×2 rectangle. The pink triangle is about half the unit square.

? "How does knowing the area of these pieces help you approximate the area of the circle?" Ask the question and let students begin their work.

- When groups have finished, have several record their work at the board. They should have Area of circle $= 100 - 4(3) - 8(1) - 4\left(\frac{1}{2}\right) = 78$ units^2.

? "The area of the large square is 100 units^2. Can you write 100 as the product of a number times 5^2?" $100 = 4(5^2)$

? "Let's do the same with the area of the circle. Can you write 78 as the product of a number times 5^2? Explain." yes; divide 78 by 25; $78 = 3.12(5^2)$

? "Thinking about the circle, what part of the circle is 5 units long?" radius

- It is very likely that at this point students may guess that 3.12 is supposed to be π and may want to know what they did wrong because they got 3.12 instead of 3.14. It's important to say that they did nothing wrong. The areas of the two triangles are approximations because they aren't even triangles—there is a curved edge.
- The conclusion you are hoping for is that the area of this circle is πr^2.

? "Do you think this is a formula that works for all circles?" open-ended

What Your Students Will Learn

- Find areas of circles and semicircles using formulas.

Previous Learning

Students should know how to find areas of parallelograms.

Technology for the Teacher

Dynamic Classroom

Lesson Plans

Complete Materials List

13.3 Record and Practice Journal

Essential Question How can you find the area of a circle?

1 ACTIVITY: Estimating the Area of a Circle

Work with a partner. Each square in the grid is 1 unit by 1 unit.

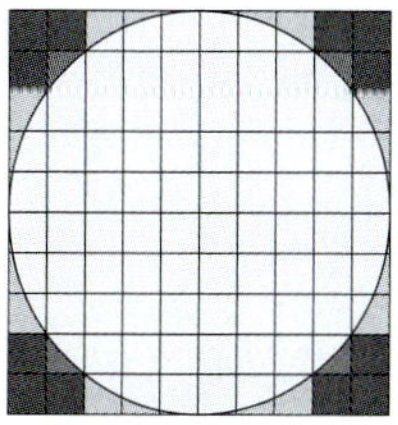

a. Find the area of the large 10-by-10 square.

100 units^2

b. Complete the table.

Region	(black square)	(triangle)	(triangle)
Area (square units)	1 unit^2	$\frac{1}{2}$ unit^2	1 unit^2

c. Use your results to estimate the area of the circle. Explain your reasoning.

about 78 units^2

Differentiated Instruction

Visual

To help students visualize the relationship between area and radius, have them use a compass and grid paper to draw a circle with a radius of 5 units. Ask them to estimate the area of the circle by counting the squares. They should get an estimate that is slightly greater than $3r^2$. Then have students draw a square on the grid paper that has a vertex at the center of the circle and two perpendicular radii as sides. Students should see that one-quarter of the area is enclosed in the square. Because the area of the square is r^2, the area of the circle is less than $4r^2$. So, the area of the circle is between $3r^2$ and $4r^2$, or more precisely πr^2.

13.3 Record and Practice Journal

d. Fill in the blanks. Explain your reasoning.

Area of large square = **4** • 5^2 square units

Area of circle = **3.12** • 5^2 square units

e. What dimension of the circle does 5 represent? What can you conclude?

radius; The area of a circle is about πr^2.

2 ACTIVITY: Approximating the Area of a Circle

Work with a partner.

a. Draw a circle. Label the radius as r.*

b. Divide the circle into 24 equal sections.

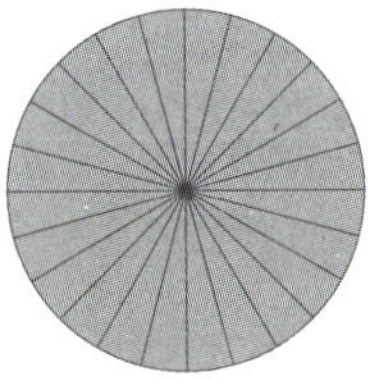

*Cut-outs are available in the back of the Record and Practice Journal.

c. Cut the sections apart. Then arrange them to approximate a parallelogram.

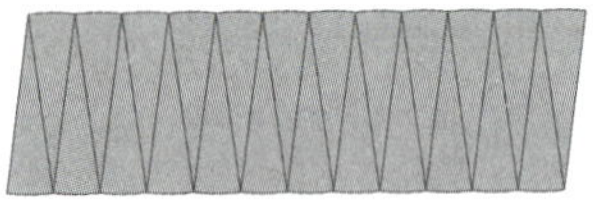

d. What is the approximate height and base of the parallelogram?

height ≈ r; base ≈ $r\pi$

e. Find the area of the parallelogram. What can you conclude?

area ≈ πr^2; The area of the circle is about πr^2.

What Is Your Answer?

3. **IN YOUR OWN WORDS** How can you find the area of a circle?

Check students' work.

4. Write a formula for the area of a circle with radius r. Find an object that is circular. Use your formula to find the area.

$A = \pi r^2$

Laurie's Notes

Activity 2

- **Preparation:** Trace and cut 2 circles from file folder weight paper. Cut one circle into 4 sectors (pie-shaped pieces) and one circle into 12 sectors.
- In this activity, a different approach is used to develop the area formula.
- To connect the two activities, tell students that you want to think about the circle in the large square.

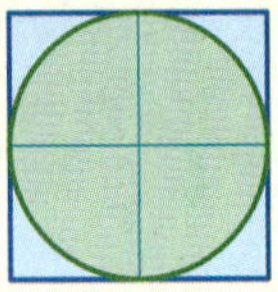

? "If you cut the circle into 4 parts and rearranged them, do you think it would make a shape that was familiar?"

- Take out the 4 prepared pieces. Arrange the 4 pieces in a circle and then rearrange them in a fashion similar to the design in the book. The curvature of the circle is still very apparent, so students will not suggest a similarity to any figure they know.

? "What if the circle were cut into smaller pieces and then rearranged?"

- Take out the 12 prepared pieces. Arrange the 12 pieces in a circle and then rearrange them in a fashion similar to the design in the book. The curvature of the circle is much less apparent.
- **Look for and Express Regularity in Repeated Reasoning** and **Big Idea:** The 24 pieces approximate a parallelogram. The curvature of the circle is not obvious. The area of a parallelogram is base × height. The base is half the circumference. (Pieces are alternated so that $\frac{1}{2}$ the circumference is on the top and $\frac{1}{2}$ is on the bottom.) The height is the radius of the original circle.

$$\begin{aligned} \text{Area} &= \text{base} \times \text{height} \\ &= \frac{1}{2}C \cdot r \\ &= \frac{1}{2}(2\pi r)r \\ &= \pi r^2 \end{aligned}$$

What Is Your Answer?

- **Neighbor Check:** Have students work independently and then have their neighbors check their work. Have students discuss any discrepancies.

Closure

- What percent of the lawn is getting watered? 78%
- **Connection:** This is an example of a geometric probability question where the area of the circle is being compared to the area of the square. In Activity 1, students found the area of the square and the circle.

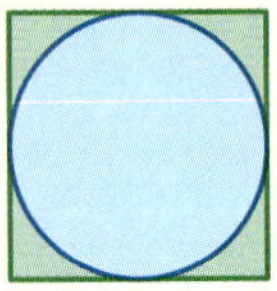

2 ACTIVITY: Approximating the Area of a Circle

Work with a partner.

a. Draw a circle. Label the radius as *r*.

b. Divide the circle into 24 equal sections.

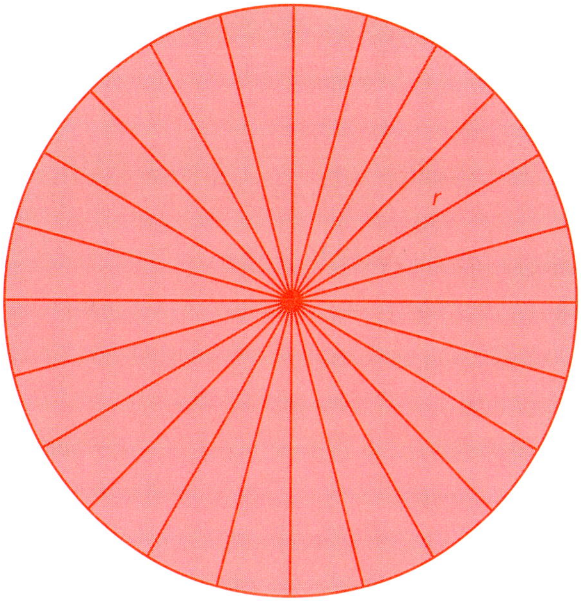

c. Cut the sections apart. Then arrange them to approximate a parallelogram.

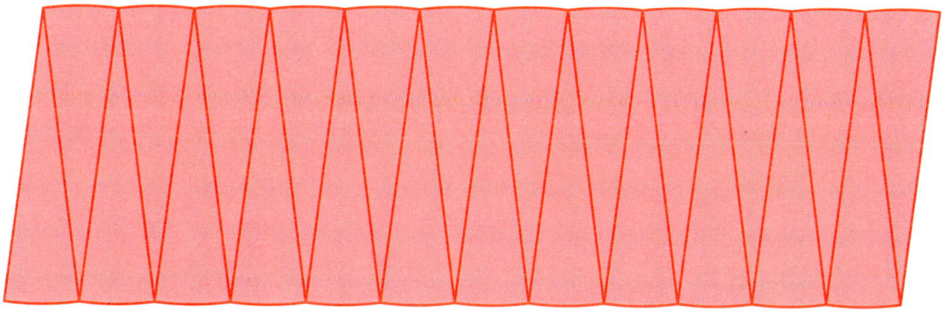

d. What is the approximate height and base of the parallelogram?

e. Find the area of the parallelogram. What can you conclude?

Math Practice

Interpret a Solution

What does the area of the parallelogram represent? Explain.

What Is Your Answer?

3. **IN YOUR OWN WORDS** How can you find the area of a circle?

4. Write a formula for the area of a circle with radius *r*. Find an object that is circular. Use your formula to find the area.

Practice Use what you learned about areas of circles to complete Exercises 3–5 on page 568.

13.3 Lesson

Check It Out
Lesson Tutorials
BigIdeasMath.com

Key Idea

Area of a Circle

Words The area A of a circle is the product of π and the square of the radius.

Algebra $A = \pi r^2$

EXAMPLE 1 Finding Areas of Circles

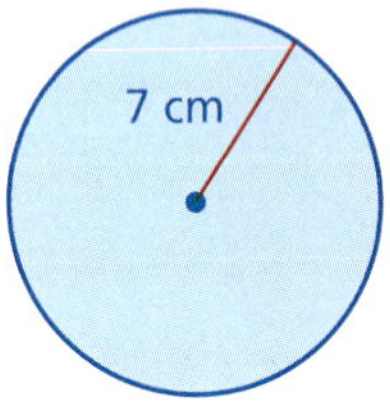

a. Find the area of the circle. Use $\frac{22}{7}$ for π.

Estimate $3 \times 7^2 \approx 3 \times 50 = 150$

$A = \pi r^2$ Write formula for area.

$\approx \frac{22}{7} \cdot 7^2$ Substitute $\frac{22}{7}$ for π and 7 for r.

$= \frac{22}{\cancel{7}_1} \cdot \cancel{49}^{7}$ Evaluate 7^2. Divide out the common factor.

$= 154$ Multiply.

The area is about 154 square centimeters.

Reasonable? $154 \approx 150$ ✓

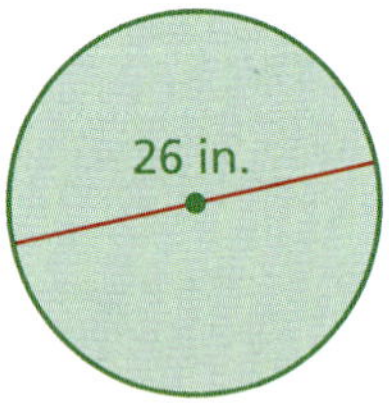

b. Find the area of the circle. Use 3.14 for π.

The radius is $26 \div 2 = 13$ inches.

Estimate $3 \times 13^2 \approx 3 \times 170 = 510$

$A = \pi r^2$ Write formula for area.

$\approx 3.14 \cdot 13^2$ Substitute 3.14 for π and 13 for r.

$= 3.14 \cdot 169$ Evaluate 13^2.

$= 530.66$ Multiply.

The area is about 530.66 square inches.

Reasonable? $530.66 \approx 510$ ✓

On Your Own

Now You're Ready
Exercises 3–10

1. Find the area of a circle with a radius of 6 feet. Use 3.14 for π.
2. Find the area of a circle with a diameter of 28 meters. Use $\frac{22}{7}$ for π.

Laurie's Notes

Introduction

Connect

- **Yesterday:** Students gained an intuitive understanding about how to find the area of a circle.
- **Today:** Students will use the formula for area to solve real-life problems.

Motivate

- ? "At a pizza restaurant, you have a choice of ordering a 16-inch pizza or a 12-inch pizza for dinner. What measurement is being used to describe the pizza?" the diameter
- ? "Both pizzas are cut into the same number of pieces. Do both pizzas give you the same amount of pizza?" No, the size of the slices is different.
- ? "How would you figure out how many times more pizza you get with the 16-inch pizza than with the 12-inch pizza?" Listen for students to mention the idea of comparing the areas of the pizzas.
- Mention to students that after today's lesson, they will be able to answer this question. You may want to return to this question at the end of class.

Lesson Notes

Key Idea

- Discuss the formula, written in words and algebraically.
- **Common Error:** Students square the product of π and r, instead of just the radius.

Example 1

- Work through each example. Remind students that pi can be rounded to 3 when estimating.
- **Common Error:** Make sure students are reading the figures correctly. A line segment from the center to the outside edge indicates a radius. A line segment across the circle through the center is a diameter.
- ? "What would a reasonable answer be?" Answers will vary, but should be similar to what is shown in the text.
- ? "In part (a), what estimate should be used for π? Why?" $\frac{22}{7}$; The radius is a multiple of 7.
- Remind students to label answers with the appropriate units.
- **Teaching Tip:** Many students view a diagram differently than a written problem: *Find the area of a circle with a radius of 7 centimeters.* The visual learner is aided by a simple diagram.
- **Common Error:** For part (b), students may forget to divide the diameter by 2 to find the radius before substituting in the area formula.
- If students are not using calculators, ask them to give a reasonable estimate for 3.14×169 before they multiply.

On Your Own

- **Neighbor Check:** Have students work independently and then have their neighbors check their work. Have students discuss any discrepancies.

Goal Today's lesson is finding the area of a circle and a semicircle.

Technology for the Teacher

Dynamic Classroom

Lesson Tutorials
Lesson Plans
Answer Presentation Tool

Extra Example 1

a. Find the area of a circle with a radius of 21 feet. Use $\frac{22}{7}$ for π.
about 1386 ft^2

b. Find the area of a circle with a diameter of 16 meters. Use 3.14 for π.
about 200.96 m^2

On Your Own

1. about 113.04 ft^2
2. about 616 m^2

Extra Example 2

The 1893 World's Fair in Chicago boasted the first ever Ferris wheel with a diameter of 250 feet. How far did a person travel in one revolution of the wheel? about 785 ft

On Your Own

3. the diameter of the tire

Extra Example 3

Find the area of a semicircle with a radius of 9 inches. about 127.17 in.^{2}

On Your Own

4. about 25.12 m^2
5. about 9.8125 yd^2
6. about 189.97 cm^2

English Language Learners

Vocabulary

Discuss with students the word *approximation*. Give and ask for examples of times they use approximations in their daily life. (e.g., "I'll meet you in ten minutes.") Discuss when using an approximation for an answer is as useful as an exact answer. Discuss the usefulness of approximations for π, 3.14 and $\frac{22}{7}$.

Laurie's Notes

Example 2

- This monster truck example will grab students' attention!
- **Model with Mathematics:** A circle is a common shape. In this question, students connect vocabulary of a circle to a real-life application of a circle.

On Your Own

- Review this together as a class.

Example 3

- Ask if anyone knows what an orchestra pit is. An orchestra pit is a lowered area in front of the stage where musicians perform.

? "How do you find the area of a semicircle?" Find $\frac{1}{2}$ the area of the circle.

? "What is the radius of the semicircle?" 15 ft

- Work through the problem.
- **Extension:** If time permits and students are using the area formula correctly, try the following problems.

  ? "If you make a mistake and use the diameter instead of the radius in the area formula, is the area doubled? Explain." No; It is quadrupled.

  ? "Find the area of the doughnut region." $8\pi \approx 25.12$ in.^{2}

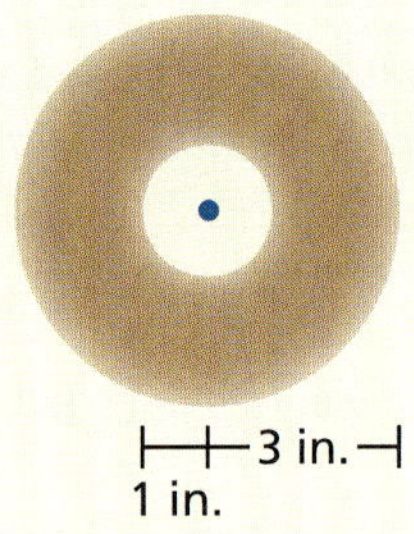

On Your Own

- Check to see that students recognize the difference in how the dimensions are labeled. The radius is marked with a line segment from the center to the outer edge. The diameter is marked with a line segment from one edge through the center to the opposite edge.

Closure

- Find the areas of a 16-inch pizza and a 12-inch pizza. How many times larger is the 16-inch pizza than the 12-inch pizza? about 1.8 times
- Students will find this easier to calculate if the area is kept in terms of π.
- This situation can also be thought of as "how many times more toppings are needed for the larger pizza?"

EXAMPLE 2 Describing a Distance

You want to find the distance the monster truck travels when the tires make one 360 degree rotation. Which best describes this distance?

Ⓐ the radius of the tire Ⓑ the diameter of the tire

Ⓒ the circumference of the tire Ⓓ the area of the tire

The distance the truck travels after one rotation is the same as the distance *around* the tire. So, the circumference of the tire best describes the distance in one rotation.

∴ The correct answer is Ⓒ.

On Your Own

3. You want to find the height of one of the tires. Which measurement would best describe the height?

EXAMPLE 3 Finding the Area of a Semicircle

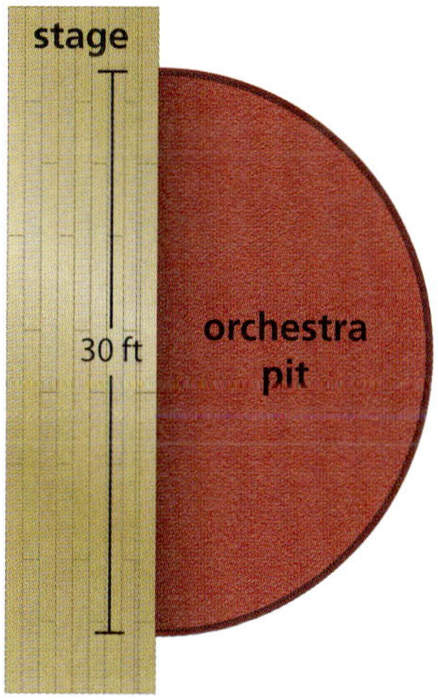

Find the area of the semicircular orchestra pit.

The area of the orchestra pit is one-half the area of a circle with a diameter of 30 feet.

The radius of the circle is 30 ÷ 2 = 15 feet.

$$\frac{A}{2} = \frac{\pi r^2}{2}$$ Divide the area by 2.

$$\approx \frac{3.14 \cdot 15^2}{2}$$ Substitute 3.14 for π and 15 for r.

$$= \frac{3.14 \cdot 225}{2}$$ Evaluate 15^2.

$$= 353.25$$ Simplify.

∴ So, the area of the orchestra pit is about 353.25 square feet.

On Your Own

Now You're Ready
Exercises 13–15

Find the area of the semicircle.

4. 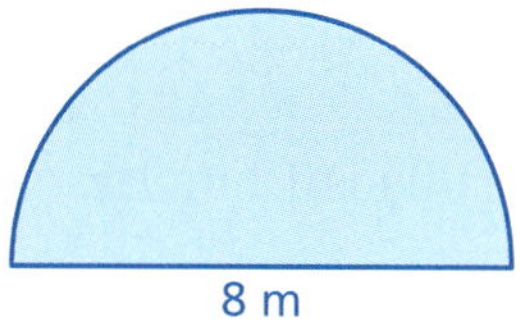

5. 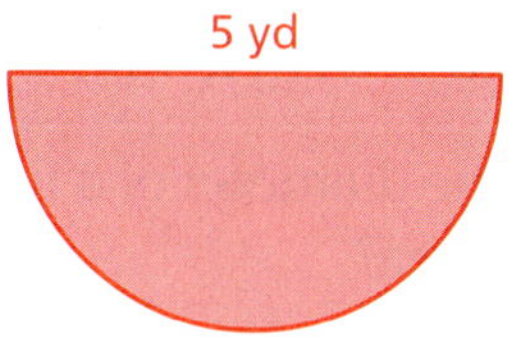

6. 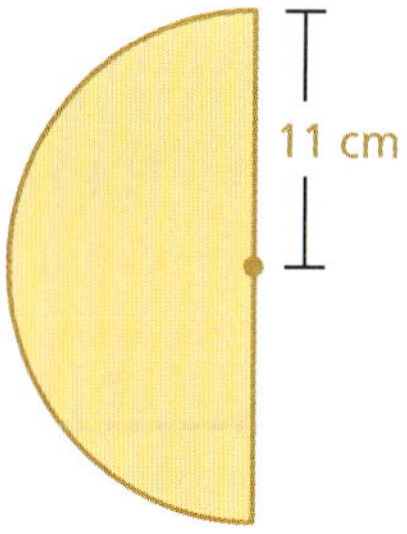

13.3 Exercises

Vocabulary and Concept Check

1. **VOCABULARY** Explain how to find the area of a circle given its diameter.

2. **DIFFERENT WORDS, SAME QUESTION** Which is different? Find "both" answers.

What is the area of a circle with a diameter of 1 m?	What is the area of a circle with a diameter of 100 cm?
What is the area of a circle with a radius of 100 cm?	What is the area of a circle with a radius of 500 mm?

Practice and Problem Solving

Find the area of the circle. Use 3.14 or $\frac{22}{7}$ for π.

1 3.

4.

5.

6.

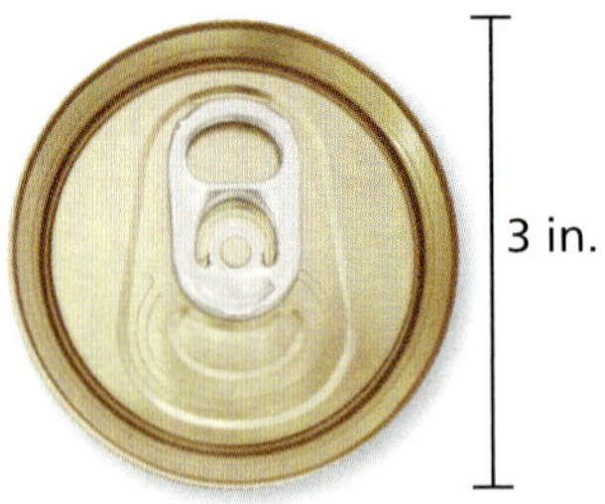

7.

8.

9. Find the area of a circle with a diameter of 56 millimeters.

10. Find the area of a circle with a radius of 5 feet.

11. **TORTILLA** The diameter of a flour tortilla is 12 inches. What is the area?

12. **LIGHTHOUSE** The Hillsboro Inlet Lighthouse lights up how much more area than the Jupiter Inlet Lighthouse?

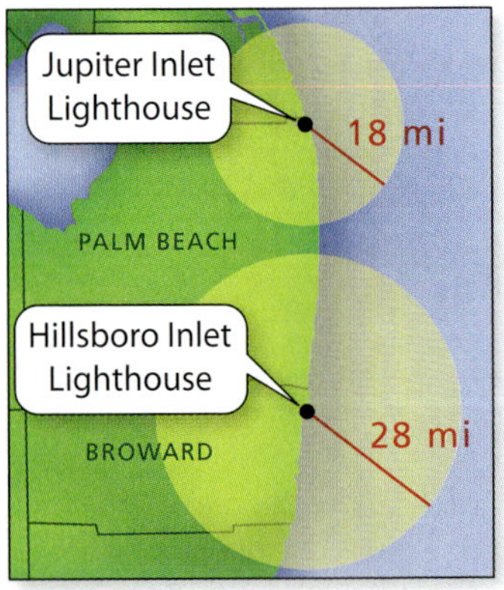

Assignment Guide and Homework Check

Level	Assignment	Homework Check
Advanced	1–12, 14, 16–25	12, 14, 16, 18, 21

For Your Information

- **Exercise 16** Tell students that *radii* is the plural form of radius.
- **Exercise 16** Have students complete the table with their answers in terms of pi. The patterns in the table will be easier for the students to see.

Common Errors

- **Exercises 3–8, 13–15** Students may forget to divide the diameter by 2 to find the area. Remind them that they need the *radius* for the area formula of a circle.
- **Exercises 3–8, 13–15** Students may write the incorrect units for the area. Remind them to carefully check the units and to square the units as well.
- **Exercises 3–8** Students may refer to the formula for circumference and forget to square the radius when finding the area. Remind them that the area formula uses the radius squared.
- **Exercises 13–15** Students may forget to divide the area in half. Remind them that they are finding the area of half of a circle so the area is half of a whole circle.

13.3 Record and Practice Journal

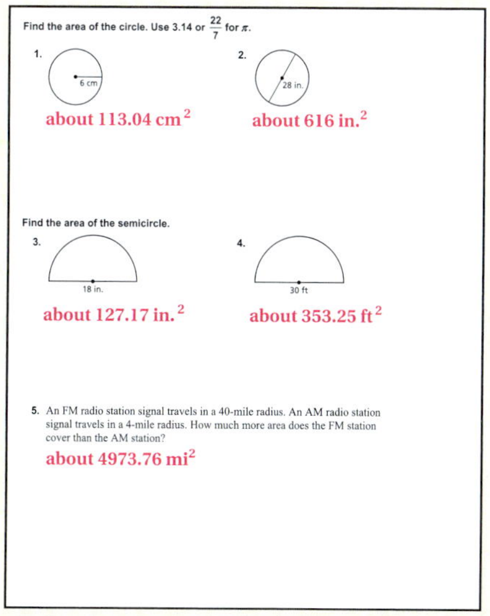

Find the area of the circle. Use 3.14 or $\frac{22}{7}$ for π.

1. (6 cm) about 113.04 cm^2
2. (28 in.) about 616 $in.^2$

Find the area of the semicircle.

3. (18 in.) about 127.17 $in.^2$
4. (30 ft) about 353.25 ft^2

5. An FM radio station signal travels in a 40-mile radius. An AM radio station signal travels in a 4-mile radius. How much more area does the FM station cover than the AM station?

about 4973.76 mi^2

Vocabulary and Concept Check

1. Divide the diameter by 2 to get the radius. Then use the formula $A = \pi r^2$ to find the area.
2. What is the area of a circle with a radius of 100 cm; about 31,400 cm^2; about 7850 cm^2?

Practice and Problem Solving

3. about 254.34 mm^2
4. about 616 cm^2
5. about 314 $in.^2$
6. about 7.065 $in.^2$
7. about 3.14 cm^2
8. about 1.76625 ft^2
9. about 2461.76 mm^2
10. about 78.5 ft^2
11. about 113.04 $in.^2$
12. about 1444.4 mi^2

Practice and Problem Solving

13–16. See Additional Answers.

17. See *Taking Math Deeper*.

18. greater than; The circle's diameter is one-half as long, so it equals the radius of the semicircle. A diagram shows that the area of the semicircle is greater.

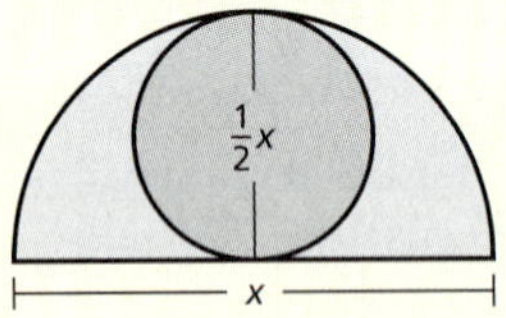

19–21. See Additional Answers.

Fair Game Review

22. 44	**23.** 53
24. 73	**25.** A

Mini-Assessment

Find the area of the circle. Use 3.14 or $\frac{22}{7}$ for π.

1.

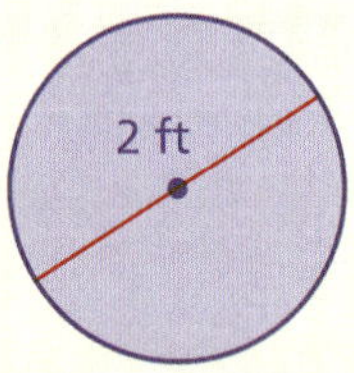

about 3.14 ft^2

2.

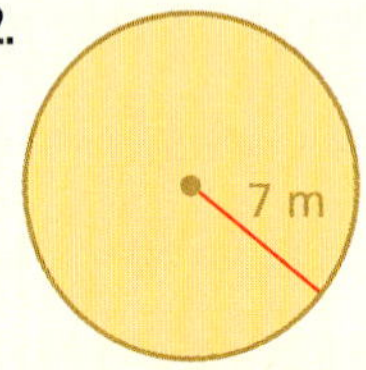

about 154 m^2

3.

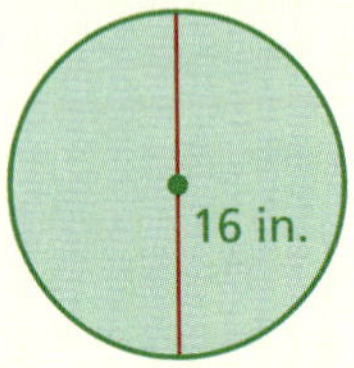

about 200.96 $in.^2$

4.

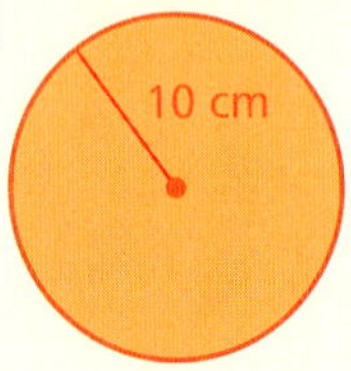

about 314 cm^2

5. Find the area of the rug.
about 28.26 ft^2

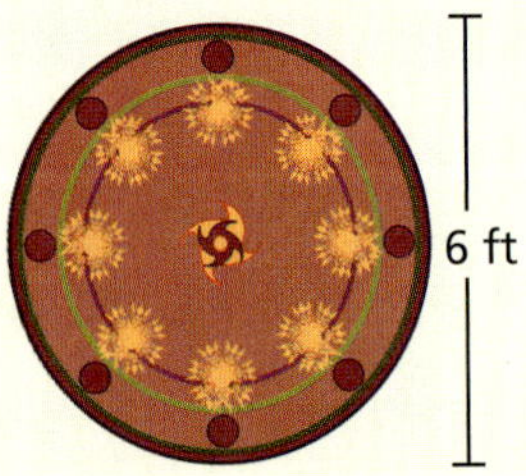

Taking Math Deeper

Exercise 17

This problem is not difficult, once a student realizes that the dog's running area is three-quarters of a circle.

1. Draw and label a diagram.

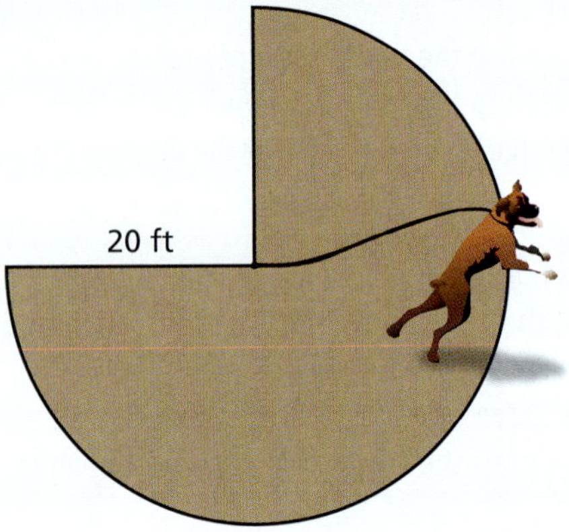

2. Find the area.

$$\text{Area} = \frac{3}{4}(\text{Area of circle})$$
$$= \frac{3}{4} \cdot \pi \cdot 20^2$$
$$\approx 942 \text{ ft}^2$$

3. Ask students to talk about the advantages and disadvantages of this type of outside exercise plan for a dog. What other ways can you allow your dog to be outside and get exercise?
   - Build a dog run.
   - Fence in a back yard.
   - Use an invisible fence.
   - Take the dog for a long walk each morning and night.

Project

Design a dog run for your back yard. Do some research to determine the cost of the run.

Reteaching and Enrichment Strategies

If students need help. . .	If students got it. . .
Resources by Chapter • Practice A and Practice B • Puzzle Time Record and Practice Journal Practice Differentiating the Lesson Lesson Tutorials Skills Review Handbook	Resources by Chapter • Enrichment and Extension • Technology Connection Start the next section

Find the area of the semicircle.

3 **13.**

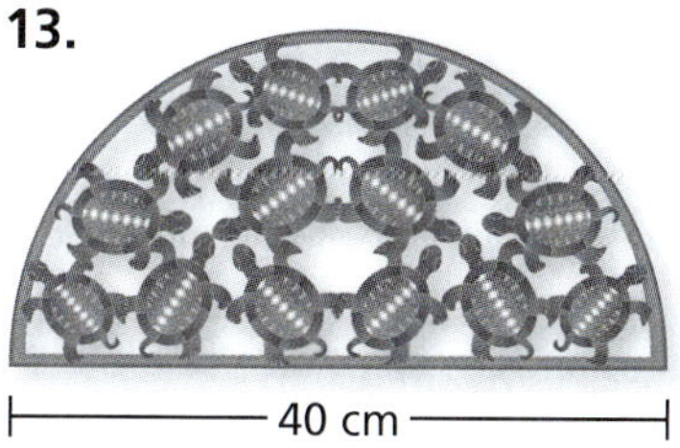

14.

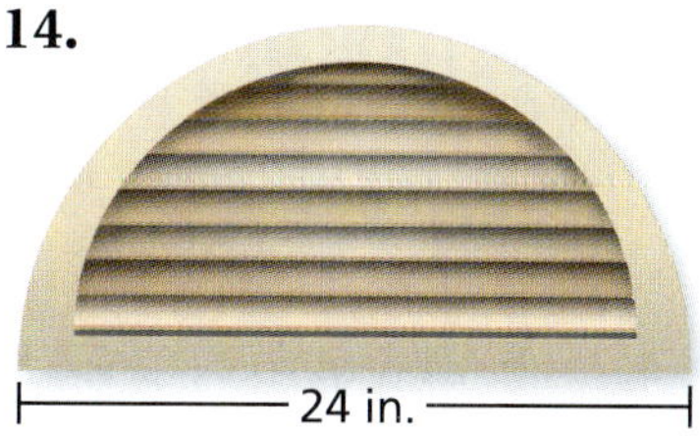

15.

16. **REPEATED REASONING** Consider five circles with radii of 1, 2, 4, 8, and 16 inches.

a. Copy and complete the table. Write your answers in terms of π.

b. Compare the areas and circumferences. What happens to the circumference of a circle when you double the radius? What happens to the area?

c. What happens when you triple the radius?

Radius	Circumference	Area
1	2π in.	π in.^{2}
2		
4		
8		
16		

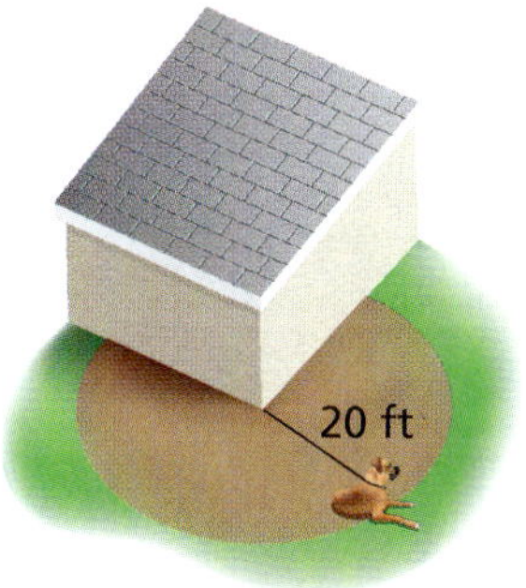

17. **DOG** A dog is leashed to the corner of a house. How much running area does the dog have? Explain how you found your answer.

18. **CRITICAL THINKING** Is the area of a semicircle with a diameter of x *greater than*, *less than*, or *equal to* the area of a circle with a diameter of $\frac{1}{2}x$? Explain.

Reasoning **Find the area of the shaded region. Explain how you found your answer.**

19.

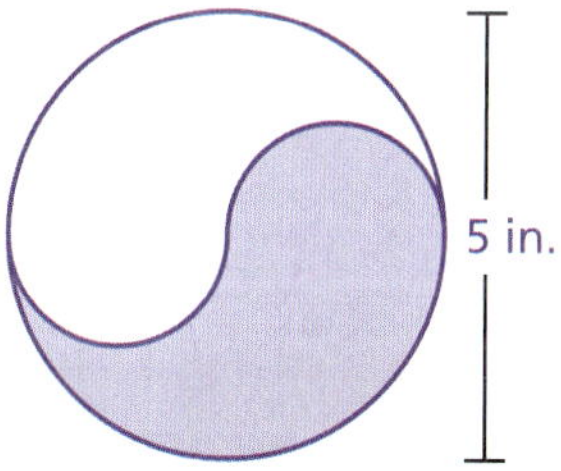

20.

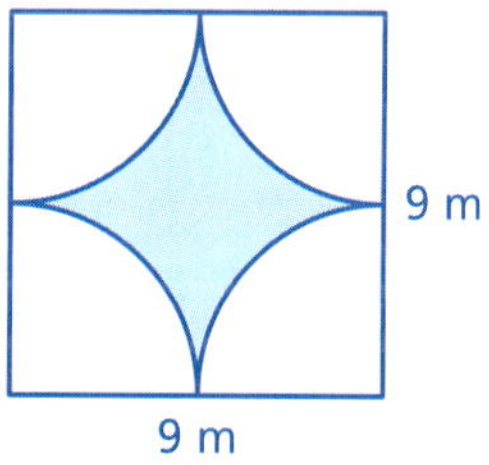

21.

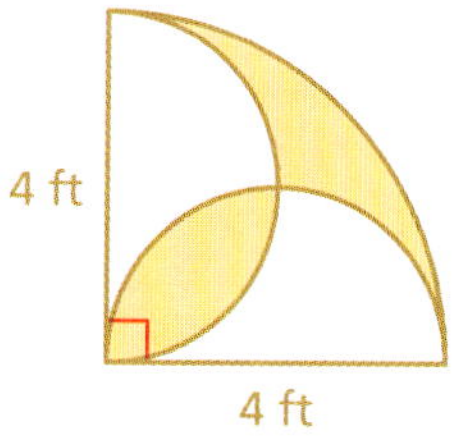

Fair Game Review What you learned in previous grades & lessons

Evaluate the expression. *(Skills Review Handbook)*

22. $\frac{1}{2}(7)(4) + 6(5)$

23. $\frac{1}{2} \cdot 8^2 + 3(7)$

24. $12(6) + \frac{1}{4} \cdot 2^2$

25. **MULTIPLE CHOICE** What is the product of $-8\frac{1}{3}$ and $3\frac{2}{5}$? *(Skills Review Handbook)*

Ⓐ $-28\frac{1}{3}$ Ⓑ $-24\frac{2}{15}$ Ⓒ $24\frac{2}{15}$ Ⓓ $28\frac{1}{3}$

13.4 Areas of Composite Figures

Essential Question How can you find the area of a composite figure?

1 ACTIVITY: Estimating Area

Work with a partner.

a. Choose a state. On grid paper, draw a larger outline of the state.

b. Use your drawing to estimate the area (in square miles) of the state.

c. Which state areas are easy to find? Which are difficult? Why?

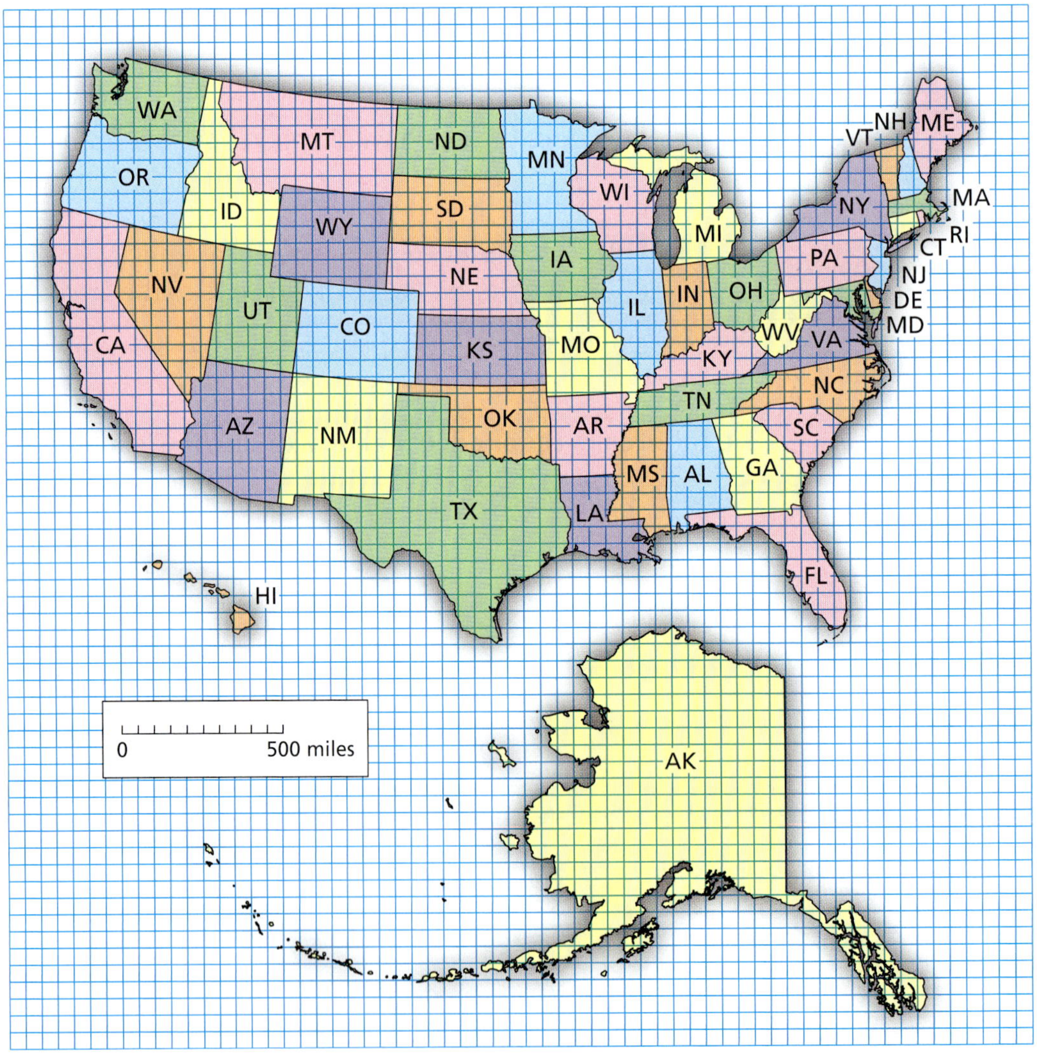

Geometry

In this lesson, you will

- find areas of composite figures by separating them into familiar figures.
- solve real-life problems.

Laurie's Notes

Introduction

Applying Mathematical Practices

- **Make Sense of Problems and Persevere in Solving Them:** Students have worked with a number of area formulas. In making sense of composite figures, they need to view the figures as composed of smaller, familiar figures.

Motivate

- Share U.S. state trivia in question/answer format.
- Largest states: Alaska, Texas, California
- Smallest states: Rhode Island, Delaware, Connecticut
- Most densely populated states: New Jersey, Rhode Island, Massachusetts

Activity Notes

For Your Information

- The three activities may likely take longer than one class period. You may want to do just Activities 1 and 3, or Activities 2 and 3. Students really like Activity 1, and it is interesting to hear their comments about different states or regions of the country.

Activity 1

- ? Ask a variety of questions about the attributes of the states: state with most coastline; states with no coastline; state furthest north, south, east, and west.
- Questions are answered by using eyesight only. The goal is to get students thinking about the states and how each might be described.
- **Make Sense of Problems and Persevere in Solving Them** and **Big Idea:** Each square in the grid is 2500 square miles. Each state is composed of whole squares and parts of whole squares. Students will find the area of the state by finding the area of composite figures.
- Here are a few strategies for implementing this activity.
  - Students are assigned states within one particular region.
  - Students use $\frac{1}{2}$-inch or 1-inch grid paper to draw the larger outline. This is *not* a lesson on scale drawings. A larger copy of the state allows students to record their areas for each part on the drawing.
  - Students are given an enlarged photocopy of the state to use.
  - Assign the same state to two students. Have each student make a drawing and then compare their estimates for the non-square parts of the state.
- Discuss part (c). States with straighter borders are easier to draw and to estimate their areas.
- **Extension:** Drawings of adjacent states can be cut out and taped together on a bulletin board. Students can do outside research to gather information about their state: population, population density; or to rank the states by area and find the mean and median of the state areas.

What Your Students Will Learn

- Find areas of composite figures using grid paper.
- Find areas of composite figures by separating them into familiar figures, finding those areas, and then adding up the areas.

Previous Learning

Students should know area formulas.

Technology for the Teacher

Dynamic Classroom

Lesson Plans
Complete Materials List

13.4 Record and Practice Journal

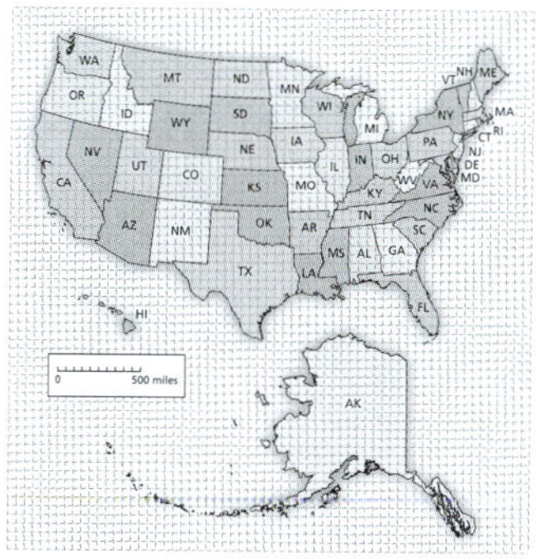

Essential Question How can you find the area of a composite figure?

1 ACTIVITY: Estimating Area

Work with a partner. Check students' work.

a. Choose a state. On grid paper, draw a larger outline of the state.

b. Use your drawing to estimate the area (in square miles) of the state.

c. Which state areas are easy to find? Which are difficult? Why?

Differentiated Instruction

Kinesthetic

Using grid paper, have students draw and cut out a square 4 units on each side. Next, have students draw and cut out a rectangle that is 3 units by 2 units. Have students place the two figures so that they are touching, but not overlapping. Ask them to find the combined area by finding the area of each piece. Have them draw other shapes that are touching on the grid paper, for example a square with a semicircle on top. Find the total area. Discuss how they could use this idea to find the area of other figures.

13.4 Record and Practice Journal

2 ACTIVITY: Estimating Areas

Work with a partner. The completed puzzle has an area of 150 square centimeters.*

a. Estimate the area of each puzzle piece.

Answers will vary depending on students' measurements.

b. Check your work by adding the six areas. Why is this a check?

They should add up to 150 cm^2 when put together.

3 ACTIVITY: Filling a Square with Circles

Work with a partner. Which pattern fills more of the square with circles? Explain.

a.

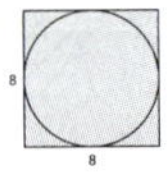

b.

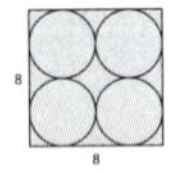

a–d. None; Each pattern of circles has the same area.

*Cut-outs are available in the back of the Record and Practice Journal.

c.

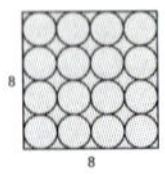

d.

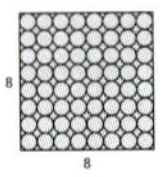

What Is Your Answer?

4. **IN YOUR OWN WORDS** How can you find the area of a composite figure?

Sample answer: Add the areas of all the pieces that form the figure.

5. Summarize the area formulas for all the basic figures you have studied. Draw a single composite figure that has each type of basic figure. Label the dimensions and find the total area.

Area formulas for square, rectangle, triangle, trapezoid, parallelogram, and circle; Check students' work.

Laurie's Notes

Activity 2

- **Make Sense of Problems and Persevere in Solving Them:** Students will work with actual-size pieces. The interlocking cuts are geometric shapes of which students should know how to find the area. Like Activity 1, to find the area, component parts are used. Unlike Activity 1, a component part may be missing so the area is subtracted off.
- Students will need to make accurate measurements to the nearest tenth of a centimeter.
- Scaled pieces are shown in the textbook.
- This activity reviews measurement, computation with decimals, and area formulas.
- Watch for computation errors.
- The completed puzzle is shown at the right.

Activity 3

- **Big Idea:** Each square has the same amount of green. Students can determine the radius of the circles. Then they can find the sum of the areas of circles in each diagram.
- You may need to suggest a table format to help guide student thinking.

Number of Circles	Radius of One Circle	Total Area
1	4	$1 \times \pi(4)^2 = 16\pi$ square units
4	2	$4 \times \pi(2)^2 = 16\pi$ square units
16	1	$16 \times \pi(1)^2 = 16\pi$ square units
64	$\frac{1}{2}$	$64 \times \pi\left(\frac{1}{2}\right)^2 = 16\pi$ square units

- Students find the results of this activity surprising.

What Is Your Answer?

- Question 5 is a nice summary problem that could become a small project.

Closure

- Does the area of a figure change when it is divided into smaller regions? Explain. No. As each region becomes smaller, the number of regions increases, but the total area remains the same.

2 ACTIVITY: Estimating Areas

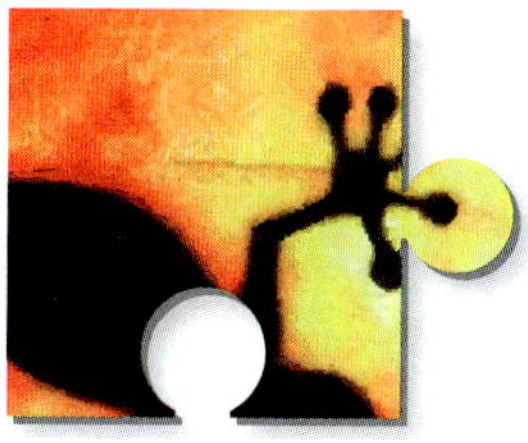

Work with a partner. The completed puzzle has an area of 150 square centimeters.

a. Estimate the area of each puzzle piece.

b. Check your work by adding the six areas. Why is this a check?

3 ACTIVITY: Filling a Square with Circles

> **Math Practice**
>
> **Make a Plan**
>
> What steps will you use to solve this problem?

Work with a partner. Which pattern fills more of the square with circles? Explain.

a.

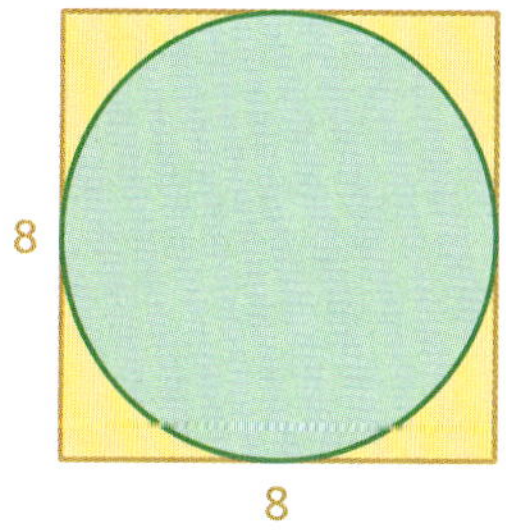

b.

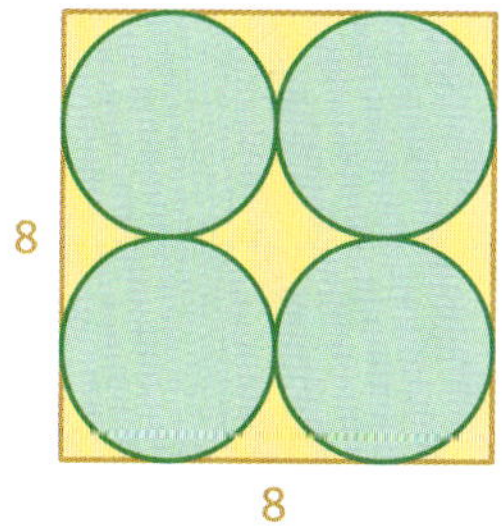

c.

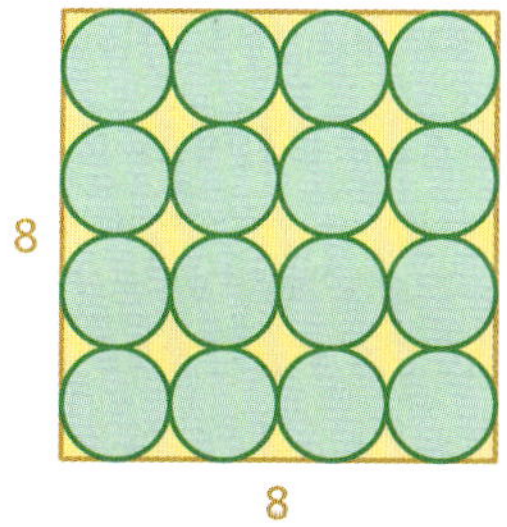

d.

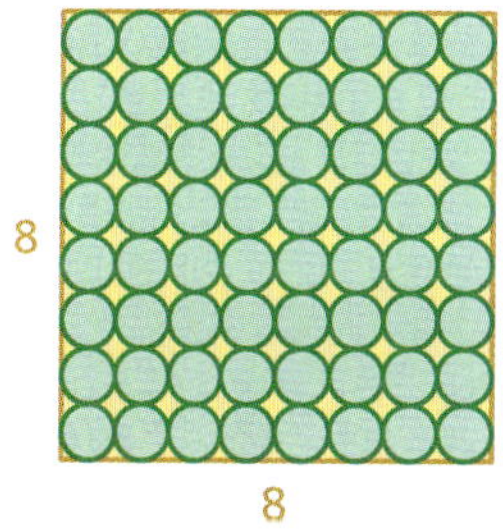

What Is Your Answer?

4. **IN YOUR OWN WORDS** How can you find the area of a composite figure?

5. Summarize the area formulas for all the basic figures you have studied. Draw a single composite figure that has each type of basic figure. Label the dimensions and find the total area.

Practice Use what you learned about areas of composite figures to complete Exercises 3–5 on page 574.

13.4 Lesson

Check It Out
Lesson Tutorials
BigIdeasMath.com

To find the area of a composite figure, separate it into figures with areas you know how to find. Then find the sum of the areas of those figures.

EXAMPLE 1 Finding an Area Using Grid Paper

Find the area of the yellow figure.

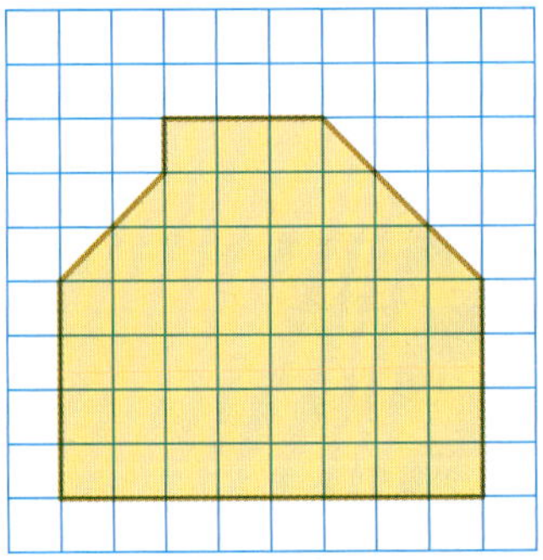

Count the number of squares that lie entirely in the figure. There are 45.

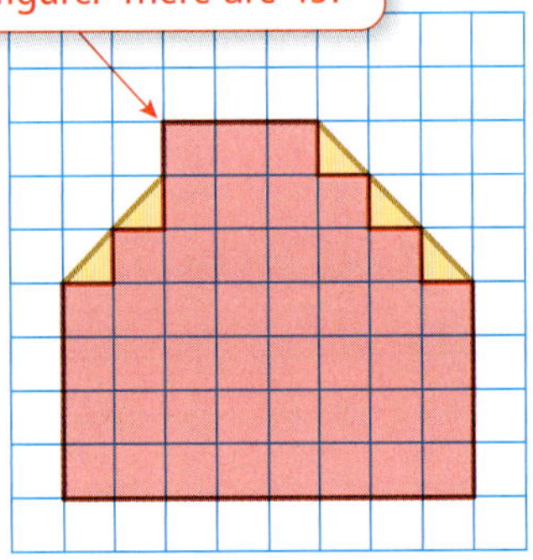

Count the number of half squares in the figure. There are 5.

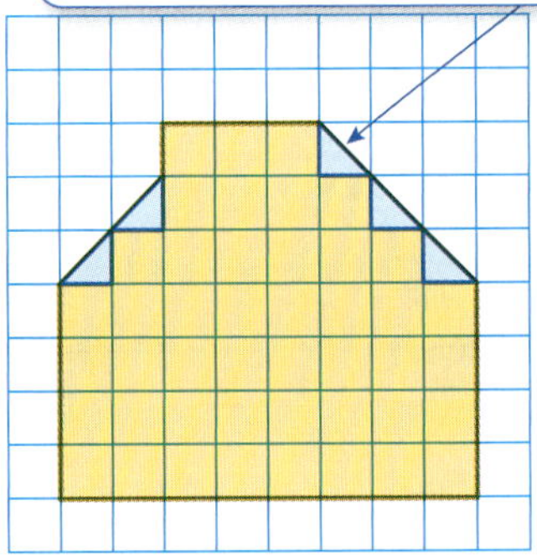

The area of a half square is $1 \div 2 = 0.5$ square unit.

Area of 45 squares: $45 \times 1 = 45$ square units

Area of 5 half squares: $5 \times 0.5 = 2.5$ square units

So, the area is $45 + 2.5 = 47.5$ square units.

On Your Own

Now You're Ready
Exercises 3–8

Find the area of the shaded figure.

1. 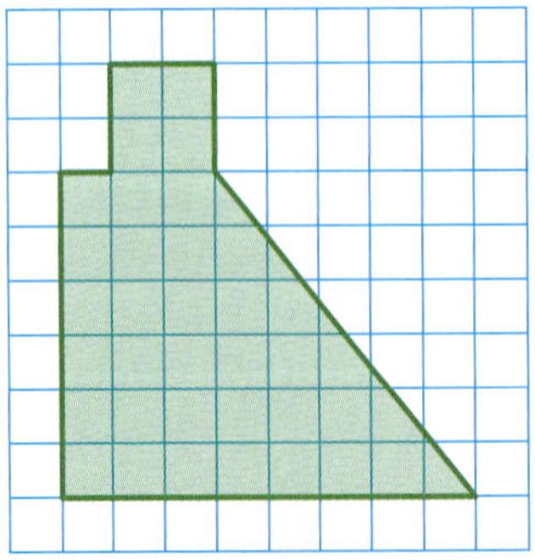

2. 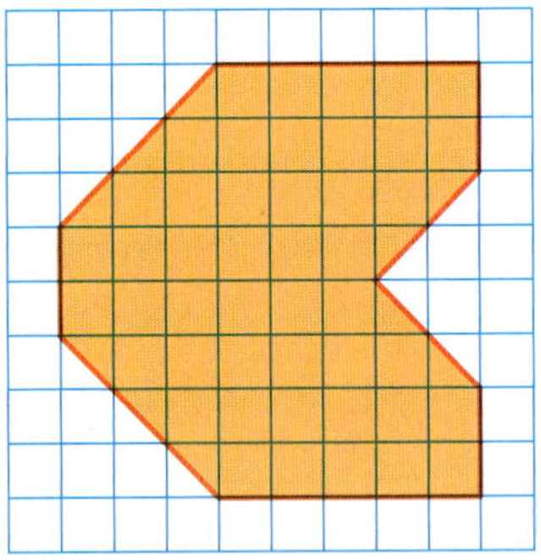

Laurie's Notes

Introduction

Connect

- **Yesterday:** Students gained an intuitive understanding about how to find the area of composite figures.
- **Today:** Students will divide composite figures into familiar geometric shapes and use known area formulas to find the total area.

Motivate

- Arrange a set of tangram pieces in a square. Tell students that the area of the square is 16 square units.

? "What are the dimensions of the square?" 4 units by 4 units

- Now rearrange all of the tangram pieces to make a new shape.

? "Can you find the area of each of these? Explain" yes, 16 square units; Each new figure is composed of the same 7 pieces that made the square.

Lesson Notes

Example 1

- Work through the example as shown and then try an alternate approach.
- **Construct Viable Arguments and Critique the Reasoning of Others**: There are several ways to find the area of this figure. It is important for students to describe the process. The needed dimensions are unknown, so student explanations are key.

? "What is the name of the yellow polygon?" heptagon

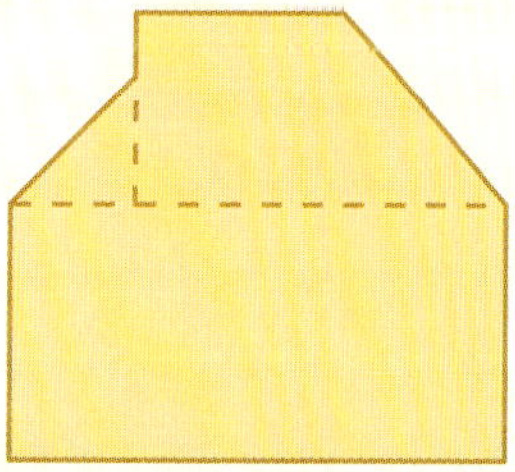

? "Can you think of another way to find the area of the heptagon? Explain." Yes. Listen for a suggestion that divides the heptagon into a rectangle, a triangle, and a trapezoid.

- Note: The rectangle has dimensions 4×8, the triangle has a base of 2 and height of 2, and the trapezoid has bases of 3 and 6 and a height of 3. The total area is 47.5 square units.

? "Are there other strategies for finding the area of the yellow heptagon?" Yes, it could be divided into a rectangle and two trapezoids using two parallel horizontal lines (or two parallel vertical lines).

- **Extension:** Ask students if the area of their classroom is more or less than 48 square meters. Hold two meter sticks perpendicular to each other as a visual clue.

On Your Own

- Encourage students to find more than one method to find the area of each.
- **Construct Viable Arguments and Critique the Reasoning of Others:** Have students share different strategies at the board.

Goal Today's lesson is finding the areas of composite figures.

Technology for the Teacher

Dynamic Classroom

Lesson Tutorials
Lesson Plans
Answer Presentation Tool

Extra Example 1

Find the area of the figure.

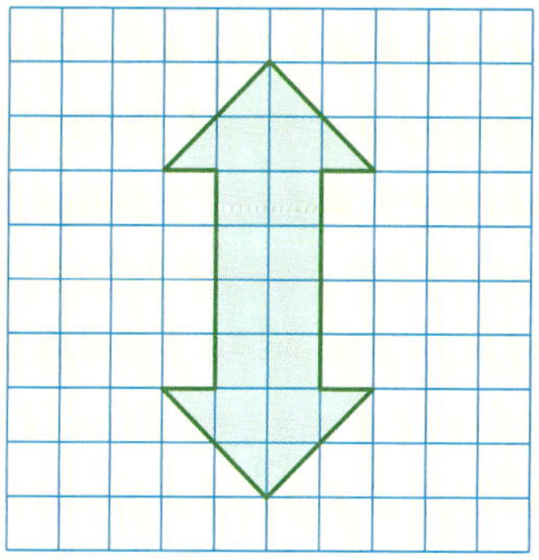

16 square units

On Your Own

1. 37 square units
2. 51 square units

Extra Example 2

Find the area of the pool and the deck. The figure is made up of a right triangle, a square, and a semicircle.

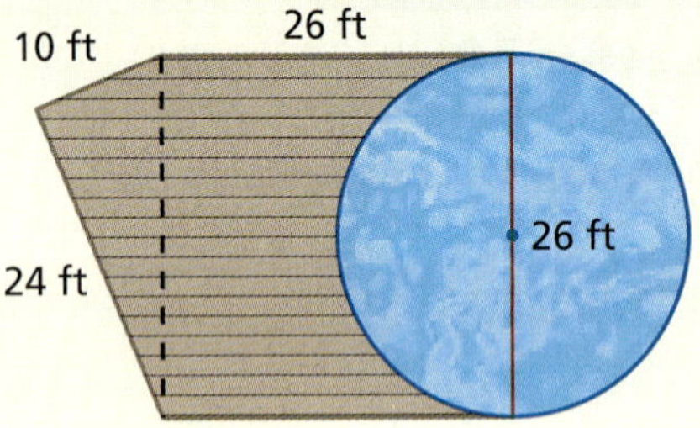

about 1061.33 ft^2

Extra Example 3

Find the area of the figure made up of a triangle, a rectangle, and a trapezoid.

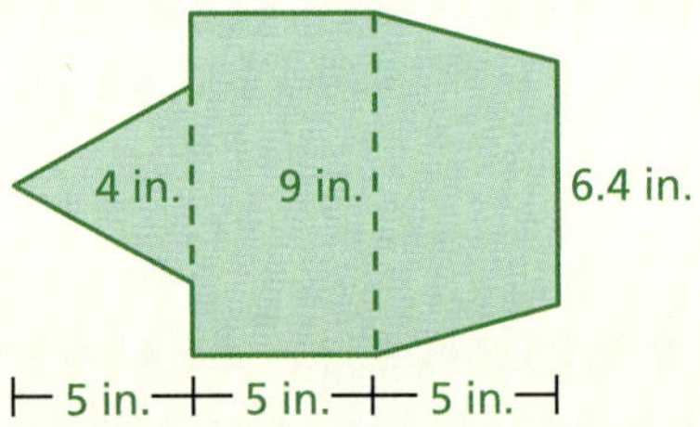

93.5 $in.^2$

On Your Own

3. $90\ m^2$
4. about $10.28\ ft^2$

English Language Learners

Vocabulary

Formulas use letters to represent specific items or measures. For instance, in the formula for the area of a rectangle, $A = bh$, A represents area, b represents the base, and h represents the height. Review with your English learners the capital and lowercase letters used in formulas.

area *(A)*	base *(b)*
circumference *(C)*	diameter *(d)*
distance *(d)*	height *(h)*
length *(ℓ)*	perimeter *(P)*
radius *(r)*	rate *(r)*
time *(t)*	volume *(V)*
width *(w)*	

Laurie's Notes

Example 2

- After Example 1, it should be clear to students that this composite figure will be divided into a semicircle and a rectangle.
- ? "What are the dimensions of the rectangle?" 19 ft by 12 ft
- ? "What is the radius of the circle?" 6 ft
- In calculating the area of the semicircle, watch for student errors in multiplying and dividing decimals.
- **Extension:** Ask students if the area of a semicircle of radius 6 is the same as the area of a circle of radius 3. The answer is no! 18π compared to 9π

Example 3

- Draw the composite figure without the interior dotted lines. Ask students to think about how the figure could be divided into regions with known area formulas. They may come up with more than three regions.
- Label the diagram with the dimensions given. I find it helpful to draw small arrows from the measurement to the segment it is associated with, especially for interior dimensions.
- Work through the problem, questioning students along the way about the process.
- ? "How do you find the area of a triangle? How do you multiply $\frac{1}{2}$ by 11.2? What units are used to label area?"
- **Common Misconception:** In a right triangle, the height is a side length. It is not an interior measurement.

On Your Own

- Have students share their strategies for these two problems. Question 4 might be done as 4 semicircles or 2 circles.

Closure

- **Exit Ticket:** Find the area. 100 $units^2$

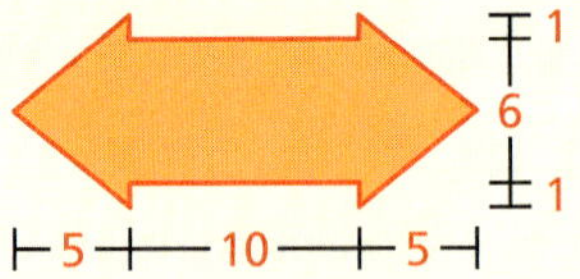

EXAMPLE 2 Finding an Area

Find the area of the portion of the basketball court shown.

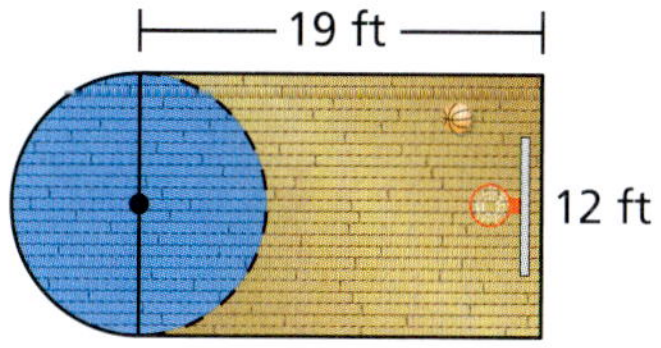

The figure is made up of a rectangle and a semicircle. Find the area of each figure.

Area of Rectangle

$A = \ell w$

$= 19(12)$

$= 228$

Area of Semicircle

$A = \dfrac{\pi r^2}{2}$

$\approx \dfrac{3.14 \cdot 6^2}{2}$

$= 56.52$

The semicircle has a radius of $\dfrac{12}{2} = 6$ feet.

So, the area is about $228 + 56.52 = 284.52$ square feet.

EXAMPLE 3 Finding an Area

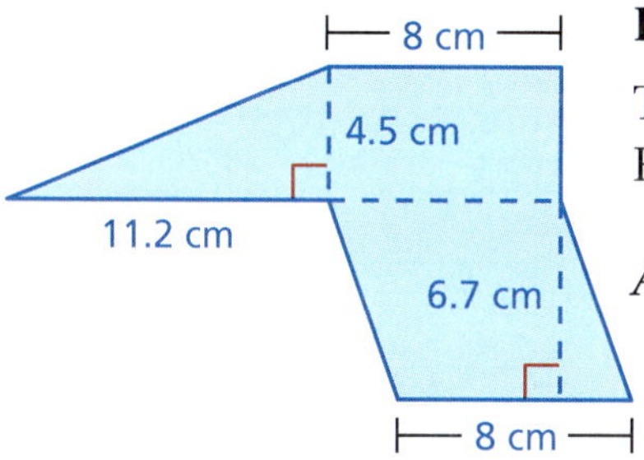

Find the area of the figure.

The figure is made up of a triangle, a rectangle, and a parallelogram. Find the area of each figure.

Area of Triangle

$A = \dfrac{1}{2}bh$

$= \dfrac{1}{2}(11.2)(4.5)$

$= 25.2$

Area of Rectangle

$A = \ell w$

$= 8(4.5)$

$= 36$

Area of Parallelogram

$A = bh$

$= 8(6.7)$

$= 53.6$

So, the area is $25.2 + 36 + 53.6 = 114.8$ square centimeters.

On Your Own

Now You're Ready
Exercises 9 and 10

Find the area of the figure.

3.

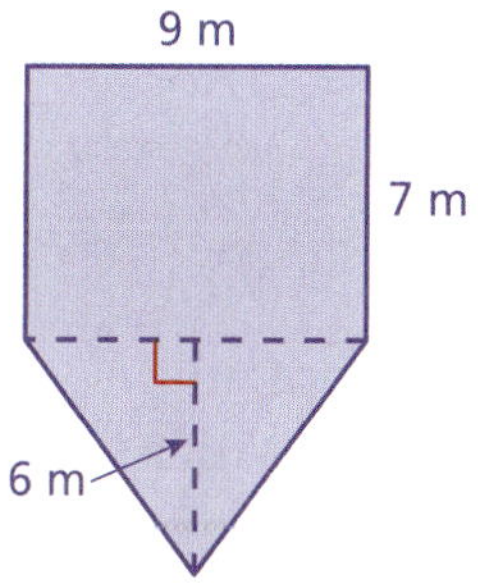

4.

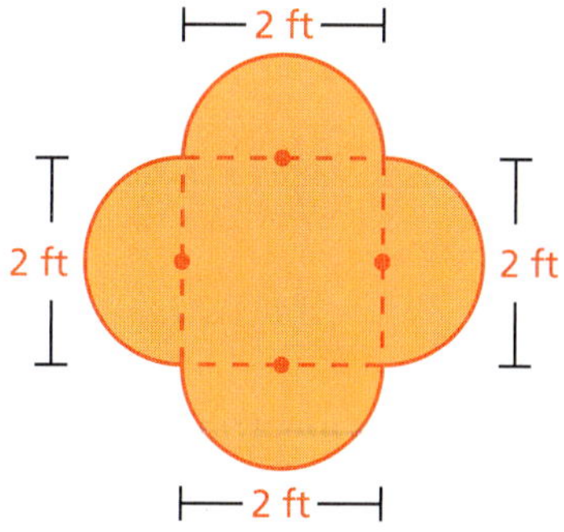

13.4 Exercises

Check It Out
Help with Homework
BigIdeasMath.com

Vocabulary and Concept Check

1. **REASONING** Describe two different ways to find the area of the figure. Name the types of figures you used and the dimensions of each.

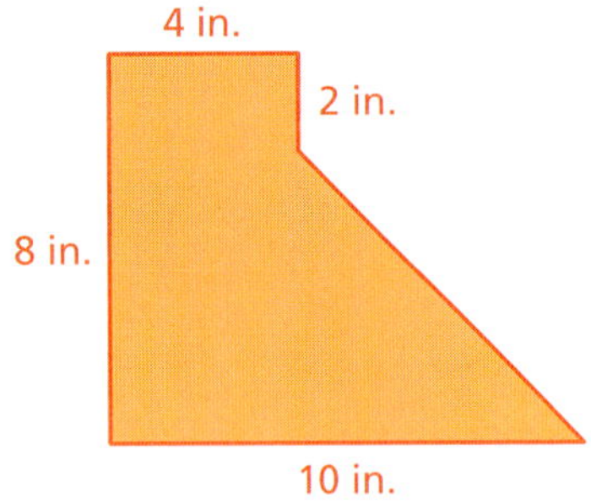

2. **REASONING** Draw a trapezoid. Explain how you can think of the trapezoid as a composite figure to find its area.

Practice and Problem Solving

Find the area of the figure.

1 3. 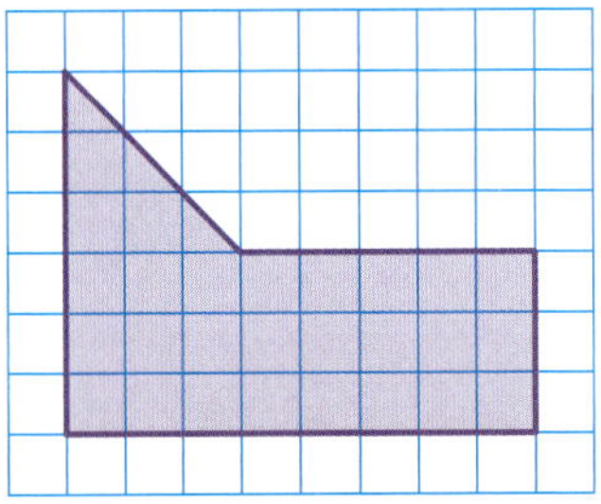

4. 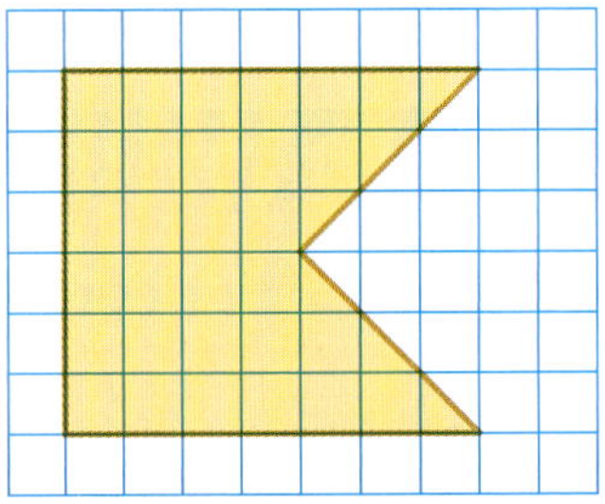

5. 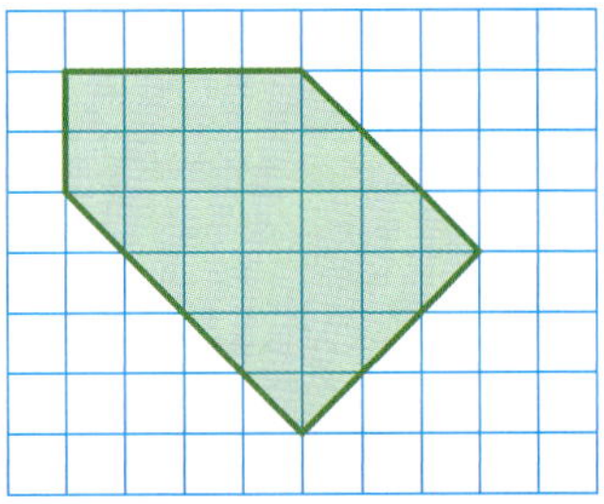

6. 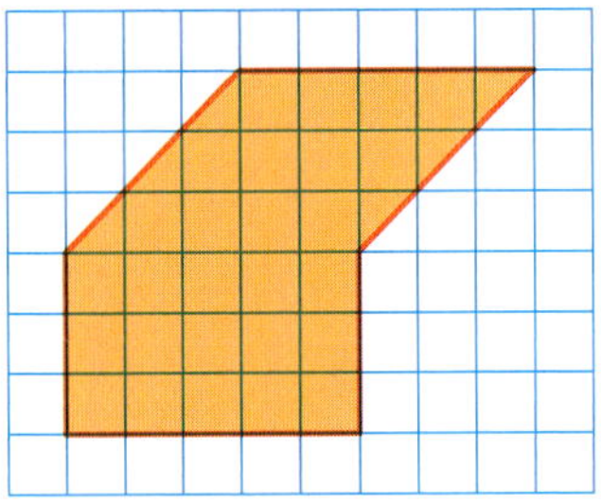

7. 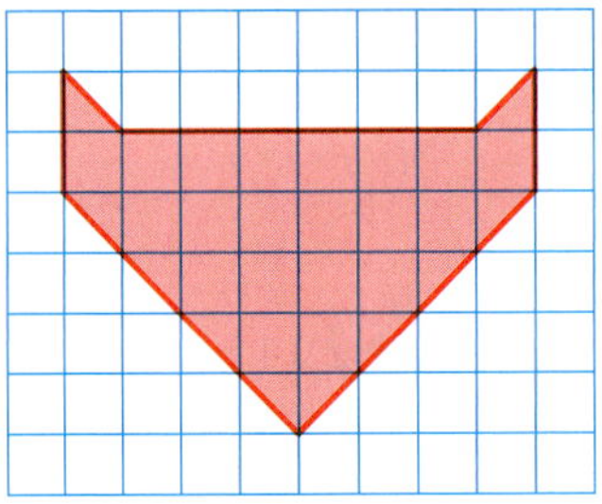

8. 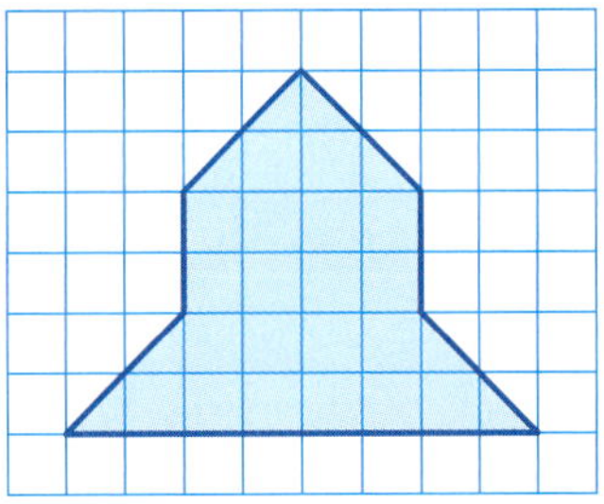

Find the area of the figure.

2 3 9. 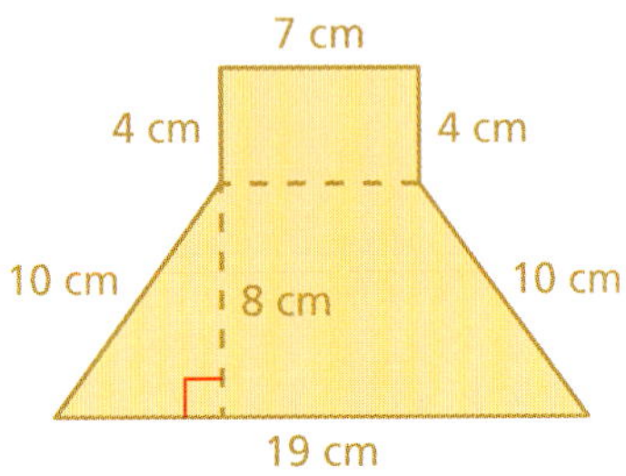

10. 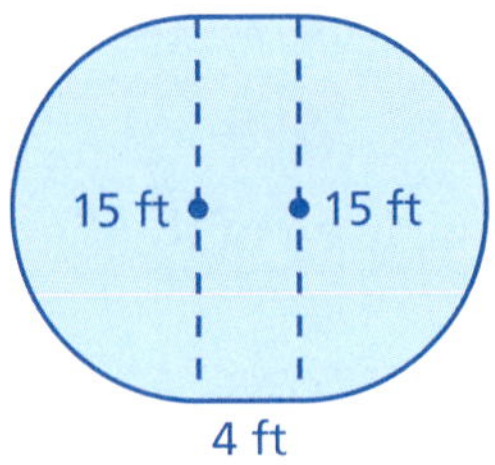

11. **OPEN-ENDED** Trace your hand and your foot on grid paper. Then estimate the area of each. Which one has the greater area?

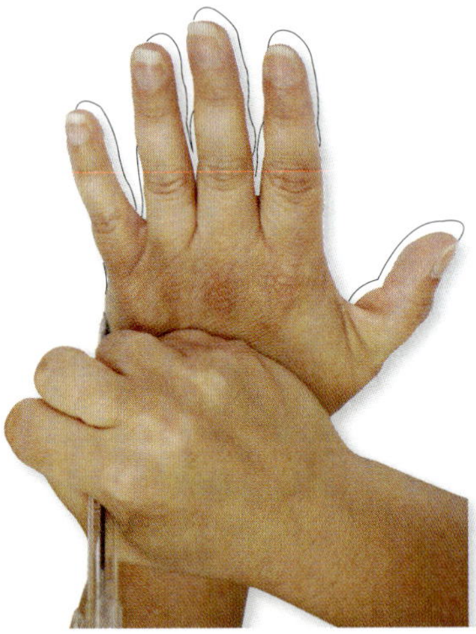

Assignment Guide and Homework Check

Level	Assignment	Homework Check
Advanced	1–5, 8–10, 12–22	8, 12, 15, 16

Common Errors

- **Exercises 3–8** Students may forget to count all the squares inside the figure and just count the ones along the border because this is what they did for perimeter. Remind them that the area includes everything inside the figure.
- **Exercises 9 and 10** Students may forget to include one of the areas of the composite figures or may count one area more than once. Tell them to break apart the figure into several figures. Draw and label each figure and then find the area of each part. Finally add the areas of each part together for the area of the whole figure.
- **Exercises 12–14** Students may forget to subtract the unshaded area from the figure. Remind them that in this situation instead of adding on the area of a figure they must subtract a portion out. Give a real-life example to help students understand, like tiling a bathroom floor but taking out the part where the sink or toilet is.

13.4 Record and Practice Journal

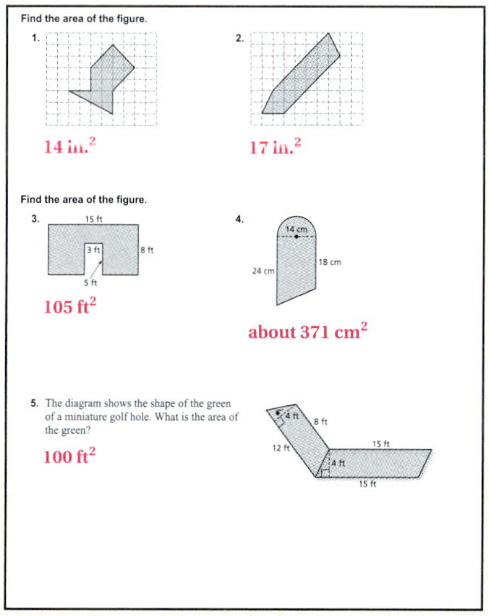
Find the area of the figure.

1. 14 in.^2

2. 17 in.^2

Find the area of the figure.

3. 105 ft^2

4. about 371 cm^2

5. The diagram shows the shape of the green of a miniature golf hole. What is the area of the green?

100 ft^2

Vocabulary and Concept Check

1. *Sample answer:* You could add the areas of an 8-inch × 4-inch rectangle and a triangle with a base of 6 inches and a height of 6 inches. Also you could add the area of a 2-inch × 4-inch rectangle to the area of a trapezoid with a height of 6 inches, and base lengths of 4 inches and 10 inches.
2. *Sample answer:* You can think of the trapezoid as a rectangle and two triangles.

Practice and Problem Solving

3. 28.5 units^2
4. 33 units^2
5. 25 units^2
6. 30 units^2
7. 25 units^2
8. 24 units^2
9. 132 cm^2
10. about 236.625 ft^2
11. *Answer should include, but is not limited to:* Tracings of a hand and foot on grid paper, estimates of the areas, and a statement of which is greater.

Practice and Problem Solving

12–15. See Additional Answers.

16. P = about 94.2 ft
A = about 628 ft^2

17. See *Taking Math Deeper*.

Fair Game Review

18. $x - 12$ **19.** $y \div 6$

20. $b + 3$ **21.** $7w$

22. A

Mini-Assessment

Find the area of the figure.

1.

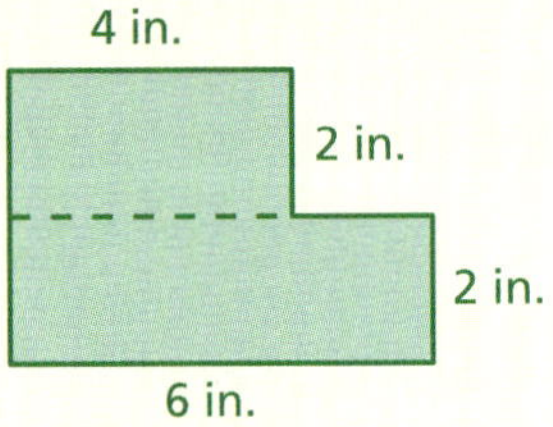

20 in.^{2}

2.

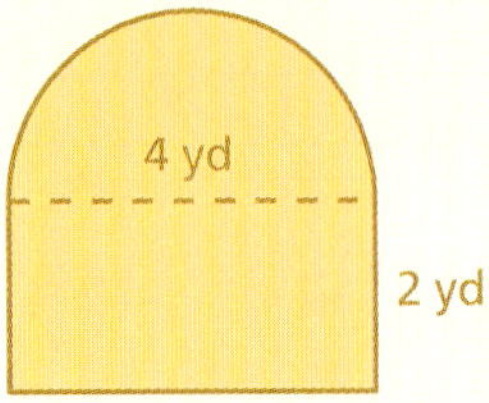

about 14.28 yd^2

3.

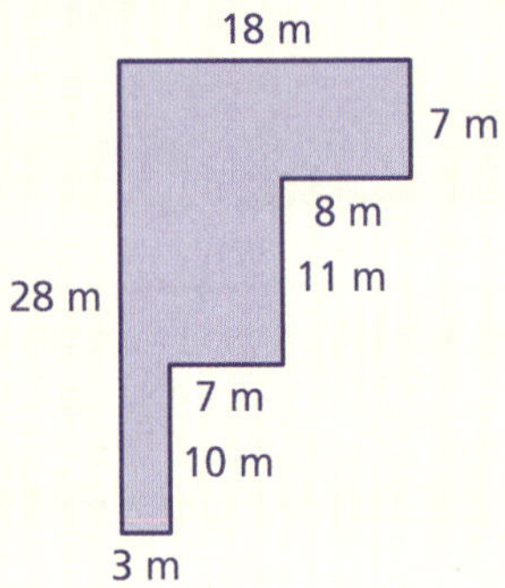

266 m^2

4. Find the area of the red region.
about 3.44 in.^{2}

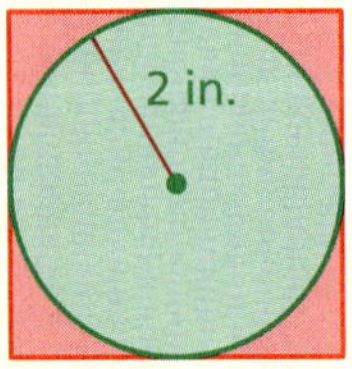

Taking Math Deeper

Exercise 17

This problem is not conceptually difficult. However, it does have a lot of calculations and it is difficult to do without a calculator. Before starting the calculations, suggest that students think about which design is more efficient. Each design will fit on an 11-inch × 17-inch sheet of paper. By cutting and folding each pattern, a student can see how much overlap each design has. The design with the greatest overlap is the least efficient.

1 Find the area of the first design.

$\frac{1}{2} \cdot 5.5 \cdot 2.5 = 6.875$

$\frac{1}{2} \cdot 4.5 \cdot 3 = 6.75$

$4.5 \cdot 5.5 = 24.75$

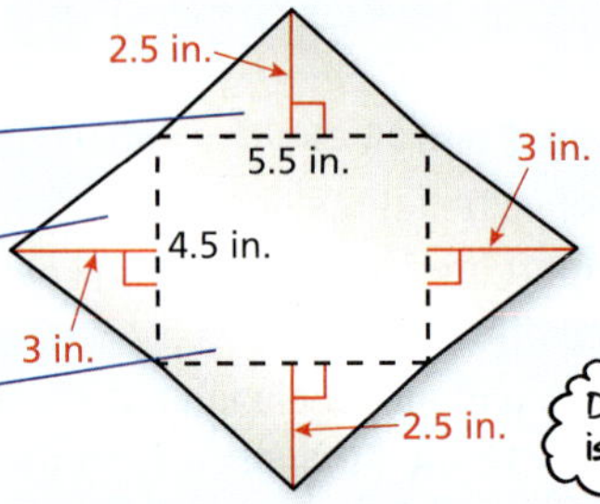

$A = 2(6.875) + 2(6.75) + 24.75$
$= 52 \text{ in.}^2$

2 Find the area of the second design.

$2 \cdot 5.5 = 11$

$\frac{1}{2}(4 + 4.5) \cdot 0.75 = 3.1875$

$4.5 \cdot 5.5 = 24.75$

$3.5 \cdot 5.5 = 19.25$

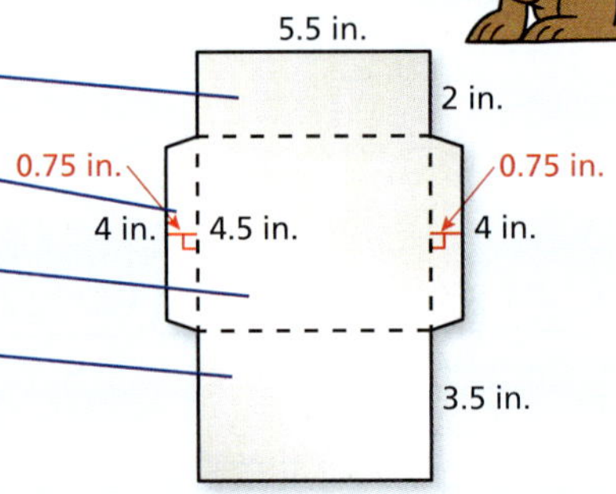

$A = 11 + 24.75 + 19.25 + 2(3.1875)$
$= 61.375 \text{ in.}^2$

3 **a.** So, the second design has the greater area. Answer the question.

$500 \cdot 61.375 = 30{,}687.5 \text{ in.}^2$ (500 envelopes using Design 2)

$\frac{30{,}687.5}{52} \approx 590.1$ (number of envelopes using Design 1)

b. You could make 90 more envelopes using Design 1.

Reteaching and Enrichment Strategies

If students need help. . .	If students got it. . .
Resources by Chapter • Practice A and Practice B • Puzzle Time Record and Practice Journal Practice Differentiating the Lesson Lesson Tutorials Skills Review Handbook	Resources by Chapter • Enrichment and Extension • Technology Connection Start the next section

Find the area of the figure.

12.

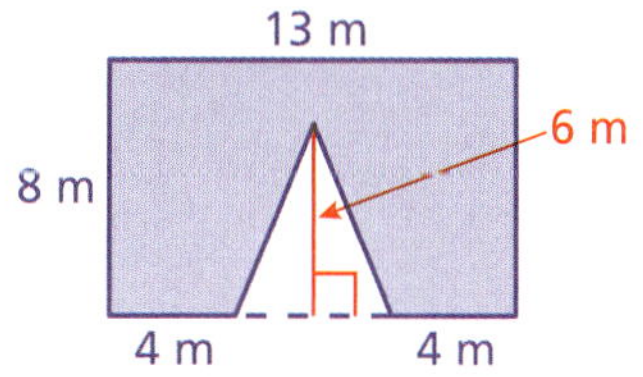

13.

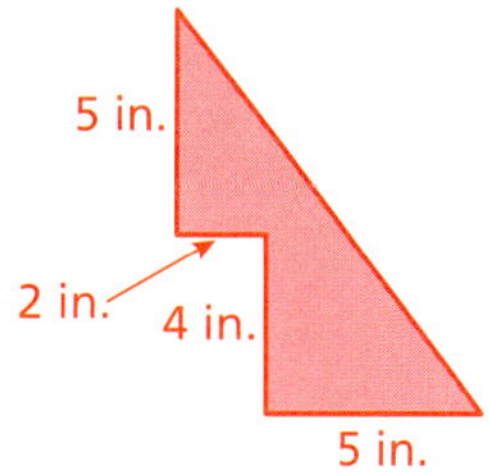

14.

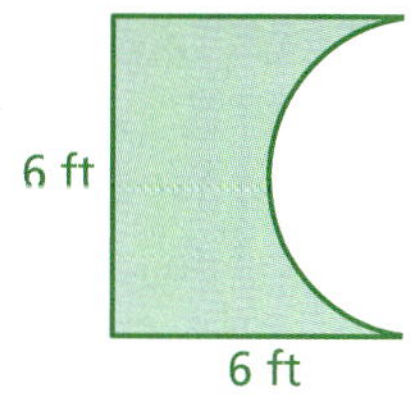

15. **STRUCTURE** The figure is made up of a square and a rectangle. Find the area of the shaded region.

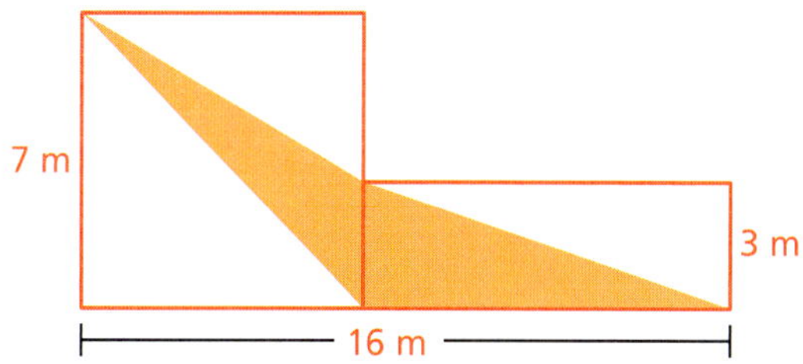

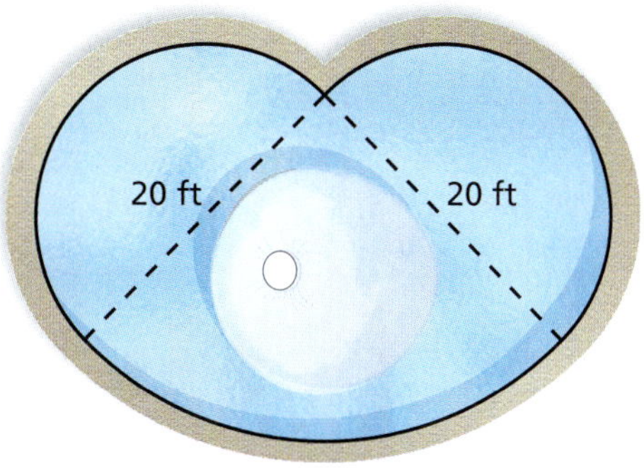

16. **FOUNTAIN** The fountain is made up of two semicircles and a quarter circle. Find the perimeter and the area of the fountain.

17. Critical Thinking You are deciding on two different designs for envelopes.

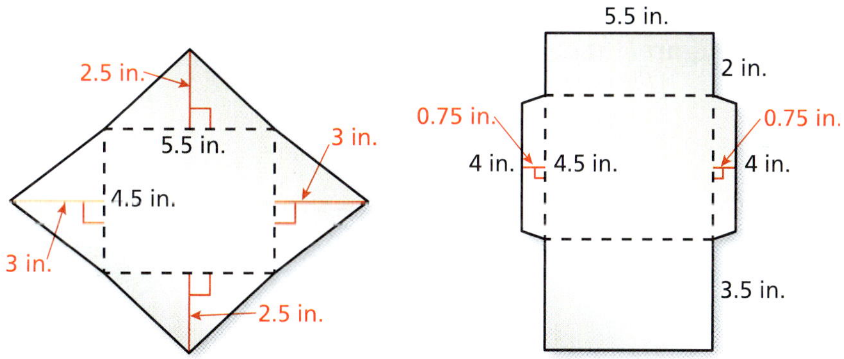

a. Which design has the greater area?

b. You make 500 envelopes using the design with the greater area. Using the same amount of paper, how many more envelopes can you make with the other design?

Fair Game Review What you learned in previous grades & lessons

Write the phrase as an expression. *(Skills Review Handbook)*

18. 12 less than a number x

19. a number y divided by 6

20. a number b increased by 3

21. the product of 7 and a number w

22. **MULTIPLE CHOICE** What number is 0.02% of 50? *(Skills Review Handbook)*

Ⓐ 0.01 Ⓑ 0.1 Ⓒ 1 Ⓓ 100

13.3–13.4 Quiz

Check It Out
Progress Check
BigIdeasMath.com

Find the area of the figure. *(Section 13.4)*

1.

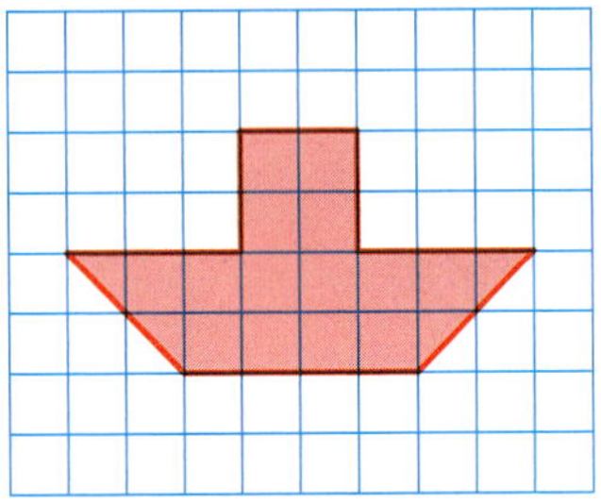

2.

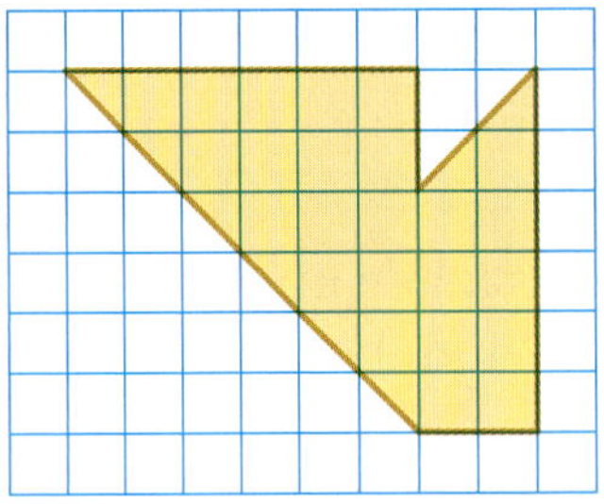

3.

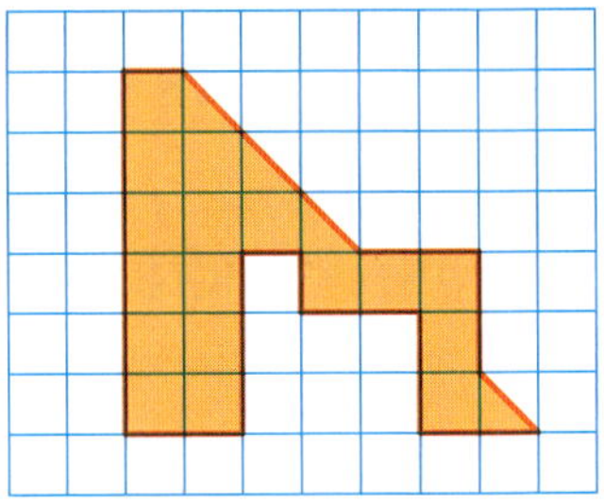

Find the area of the circle. Use 3.14 or $\frac{22}{7}$ for π. *(Section 13.3)*

4.

5.

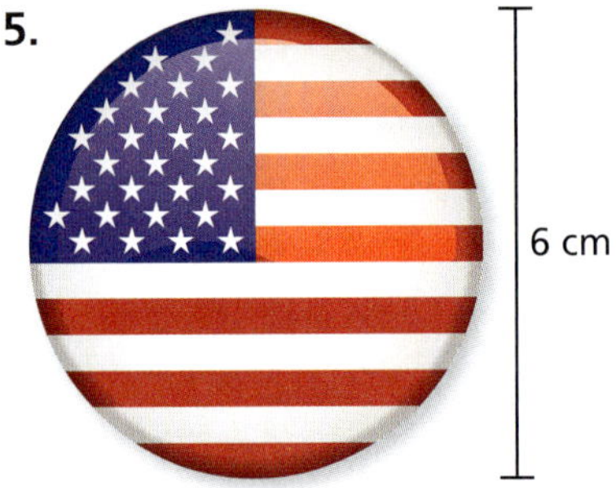

6.

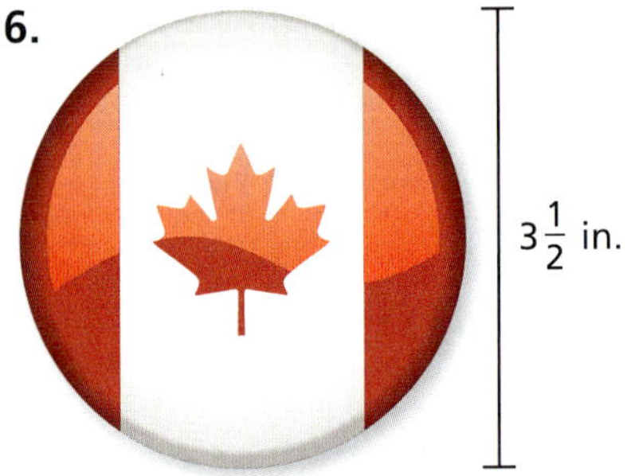

Find the area of the figure. *(Section 13.4)*

7.

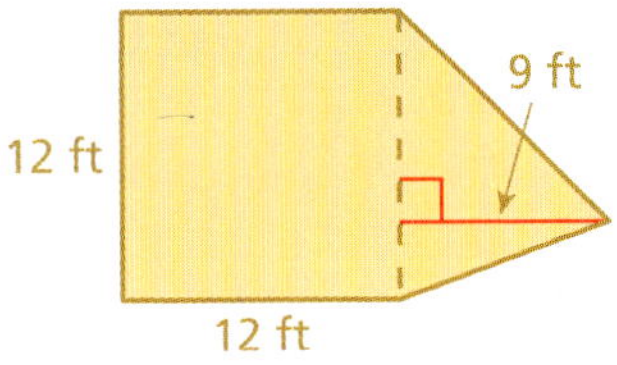

8.

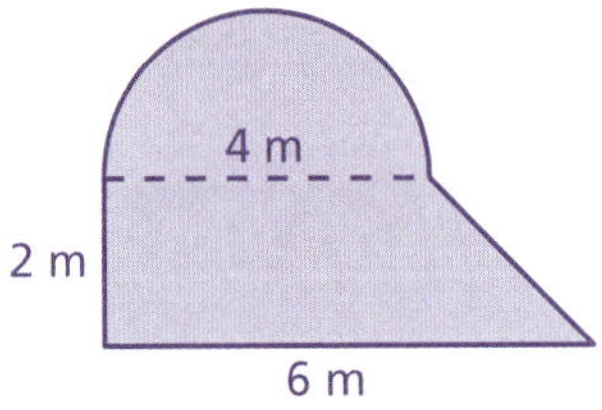

9.

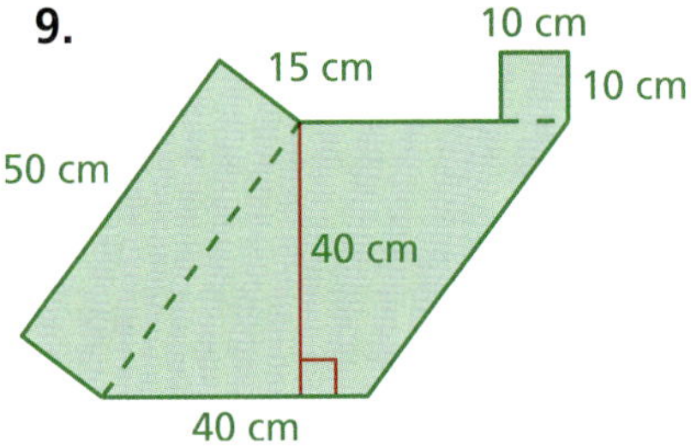

10. POT HOLDER A knitted pot holder is shaped like a circle. Its radius is 3.5 inches. What is its area? *(Section 13.3)*

11. CARD The heart-shaped card is made up of a square and two semicircles. What is the area of the card? *(Section 13.4)*

12. DESK A desktop is shaped like a semicircle with a diameter of 28 inches. What is the area of the desktop? *(Section 13.3)*

13. RUG The circular rug is placed on a square floor. The rug touches all four walls. How much of the floor space is *not* covered by the rug? *(Section 13.4)*

Alternative Assessment Options

Math Chat	Student Reflective Focus Question
Structured Interview	Writing Prompt

Math Chat

- Have students work in pairs. Assign Exercises 10–13 from the Quiz to each pair. Each student works through all four problems. After the students have worked through the problems, they take turns talking through the process that they used to get the answer. Students analyze and evaluate the mathematical thinking and strategies used.
- The teacher should walk around the classroom listening to the pairs and asking questions to ensure understanding.

Study Help Sample Answers

Remind students to complete Graphic Organizers for the rest of the chapter.

4.

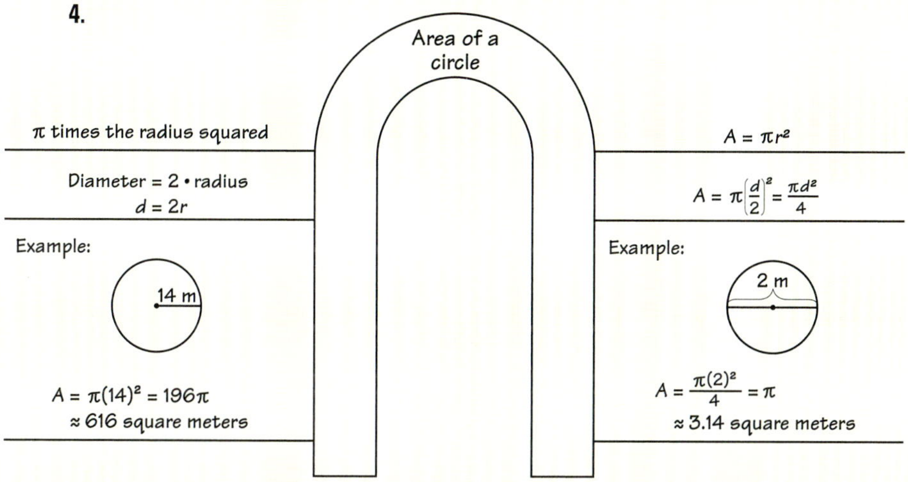

5.

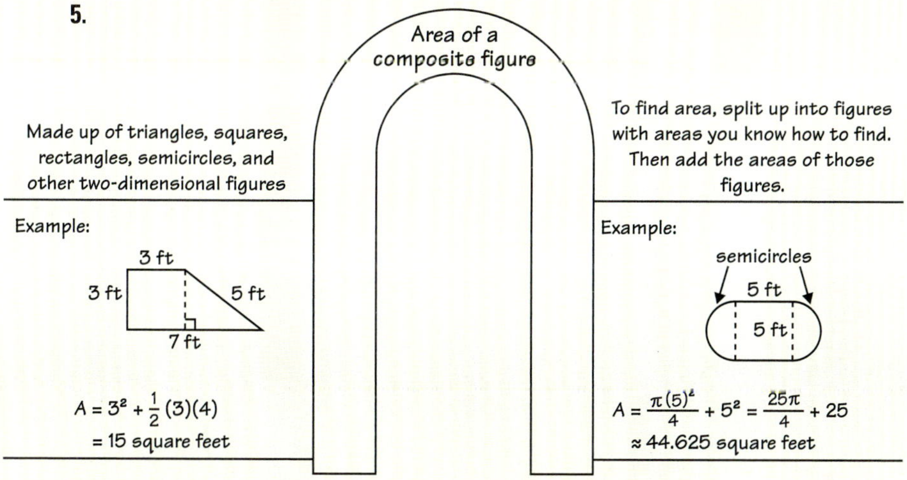

Reteaching and Enrichment Strategies

If students need help. . .	If students got it. . .
Resources by Chapter • Practice A and Practice B • Puzzle Time Lesson Tutorials *BigIdeasMath.com*	Resources by Chapter • Enrichment and Extension • Technology Connection Game Closet at *BigIdeasMath.com* Start the Chapter Review

Answers

1. 16 units^2
2. 28 units^2
3. 19 units^2
4. about 452.16 in.^2
5. about 28.26 cm^2
6. about 9.625 in.^2
7. 198 ft^2
8. about 16.28 m^2
9. 2450 cm^2
10. about 38.465 in.^2
11. about 114.24 cm^2
12. about 308 in.^2
13. about 42 ft^2

Technology for the Teacher

Online Assessment
Assessment Book
ExamView® Assessment Suite

For the Teacher
Additional Review Options

- *BigIdeasMath.com*
- Online Assessment
- Game Closet at *BigIdeasMath.com*
- Vocabulary Help
- Resources by Chapter

Answers

1. 4 in.
2. 30 mm
3. 50 m
4. 1.5 yd
5. 40 ft
6. 10 m
7. 2 in.
8. 50 mm
9. about 18.84 ft
10. about 66 cm
11. about 132 in.

Review of Common Errors

Exercises 1–8

- Students may confuse what they are finding and double the diameter or halve the radius. Remind them that the radius is half the diameter.

Exercises 9–11

- Students may use the radius in the circumference formula that calls for diameter or the diameter in the circumference formula that calls for radius. Remind them of the different formulas.

Exercises 12–17

- Students may include the dashed lines in their calculation of the perimeter. Remind them that the dashed lines are for reference and sometimes give information to find another length. Only the outside lengths are counted.

Exercises 18–20

- Students may forget to divide the diameter by 2 to find the area. Remind them that they need the *radius* in the given formula for area of a circle.
- Students may use the formula for circumference or forget to square the radius when finding the area.

Exercises 21–23

- Students may forget to include the area of one or more parts of a composite figure or count a part more than once. Tell them to draw and label each part, find the area of each part, and add the areas of the parts to find the area of the composite figure.

13 Chapter Review

Check It Out
Vocabulary Help
BigIdeasMath.com

Review Key Vocabulary

circle, *p. 550*	circumference, *p. 551*
center, *p. 550*	pi, *p. 551*
radius, *p. 550*	semicircle, *p. 552*
diameter, *p. 550*	composite figure, *p. 558*

Review Examples and Exercises

13.1 Circles and Circumference *(pp. 548–555)*

Find the circumference of the circle. Use 3.14 for π.

The radius is 4 millimeters.

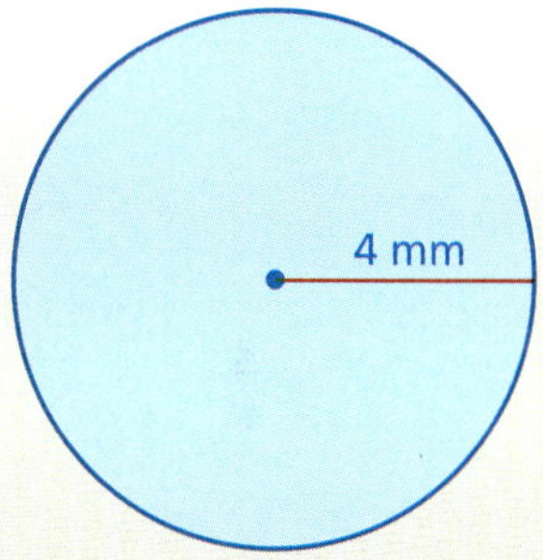

$C = 2\pi r$	Write formula for circumference.
$\approx 2 \cdot 3.14 \cdot 4$	Substitute 3.14 for π and 4 for r.
$= 25.12$	Multiply.

The circumference is about 25.12 millimeters.

Exercises

Find the radius of the circle with the given diameter.

1. 8 inches

2. 60 millimeters

3. 100 meters

4. 3 yards

Find the diameter of the circle with the given radius.

5. 20 feet

6. 5 meters

7. 1 inch

8. 25 millimeters

Find the circumference of the circle. Use 3.14 or $\frac{22}{7}$ for π.

9.

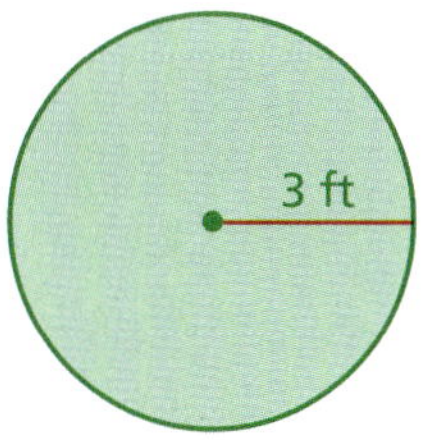

10.

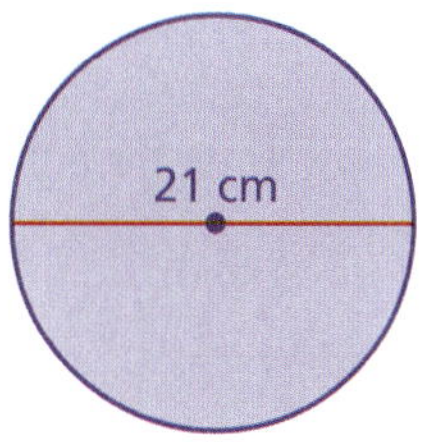

11.

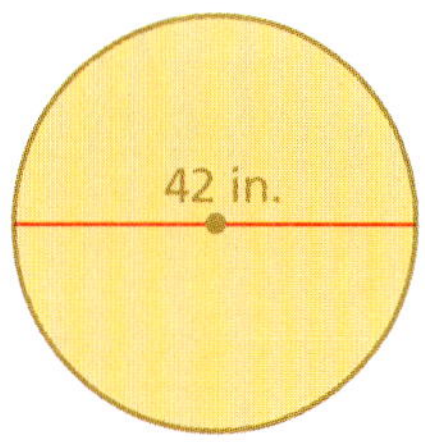

13.2 Perimeters of Composite Figures (pp. 556–561)

The figure is made up of a semicircle and a square. Find the perimeter.

The distance around the square part is 6 + 6 + 6 = 18 meters. The distance around the semicircle is one-half the circumference of a circle with $d = 6$ meters.

$\frac{C}{2} = \frac{\pi d}{2}$ Divide the circumference by 2.

$\approx \frac{3.14 \cdot 6}{2}$ Substitute 3.14 for π and 6 for d.

$= 9.42$ Simplify.

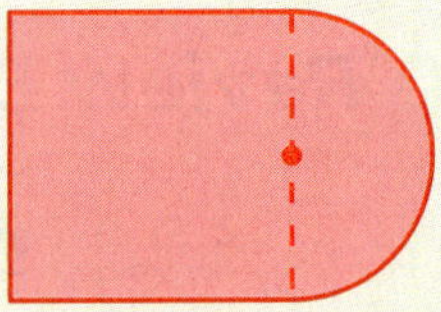

So, the perimeter is about 18 + 9.42 = 27.42 meters.

Exercises

Find the perimeter of the figure.

12.

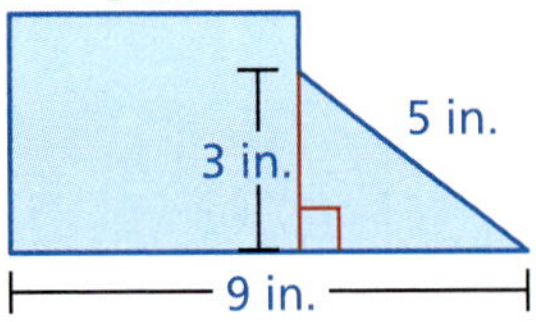

13.

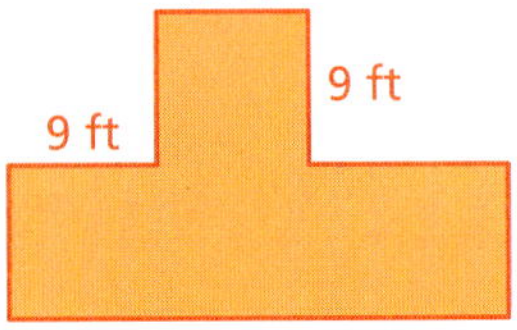

14.

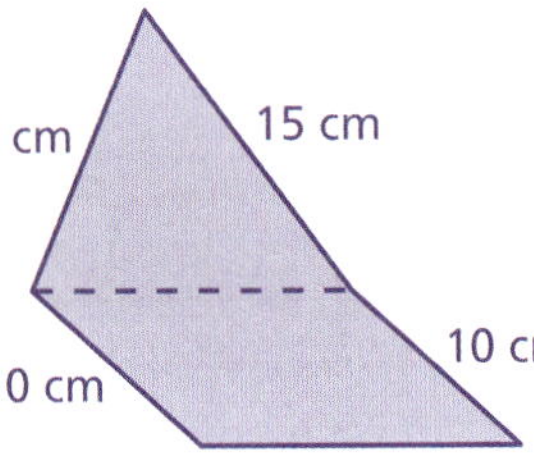

15.

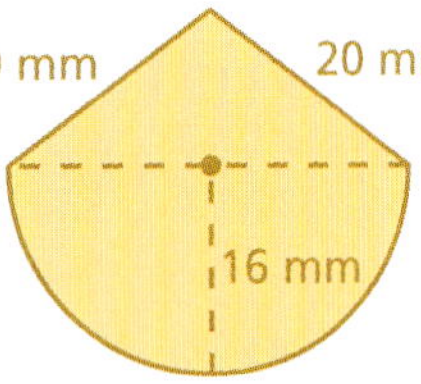

16.

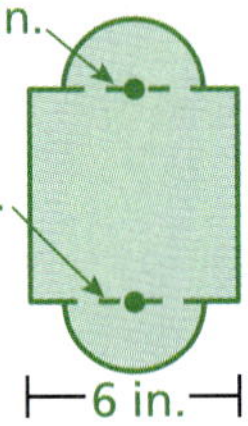

17.

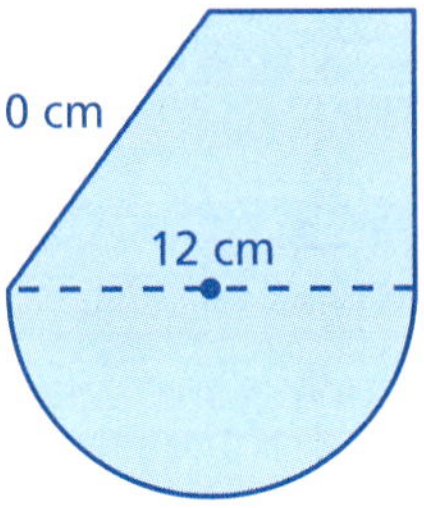

13.3 Areas of Circles (pp. 564–569)

Find the area of the circle. Use 3.14 for π.

$A = \pi r^2$ Write formula for area.

$\approx 3.14 \cdot 20^2$ Substitute 3.14 for π and 20 for r.

$= 1256$ Multiply.

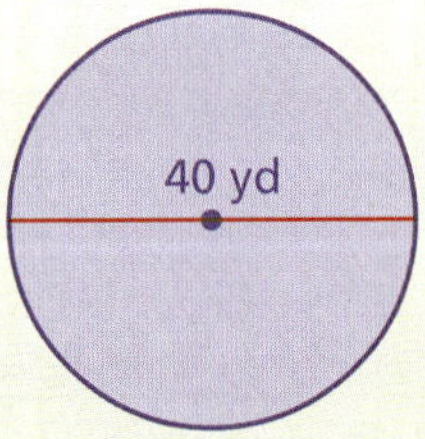

The area is about 1256 square yards.

Review Game

Area, Perimeter, and Circumference

Materials per Pair:

- paper
- pencils (colored pencils optional)

Directions:

Each group of four divides into teams of two. Each team competes with the other team. The class is directed to draw and label a composite figure using selected shapes and a predetermined area. For example, the teacher could say, "Using squares, rectangles, and circles, draw and label a composite figure with an area of 50 square units." Each team then draws their figure and calculates the area. Papers are exchanged within the group and teams check each other's work. The team whose correctly-drawn figure has an area that is closest to the area specified receives 1 point. If both teams are equally close, both teams receive 1 point. Play for a set amount of time.

Who Wins?

When time is up, the team with the most points wins.

For the Student Additional Practice

- Lesson Tutorials
- Multi-Language Glossary
- Self-Grading Progress Check
- *BigIdeasMath.com*
  Dynamic Student Edition
  Student Resources

Answers

12. 24 in.

13. 96 ft

14. 62 cm

15. about 90.24 mm

16. about 28.56 in.

17. about 42.84 cm

18. about 50.24 in.^{2}

19. about 379.94 cm^2

20. about 1386 mm^2

21. about 79.25 in.^{2}

22. 29 in.^{2}

23. about 31.625 ft^2

My Thoughts on the Chapter

Teacher Tip

Not allowed to write in your teaching edition? Use sticky notes to record your thoughts.

What worked. . .

What did not work. . .

What I would do differently. . .

Exercises

Find the area of the circle. Use 3.14 or $\frac{22}{7}$ for π.

18.

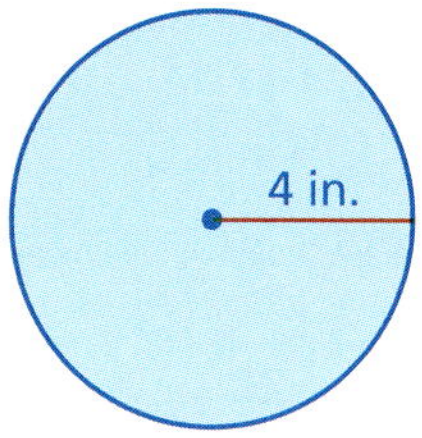

19.

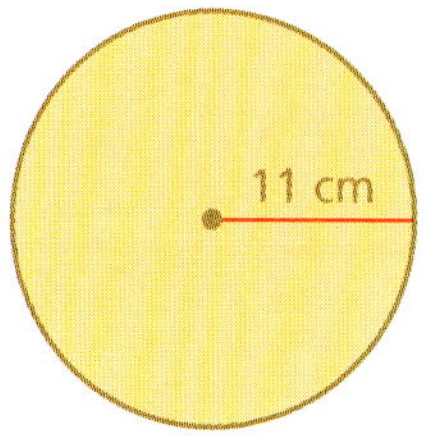

20.

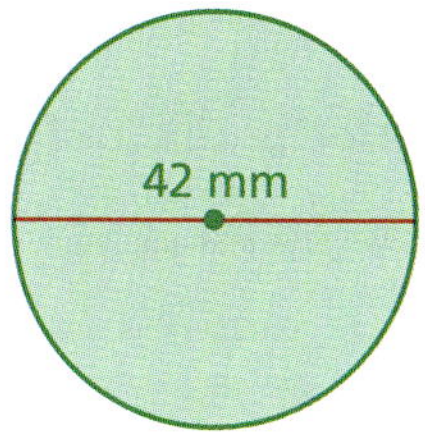

13.4 Areas of Composite Figures *(pp. 570–575)*

Find the area of the figure.

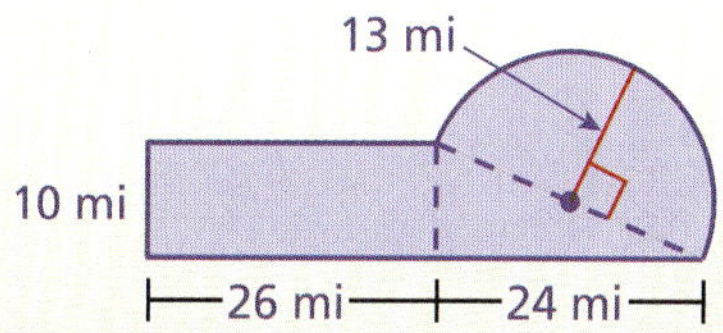

The figure is made up of a rectangle, a triangle and a semicircle. Find the area of each figure.

Area of Rectangle	***Area of Triangle***	***Area of Semicircle***
$A = \ell w$	$A = \frac{1}{2}bh$	$A = \frac{\pi r^2}{2}$
$= 26(10)$	$= \frac{1}{2}(10)(24)$	$\approx \frac{3.14 \cdot 13^2}{2}$
$= 260$	$= 120$	$= 265.33$

So, the area is about $260 + 120 + 265.33 = 645.33$ square miles.

Exercises

Find the area of the figure.

21.

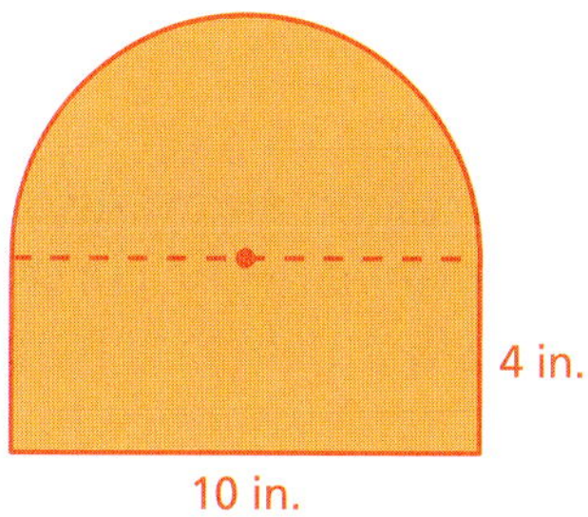

22.

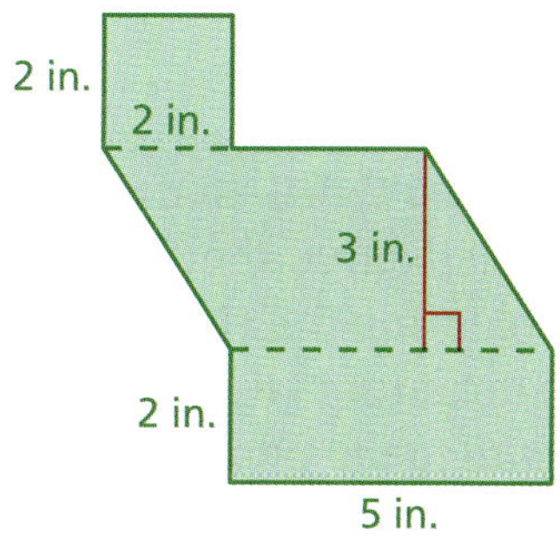

23.

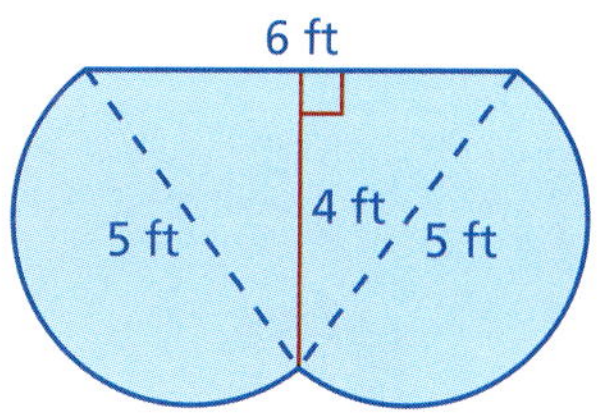

13 Chapter Test

Check It Out
Test Practice
BigIdeasMath.com

Find the radius of the circle with the given diameter.

1. 10 inches

2. 5 yards

Find the diameter of the circle with the given radius.

3. 34 feet

4. 19 meters

Find the circumference and the area of the circle. Use 3.14 or $\frac{22}{7}$ for π.

5.

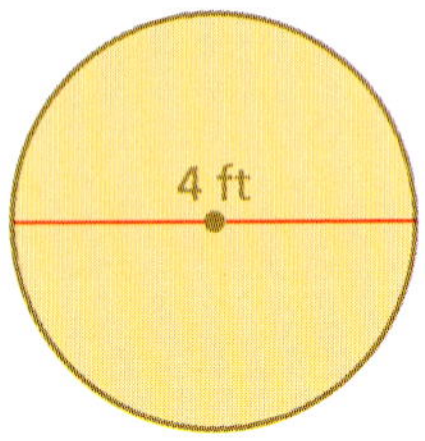

6.

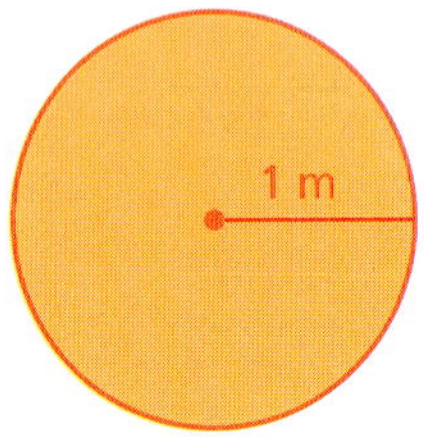

7.

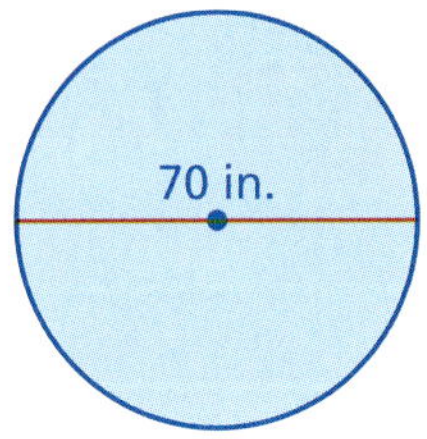

8. Estimate the perimeter of the figure. Then find the area.

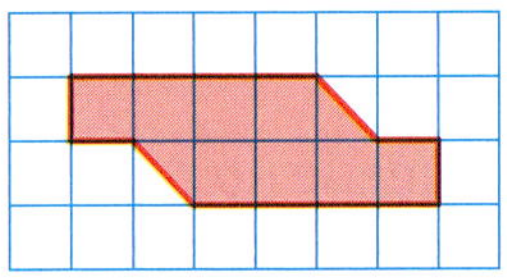

Find the perimeter and the area of the figure. Use 3.14 or $\frac{22}{7}$ for π.

9.

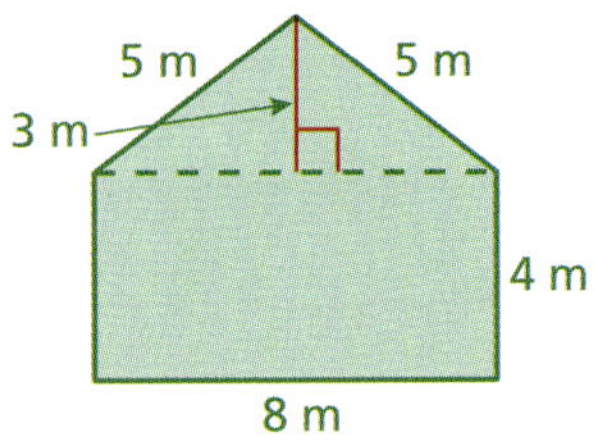

10.

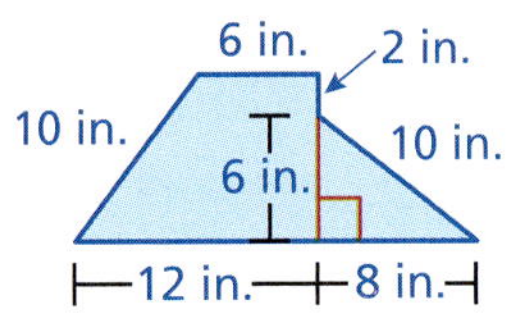

11.

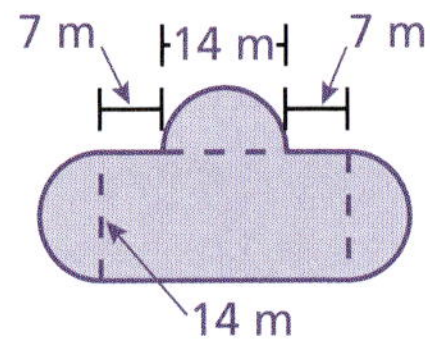

12. **MUSEUM** A museum plans to rope off the perimeter of the L-shaped exhibit. How much rope does it need?

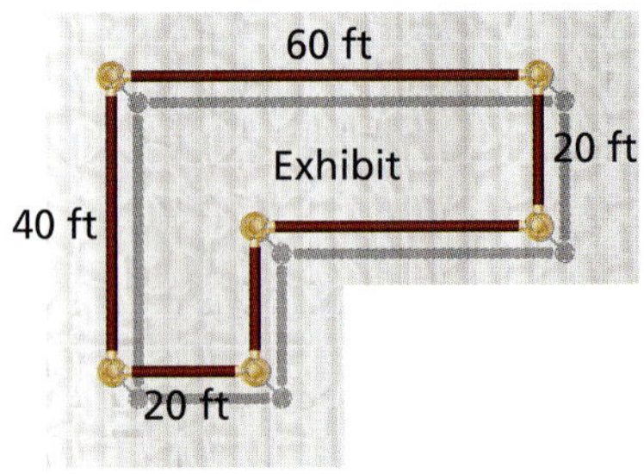

13. **ANIMAL PEN** You unfold chicken wire to make a circular pen with a diameter of 2.9 meters. How many meters of chicken wire do you need?

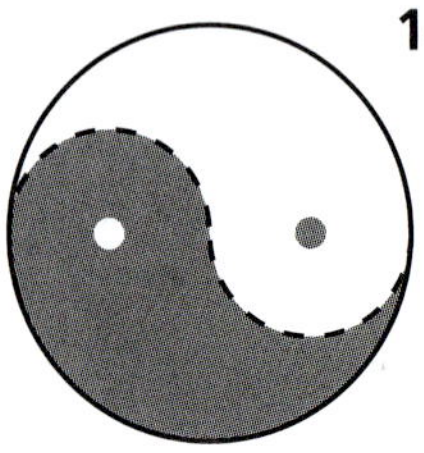

14. **YIN AND YANG** In the Chinese symbol for yin and yang, the dashed curve shows two semicircles formed by the curve separating the yin (dark) and the yang (light). Is the circumference of the entire yin and yang symbol *less than*, *greater than*, or *equal to* the perimeter of the yin?

Test Item References

Chapter Test Questions	Section to Review
1–7, 13, 14	13.1
8–12, 14	13.2
5–7	13.3
8–11	13.4

Test-Taking Strategies

Remind students to quickly look over the entire test before they start so that they can budget their time. There is a lot of vocabulary in this chapter, so students should have been making flash cards as they worked through the chapter. Words that get mixed up should be jotted on the back of the test before they start. (For instance, they should write down *radius* and *circumference*.) Students need to use the **Stop** and **Think** strategy before they answer the questions.

Common Errors

- **Exercises 1–4** Students may confuse what they are finding and double the diameter or halve the radius. Remind them that the radius is half the diameter. Encourage students to draw a line segment representing each radius or diameter so that they have a visual for reference.
- **Exercises 5–7** Students may use the radius in a formula that calls for diameter, the diameter in a formula that calls for radius, the area formula when calculating circumference, or the circumference formula when calculating area. Again, remind students that the radius is half the diameter. Also, remind them of the circumference and area formulas.
- **Exercise 8** Students may count the sides of the squares inside the figure when finding the perimeter or confuse area and perimeter. Remind students of how to find perimeter and area.
- **Exercises 9–11** When calculating the perimeter of the figure, students may count the dashed lines. Remind them that the dashed lines are for reference and that only the outside lengths are counted. When calculating the area of the figure, students may forget to include the area of one or more parts or count part(s) more than once. Tell them to draw and label each part, find the area of each part, and add the areas of the parts.

Reteaching and Enrichment Strategies

If students need help. . .	If students got it. . .
Resources by Chapter • Practice A and Practice B • Puzzle Time Record and Practice Journal Practice Differentiating the Lesson Lesson Tutorials *BigIdeasMath.com* Skills Review Handbook	Resources by Chapter • Enrichment and Extension • Technology Connection Game Closet at *BigIdeasMath.com* Start Cumulative Assessment

Answers

1. 5 in.
2. 2.5 yd
3. 68 ft
4. 38 m
5. $C \approx 12.56$ ft, $A \approx 12.56$ ft^2
6. $C \approx 6.28$ m, $A \approx 3.14$ m^2
7. $C \approx 220$ in., $A \approx 3850$ in.^{2}
8. $P \approx 15$ units, $A = 9$ units^2
9. $P = 26$ m, $A = 44$ m^2
10. $P = 48$ in., $A = 96$ in.^{2}
11. $P \approx 108$ m, $A \approx 623$ m^2
12. 200 ft
13. about 9.106 m
14. *equal to*

Technology for the Teacher

Online Assessment
Assessment Book
ExamView® Assessment Suite

Test-Taking Strategies

Available at *BigIdeasMath.com*

After Answering Easy Questions, Relax
Answer Easy Questions First
Estimate the Answer
Read All Choices before Answering
Read Question before Answering
Solve Directly or Eliminate Choices
Solve Problem before Looking at Choices
Use Intelligent Guessing
Work Backwards

About this Strategy

When taking a multiple choice test, be sure to read each question carefully and thoroughly. When taking a timed test, it is often best to skim the test and answer the easy questions first. Be careful that you record your answer in the correct position on the answer sheet.

Answers

1. C
2. 42
3. I
4. C

Technology for the Teacher

Performance Tasks
Online Assessment
Assessment Book
ExamView® Assessment Suite

Item Analysis

1. **A.** The student does not convert to the same units (either 2 quarts to 8 cups or 5 cups to 1.25 quarts) and inverts one of the ratios.

   B. The student does not convert to the same units (either 2 quarts to 8 cups or 5 cups to 1.25 quarts).

   C. Correct answer

   D. The student uses the correct numbers but inverts one of the ratios.

2. **Gridded Response:** Correct answer: 42

   Common Error: The student sets up the equation incorrectly. For instance, $2x + 1 + 85 = 180$ and finds $x = 47$.

3. **F.** On the left side of the equation, the student writes an expression that represents the product of n and a number that is 7 less than 5.

   G. On the left side of the equation, the student writes an expression that represents the product of n and a number that is 5 less than 7.

   H. On the left side of the equation, the student misinterprets the use of "less than" and reverses the order of the subtraction.

   I. Correct answer

4. **A.** The student did not square the radius.

   B. The student multiplied the radius by 2 instead of squaring it.

   C. Correct answer

   D. The student squared the diameter instead of the radius.

13 Cumulative Assessment

Test-Taking Strategy
Answer Easy Questions First

"Scan the test and answer the easy questions first. You know that the radius is half the diameter."

1. To make 6 servings of soup, you need 5 cups of chicken broth. You want to know how many servings you can make with 2 quarts of chicken broth. Which proportion should you use?

   A. $\frac{6}{5} = \frac{2}{x}$ **C.** $\frac{6}{5} = \frac{x}{8}$

   B. $\frac{6}{5} = \frac{x}{2}$ **D.** $\frac{5}{6} = \frac{x}{8}$

2. What is the value of x?

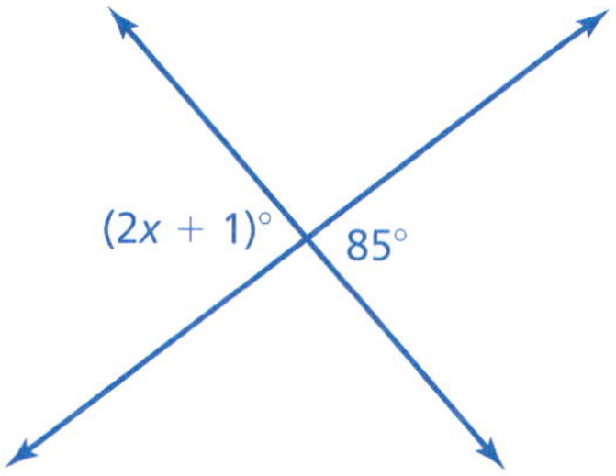

3. Your mathematics teacher described an equation in words. Her description is in the box below.

   "5 less than the product of 7 and an unknown number is equal to 42."

   Which equation matches your mathematics teacher's description?

   F. $(5 - 7)n = 42$ **H.** $5 - 7n = 42$

   G. $(7 - 5)n = 42$ **I.** $7n - 5 = 42$

4. What is the area of the circle below? $\left(\text{Use } \frac{22}{7} \text{ for } \pi.\right)$

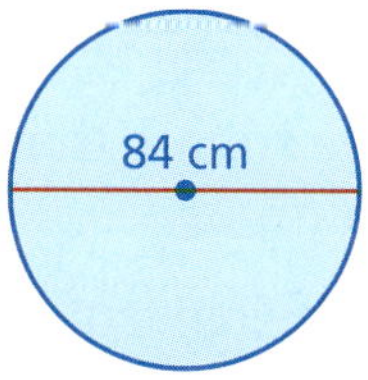

   A. 132 cm^2 **C.** 5544 cm^2

   B. 264 cm^2 **D.** $22{,}176 \text{ cm}^2$

5. John was finding the area of the figure below.

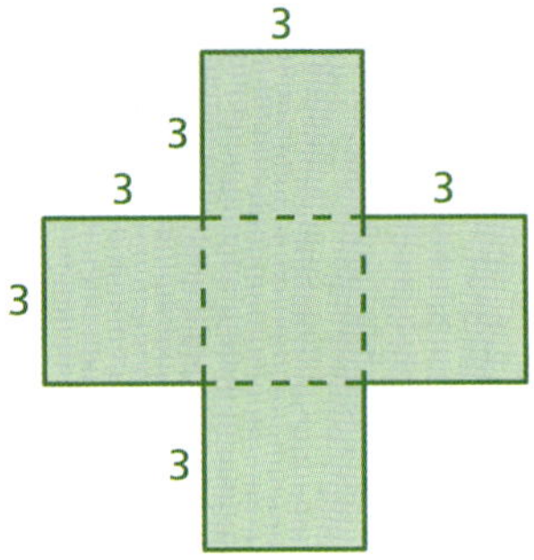

John's work is in the box below.

area of horizontal rectangle

$A = 3 \times (3 + 3 + 3)$

$= 3 \times 9$

$= 27$ square units

area of vertical rectangle

$A = (3 + 3 + 3) \times 3$

$= 9 \times 3$

$= 27$ square units

total area of figure

$A = 27 + 27$

$= 54$ square units

What should John do to correct the error that he made?

F. Add the area of the center square to the 54 square units.

G. Find the area of one square and multiply this number by 4.

H. Subtract the area of the center square from the 54 square units.

I. Subtract 54 from the area of a large square that is 9 units on each side.

6. Which value of x makes the equation below true?

$$5x - 3 = 11$$

A. 1.6

B. 2.8

C. 40

D. 70

Item Analysis (continued)

5. **F.** The student adds the area of the center square that was included twice, instead of subtracting it.

   G. The student does not include the area of the center square at all.

   H. Correct answer

   I. The student employs a plausible strategy but makes the same error as in John's work of counting the center square twice.

6. **A.** The student subtracts 3 instead of adding 3.

   B. Correct answer

   C. The student subtracts 3 instead of adding and then multiplies instead of dividing.

   D. The student multiplies instead of dividing.

Answers

5. H

6. B

Answers

7. 29.42

8. G

9. A

10. *Part A* 942 ft^2

Part B 134.2 ft

Item Analysis (continued)

7. **Gridded Response:** Correct answer: 29.42

Common Error: The student includes the circumference of the circle instead of just the circumference of the semicircle.

8. **F.** The student does not reverse the inequality sign, leading to $x \geq 1$.

G. Correct answer

H. The student does not realize that $x < 5$ does not include 5.

I. The student does not realize that $x > 5$ does not include 5.

9. **A.** Correct answer

B. The student chooses the store offering the greatest percent off but ignores the fact that this store also had the greatest original price.

C. The student chooses the store that had the least original price but ignores the fact that this store also offered the least percent off.

D. The student chooses the store with the highest sale price.

10. **4 points** The student demonstrates a thorough understanding of solving problems involving area and perimeter of composite figures. In Part A, the student correctly finds the area of the sprayed region to be approximately 942 square feet. In Part B, the student correctly finds the perimeter of the sprayed region to be approximately 134.2 feet. The student shows accurate, complete work for both parts and provides clear and complete explanations.

3 points The student demonstrates an understanding of solving problems involving area and perimeter of composite figures, but the student's work and explanations demonstrate an essential but less than thorough understanding.

2 points The student demonstrates a partial understanding of solving problems involving area and perimeter of composite figures. The student's work and explanations demonstrate a lack of essential understanding.

1 point The student demonstrates very limited understanding of solving problems involving area and perimeter of composite figures. The student's response is incomplete and exhibits many flaws.

0 points The student provided no response, a completely incorrect or incomprehensible response, or a response that demonstrates insufficient understanding of solving problems involving area and perimeter of composite figures.

7. What is the perimeter of the figure below? (Use 3.14 for π.)

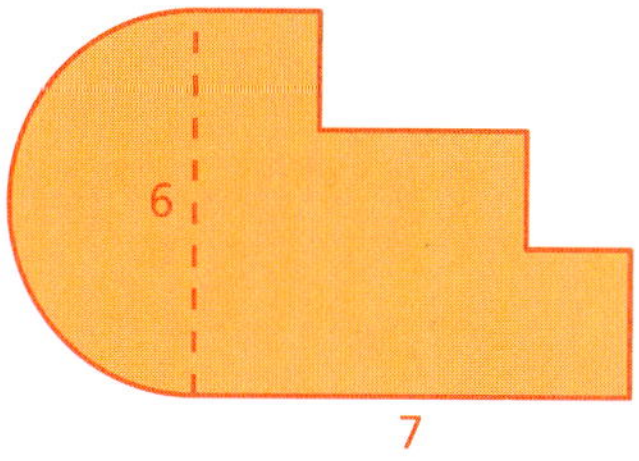

8. Which inequality has 5 in its solution set?

F. $5 - 2x \geq 3$

G. $3x - 4 \geq 8$

H. $8 - 3x > -7$

I. $4 - 2x < -6$

9. Four jewelry stores are selling an identical pair of earrings.

- Store A: original price of \$75; 20% off during sale
- Store B: original price of \$100; 35% off during sale
- Store C: original price of \$70; 10% off during sale
- Store D: original price of \$95; 30% off during sale

Which store has the least sale price for the pair of earrings?

A. Store A

B. Store B

C. Store C

D. Store D

10. A lawn sprinkler sprays water onto part of a circular region, as shown below.

Think Solve Explain

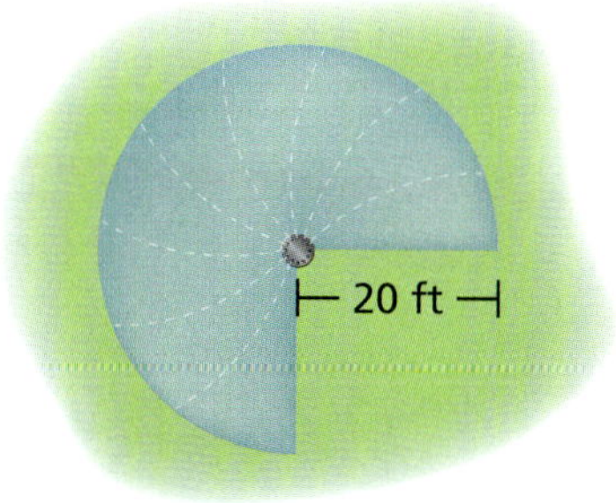

Part A What is the area, in square feet, of the region that the sprinkler sprays with water? Show your work and explain your reasoning. (Use 3.14 for π.)

Part B What is the perimeter, in feet, of the region that the sprinkler sprays with water? Show your work and explain your reasoning. (Use 3.14 for π.)

14 Surface Area and Volume

14.1 Surface Areas of Prisms

14.2 Surface Areas of Pyramids

14.3 Surface Areas of Cylinders

14.4 Volumes of Prisms

14.5 Volumes of Pyramids

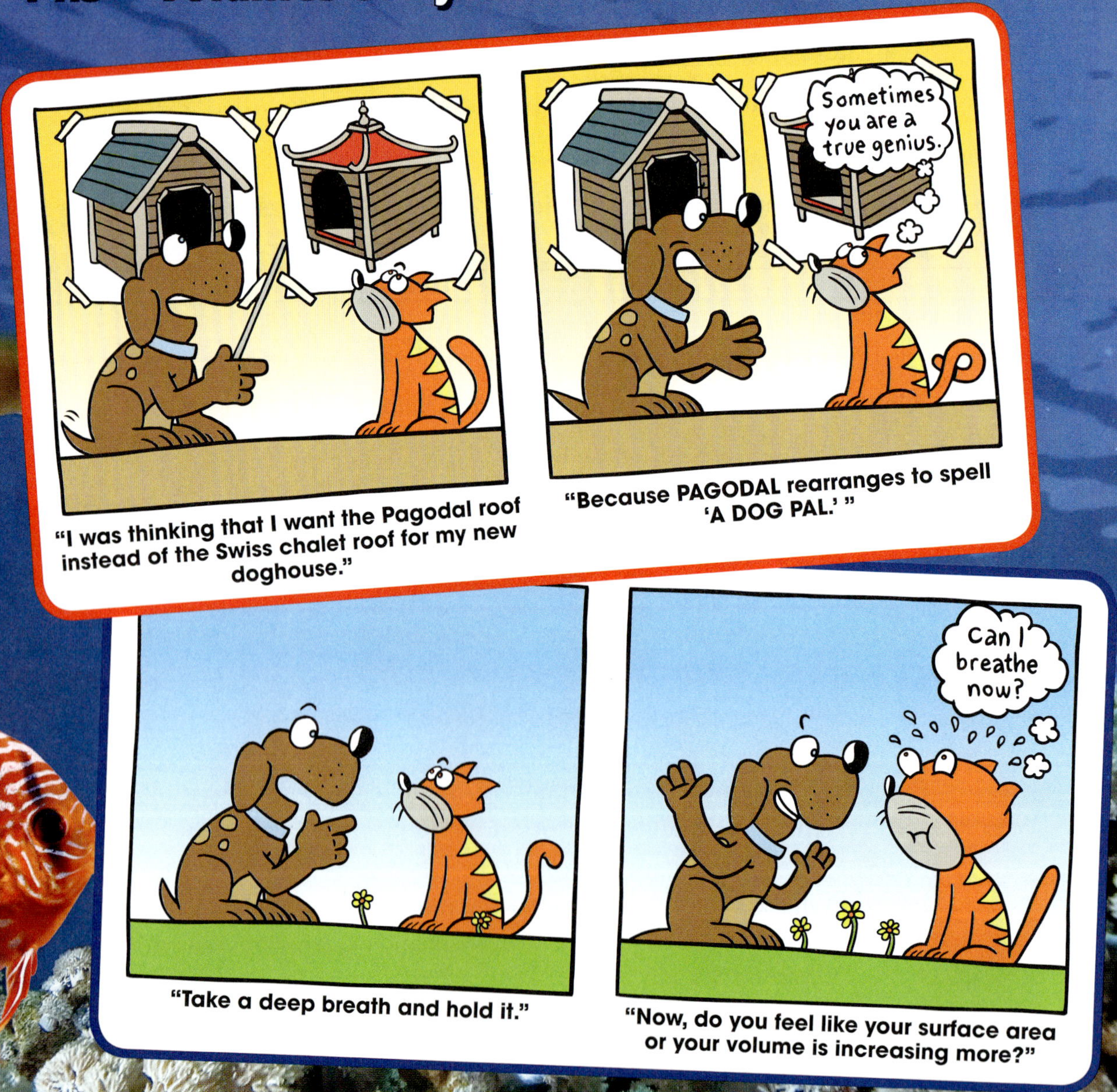

What Your Students Have Learned

- Find the areas of rectangles with fractional side lengths.
- Classify two-dimensional figures into categories based on properties.
- Understand volume, and measure it by counting unit cubes.
- Find the volumes of rectangular prisms using a formula.
- Find the areas of triangles, special quadrilaterals, and polygons.
- Use nets made up of rectangles and triangles to find surface areas.
- Find the volumes of prisms with fractional edge lengths.

What Your Students Will Learn

- Use two-dimensional nets to represent three-dimensional solids.
- Solve real-world problems involving surface areas and volumes of objects composed of prisms, pyramids, and cylinders.
- Describe the cross sections that result from slicing three-dimensional figures.

Pacing Guide for Chapter 14

Chapter Opener Advanced	1 Day
Section 1 Advanced	1 Day
Section 2 Advanced	2 Days
Section 3 Advanced	1 Day
Study Help / Quiz Advanced	1 Day
Section 4 Advanced	1 Day
Section 5 Advanced	2 Days
Chapter Review/ Chapter Tests Advanced	2 Days
Total Chapter 14 Advanced	11 Days
Year-to-Date Advanced	129 Days

Technology for the Teacher

BigIdeasMath.com
Chapter at a Glance
Complete Materials List
Parent Letters: English and Spanish

What Your Students Have Learned

- Find areas of rectangles, squares, and triangles using formulas.

Additional Topics for Review

- Three-Dimensional Figures
- Using Nets to Find Surface Areas
- Finding the Volumes of Rectangular Prisms

Try It Yourself

1. 99 m^2
2. 35.7 ft^2
3. $\frac{4}{9} \text{ in.}^2$
4. 39 ft^2
5. 140 m^2
6. 225 cm^2

Record and Practice Journal Fair Game Review

1. 64 cm^2
2. 84 yd^2
3. 58.88 in.^2
4. $\frac{25}{36} \text{ m}^2$
5. $3\frac{1}{9} \text{ mm}^2$
6. 321.63 ft^2
7. 6.25 ft^2
8. 20 cm^2
9. 12 ft^2
10. 21 m^2
11. 30 yd^2
12. 10 in.^2
13. 9 mm^2
14. 24 ft^2

Math Background Notes

Vocabulary Review

- Area
- Square Units
- Base of a Triangle
- Height of a Triangle

Finding Areas of Squares and Rectangles

- Students should be familiar with area formulas for squares and rectangles.
- You may want to review units with students. A complete answer will include a numeric solution and the correct units. Remind students that area is always measured in square units.
- The length and width of a rectangle are interchangeable. Some students may think that the length of the rectangle must always be the side that sits on the ground. Turn the rectangle vertically to demonstrate that the length and width are relative to interpretation.

Finding Areas of Triangles

- Students should be familiar with area formulas for triangles.
- Review how to identify a base and the corresponding height of the triangle.
- **Common Error:** Students may want to take $\frac{1}{2}$ of 6 and $\frac{1}{2}$ of 7 because they think it is the Distributive Property.
- Note that the product of 6 and 7 was found (Associative Property), and then $\frac{1}{2}$ of that product was computed. It would also be correct to multiply from left to right according to the order of operations.

Reteaching and Enrichment Strategies

If students need help. . .	If students got it. . .
Record and Practice Journal • Fair Game Review Skills Review Handbook Lesson Tutorials	Game Closet at *BigIdeasMath.com* Start the next section

What You Learned Before

"Descartes, how would you like it if I could double the height of your cat food can?"

Finding Areas of Squares and Rectangles

Example 1 Find the area of the rectangle.

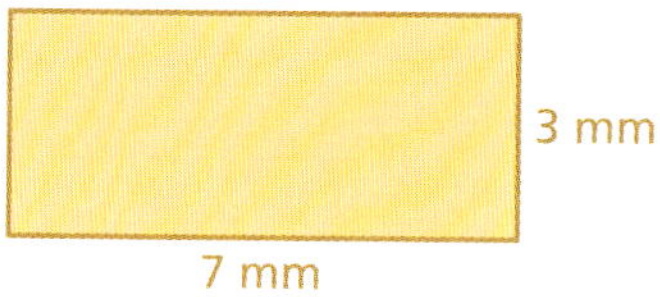

Area $= \ell w$	Write formula for area.
$= 7(3)$	Substitute 7 for ℓ and 3 for w.
$= 21$	Multiply.

∴ The area of the rectangle is 21 square millimeters.

Try It Yourself

Find the area of the square or rectangle.

1.

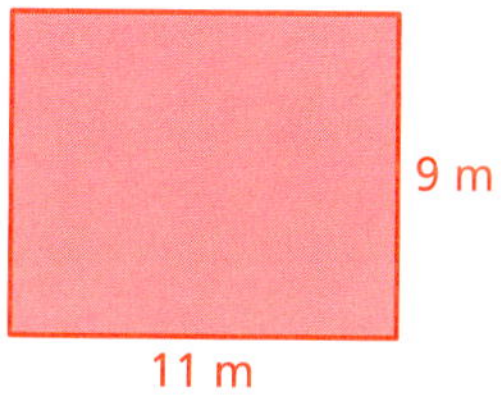

2.

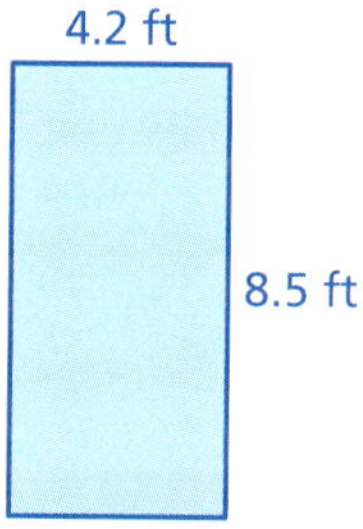

3.

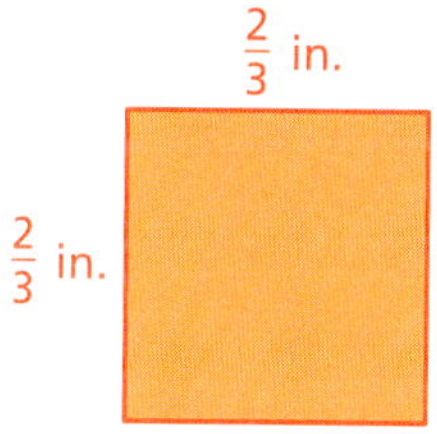

Finding Areas of Triangles

Example 2 Find the area of the triangle.

$A = \frac{1}{2}bh$	Write formula.
$= \frac{1}{2}(6)(7)$	Substitute 6 for b and 7 for h.
$= \frac{1}{2}(42)$	Multiply 6 and 7.
$= 21$	Multiply $\frac{1}{2}$ and 42.

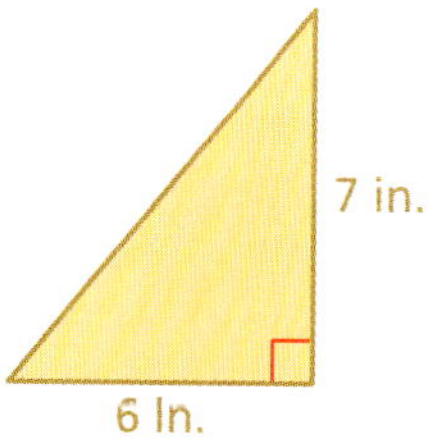

∴ The area of the triangle is 21 square inches.

Try It Yourself

Find the area of the triangle.

4.

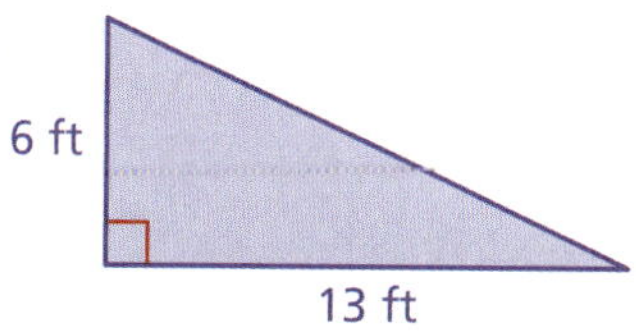

5.

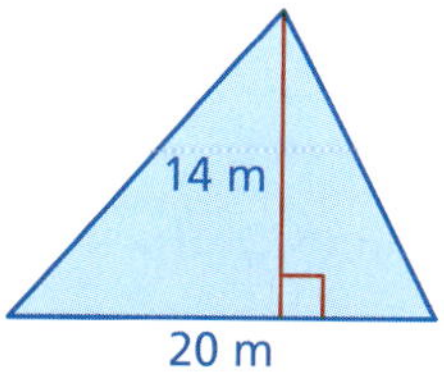

6.

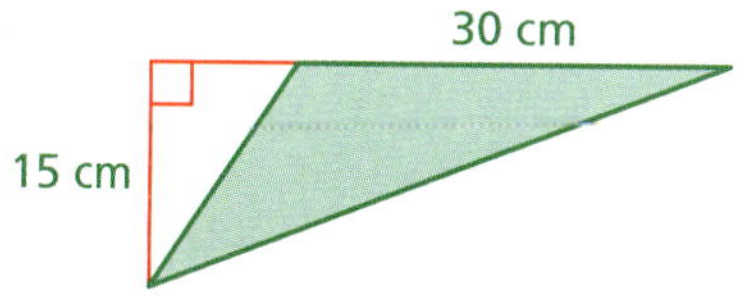

14.1 Surface Areas of Prisms

Essential Question How can you find the surface area of a prism?

1 ACTIVITY: Surface Area of a Rectangular Prism

Work with a partner. Copy the net for a rectangular prism. Label each side as h, w, or ℓ. Then use your drawing to write a formula for the surface area of a rectangular prism.

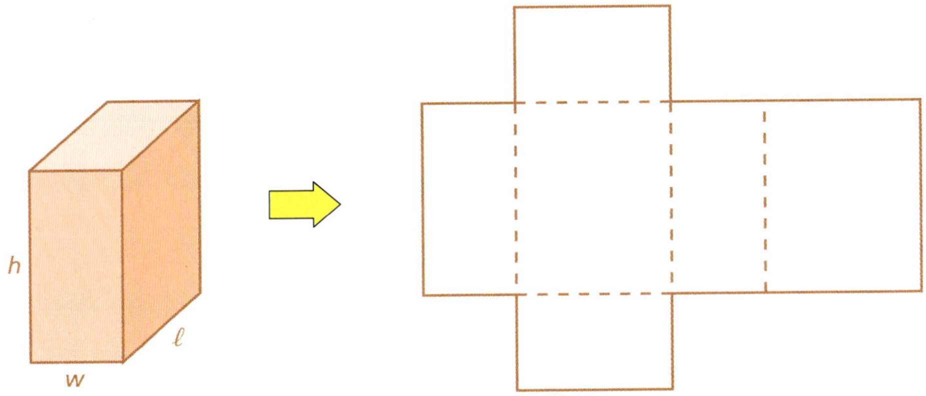

2 ACTIVITY: Surface Area of a Triangular Prism

Work with a partner.

a. Find the surface area of the solid shown by the net. Copy the net, cut it out, and fold it to form a solid. Identify the solid.

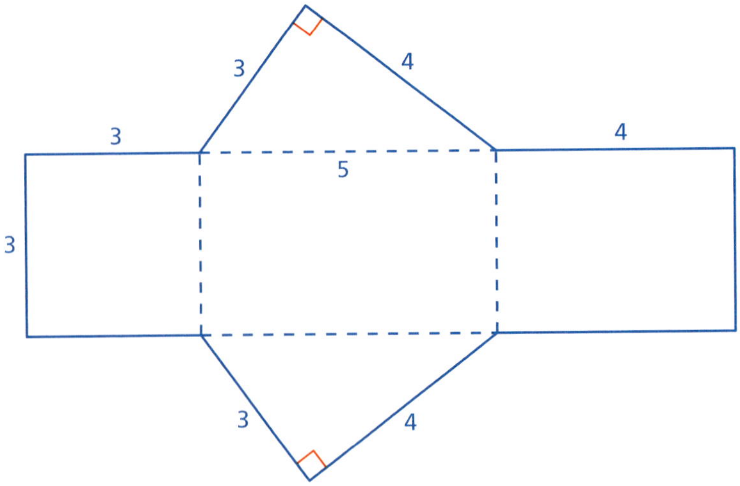

b. Which of the surfaces of the solid are bases? Why?

Geometry

In this lesson, you will

- use two-dimensional nets to represent three-dimensional solids.
- find surface areas of rectangular and triangular prisms.
- solve real-life problems.

Laurie's Notes

Introduction

Applying Mathematical Practices

- **Model with Mathematics:** Finding the surface area of a solid can be challenging when the faces are not all visible. Drawing a net allows students to see all of the faces of the solid.

Motivate

- Begin by standing on a solid box. Students will naturally be curious.
- ? "I'm standing on a face of this rectangular prism. Is there another face that is congruent to the one I'm standing on? Explain." The opposite face, on the bottom, is congruent to the top.
- Rotate the box and stand on a different face. Repeat the question.
- ? "How many pairs of congruent faces are there?" three

Write

- Write the definition of surface area. Describe and show examples of nets. Real-life models of nets (donut box, pizza box) are helpful.

Activity Notes

Activity 1

- **Classroom Management:** Collect cardboard samples of prisms such as donut, pizza, and tissue boxes. These can be taken apart to demonstrate nets.
- Explain that any pair of opposite faces of a rectangular prism can be identified as the bases, and the other four faces are the lateral faces.
- **Common Misconception:** The bases do not have to be on the top and bottom, although they are often identified that way.
- **Model with Mathematics:** A net helps students to see how to find the surface area of a prism.
- Students may be challenged to visualize how the prism unfolds as the net.
- Students will likely label the vertical segments to the far left and right as h.
- Encourage students to label all of the segments.
- **Construct Viable Arguments and Critique the Reasoning of Others:** Have students discuss ways to write the formula. Here are a few.
  - $S = \ell h + wh + \ell h + wh + \ell w + \ell w$
  - $S = 2\ell h + 2wh + 2\ell w$
  - $S = 2(\ell h + wh + \ell w)$
- Have students verify suggested formulas and discuss their usability.
- **Extension:** Find how many different nets there are for a cube. Use standard grid paper or investigate at *illuminations.nctm.org*. There are 11.

Activity 2

- **Management Tip:** To save time, you can precut the nets before class.
- ? "What are the lateral faces of this triangular prism?" The 3 rectangles
- A prism is identified by the type of base it has, such as a triangular prism and a pentagonal prism. The lateral faces are all rectangles.

What Your Students Will Learn

- Use two-dimensional nets to represent three-dimensional solids.
- Find surface areas of rectangular prisms, triangular prisms, and cubes using formulas.

Previous Learning

Students should know how to make a net for a prism and how to find the surface area of a prism by adding unit squares.

Technology for the Teacher

Dynamic Classroom

Lesson Plans
Complete Materials List

14.1 Record and Practice Journal

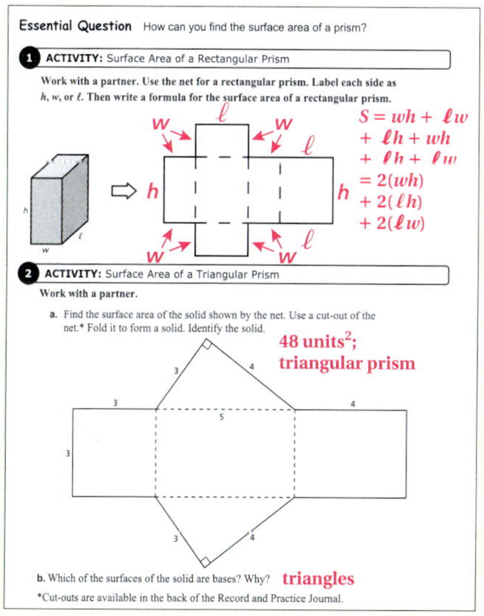

Essential Question How can you find the surface area of a prism?

1 ACTIVITY: Surface Area of a Rectangular Prism

Work with a partner. Use the net for a rectangular prism. Label each side as h, w, or ℓ. Then write a formula for the surface area of a rectangular prism.

2 ACTIVITY: Surface Area of a Triangular Prism

Work with a partner.

a. Find the surface area of the solid shown by the net. Use a cut-out of the net.* Fold it to form a solid. Identify the solid. 48 units^2; triangular prism

b. Which of the surfaces of the solid are bases? Why? triangles

*Cut-outs are available in the back of the Record and Practice Journal.

Differentiated Instruction

Kinesthetic

After students draw a net and find the surface area, have them cut out the net, fold it, and tape it together to create the prism.

14.1 Record and Practice Journal

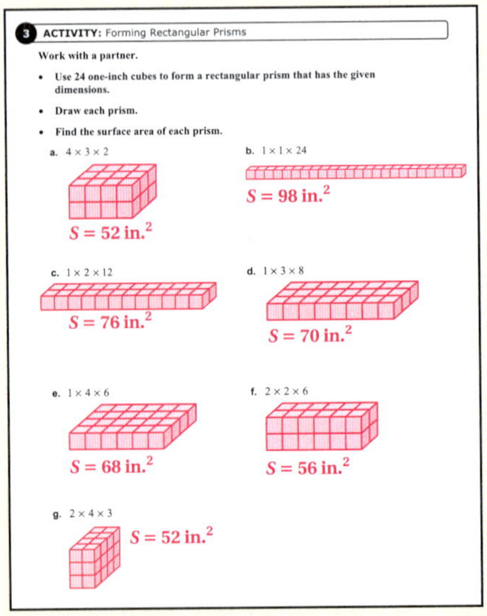

3 ACTIVITY: Forming Rectangular Prisms

Work with a partner.

- Use 24 one-inch cubes to form a rectangular prism that has the given dimensions.
- Draw each prism.
- Find the surface area of each prism.

a. $4 \times 3 \times 2$ $S = 52$ in.^{2}

b. $1 \times 1 \times 24$ $S = 98$ in.^{2}

c. $1 \times 2 \times 12$ $S = 76$ in.^{2}

d. $1 \times 3 \times 8$ $S = 70$ in.^{2}

e. $1 \times 4 \times 6$ $S = 68$ in.^{2}

f. $2 \times 2 \times 6$ $S = 56$ in.^{2}

g. $2 \times 4 \times 3$ $S = 52$ in.^{2}

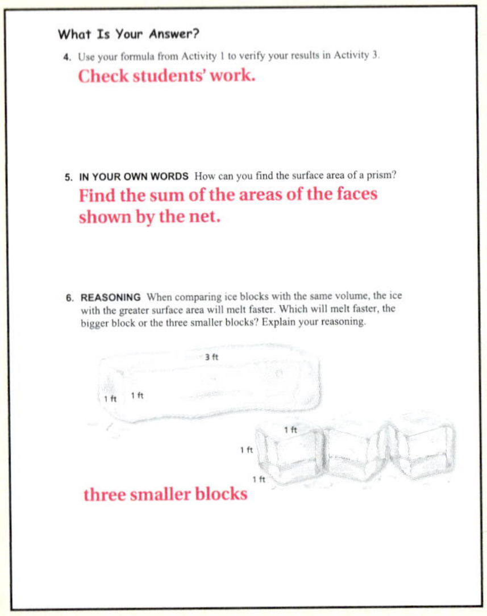

What Is Your Answer?

4. Use your formula from Activity 1 to verify your results in Activity 3.
Check students' work.

5. IN YOUR OWN WORDS How can you find the surface area of a prism?
Find the sum of the areas of the faces shown by the net.

6. REASONING When comparing ice blocks with the same volume, the ice with the greater surface area will melt faster. Which will melt faster, the bigger block or the three smaller blocks? Explain your reasoning.
three smaller blocks

Laurie's Notes

Activity 3

- To help facilitate this activity, you can package 24 one-inch cubes in plastic bags before class.
- ? "How does the $4 \times 3 \times 2$ prism compare to the $2 \times 4 \times 3$ prism?"
The volumes are the same. One has a base of 4×3 and a height of 2. The other has a base of 2×4 and a height of 3.
- ? "Was the surface area always the same?" no
- **Common Misconception:** Students often think that because the number of wooden cubes stays the same (volume), the surface area should stay the same as well. This activity should help students recognize that the surface areas vary.

What Is Your Answer?

- Review the answers to the questions together as a class. Have volunteers share their answers.

Closure

- Draw the net for a pizza box. Label with approximate dimensions and find the surface area.

$S = 640$ in.^{2}

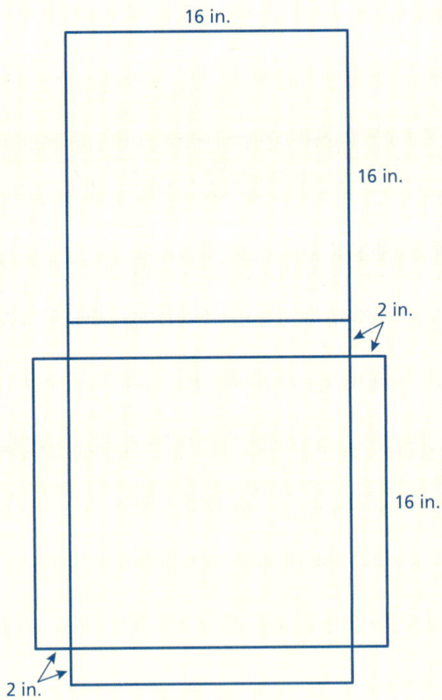

3 ACTIVITY: Forming Rectangular Prisms

Math Practice

Construct Arguments

What method did you use to find the surface area of the rectangular prism? Explain.

Work with a partner.

- **Use 24 one-inch cubes to form a rectangular prism that has the given dimensions.**
- **Draw each prism.**
- **Find the surface area of each prism.**

a. $4 \times 3 \times 2$ ***Drawing***   ***Surface Area***

▭ in.^2

b. $1 \times 1 \times 24$ **c.** $1 \times 2 \times 12$ **d.** $1 \times 3 \times 8$

e. $1 \times 4 \times 6$ **f.** $2 \times 2 \times 6$ **g.** $2 \times 4 \times 3$

What Is Your Answer?

4. Use your formula from Activity 1 to verify your results in Activity 3.

5. **IN YOUR OWN WORDS** How can you find the surface area of a prism?

6. **REASONING** When comparing ice blocks with the same volume, the ice with the greater surface area will melt faster. Which will melt faster, the bigger block or the three smaller blocks? Explain your reasoning.

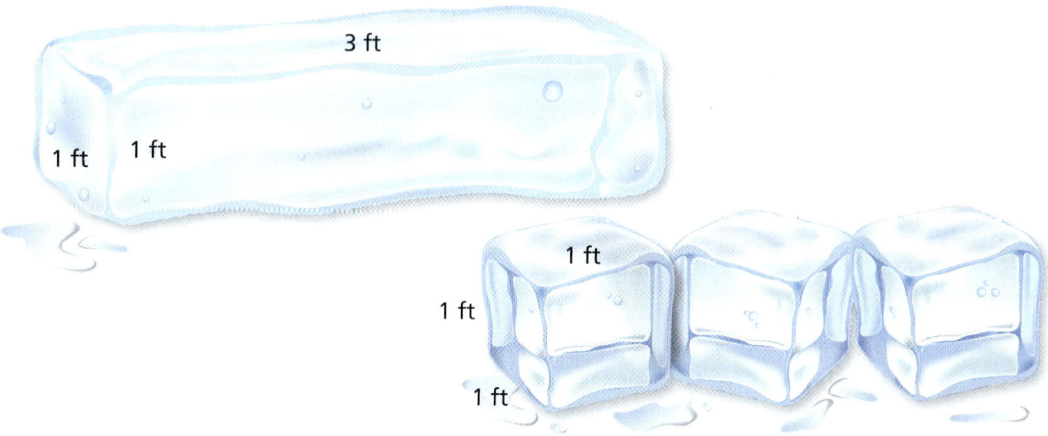

Practice Use what you learned about the surface areas of rectangular prisms to complete Exercises 4–6 on page 591.

14.1 Lesson

Check It Out
Lesson Tutorials
BigIdeasMath.com

Key Vocabulary
lateral surface area, *p. 590*

Key Idea

Surface Area of a Rectangular Prism

Words The surface area S of a rectangular prism is the sum of the areas of the bases and the lateral faces.

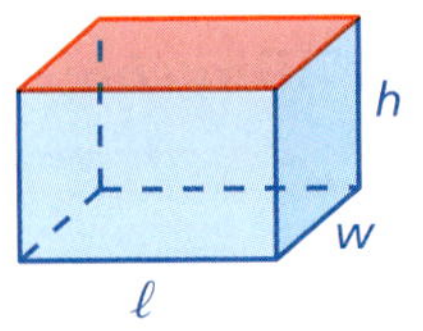

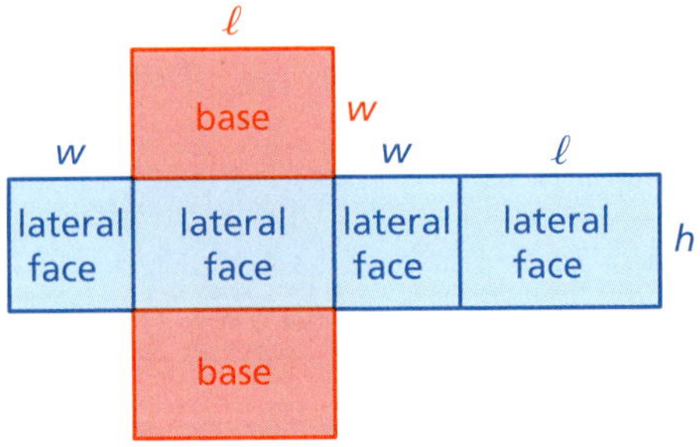

Algebra $S = 2\ell w + 2\ell h + 2wh$

- $2\ell w$: Areas of bases
- $2\ell h + 2wh$: Areas of lateral faces

EXAMPLE 1 Finding the Surface Area of a Rectangular Prism

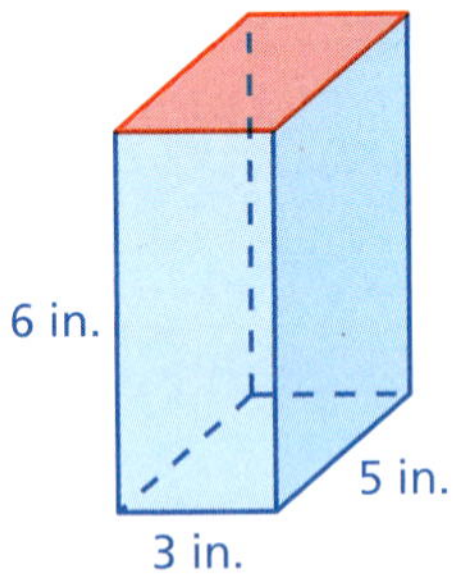

Find the surface area of the prism.

Draw a net.

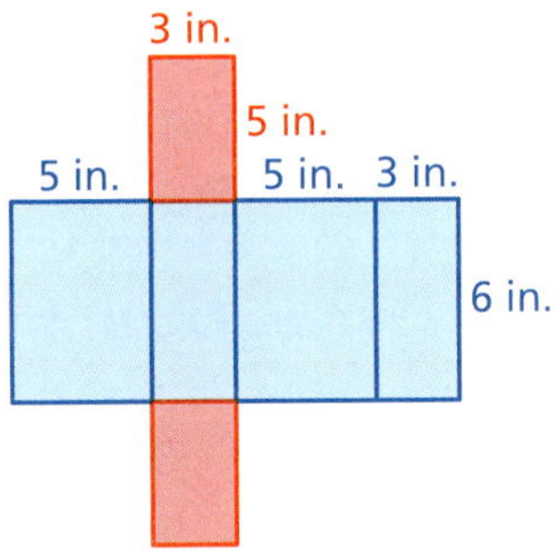

$$\begin{aligned} S &= 2\ell w + 2\ell h + 2wh \\ &= 2(3)(5) + 2(3)(6) + 2(5)(6) \\ &= 30 + 36 + 60 \\ &= 126 \end{aligned}$$

The surface area is 126 square inches.

On Your Own

Now You're Ready
Exercises 7–9

Find the surface area of the prism.

1.

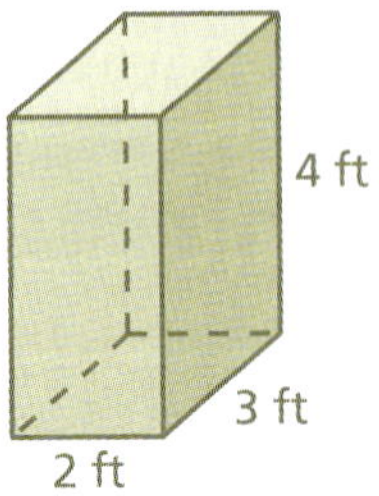

2.

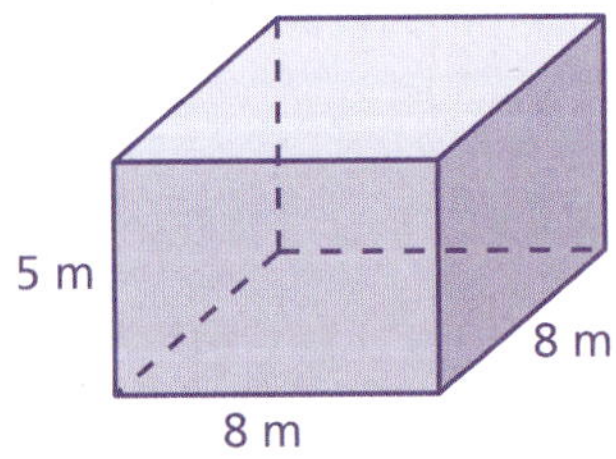

Multi-Language Glossary at BigIdeasMath.com

Laurie's Notes

Introduction

Connect

- **Yesterday:** Students explored surface area using the nets of two prisms.
- **Today:** Students will work with a formula for the surface area of a prism.

Motivate

- Hold a prism, perhaps a box that has a wrapping on it. It could be wrapping paper or simply clear shrink wrapping.
- ? "Why would I need to know the surface area of this prism?" Listen for the concept of wrapping the box in some type of paper.
- Explain to students that when items are mass produced, someone needs to calculate the amount of wrapping needed to cover the prism. The wrapping is the surface area.

Lesson Notes

Key Idea

- Have a net available as a visual, whether it is a cardboard box or a rectangular prism made from the snap together polygon frames.
- Note the color coding of the formula, prism, and net.
- **FYI:** Remind students that any two opposite faces can be called the bases. Once the bases are identified, the remaining 4 faces form the lateral portion.
- ? "Can you explain why there are three parts to finding the surface area of the rectangular prism?" Students should recognize that there are 3 pairs of congruent faces.

Example 1

- The challenge for students is not to get bogged down in symbols. Students need to remember that there are 3 pairs of congruent faces. They need to make sure that they calculate the area of each one of the 3 different faces, then double the answer to account for the pair.
- **Common Error:** When students multiply (2)(3)(5), they sometimes multiply $2 \times 3 \times 2 \times 5$, similar to using the Distributive Property. Remind them that multiplication is both commutative and associative, and they can multiply in order $(2 \times 3 \times 5)$ or in a different order $(2 \times 5 \times 3)$.
- **Teaching Tip:** Write the equation for surface area as: $S =$ bases + sides + (front and back). Students follow the words and find the area of each pair without thinking about the variables.

On Your Own

- Students need to record their work neatly so they, and you, can look back and see what corrections are needed.
- **Think-Pair-Share:** Students should read each question independently and then work in pairs to answer the questions. When they have answered the questions, the pair should compare their answers with another group and discuss any discrepancies.

Goal Today's lesson is finding the **surface area** of a prism using a formula.

Technology for the Teacher

Dynamic Classroom

Lesson Tutorials
Lesson Plans
Answer Presentation Tool

Extra Example 1

Find the surface area of a rectangular prism with a length of 6 yards, a width of 4 yards, and a height of 9 yards.
228 yd^2

On Your Own

1. 52 ft^2
2. 288 m^2

Differentiated Instruction

Visual

Use a rectangular box to demonstrate three ways of finding surface area. The first method is to find the area of each face and then add areas. The second method is to open the box into a net and find the area of the net. The third method is to use the formula for finding surface area. Students should see that the three methods have the same result.

Extra Example 2

Find the surface area of the triangular prism.

216 in.^{2}

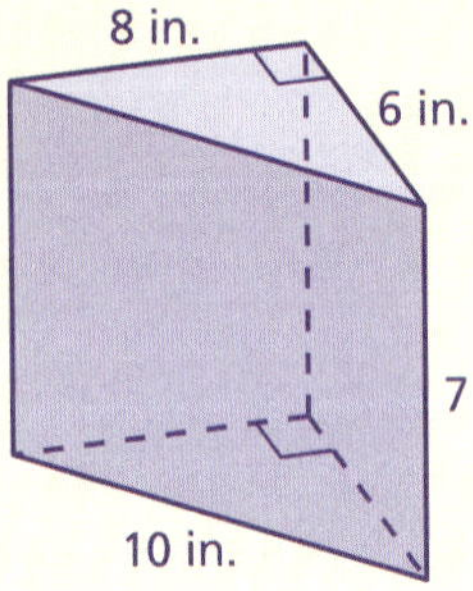

On Your Own

3. 150 m^2

4. 60 cm^2

Laurie's Notes

Key Idea

- This is the general formula for a prism without variables. Most students are comfortable with this form.
- ? "How many faces are there that make up the bases?" two
- ? "How many faces are there that make up the lateral faces?" It depends on how many sides the bases have.

Example 2

- Note that the net is a visual reminder of each face whose area must be found. Color coding the faces should help students keep track of their work.
- **Model with Mathematics**: Encourage students to write the formula in words for each new problem: $S =$ area of bases $+$ areas of lateral faces.

On Your Own

- Give students sufficient time to do their work before asking volunteers to share their work *and* sketch at the board.
- Students having difficulty with Question 4 may want to redraw the triangular prism with the base on the bottom.

Key Idea

Surface Area of a Prism

The surface area S of any prism is the sum of the areas of the bases and the lateral faces.

$$S = \text{areas of bases} + \text{areas of lateral faces}$$

EXAMPLE 2 Finding the Surface Area of a Triangular Prism

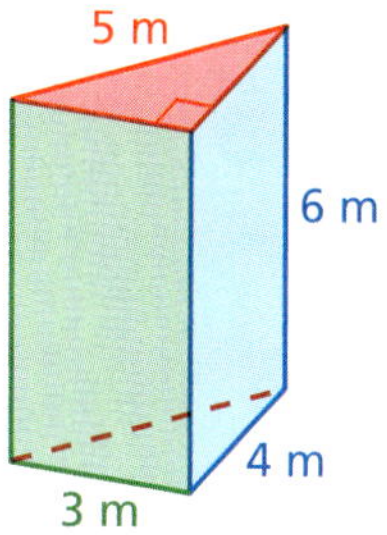

Find the surface area of the prism.

Draw a net.

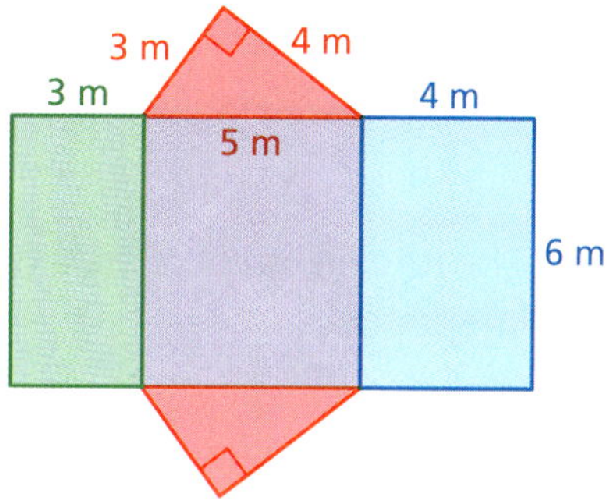

> **Remember**
> The area A of a triangle with base b and height h is $A = \frac{1}{2}bh$.

Area of a Base	*Areas of Lateral Faces*
Red base: $\frac{1}{2} \cdot 3 \cdot 4 = 6$	Green lateral face: $3 \cdot 6 = 18$
	Purple lateral face: $5 \cdot 6 = 30$
	Blue lateral face: $4 \cdot 6 = 24$

Add the areas of the bases and the lateral faces.

$$\begin{aligned} S &= \text{areas of bases} + \text{areas of lateral faces} \\ &= 6 + 6 + 18 + 30 + 24 \\ &= 84 \end{aligned}$$

There are two identical bases. Count the area twice.

The surface area is 84 square meters.

On Your Own

Now You're Ready
Exercises 10–12

Find the surface area of the prism.

3.

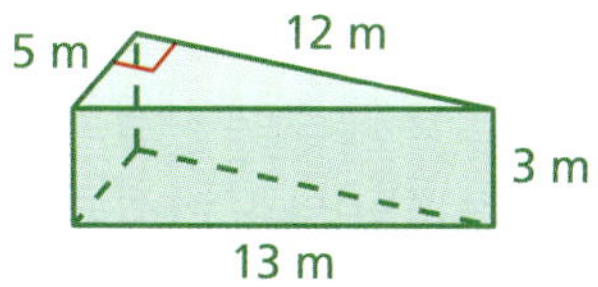

4.

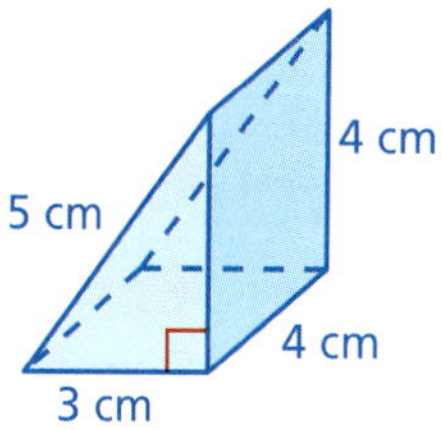

Remember

A cube has 6 congruent square faces.

When all the edges of a rectangular prism have the same length s, the rectangular prism is a cube. The formula for the surface area of a cube is

$S = 6s^2$. Formula for surface area of a cube

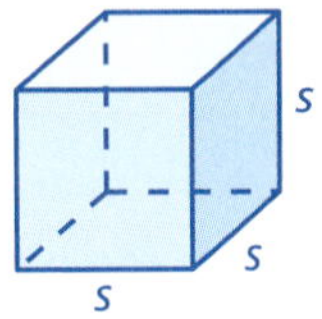

EXAMPLE 3 Finding the Surface Area of a Cube

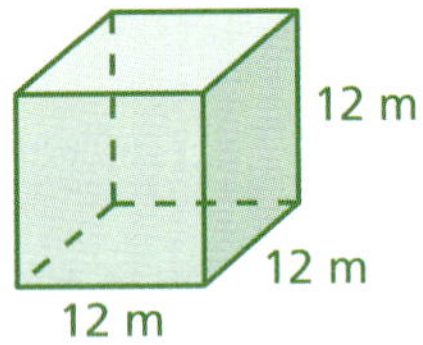

Find the surface area of the cube.

$S = 6s^2$ Write formula for surface area of a cube.

$= 6(12)^2$ Substitute 12 for s.

$= 864$ Simplify.

The surface area of the cube is 864 square meters.

The **lateral surface area** of a prism is the sum of the areas of the lateral faces.

EXAMPLE 4 Real-Life Application

The outsides of purple traps are coated with glue to catch emerald ash borers. You make your own trap in the shape of a rectangular prism with an open top and bottom. What is the surface area that you need to coat with glue?

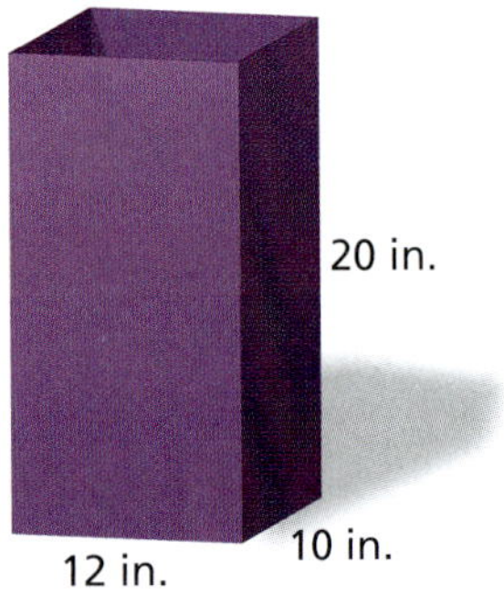

Find the lateral surface area.

$S = 2\ell h + 2wh$ ← Do not include the areas of the bases in the formula.

$= 2(12)(20) + 2(10)(20)$ Substitute.

$= 480 + 400$ Multiply.

$= 880$ Add.

So, you need to coat 880 square inches with glue.

On Your Own

Now You're Ready
Exercises 13–15

5. Which prism has the greater surface area?

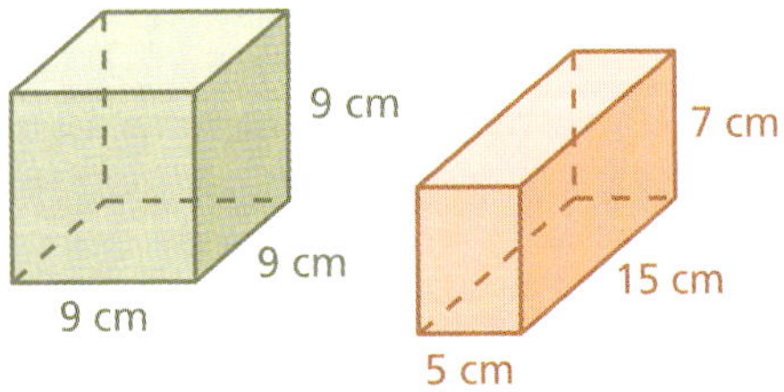

6. **WHAT IF?** In Example 4, both the length and the width of your trap are 12 inches. What is the surface area that you need to coat with glue?

Laurie's Notes

Discuss

- Ask students to discuss how they would find the surface area of a cube. Students should recognize that all six faces are congruent, so finding the area of one face and multiplying by 6 will give you the surface area of the cube.

Example 3

- Write the formula for the surface area of a cube.
- Substitute for the variable and work through the problem.

Example 4

- **FYI:** The emerald ash borer is an exotic beetle that was discovered in southeastern Michigan near Detroit in the summer of 2002. The adult beetles nibble on ash foliage, but cause little damage. The larvae (the immature stage) feed on the inner bark of ash trees, disrupting the tree's ability to transport water and nutrients. The emerald ash borer probably arrived in the United States on solid wood packing material carried in cargo ships or airplanes originating in its native Asia. Since its discovery in Michigan, the emerald ash borer has killed millions of ash trees.

? "How many faces does the trap have?" four

? "How will you find the surface area of the trap?" Only find the area of the lateral faces.

- Write the modified surface area formula and work through the problem.
- Encourage students to write the formula in words.

  $S =$ area of lateral faces

On Your Own

- These two problems are a good summary of the lesson. Students must find the surface area of each prism and then compare the answers. The surface areas are not that different so students cannot simply guess.

Closure

- **Writing:** Hold the prism from the Motivate section of today's lesson and ask students to write about how they would find the surface area of the prism. Look for: finding the areas of the bases and lateral faces and adding those together, *or* unwrapping the prism and finding the area of the wrapping.

Extra Example 3

Find the surface area of the cube.

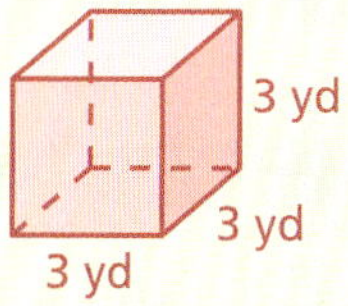

54 yd^2

Extra Example 4

In Example 4, the height of the trap is 30 inches. What is the surface area that you need to coat with glue? 1320 in.^2

On Your Own

5. the cube
6. 960 in.^2

Vocabulary and Concept Check

1. *Sample answer:* 1) Use a net. 2) Use the formula $S = 2\ell w + 2\ell h + 2wh$.
2. same number of faces, vertices and edges; A cube is made of 6 square faces. A rectangular prism is not.
3. Find the area of the bases of the prism; 24 in.^{2}; 122 in.^{2}

Practice and Problem Solving

4.

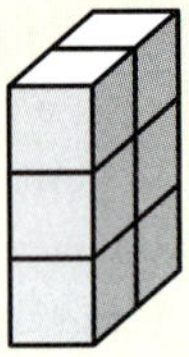

22 in.^{2}

5.

38 in.^{2}

6.

32 in.^{2}

7. 324 m^2
8. 166 mm^2
9. 49.2 yd^2
10. 920 ft^2
11. 136 m^2
12. 382.5 in.^{2}
13. 294 yd^2
14. 1.5 cm^2
15. $2\frac{2}{3}$ ft^2

Assignment Guide and Homework Check

Level	Assignment	Homework Check
Advanced	1–6, 8–26 even, 27–32	8, 12, 14, 18, 26

Common Errors

- **Exercises 7–9** Students may find the area of only three of the faces instead of all six. Remind them that each face is paired with another. Show students the net of a rectangular solid to remind them of the six faces.
- **Exercises 7–9** Some students may multiply length by width by height to find the surface area. Show them that the surface area is the sum of the areas of all six faces, so they must multiply and add to find the solution.
- **Exercises 10–12** Students may try to use the formula for a rectangular prism to find the surface area of a triangular prism. Show them that this will not work by focusing on the area of the triangular base. For students who are struggling to identify all the faces, draw a net of the prism and tell them to label the length, width, and height of each part before finding the surface area.

14.1 Record and Practice Journal

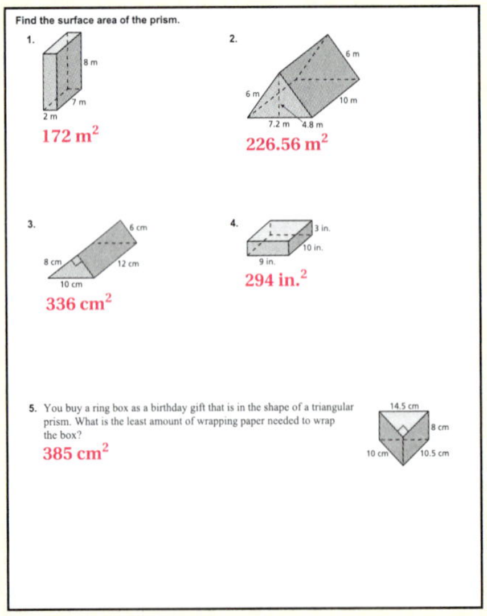

Find the surface area of the prism.

1. 172 m^2

2. 226.56 m^2

3. 336 cm^2

4. 294 in.^{2}

5. You buy a ring box as a birthday gift that is in the shape of a triangular prism. What is the least amount of wrapping paper needed to wrap the box?
385 cm^2

14.1 Exercises

Check It Out
Help with Homework
BigIdeasMath.com

Vocabulary and Concept Check

1. **VOCABULARY** Describe two ways to find the surface area of a rectangular prism.
2. **WRITING** Compare and contrast a rectangular prism to a cube.
3. **DIFFERENT WORDS, SAME QUESTION** Which is different? Find "both" answers.

Find the surface area of the prism.	Find the area of the bases of the prism.
Find the area of the net of the prism.	Find the sum of the areas of the bases and the lateral faces of the prism.

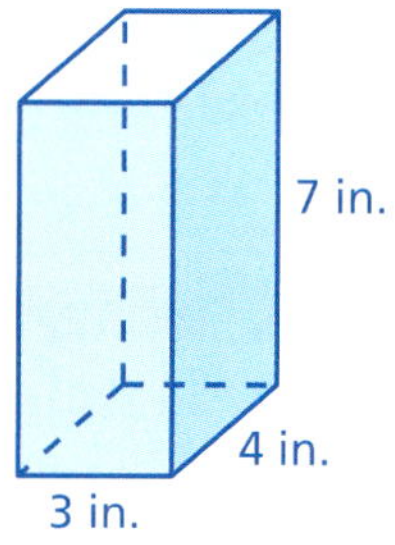

Practice and Problem Solving

Use one-inch cubes to form a rectangular prism that has the given dimensions. Then find the surface area of the prism.

4. $1 \times 2 \times 3$
5. $3 \times 4 \times 1$
6. $2 \times 3 \times 2$

Find the surface area of the prism.

1 7.

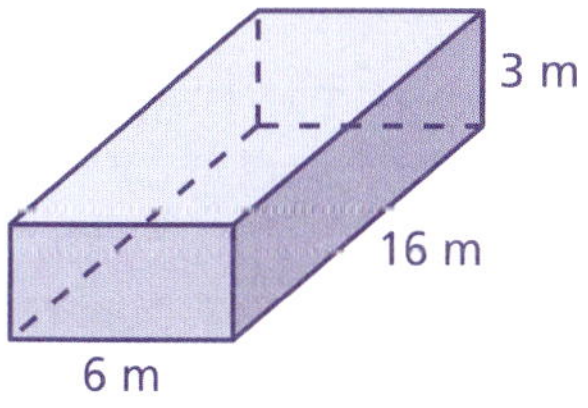

8.

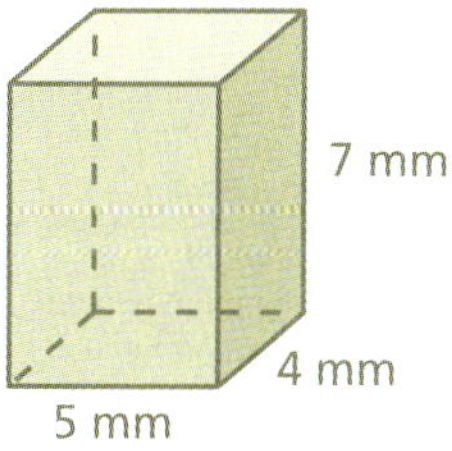

9.

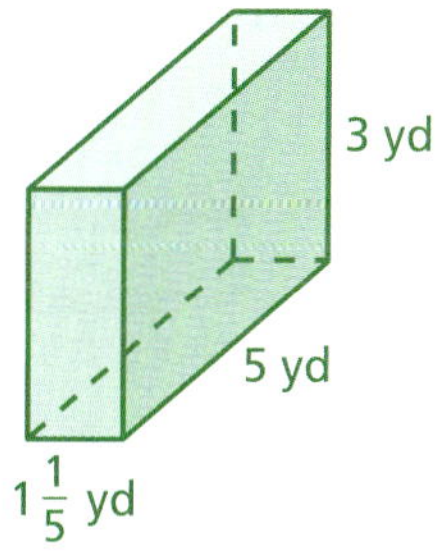

2 10.

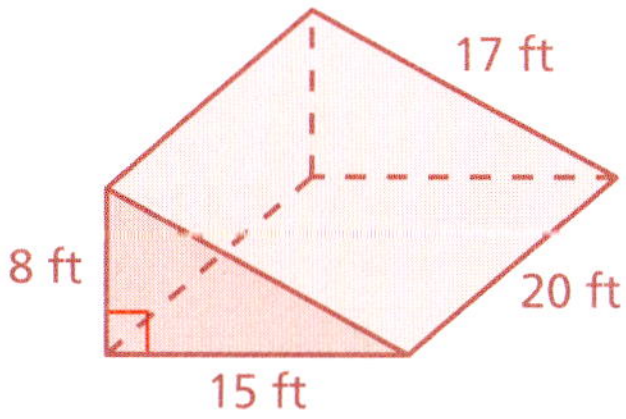

11.

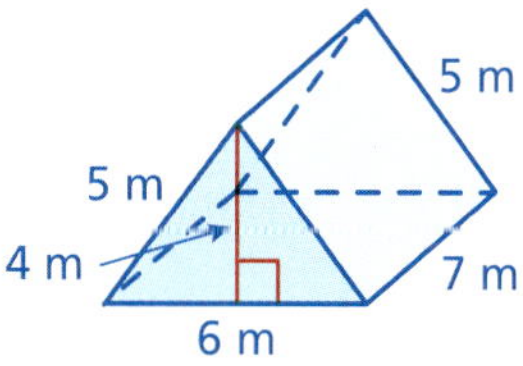

12.

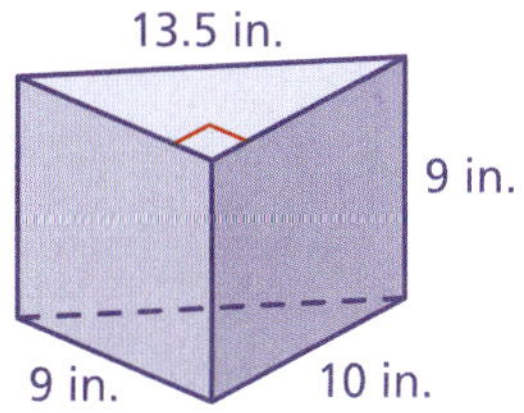

3 13.

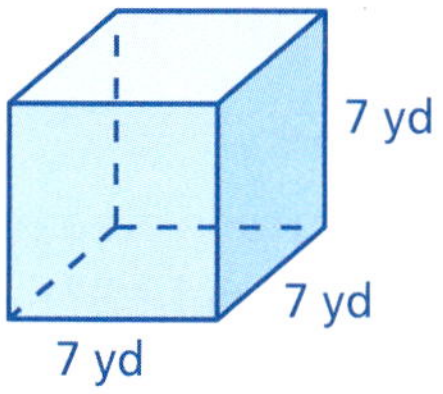

14.

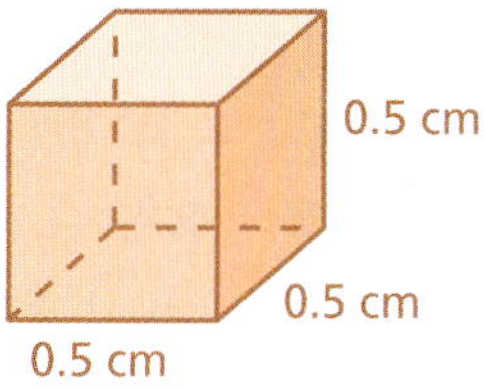

15.

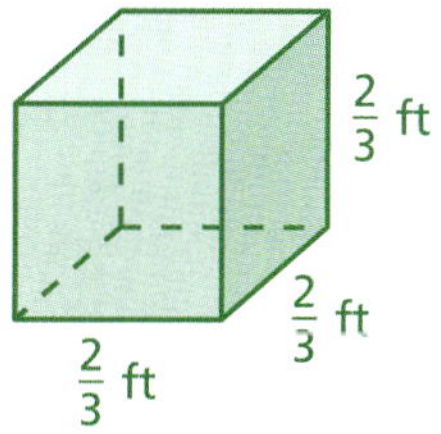

16. **ERROR ANALYSIS** Describe and correct the error in finding the surface area of the prism.

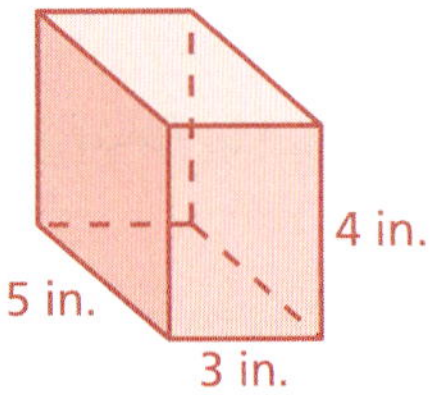

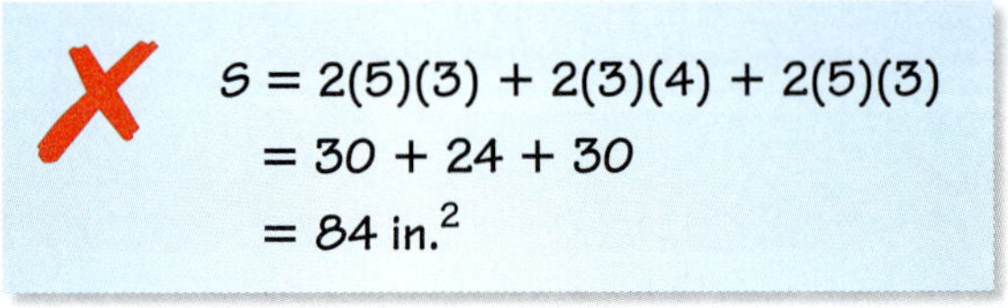

17. **GAME** Find the surface area of the tin game case.

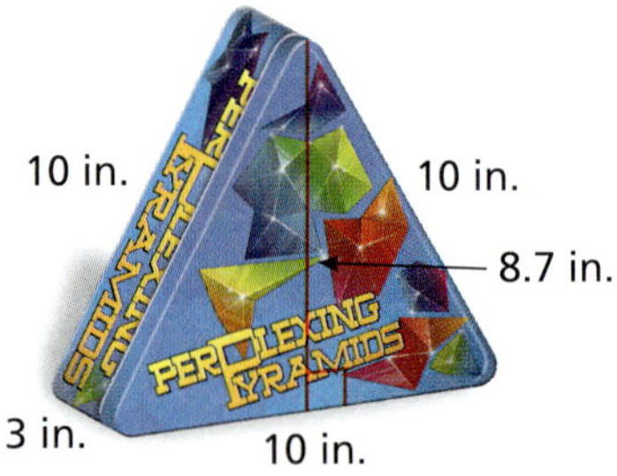

18. **WRAPPING PAPER** A cube-shaped gift is 11 centimeters long. What is the least amount of wrapping paper you need to wrap the gift?

19. **FROSTING** One can of frosting covers about 280 square inches. Is one can of frosting enough to frost the cake? Explain.

Find the surface area of the prism.

20.

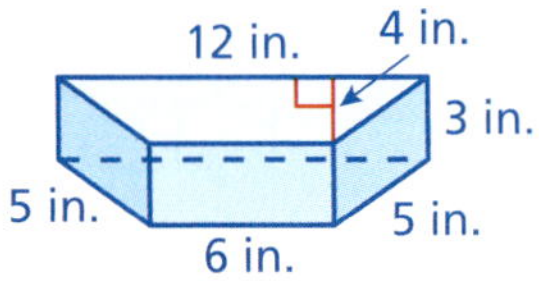

21.

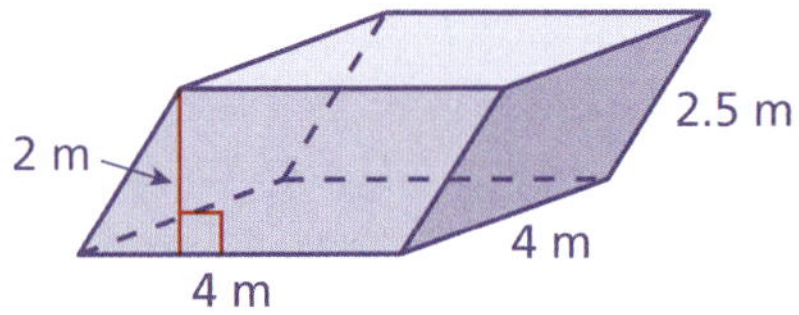

22. **OPEN-ENDED** Draw and label a rectangular prism that has a surface area of 158 square yards.

23. **LABEL** A label that wraps around a box of golf balls covers 75% of its lateral surface area. What is the value of x?

24. **BREAD** Fifty percent of the surface area of the bread is crust. What is the height h?

Common Errors

- **Exercise 19** Students may include the bottom of the cake as part of the area that needs to be frosted. Remind them to pay attention to the context of the problem.
- **Exercises 25 and 26** Students may use the ratio of a side length of the red prism to the corresponding side length of the blue prism. Remind them that the areas are not necessarily in the same proportions as the side lengths.

English Language Learners

Vocabulary

The term *net* means a two-dimensional representation of a solid figure. This definition is very different from its everyday meanings such as a device used to catch birds and fish or a fabric barricade dividing a tennis or volleyball court in half.

Practice and Problem Solving

16. The area of the 3×5 face is used 4 times rather than just twice.

$$S = 2(3)(5) + 2(3)(4) + 2(5)(4)$$
$$= 30 + 24 + 40$$
$$= 94 \text{ in.}^2$$

17. 177 in.^2

18. 726 cm^2

19. yes; Because you do not need to frost the bottom of the cake, you only need 249 square inches of frosting.

20. 156 in.^2

21. 68 m^2

22. Answers will vary.

23. $x = 4$

Practice and Problem Solving

24. See *Taking Math Deeper.*

25. The dimensions of the red prism are three times the dimensions of the blue prism. The surface area of the red prism is 9 times the surface area of the blue prism.

26–28. See Additional Answers.

Fair Game Review

29. 160 ft^2	**30.** 54 m^2
31. 28 ft^2	**32.** B

Mini-Assessment

Find the surface area of the prism.

1.

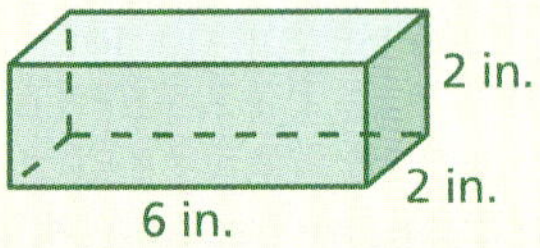

56 in.^2

2.

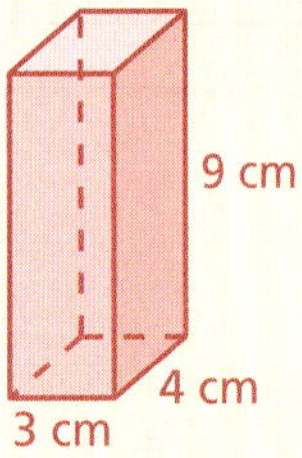

150 cm^2

3. Find the least amount of wrapping paper needed to cover the box. 22 ft^2

4. Find the least amount of fabric needed to make the tent. 152 ft^2

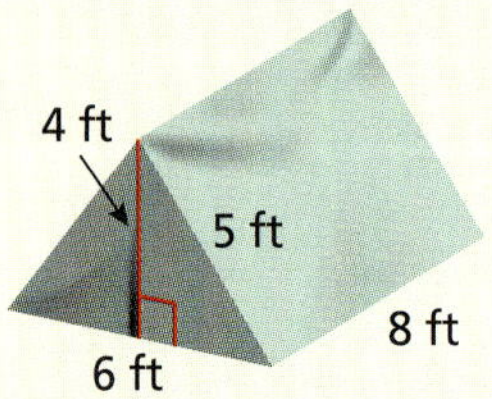

Taking Math Deeper

Exercise 24

To complete this problem, students may use a percent. Remind them to write the percent as a decimal before multiplying.

1 Find the total surface area.

$$S = 2(10^2) + 2(10h) + 2(10h)$$
$$= 200 + 40h$$

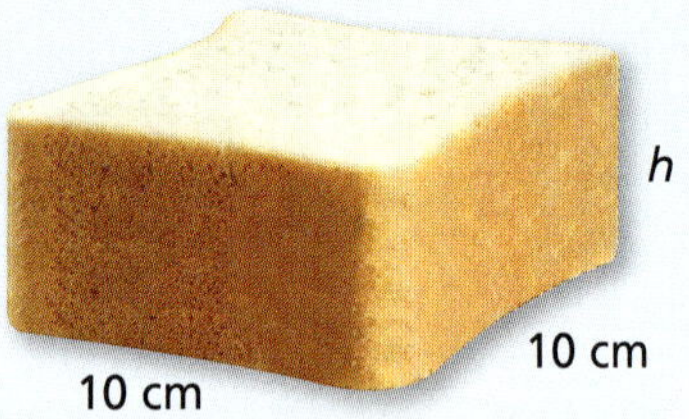

2 Find the surface area of the crust.

$$C = 2(10h) + 2(10h)$$
$$= 40h$$

3 In part 1, notice that 200 is the area of the top and bottom, and $40h$ is the area of the crust. These areas are equal, so $40h = 200$, or $h = 5$. Otherwise you can write and solve an equation with *variables on both sides.*

$$0.5(200 + 40h) = 40h$$
$$100 + 20h = 40h$$
$$100 = 20h$$
$$5 = h$$

So, the height of the bread is 5 centimeters.

Fun Bread Facts:

- It takes 1 second for a combine to harvest enough wheat to make about 8 loaves of bread.
- On average, each American consumes 53 pounds of bread per year.
- An average slice of packaged bread contains only 1 gram of fat and about 70 calories.

Reteaching and Enrichment Strategies

If students need help. . .	If students got it. . .
Resources by Chapter • Practice A and Practice B • Puzzle Time Record and Practice Journal Practice Differentiating the Lesson Lesson Tutorials Skills Review Handbook	Resources by Chapter • Enrichment and Extension • Technology Connection Start the next section

Compare the dimensions of the prisms. How many times greater is the surface area of the red prism than the surface area of the blue prism?

25.

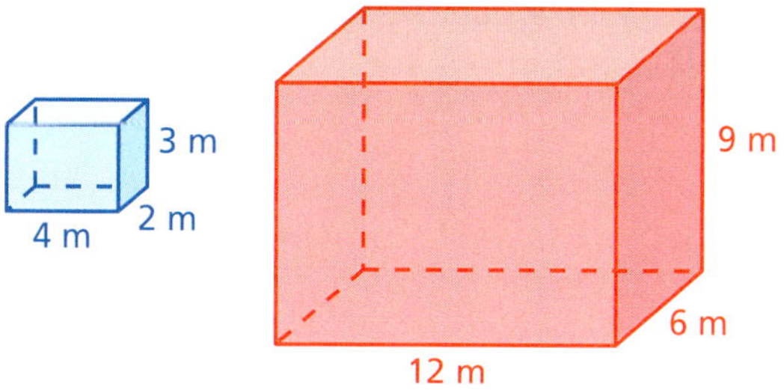

26.

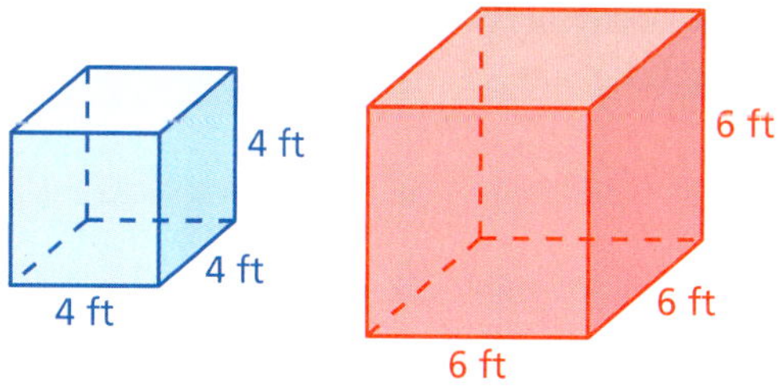

27. **STRUCTURE** You are painting the prize pedestals shown (including the bottoms). You need 0.5 pint of paint to paint the red pedestal.

a. The side lengths of the green pedestal are one-half the side lengths of the red pedestal. How much paint do you need to paint the green pedestal?

b. The side lengths of the blue pedestal are triple the side lengths of the green pedestal. How much paint do you need to paint the blue pedestal?

c. Compare the ratio of paint amounts to the ratio of side lengths for the green and red pedestals. Repeat for the green and blue pedestals. What do you notice?

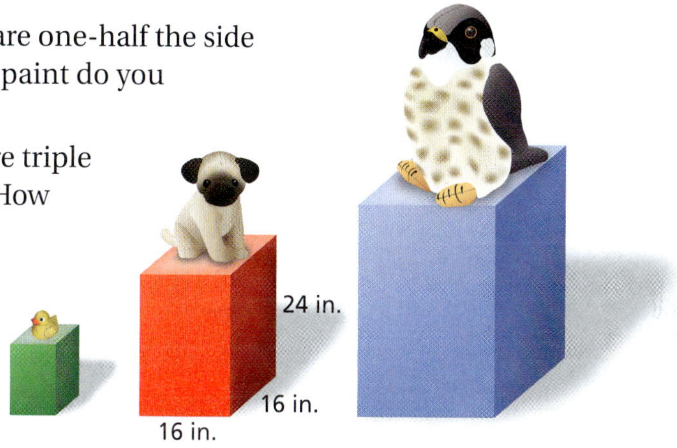

28. Number Sense A keychain-sized Rubik's Cube® is made up of small cubes. Each small cube has a surface area of 1.5 square inches.

a. What is the side length of each small cube?

b. What is the surface area of the entire Rubik's Cube®?

Fair Game Review What you learned in previous grades & lessons

Find the area of the triangle *(Skills Review Handbook)*

29.

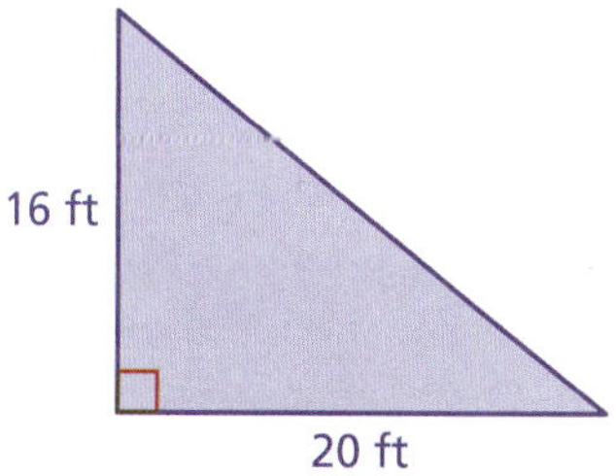

30.

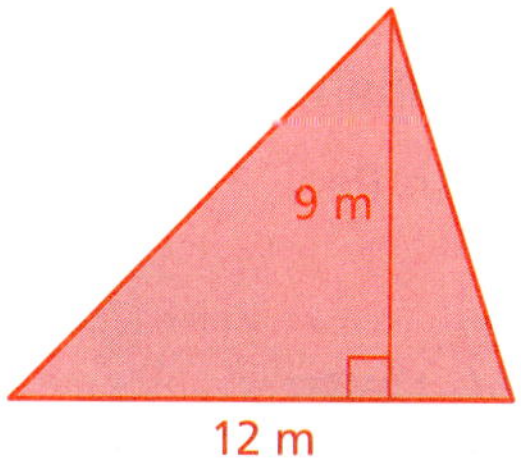

31.

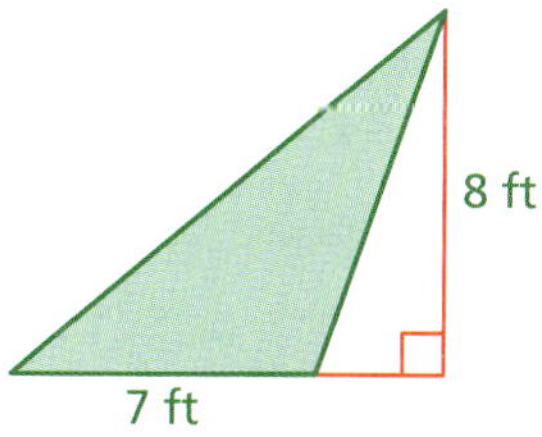

32. **MULTIPLE CHOICE** What is the circumference of the basketball? Use 3.14 for π. *(Section 13.1)*

Ⓐ 14.13 in. Ⓑ 28.26 in. Ⓒ 56.52 in. Ⓓ 254.34 in.

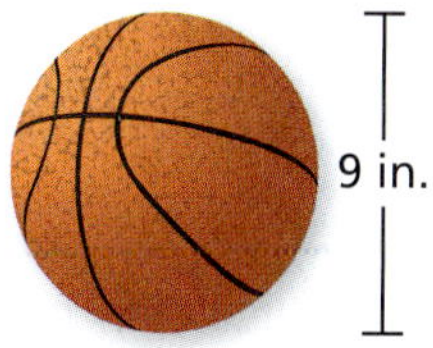

14.2 Surface Areas of Pyramids

Essential Question How can you find the surface area of a pyramid?

Even though many well-known pyramids have square bases, the base of a pyramid can be any polygon.

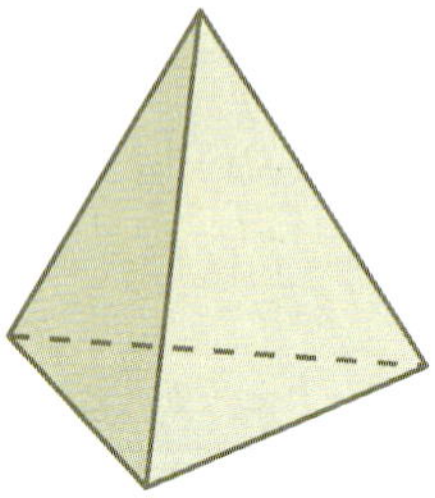

Triangular Base

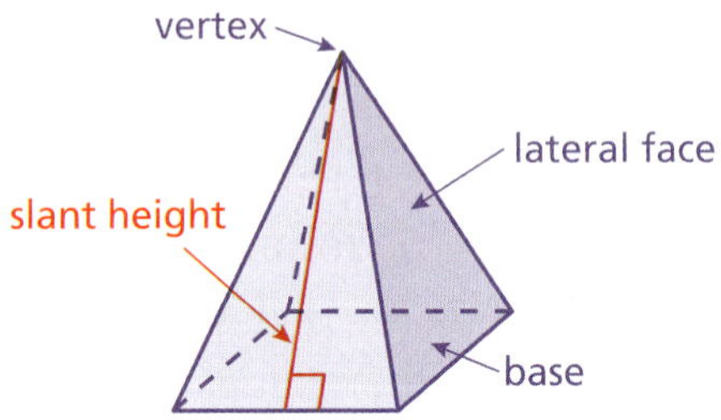

Square Base

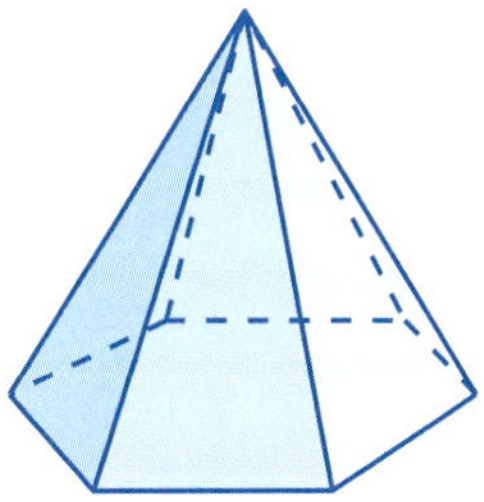

Hexagonal Base

1 ACTIVITY: Making a Scale Model

Work with a partner. Each pyramid has a square base.

- **Draw a net for a scale model of one of the pyramids. Describe your scale.**
- **Cut out the net and fold it to form a pyramid.**
- **Find the lateral surface area of the real-life pyramid.**

a. Cheops Pyramid in Egypt

Side = 230 m, Slant height ≈ 186 m

b. Muttart Conservatory in Edmonton

Side = 26 m, Slant height ≈ 27 m

c. Louvre Pyramid in Paris

Side = 35 m, Slant height ≈ 28 m

d. Pyramid of Caius Cestius in Rome

Side = 22 m, Slant height ≈ 29 m

Geometry

In this lesson, you will
- find surface areas of regular pyramids.
- solve real-life problems.

Laurie's Notes

Introduction

Applying Mathematical Practices

- **Model with Mathematics:** Finding the surface area of a solid can be challenging when the faces are not all visible. Drawing a net allows students to see all the faces of the pyramid.

Motivate

- Share information about the Great Pyramid of Egypt, also known as Cheops Pyramid.
- The Great Pyramid is the largest of the original *Seven Wonders of the World.* It was built in the 5th century B.C. and is estimated to have taken 100,000 men over 20 years to build it.
- The Great Pyramid is a square pyramid. It covers an area of 13 acres. The original height of the Great Pyramid was 485 feet, but due to erosion its height has declined to 450 feet. Each side of the square base is 755.5 feet in length (about 2.5 football field lengths).
- The Great Pyramid consists of approximately 2.5 million blocks that weigh from 2 tons to over 70 tons. The stones are cut so precisely that a credit card cannot fit between them.

Activity Notes

Discuss

- Discuss the vocabulary of pyramids and how they are named according to the base. Make a distinction between the slant height and the height.

Activity 1

- This activity connects scale drawings with the study of pyramids.
- This activity may take up most of the class period, but there may be time to complete Activity 2 or 3 as well.
- To ensure a variety, assign one pyramid to each pair of students and make sure about $\frac{1}{4}$ of the class makes each pyramid.
- Students will need to decide on the scale they will use.

  Example: To make a scale model for pyramid A, assume the scale selected is 1 cm = 20 m.

  $\frac{1 \text{ cm}}{20 \text{ m}} = \frac{x \text{ cm}}{230 \text{ m}} \rightarrow x = 11.5$ $\quad$ $\frac{1 \text{ cm}}{20 \text{ m}} = \frac{x \text{ cm}}{186 \text{ m}} \rightarrow x = 9.3$
- Students will use their eyesight and knowledge of squares and isosceles triangles to construct the square and four isosceles triangles.
- **Attend to Precision:** Have groups discuss the scale they used and how they found the lateral surface area. Listen for how clearly they communicate the process. Multiply the area of one triangular face by 4.
- **Model with Mathematics:** Making a pyramid from a net and finding the lateral surface area helps students remember the process and understand the formula.

What Your Students Will Learn

- Find surface areas of squares or triangular pyramids using formulas.

Previous Learning

Students should know how to find the area of a triangle and should know the general properties of squares and isosceles triangles.

Technology for the Teacher

Dynamic Classroom

Lesson Plans
Complete Materials List

14.2 Record and Practice Journal

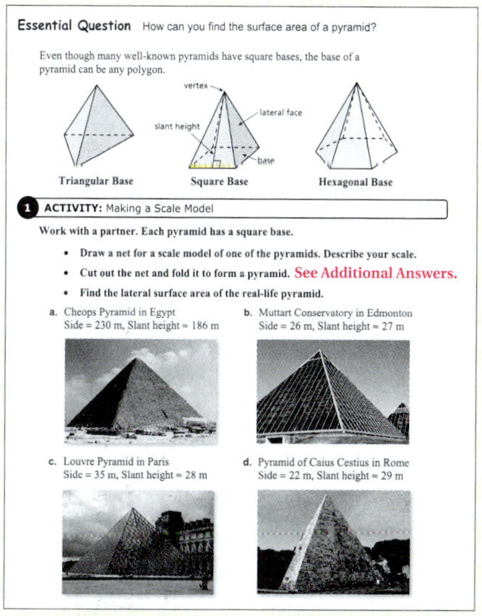

Essential Question How can you find the surface area of a pyramid?

Even though many well-known pyramids have square bases, the base of a pyramid can be any polygon.

Triangular Base — Square Base — Hexagonal Base

(labels: vertex, lateral face, slant height, base)

1 ACTIVITY: Making a Scale Model

Work with a partner. Each pyramid has a square base.

- Draw a net for a scale model of one of the pyramids. Describe your scale.
- Cut out the net and fold it to form a pyramid. See Additional Answers.
- Find the lateral surface area of the real-life pyramid.

a. Cheops Pyramid in Egypt
Side = 230 m, Slant height ≈ 186 m

b. Muttart Conservatory in Edmonton
Side = 26 m, Slant height ≈ 27 m

c. Louvre Pyramid in Paris
Side = 35 m, Slant height ≈ 28 m

d. Pyramid of Caius Cestius in Rome
Side = 22 m, Slant height ≈ 29 m

English Language Learners

Vocabulary

English learners may struggle with understanding the *slant height* of a pyramid. Use a skateboard ramp as an example. Ask students to find the length of the ramp. Most likely students will find the length of the slanted portion of the ramp. Compare this length to the slant height of a pyramid.

14.2 Record and Practice Journal

2 ACTIVITY: Estimation

Work with a partner. There are many different types of gemstone cuts. Here is one called a brilliant cut.

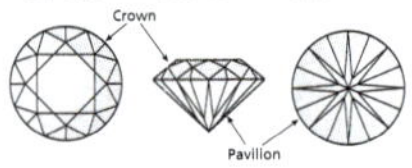

The size and shape of the pavilion can be approximated by an octagonal pyramid.

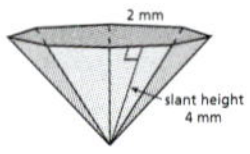

a. What does *octagonal* mean?

8 sided

b. Draw a net for the pyramid.

See Additional Answers.

c. Find the lateral surface area of the pyramid.

32 mm^2

3 ACTIVITY: Comparing Surface Areas

Work with a partner. Both pyramids have the same side lengths of the base and the same slant heights.

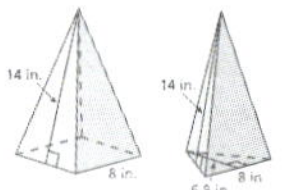

a. **REASONING** Without calculating, which pyramid has the greater surface area? Explain.

square pyramid

b. Verify your answer to part (a) by finding the surface area of each pyramid.

square pyramid: 288 $in.^2$
triangular pyramid: 195.6 $in.^2$

What Is Your Answer?

4. **IN YOUR OWN WORDS** How can you find the surface area of a pyramid? Draw a diagram with your explanation.

Find area of base and add area of lateral faces.

Laurie's Notes

Activity 2

- Note that students are only asked to find the lateral surface area. This means that they are finding the surface area of 8 congruent isosceles triangles.
- **FYI:** The prefix octa- means eight. An octopus has 8 arms; when October was named in the Roman calendar, it was the 8th month; an octave on the piano has 8 notes; an octad is a group of 8 things.

? "What common road sign is an octagon?" Stop sign

- The net includes the octagonal base, but the surface area of the base is not needed for this problem.
- Ask a volunteer to sketch his or her net at the board. If you have the appropriate snap together polygon frames, make the net.

Activity 3

- Students can reason that the area of a triangle with 8-inch sides must be less than the area of a square with 8-inch sides. By reasoning that the lateral surface area of the triangular pyramid is also less than that of the square pyramid, they can answer part (a) without any computations.

? "What is true about the lateral faces of both pyramids?" All the lateral faces of the two pyramids are congruent triangles.

? "Which pyramid has the greater lateral surface area? Explain." The square pyramid has a greater lateral surface area, because there is one more isosceles triangle.

What Is Your Answer?

- Have students work in pairs to answer the question.

Closure

- **Exit Ticket:** Sketch a net for a hexagonal pyramid and describe how to find the lateral surface area. Find the area of one of the lateral faces and multiply by 6.

2 ACTIVITY: Estimation

Math Practice

Calculate Accurately

How can you verify that you have calculated the lateral surface area accurately?

Work with a partner. There are many different types of gemstone cuts. Here is one called a brilliant cut.

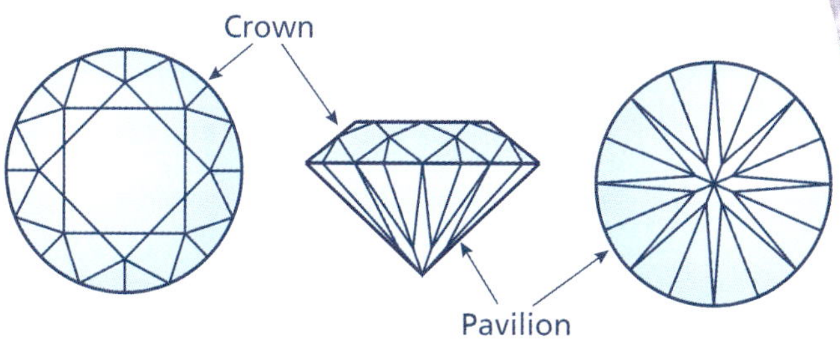

The size and shape of the pavilion can be approximated by an octagonal pyramid.

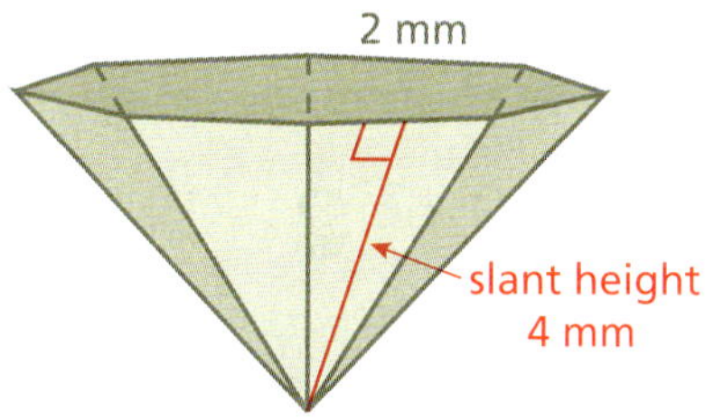

a. What does *octagonal* mean?

b. Draw a net for the pyramid.

c. Find the lateral surface area of the pyramid.

3 ACTIVITY: Comparing Surface Areas

Work with a partner. Both pyramids have the same side lengths of the base and the same slant heights.

a. **REASONING** Without calculating, which pyramid has the greater surface area? Explain.

b. Verify your answer to part (a) by finding the surface area of each pyramid.

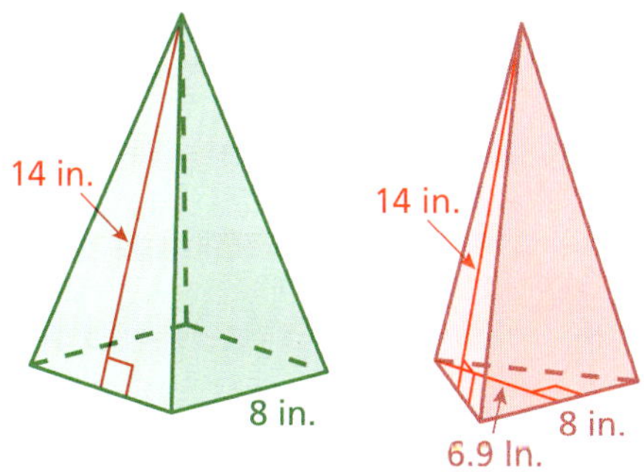

What Is Your Answer?

4. **IN YOUR OWN WORDS** How can you find the surface area of a pyramid? Draw a diagram with your explanation.

Practice Use what you learned about the surface area of a pyramid to complete Exercises 4–6 on page 598.

14.2 Lesson

Check It Out
Lesson Tutorials
BigIdeasMath.com

Key Vocabulary

regular pyramid, *p. 596*

slant height, *p. 596*

A **regular pyramid** is a pyramid whose base is a regular polygon. The lateral faces are triangles. The height of each triangle is the **slant height** of the pyramid.

Remember

In a regular polygon, all the sides are congruent and all the angles are congruent.

Key Idea

Surface Area of a Pyramid

The surface area S of a pyramid is the sum of the areas of the base and the lateral faces.

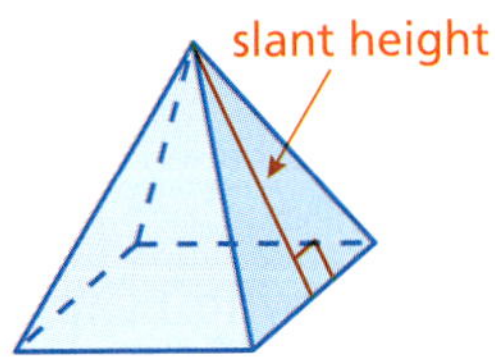

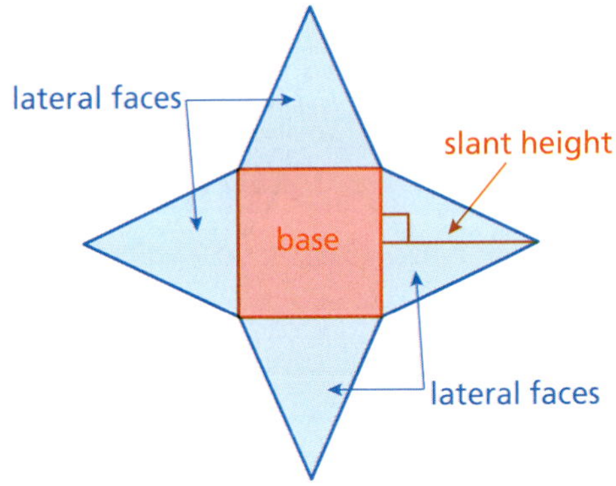

S = area of base + areas of lateral faces

EXAMPLE 1 Finding the Surface Area of a Square Pyramid

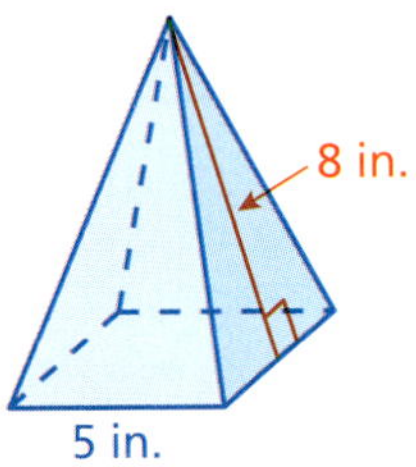

Find the surface area of the regular pyramid.

Draw a net.

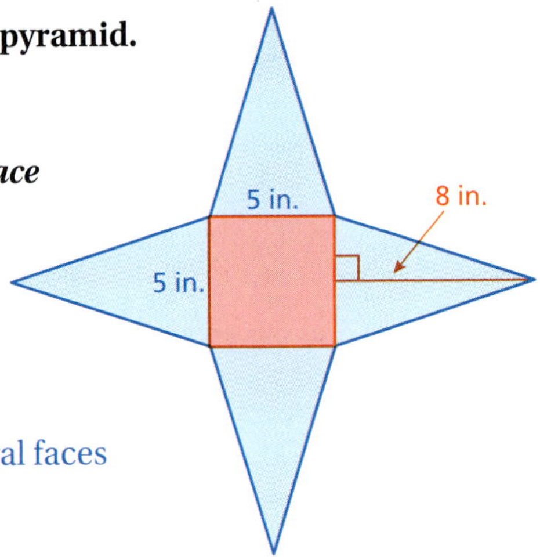

Area of Base	*Area of a Lateral Face*
$5 \cdot 5 = 25$	$\frac{1}{2} \cdot 5 \cdot 8 = 20$

Find the sum of the areas of the base and the lateral faces.

$$S = \text{area of base} + \text{areas of lateral faces}$$
$$= 25 + \underbrace{20 + 20 + 20 + 20}$$
$$= 105$$

There are 4 identical lateral faces. Count the area 4 times.

The surface area is 105 square inches.

On Your Own

1. What is the surface area of a square pyramid with a base side length of 9 centimeters and a slant height of 7 centimeters?

Laurie's Notes

Introduction

Connect

- **Yesterday:** Students discovered how to find the surface area of a pyramid by examining the net that makes up a pyramid.
- **Today:** Students will work with a formula for the surface area of a pyramid.

Motivate

- Ask students where they have heard about pyramids or have seen them before. Give groups of students 3–4 minutes to brainstorm a list. They may mention the pyramid on the back of U.S. dollar bills, camping tents, roof designs, tetrahedral dice, and of course, Egyptian pyramids.

Lesson Notes

Key Idea

- Introduce the vocabulary: regular pyramid, regular polygon, slant height.
- ? "What information does the type of base give you about the lateral faces?" number of sides in the base = number of congruent isosceles triangles for the lateral surface area
- ? "If you know the length of each side of the base, what else do you know?" the length of the base of the triangular lateral faces

Example 1

- **Model with Mathematics:** Draw the net and label the known information. This should remind students of the work they did yesterday making a scale model of a pyramid.
- Write the formula in words first to model good problem-solving techniques.
- Continue to ask questions as you find the total surface area: "How do you find the area of the base? How many lateral faces are there? What is the area of just one lateral face? How do you find the area of a triangle?"
- **Common Error:** In using the area formula for a triangle, the $\frac{1}{2}$ often produces a computation mistake. In this instance, students must multiply $\frac{1}{2} \times 5 \times 8$. Remind students that it's okay to change the order of the factors (Commutative Property). Rewriting the problem as $\frac{1}{2} \times 8 \times 5$ means that you can work with whole numbers: $\frac{1}{2} \times 8 \times 5 = 4 \times 5 = 20$.

On Your Own

- **Model with Mathematics:** Encourage students to sketch a three-dimensional model of the pyramid and the net for the pyramid. Label the net with the known information.
- Ask a volunteer to share his or her work at the board.

Goal Today's lesson is finding the surface area of a pyramid using a formula.

Technology for the Teacher

Dynamic Classroom

Lesson Tutorials
Lesson Plans
Answer Presentation Tool

Extra Example 1

What is the surface area of a square pyramid with a base side length of 3 meters and a slant height of 6 meters? $45\ \text{m}^2$

On Your Own

1. $207\ \text{cm}^2$

Extra Example 2

Find the surface area of the regular pyramid.

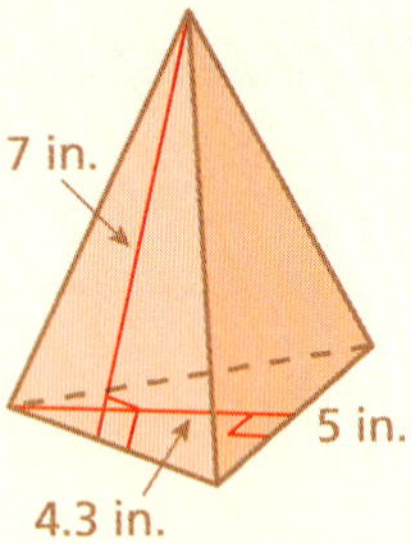

63.25 in.^2

Extra Example 3

The slant height of the roof in Example 3 is 13 feet. One bundle of shingles covers 30 square feet. How many bundles of shingles should you buy to cover the roof? 16 bundles of shingles

On Your Own

2. 105.6 ft^2
3. 17 bundles

Differentiated Instruction

Kinesthetic

Photocopy nets of solids for students to cut out and assemble. Then have students draw their own nets to cut out and assemble.

Laurie's Notes

Example 2

- Remind students of the definition of a regular pyramid. This is important because the base, as drawn, doesn't look like an equilateral triangle. This is the challenge of representing a 3-dimensional figure on a flat 2-dimensional sheet of paper.
- Drawing the net is an important step. It allows the key dimensions to be labeled in a way that can be seen.
- Encourage mental math when multiplying $\frac{1}{2} \times 10 \times 8.7$ and $\frac{1}{2} \times 10 \times 14$. Ask students to share their strategies with other students.

Example 3

? "How does the lateral surface area of the roof relate to the bundles of shingles needed?" The lateral surface area divided by the area covered per bundle gives the number of bundles needed.

- Have students compute the lateral surface area. Some students may need to draw the triangular lateral face first before performing the computation.

? **Reason Abstractly and Quantitatively:** "Suppose you compute the number of bundles needed on another roof, and you get an exact answer of 25.2. Is it okay to round down to 25 bundles? Explain." No; Always round up to the next whole number, so you do not run short of shingles.

- **FYI:** When shingles are placed on a roof, they need to overlap the shingle below. The coverage given per bundle takes into account the overlap.
- **Extension:** Suppose a bundle of shingles sells for \$34.75. What will the total cost be for the shingles? \$764.50

On Your Own

- Give students sufficient time to do their work for each problem before asking volunteers to share their work at the board.

Closure

- **Exit Ticket:** Sketch a square pyramid with a slant height of 4 centimeters and a base side length of 3 centimeters. Sketch the net and find the surface area. 33 cm^2

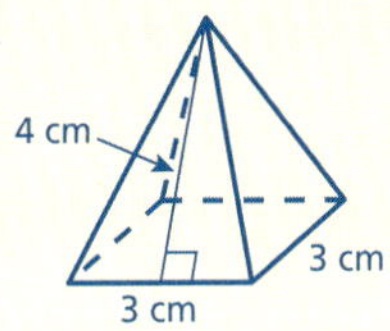

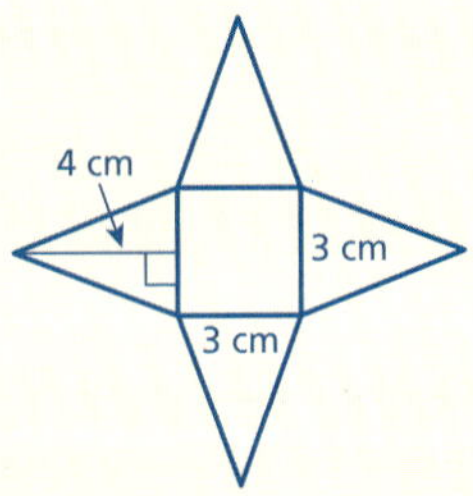

EXAMPLE 2 Finding the Surface Area of a Triangular Pyramid

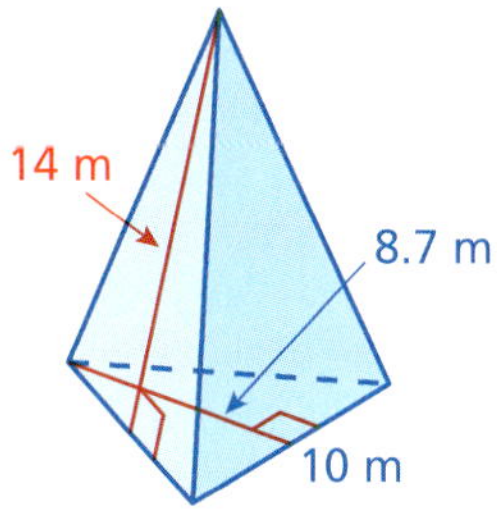

Find the surface area of the regular pyramid.

Draw a net.

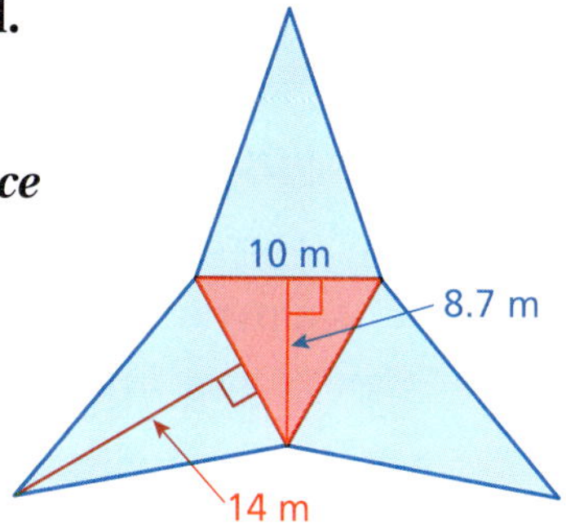

Area of Base

$\frac{1}{2} \cdot 10 \cdot 8.7 = 43.5$

Area of a Lateral Face

$\frac{1}{2} \cdot 10 \cdot 14 = 70$

Find the sum of the areas of the base and the lateral faces.

S = area of base + areas of lateral faces

$= 43.5 + \underbrace{70 + 70 + 70}$

$= 253.5$

There are 3 identical lateral faces. Count the area 3 times.

The surface area is 253.5 square meters.

EXAMPLE 3 Real-Life Application

A roof is shaped like a square pyramid. One bundle of shingles covers 25 square feet. How many bundles should you buy to cover the roof?

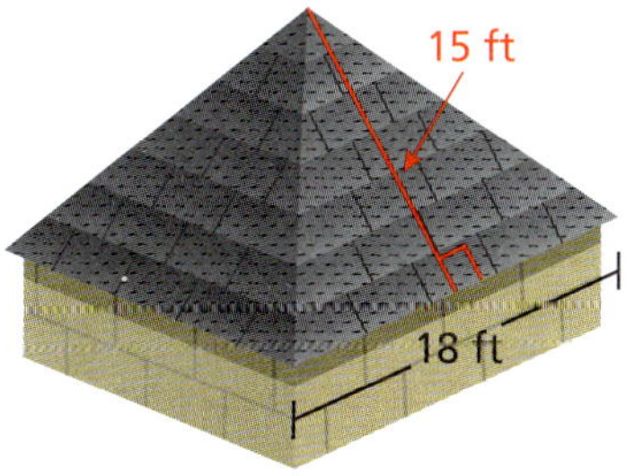

The base of the roof does not need shingles. So, find the sum of the areas of the lateral faces of the pyramid.

Area of a Lateral Face

$\frac{1}{2} \cdot 18 \cdot 15 = 135$

There are four identical lateral faces. So, the lateral surface area is

$135 + 135 + 135 + 135 = 540.$

Because one bundle of shingles covers 25 square feet, it will take $540 \div 25 = 21.6$ bundles to cover the roof.

So, you should buy 22 bundles of shingles.

On Your Own

Now You're Ready
Exercises 7–12

2. What is the surface area of the regular pyramid at the right?

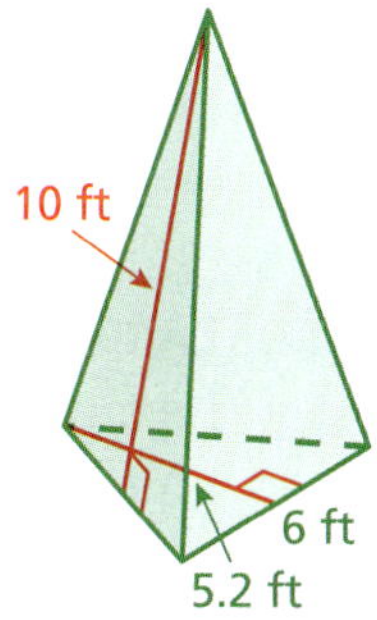

3. **WHAT IF?** In Example 3, one bundle of shingles covers 32 square feet. How many bundles should you buy to cover the roof?

14.2 Exercises

Check It Out
Help with Homework
BigIdeasMath.com

Vocabulary and Concept Check

1. **VOCABULARY** Can a pyramid have rectangles as lateral faces? Explain.

2. **CRITICAL THINKING** Why is it helpful to know the slant height of a pyramid to find its surface area?

3. **WHICH ONE DOESN'T BELONG?** Which description of the solid does *not* belong with the other three? Explain your answer.

square pyramid	regular pyramid
rectangular pyramid	triangular pyramid

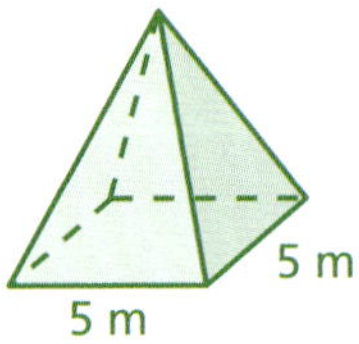

Practice and Problem Solving

Use the net to find the surface area of the regular pyramid.

4. 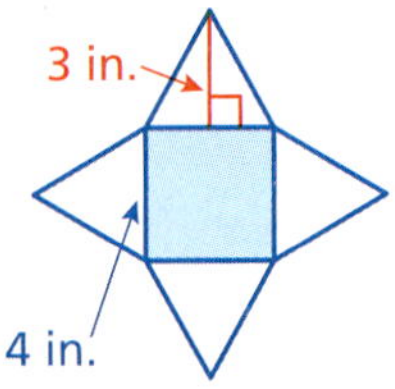

5. 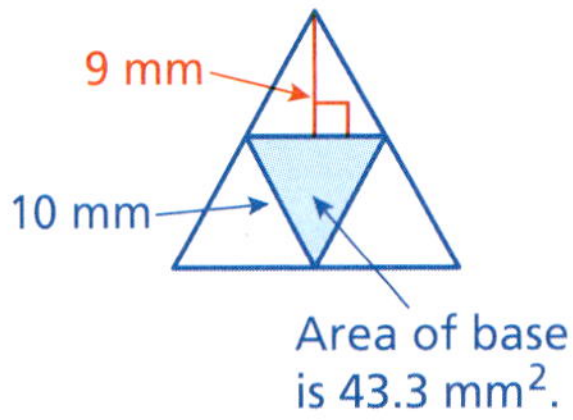

6. 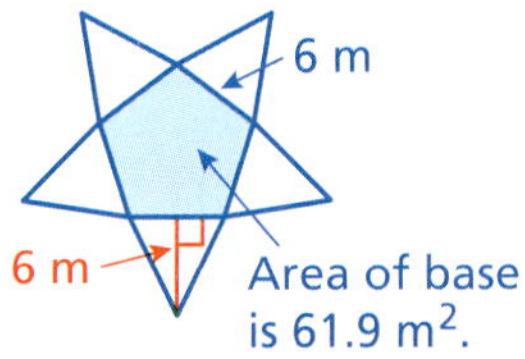

In Exercises 7–11, find the surface area of the regular pyramid.

1 2 7. 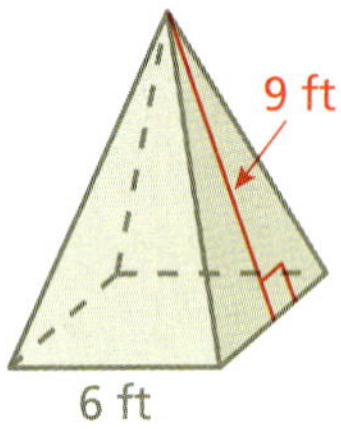

8. 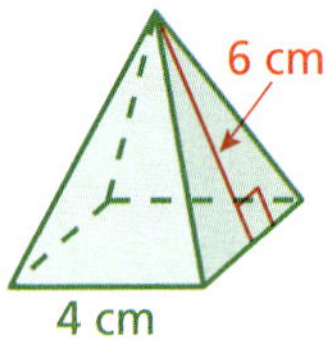

9. 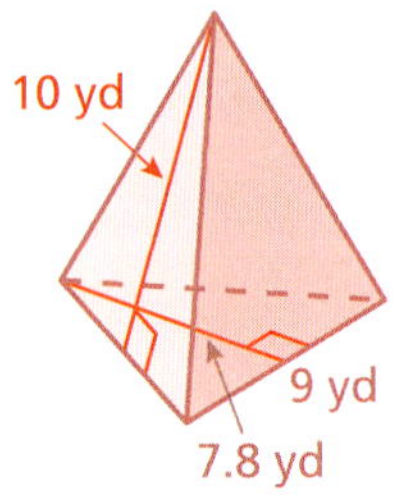

10. 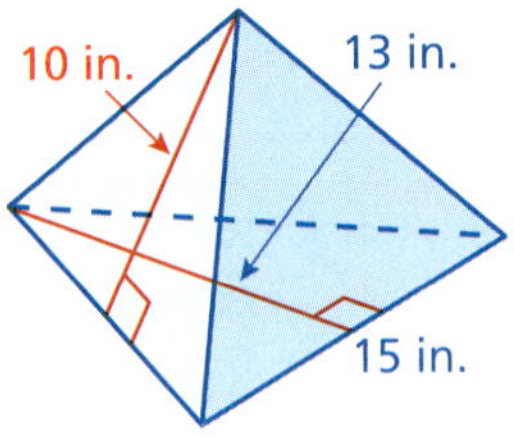

11. 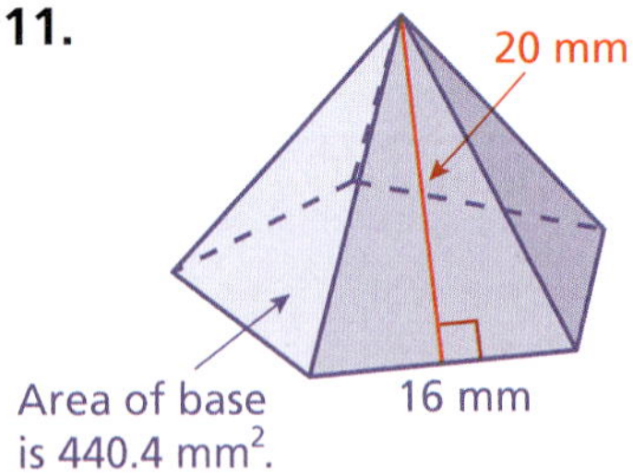

3 12. **LAMPSHADE** The base of the lampshade is a regular hexagon with a side length of 8 inches. Estimate the amount of glass needed to make the lampshade.

13. **GEOMETRY** The surface area of a square pyramid is 85 square meters. The base length is 5 meters. What is the slant height?

Assignment Guide and Homework Check

Level	Assignment	Homework Check
Advanced	1–6, 8–20 even, 21–24	10, 12, 16, 18, 20

Common Errors

- **Exercises 7–11** Students may forget to add on the area of the base when finding the surface area. Remind them that when asked to find the surface area, the base is included.
- **Exercises 7–11** Students may add the wrong number of lateral face areas to the area of the base. Examine several different pyramids with different bases and ask if they can find a relationship between the number of sides of the base and the number of lateral faces. (They are the same.) Remind students that the number of sides on the base determines how many triangles make up the lateral surface area.
- **Exercise 12** Students may think that there is not enough information to solve the problem because it is not all labeled in the picture. Tell them to use the information in the word problem to finish labeling the picture. Also ask students to identify how many lateral faces are part of the lamp before they find the area of one face.

14.2 Record and Practice Journal

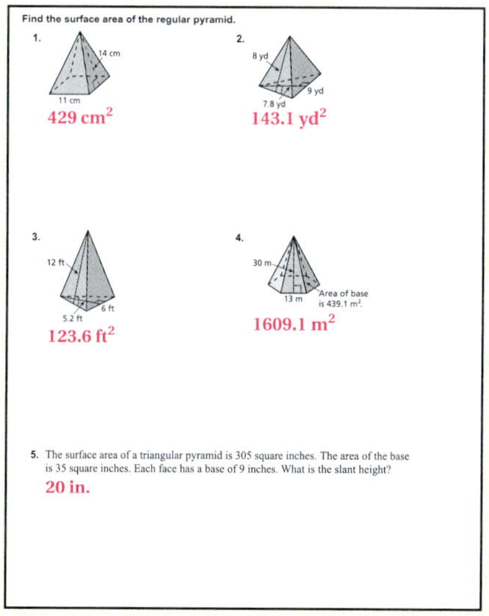
Find the surface area of the regular pyramid.

1. 14 cm, 11 cm
429 cm²

2. 8 yd, 9 yd, 7.8 yd
143.1 yd²

3. 12 ft, 6 ft, 5.2 ft
123.6 ft²

4. 30 m, 13 m, Area of base is 439.1 m².
1609.1 m²

5. The surface area of a triangular pyramid is 305 square inches. The area of the base is 35 square inches. Each face has a base of 9 inches. What is the slant height?
20 in.

Vocabulary and Concept Check

1. no; The lateral faces of a pyramid are triangles.
2. Knowing the slant height helps because it represents the height of the triangle that makes up each lateral face. So, the slant height helps you to find the area of each lateral face.
3. triangular pyramid; The other three are names for the pyramid.

Practice and Problem Solving

4. 40 in.^{2}
5. 178.3 mm^2
6. 151.9 m^2
7. 144 ft^2
8. 64 cm^2
9. 170.1 yd^2
10. 322.5 in.^{2}
11. 1240.4 mm^2
12. 240 in.^{2}
13. 6 m

Practice and Problem Solving

14. 165 ft^2 **15.** 283.5 cm^2

16. 281 ft^2

17. See *Taking Math Deeper.*

18–20. See Additional Answers.

Fair Game Review

21. $A \approx 452.16$ units^2; $C \approx 75.36$ units

22. $A \approx 200.96$ units^2; $C \approx 50.24$ units

23. $A \approx 572.265$ units^2; $C \approx 84.78$ units

24. B

Mini-Assessment

Find the surface area of the regular pyramid.

1.

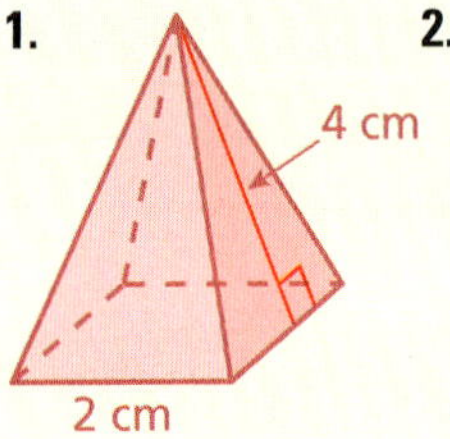

20 cm^2

2.

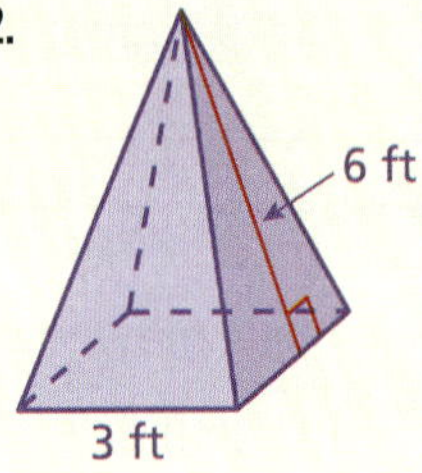

45 ft^2

3.

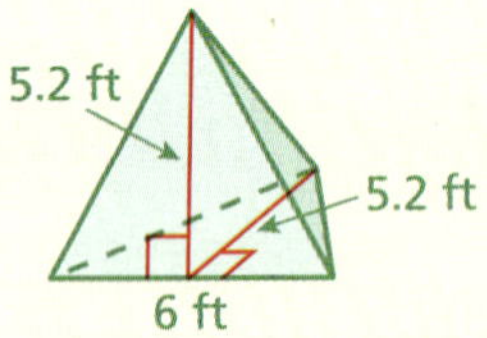

62.4 ft^2

4. Find the surface area of the roof of the doll house. 480 in.^{2}

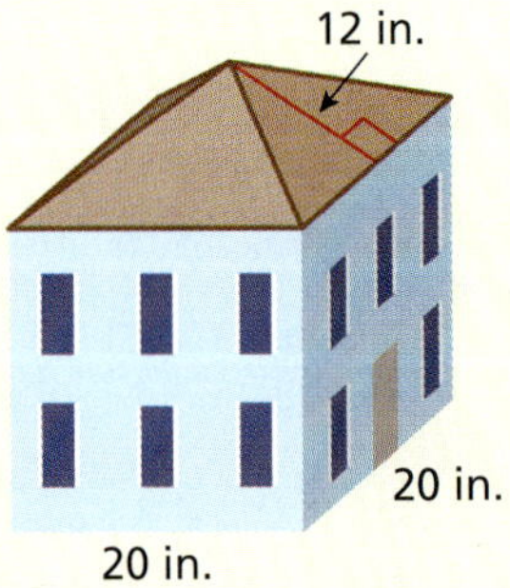

Taking Math Deeper

Exercise 17

Creating something with fabric takes precision. You need to be precise with your measurements and pattern placement. Once you cut the fabric, there is no way to undo it.

① **a.** Find the area of the 8 pieces.

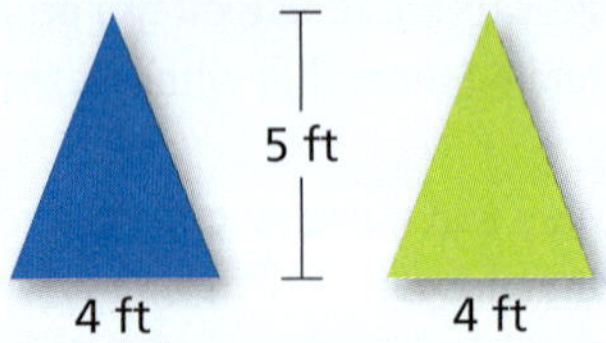

$$\text{Area} = 8\left(\frac{1}{2} \cdot 4 \cdot 5\right)$$
$$= 80 \text{ ft}^2$$

② **b.** Draw a diagram.

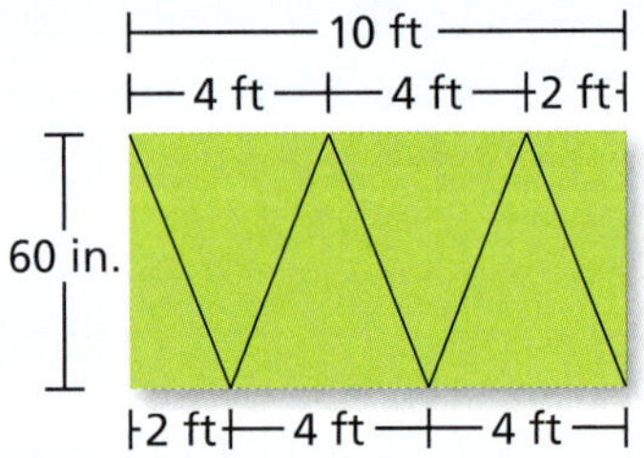

③ Answer the question.

For each color, you cut the four pieces from fabric that is 5 feet wide and 10 feet long.

Fabric Area $= 2(5 \bullet 10) = 100$ ft^2

Area of 8 pieces $= 80$ ft^2

c. Area of waste $= 100 - 80 = 20$ ft^2

Project

Use construction paper and a pencil to create an "umbrella" using the least possible amount of paper. The umbrella should be a scale model of the one in the exercise, using a scale of 1 inch to 1 foot.

Reteaching and Enrichment Strategies

If students need help. . .	If students got it. . .
Resources by Chapter • Practice A and Practice B • Puzzle Time Record and Practice Journal Practice Differentiating the Lesson Lesson Tutorials Skills Review Handbook	Resources by Chapter • Enrichment and Extension • Technology Connection Start the next section

Find the surface area of the composite solid.

14.

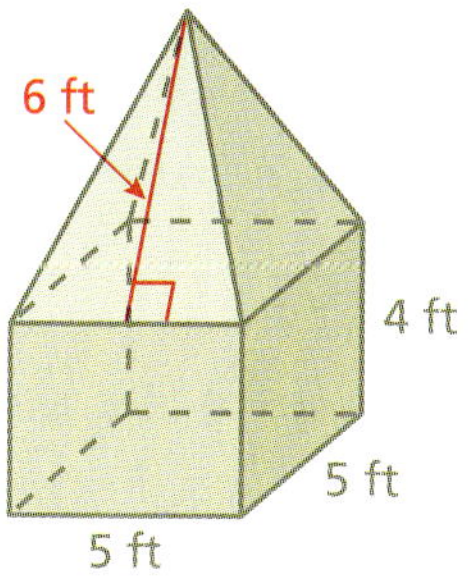

15.

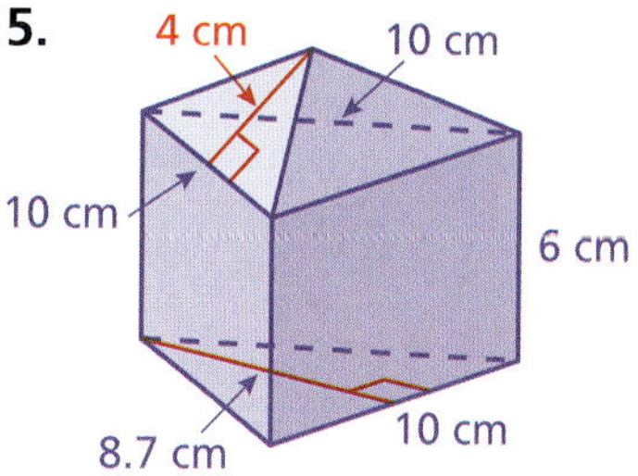

16.

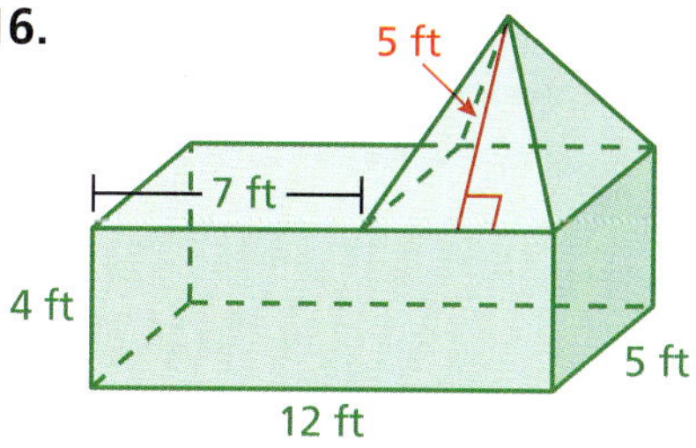

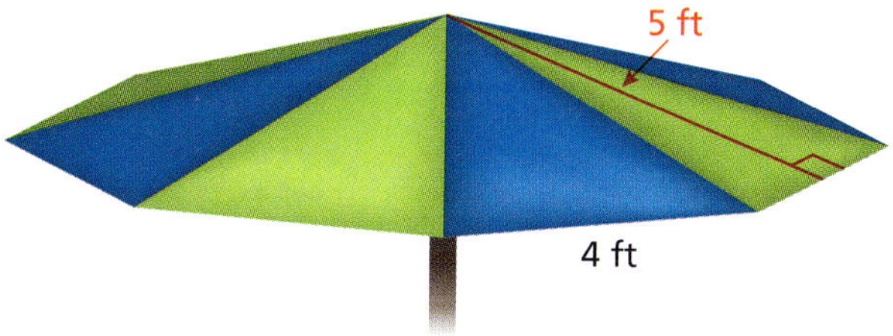

17. **PROBLEM SOLVING** You are making an umbrella that is shaped like a regular octagonal pyramid.

a. Estimate the amount of fabric that you need to make the umbrella.

b. The fabric comes in rolls that are 60 inches wide. Draw a diagram of how you can cut the fabric from a roll that is 10 feet long.

c. How much fabric is wasted?

18. **REASONING** The *height* of a pyramid is the perpendicular distance between the base and the top of the pyramid. Which is greater, the height of a pyramid or the slant height? Explain your reasoning.

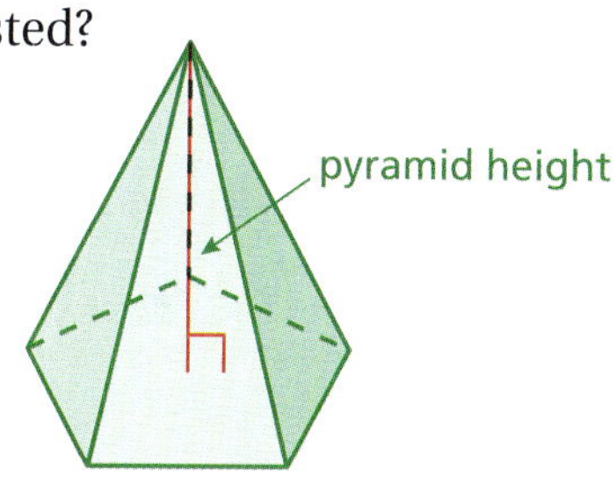

19. **TETRAHEDRON** A tetrahedron is a triangular pyramid whose four faces are identical equilateral triangles. The total lateral surface area is 93 square centimeters. Find the surface area of the tetrahedron.

20. **Reasoning** Is the total area of the lateral faces of a pyramid *greater than*, *less than*, or *equal to* the area of the base? Explain.

Fair Game Review What you learned in previous grades & lessons

Find the area and the circumference of the circle. Use 3.14 for π.
(Section 13.1 and Section 13.3)

21.

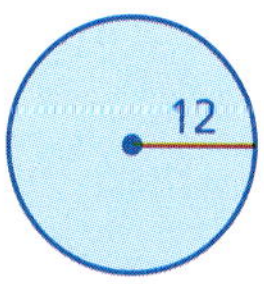

22.

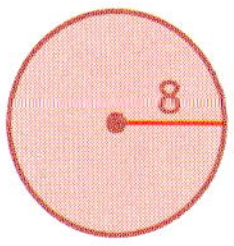

23.

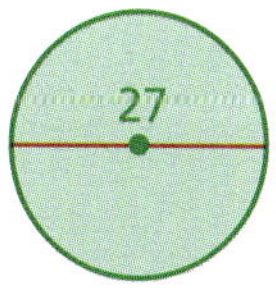

24. **MULTIPLE CHOICE** The distance between bases on a youth baseball field is proportional to the distance between bases on a professional baseball field. The ratio of the youth distance to the professional distance is 2 : 3. Bases on a youth baseball field are 60 feet apart. What is the distance between bases on a professional baseball field? *(Skills Review Handbook)*

Ⓐ 40 ft　　Ⓑ 90 ft　　Ⓒ 120 ft　　Ⓓ 180 ft

14.3 Surface Areas of Cylinders

Essential Question How can you find the surface area of a cylinder?

A *cylinder* is a solid that has two parallel, identical circular bases.

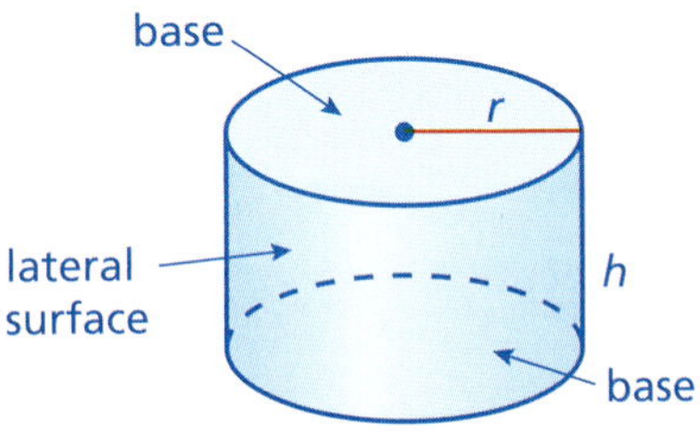

1 ACTIVITY: Finding Area

Work with a partner. Use a cardboard cylinder.

- **Talk about how you can find the area of the outside of the roll.**
- **Estimate the area using the methods you discussed.**
- **Use the roll and the scissors to find the actual area of the cardboard.**
- **Compare the actual area to your estimates.**

2 ACTIVITY: Finding Surface Area

Work with a partner.

- **Make a net for the can. Name the shapes in the net.**
- **Find the surface area of the can.**
- **How are the dimensions of the rectangle related to the dimensions of the can?**

Geometry
In this lesson, you will
- find surface areas of cylinders.

Laurie's Notes

Introduction

Applying Mathematical Practices

- **Model with Mathematics:** Drawing, cutting, and measuring the components of a cylinder allow students to make sense of the formula for the surface area of a cylinder.

Motivate

- Use two different cans (cylinders), where the taller can has a lesser radius. A tuna can and a 6-ounce vegetable can work well.
- ? Hold both cans. "Which can required more metal to make?" Answers will vary depending on can sizes.
- This question focuses attention on the surface areas of the cans, the need to consider their components, and how they were made.

Activity Notes

Activity 1

- **Big Idea:** Recall the connection between a prism and a cylinder—*structurally they are the same*. Unlike the prism, you don't have cardboard models to unfold. Use cardboard rolls from paper towels or toilet paper, or make rolls from strips of file folder paper.
- By estimating the area before they cut, students are engaging their spatial skills to think about the area of a curved surface.
- ? "What shape do you have when the roll is flattened out?" rectangle
- ? "How are the dimensions of the rectangle related to the dimensions of the original cylinder?" The height of the rectangle is the height of the cylinder. The length of the rectangle is the circumference of the cylinder.
- ? "What units do you use to measure the dimensions of your rectangle?" depending upon the ruler, centimeters or inches
- ? "What units do you use to label your answer?" cm^2 or $in.^2$

Activity 2

- Each pair of students will need scrap paper, tape, scissors, and a can. It is more interesting when the cans around the room are of different sizes.
- **Teaching Tip:** Recycle plastic bags and tape to desks for trash disposal.
- Students wrap paper around the can to make the net for the lateral surface. This helps them relate the cylinder and rectangle dimensions.
- **Discuss Results:** Students should describe the parts of a cylinder, how to find the area of each part, and how the dimensions of each part are related to the dimensions of the cylinder.
- **Make Sense of Problems and Persevere in Solving Them:** Making the cylinder and calculating its surface area help students to remember the process and understand the formula.

What Your Students Will Learn

- Find surface areas of cylinders using a formula.

Previous Learning

Students should know the formulas for the area of a rectangle and the area of a circle.

Technology for the Teacher

Dynamic Classroom

Lesson Plans
Complete Materials List

14.3 Record and Practice Journal

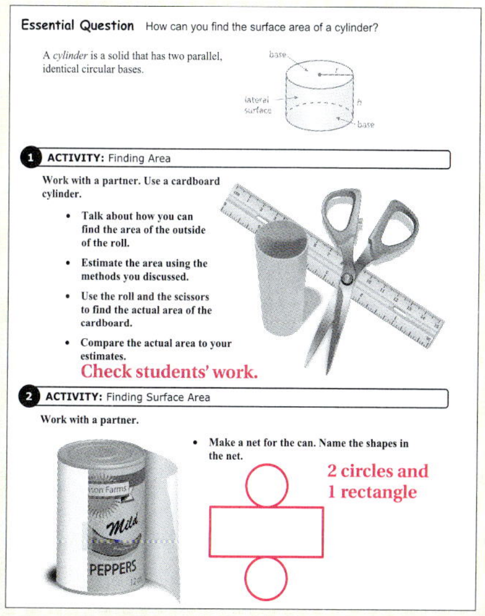

Essential Question How can you find the surface area of a cylinder?

A *cylinder* is a solid that has two parallel, identical circular bases.

1 ACTIVITY: Finding Area

Work with a partner. Use a cardboard cylinder.

- Talk about how you can find the area of the outside of the roll.
- Estimate the area using the methods you discussed.
- Use the roll and the scissors to find the actual area of the cardboard.
- Compare the actual area to your estimates.

Check students' work.

2 ACTIVITY: Finding Surface Area

Work with a partner.

- Make a net for the can. Name the shapes in the net.

2 circles and 1 rectangle

English Language Learners

Vocabulary

Have students work in pairs, one English learner and one English speaker. Have each pair write a problem involving the surface area of a cylinder. On a separate piece of paper, students should solve their own problem. Then have students exchange their problem with another pair of students. Students solve the new problem. After solving the problem, the four students discuss the problems and solutions.

14.3 Record and Practice Journal

- Find the surface area of the can
  Check students' work.

- How are the dimensions of the rectangle related to the dimensions of the can?
  rectangle length = can circumference
  rectangle width = can height

3 ACTIVITY: Estimation

Work with a partner. From memory, estimate the dimensions of the real-life item in inches. Then use the dimensions to estimate the surface area of the item in square inches.

a.

$r \approx 1.5$ in.
$h \approx 4$ in.
$s \approx 52$ in.^{2}

b.

$r \approx 1.25$ in.
$h \approx 4.5$ in.
$s \approx 45$ in.^{2}

c.

$r \approx 2.75$ in.
$h \approx 6$ in.
$s \approx 151$ in.^{2}

d.

$r \approx 1.75$ in.
$h \approx 1.5$ in.
$s \approx 36$ in.^{2}

What Is Your Answer?

4. IN YOUR OWN WORDS How can you find the surface area of a cylinder? Give an example with your description. Include a drawing of the cylinder.
   Add twice the area of the circular base to the product of the circumference and the height of the cylinder.

5. To eight decimal places, $\pi \approx 3.14159265$. Which of the following is closest to π? c

   a. 3.14 b. $\frac{22}{7}$ c. $\frac{355}{113}$

"To approximate $\pi \approx 3.141593$, I simply remember 1, 1, 3, 3, 5, 5." "Then I compute $\frac{355}{113} \approx 3.141593$."

Laurie's Notes

Activity 3

- When estimating the dimensions of the common cylinders, encourage students to use their hands to visualize the size of the cylinder.
- If time permits, set up stations in the room with a different cylinder at each. Provide rulers. In small groups, students move from one station to the next. Make sure you have a good variety of common cylinders: soup can, soft drink can, tuna can, AA battery, etc.
- ? "What dimensions did you measure for each cylinder?" Students will often say diameter and height.
- ? "Could the radius be measured?" yes; Take $\frac{1}{2}$ of the diameter to find the radius.
- **Extension:** Gather the results of the activity, and then find the mean for several of the cylinders.

What Is Your Answer?

- Have students work in pairs. Review answers as a class.

Closure

- Hold the two cans from the Motivate section and ask, "Which of the two cans from the beginning of today's lesson required more metal to make?" Answers will vary depending on can sizes.

3 ACTIVITY: Estimation

Math Practice

View as Components

How can you use the results of Activity 2 to help you identify the components of the surface area?

Work with a partner. From memory, estimate the dimensions of the real-life item in inches. Then use the dimensions to estimate the surface area of the item in square inches.

a.

b.

c.

d.

What Is Your Answer?

4. **IN YOUR OWN WORDS** How can you find the surface area of a cylinder? Give an example with your description. Include a drawing of the cylinder.

5. To eight decimal places, $\pi \approx 3.14159265$. Which of the following is closest to π?

   a. 3.14 **b.** $\frac{22}{7}$ **c.** $\frac{355}{113}$

"To approximate $\pi \approx 3.141593$, I simply remember 1, 1, 3, 3, 5, 5."

"Then I compute $\frac{355}{113} \approx 3.141593$."

Practice Use what you learned about the surface area of a cylinder to complete Exercises 3–5 on page 604.

14.3 Lesson

Check It Out
Lesson Tutorials
BigIdeasMath.com

Key Idea

Surface Area of a Cylinder

Words The surface area S of a cylinder is the sum of the areas of the bases and the lateral surface.

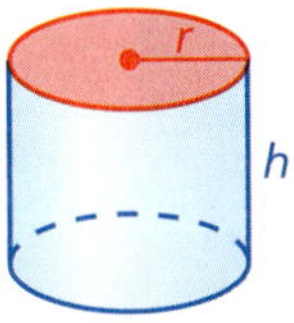

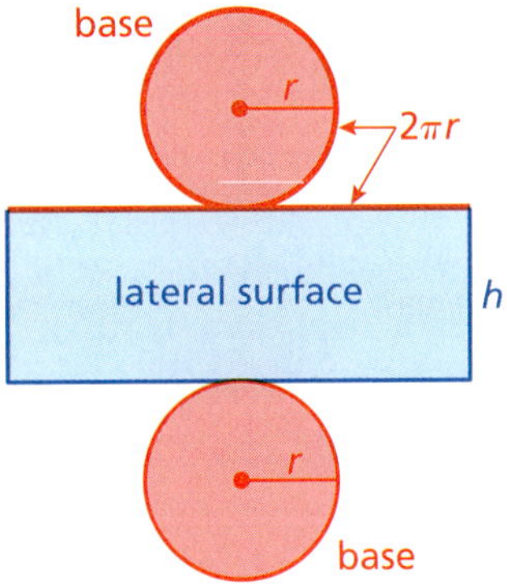

Algebra $S = 2\pi r^2 + 2\pi rh$

Areas of bases ($2\pi r^2$)

Area of lateral surface ($2\pi rh$)

Remember

Pi can be approximated as 3.14 or $\frac{22}{7}$.

EXAMPLE 1 Finding the Surface Area of a Cylinder

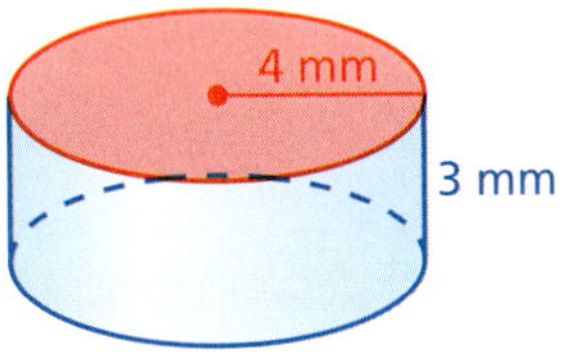

Find the surface area of the cylinder. Round your answer to the nearest tenth.

Draw a net.

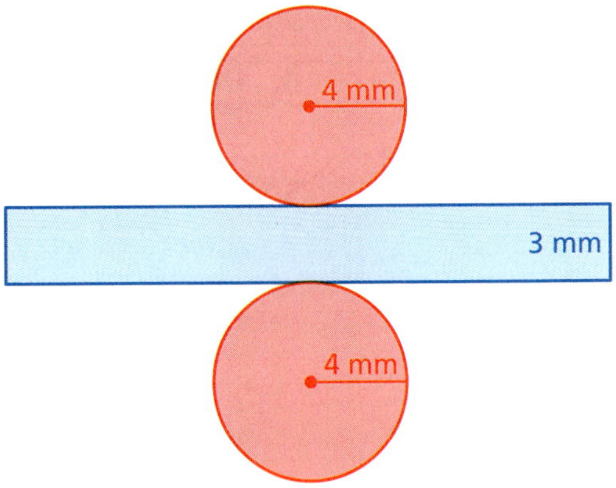

$$
\begin{aligned}
S &= 2\pi r^2 + 2\pi rh \\
&= 2\pi(4)^2 + 2\pi(4)(3) \\
&= 32\pi + 24\pi \\
&= 56\pi \\
&\approx 175.8
\end{aligned}
$$

∴ The surface area is about 175.8 square millimeters.

On Your Own

Now You're Ready
Exercises 6–8

Find the surface area of the cylinder. Round your answer to the nearest tenth.

1.

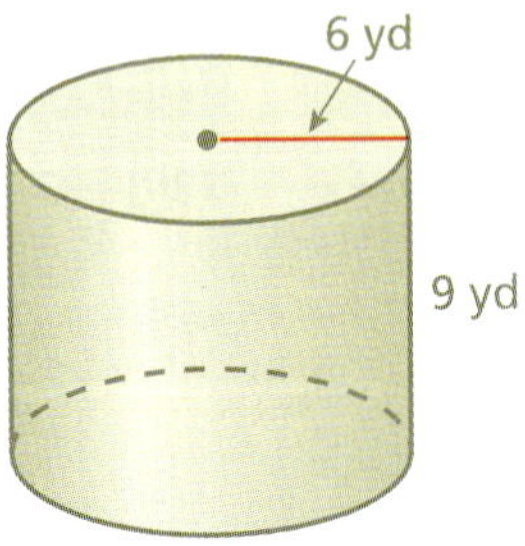

2.

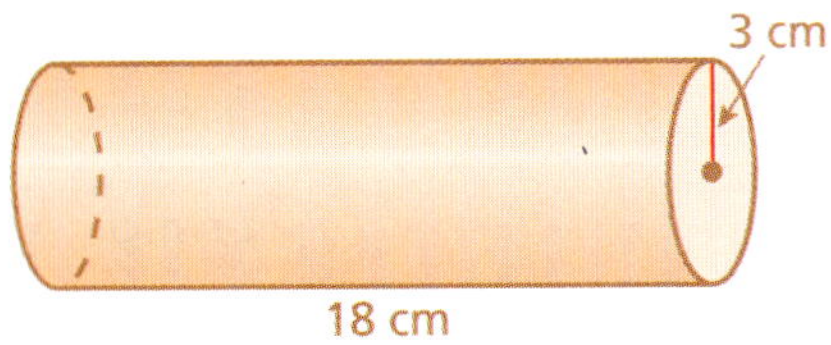

Laurie's Notes

Introduction

Connect

- **Yesterday:** Students discovered how to find the surface area of a cylinder by examining the net that makes up a cylinder.
- **Today:** Students will work with a formula for the surface area of a cylinder.

Motivate

- Find two cans (cylinders) that have volumes in the ratio of 1 : 2 (i.e., 10 fl oz and 20 fl oz).
- ? "The larger can has twice the volume of the smaller can. Do you think the surface area is twice as much?" Answers may differ, but most students believe this is true. Return to this question at the end of the lesson.

Lesson Notes

Key Idea

- ? "How are cylinders and rectangular prisms alike?" Both have 2 congruent bases and a lateral portion. "Different?" Cylinders have circular bases, while the rectangular prism has a rectangular base.
- Refer to the diagram with the radius marked. Review the formulas for area and circumference.
- Write the formula in words first. Before writing the formula in symbols, ask direct questions to help students make the connection between the words and the symbols.
- ? "How do you find the area of the bases?" Find the area of one base, πr^2, and then multiply by 2.
- ? "How do you find the area of the lateral portion?" The dimensions of its rectangular net are the height and circumference of the cylinder, so the area of the lateral portion is $2\pi rh$.
- Write the formula in symbols with each part identified (area of bases + lateral surface area).

Example 1

- Write the formula first to model good problem-solving techniques.
- Notice that the values of the variables are substituted, with each term being left in terms of π.
- **Common Misconception:** Students are unsure of how to perform the multiplication with π in the middle of the term. Remind students that π is a number, a factor in this case, just like the other numbers. Because of the Commutative and Associative Properties, the whole numbers can be multiplied first. Then the two like terms, 32π and 24π, are combined. The last step is to substitute 3.14 for π.
- ? Review $\approx$. "What does this symbol mean? Why is it used?" approximately equal to; π is an irrational number and an estimate for pi is used in the calculation.

On Your Own

- Ask volunteers to share their answers.

Goal Today's lesson is finding the surface area of a cylinder using a formula.

Technology for the Teacher

Dynamic Classroom

Lesson Tutorials
Lesson Plans
Answer Presentation Tool

Extra Example 1

Find the surface area of a cylinder with a radius of 3 inches and a height of 4 inches. Round your answer to the nearest tenth. $42\pi \approx 131.9 \text{ in.}^2$

On Your Own

1. $180\pi \approx 565.2 \text{ yd}^2$
2. $126\pi \approx 395.6 \text{ cm}^2$

Extra Example 2

Find the lateral surface area of a cylinder with a radius of 2 inches and a height of 6 inches. Round your answer to the nearest tenth. $24\pi \approx 75.4$ in.^{2}

Extra Example 3

You earn \$0.07 for recycling a can with a radius of 3 inches and a height of 4 inches (Extra Example 1). How much can you expect to earn for recycling a can with a radius of 2 inches and a height of 6 inches (Extra Example 2)? Assume that the recycle value is proportional to the surface area. \$0.05

On Your Own

3. a. yes

   b. no; Only the lateral surface area doubled. Because the surface area of the can does not double, the recycle value does not double.

Differentiated Instruction

Visual

Encourage students to estimate their answers for reasonableness. For the surface area of a cylinder, a common error is using the diameter in the formula instead of the radius. Have students imagine the cylinder inside of a rectangular prism. By calculating the surface area of the prism, the student has an overestimate of the surface area of the cylinder.

Laurie's Notes

Example 2

- Note that only the lateral surface area is asked for in this example.
- If you have a small can with a label on it, use it as a model for this problem.
- Note again that the answer is left in terms of π until the last step.

Example 3

- **Make Sense of Problems and Persevere in Solving Them:** Ask a volunteer to read the problem. Check to see if students understand what is being asked in this problem.
- ? "How does the surface area of the can relate to the recycling value?" The value of the recycled can is proportional to the surface area of the can.
- Have students compute the surface area of each can.
- ? "Approximately how much more metal is there in the larger can?" 24π in.$^2 \approx 75$ in.^{2}
- ? "How many times more metal is there in the larger can?" 5 times
- This is a good review of proportions. Calculators are helpful.

On Your Own

- Give students sufficient time to do their work before asking volunteers to share their work at the board.

Closure

- Hold the two cans from the Motivate section of today's lesson and ask students to find the surface area of each. Answers will vary depending on can sizes.

EXAMPLE 2 Finding Surface Area

How much paper is used for the label on the can of peas?

Find the lateral surface area of the cylinder.

$S = 2\pi rh$ ← Do not include the areas of the bases in the formula.

$= 2\pi(1)(2)$ Substitute.

$= 4\pi \approx 12.56$ Multiply.

∴ About 12.56 square inches of paper is used for the label.

EXAMPLE 3 Real-Life Application

You earn \$0.01 for recycling the can in Example 2. How much can you expect to earn for recycling the tomato can? Assume that the recycle value is proportional to the surface area.

Find the surface area of each can.

Tomatoes	***Peas***
$S = 2\pi r^2 + 2\pi rh$	$S = 2\pi r^2 + 2\pi rh$
$= 2\pi(2)^2 + 2\pi(2)(5.5)$	$= 2\pi(1)^2 + 2\pi(1)(2)$
$= 8\pi + 22\pi$	$= 2\pi + 4\pi$
$= 30\pi$	$= 6\pi$

Use a proportion to find the recycle value x of the tomato can.

$$\frac{30\pi \text{ in.}^2}{x} = \frac{6\pi \text{ in.}^2}{\$0.01}$$

← surface area
← recycle value

$30\pi \cdot 0.01 = x \cdot 6\pi$ Cross Products Property

$5 \cdot 0.01 = x$ Divide each side by 6π.

$0.05 = x$ Simplify.

∴ You can expect to earn \$0.05 for recycling the tomato can.

On Your Own

Now You're Ready
Exercises 9–11

3. **WHAT IF?** In Examples 2 and 3, the height of the can of peas is doubled.
   a. Does the amount of paper used in the label double?
   b. Does the recycle value double? Explain.

14.3 Exercises

Check It Out
Help with Homework
BigIdeasMath.com

Vocabulary and Concept Check

1. **CRITICAL THINKING** Which part of the formula $S = 2\pi r^2 + 2\pi rh$ represents the lateral surface area of a cylinder?

2. **CRITICAL THINKING** You are given the height and the circumference of the base of a cylinder. Describe how to find the surface area of the entire cylinder.

Practice and Problem Solving

Make a net for the cylinder. Then find the surface area of the cylinder. Round your answer to the nearest tenth.

3. 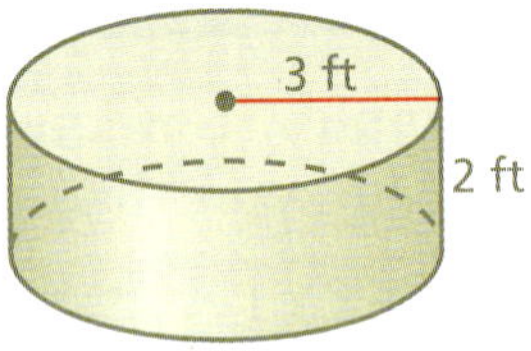

4. 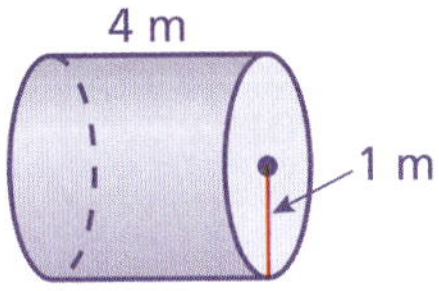

5. 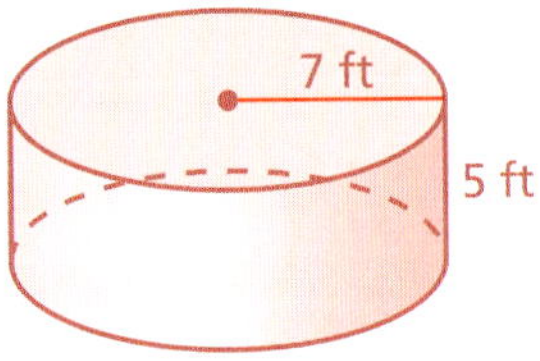

Find the surface area of the cylinder. Round your answer to the nearest tenth.

1 6. 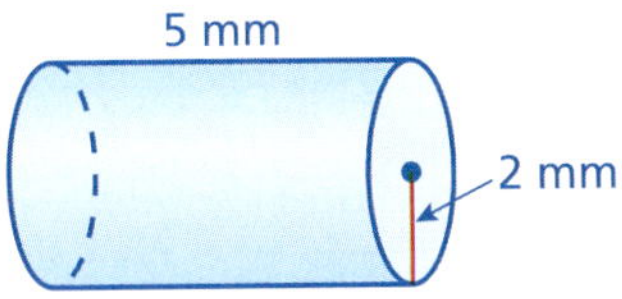

7. 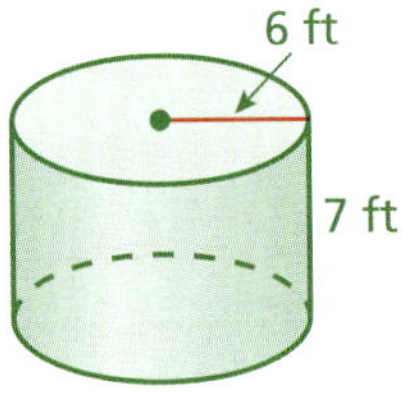

8. 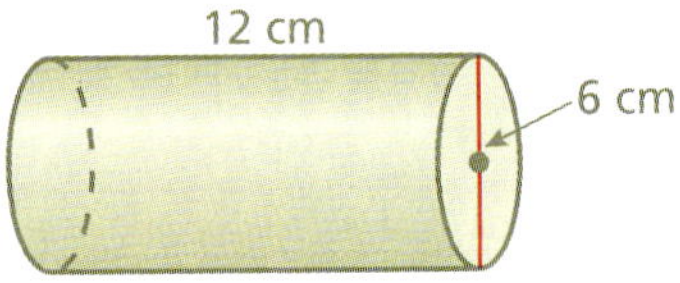

Find the lateral surface area of the cylinder. Round your answer to the nearest tenth.

2 9. 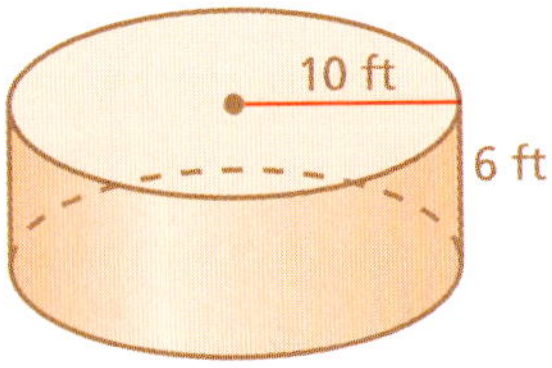

10. 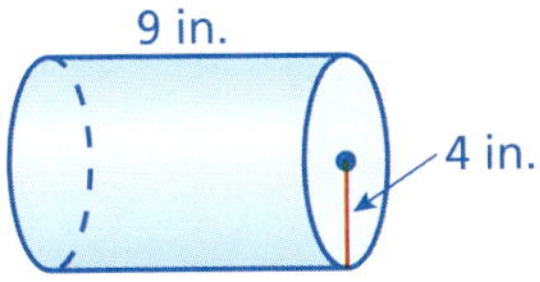

11. 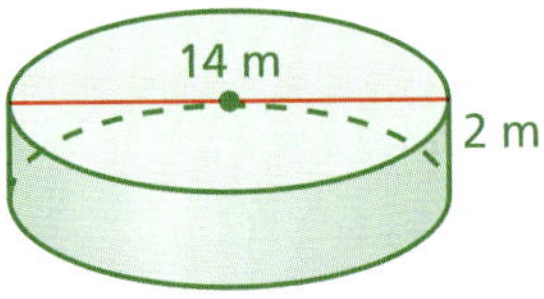

12. **ERROR ANALYSIS** Describe and correct the error in finding the surface area of the cylinder.

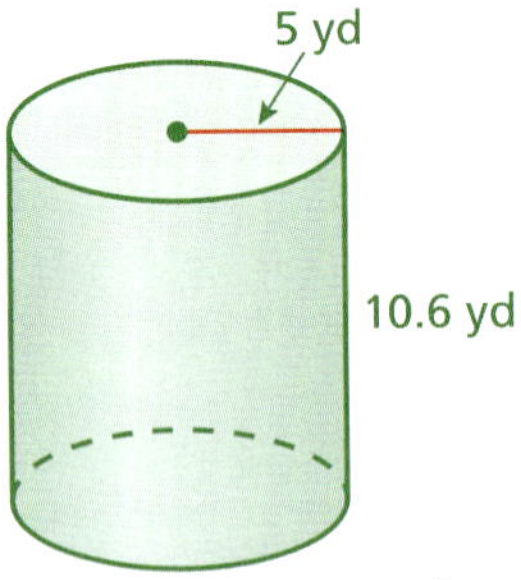

✗
$S = \pi r^2 + 2\pi rh$
$= \pi(5)^2 + 2\pi(5)(10.6)$
$= 25\pi + 106\pi$
$= 131\pi \approx 411.3 \text{ yd}^2$

13. **TANKER** The truck's tank is a stainless steel cylinder. Find the surface area of the tank.

Assignment Guide and Homework Check

Level	Assignment	Homework Check
Advanced	1–5, 6–18 even, 19–22	8, 10, 14, 16, 18

Common Errors

- **Exercises 6–8** Students may add the area of only one base. Remind them of the net for a cylinder and that there are two circles as bases.
- **Exercises 6–8** Students may double the radius instead of squaring it. Remind them of the area of a circle and also the order of operations.
- **Exercise 8** Students may use the diameter instead of the radius. Remind them that the radius is in the formula, so they should find the radius before finding the surface area.
- **Exercises 9–11** Students may multiply the height by the area of the circular base instead of the circumference. Review with them how the lateral surface is created to show that the length of the rectangle is the circumference of the circular bases.

14.3 Record and Practice Journal

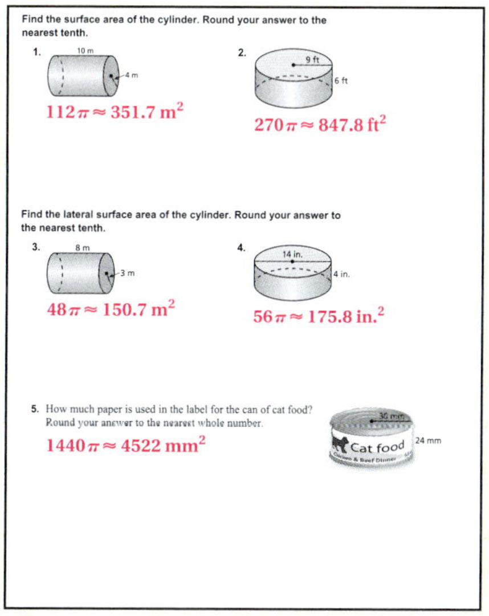
Find the surface area of the cylinder. Round your answer to the nearest tenth.

1. 10 m, 4 m — $112\pi \approx 351.7\text{ m}^2$
2. 9 ft, 6 ft — $270\pi \approx 847.8\text{ ft}^2$

Find the lateral surface area of the cylinder. Round your answer to the nearest tenth.

3. 8 m, 3 m — $48\pi \approx 150.7\text{ m}^2$
4. 14 in., 4 in. — $56\pi \approx 175.8\text{ in.}^2$

5. How much paper is used in the label for the can of cat food? Round your answer to the nearest whole number. (30 mm, 24 mm, Cat food)
$1440\pi \approx 4522\text{ mm}^2$

Vocabulary and Concept Check

1. $2\pi rh$
2. Use the given circumference to find the radius by solving $C = 2\pi r$ for r. Then use the formula for the surface area of a cylinder.

Practice and Problem Solving

3. 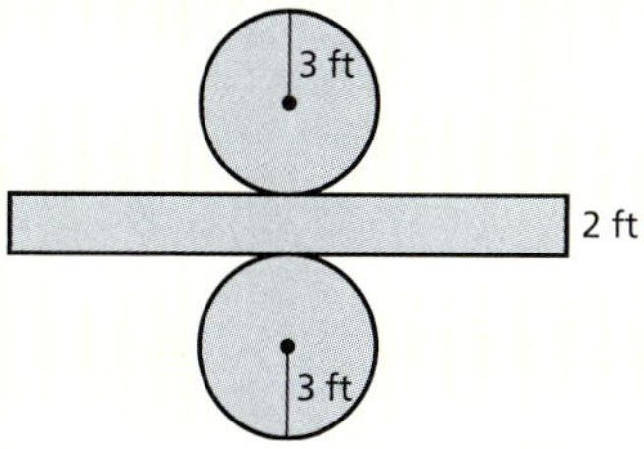

$30\pi \approx 94.2\text{ ft}^2$

4. 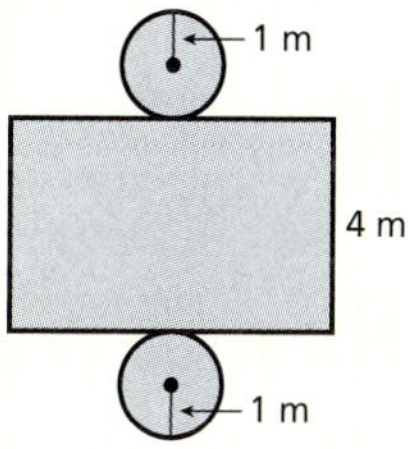

$10\pi \approx 31.4\text{ m}^2$

5. 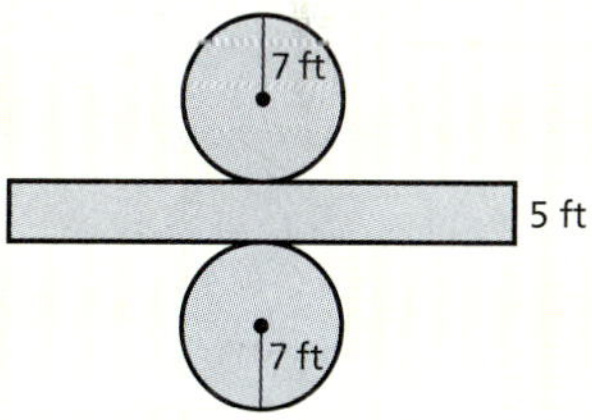

$168\pi \approx 527.5\text{ ft}^2$

6. $28\pi \approx 87.9\text{ mm}^2$
7. $156\pi \approx 489.8\text{ ft}^2$
8. $90\pi \approx 282.6\text{ cm}^2$
9. $120\pi \approx 376.8\text{ ft}^2$
10. $72\pi \approx 226.1\text{ in.}^2$
11. $28\pi \approx 87.9\text{ m}^2$
12. See Additional Answers.
13. $432\pi \approx 1356.48\text{ ft}^2$

Practice and Problem Solving

14. about 36.4%

15. The surface area of the cylinder with the height of 8.5 inches is greater than the surface area of the cylinder with the height of 11 inches.

16. **a.** $S = 41.125\pi \approx 129.1\ \text{cm}^2$, $S = 149.875\pi \approx 470.6\ \text{cm}^2$

b. about 4.0 lb

17. See *Taking Math Deeper.*

18. See Additional Answers.

Fair Game Review

19. $10\ \text{ft}^2$ **20.** $16\ \text{cm}^2$

21. $47.5\ \text{in.}^2$ **22.** C

Mini-Assessment

Find the surface area of the cylinder. Round your answer to the nearest tenth.

1.

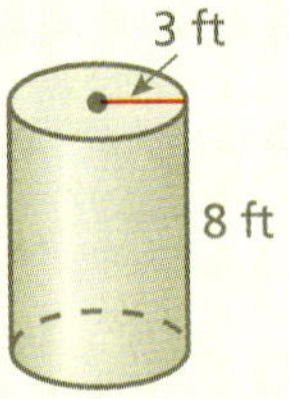

$66\pi \approx 207.2\ \text{ft}^2$

2.

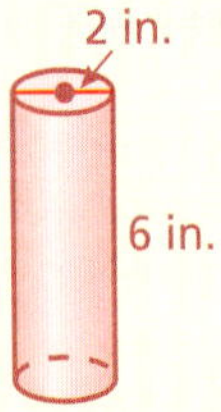

$14\pi \approx 44.0\ \text{in.}^2$

3. Find the surface area of the roll of paper towels.

$67.5\pi \approx 212.0\ \text{in.}^2$

4. How much paper is used for the label on the can of tuna?

$12.56\ \text{in.}^2$

Taking Math Deeper

Exercise 17

This is a real-life problem. That is, when you leave cheese in the refrigerator without covering it, the amount that dries out is proportional to the surface area.

1 Find the surface area of the uncut cheese.

$$S = 2\pi r^2 + 2\pi rh$$
$$= 2\pi \cdot 3^2 + 2\pi \cdot 3 \cdot 1$$
$$= 24\pi$$

a. $\approx 75.36\ \text{in.}^2$

2 Find the surface area of the remaining cheese.
One-eighth of the surface area is removed.
But, two 3-by-1 rectangular regions are added.

$$S = \frac{7}{8} \cdot 24\pi + 2(3 \cdot 1)$$
$$= 21\pi + 6$$

b. $\approx 71.94\ \text{in.}^2$

3 Answer the question.
b. The surface area decreased.

Reteaching and Enrichment Strategies

If students need help. . .	If students got it. . .
Resources by Chapter • Practice A and Practice B • Puzzle Time Record and Practice Journal Practice Differentiating the Lesson Lesson Tutorials Skills Review Handbook	Resources by Chapter • Enrichment and Extension • Technology Connection Start the next section

14. **OTTOMAN** What percent of the surface area of the ottoman is green (not including the bottom)?

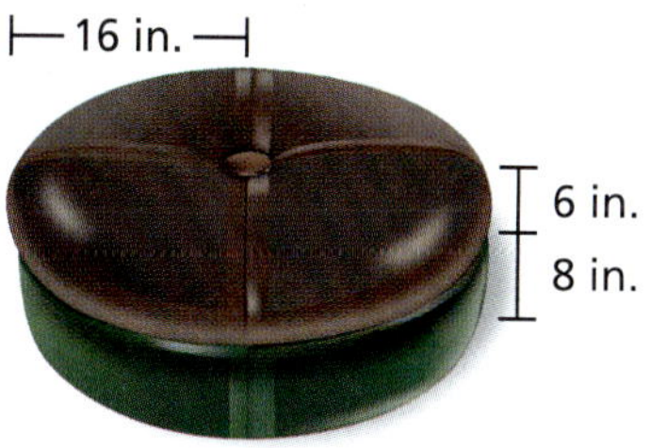

15. **REASONING** You make two cylinders using 8.5-by-11-inch pieces of paper. One has a height of 8.5 inches, and the other has a height of 11 inches. Without calculating, compare the surface areas of the cylinders.

16. **INSTRUMENT** A *ganza* is a percussion instrument used in samba music.

    a. Find the surface area of each of the two labeled ganzas.

    b. The weight of the smaller ganza is 1.1 pounds. Assume that the surface area is proportional to the weight. What is the weight of the larger ganza?

17. **BRIE CHEESE** The cut wedge represents one-eighth of the cheese.

    a. Find the surface area of the cheese before it is cut.

    b. Find the surface area of the remaining cheese after the wedge is removed. Did the surface area increase, decrease, or remain the same?

18. Repeated Reasoning A cylinder has radius r and height h.

    a. How many times greater is the surface area of a cylinder when both dimensions are multiplied by a factor of 2? 3? 5? 10?

    b. Describe the pattern in part (a). How many times greater is the surface area of a cylinder when both dimensions are multiplied by a factor of 20?

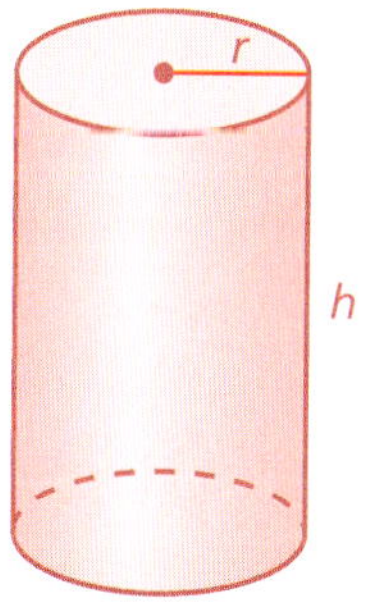

Fair Game Review What you learned in previous grades & lessons

Find the area. *(Skills Review Handbook)*

19. 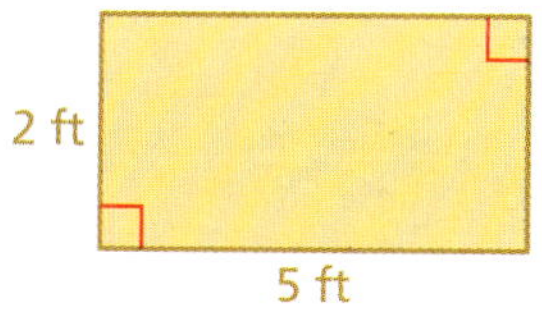

20. 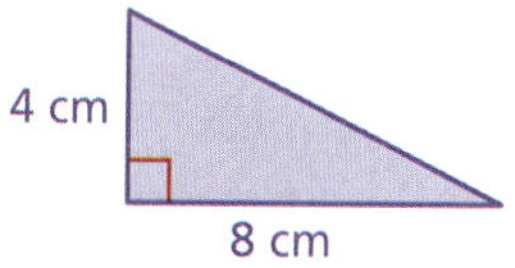

21. 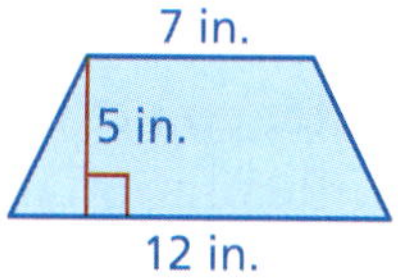

22. **MULTIPLE CHOICE** 40% of what number is 80? *(Skills Review Handbook)*

Ⓐ 32 Ⓑ 48 Ⓒ 200 Ⓓ 320

14 Study Help

Check It Out
Graphic Organizer
BigIdeasMath.com

You can use an **information frame** to help you organize and remember concepts. Here is an example of an information frame for surface areas of rectangular prisms.

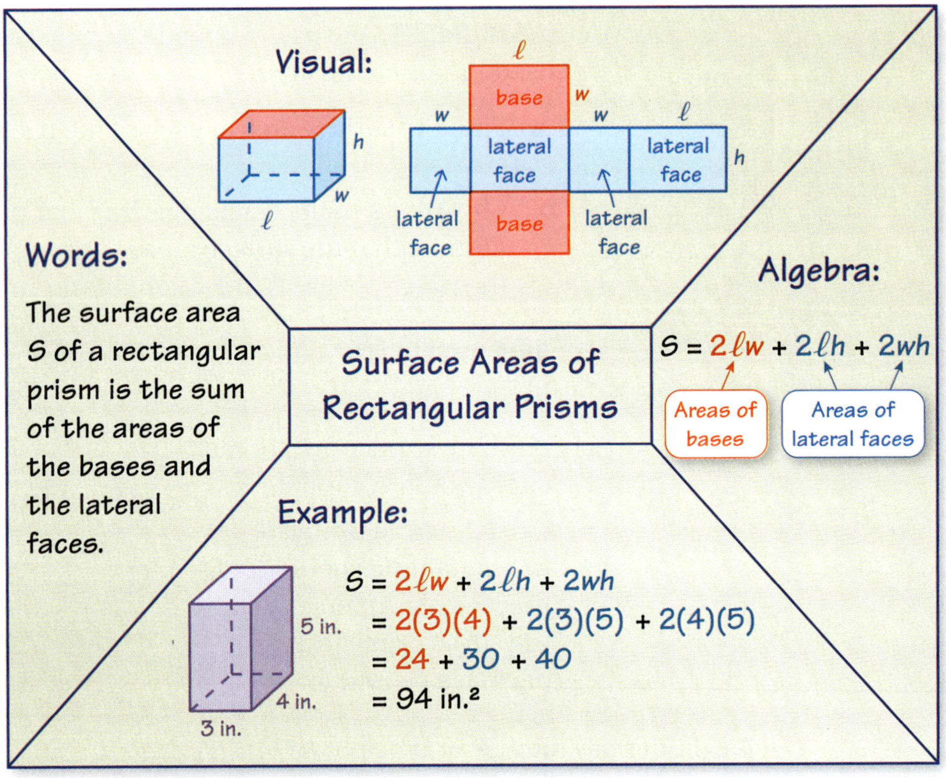

On Your Own

Make information frames to help you study the topics.

1. surface areas of prisms
2. surface areas of pyramids
3. surface areas of cylinders

After you complete this chapter, make information frames for the following topics.

4. volumes of prisms
5. volumes of pyramids

"I'm having trouble thinking of a good title for my information frame."

Sample Answers

1.

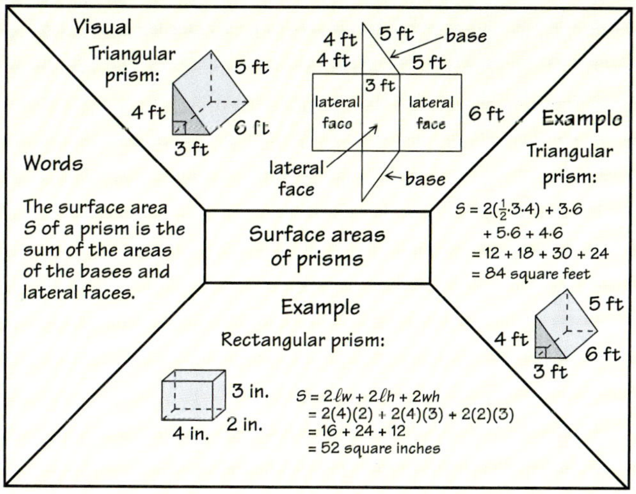

2.

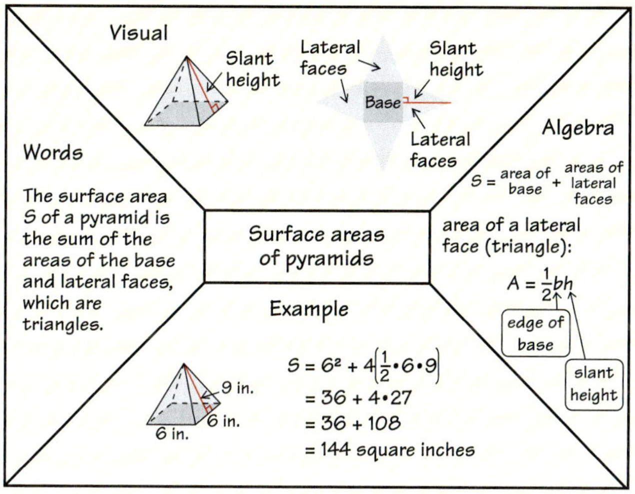

3.

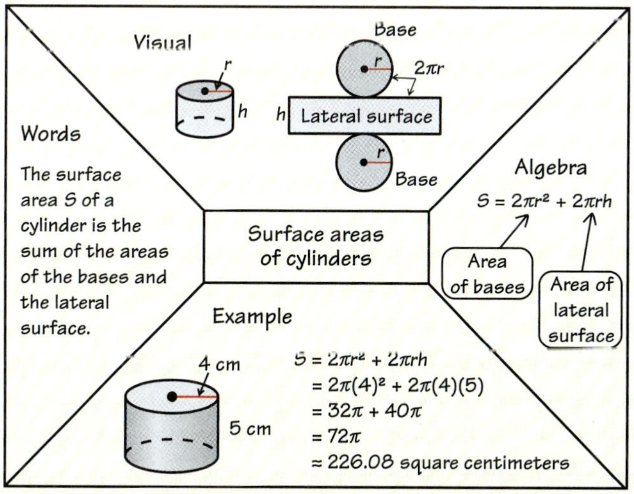

List of Organizers

Available at *BigIdeasMath.com*

Comparison Chart
Concept Circle
Example and Non-Example Chart
Formula Triangle
Four Square
Idea (Definition) and Examples Chart
Information Frame
Information Wheel
Notetaking Organizer
Process Diagram
Summary Triangle
Word Magnet
Y Chart

About this Organizer

An **Information Frame** can be used to help students organize and remember concepts. Students write the topic in the middle rectangle. Then students write related concepts in the spaces around the rectangle. Related concepts can include *Words*, *Numbers*, *Algebra*, *Example*, *Definition*, *Non-Example*, *Visual*, *Procedure*, *Details*, and *Vocabulary*. Students can place their information frames on note cards to use as a quick study reference.

Technology for the Teacher

Editable Graphic Organizer

Answers

1. 132 cm^2
2. 100 mm^2
3. 245 m^2
4. 28 cm^2
5. $78\pi \approx 244.9\text{ ft}^2$
6. $110\pi \approx 345.4\text{ m}^2$
7. $126\pi \approx 395.6\text{ cm}^2$
8. $97.6\pi \approx 306.5\text{ mm}^2$
9. **a.** 18 ft^2

   b. yes
10. $108\pi \approx 339.12\text{ in.}^2$
11. $11{,}520\text{ in.}^2$

Technology for the Teacher

Online Assessment
Assessment Book
ExamView® Assessment Suite

Alternative Quiz Ideas

100% Quiz
Error Notebook
Group Quiz
Homework Quiz
Math Log
Notebook Quiz
Partner Quiz
Pass the Paper

100% Quiz

This is a quiz where students are given the answers and then they have to explain and justify each answer.

Reteaching and Enrichment Strategies

If students need help. . .	If students got it. . .
Resources by Chapter • Practice A and Practice B • Puzzle Time Lesson Tutorials *BigIdeasMath.com*	Resources by Chapter • Enrichment and Extension • Technology Connection Game Closet at *BigIdeasMath.com* Start the next section

14.1–14.3 Quiz

Check It Out
Progress Check
BigIdeasMath.com

Find the surface area of the prism. *(Section 14.1)*

1. 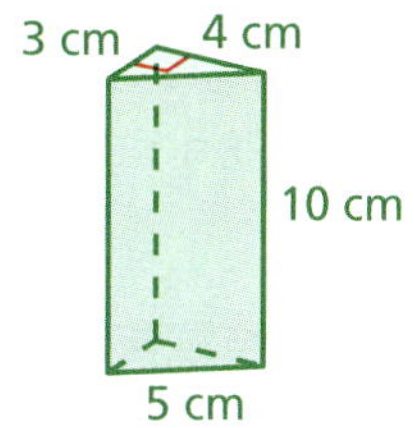

2. 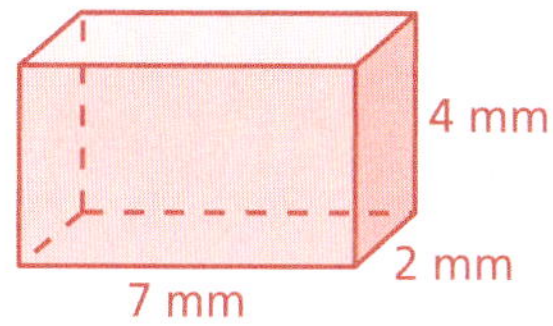

Find the surface area of the regular pyramid. *(Section 14.2)*

3. 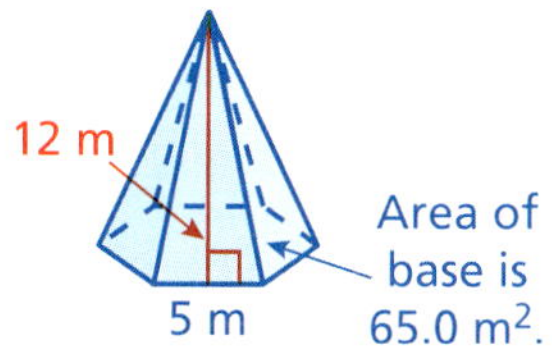

4. 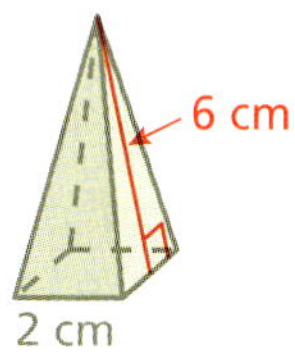

Find the surface area of the cylinder. Round your answer to the nearest tenth. *(Section 14.3)*

5. 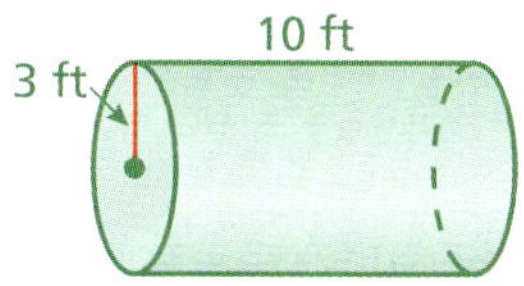

6. 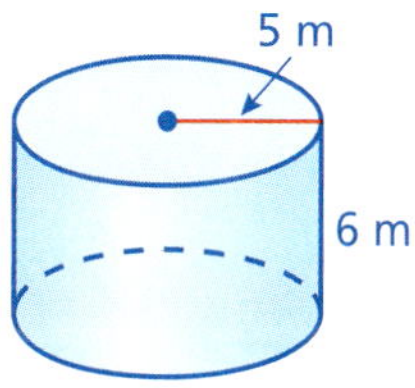

Find the lateral surface area of the cylinder. Round your answer to the nearest tenth. *(Section 14.3)*

7. 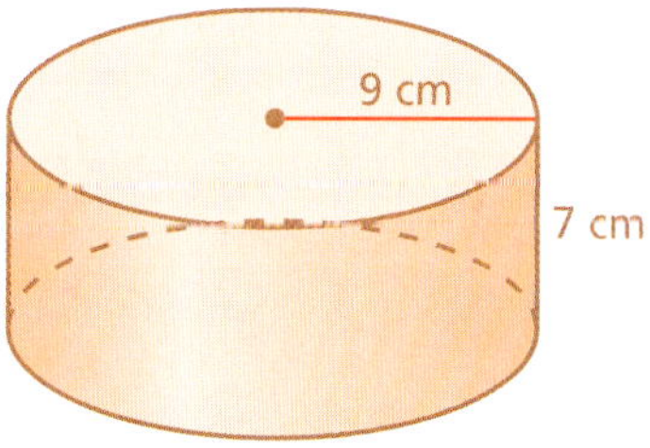

8. 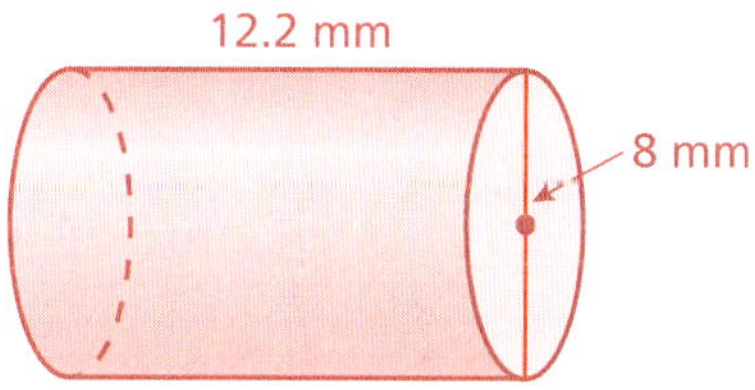

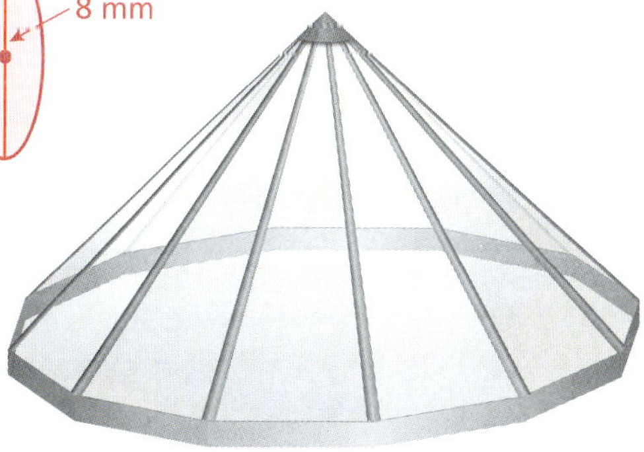

9. SKYLIGHT You are making a skylight that has 12 triangular pieces of glass and a slant height of 3 feet. Each triangular piece has a base of 1 foot. *(Section 14.2)*

a. How much glass will you need to make the skylight?

b. Can you cut the 12 glass triangles from a sheet of glass that is 4 feet by 8 feet? If so, draw a diagram showing how this can be done.

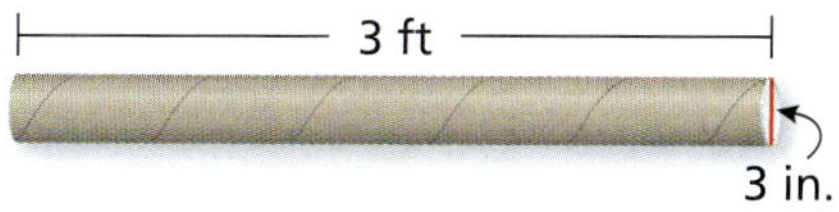

10. MAILING TUBE What is the least amount of material needed to make the mailing tube? *(Section 14.3)*

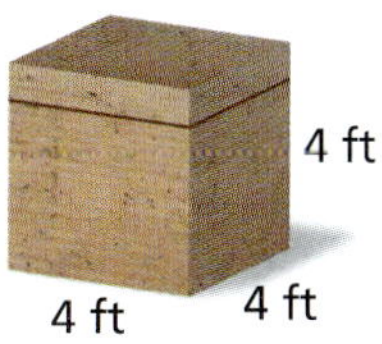

11. WOODEN CHEST All the faces of the wooden chest will be painted except for the bottom. Find the area to be painted, in *square inches*. *(Section 14.1)*

14.4 Volumes of Prisms

Essential Question How can you find the volume of a prism?

1 ACTIVITY: Pearls in a Treasure Chest

Work with a partner. A treasure chest is filled with valuable pearls. Each pearl is about 1 centimeter in diameter and is worth about $80.

Use the diagrams below to describe two ways that you can estimate the number of pearls in the treasure chest.

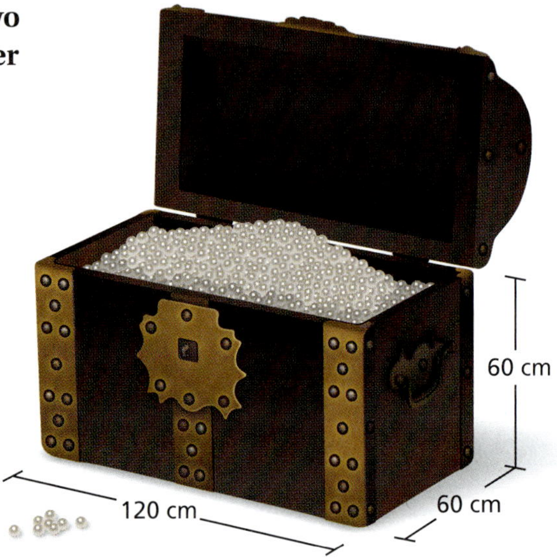

a.

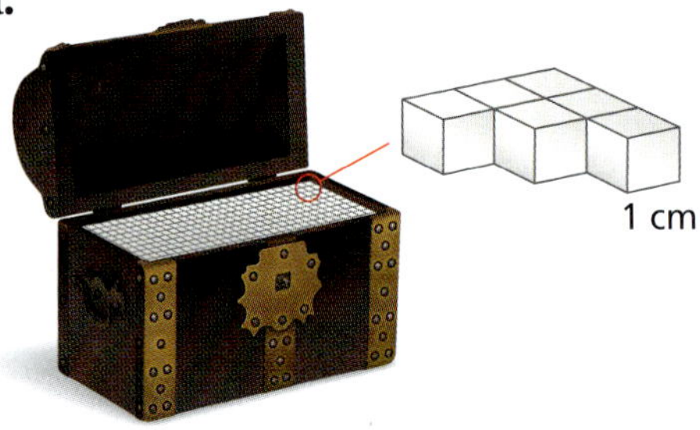

b.

c. Use the method in part (a) to estimate the value of the pearls in the chest.

Geometry

In this lesson, you will

- find volumes of prisms.
- solve real-life problems.

2 ACTIVITY: Finding a Formula for Volume

Work with a partner. You know that the formula for the volume of a rectangular prism is $V = \ell wh$.

a. Write a formula that gives the volume in terms of the area of the base B and the height h.

b. Use both formulas to find the volume of each prism. Do both formulas give you the same volume?

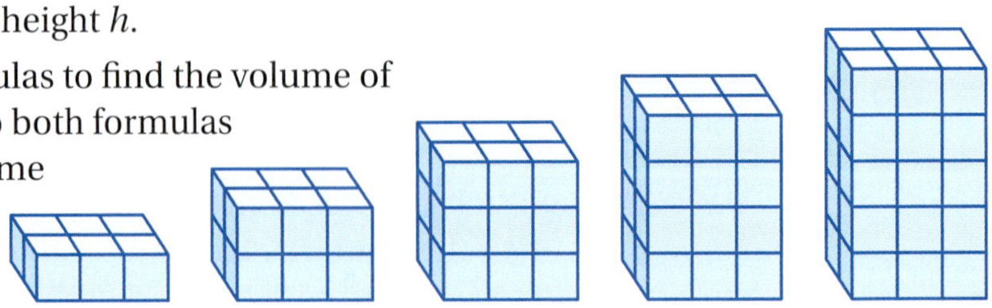

Laurie's Notes

Introduction

Applying Mathematical Practices

- **Look for and Express Regularity in Repeated Reasoning:** The approach used in developing a formula for the volume of a prism is to consider repeated layers with the same base. Mathematically proficient students notice that each layer increases the volume by the number of units of the area of the base.

Motivate

- Hold up a variety of common containers and ask what is commonly found inside. Examples: egg carton (12 eggs); playing cards box (52 cards); crayon box (8 crayons)
- Discuss with students these examples of volume. Each container is filled with objects of the same size. How many eggs fit in the egg carton, or how many crayons fit in the crayon box? Because the units are different (eggs, cards, crayons), you can't compare the volumes.

Activity Notes

Activity 1

- If you have beads (marbles), use them to model this activity. "I've filled this box with beads. How would you estimate the number of beads in the box?"
- ? "How big is the treasure chest? Compare it to an object in this room." Students should recognize that 120 centimeters is more than 3 feet long.
- ? "Do you think there is a thousand dollars worth of pearls in the treasure chest? a million dollars? a billion dollars?" Students will likely have only a wild guess about the value of the pearls in the chest at this point.
- In part (a), listen for students to say you can fit a layer of centimeter cubes on the bottom and a total of 60 layers in the chest.
- ? "How did you estimate the number of pearls using the method in part (a)?" The bottom layer holds about 7200 pearls. Times 60 layers is 432,000 pearls.
- **Attend to Precision:** Discuss how the estimate compares to the actual number of pearls. It will be less because a 1-centimeter pearl has less volume than a cubic centimeter, so more will fit in the chest.
- ? Explain how the method in part (b) could be used to estimate the number of pearls in the chest. *Sample answer:* Weigh the chest full, and then empty to find the weight of the pearls. Then weigh 10 pearls and use this information to estimate the total number of pearls.

Activity 2

- From the figure, students should see that the bottom layer has 6 cubes, the second layer has 6 cubes, the third layer has 6 cubes, and so on.
- **Look for and Express Regularity in Repeated Reasoning:** If students are not thinking about layers (height), suggest that writing the volume of each prism would be helpful: 6, 12, 18, 24, 30.
- **Big Idea:** The area of the base (denoted B) is 6. The height (denoted h) is how many layers?

What Your Students Will Learn

- Find volumes of prisms using formulas.

Previous Learning

Students should know how to find areas of two-dimensional figures, surface areas of three-dimensional figures, and volumes of rectangular prisms using $V = \ell wh$ or by counting unit cubes.

Technology for the Teacher

Dynamic Classroom

Lesson Plans
Complete Materials List

14.4 Record and Practice Journal

Essential Question How can you find the volume of a prism?

1 ACTIVITY: Pearls in a Treasure Chest

Work with a partner. A treasure chest is filled with valuable pearls. Each pearl is about 1 centimeter in diameter and is worth about $80.

Use the diagrams below to describe two ways that you can estimate the number of pearls in the treasure chest.

a. 1 cm; 60 cm; 120 cm; 60 cm

Because each pearl is about 1 cubic centimeter, the number of pearls in the chest can be estimated by multiplying the length, width, and height of the treasure chest.

b. Weigh one pearl then weigh all the pearls.

c. Use the method in part (a) to estimate the value of the pearls in the chest.

$34,560,000

Differentiated Instruction

Visual

Students may think that prisms with the same volume have the same surface area. Have them work together to find prisms with the same volume, but different dimensions. Then direct the students to find the surface areas of each of the prisms. Ask them to share their results.

14.4 Record and Practice Journal

2 ACTIVITY: Finding a Formula for Volume

Work with a partner. You know that the formula for the volume of a rectangular prism is $V = \ell wh$.

a. Write a formula that gives the volume in terms of the area of the base B and the height h.

$V = Bh$

b. Use both formulas to find the volume of each prism. Do both formulas give you the same volume?

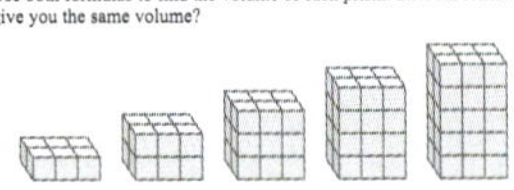

6 12 18 24 30

yes

3 ACTIVITY: Finding a Formula for Volume

Work with a partner. Use the concept in Activity 2 to find a formula that gives the volume of any prism. $V = Bh$; B = area of the base; h = height of the prism

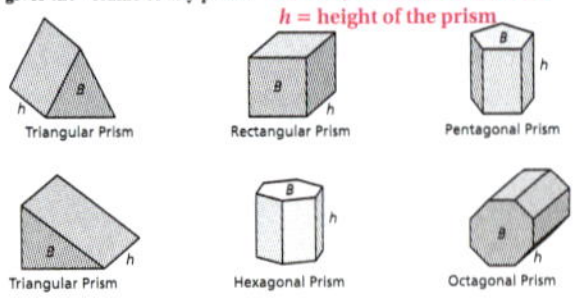

4 ACTIVITY: Using a Formula

Work with a partner. A ream of paper has 500 sheets.

a. Does a single sheet of paper have a volume? Why or why not?

yes

b. If so, explain how you can find the volume of a single piece of paper.

Find the volume of a ream of paper and divide by 500.

What Is Your Answer?

5. **IN YOUR OWN WORDS** How can you find the volume of a prism?

Find the area of the base and multiply it by the height of the prism.

6. **STRUCTURE** Draw a prism that has a trapezoid as its base. Use your formula to find the volume of the prism.

Check students' work.

Laurie's Notes

Activity 3

- The formula discovered in Activity 2 is now used in Activity 3.
- **Make Sense of Problems and Persevere in Solving Them:** Students can memorize formulas and have little understanding of why the formula makes sense. It is important throughout this chapter that students see that the formulas are all similar. The volume is found by finding the area of the base (B) and then multiplying by the number of layers (h).
- **Model with Mathematics:** Having models of these prisms is very helpful.
- **Common Misconception:** The height of a prism does not need to be the vertical measure. Demonstrate this by holding a rectangular prism (a tissue box is fine). Ask students to identify the base (a face of the prism) and the height (an edge). Chances are students will identify the (standard) bottom of the box as the base. Now, rotate the tissue box so that the base is vertical. Again ask students to identify the base and height. Students may stick with their first answers or may now switch to the "bottom face" as the base.
- A prism is named by its base. A triangular prism has 2 triangular bases.
- **Attend to Precision:** Give students time to discuss the solids in this activity. If you have physical models of each of these, ask six volunteers to describe how to find the volume of the solid. Expect the student volunteer to point to the base and the height as they explain how to find the volume.

Activity 4

- **Common Misconception:** Students may believe a sheet of paper has no height and so, no volume, only area. It may be difficult to measure the height with tools available to us, but a sheet of paper does have a height.
- **Use Appropriate Tools Strategically:** In part (b), students should deduce that they can measure one sheet of paper indirectly by first measuring a whole ream of paper.
- The ream of copy paper is a good visual model.

What Is Your Answer?

- **Think-Pair-Share:** Students should read each question independently and then work in pairs to answer the questions. When they have answered the questions, the pair should compare their answers with another group and discuss any discrepancies.

Closure

- **Writing Prompt:** To find the volume of a tissue box ...

3 ACTIVITY: Finding a Formula for Volume

Math Practice

Use a Formula

What are the given quantities? How can you use the quantities to write a formula?

Work with a partner. Use the concept in Activity 2 to find a formula that gives the volume of any prism.

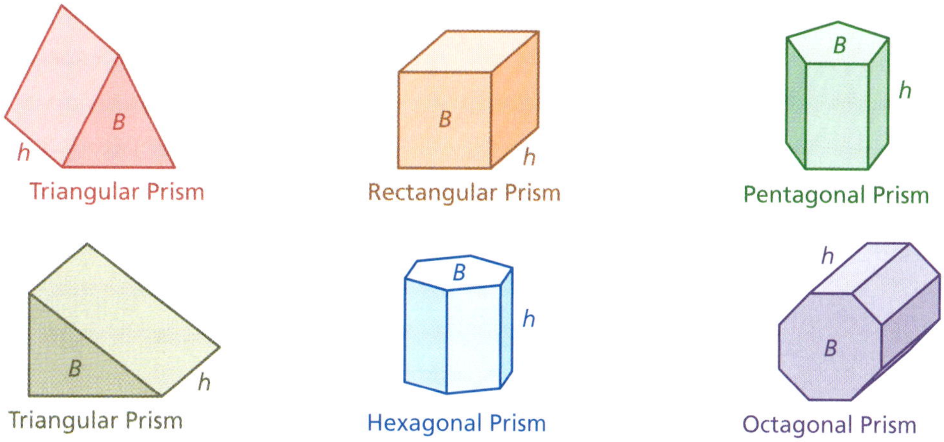

4 ACTIVITY: Using a Formula

Work with a partner. A ream of paper has 500 sheets.

a. Does a single sheet of paper have a volume? Why or why not?

b. If so, explain how you can find the volume of a single sheet of paper.

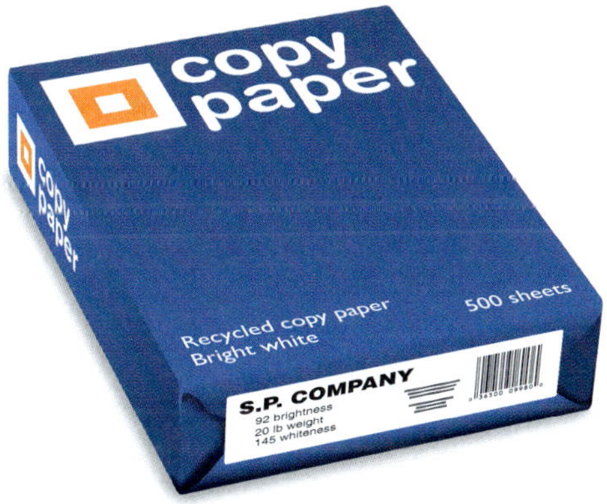

What Is Your Answer?

5. **IN YOUR OWN WORDS** How can you find the volume of a prism?

6. **STRUCTURE** Draw a prism that has a trapezoid as its base. Use your formula to find the volume of the prism.

Practice Use what you learned about the volumes of prisms to complete Exercises 4–6 on page 612.

14.4 Lesson

Check It Out
Lesson Tutorials
BigIdeasMath.com

The *volume* of a three-dimensional figure is a measure of the amount of space that it occupies. Volume is measured in cubic units.

Key Idea

Volume of a Prism

Words The volume V of a prism is the product of the area of the base and the height of the prism.

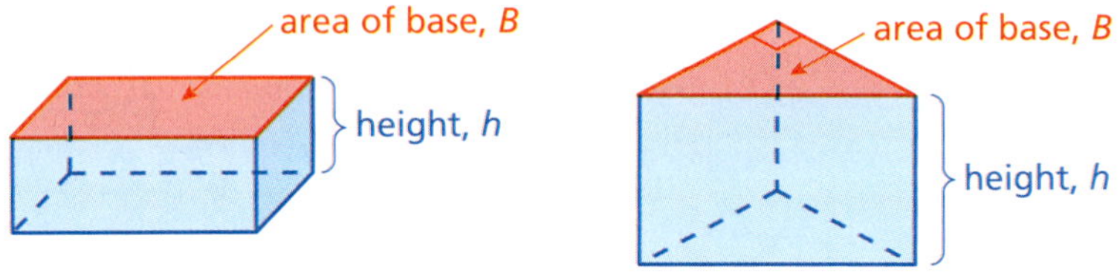

Algebra $V = Bh$

Area of base → B; Height of prism → h

Remember

The volume V of a cube with an edge length of s is $V = s^3$.

EXAMPLE 1 Finding the Volume of a Prism

Find the volume of the prism.

$V = Bh$	Write formula for volume.
$= 6(8) \cdot 15$	Substitute.
$= 48 \cdot 15$	Simplify.
$= 720$	Multiply.

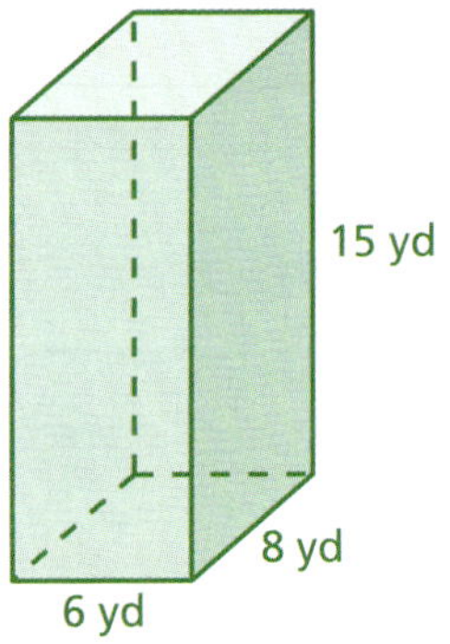

The volume is 720 cubic yards.

Study Tip

The area of the base of a rectangular prism is the product of the length ℓ and the width w.

You can use $V = \ell wh$ to find the volume of a rectangular prism.

EXAMPLE 2 Finding the Volume of a Prism

Find the volume of the prism.

$V = Bh$	Write formula for volume.
$= \frac{1}{2}(5.5)(2) \cdot 4$	Substitute.
$= 5.5 \cdot 4$	Simplify.
$= 22$	Multiply.

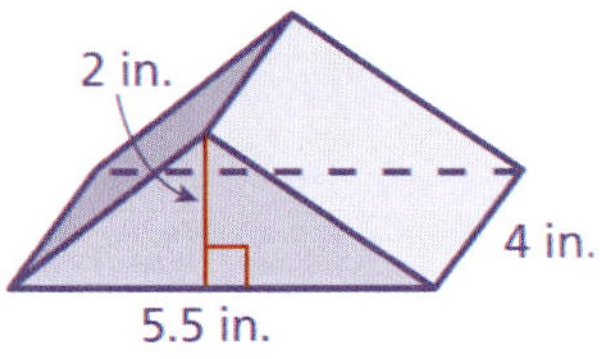

The volume is 22 cubic inches.

Laurie's Notes

Introduction

Connect

- **Yesterday:** Students explored how to find the volume of a prism.
- **Today:** Students will use the formula for the volume of a prism to solve problems.

Motivate

- **True Story:** Baseball legend Ken Griffey Jr. owed teammate Josh Fogg some money and paid him back in pennies. Griffey stacked 60 cartons, each holding $25 worth of pennies, in Fogg's locker.

? Ask the following questions.
  - "How does this story relate to the volume of a prism?" The volume of the carton is being measured in pennies.
  - "How big is a carton that can hold $25 worth of pennies?" open-ended
  - "How many pennies were in each carton?" 2500
  - "How much did Griffey owe Fogg?" $1500
  - "How much do you think each carton weighed?" A $25 carton of pennies weighs about 16 pounds.

Lesson Notes

Key Idea

? "What is a prism?" three-dimensional solid with two congruent bases and lateral faces that are rectangles

? "What are cubic units? Give an example." Cubic units are cubes which fill a space completely without overlapping or leaving gaps. Cubic inches and cubic centimeters are common examples.

- Point out to students that the bases of the prisms are shaded red. The height will be perpendicular to the two congruent bases. The dotted lines are edges that would not be visible through the solid prism.
- **Teaching Tip:** Use words (area of base, height) and symbols (B, h) when writing the formula.
- **Review Vocabulary:** *Product* is the answer to a multiplication problem.

Example 1

- Discuss the *Study Tip* with students.

? "Could the face measuring 8 yards by 15 yards be the base?" Yes, the height would then be 6 yards.

- **Extension:** Point out to students that all of the measurements are in terms of yards. "What if the 6 yard edge had been labeled 18 feet. Now how would you find the volume?" Convert all 3 dimensions to yards or to feet.

Example 2

- Ask a volunteer to describe the base of this triangular prism.

? "What property is used to simplify the area of the base?" Commutative Property of Multiplication

- Caution students to distinguish between the height of the base and the height of the prism.

Goal Today's lesson is finding the volumes of prisms.

Technology for the Teacher

Dynamic Classroom

Lesson Tutorials
Lesson Plans
Answer Presentation Tool

Extra Example 1

Find the volume of a rectangular prism with a length of 2 meters, a width of 6 meters, and a height of 3 meters. $36\ m^3$

Extra Example 2

Find the volume of the prism.

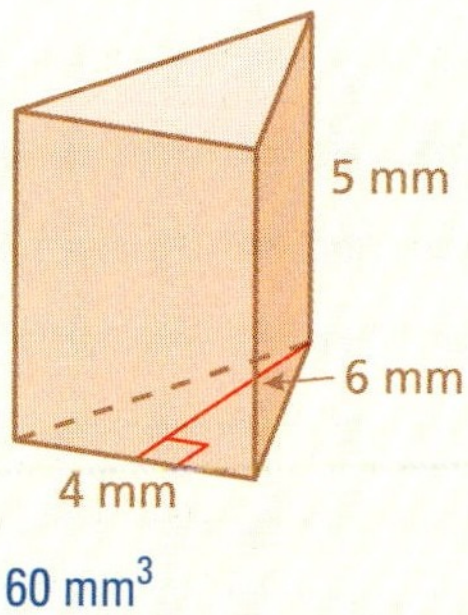

$60\ mm^3$

On Your Own

1. 64 ft^3
2. 270 m^3

Extra Example 3

Two rectangular prisms each have a volume of 120 cubic centimeters. The base of Prism A is 2 centimeters by 4 centimeters. The base of Prism B is 4 centimeters by 6 centimeters.

a. Find the height of each prism. Prism A: 15 cm, Prism B: 5 cm

b. Which prism has the lesser surface area? Prism B

On Your Own

3. yes; Because it has the same volume as the other two bags, but its surface area is 107.2 square inches which is less than both Bag A and Bag B.

English Language Learners

Vocabulary

Discuss the meaning of the words *volume* and *cubic units*. Have students add these words to their notebooks.

Laurie's Notes

On Your Own

- Ask volunteers to share their work at the board.

Example 3

- This example connects volume, surface area, and solving equations.
- Ask a student to read the example.
- **Make Sense of Problems and Persevere in Solving Them:** Ask probing questions to make sure that students understand the problem.

? "What type of measurement is 96 cubic inches?" volume

? "What type of measurement is part (b) concerned with?" surface area

- Work through part (a).
- Before beginning part (b), ask students to review the formula for surface area of a rectangular prism. Note that only five of the six faces are considered.
- Work through part (b).

? "Both bags hold the same amount of popcorn. Are there any practical advantages of one bag over the other?" Bag A: can grip it in your hand more easily; Bag B: less likely to tip over and uses less paper

On Your Own

- Given the discussion of the practical features of Bags A and B, students should have a sense of the problems in the design of Bag C.

Closure

- Sketch a rectangular prism. Label the dimensions 4 centimeters, 6 centimeters, and 10 centimeters. Find the volume of the prism. $V = 4 \cdot 6 \cdot 10 = 240 \text{ cm}^3$

On Your Own

Now You're Ready
Exercises 4–12

Find the volume of the prism.

1. 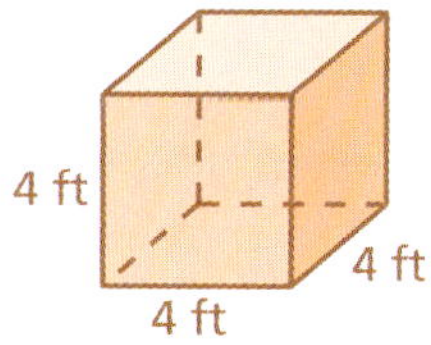

2. 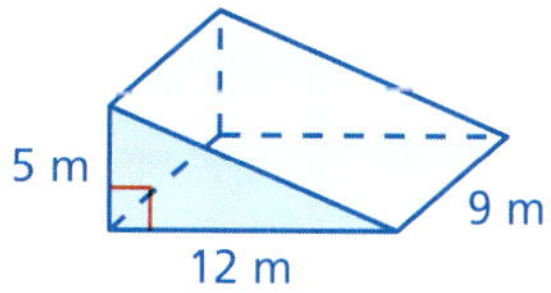

EXAMPLE 3 Real-Life Application

A movie theater designs two bags to hold 96 cubic inches of popcorn. (a) Find the height of each bag. (b) Which bag should the theater choose to reduce the amount of paper needed? Explain.

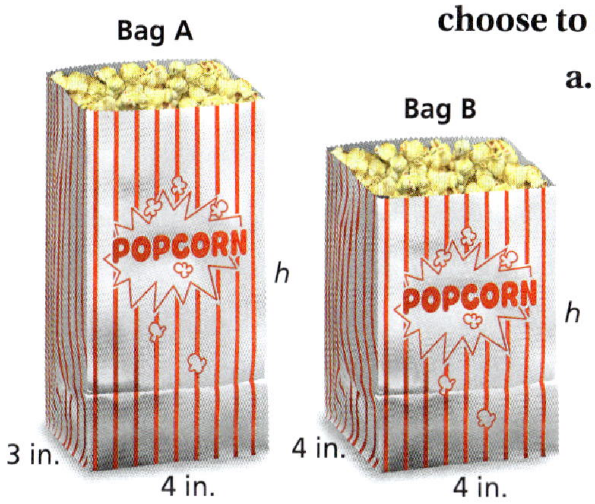

a. Find the height of each bag.

Bag A	***Bag B***
$V = Bh$	$V = Bh$
$96 = 4(3)(h)$	$96 = 4(4)(h)$
$96 = 12h$	$96 = 16h$
$8 = h$	$6 = h$
The height is 8 inches.	The height is 6 inches.

b. To determine the amount of paper needed, find the surface area of each bag. Do not include the top base.

Bag A	***Bag B***
$S = \ell w + 2\ell h + 2wh$	$S = \ell w + 2\ell h + 2wh$
$= 4(3) + 2(4)(8) + 2(3)(8)$	$= 4(4) + 2(4)(6) + 2(4)(6)$
$= 12 + 64 + 48$	$= 16 + 48 + 48$
$= 124 \text{ in.}^2$	$= 112 \text{ in.}^2$

The surface area of Bag B is less than the surface area of Bag A. So, the theater should choose Bag B.

On Your Own

3. You design Bag C that has a volume of 96 cubic inches. Should the theater in Example 3 choose your bag? Explain.

14.4 Exercises

Check It Out
Help with Homework
BigIdeasMath.com

Vocabulary and Concept Check

1. **VOCABULARY** What types of units are used to describe volume?
2. **VOCABULARY** Explain how to find the volume of a prism.
3. **CRITICAL THINKING** How are volume and surface area different?

Practice and Problem Solving

Find the volume of the prism.

1 2 **4.**

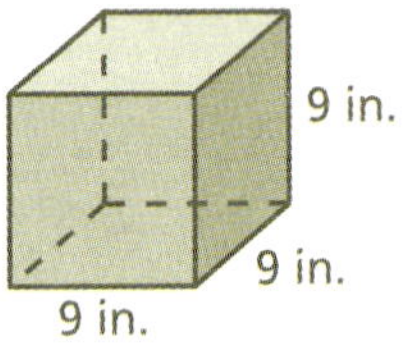

5.

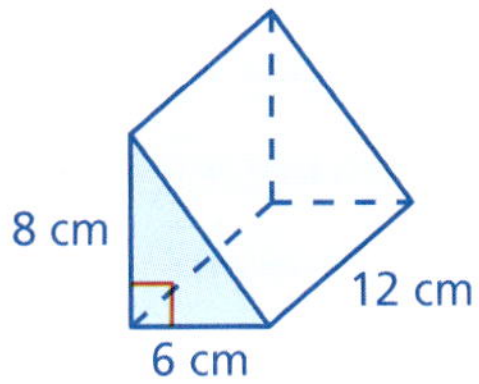

6.

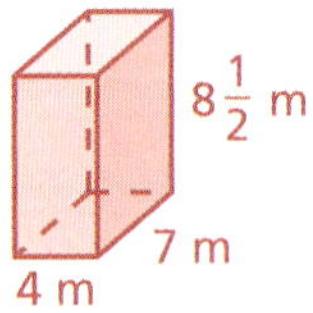

7.

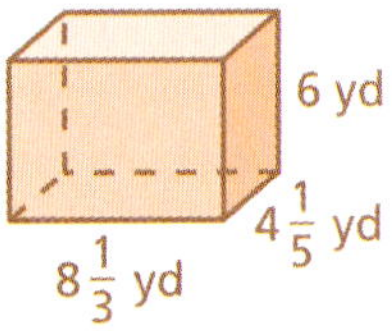

8.

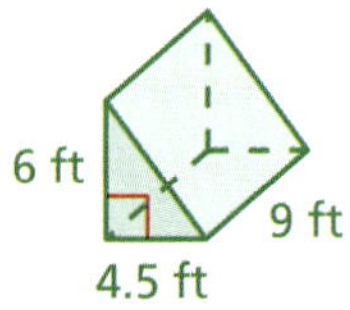

9.

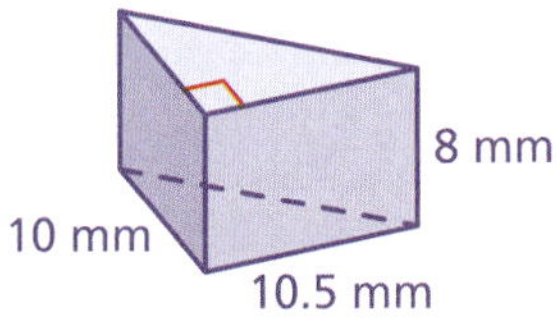

10.

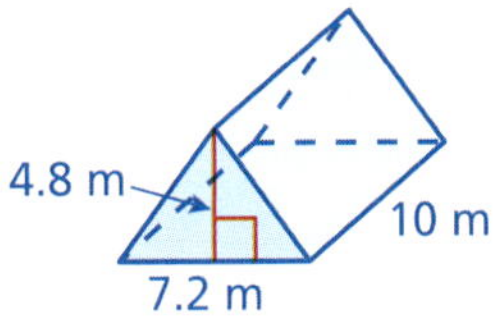

11.

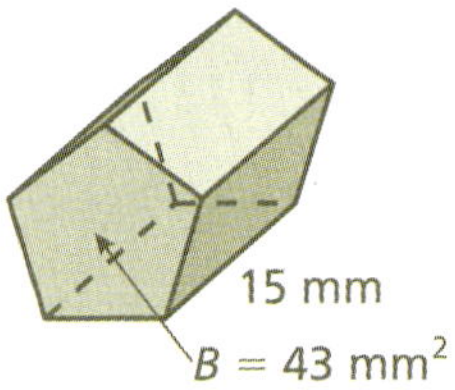

12.

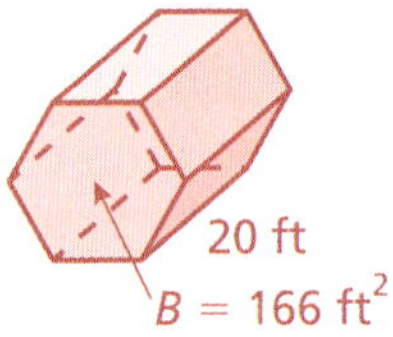

13. **ERROR ANALYSIS** Describe and correct the error in finding the volume of the triangular prism.

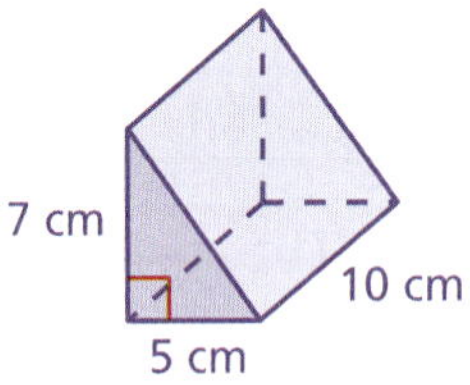

✗
$$\begin{aligned} V &= Bh \\ &= 10(5)(7) \\ &= 50 \cdot 7 \\ &= 350 \text{ cm}^3 \end{aligned}$$

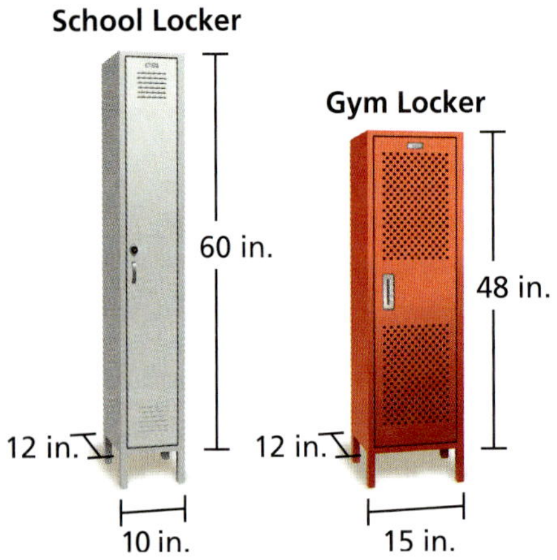

14. **LOCKER** Each locker is shaped like a rectangular prism. Which has more storage space? Explain.

15. **CEREAL BOX** A cereal box is 9 inches by 2.5 inches by 10 inches. What is the volume of the box?

Assignment Guide and Homework Check

Level	Assignment	Homework Check
Advanced	1–6, 8–12 even, 13, 14–24 even, 25–28	16, 18, 20, 22

Common Errors

- **Exercises 4–12** Students may write the units incorrectly, often writing square units instead of cubic units. Remind them that they are working in three dimensions, so the units are cubed. Give an example showing the formula for the base as three units multiplied together. For example, write the volume of Exercise 5 as $V = \frac{1}{2}(6\text{ cm})(8\text{ cm})(12\text{ cm})$.

14.4 Record and Practice Journal

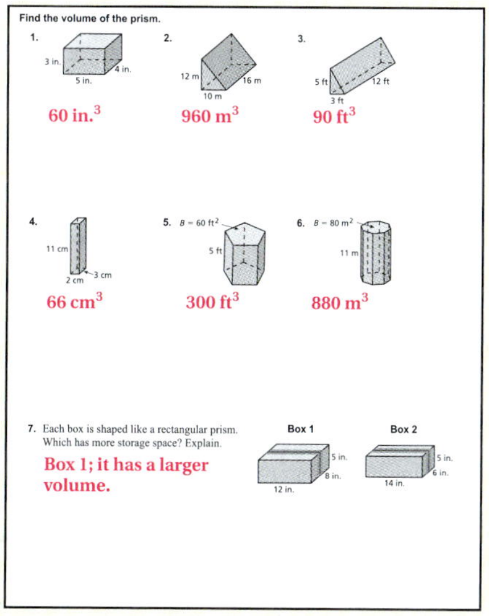

Find the volume of the prism.

1. 60 in.^3
2. 960 m^3
3. 90 ft^3
4. 66 cm^3
5. 300 ft^3
6. 880 m^3
7. Each box is shaped like a rectangular prism. Which has more storage space? Explain.

   Box 1; it has a larger volume.

Vocabulary and Concept Check

1. cubic units
2. Find the area of the base and multiply it by the height.
3. The volume of an object is the amount of space it occupies. The surface area of an object is the sum of the areas of all its faces.

Practice and Problem Solving

4. 729 in.^3
5. 288 cm^3
6. 238 m^3
7. 210 yd^3
8. 121.5 ft^3
9. 420 mm^3
10. 172.8 m^3
11. 645 mm^3
12. 3320 ft^3
13. The area of the base is wrong.

    $V = \frac{1}{2}(7)(5) \cdot 10$

    $= 175\text{ cm}^3$
14. The gym locker has more storage space because it has a greater volume.
15. 225 in.^3

Practice and Problem Solving

16. 1440 in.^{3} **17.** 7200 ft^3

18. sometimes; The prisms in Example 3 have different surface areas, but the same volume. Two prisms that are exactly the same will have the same surface area.

19. See Additional Answers.

20. 48 packets

21. 20 cm

22. *Sample answer:* gas about \$3 per gallon; \$36

23. See *Taking Math Deeper.*

24. The volume is 2 times greater; The volume is 8 times greater.

Fair Game Review

25. \$90 **26.** \$144

27. \$240.50 **28.** D

Mini-Assessment

Find the volume of the prism.

1.

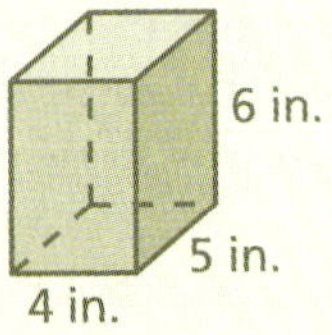

120 in.^{3}

2.

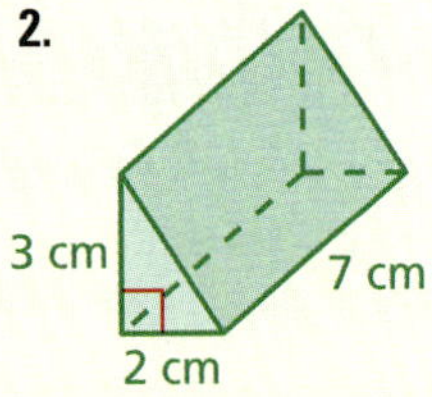

21 cm^3

3.

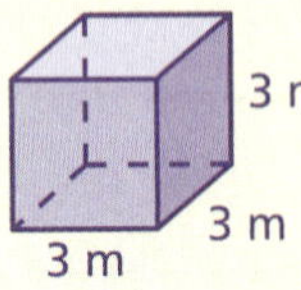

27 m^2

4. Find the volume of the fish tank.

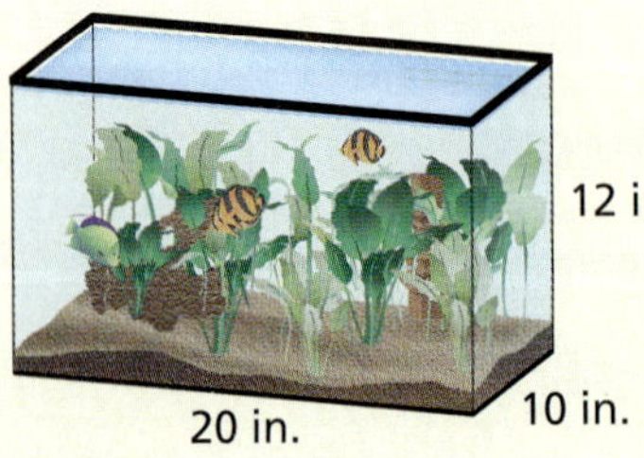

2400 in.^{3}

Taking Math Deeper

Exercise 23

This problem gives students a chance to relate dimensions of a solid with the volume of a solid. It also gives students an opportunity to work with prime factorization. Although aquariums are traditionally the shape of a rectangular prism, remember that other shapes are also possible.

1 Find the volume of the aquarium in cubic inches.

$$\text{Volume} = (450 \cancel{\text{gal}})\left(231 \frac{\text{in.}^3}{\cancel{\text{gal}}}\right) = 103{,}950 \text{ in.}^3$$

2 You could choose two of the dimensions and solve for the third, or you can use prime factorization to find whole number dimensions.

Find the prime factorization of 103,950.

103,950
150 — 693
150: 10, 15; 10: 2, 5; 15: 3, 5
693: 9, 77; 9: 3, 3; 77: 7, 11

The prime factorization is $2 \times 3 \times 3 \times 3 \times 5 \times 5 \times 7 \times 11$.

3 Rearrange the factors to find one set of possible dimensions.

Length: $2 \times 3 \times 5 \times 5 = 150$ in.
Width: $3 \times 7 = 21$ in.
Height: $3 \times 11 = 33$ in.

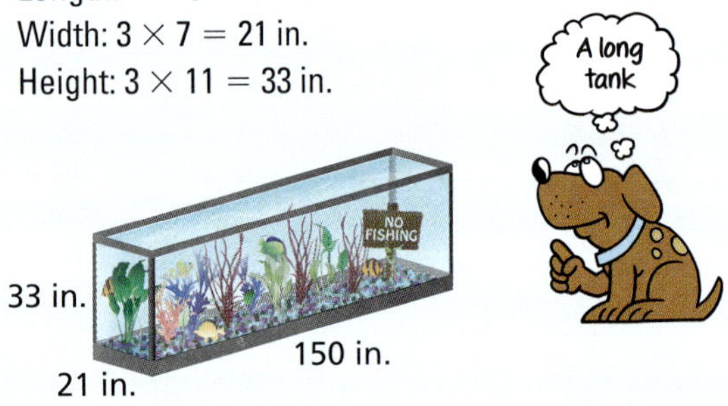

Project

Research a local aquarium. Select an exhibit and draw a picture of the exhibit you selected. Include a short report about the details of the exhibit.

Reteaching and Enrichment Strategies

If students need help. . .	If students got it. . .
Resources by Chapter • Practice A and Practice B • Puzzle Time Record and Practice Journal Practice Differentiating the Lesson Lesson Tutorials Skills Review Handbook	Resources by Chapter • Enrichment and Extension • Technology Connection Start the next section

Find the volume of the prism.

16.

17.

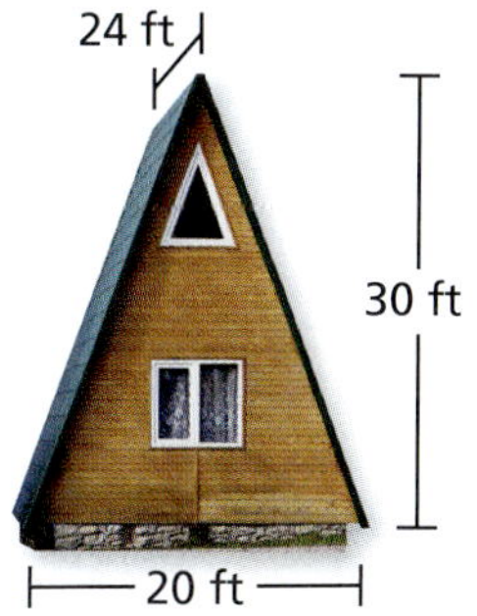

18. **LOGIC** Two prisms have the same volume. Do they *always*, *sometimes*, or *never* have the same surface area? Explain.

19. **CUBIC UNITS** How many cubic inches are in a cubic foot? Use a sketch to explain your reasoning.

20. **CAPACITY** As a gift, you fill the calendar with packets of chocolate candy. Each packet has a volume of 2 cubic inches. Find the maximum number of packets you can fit inside the calendar.

21. **PRECISION** Two liters of water are poured into an empty vase shaped like an octagonal prism. The base area is 100 square centimeters. What is the height of the water? ($1 \text{ L} = 1000 \text{ cm}^3$)

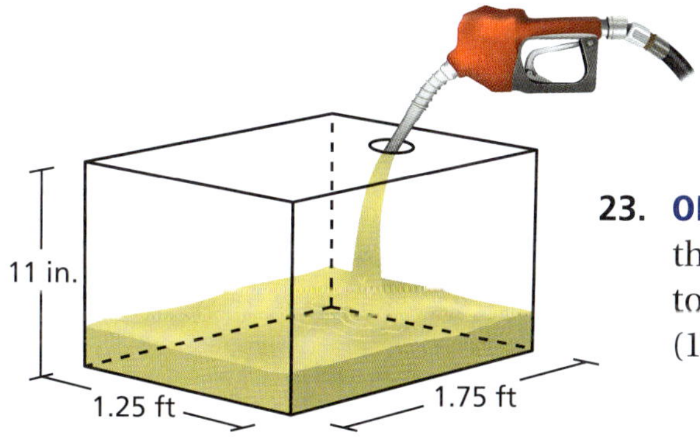

22. **GAS TANK** The gas tank is 20% full. Use the current price of regular gasoline in your community to find the cost to fill the tank. ($1 \text{ gal} = 231 \text{ in.}^3$)

23. **OPEN-ENDED** You visit an aquarium. One of the tanks at the aquarium holds 450 gallons of water. Draw a diagram to show one possible set of dimensions of the tank. ($1 \text{ gal} = 231 \text{ in.}^3$)

24. **Critical Thinking** How many times greater is the volume of a triangular prism when one of its dimensions is doubled? when all three dimensions are doubled?

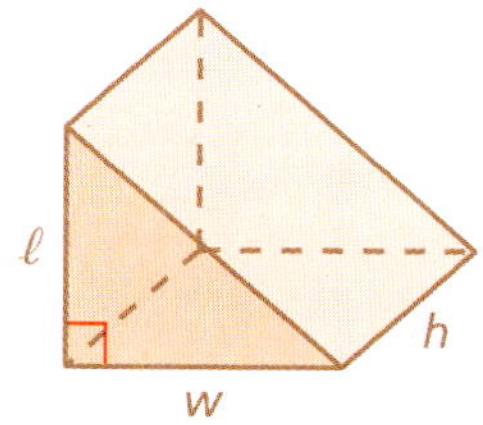

Fair Game Review What you learned in previous grades & lessons

Find the selling price. *(Skills Review Handbook)*

25. Cost to store: \$75
Markup: 20%

26. Cost to store: \$90
Markup: 60%

27. Cost to store: \$130
Markup: 85%

28. **MULTIPLE CHOICE** What is the approximate surface area of a cylinder with a radius of 3 inches and a height of 10 inches? *(Section 14.3)*

Ⓐ 30 in.^2 Ⓑ 87 in.^2 Ⓒ 217 in.^2 Ⓓ 245 in.^2

14.5 Volumes of Pyramids

Essential Question How can you find the volume of a pyramid?

1 ACTIVITY: Finding a Formula Experimentally

Work with a partner.

- **Draw the two nets on cardboard and cut them out.**

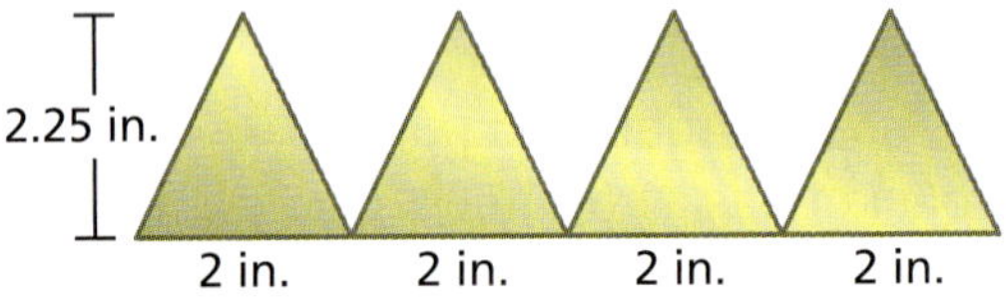

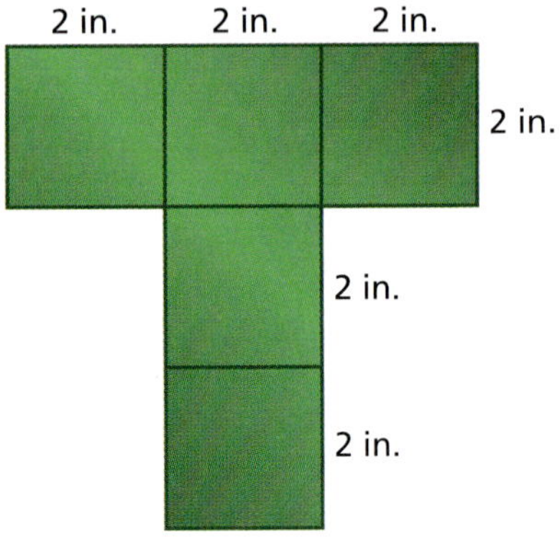

- **Fold and tape the nets to form an open square box and an open pyramid.**
- **Both figures should have the same size square base and the same height.**
- **Fill the pyramid with pebbles. Then pour the pebbles into the box. Repeat this until the box is full. How many pyramids does it take to fill the box?**
- **Use your result to find a formula for the volume of a pyramid.**

2 ACTIVITY: Comparing Volumes

Work with a partner. You are an archaeologist studying two ancient pyramids. What factors would affect how long it took to build each pyramid? Given similar conditions, which pyramid took longer to build? Explain your reasoning.

Geometry

In this lesson, you will

- find volumes of pyramids.
- solve real-life problems.

The Sun Pyramid in Mexico
Height: about 246 ft
Base: about 738 ft by 738 ft

Cheops Pyramid in Egypt
Height: about 480 ft
Base: about 755 ft by 755 ft

Laurie's Notes

Introduction

Applying Mathematical Practices

- **Reason Abstractly and Quantitatively** and **Model with Mathematics:** In constructing physical models of a prism and a pyramid that have the same base area and height and then comparing their volumes, students make sense of the formula for the volume of a pyramid.

Motivate

- Share information about two well known pyramid-shaped buildings.
- The Luxor Resort and Casino in Las Vegas reaches 350 feet into the sky and is 36 stories tall. Luxor has over 4400 guest rooms, making it one of the top ten largest hotels in the world.
- The Rock and Roll Hall of Fame in Cleveland was designed by architect I. M. Pei. The design of the glass-faced main building uses a pyramid shape to invoke the image of a guitar neck rising to the sky.

Activity Notes

Activity 1

- ? "How are the two shapes alike?" same base, same height "How are they different?" One is a square pyramid. The other is a square prism.
- ? "How do you think their volumes compare?" Most students guess that the prism has twice the volume as the pyramid.
- ? "How can you test your hunch about the volumes?" If students have looked at the activity, they'll want to fill the pyramid.
- After the first pour, students should start to suspect that their guess might be off.
- After the second pour, students are pretty sure the relationship is 3 to 1.
- **Reason Abstractly and Quantitatively** and **Model with Mathematics:** This hands-on experience of making and filling the prism will help students remember the factor of $\frac{1}{3}$. The formula should now make sense to them. The volume of a pyramid should be $\frac{1}{3}$ the volume of a prism with the same base and height as the pyramid.

Activity 2

- ? **Attend to Precision:** "What do you know about the pyramids from looking only at the pictures?" They look like square pyramids.
- ? "What do you know about the pyramids from looking at their dimensions?" Cheops has a larger base and is nearly twice as tall.
- Give time for students to calculate the volume. From the first activity, they should feel comfortable finding the area of the base and multiplying by the height (this would be the prism's volume), and then taking $\frac{1}{3}$ of this answer.
- ? "Which pyramid has the greater volume, and about how many times greater is its volume than the other pyramid?" Cheops Pyramid has about twice the volume of The Sun Pyramid.

What Your Students Will Learn

- Find volumes of pyramids using formulas.

Previous Learning

Students should know how to find the volumes of prisms and how to perform operations on rational numbers.

Technology for the Teacher

Dynamic Classroom

Lesson Plans
Complete Materials List

14.5 Record and Practice Journal

Essential Question How can you find the volume of a pyramid?

1 ACTIVITY: Finding a Formula Experimentally

Work with a partner.

- **Draw the two nets on cardboard and cut them out.***

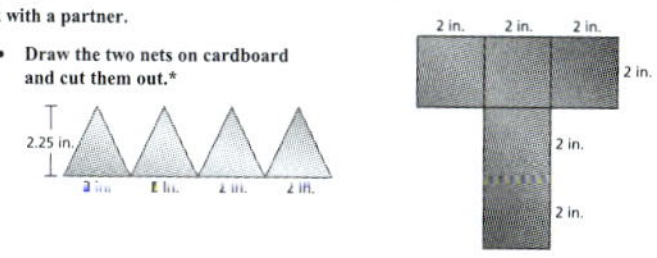

- **Fold and tape the nets to form an open square box and an open pyramid.**
- **Both figures should have the same size square base and the same height.**
- **Fill the pyramid with pebbles. Then pour the pebbles into the box. Repeat this until the box is full. How many pyramids does it take to fill the box?** 3

- **Use your result to find a formula for the volume of a pyramid.** $V = \frac{1}{3}Bh$

2 ACTIVITY: Comparing Volumes

Work with a partner. You are an archaeologist studying two ancient pyramids. What factors would affect how long it took to build each pyramid? Given similar conditions, which pyramid took longer to build? Explain your reasoning.

The Sun Pyramid in Mexico	Cheops Pyramid in Egypt
Height: 246 ft	Height: about 480 ft
Base: 738 ft by 738 ft	Base: about 755 ft by 755 ft

Cheops Pyramid in Egypt

*Cut-outs are available in the back of the Record and Practice Journal.

Differentiated Instruction

Visual

Students may confuse pyramids and triangular prisms. Show the students models and point out the following characteristics. A pyramid has one base, which can be any polygon. The remaining faces are triangles. A triangular prism has two bases, which are triangles. The remaining faces are rectangles.

14.5 Record and Practice Journal

3 ACTIVITY: Finding and Using a Pattern

Work with a partner.

- Find the volumes of the pyramids.

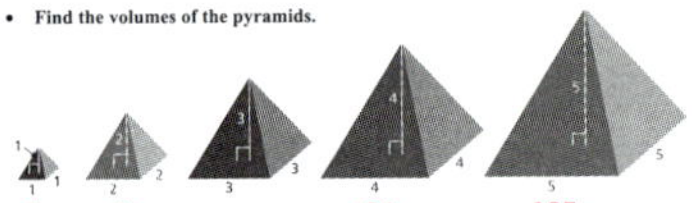

$\frac{1}{3}$ $\frac{8}{3}$ 9 $\frac{64}{3}$ $\frac{125}{3}$

- Organize your results in a table.

Pyramid	Volume (cubic units)
1	$\frac{1}{3}$
2	$\frac{8}{3}$
3	9
4	$\frac{64}{3}$
5	$\frac{125}{3}$

- Describe the pattern.
  side length cubed divided by 3

- Use your pattern to find the volume of a pyramid with a side length and a height of 20.
  $\frac{8000}{3}$ cubic units

4 ACTIVITY: Breaking a Prism into Pyramids

Work with a partner. The rectangular prism can be cut to form three pyramids. Show that the sum of the volumes of the three pyramids is equal to the volume of the prism.

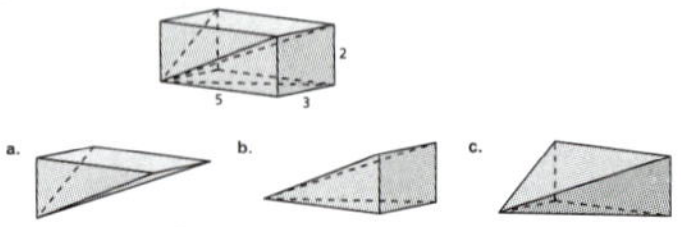

Volume of Prism $= 2 \cdot 3 \cdot 5 = 30$

a. $V = \frac{1}{3}(3 \cdot 5)(2) = 10$

b. $V = \frac{1}{3}(2 \cdot 3)(5) = 10$

c. $V = \frac{1}{3}(2 \cdot 5)(3) = 10$

What Is Your Answer?

5. **IN YOUR OWN WORDS** How can you find the volume of a pyramid?
   Find the area of the base and multiply it by the height of the pyramid. Then divide by 3.

6. **STRUCTURE** Write a general formula for the volume of a pyramid.
   $V = \frac{1}{3}Bh$

Laurie's Notes

Activity 3

- **Look for and Express Regularity in Repeated Reasoning:** Ask a student to describe the five pyramids shown. You want to make sure that students recognize that the bases are all squares, and the height of the pyramid is the same as the length of the base edge.
- **Use Appropriate Tools Strategically:** Reinforce good problem solving by having students organize their data in a table.
- Allow time for students to record the volume of each pyramid.

? "What was the volume of the smallest pyramid and how did you find it?"
$V = \frac{1}{3} \cdot 1^3 = \frac{1}{3}$

? "What was the volume of the next pyramid and how did you find it?"
$V = \frac{1}{3} \cdot 2^3 = \frac{8}{3}$

- **Look for and Express Regularity in Repeated Reasoning:** Repeat the question for the remaining pyramids, and then ask a volunteer to clearly summarize the pattern of the pyramids and their volumes. The height of each pyramid is equal to its side length s. The volume of each pyramid is given by $V = \frac{1}{3} \cdot s^3$.

Activity 4

- Discuss with students how to follow the color coding so that correct dimensions can be matched up.

What Is Your Answer?

- **Neighbor Check:** Have students work independently and then have their neighbors check their work. Have students discuss any discrepancies.

Closure

- Does the volume formula you wrote for Question 5 need to have a square base? Explain your thinking. No; Students should try to sketch or make pyramids with other polygonal bases.

3 ACTIVITY: Finding and Using a Pattern

Math Practice

Look for Patterns

As the height and the base lengths increase, how does this pattern affect the volume? Explain.

Work with a partner.

- **Find the volumes of the pyramids.**
- **Organize your results in a table.**
- **Describe the pattern.**
- **Use your pattern to find the volume of a pyramid with a base length and a height of 20.**

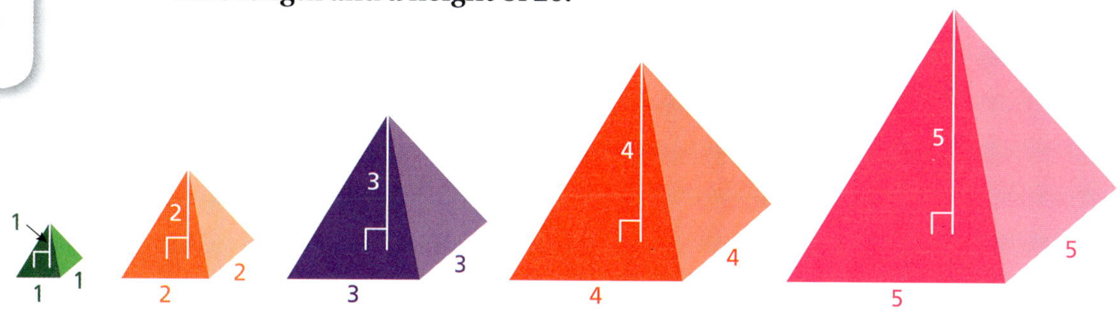

4 ACTIVITY: Breaking a Prism into Pyramids

Work with a partner. The rectangular prism can be cut to form three pyramids. Show that the sum of the volumes of the three pyramids is equal to the volume of the prism.

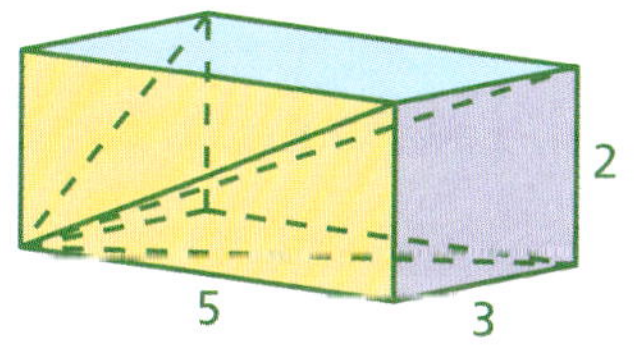

a.

b.

c.

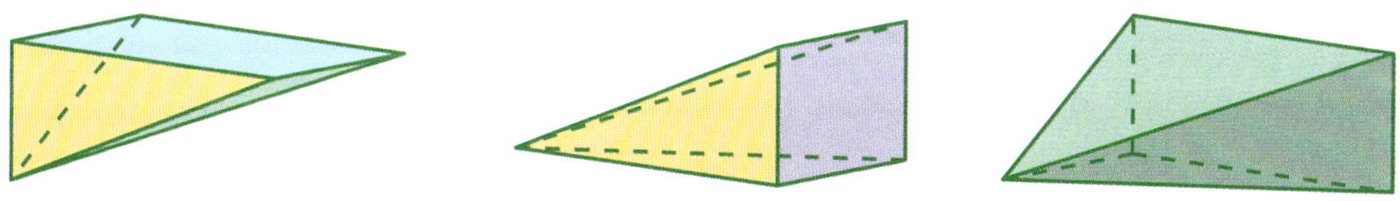

What Is Your Answer?

5. **IN YOUR OWN WORDS** How can you find the volume of a pyramid?
6. **STRUCTURE** Write a general formula for the volume of a pyramid.

Practice Use what you learned about the volumes of pyramids to complete Exercises 4–6 on page 618.

14.5 Lesson

Check It Out
Lesson Tutorials
BigIdeasMath.com

Key Idea

Volume of a Pyramid

Words The volume V of a pyramid is one-third the product of the area of the base and the height of the pyramid.

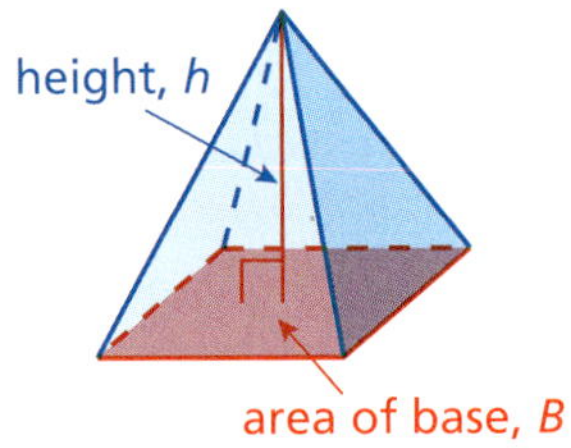

Algebra $V = \frac{1}{3}Bh$

Area of base (B)

Height of pyramid (h)

> **Study Tip**
>
> The *height* of a pyramid is the perpendicular distance from the base to the vertex.

EXAMPLE 1 Finding the Volume of a Pyramid

Find the volume of the pyramid.

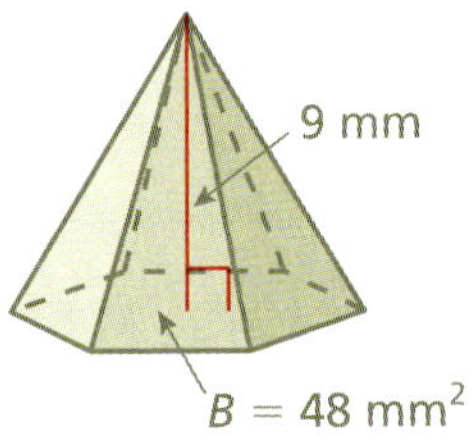

$V = \frac{1}{3}Bh$ Write formula for volume.

$= \frac{1}{3}(48)(9)$ Substitute.

$= 144$ Multiply.

The volume is 144 cubic millimeters.

EXAMPLE 2 Finding the Volume of a Pyramid

Find the volume of the pyramid.

a.

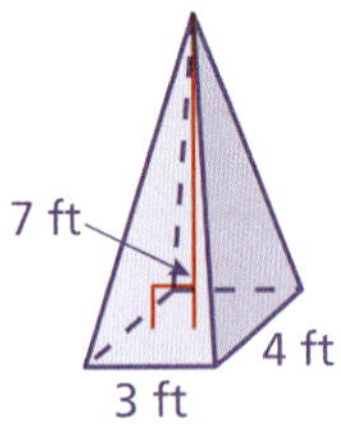

$V = \frac{1}{3}Bh$

$= \frac{1}{3}(4)(3)(7)$

$= 28$

The volume is 28 cubic feet.

b.

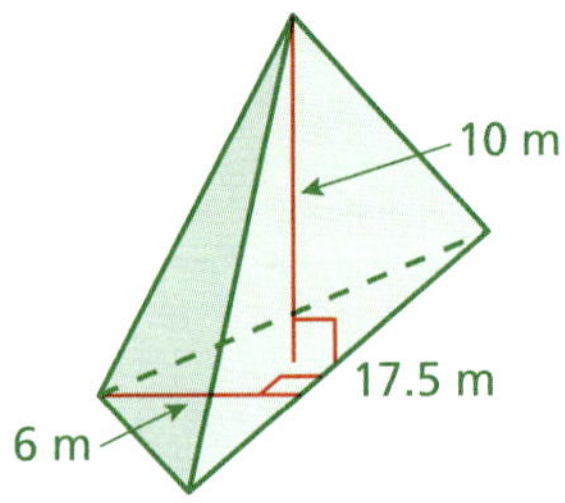

$V = \frac{1}{3}Bh$

$= \frac{1}{3}\left(\frac{1}{2}\right)(17.5)(6)(10)$

$= 175$

The volume is 175 cubic meters.

> **Study Tip**
>
> The area of the base of a rectangular pyramid is the product of the length ℓ and the width w.
>
> You can use $V = \frac{1}{3}\ell wh$ to find the volume of a rectangular pyramid.

Laurie's Notes

Introduction

Connect

- **Yesterday:** Students discovered how to find the volume of a pyramid by comparing it to the volume of a prism with the same base area and same height.
- **Today:** Students will work with a formula for the volume of a pyramid.

Motivate

- Show and discuss a picture of the Transamerica Pyramid building.
- It is the tallest building in San Francisco.
- The tapered design reduces the building's shadow to let more light reach the streets below.
- A San Francisco regulation limits the ratio of surface area to height for a building.

Lesson Notes

Key Idea

- Write the formula in words, and then in symbols.
- ? "How will you find the area of the base?" It depends on what type of polygon the base is.
- Discuss the dotted lines and the shaded base. Pyramids are difficult to draw in two dimensions. Have students sketch a triangular pyramid and a square pyramid to practice.

Example 1

- Model good problem solving. Write the formula in words. Write the symbols underneath the words. Substitute the values for the symbols.
- **Common Error:** In using the volume formula, students often find $\frac{1}{3}$ of both B and h $\left(\frac{1}{3} \text{ of } 48 \text{ and } \frac{1}{3} \text{ of } 9\right)$ as though they are using the Distributive Property. Remind them that the Distributive Property is used when there is addition or subtraction involved. The correct steps for this problem are to multiply from left to right.

Example 2

- ? "Describe the base of each pyramid." part (a): rectangular base; part (b): triangular base
- ? "How do you find the area of the base in part (b)?" $\frac{1}{2} \times 17.5 \times 6$
- **FYI:** A statement like "one-half of the base times the height" will confuse students in part (b). Specify when you are speaking of parts of the base triangle and when you are speaking of parts of the pyramid.
- Students may need help in multiplying the fractions in each problem. They can apply the Commutative Property to get products of reciprocals in each problem. In part (a), $\frac{1}{3} \times 3 = 1$ and in part (b), $\frac{1}{3} \times \frac{1}{2} \times 6 = 1$.

Goal Today's lesson is finding the volumes of pyramids.

Technology for the Teacher

Dynamic Classroom

Lesson Tutorials
Lesson Plans
Answer Presentation Tool

Extra Example 1

Find the volume of a pentagonal pyramid with a base area of 24 square feet and a height of 8 feet. 64 ft^3

Extra Example 2

a. Find the volume of a rectangular pyramid with a base of 2 meters by 6 meters and a height of 3 meters. 12 m^3

b. Find the volume of a triangular pyramid with a height of 8 inches and where the triangular base has a width of 4 inches and an altitude of 9 inches. 48 in.^3

On Your Own

1. 42 ft^3
2. $186\frac{2}{3}\text{ in.}^3$
3. 231 cm^3

Extra Example 3

a. The volume of lotion in Bottle B is how many times the volume in Bottle A? $1\frac{2}{3}$

b. Which is the better buy? Bottle B

On Your Own

4. yes; Bottles B and C have the same volume, but Bottle C has a unit cost of $2.20.

English Language Learners

Forming Answers

Encourage English learners to form complete sentences in their responses. Students can use the question to help them form the answer.

Question: If you know the area of the base of a pyramid, what else do you need to know to find the volume?

Response: If you know the area of the base of a pyramid, you need to know the height of the pyramid to find the volume.

Laurie's Notes

On Your Own

- Have students name each pyramid and describe what they know about each base. Note that for the pentagonal pyramid, the area of the base has already been computed.
- In Question 2, none of the dimensions contain factors of 3. In computing the volume, $V = \frac{1}{3} \times 10 \times 8 \times 7$, suggest to students that they multiply the whole numbers for a product of 560 and then multiply by $\frac{1}{3}$. Remind students how to rewrite the improper fraction $\frac{560}{3}$ as a mixed number, $186\frac{2}{3}$.

Example 3

- If you have any lotion or shampoo that is in a pyramidal bottle, use it as a model.
- Work through the computation of volume for each bottle.
- **Reason Abstractly and Quantitatively:** Explain different approaches to multiplying the factors in Bottle A: (1) multiply in order from left to right or (2) use the Commutative Property to multiply the whole numbers, and then multiply by $\frac{1}{3}$.
- Discuss the phrase "how many times." Because the volume of Bottle B is not a multiple of the volume of Bottle A, students are uncertain how to compare the volumes.

? "How do you decide which bottle is the better buy?" Students will try to describe how to find the cost for one cubic inch. This is the unit price.

- Use the language, cost per volume or cost per cubic inch.

On Your Own

- Give students time to complete this problem. Ask volunteers to share their work at the board.

Closure

- **Exit Ticket:** Sketch a rectangular pyramid with base 3 units by 4 units, and a height of 5 units. What is the volume? 20 units^3

On Your Own

Now You're Ready
Exercises 4–11

Find the volume of the pyramid.

1.

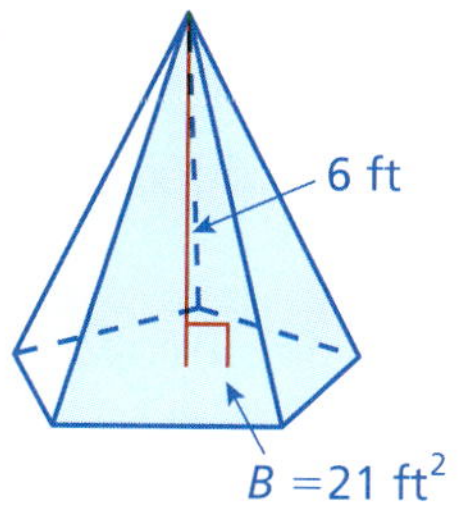

2.

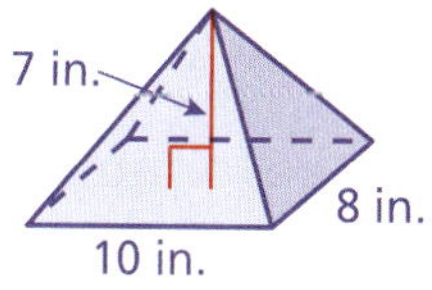

3.

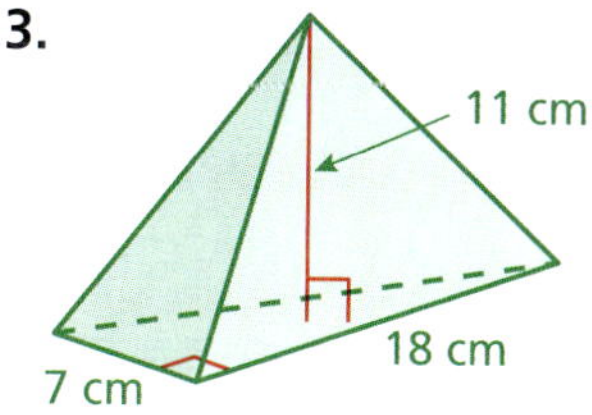

EXAMPLE 3 Real-Life Application

a. The volume of sunscreen in Bottle B is about how many times the volume in Bottle A?

b. Which is the better buy?

Bottle A
$9.96

Bottle B
$14.40

a. Use the formula for the volume of a pyramid to estimate the amount of sunscreen in each bottle.

Bottle A	***Bottle B***
$V = \frac{1}{3}Bh$	$V = \frac{1}{3}Bh$
$= \frac{1}{3}(2)(1)(6)$	$= \frac{1}{3}(3)(1.5)(4)$
$= 4 \text{ in.}^3$	$= 6 \text{ in.}^3$

So, the volume of sunscreen in Bottle B is about $\frac{6}{4} = 1.5$ times the volume in Bottle A.

b. Find the unit cost for each bottle.

Bottle A	***Bottle B***
$\frac{\text{cost}}{\text{volume}} = \frac{\$9.96}{4 \text{ in.}^3}$	$\frac{\text{cost}}{\text{volume}} = \frac{\$14.40}{6 \text{ in.}^3}$
$= \frac{\$2.49}{1 \text{ in.}^3}$	$= \frac{\$2.40}{1 \text{ in.}^3}$

The unit cost of Bottle B is less than the unit cost of Bottle A. So, Bottle B is the better buy.

On Your Own

Now You're Ready
Exercise 16

4. Bottle C is on sale for $13.20. Is Bottle C a better buy than Bottle B in Example 3? Explain.

Bottle C

14.5 Exercises

Vocabulary and Concept Check

1. **WRITING** How is the formula for the volume of a pyramid different from the formula for the volume of a prism?
2. **OPEN-ENDED** Describe a real-life situation that involves finding the volume of a pyramid.
3. **REASONING** A triangular pyramid and a triangular prism have the same base and height. The volume of the prism is how many times the volume of the pyramid?

Practice and Problem Solving

Find the volume of the pyramid.

4.
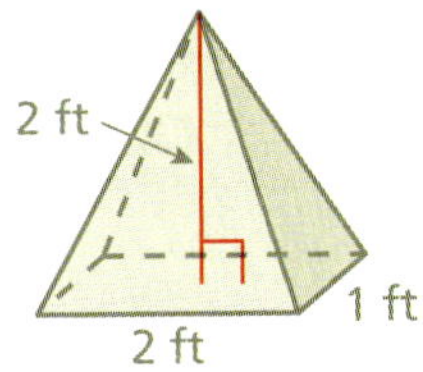

5.
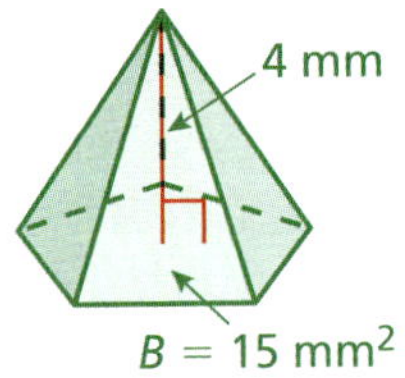

6.
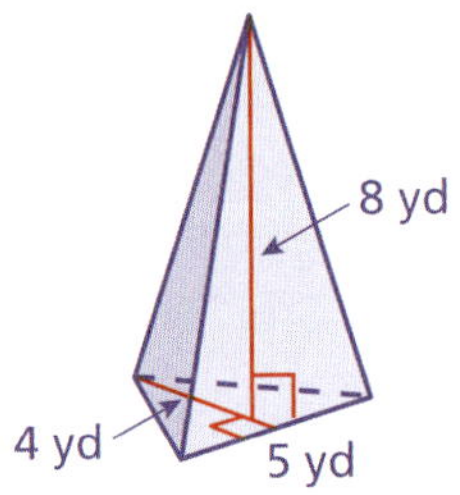

7.
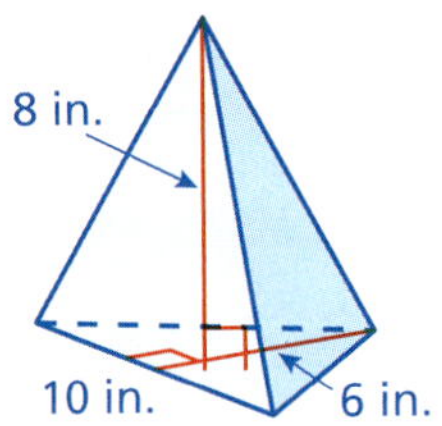

8.
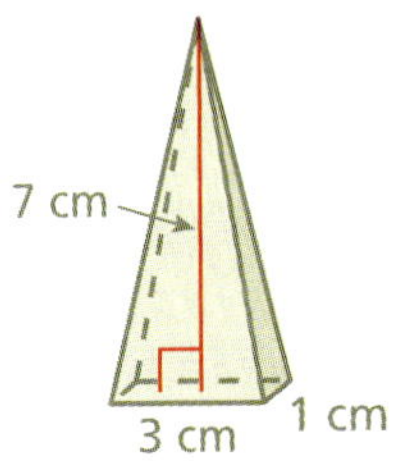

9.
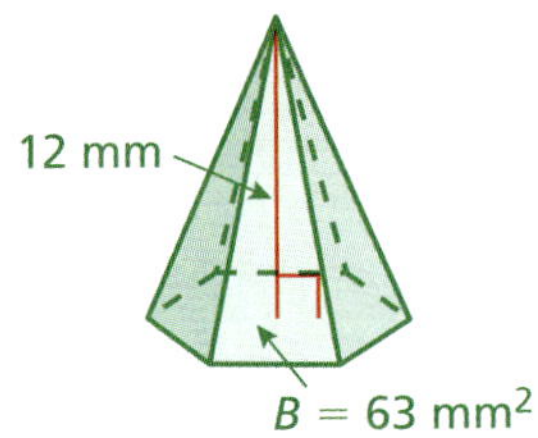

10.
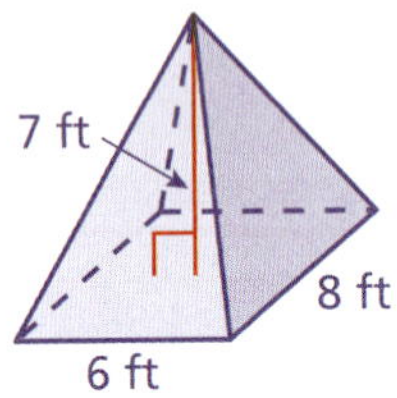

11.
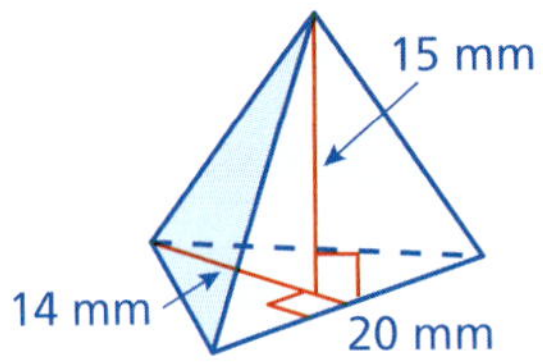

12. **PARACHUTE** In 1483, Leonardo da Vinci designed a parachute. It is believed that this was the first parachute ever designed. In a notebook, he wrote, "If a man is provided with a length of gummed linen cloth with a length of 12 yards on each side and 12 yards high, he can jump from any great height whatsoever without injury." Find the volume of air inside Leonardo's parachute.

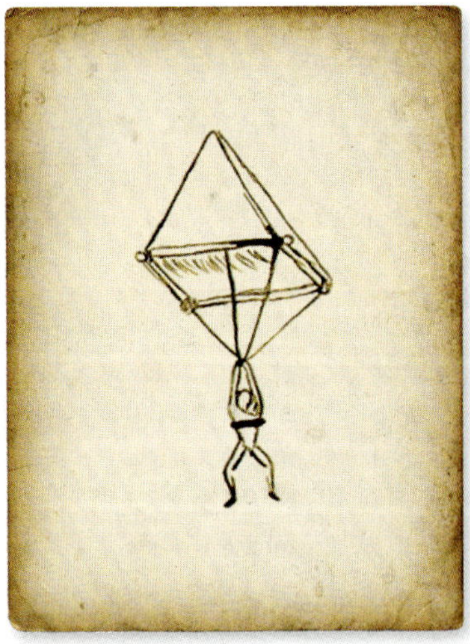
Not drawn to scale

Assignment Guide and Homework Check

Level	Assignment	Homework Check
Advanced	1–7, 8–20 even, 21–24	7, 10, 12, 18, 20

For Your Information

- **Exercise 12** Skydiver Adrian Nicholas tested the design, jumping from a hot-air balloon at 3000 meters. The parachute weighed over 90 kilograms.

Common Errors

- **Exercises 4–11** Students may write the units incorrectly, often writing square units instead of cubic units. This is especially true when the area of the base is given. Remind them that the units are cubed because there are three dimensions.
- **Exercises 4–11** Students may forget to multiply by one of the measurements, especially when finding the area of the base. Encourage them to find the area of the base separately and then substitute it into the equation. Using colored pencils for each part can also assist students. Tell them to write the formula using different colors for the base and height, as in the lesson. When they substitute values into the equation for volume, they will be able to clearly see that they have accounted for all of the dimensions.

14.5 Record and Practice Journal

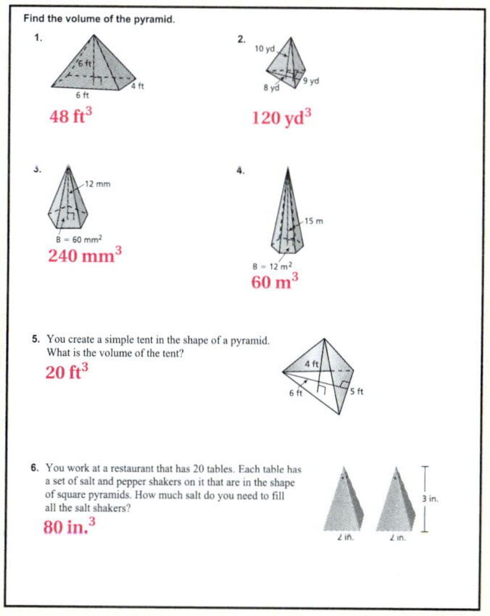
Find the volume of the pyramid.

1. $48\ \text{ft}^3$
2. $120\ \text{yd}^3$
3. $240\ \text{mm}^3$
4. $60\ \text{m}^3$
5. You create a simple tent in the shape of a pyramid. What is the volume of the tent? $20\ \text{ft}^3$
6. You work at a restaurant that has 20 tables. Each table has a set of salt and pepper shakers on it that are in the shape of square pyramids. How much salt do you need to fill all the salt shakers? $80\ \text{in.}^3$

Vocabulary and Concept Check

1. The volume of a pyramid is $\frac{1}{3}$ times the area of the base times the height. The volume of a prism is the area of the base times the height.
2. *Sample answer:* You are comparing the sizes of two tents and want to know which one has more space inside of it.
3. 3 times

Practice and Problem Solving

4. $1\frac{1}{3}\ \text{ft}^3$
5. $20\ \text{mm}^3$
6. $26\frac{2}{3}\ \text{yd}^3$
7. $80\ \text{in.}^3$
8. $7\ \text{cm}^3$
9. $252\ \text{mm}^3$
10. $112\ \text{ft}^3$
11. $700\ \text{mm}^3$
12. $576\ \text{yd}^3$

Practice and Problem Solving

13. $156\ \text{ft}^3$

14. $240\ \text{m}^3$

15. $340.4\ \text{in.}^3$

16. Spire B; $4\ \text{in.}^3$

17. See Additional Answers.

18. See *Taking Math Deeper.*

19. *Sample answer:* 5 ft by 4 ft

20. yes; Prism: $V = xyz$

Pyramid: $V = \frac{1}{3}(xy)(3z) = xyz$

Fair Game Review

21. 153°; 63° **22.** 98°; 8°

23. 60°; none **24.** C

Mini-Assessment

Find the volume of the pyramid.

1.

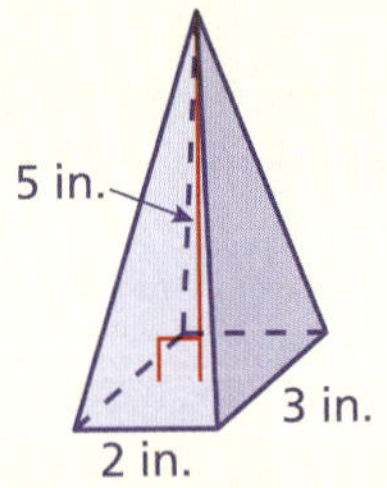

$10\ \text{in.}^3$

2.

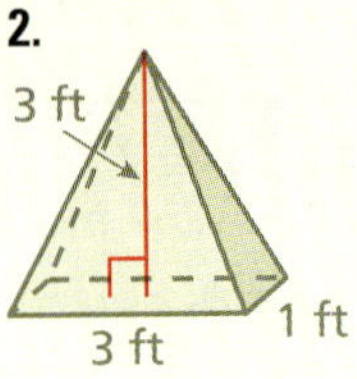

$3\ \text{ft}^3$

3.

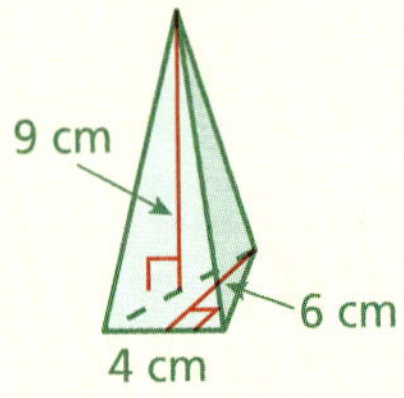

$36\ \text{cm}^3$

4. Find the volume of the paper weight.

$4\ \text{in.}^3$

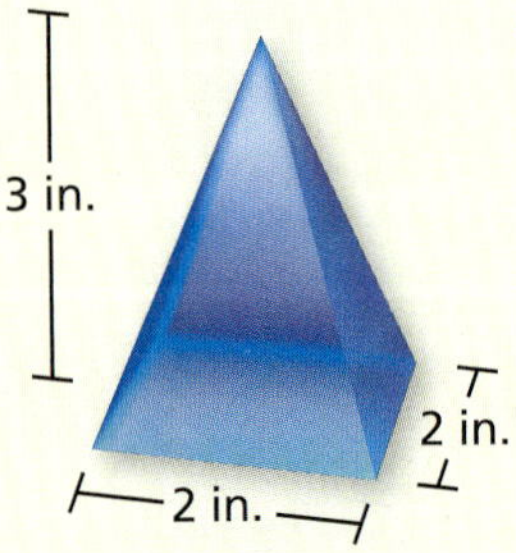

Taking Math Deeper

Exercise 18

Students have to think a bit about this question. At first it seems like you cannot tell the shape of the base. However, you can count the number of support sticks to find the shape of the base.

1 Count the supports.

a. There are 12. So, the base is a dodecagon (a 12-sided polygon).

2 Using a ruler, the base of the teepee appears to be about the same as its height. So, estimate the width of the base to be 10 feet.

3 Use a 10-by-10 grid to estimate the area of the base. It appears to have an area of about 80 square feet.

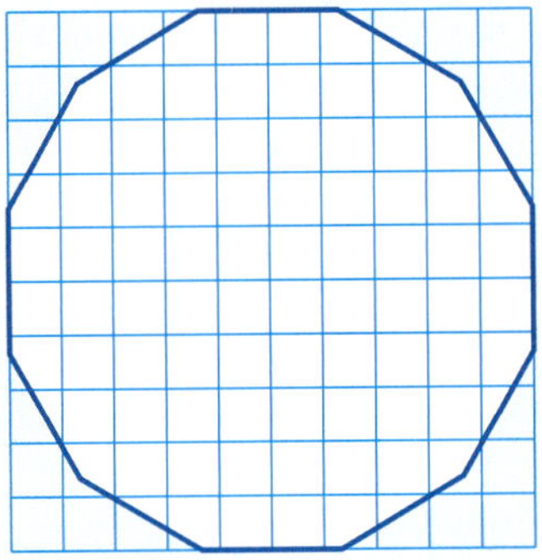

b. $V = \frac{1}{3}Bh$

$= \frac{1}{3} \cdot 80 \cdot 10$

$\approx 267\ \text{ft}^3$

You need to give some leeway in the answers. Anything from 250 cubic feet to 300 cubic feet is a reasonable answer.

Reteaching and Enrichment Strategies

If students need help. . .	If students got it. . .
Resources by Chapter • Practice A and Practice B • Puzzle Time Record and Practice Journal Practice Differentiating the Lesson Lesson Tutorials Skills Review Handbook	Resources by Chapter • Enrichment and Extension • Technology Connection Start the next section

Find the volume of the composite solid.

13.

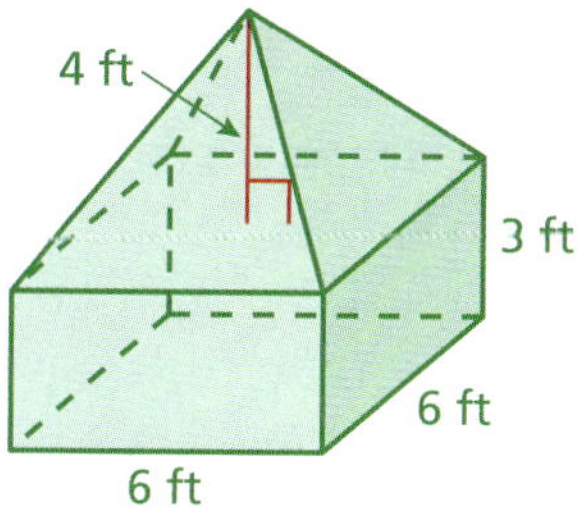

14.

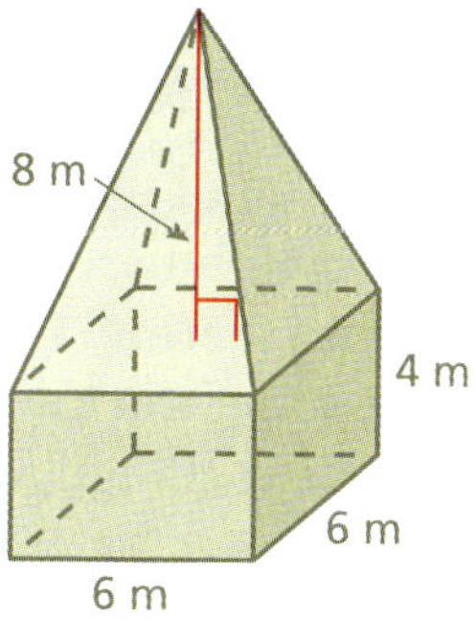

15.

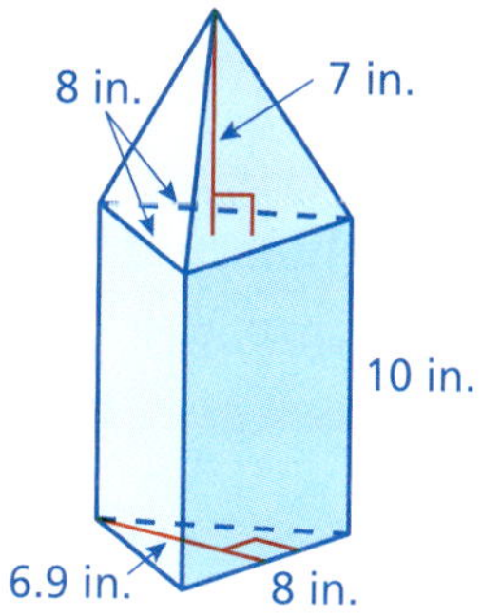

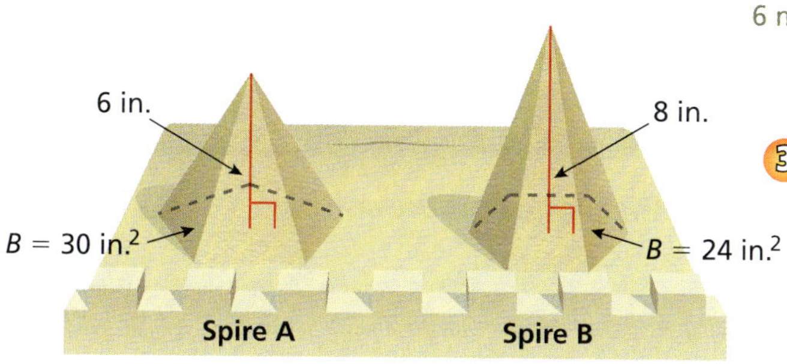

16. SPIRE Which sand-castle spire has a greater volume? How much more sand do you need to make the spire with the greater volume?

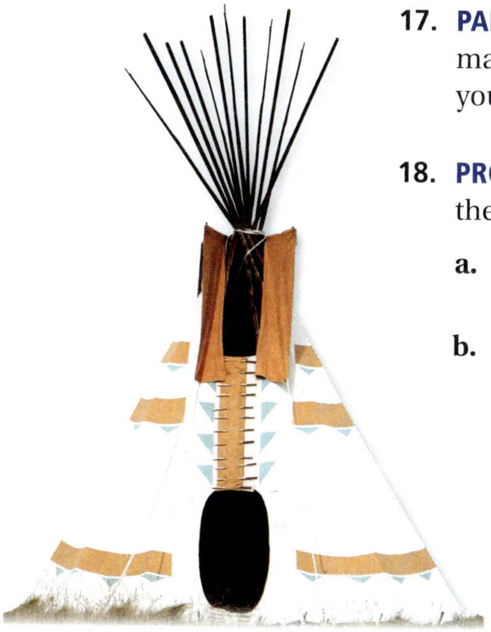

17. PAPERWEIGHT How much glass is needed to manufacture 1000 paperweights? Explain your reasoning.

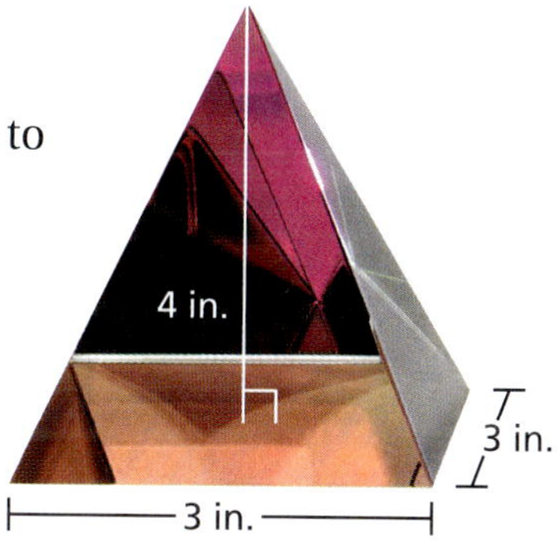

Paperweight

18. PROBLEM SOLVING Use the photo of the tepee.

a. What is the shape of the base? How can you tell?

b. The tepee's height is about 10 feet. Estimate the volume of the tepee.

19. OPEN-ENDED A pyramid has a volume of 40 cubic feet and a height of 6 feet. Find one possible set of dimensions of the rectangular base.

20. Reasoning Do the two solids have the same volume? Explain.

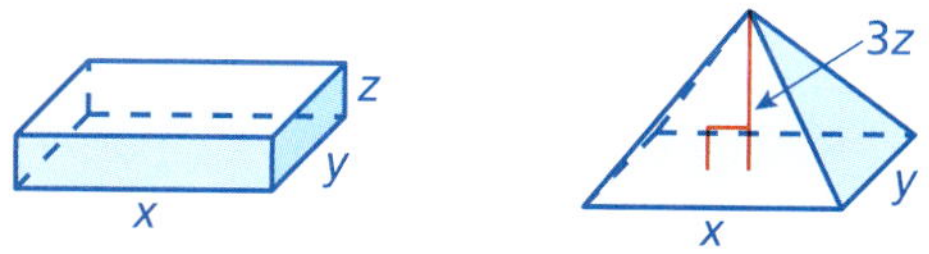

Fair Game Review What you learned in previous grades & lessons

For the given angle measure, find the measure of a supplementary angle and the measure of a complementary angle, if possible. *(Section 12.2)*

21. 27° **22.** 82° **23.** 120°

24. MULTIPLE CHOICE The circumference of a circle is 44 inches. Which estimate is closest to the area of the circle? *(Section 13.3)*

Ⓐ 7 in.^{2} Ⓑ 14 in.^{2} Ⓒ 154 in.^{2} Ⓓ 484 in.^{2}

Extension 14.5 Cross Sections of Three-Dimensional Figures

Check It Out
Lesson Tutorials
BigIdeasMath.com

Key Vocabulary
cross section, *p. 620*

Consider a plane "slicing" through a solid. The intersection of the plane and the solid is a two-dimensional shape called a **cross section**. For example, the diagram shows that the intersection of the plane and the rectangular prism is a rectangle.

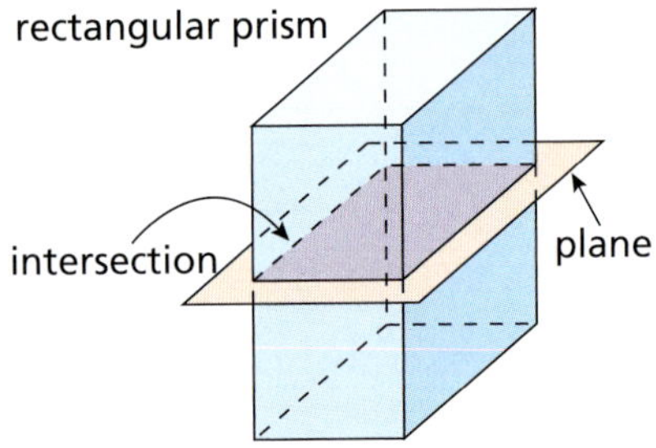

Geometry
In this extension, you will
- describe the intersections of planes and solids.

EXAMPLE 1 Describing the Intersection of a Plane and a Solid

Describe the intersection of the plane and the solid.

a.

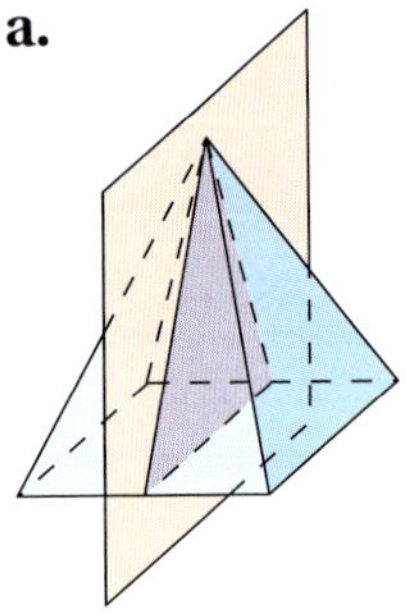

b.

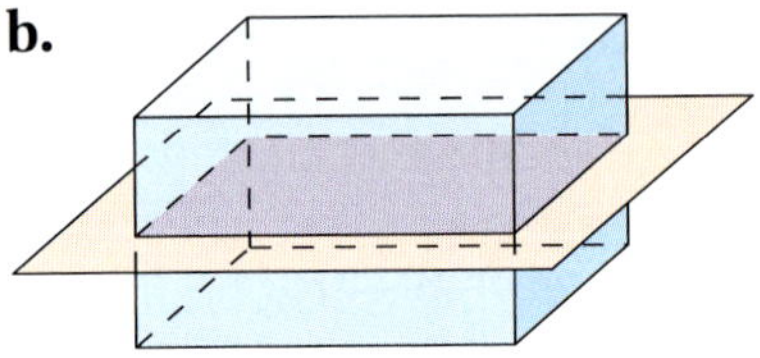

c.

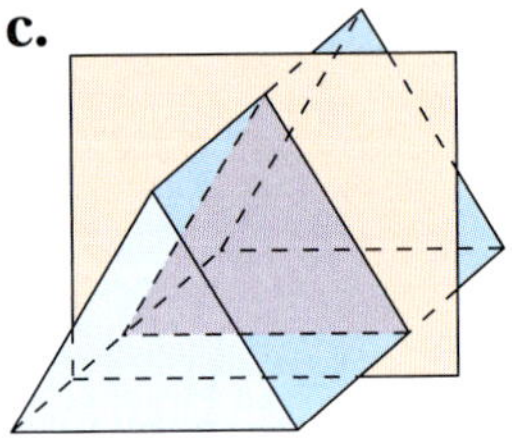

a. The intersection is a triangle.

b. The intersection is a rectangle.

c. The intersection is a triangle.

Practice

Describe the intersection of the plane and the solid.

1.

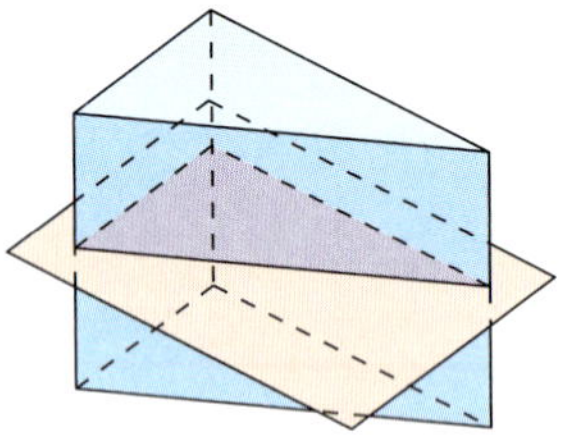

2.

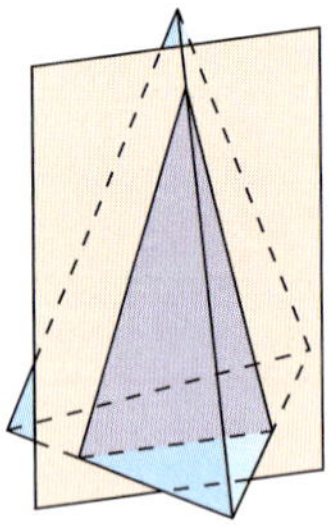

3.

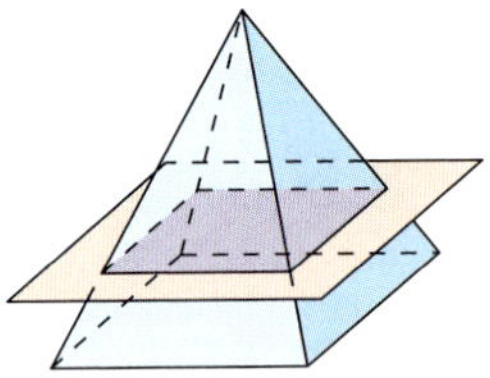

4.

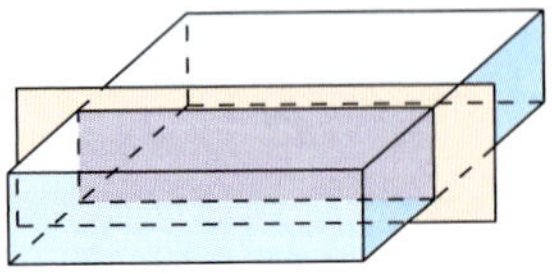

5.

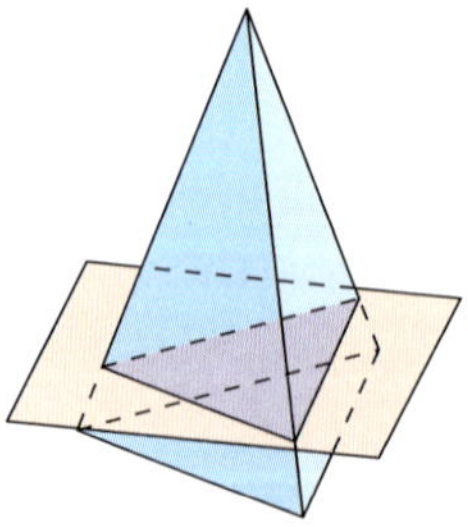

6.

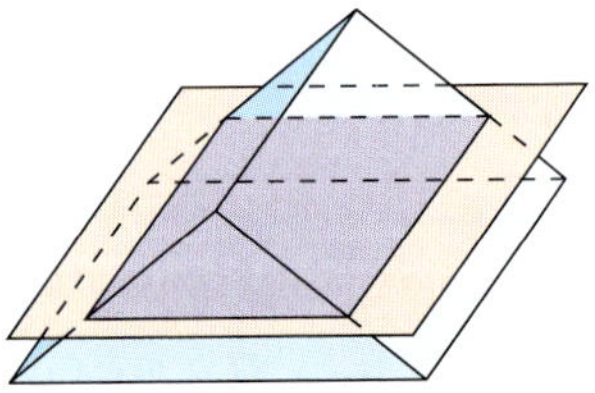

7. **REASONING** A plane that intersects a prism is parallel to the bases of the prism. Describe the intersection of the plane and the prism.

Multi-Language Glossary at BigIdeasMath.com

Laurie's Notes

Introduction

Connect

- **Yesterday:** Students found the volumes of pyramids.
- **Today:** Students will describe cross sections of three-dimensional figures.

Motivate

- Use a flashlight, overhead projector, or document camera to show two-dimensional projections of three-dimensional objects. For instance, hold a rectangular prism in different orientations to show different projections. Do the same for a triangular prism (a door stop works well) and a cylinder.
- The goal is for students to visualize different cross sections of solids.
- **Model with Mathematics:** If cross sections are difficult for students to visualize, tell them to consider a cheese slicer cutting through a wedge of cheese. The face exposed when the cheese is cut is a cross section.

Lesson Notes

Example 1

- It is helpful for students to see and hold models of solids when thinking about how a plane intersects a solid. You can stretch a rubber band around the model of a solid to help students visualize the intersection of a plane with the solid.
- Consider using technology that displays the animation of a plane cutting through a solid.

Practice

- **Common Error:** Students may incorrectly describe the intersection because of the orientation of the drawing.

What Your Students Will Learn

- Describe the intersections of planes and solids.

Goal Today's lesson is describing **cross sections** of three-dimensional figures.

Technology for the Teacher

Dynamic Classroom

Lesson Tutorials
Lesson Plans
Answer Presentation Tool

Extra Example 1

Describe the intersection of the plane and the solid.

a. 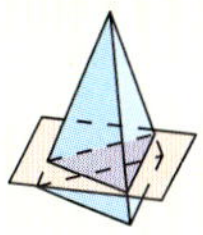

a triangle

b. 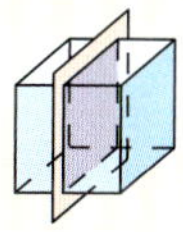

a square

c. 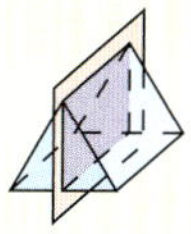

a rectangle

Practice

1. triangle
2. triangle
3. rectangle
4. rectangle
5. triangle
6. rectangle
7. The intersection is the shape of the base.

Record and Practice Journal Extension 14.5 Practice

1. triangle
2. rectangle
3. rectangle
4. rectangle
5. triangle
6. rectangle
7. triangle
8. circle
9. circle
10. rectangle
11. circle
12. circle

Extra Example 2

Describe the intersection of the plane and the solid.

a.

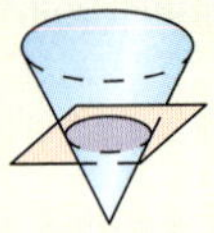

a circle

b.

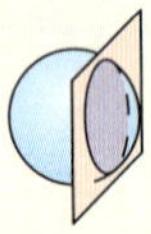

a circle

Practice

8. rectangle

9. circle

10. line segment

11. circle

12. circle

13. rectangle

14. circle

15. The intersection occurs at the vertex of the cone.

Mini-Assessment

Describe the intersection of the plane and the solid.

1.

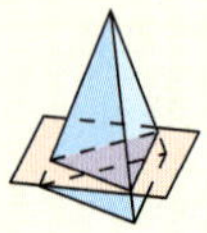

a triangle

2.

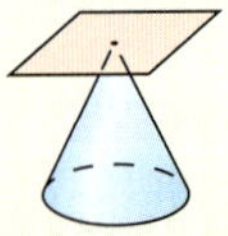

a point

3.

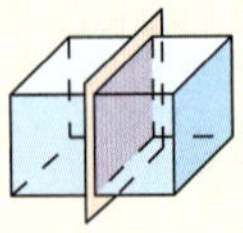

a rectangle

Laurie's Notes

Discuss

- Describe the features of a cone. Structurally it is the same as a pyramid. There is one base and a lateral surface with one vertex not on the base.

Example 2

- You can investigate these cross sections for a cylinder and cone using a flashlight as described in the Motivate activity. A rubber band stretched around the solid will also reveal the perimeter or circumference of the cross section.

Practice

- The intersections are shaded and should be clear to students.
- **Extension:** Ask students how the size of the cross section changes as the plane moves through the cylinder in Question 8.

Closure

- What are the possible cross sections of a cube? point, line, triangle, square, rectangle, parallelogram, quadrilateral, trapezoid, pentagon, hexagon

Example 1 shows how a plane intersects a polyhedron. Now consider the intersection of a plane and a solid having a curved surface, such as a cylinder or cone. As shown, a *cone* is a solid that has one circular base and one vertex.

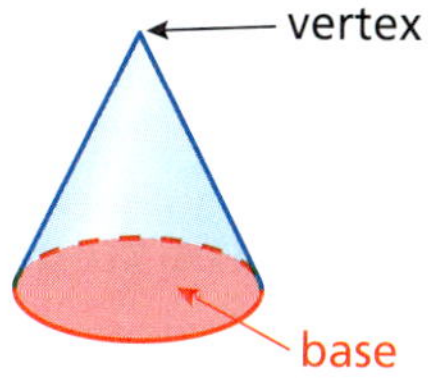

EXAMPLE 2 Describing the Intersection of a Plane and a Solid

Math Practice

Analyze Givens

What solid is shown? What are you trying to find? Explain.

Describe the intersection of the plane and the solid.

a.

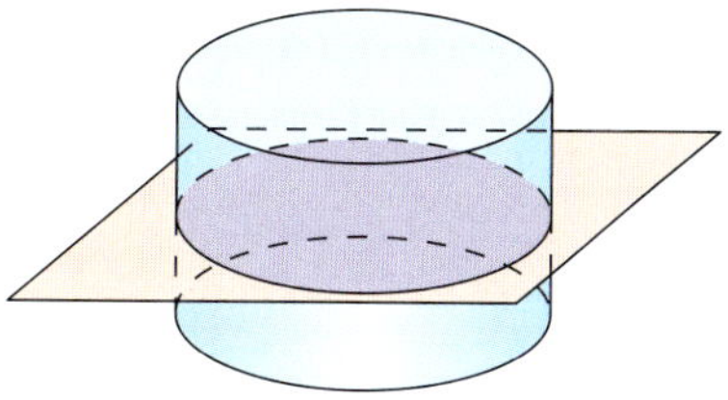

b.

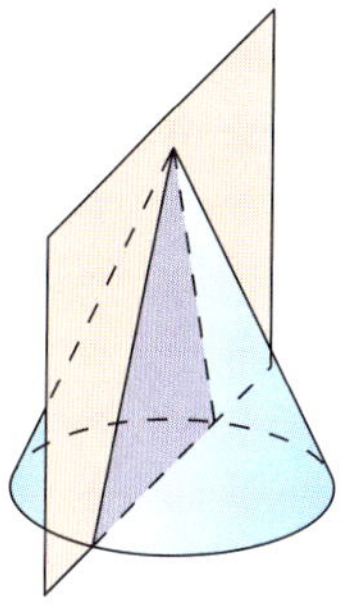

a. The intersection is a circle.

b. The intersection is a triangle.

Practice

Describe the intersection of the plane and the solid.

8.

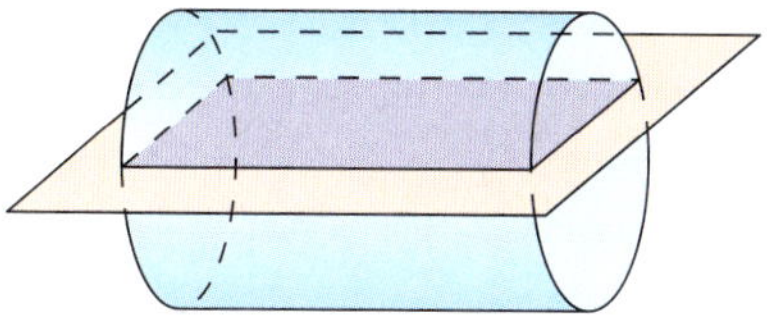

9.

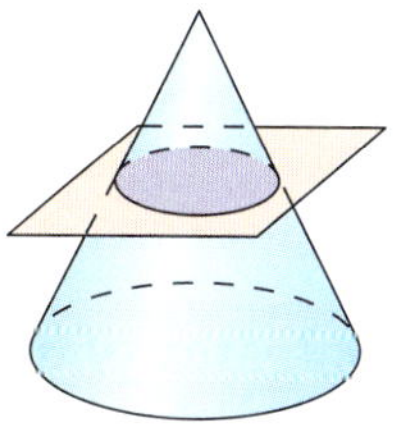

10.

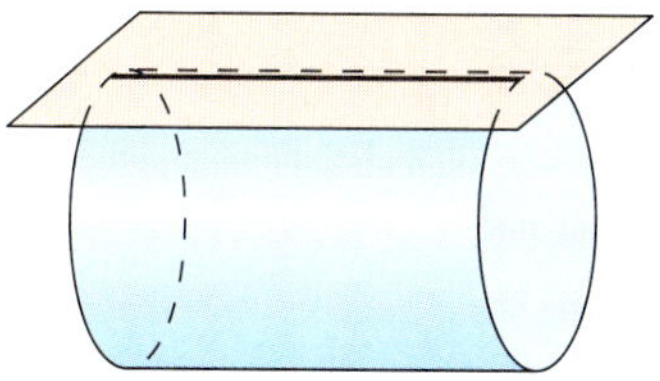

11.

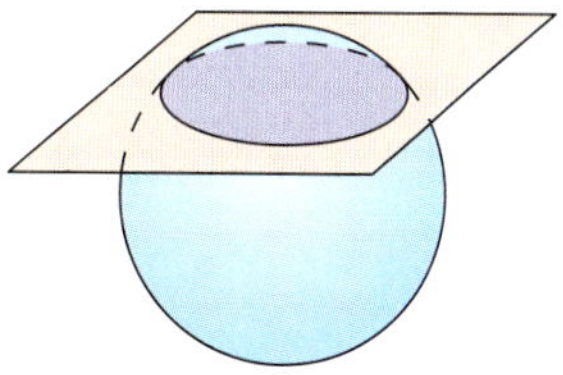

Describe the shape that is formed by the cut made in the food shown.

12.

13.

14.

15. **REASONING** Explain how a plane can be parallel to the base of a cone and intersect the cone at exactly one point.

14.4–14.5 Quiz

Check It Out
Progress Check
BigIdeasMath.com

Find the volume of the prism. *(Section 14.4)*

1.

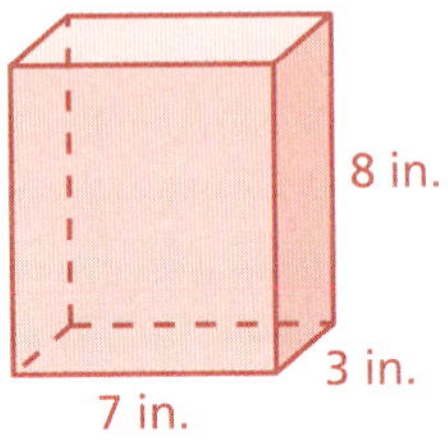

2.

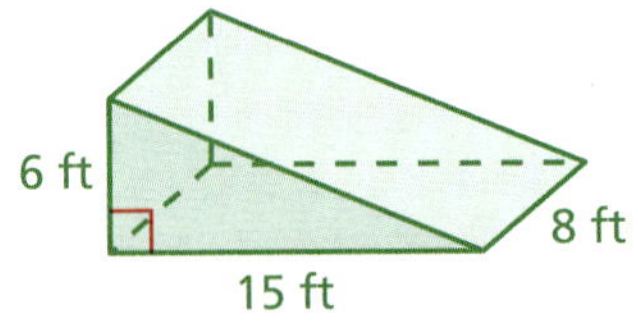

3.

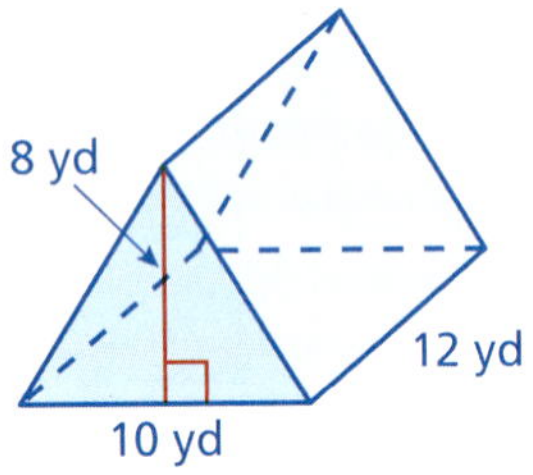

4.

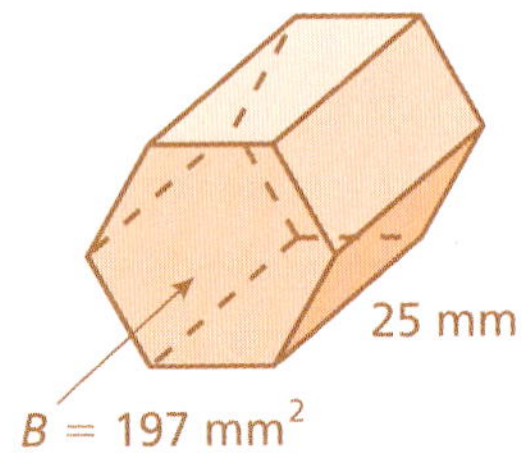

Find the volume of the solid. Round your answer to the nearest tenth if necessary. *(Section 14.5)*

5.

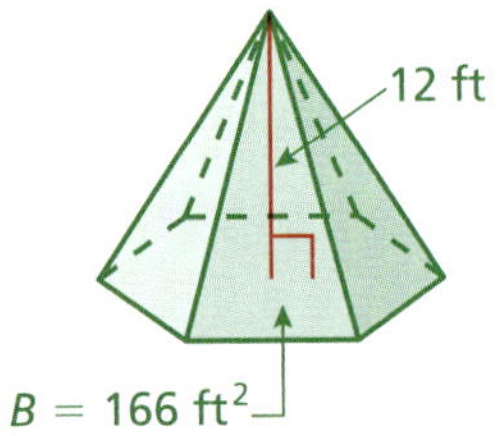

6.

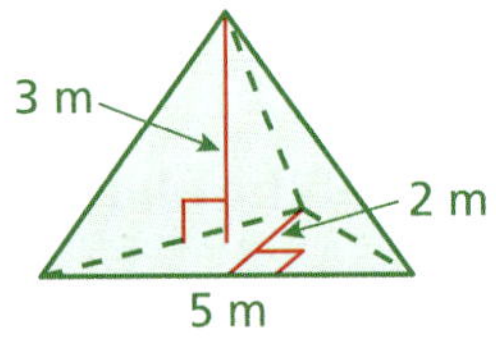

Describe the intersection of the plane and the solid. *(Section 14.5)*

7.

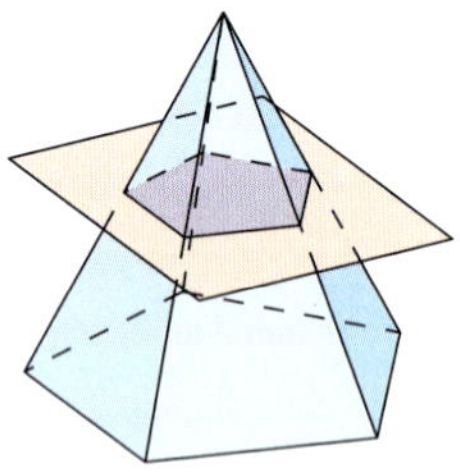

8.

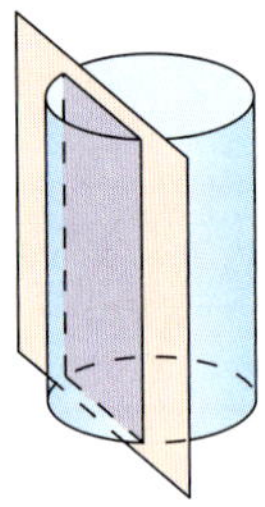

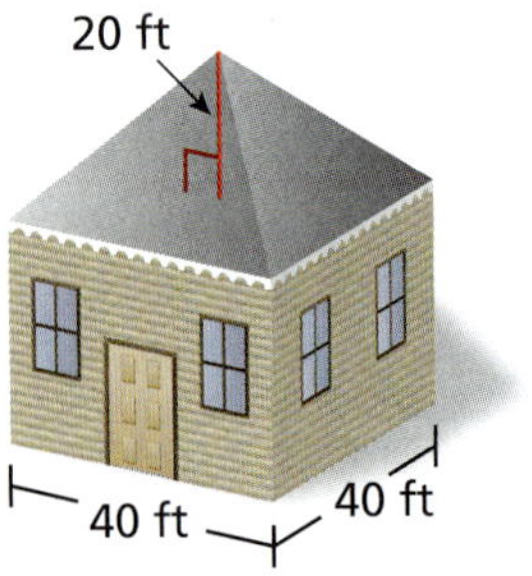

9. **ROOF** A pyramid hip roof is a good choice for a house in a hurricane area. What is the volume of the roof to the nearest tenth? *(Section 14.5)*

10. **CUBIC UNITS** How many cubic feet are in a cubic yard? Use a sketch to explain your reasoning. *(Section 14.4)*

Alternative Assessment Options

Math Chat | **Student Reflective Focus Question**
Structured Interview | Writing Prompt

Student Reflective Focus Question
Ask students to summarize the similarities and differences between volumes and cross sections of prisms and pyramids. Be sure that they include examples. Select students at random to present their summaries to the class.

Study Help Sample Answers

Remind students to complete Graphic Organizers for the rest of the chapter.

4.

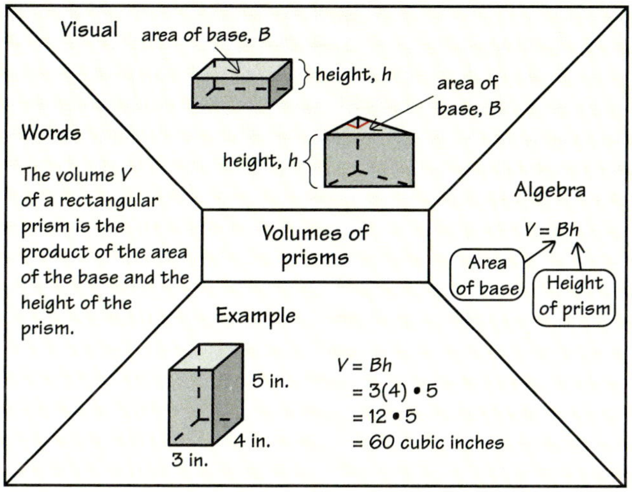

5.

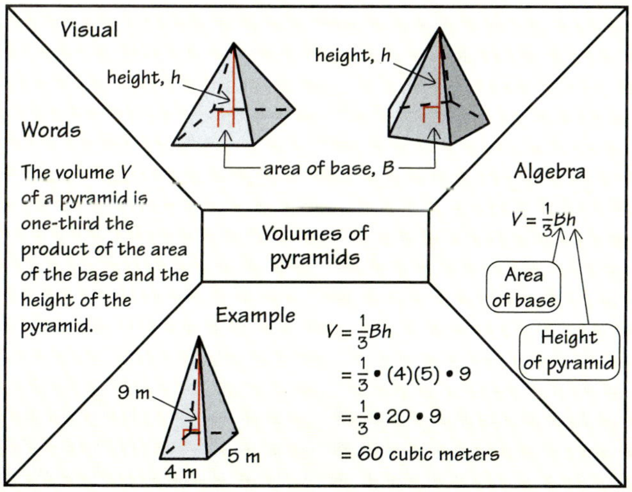

Reteaching and Enrichment Strategies

If students need help. . .	If students got it. . .
Resources by Chapter • Practice A and Practice B • Puzzle Time Lesson Tutorials *BigIdeasMath.com*	Resources by Chapter • Enrichment and Extension • Technology Connection Game Closet at *BigIdeasMath.com* Start the Chapter Review

Answers

1. 168 in.^{3}
2. 360 ft^3
3. 480 yd^3
4. 4925 mm^3
5. 664 ft^3
6. 5 m^3
7. pentagon
8. rectangle
9. 10,666.7 ft^3
10. 27 ft^3

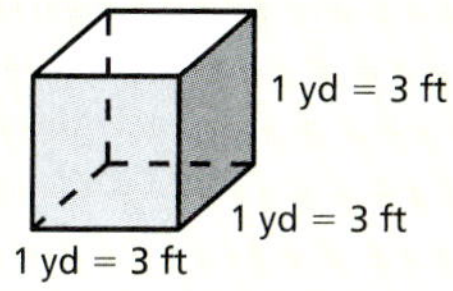

$3 \times 3 \times 3 = 27$ ft^3

Technology for the Teacher

Online Assessment
Assessment Book
ExamView® Assessment Suite

For the Teacher
Additional Review Options

- *BigIdeasMath.com*
- Online Assessment
- Game Closet at *BigIdeasMath.com*
- Vocabulary Help
- Resources by Chapter

Answers

1. 158 in.^2
2. 400 cm^2
3. 108 m^2

Review of Common Errors

Exercises 1–3

- Students may sum the areas of only 3 of the faces of the rectangular prism instead of all 6. Remind them that a rectangular prism has 6 faces.
- Students may try to use the formula for the surface area of a rectangular prism to find the surface area of a triangular prism. Show them that this will not work by comparing the nets of the two types of prisms.

Exercises 4–6

- Students may forget to add the area of the base when finding the surface area. Remind them that when asked to find the surface area, the base is included.

Exercises 4–6

- Students may add the wrong number of lateral face areas to the area of the base. Examine several different pyramids with different bases and ask if they can find a relationship between the number of sides of the base and the number of lateral faces. (They are the same.) Remind students that the number of sides on the base determines how many triangles make up the lateral surface area.

Exercises 7 and 8

- Students may add the area of only one base. Remind them that a cylinder has *two* bases.
- Students may double the radius instead of squaring it, or forget the correct order of operations when using the formula for the surface area of a cylinder. Remind them of the formula and remind them of the order of operations.

Exercises 10–15

- Students may write the units incorrectly, often writing square units instead of cubic units. Remind them that they are working in three dimensions, so the units are cubed.

14 Chapter Review

Check It Out
Vocabulary Help
BigIdeasMath.com

Review Key Vocabulary

lateral surface area, *p. 590*
regular pyramid, *p. 596*
slant height, *p. 596*
cross section, *p. 620*

Review Examples and Exercises

14.1 Surface Areas of Prisms (pp. 586–593)

Find the surface area of the prism.

Draw a net.

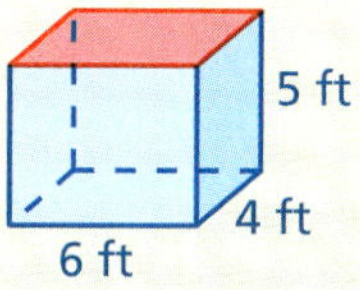

$$S = 2\ell w + 2\ell h + 2wh$$
$$= 2(6)(4) + 2(6)(5) + 2(4)(5)$$
$$= 48 + 60 + 40$$
$$= 148$$

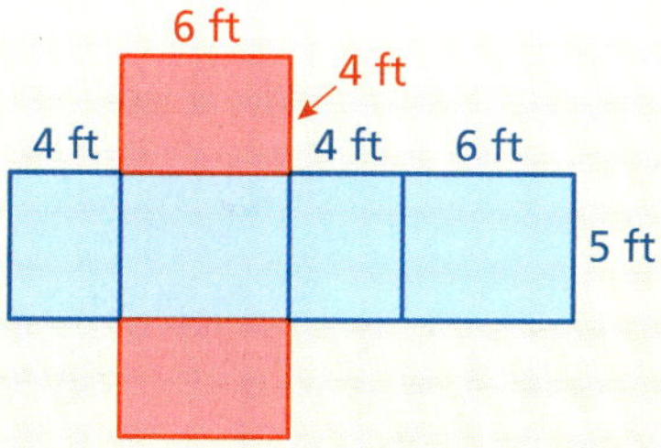

The surface area is 148 square feet.

Exercises

Find the surface area of the prism.

1. 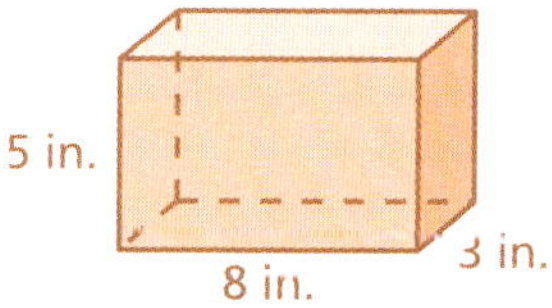

2. 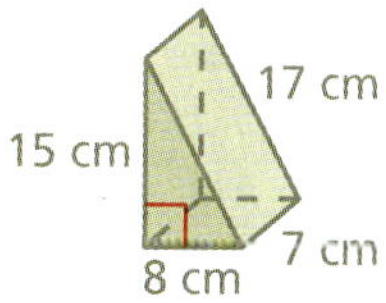

3. 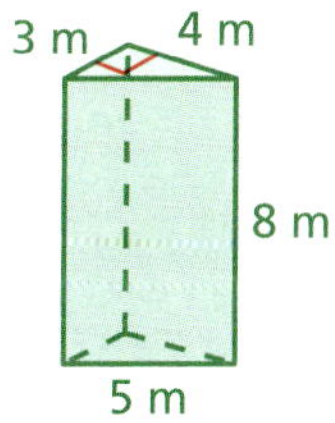

14.2 Surface Areas of Pyramids (pp. 594–599)

Find the surface area of the regular pyramid.

Draw a net.

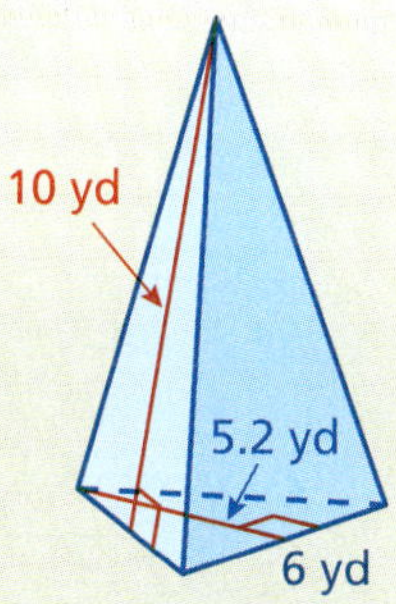

Area of Base

$\frac{1}{2} \cdot 6 \cdot 5.2 = 15.6$

Area of a Lateral Face

$\frac{1}{2} \cdot 6 \cdot 10 = 30$

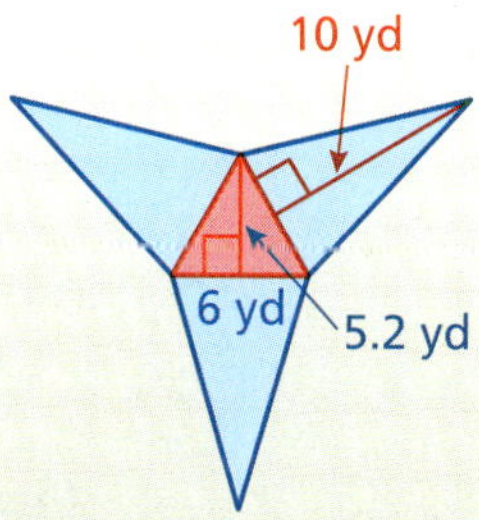

Find the sum of the areas of the base and all three lateral faces.

$$S = 15.6 + 30 + 30 + 30$$
$$= 105.6$$

There are 3 identical lateral faces. Count the area 3 times.

The surface area is 105.6 square yards.

Exercises

Find the surface area of the regular pyramid.

4.

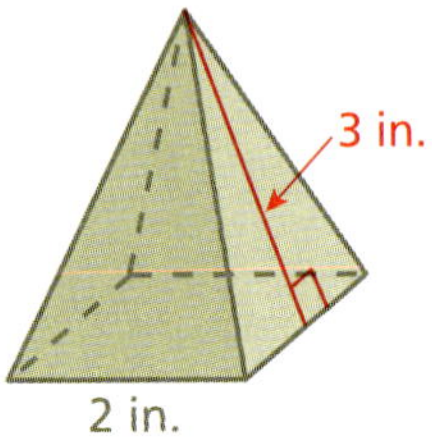

5.

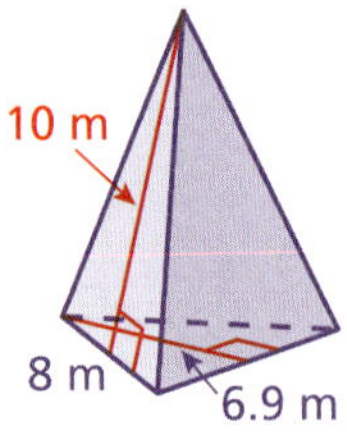

6.

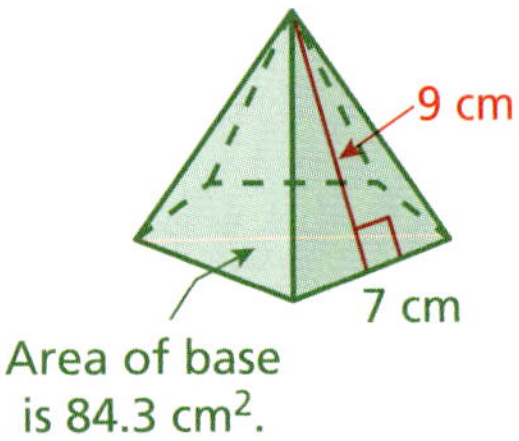

14.3 Surface Areas of Cylinders *(pp. 600–605)*

Find the surface area of the cylinder. Round your answer to the nearest tenth.

Draw a net.

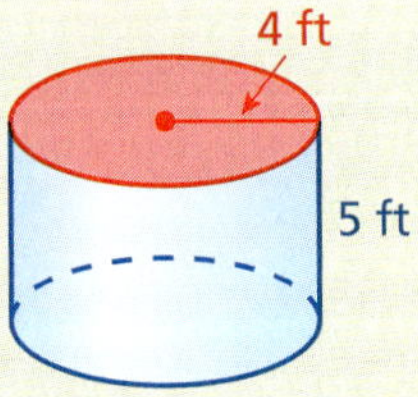

$$S = 2\pi r^2 + 2\pi rh$$
$$= 2\pi(4)^2 + 2\pi(4)(5)$$
$$= 32\pi + 40\pi$$
$$= 72\pi \approx 226.1$$

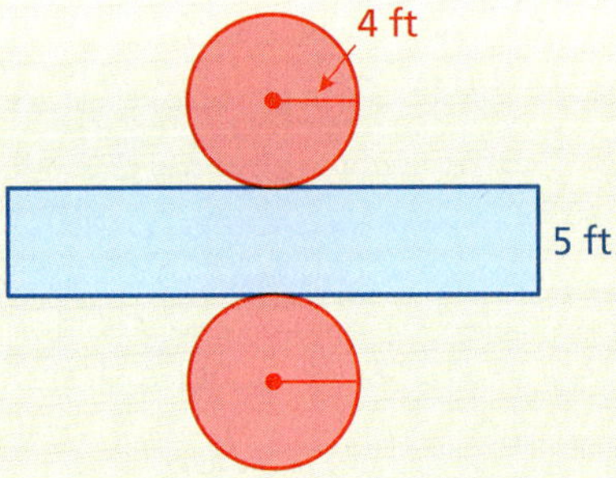

∴ The surface area is about 226.1 square millimeters.

Exercises

Find the surface area of the cylinder. Round your answer to the nearest tenth.

7.

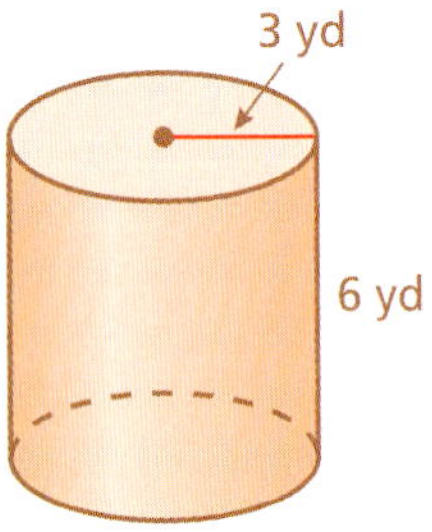

8.

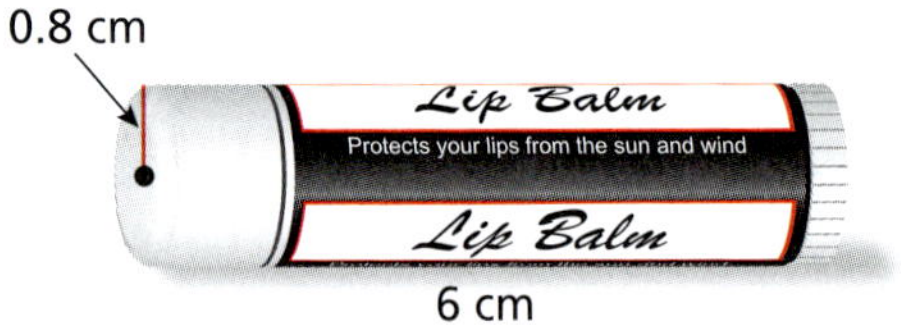

9. ORANGES Find the lateral surface area of the can of mandarin oranges.

14.4 Volumes of Prisms *(pp. 608–613)*

Find the volume of the prism.

$V = Bh$	Write formula for volume.
$= \frac{1}{2}(7)(3) \cdot 5$	Substitute.
$= 52.5$	Multiply.

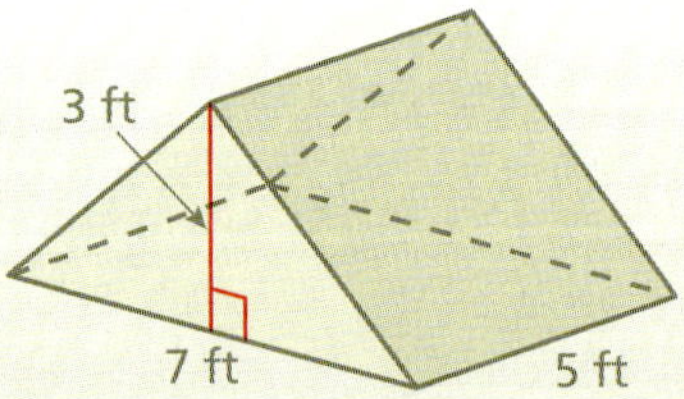

∴ The volume is 52.5 cubic feet.

Review Game

3-Dingo

Materials per Group:

- one 3-Dingo card*
- objects to cover the 3-Dingo card squares, such as bingo chips

Directions:

This activity is played like bingo. Divide the class into groups. Call out a three-dimensional figure and all of the dimensions necessary to calculate the surface area or volume. The group calculates the indicated surface area or volume. If that value is in a square on their card under the correct figure, they cover it. Keep calling out figures and their dimensions until a group wins.

Who Wins?

Just like bingo, the first group to get a row, column, or diagonal of covered squares wins. The winning team yells 3-Dingo!

*A 3-Dingo card has 5 columns of 5 squares and a three-dimensional figure (rectangular prism, triangular prism, square pyramid, triangular pyramid, and cylinder) with variable dimensions shown at the top of each column of squares. Different values for the surface area or volume of each figure are shown in each of the 5 squares below the figure. Different cards show the values in different orders, so no two 3-Dingo cards in a set are the same. A 3-Dingo card set is available at *BigIdeasMath.com*.

For the Student Additional Practice

- Lesson Tutorials
- Multi-Language Glossary
- Self-Grading Progress Check
- *BigIdeasMath.com*
  Dynamic Student Edition
  Student Resources

Answers

4. 16 in.^2
5. 147.6 m^2
6. 241.8 cm^2
7. $54\pi \approx 169.6 \text{ yd}^2$
8. $10.88\pi \approx 34.2 \text{ cm}^2$
9. $88\pi \approx 276.3 \text{ cm}^2$
10. 96 in.^3
11. 120 m^3
12. 607.5 mm^3
13. 850 ft^3
14. 2100 in.^3
15. 192 mm^3
16. rectangle
17. trianglc

My Thoughts on the Chapter

Teacher Tip

Not allowed to write in your teaching edition? Use sticky notes to record your thoughts.

What worked. . .

What did not work. . .

What I would do differently. . .

Exercises

Find the volume of the prism.

10.

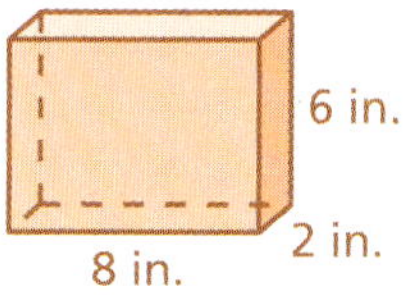

11.

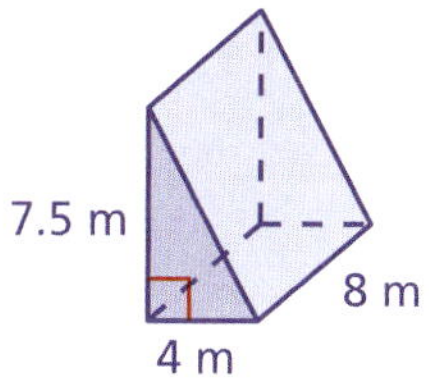

12.

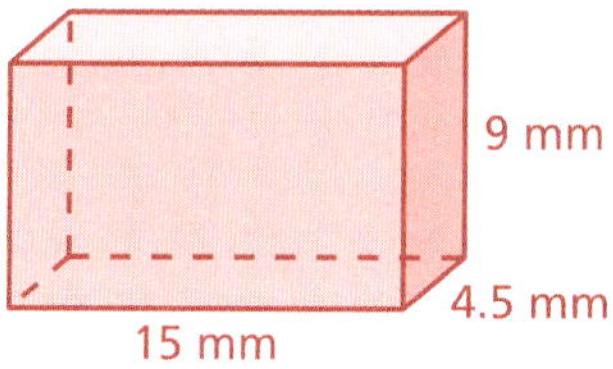

14.5 Volumes of Pyramids *(pp. 614–621)*

a. Find the volume of the pyramid.

$V = \frac{1}{3}Bh$	Write formula for volume.
$= \frac{1}{3}(6)(5)(10)$	Substitute.
$= 100$	Multiply.

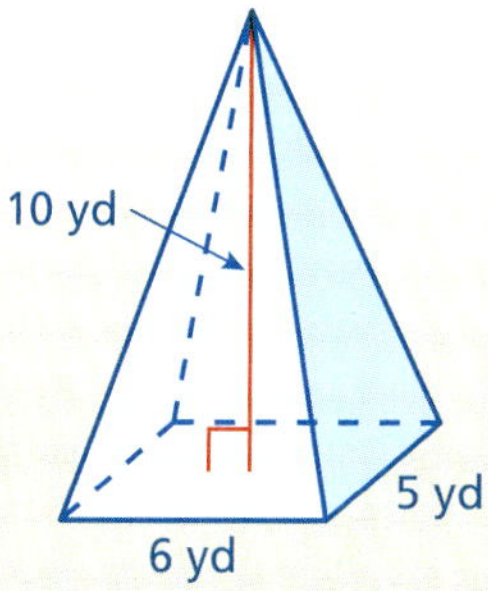

The volume is 100 cubic yards.

b. Describe the intersection of the plane and the solid.

i.

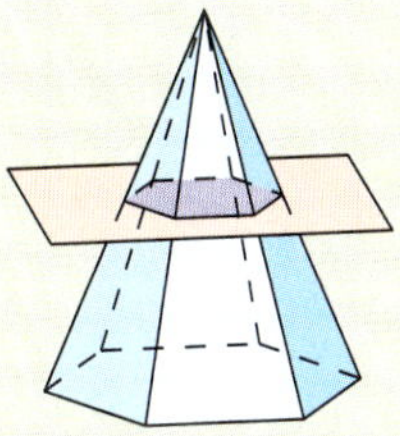

The intersection is a hexagon.

ii.

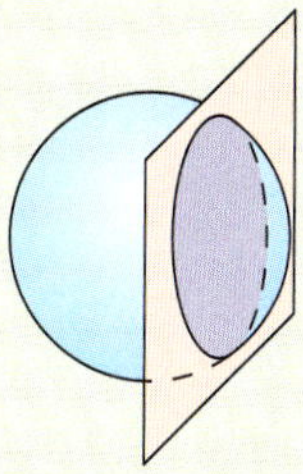

The intersection is a circle.

Exercises

Find the volume of the pyramid.

13.

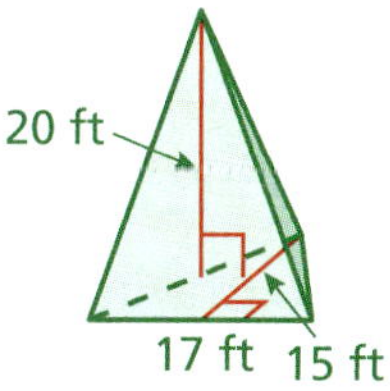

14.

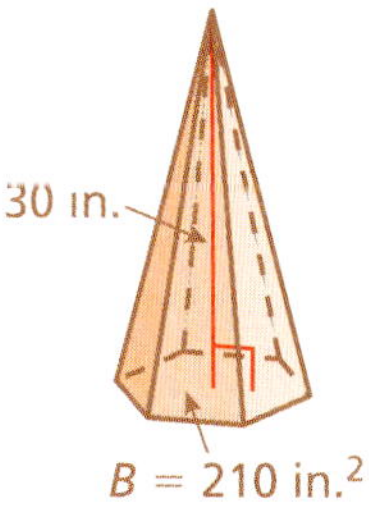

15.

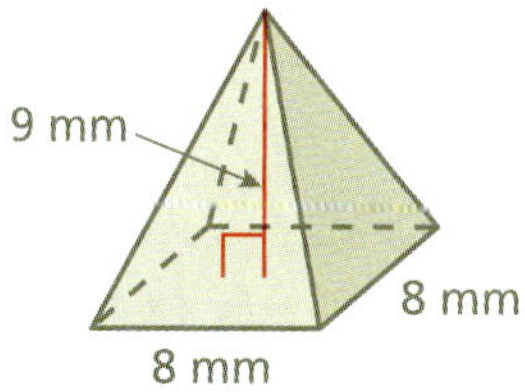

Describe the intersection of the plane and the solid.

16.

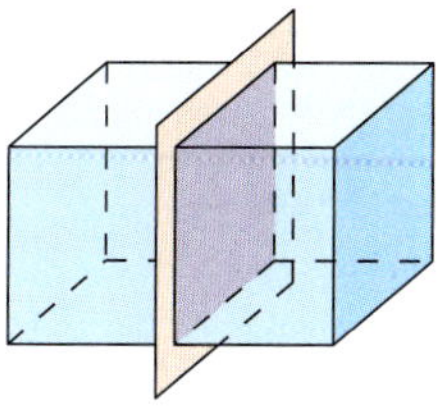

17.

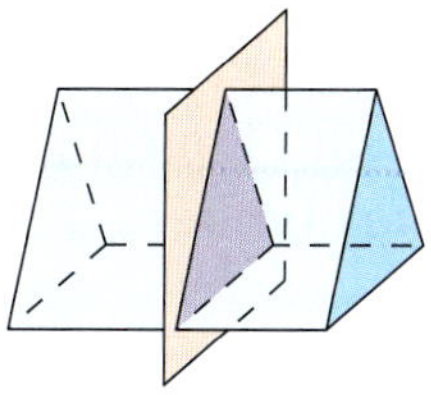

14 Chapter Test

Check It Out
Test Practice
BigIdeasMath.com

Find the surface area of the prism or regular pyramid.

1.

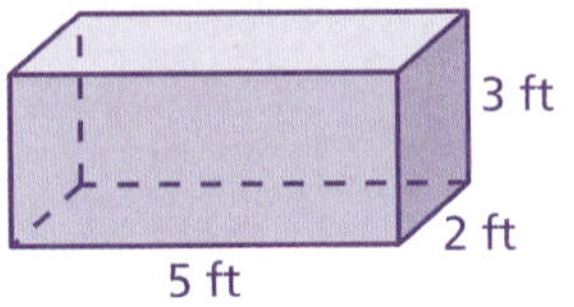

2.

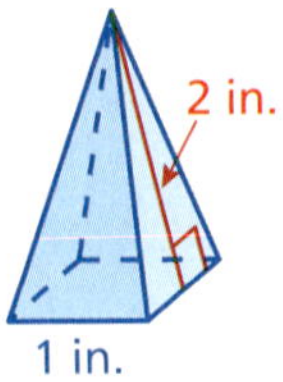

3.

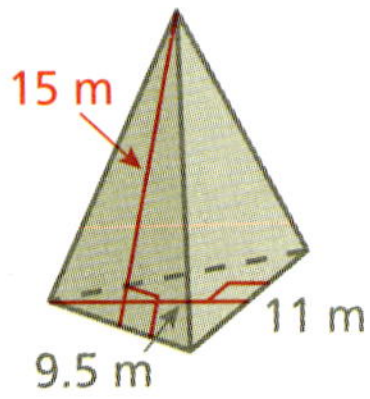

Find the surface area of the cylinder. Round your answer to the nearest tenth.

4.

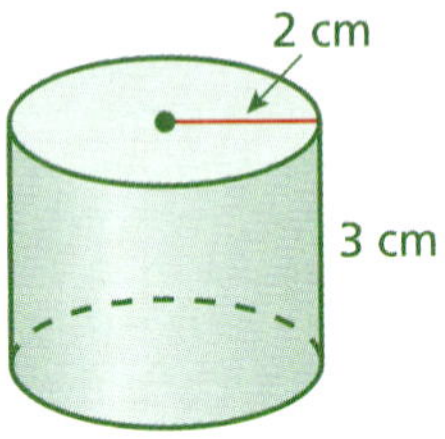

5.

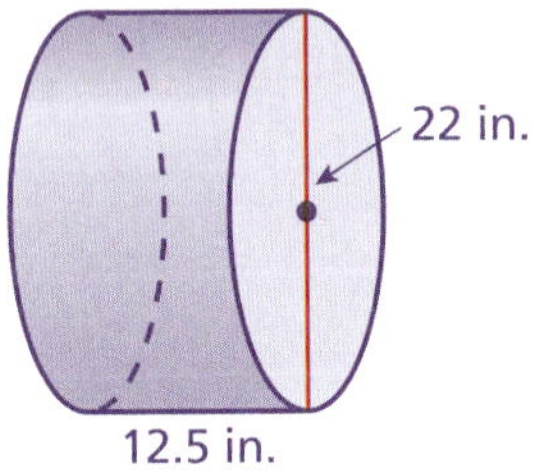

Find the volume of the solid.

6.

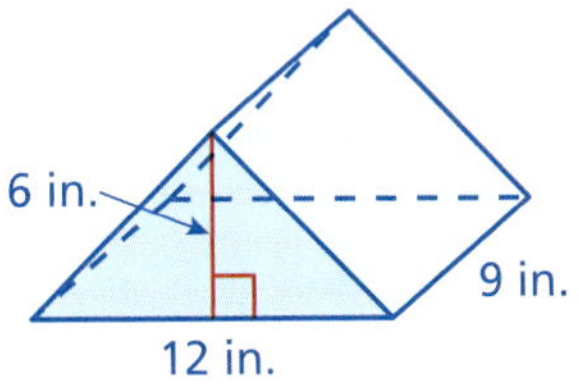

7.

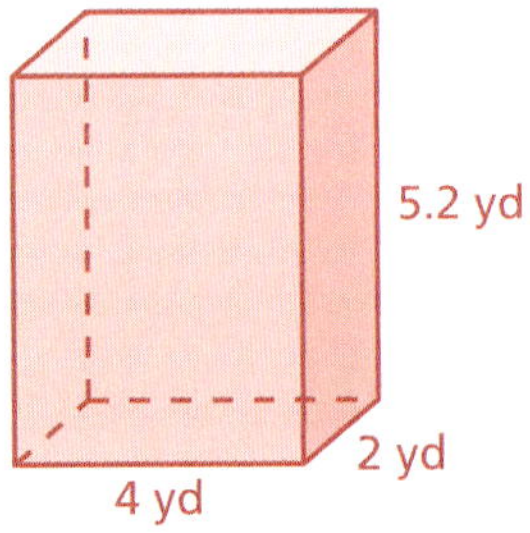

8.

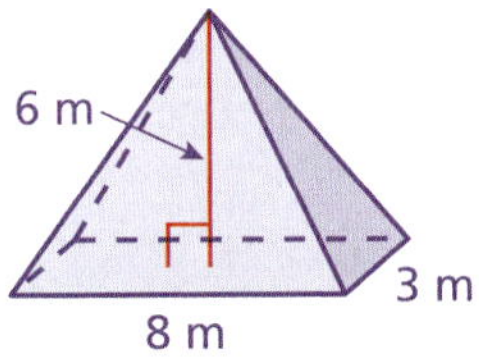

9. SKATEBOARD RAMP A quart of paint covers 80 square feet. How many quarts should you buy to paint the ramp with two coats? (Assume you will not paint the bottom of the ramp.)

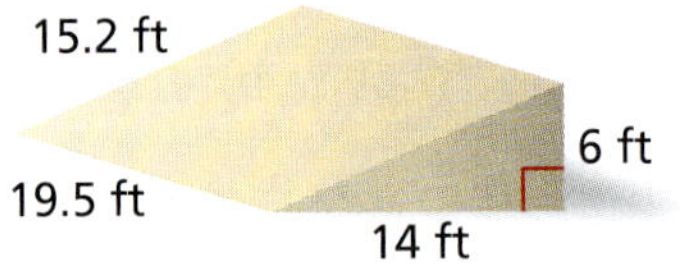

10. GRAHAM CRACKERS A manufacturer wants to double the volume of the graham cracker box. The manufacturer will either double the height or double the width.

a. Which option uses less cardboard? Justify your answer.

b. What is the volume of the new graham cracker box?

11. SOUP The label on the can of soup covers about 354.2 square centimeters. What is the height of the can? Round your answer to the nearest whole number.

Test Item References

Chapter Test Questions	Section to Review
1, 9, 10	14.1
2, 3	14.2
4, 5, 11	14.3
6, 7, 10	14.4
8	14.5

Test-Taking Strategies

Remind students to quickly look over the entire test before they start so that they can budget their time. This test is very visual and requires that students remember many terms. It might be helpful for them to jot down some of the terms on the back of the test before they start.

Common Errors

- **Exercise 1** Students may sum the areas of only 3 of the faces of the rectangular prism instead of all 6. Remind them that a rectangular prism has 6 faces.
- **Exercises 2 and 3** Students may add the wrong number of lateral face areas to the area of the base. Remind students that the number of sides on the base determines how many triangles make up the lateral surface area.
- **Exercises 4 and 5** Students may double the radius instead of squaring it or forget the correct order of operations when using the formula for the surface area of a cylinder. Remind them of the formula, and remind them of the order of operations.
- **Exercises 6–8** Students may write the units incorrectly, often writing square units instead of cubic units. Remind them that they are working in three dimensions, so the units are cubed.

Reteaching and Enrichment Strategies

If students need help. . .	If students got it. . .
Resources by Chapter • Practice A and Practice B • Puzzle Time Record and Practice Journal Practice Differentiating the Lesson Lesson Tutorials *BigIdeasMath.com* Skills Review Handbook	Resources by Chapter • Enrichment and Extension • Technology Connection Game Closet at *BigIdeasMath.com* Start Cumulative Assessment

Answers

1. 62 ft^2
2. 5 in.^2
3. 299.8 m^2
4. 62.8 cm^2
5. 1623.4 in.^2
6. 324 in.^3
7. 41.6 yd^3
8. 48 m^3
9. 13 quarts of paint
10. **a.** doubling the width

    b. 216 in.^3
11. 12 cm

Technology for the Teacher

Online Assessment
Assessment Book
ExamView® Assessment Suite

Test-Taking Strategies

Available at *BigIdeasMath.com*

After Answering Easy Questions, Relax

Answer Easy Questions First

Estimate the Answer

Read All Choices before Answering

Read Question before Answering

Solve Directly or Eliminate Choices

Solve Problem before Looking at Choices

Use Intelligent Guessing

Work Backwards

About this Strategy

When taking a multiple choice test, be sure to read each question carefully and thoroughly. After skimming the test and answering the easy questions, stop for a few seconds, take a deep breath, and relax. Work through the remaining questions carefully, using your knowledge and test-taking strategies. Remember, you already completed many of the questions on the test!

Answers

1. D
2. G
3. B

Technology for the Teacher

Performance Tasks

Online Assessment

Assessment Book

ExamView® Assessment Suite

Item Analysis

1. **A.** The student multiplies the length and height and adds the width.

   B. The student adds the areas of only three unique faces.

   C. The student multiplies the length, width, and height.

   D. Correct answer

2. **F.** The student finds what percent 60 is of 660.

   G. Correct answer

   H. The student finds 60% of 660 and misplaces the decimal point.

   I. The student thinks that the difference of the scores is equivalent to the percent.

3. **A.** The student incorrectly applies the Cross Products Property to get $3(x - 3) = 8(24)$

   B. Correct answer

   C. The student performs an order of operations error by not multiplying each side by 24 first.

   D. The student thinks that dividing both sides by 24 is the first step to isolating the variable on one side. The correct first step is to multiply both sides by 24.

14 Cumulative Assessment

Test-Taking Strategy
After Answering Easy Questions, Relax

"After answering the easy questions, relax and try the harder ones. For this, you know area is measured in square units."

1. A gift box and its dimensions are shown below.

What is the least amount of wrapping paper that you could have used to wrap the box?

A. 20 in.^{2} **C.** 64 in.^{2}

B. 56 in.^{2} **D.** 112 in.^{2}

2. A student scored 600 the first time she took the mathematics portion of her college entrance exam. The next time she took the exam, she scored 660. Her second score represents what percent increase over her first score?

F. 9.1% **H.** 39.6%

G. 10% **I.** 60%

3. Raj was solving the proportion in the box below.

$$\frac{3}{8} = \frac{x - 3}{24}$$

$$3 \cdot 24 = (x - 3) \cdot 8$$

$$72 = x - 24$$

$$96 = x$$

What should Raj do to correct the error that he made?

A. Set the product of the numerators equal to the product of the denominators.

B. Distribute 8 to get $8x - 24$.

C. Add 3 to each side to get $\frac{3}{8} + 3 = \frac{x}{24}$.

D. Divide both sides by 24 to get $\frac{3}{8} \div 24 = x - 3$.

4. A line contains the two points plotted in the coordinate plane below.

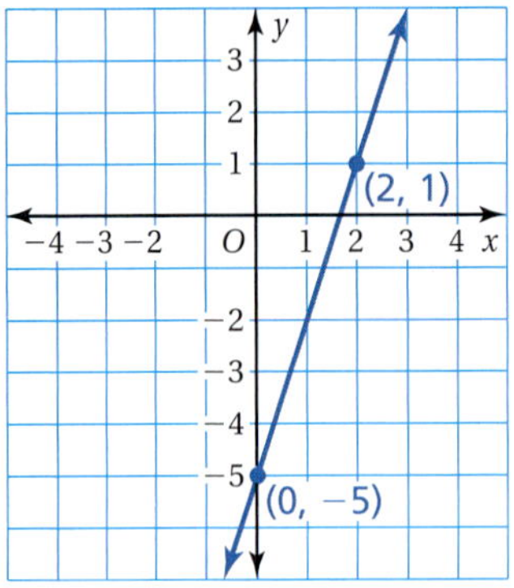

What is the slope of the line?

F. $\frac{1}{3}$

G. 2

H. 3

I. 6

5. James is getting ready for wrestling season. As part of his preparation, he plans to lose 5% of his body weight. James currently weighs 160 pounds. How much will he weigh, in pounds, after he loses 5% of his weight?

6. How much material is needed to make the popcorn container?

A. 76π in.^{2}

B. 84π in.^{2}

C. 92π in.^{2}

D. 108π in.^{2}

7. To make 10 servings of soup you need 4 cups of broth. You want to know how many servings you can make with 8 pints of broth. Which proportion should you use?

F. $\frac{10}{4} = \frac{x}{8}$

G. $\frac{4}{10} = \frac{x}{16}$

H. $\frac{10}{4} = \frac{8}{x}$

I. $\frac{10}{4} = \frac{x}{16}$

Item Analysis (continued)

4. **F.** The student finds the change in *x* divided by the change in *y*.

G. The student makes an error in subtracting the integers to find the change in *y*.

H. Correct answer

I. The student finds the change in *y* but forgets to divide by the change in *x*.

5. **Gridded Response:** Correct answer: 152 lb

Common Error: The student finds only the loss, getting an answer of 8.

6. **A.** The student only calculates the lateral surface area. They do not include the area of the base.

B. The student doubles the radius instead of squaring it when finding the area of the base.

C. Correct answer

D. The student includes the area of both bases instead of just one base.

7. **F.** The student does not convert to the same units (either 8 pints to 16 cups or 4 cups to 2 pints).

G. The student uses the correct numbers but inverts one of the ratios.

H. The student does not convert to the same units (either 8 pints to 16 cups or 4 cups to 2 pints) and inverts one of the ratios.

I. Correct answer

Answers

4. H

5. 152 lb

6. C

7. I

Answers

8. 648 $in.^3$

9. A

10. H

11. *Part A*

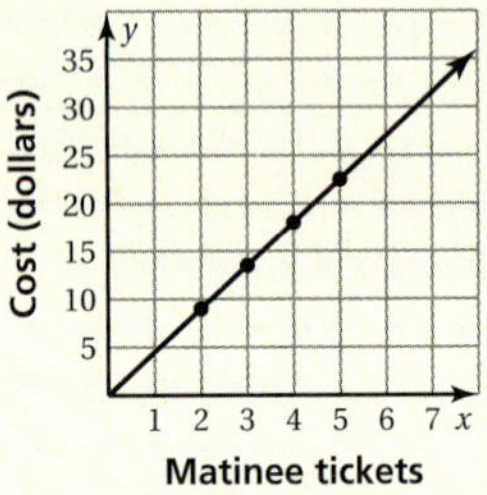

Part B 4.5; the cost of 1 movie ticket is \$4.50.

Part C \$36

Item Analysis (continued)

8. Gridded Response: Correct answer: 648 $in.^3$

Common Error: The student does not read the entire question and gets an answer of 24 cubic inches.

9. **A.** Correct answer

B. The student solves $2x + 4 = 90$.

C. The student subtracts 46 from 90.

D. The student solves $2x + 4 + 46 = 180$.

10. **F.** The student does not realize that the sum of the angle measures cannot be less than 180°.

G. The student does not realize that the sum of the angle measures cannot be greater than 180°.

H. Correct answer

I. The student does not realize that a triangle cannot have an angle with a measure of 0°.

11. **2 points** The student demonstrates a thorough understanding of graphing data points and finding the slope. In Part A, the student correctly plots the points and graphs the function $y = 4.5x$, for $x \geq 0$. In Part B, the student correctly determines that the slope is 4.5 and that it costs \$4.50 per movie ticket. In Part C, the student correctly determines that it costs \$36 to buy 8 movie tickets. The student provides clear and complete work and explanations.

1 point The student demonstrates a partial understanding of graphing data points and finding the slope. The student provides some correct work and explanation.

0 points The student demonstrates insufficient understanding of graphing data points and finding the slope. The student is unable to make any meaningful progress toward a correct answer.

8. A rectangular prism and its dimensions are shown below.

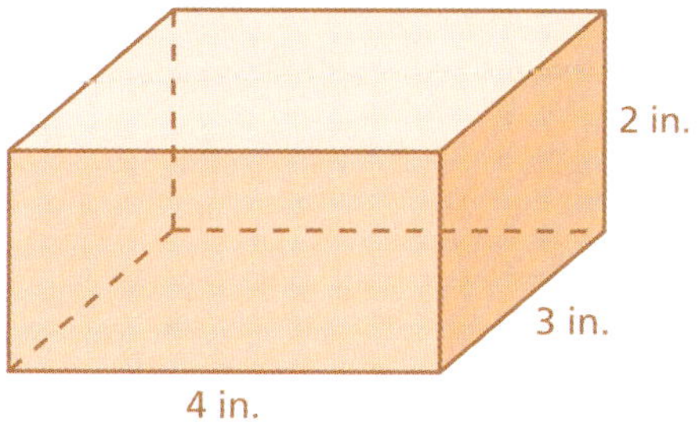

What is the volume, in cubic inches, of a rectangular prism whose dimensions are three times greater?

9. What is the value of x?

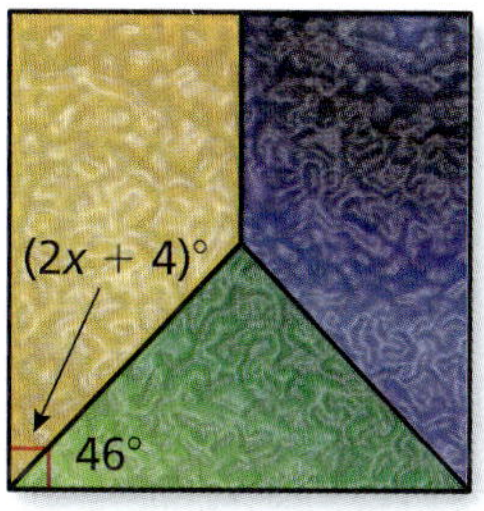

A. 20

B. 43

C. 44

D. 65

10. Which of the following could be the angle measures of a triangle?

F. 60°, 50°, 20°

G. 40°, 80°, 90°

H. 30°, 60°, 90°

I. 0°, 90°, 90°

11. The table below shows the costs of buying matinee movie tickets.

Think Solve Explain

Matinee Tickets, x	2	3	4	5
Cost, y	\$9	\$13.50	\$18	\$22.50

Part A Graph the data.

Part B Find and interpret the slope of the line through the points.

Part C How much does it cost to buy 8 matinee movie tickets?

15 Probability and Statistics

15.1 Outcomes and Events

15.2 Probability

15.3 Experimental and Theoretical Probability

15.4 Compound Events

15.5 Independent and Dependent Events

15.6 Samples and Populations

15.7 Comparing Populations

What Your Students Have Learned

- Make a line plot with data in fractions of a unit.
- Understand that a measure of center summarizes all of the values in a data set with a single number.
- Understand that a measure of variation summarizes how all of the values in a data set vary with a single number.
- Display data on a number line in dot plots and box-and-whisker plots.
- Choose measures of center and variation based on shape.

What Your Students Will Learn

- Understand representative samples (random sampling) and populations.
- Use samples to draw inferences about populations.
- Compare two populations from random samples using measures of center and variability.
- Understand that probability is the likelihood of an event occurring, expressed as a number from 0 to 1.
- Use simulations to find experimental probabilities.
- Develop probability models and use them to find probabilities.
- Find the probabilities of compound events.

Pacing Guide for Chapter 15

Chapter Opener Advanced	1 Day
Section 1 Advanced	1 Day
Section 2 Advanced	1 Day
Section 3 Advanced	1 Day
Section 4 Advanced	1 Day
Section 5 Advanced	2 Days
Study Help / Quiz Advanced	1 Day
Section 6 Advanced	2 Days
Section 7 Advanced	1 Day
Chapter Review/ Chapter Tests Advanced	2 Days
Total Chapter 15 Advanced	13 Days
Year-to-Date Advanced	142 Days

Technology for the Teacher

BigIdeasMath.com
Chapter at a Glance
Complete Materials List
Parent Letters: English and Spanish

What Your Students Have Learned

- Write ratios in simplest form.

Additional Topics for Review

- Multiplying Fractions
- Converting Between Fractions, Decimals, and Percents
- Box-and-Whisker Plots
- Dot Plots
- Stem-and-Leaf Plots

Try It Yourself

1. 1 : 3 **2.** 3 : 4
3. 1 : 2 **4.** 1 : 3
5. 1 : 2 **6.** 4 : 3
7. 1 : 5

Record and Practice Journal Fair Game Review

1. 2 : 3 **2.** 5 : 3
3. 1 : 2 **4.** 1 : 4
5. 3 : 20 **6.** 2 : 3
7. 1 : 5 **8.** 7 : 17
9. 2 : 3 **10.** 3 : 7

Math Background Notes

Vocabulary Review

- Fraction
- Simplest Form
- Ratio

Writing Ratios

- Students should know how to write ratios.
- Remind students that a ratio is a comparison between two quantities.
- Ratios can be written in three different ways. They can be expressed as a fraction, using a colon, or using the word "to."

  Example: $\frac{3}{4}$, 3 : 4, 3 to 4
- **Common Error:** Order matters when writing ratios. A ratio of 3 : 4 carries a different meaning than a ratio of 4 : 3. Remind students to write the ratio in the same order that the problem asks for it.
- It is best to express the final ratio in simplest form. Writing a ratio in simplest form is similar to simplifying fractions.

Reteaching and Enrichment Strategies

If students need help. . .	If students got it. . .
Record and Practice Journal • Fair Game Review Skills Review Handbook Lesson Tutorials	Game Closet at *BigIdeasMath.com* Start the next section

What You Learned Before

"Let's spin to decide what we will have for lunch."

Writing Ratios

Example 1 There are 32 football players and 16 cheerleaders at your school. Write the ratio of cheerleaders to football players.

cheerleaders → 16; football players → 32

$$\frac{16}{32} = \frac{1}{2}$$ Write in simplest form.

So, the ratio of cheerleaders to football players is $\frac{1}{2}$.

Example 2

a. Write the ratio of girls to boys in Classroom A.

$$\frac{\text{Girls in Classroom A}}{\text{Boys in Classroom A}} = \frac{11}{14}$$

	Boys	Girls
Classroom A	14	11
Classroom B	12	8

So, the ratio of girls to boys in Classroom A is $\frac{11}{14}$.

b. Write the ratio of boys in Classroom B to the total number of students in both classes.

$$\frac{\text{Boys in Classroom B}}{\text{Total number of students}} = \frac{12}{14 + 11 + 12 + 8} = \frac{12}{45} = \frac{4}{15}$$ Write in simplest form.

So, the ratio of boys in Classroom B to the total number of students is $\frac{4}{15}$.

Try It Yourself

Write the ratio in simplest form.

1. baseballs to footballs
2. footballs to total pieces of equipment
3. sneakers to ballet slippers
4. sneakers to total number of shoes

5. green beads to blue beads
6. red beads : green beads
7. green beads : total number of beads

15.1 Outcomes and Events

Essential Question In an experiment, how can you determine the number of possible results?

An *experiment* is an investigation or a procedure that has varying results. Flipping a coin, rolling a number cube, and spinning a spinner are all examples of experiments.

1 ACTIVITY: Conducting Experiments

Work with a partner.

a. You flip a dime.

There are ____ possible results.

Out of 20 flips, you think you will flip heads ____ times.

Flip a dime 20 times. Tally your results in a table. How close was your guess?

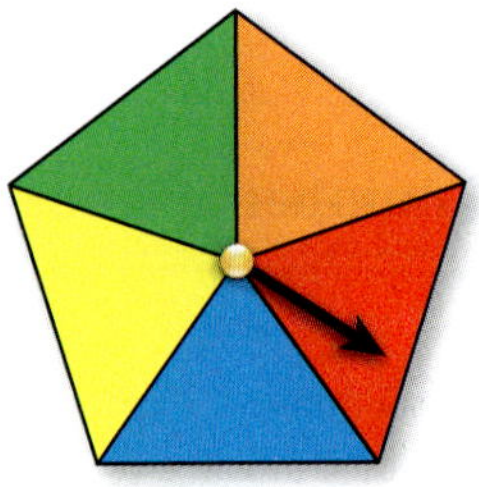

b. You spin the spinner shown.

There are ____ possible results.

Out of 20 spins, you think you will spin orange ____ times.

Spin the spinner 20 times. Tally your results in a table. How close was your guess?

c. You spin the spinner shown.

There are ____ possible results.

Out of 20 spins, you think you will spin a 4 ____ times.

Spin the spinner 20 times. Tally your results in a table. How close was your guess?

Probability and Statistics
In this lesson, you will
- identify and count the outcomes of experiments.

2 ACTIVITY: Comparing Different Results

Work with a partner. Use the spinner in Activity 1(c).

a. Do you have a better chance of spinning an even number or a multiple of 4? Explain your reasoning.

b. Do you have a better chance of spinning an even number or an odd number? Explain your reasoning.

Laurie's Notes

Introduction

Applying Mathematical Practices

- **Model with Mathematics:** Students gain a conceptual sense of probability by performing these activities. The concept of *possible outcomes* is also developed.

Motivate

? "How many of you have ever played Rock Paper Scissors?

- Different versions of Rock Paper Scissors are played in different countries. In Japan it is called *Jan-ken-pon.* In Indonesia the objects are elephant (one thumb up out of a clapped hand), person (showing one index finger), and ant (showing one little finger). The elephant beats the person. The person beats the ant. The ant beats the elephant, because if an ant gets into the elephant's ear, the elephant cannot do anything about the itchiness!
- Students will be eager to play with their partners. They will play the game as part of today's activity.

Activity Notes

Activity 1

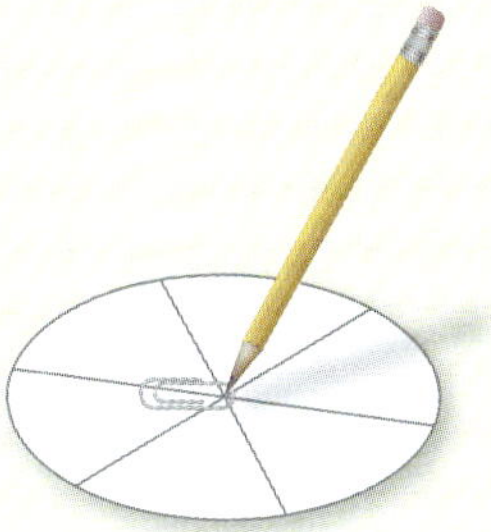

- It is important to have manipulatives for this activity. If dimes are not available, substitute pennies or two-colored counters. You can make spinners using a paper clip.
- Remind students to make a prediction about the outcomes before they perform the experiment.
- If time is short, one partner can record while the other performs the experiment, or have enough materials so that each partner can do the experiment at the same time.
- When students are finished, try to gather class data quickly for one outcome in each part of the activity. Below is an example.

Result	Total number of repetitions	Total number of occurrences
Flipping heads	20 × Number of students	?
Spinning red	20 × Number of students	?
Spinning 4	20 × Number of students	?

- Ask for volunteers to explain how they came up with their guesses in each part. This will help students prepare for Activity 2.

Activity 2

- **Construct Viable Arguments and Critique the Reasoning of Others:** Ask for volunteers to share their explanations with the class.
- **Extension:** Redraw the spinner in part (c) so the numbers are not consecutive. For example, the even numbers are consecutive and the odd numbers are consecutive. Ask students if this new spinner changes how they think about the problem. Some students may say that a particular result is more likely because the numbers are clustered together.

What Your Students Will Learn

- Identify and count the outcomes of experiments.

Previous Learning

Students should be familiar with organizing results of an experiment in a table.

Technology for the Teacher

Dynamic Classroom

Lesson Plans
Complete Materials List

15.1 Record and Practice Journal

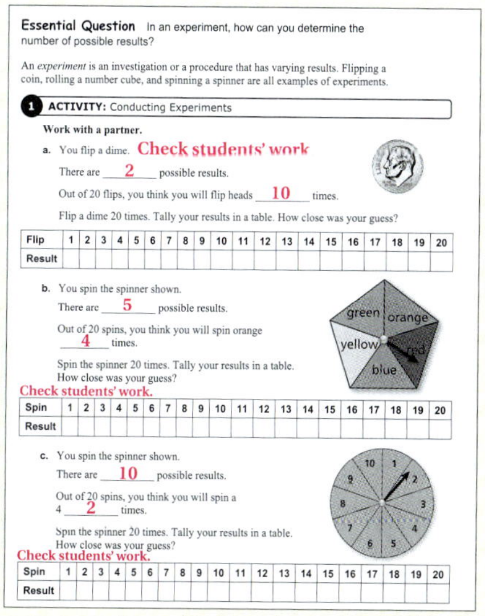

Essential Question In an experiment, how can you determine the number of possible results?

An *experiment* is an investigation or a procedure that has varying results. Flipping a coin, rolling a number cube, and spinning a spinner are all examples of experiments.

1 ACTIVITY: Conducting Experiments

Work with a partner.

a. You flip a dime. Check students' work

There are 2 possible results.

Out of 20 flips, you think you will flip heads 10 times.

Flip a dime 20 times. Tally your results in a table. How close was your guess?

Flip	1	2	3	4	5	6	7	8	9	10	11	12	13	14	15	16	17	18	19	20
Result																				

b. You spin the spinner shown.

There are 5 possible results.

Out of 20 spins, you think you will spin orange 4 times.

Spin the spinner 20 times. Tally your results in a table. How close was your guess?

Check students' work.

Spin	1	2	3	4	5	6	7	8	9	10	11	12	13	14	15	16	17	18	19	20
Result																				

c. You spin the spinner shown.

There are 10 possible results.

Out of 20 spins, you think you will spin a 4 2 times.

Spin the spinner 20 times. Tally your results in a table. How close was your guess?

Check students' work.

Spin	1	2	3	4	5	6	7	8	9	10	11	12	13	14	15	16	17	18	19	20
Result																				

Differentiated Instruction

Auditory

Discuss with students what it means to say that something happens by chance. Ask students to describe situations when a result is determined by chance, such as rolling a number cube or playing Rock Paper Scissors. Then ask students to describe an event that is impossible and an event that is certain.

15.1 Record and Practice Journal

2 ACTIVITY: Comparing Different Results

Work with a partner. Use the spinner in Activity 1(c).

a. Do you have a better chance of spinning an even number or a multiple of 4? Explain your reasoning.
even number

b. Do you have a better chance of spinning an even number or an odd number? Explain your reasoning.
neither

3 ACTIVITY: Rock Paper Scissors

Work with a partner. Check students' work.

a. Play Rock Paper Scissors 30 times. Tally your results in the table.

Rock *breaks* scissors.
Paper *covers* rock.
Scissors *cut* paper.

		Player A		
		Rock	Paper	Scissors
Player B	Rock			
	Paper			
	Scissors			

b. How many possible results are there?
9

c. Of the possible results, in how many ways can Player A win? Player B win? the players tie?
3 ways; 3 ways; 3 ways

d. Does one of the players have a better chance of winning than the other player? Explain your reasoning.
No, both players have 3 out of 9 possible ways to win.

What Is Your Answer?

4. **IN YOUR OWN WORDS** In an experiment, how can you determine the number of possible results?
Sample answer: Make a list or table that shows all possible results.

Laurie's Notes

Activity 3

- Review with students the rules for playing Rock Paper Scissors.
- Recording the results means making a tally mark for each game played.
- Students will get caught up in playing and may forget to record the results. You may want to modify the activity and have groups of 3 where the third person is the recorder.
- When students have finished, discuss student results as a whole class.

What Is Your Answer?

- **Neighbor Check:** Have students work independently and then have their neighbors check their work. Have students discuss any discrepancies.

Closure

- **Exit Ticket:** If you played a game similar to Rock Paper Scissors that had a fourth object such as a glove, how many possible results do you think there would be? Explain. 16; Each person has 4 possible results. So, two people have $4 \times 4 = 16$ possible results.

3 ACTIVITY: Rock Paper Scissors

Work with a partner.

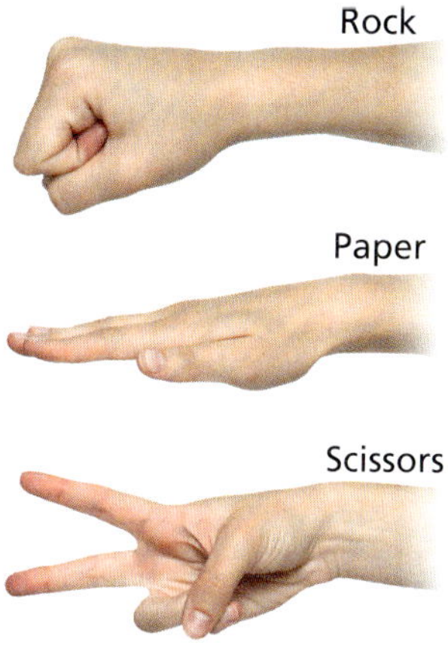

a. Play Rock Paper Scissors 30 times. Tally your results in the table.

b. How many possible results are there?

c. Of the possible results, in how many ways can Player A win? Player B win? the players tie?

d. Does one of the players have a better chance of winning than the other player? Explain your reasoning.

> **Math Practice**
>
> **Interpret a Solution**
>
> How do your results compare to the possible results? Explain.

> Rock *breaks* scissors.
> Paper *covers* rock.
> Scissors *cut* paper.

		Player A		
		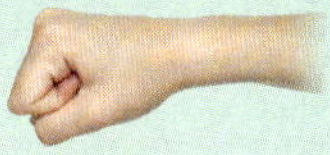		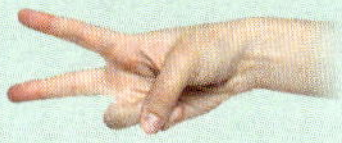
Player B	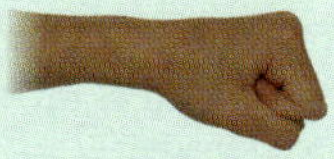			

What Is Your Answer?

4. **IN YOUR OWN WORDS** In an experiment, how can you determine the number of possible results?

Practice Use what you learned about experiments to complete Exercises 3 and 4 on page 636.

15.1 Lesson

Check It Out
Lesson Tutorials
BigIdeasMath.com

Key Vocabulary
experiment, *p. 634*
outcomes, *p. 634*
event, *p. 634*
favorable outcomes, *p. 634*

Key Ideas

Outcomes and Events

An **experiment** is an investigation or a procedure that has varying results. The possible results of an experiment are called **outcomes**. A collection of one or more outcomes is an **event**. The outcomes of a specific event are called **favorable outcomes**.

For example, randomly selecting a marble from a group of marbles is an experiment. Each marble in the group is an outcome. Selecting a green marble from the group is an event.

Possible outcomes

Event: Choosing a green marble

Number of favorable outcomes: 2

Reading

When an experiment is performed *at random* or *randomly*, all of the possible outcomes are equally likely.

EXAMPLE 1 Identifying Outcomes

You roll the number cube.

a. What are the possible outcomes?

The six possible outcomes are rolling a 1, 2, 3, 4, 5, and 6.

b. What are the favorable outcomes of rolling an even number?

even	*not* even
2, 4, 6	1, 3, 5

The favorable outcomes of the event are rolling a 2, 4, and 6.

c. What are the favorable outcomes of rolling a number greater than 5?

greater than 5	*not* greater than 5
6	1, 2, 3, 4, 5

The favorable outcome of the event is rolling a 6.

Multi-Language Glossary at BigIdeasMath.com

Laurie's Notes

Introduction

Connect

- **Yesterday:** Students counted the number of possible outcomes of an experiment.
- **Today:** Students will describe the outcomes of an experiment.

Motivate

- Give students a chance to have no homework. Hold a standard deck of cards and tell them one student is going to select a card from the deck. If it is a 7, there is no homework. If it is not a 7, there is homework.
- Students immediately say it is not fair. Act indignant, and suggest that if they do not draw a card, then there definitely will be homework. This will change their attitude.
- If you are willing to accept the outcome, let one student draw a card. The *odds* are in your favor!

Lesson Notes

Key Idea

- Discuss the vocabulary words: experiment, outcomes, event, and favorable outcomes. You can relate the vocabulary to the various activities yesterday and to tossing two number cubes.
- ? Ask students to identify the favorable outcomes for the events of choosing each color of marble. green (2), blue (1), red (1), yellow (1), purple (1)
- Discuss the Study Tip. This tip provides the statistical meaning. Go over the "meaning of the word" in everyday life; *having no specific pattern, purpose, or objective.*

Example 1

- Make sure that students understand that there can be more than one favorable outcome.
- ? "What are some other examples of experiments and events? What are the favorable outcomes for these events?" *Sample answer:* An experiment is spinning a spinner with the numbers 1–12. An event is spinning a number greater than 10, with the favorable outcomes 11 and 12.
- ? "What are the favorable outcomes of rolling a prime number?" 2, 3, and 5
- ? "What are the favorable outcomes of rolling a number divisible by 3?" 3 and 6
- ? "What are the favorable outcomes of rolling a number that is a multiple of 8?" There are none.

Goal Today's lesson is identifying the **favorable outcomes** of an **event**.

Technology for the Teacher

Dynamic Classroom

Lesson Tutorials
Lesson Plans
Answer Presentation Tool

Extra Example 1

You roll a number cube.

a. What are the possible outcomes? 1, 2, 3, 4, 5, 6

b. What are the favorable outcomes of rolling an odd number? 1, 3, 5

c. What are the favorable outcomes of rolling a number greater than 4? 5, 6

On Your Own

1. a. A, B, C, D, E, F, G, H, I, J, K
   b. A, E, I

Extra Example 2

You spin the spinner shown in Example 2.

a. How many ways can spinning blue occur? 1 way
b. How many ways can spinning *not* green occur? 5 ways
c. What are the favorable outcomes of spinning *not* green? red, red, red, purple, blue

On Your Own

2. a. 8 outcomes
   b. 2 ways
   c. 5 ways; blue, blue, red, green, purple

English Language Learners

Vocabulary

Some English learners may confuse the words *outcome* and *event*. Help students see how the two words are related using Example 2 parts (a) and (b).

Possible Outcomes			
red	red	red	← *Favorable Outcomes of event choosing red*
purple	blue	green	

Laurie's Notes

On Your Own

- **Think-Pair-Share:** Students should read each question independently and then work in pairs to answer the questions. When they have answered the questions, the pair should compare their answers with another group and discuss any discrepancies.

Example 2

- **Make Sense of Problems and Persevere in Solving Them:** Discuss the difference between outcomes and favorable outcomes. Refer back to the marbles in the Key Ideas.
- **Common Error:** When answering part (a), many students will say that there are only 4 possible outcomes, not 6, because there are only four colors in the spinner. Explain to students that they need to count every occurrence of a color as a possible outcome. Students may be able to understand this concept better after part (c).

On Your Own

- Discuss answers as a class.

Closure

- **Exit Ticket:** What are the favorable outcomes of drawing a face card from a deck of cards? There are three face cards (jack, queen, and king) and four suits, so there are 12 favorable outcomes.

On Your Own

Now You're Ready
Exercises 5–11

1. You randomly choose a letter from a hat that contains the letters A through K.

   a. What are the possible outcomes?

   b. What are the favorable outcomes of choosing a vowel?

EXAMPLE 2 Counting Outcomes

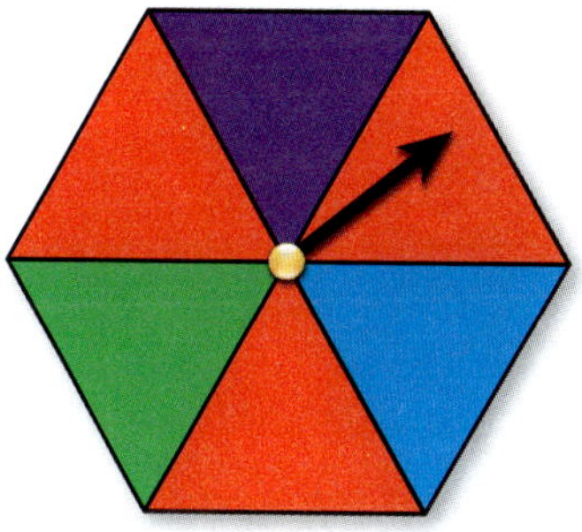

You spin the spinner.

a. How many possible outcomes are there?

The spinner has 6 sections. So, there are 6 possible outcomes.

b. In how many ways can spinning red occur?

The spinner has 3 red sections. So, spinning red can occur in 3 ways.

c. In how many ways can spinning *not* purple occur? What are the favorable outcomes of spinning *not* purple?

The spinner has 5 sections that are *not* purple. So, spinning *not* purple can occur in 5 ways.

purple	*not* purple
purple	red, red, red, green, blue

The favorable outcomes of the event are red, red, red, green, and blue.

On Your Own

Now You're Ready
Exercises 12–17

2. You randomly choose a marble.

   a. How many possible outcomes are there?

   b. In how many ways can choosing blue occur?

   c. In how many ways can choosing *not* yellow occur? What are the favorable outcomes of choosing *not* yellow?

15.1 Exercises

Check It Out
Help with Homework
BigIdeasMath.com

Vocabulary and Concept Check

1. **VOCABULARY** Is rolling an even number on a number cube an *outcome* or an *event*? Explain.

2. **WRITING** Describe how an outcome and a favorable outcome are different.

Practice and Problem Solving

You spin the spinner shown.

3. How many possible results are there?

4. Of the possible results, in how many ways can you spin an even number? an odd number?

1 5. **TILES** What are the possible outcomes of randomly choosing one of the tiles shown?

You randomly choose one of the tiles shown above. Find the favorable outcomes of the event.

6. Choosing a 6
7. Choosing an odd number
8. Choosing a number greater than 5
9. Choosing an odd number less than 5
10. Choosing a number less than 3
11. Choosing a number divisible by 3

You randomly choose one marble from the bag. (a) Find the number of ways the event can occur. (b) Find the favorable outcomes of the event.

2 12. Choosing blue
13. Choosing green
14. Choosing purple
15. Choosing yellow
16. Choosing *not* red
17. Choosing *not* blue

18. **ERROR ANALYSIS** Describe and correct the error in finding the number of ways that choosing *not* purple can occur.

✗

purple	*not* purple
purple	red, blue, green, yellow

Choosing *not* purple can occur in 4 ways.

Assignment Guide and Homework Check

Level	Assignment	Homework Check
Advanced	1–4, 6–26 even, 27–31	10, 16, 24, 26

Common Errors

- **Exercises 6–11** Students may forget to include, or include too many, favorable outcomes. Encourage them to write out all of the possible outcomes and then circle the favorable outcomes for the given event.
- **Exercises 12–17** Students may forget to include the repeats of a color when describing the favorable outcomes and how many ways the event can occur. Ask students to describe how many times you could pull out a specific color from the bag if you happened to pull out the same color each time (without replacing). For example, if the event is choosing a red marble, you can pull out a red marble three times before there are no more red marbles in the bag. So, the event can occur three times.

15.1 Record and Practice Journal

A bag is filled with 4 red marbles, 3 blue marbles, 3 yellow marbles, and 2 green marbles. You randomly choose one marble from the bag. (a) Find the number of ways the event can occur. (b) Find the favorable outcomes of the event.

1. Choosing red
   a. 4
   b. red, red, red, red
2. Choosing green
   a. 2
   b. green, green
3. Choosing yellow
   a. 3
   b. yellow, yellow, yellow
4. Choosing *not* blue
   a. 9
   b. red, red, red, red, yellow, yellow, yellow, green, green
5. In order to figure out who will go first in a game, your friend asks you to pick a number between 1 and 25.
   a. What are the possible outcomes?
      1, 2, 3, 4, 5, 6, 7, 8, 9, 10, 11, 12, 13, 14, 15, 16, 17, 18, 19, 20, 21, 22, 23, 24, 25
   b. What are the favorable outcomes of choosing an even number?
      2, 4, 6, 8, 10, 12, 14, 16, 18, 20, 22, 24
   c. What are the favorable outcomes of choosing a number less than 20?
      1, 2, 3, 4, 5, 6, 7, 8, 9, 10, 11, 12, 13, 14, 15, 16, 17, 18, 19

Vocabulary and Concept Check

1. event; It is a collection of several outcomes.
2. An outcome is one possible result of an experiment. A favorable outcome is an outcome of a specific event.

Practice and Problem Solving

3. 8
4. 4 ways; 4 ways
5. 1, 2, 3, 4, 5, 6, 7, 8, 9
6. 6
7. 1, 3, 5, 7, 9
8. 6, 7, 8, 9
9. 1, 3
10. 1, 2
11. 3, 6, 9
12. **a.** 2 ways **b.** blue, blue
13. **a.** 1 way **b.** green
14. **a.** 2 ways
    b. purple, purple
15. **a.** 1 way **b.** yellow
16. **a.** 6 ways
    b. yellow, green, blue, blue, purple, purple
17. **a.** 7 ways
    b. red, red, red, purple, purple, green, yellow
18. There are 7 marbles that are *not* purple, even though there are only 4 colors. Choosing *not* purple could be red, red, red, blue, blue, green, or yellow.

Practice and Problem Solving

19. 7 ways

20. false; red

21. true

22. false; five

23. true

24. false; eight

25. 30 rock CDs

26. See *Taking Math Deeper*.

Fair Game Review

27. $x = 2$

28. $n = 21$

29. $w = 12$

30. $b = 68$

31. C

Mini-Assessment

You randomly choose one number below. Find the favorable outcomes of the event.

10, 11, 12, 13, 14, 15, 16, 17, 18, 19

1. Choosing a 14 14
2. Choosing an even number 10, 12, 14, 16, 18
3. Choosing an odd number less than 15 11, 13
4. Choosing a number greater than 16 17, 18, 19
5. Choosing a number divisible by 2 10, 12, 14, 16, 18

Taking Math Deeper

Exercise 26

This problem previews the concept of dependent events.

1. With all five cards available, the number of possible outcomes is 5.

Choose 1 at random.

2. With only four cards left, the number of possible outcomes is reduced to 4.

Choose 1 at random.

3. Answer the question.

After the baker is chosen, the number of possible outcomes decreases.

Project

Create a game that uses picture cards and the changing probabilities indicated in the problem. Create the cards. Play the game with a partner.

Reteaching and Enrichment Strategies

If students need help...	If students got it...
Resources by Chapter • Practice A and Practice B • Puzzle Time Record and Practice Journal Practice Differentiating the Lesson Lesson Tutorials Skills Review Handbook	Resources by Chapter • Enrichment and Extension • Technology Connection Start the next section

19. **COINS** You have 10 coins in your pocket. Five are Susan B. Anthony dollars, two are Kennedy half-dollars, and three are presidential dollars. You randomly choose a coin. In how many ways can choosing *not* a presidential dollar occur?

Tell whether the statement is *true* or *false*. If it is false, change the italicized word to make the statement true.

Spinner A

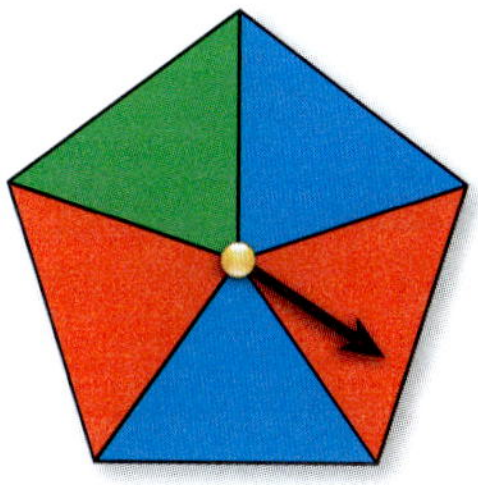

Spinner B

20. Spinning blue and spinning *green* have the same number of favorable outcomes on Spinner A.

21. Spinning blue has one *more* favorable outcome than spinning green on Spinner B.

22. There are *three* possible outcomes of spinning Spinner A.

23. Spinning *red* can occur in four ways on Spinner B.

24. Spinning not green can occur in *three* ways on Spinner B.

25. **MUSIC** A bargain bin contains classical and rock CDs. There are 60 CDs in the bin. Choosing a rock CD and *not* choosing a rock CD have the same number of favorable outcomes. How many rock CDs are in the bin?

26. **Precision** You randomly choose one of the cards and set it aside. Then you randomly choose a second card. Describe how the number of possible outcomes changes after the first card is chosen.

Fair Game Review What you learned in previous grades & lessons

Solve the proportion. *(Skills Review Handbook)*

27. $\frac{x}{10} = \frac{1}{5}$

28. $\frac{60}{n} = \frac{20}{7}$

29. $\frac{1}{3} = \frac{w}{36}$

30. $\frac{25}{17} = \frac{100}{b}$

31. **MULTIPLE CHOICE** What is the surface area of the rectangular prism? *(Section 14.1)*

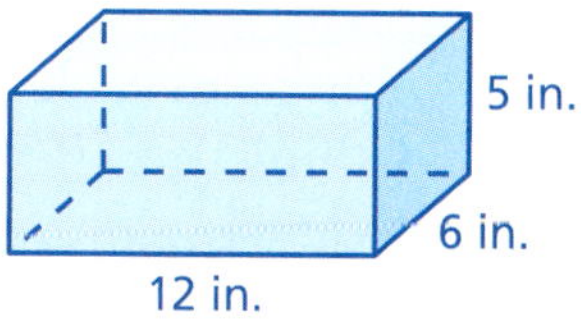

Ⓐ 162 in.^2 Ⓑ 264 in.^2

Ⓒ 324 in.^2 Ⓓ 360 in.^2

15.2 Probability

Essential Question How can you describe the likelihood of an event?

1 ACTIVITY: Black-and-White Spinner Game

Work with a partner. You work for a game company. You need to create a game that uses the spinner below.

a. Write rules for a game that uses the spinner. Then play it.

b. After playing the game, do you want to revise the rules? Explain.

Probability and Statistics

In this lesson, you will

- understand the concept of probability and the relationship between probability and likelihood.
- find probabilities of events.

c. **CHOOSE TOOLS** Using the center of the spinner as the vertex, measure the angle of each pie-shaped section. Is each section the same size? How do you think this affects the likelihood of spinning a given number?

d. Your friend is about to spin the spinner and wants to know how likely it is to spin a 3. How would you describe the likelihood of this event to your friend?

Laurie's Notes

Introduction

Applying Mathematical Practices

- **Attend to Precision:** Mathematically proficient students try to communicate precisely to others. In the activities today, students should pay attention to whether outcomes are equally likely based upon angle measures of sections in spinners.

Motivate

? Ask a few trivia questions about mammals.

- "What is the world's largest mammal?" blue whale; over 200 tons
- "What is the world's largest land mammal?" African elephant; about 16,500 pounds
- "What is the world's smallest mammal?" Thailand bumblebee bat; weighs less than a penny
- "What is the world's smallest mammal in terms of length?" Etruscan Pygmy Shrew; about 1.5 inches long for body and head

Activity Notes

Activity 1

- Students will have fun with this activity. The six mammals add interest and increase student creativity.

? "Name the mammals from 1 to 6." skunk, dog, whale, zebra, penguin, cow

- It is hard to predict what rules students might develop and how simple or involved the rules might be. Here are a few things that students may focus on:
  - 5-letter mammals skunk, whale, zebra
  - mammals that spend most of their time in or around water whale, penguin
  - 4-legged mammals skunk, dog, zebra, cow
  - neighborhood animals dog, skunk, cow
  - ignore the mammals and look at prime numbers 2, 3, 5
- The rules are open-ended. Sometimes students write simple rules—you spin *x* and you win. Other times students will create a game board, or they might do a point accumulation, or have cards that they distribute. This activity can become a multi-day project if you have time.
- The opportunity to revise the rules is included in case students think of a new or necessary change while playing the game.
- A *central angle* is an angle with its vertex at the center of a circle and whose sides are radii of the circle.

? "What is the measure of the central angle for section 1 (the skunk)?" 60°

- **Attend to Precision:** Discuss the students' answers for part (d). Listen for precision in their reasoning. Do students simply say that it is unlikely, or do they explain that each outcome is equally likely because the measures of the central angles are the same?

What Your Students Will Learn

- Understand the concept of probability and the relationship between probability and likelihood.
- Find probabilities of events using a formula.

Previous Learning

Students should know how many degrees are in a circle and how to measure angles.

Technology for the Teacher

Dynamic Classroom

Lesson Plans
Complete Materials List

15.2 Record and Practice Journal

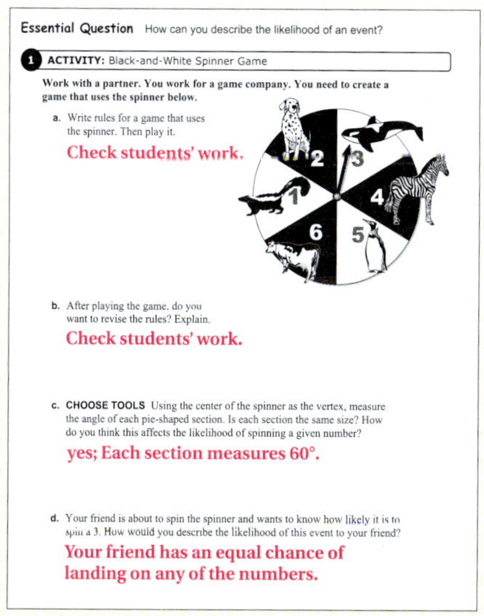

Essential Question How can you describe the likelihood of an event?

1 ACTIVITY: Black-and-White Spinner Game

Work with a partner. You work for a game company. You need to create a game that uses the spinner below.

a. Write rules for a game that uses the spinner. Then play it.

Check students' work.

b. After playing the game, do you want to revise the rules? Explain.

Check students' work.

c. **CHOOSE TOOLS** Using the center of the spinner as the vertex, measure the angle of each pie-shaped section. Is each section the same size? How do you think this affects the likelihood of spinning a given number?

yes; Each section measures 60°.

d. Your friend is about to spin the spinner and wants to know how likely it is to spin a 3. How would you describe the likelihood of this event to your friend?

Your friend has an equal chance of landing on any of the numbers.

Differentiated Instruction

Auditory

Ask students what it means when something happens by chance. Discuss with students situations in which the outcomes happen by chance, such as flipping a coin, rolling a number cube, or spinning a spinner. Ask students what it means for an event to be impossible.

15.2 Record and Practice Journal

2 ACTIVITY: Changing the Spinner

Work with a partner. For each spinner, do the following.

- Measure the angle of each pie-shaped section.
- Tell whether you are more likely to spin a particular number. Explain your reasoning.
- Tell whether your rules from Activity 1 make sense for these spinners. Explain your reasoning.

a.

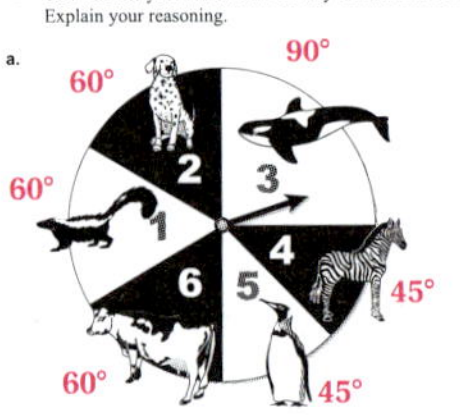

Chance of landing on 3 is greater; yes, but the game is not fair.

b.

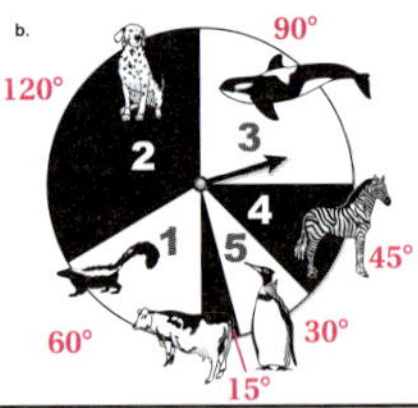

Chance of landing on 2 is greater; yes, but the game is not fair.

3 ACTIVITY: Is This Game Fair?

Work with a partner. Apply the following rules to each spinner in Activities 1 and 2. Is the game fair? Why or why not? If not, who has the better chance of winning?

- **Take turns spinning the spinner.**
- **If you spin an odd number, Player 1 wins.**
- **If you spin an even number, Player 2 wins.**

Spinner 1 is fair. Spinner 2a is not fair and Player 1 has a better chance of winning. Spinner 2b is fair.

What Is Your Answer?

4. IN YOUR OWN WORDS How can you describe the likelihood of an event?
Sample answer: You can describe an event as *impossible, unlikely to occur, likely to occur, certain,* or *equally likely to happen or not happen.*

5. Describe the likelihood of spinning an 8 in Activity 1.
not possible

6. Describe a career in which it is important to know the likelihood of an event.
Sample answer: Meteorologists use likelihood when predicting whether it will rain or not.

Laurie's Notes

Activity 2

- In this activity, students must consider spinners with different central angle measures. The Black-and-White Spinner Game students created for Activity 1 will be played using these new spinners. First, they need to measure the central angles.
- ? "How can you tell whether you are more likely to spin a particular number?" Listen for comparisons of the central angle measures.
- **Big Idea:** The outcomes are not equally likely. Students should recognize that not all of the central angle measures are the same, so the likelihood of spinning a particular number varies.
- Ask students to compare and contrast the spinners. Note that for the second spinner, spinning black and spinning white are equally likely, which is a key observation in Activity 3.

Activity 3

- ? "What would make the game fair?" if each player has an equal chance of winning
- Give time for students to play 10 games on each spinner. Have each pair of students gather data in a table.
- Collect class data for wins and losses on each spinner.
- ? "What does the data tell you about the fairness of the three spinners?" The data suggests that using these rules with the spinners in Activity 1 and Activity 2(b) results in fair games.

What Is Your Answer?

- Discuss Question 5 as a class. Try to lead them to describe the likelihood as *impossible* by first asking them if it is possible to spin an 8, and then having them complete the statement "There is not an 8 on the spinner, so spinning an 8 is _______."

Closure

- Draw the spinners below on the board. You might offer students an opportunity to spin for no homework if they can come to a consensus as to which of the spinners to use (N = no homework and Y = homework).

2 ACTIVITY: Changing the Spinner

Work with a partner. For each spinner, do the following.

- **Measure the angle of each pie-shaped section.**
- **Tell whether you are more likely to spin a particular number. Explain your reasoning.**
- **Tell whether your rules from Activity 1 make sense for these spinners. Explain your reasoning.**

a.

b.

3 ACTIVITY: Is This Game Fair?

Math Practice

Use Prior Results

How can you use the results of the previous activities to determine whether the game is fair?

Work with a partner. Apply the following rules to each spinner in Activities 1 and 2. Is the game fair? Why or why not? If not, who has the better chance of winning?

- **Take turns spinning the spinner.**
- **If you spin an odd number, Player 1 wins.**
- **If you spin an even number, Player 2 wins.**

What Is Your Answer?

4. **IN YOUR OWN WORDS** How can you describe the likelihood of an event?
5. Describe the likelihood of spinning an 8 in Activity 1.
6. Describe a career in which it is important to know the likelihood of an event.

Practice Use what you learned about the likelihood of an event to complete Exercises 4 and 5 on page 642.

15.2 Lesson

Check It Out
Lesson Tutorials
BigIdeasMath.com

Key Vocabulary
probability, *p. 640*

Key Idea

Probability

The **probability** of an event is a number that measures the likelihood that the event will occur. Probabilities are between 0 and 1, including 0 and 1. The diagram relates likelihoods (above the diagram) and probabilities (below the diagram).

Study Tip

Probabilities can be written as fractions, decimals, or percents.

EXAMPLE 1 Describing the Likelihood of an Event

There is an 80% chance of thunderstorms tomorrow. Describe the likelihood of the event.

The probability of thunderstorms tomorrow is 80%.

∴ Because 80% is close to 75%, it is *likely* that there will be thunderstorms tomorrow.

On Your Own

Now You're Ready
Exercises 6–9

Describe the likelihood of the event given its probability.

1. The probability that you land a jump on a snowboard is $\frac{1}{2}$.
2. There is a 100% chance that the temperature will be less than 120°F tomorrow.

Key Idea

Finding the Probability of an Event

When all possible outcomes are equally likely, the probability of an event is the ratio of the number of favorable outcomes to the number of possible outcomes. The probability of an event is written as $P(\text{event})$.

$$P(\text{event}) = \frac{\text{number of favorable outcomes}}{\text{number of possible outcomes}}$$

Multi-Language Glossary at BigIdeasMath.com

Laurie's Notes

Introduction

Connect

- **Yesterday:** Students developed an intuitive understanding of how to predict the likelihood of the results of a spinner.
- **Today:** Students will compute the probability of an event.

Motivate

- The name of a county in a U.S. state is often named after a person, such as a president. In fact, there are 31 states in the U.S. with a county named Washington. Sometimes the county is named after a historical figure. Madison is a county name in 20 states and Calhoun is a county name in 11 states.
- Draw a spinner representing this information.

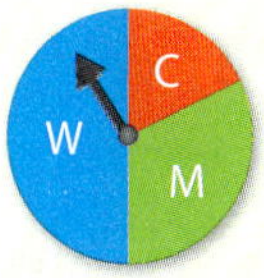

Key:
C – Calhoun M – Madison W – Washington

? "If you spin the spinner 100 times, how many times would you expect it to land on Washington? Explain." 50; There are 62 different counties represented and half of them are named Washington.

Lesson Notes

Key Idea

- Discuss possible events which have probabilities near each benchmark. Make them personal for your situation, if possible. Examples: the sun rising tomorrow = 1, math homework = 0.75, winning the softball game = 0.50, skipping breakfast = 0.25, a winter in Vermont with no snow = 0
- Spend time discussing what *equally likely* means. Give examples of events that are equally likely and not equally likely. Equally likely: number cube; Not equally likely: spinner with sections that are not all the same size

Example 1

? "Has anyone heard a weather report for tomorrow?" Try to turn student responses into a percent or fraction. For example, you could translate "it's supposed to be nice tomorrow" into "there's a 90% chance of sunshine."

- Be sure that students understand that probabilities can be represented as fractions, decimals, or percents.

On Your Own

- Discuss answers as a class.

Key Idea

- Write the Key Idea.
- Discuss probability and give several examples with which students would be familiar, such as cards, number cubes, and marbles in a bag. Stress that the outcomes must be equally likely to use this ratio.

Goal Today's lesson is finding the **probability** of an event.

Technology for the Teacher

Dynamic Classroom

Lesson Tutorials
Lesson Plans
Answer Presentation Tool

Extra Example 1

There is a 20% chance of snow flurries tomorrow. Describe the likelihood of the event. Because 20% is close to 25%, it is *unlikely* that there will be snow flurries tomorrow.

On Your Own

1. equally likely to happen or not happen
2. certain

English Language Learners

Comprehension

This chapter has more word problems and fewer skill problems than most chapters. It may be more difficult for some students. You can present problems in a predictable format so the students will not get stuck on reading and be able to focus on the mathematics.

Extra Example 2

In Example 2, what is the probability of rolling a number greater than 4? $\frac{1}{3}$

Extra Example 3

The probability that you randomly draw a short straw from a group of 50 straws is $\frac{9}{25}$. How many are short straws? 18

On Your Own

3. $\frac{2}{3}$
4. 0
5. 5

Laurie's Notes

Example 2

? "How many possible outcomes are there?" 6 "How many are favorable outcomes?" 3

? "Can you use the ratio in the previous Key Idea to find the probability of rolling an odd number? Explain." Yes, because the outcomes are equally likely.

- **Attend to Precision:** Students should communicate precisely, explaining why the outcomes are equally likely when providing their explanations.
- **Extension:** Ask about the probability of rolling an even number, rolling a prime number, or rolling a number greater than 8. $\frac{1}{2}$; $\frac{1}{2}$; 0

Example 3

- Work through the example by writing and solving the proportion.
- **Alternate Solution:** Because $\frac{3}{20} = \frac{15}{100} = 15\%$, this problem could be solved by finding the percent of a number.

$$\begin{aligned} 15\% \text{ of } 40 &= 0.15 \times 40 \\ &= 6 \end{aligned}$$

It is important for students to see this connection.

On Your Own

- **Think-Pair-Share:** Students should read each question independently and then work in pairs to answer the questions. When they have answered the questions, the pair should compare their answers with another group and discuss any discrepancies.

Closure

- Write an example of an event that has the following probabilities.
  - **a.** close to 1
  - **b.** exactly $\frac{1}{2}$
  - **c.** close to 0
  - **d.** exactly $\frac{2}{5}$

EXAMPLE 2 Finding a Probability

You roll the number cube. What is the probability of rolling an odd number?

$$P(\text{event}) = \frac{\text{number of favorable outcomes}}{\text{number of possible outcomes}}$$

$$P(\text{odd}) = \frac{3}{6}$$ There are 3 odd numbers (1, 3, and 5). There is a total of 6 numbers.

$$= \frac{1}{2}$$ Simplify.

∴ The probability of rolling an odd number is $\frac{1}{2}$, or 50%.

EXAMPLE 3 Using a Probability

The probability that you randomly draw a short straw from a group of 40 straws is $\frac{3}{20}$. How many are short straws?

Ⓐ 4 Ⓑ 6

Ⓒ 15 Ⓓ 34

$$P(\text{short}) = \frac{\text{number of short straws}}{\text{total number of straws}}$$

$$\frac{3}{20} = \frac{n}{40}$$ Substitute. Let n be the number of short straws.

$$6 = n$$ Solve for n.

There are 6 short straws.

∴ So, the correct answer is Ⓑ.

On Your Own

Now You're Ready Exercises 11–15

3. In Example 2, what is the probability of rolling a number greater than 2?
4. In Example 2, what is the probability of rolling a 7?
5. The probability that you randomly draw a short straw from a group of 75 straws is $\frac{1}{15}$. How many are short straws?

15.2 Exercises

Check It Out
Help with Homework
BigIdeasMath.com

Vocabulary and Concept Check

1. **VOCABULARY** Explain how to find the probability of an event.
2. **REASONING** Can the probability of an event be 1.5? Explain.
3. **OPEN-ENDED** Give a real-life example of an event that is impossible. Give a real-life example of an event that is certain.

Practice and Problem Solving

You are playing a game using the spinners shown.

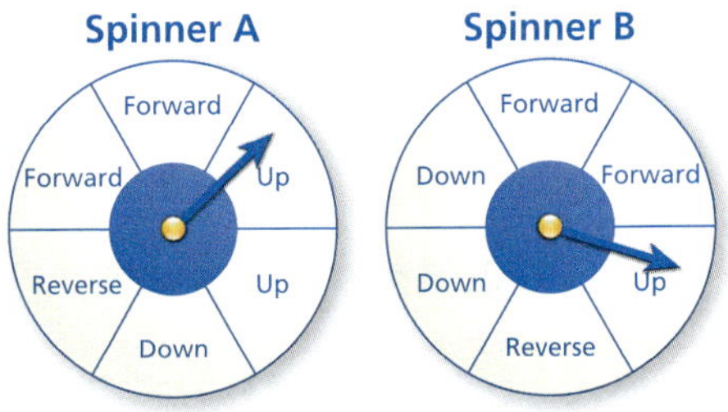

4. You want to move down. On which spinner are you more likely to spin "Down"? Explain.
5. You want to move forward. Which spinner would you spin? Explain.

Describe the likelihood of the event given its probability.

1 6. Your soccer team wins $\frac{3}{4}$ of the time.
7. There is a 0% chance that you will grow 12 more feet.
8. The probability that the sun rises tomorrow is 1.
9. It rains on $\frac{1}{5}$ of the days in July.

10. **VIOLIN** You have a 50% chance of playing the correct note on a violin. Describe the likelihood of playing the correct note.

You randomly choose one shirt from the shelves. Find the probability of the event.

2 11. Choosing a red shirt
12. Choosing a green shirt
13. *Not* choosing a white shirt
14. *Not* choosing a black shirt
15. Choosing an orange shirt

16. **ERROR ANALYSIS** Describe and correct the error in finding the probability of *not* choosing a blue shirt from the shelves above.

✗ $P(\text{not blue}) = \frac{4}{10} = \frac{2}{5}$

Assignment Guide and Homework Check

Level	Assignment	Homework Check
Advanced	1–5, 6–22 even, 23–27	8, 14, 18, 20, 22

Common Errors

- **Exercises 11–15** Students may write the probability as the ratio of the number of favorable outcomes to the number of unfavorable outcomes. Remind them that the probability of an event is the ratio of the number of favorable outcomes to the number of possible outcomes.
- **Exercise 18** In part (a), students may write and solve the proportion to find the number of winning ducks correctly. But then they forget to subtract their result from 25 to find the number of *not* winning ducks.

15.2 Record and Practice Journal

Describe the likelihood of the event given its probability.

1. There is a 30% chance of snow tomorrow.
   unlikely
2. You solve a brain teaser 0.75 of the time.
   likely

You randomly choose one hat from 3 green hats, 4 black hats, 2 white hats, 2 red hats, and 1 blue hat. Find the probability of the event.

3. Choosing a red hat
   $\frac{1}{6}$
4. Choosing a black hat
   $\frac{1}{3}$
5. *Not* choosing a white hat
   $\frac{5}{6}$
6. Choosing a blue hat
   $\frac{1}{12}$
7. *Not* choosing a black hat
   $\frac{2}{3}$
8. *Not* choosing a green hat
   $\frac{3}{4}$
9. The probability that you draw a mechanical pencil from a group of 25 mechanical and wooden pencils is $\frac{3}{5}$. How many are mechanical pencils?
   15 mechanical pencils

Vocabulary and Concept Check

1. The probability of an event is the ratio of the number of favorable outcomes to the number of possible outcomes.
2. no; Probabilities are between 0 and 1, including 0 and 1.
3. *Sample answer:* You will not have any homework this week.; You will fall asleep tonight.

Practice and Problem Solving

4. Spinner B; There are more chances to land on "Down" with Spinner B.
5. either; Both spinners have the same number of chances to land on "Forward."
6. likely
7. impossible
8. certain
9. unlikely
10. equally likely to happen or not happen
11. $\frac{1}{10}$
12. $\frac{1}{5}$
13. $\frac{9}{10}$
14. $\frac{4}{5}$
15. 0
16. The student found the probability of choosing a blue shirt.;
    $P(\textit{not}\ \text{blue}) = \frac{6}{10} = \frac{3}{5}$
17. 20
18. See *Taking Math Deeper*.
19. **a.** $\frac{2}{3}$; likely
    b. $\frac{1}{3}$; unlikely
    c. $\frac{1}{2}$; equally likely to happen or not happen

Practice and Problem Solving

20.

Father's Genes \ Mother's Genes	X	X
X	XX	XX
Y	XY	XY

21. There are 2 combinations for each.

22. a. $\frac{1}{4}$, or 25%

Parent 2 \ Parent 1	C	s
C	CC	Cs
s	Cs	ss

b. $\frac{3}{4}$, or 75%

Fair Game Review

23. $x < 4$;

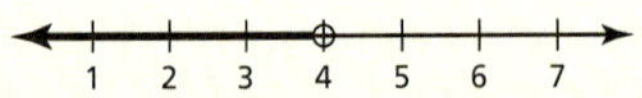

24. $b \geq -5$;

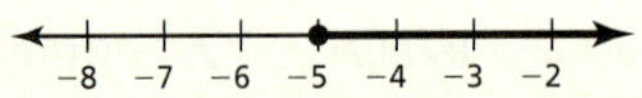

25. $w > -3$;

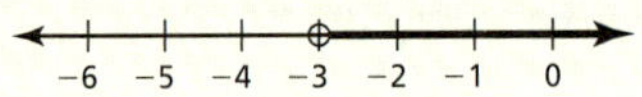

26. $g \leq -3$;

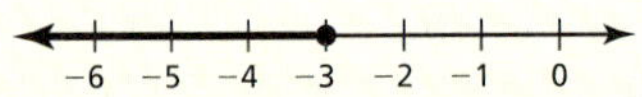

27. C

Mini-Assessment

You randomly choose one number below. Find the probability of the event.

2, 5, 6, 9, 13, 16, 22, 25, 27, 31

1. Choosing an even number $\frac{2}{5}$
2. Choosing an odd number $\frac{3}{5}$
3. Choosing a prime number $\frac{2}{5}$
4. Choosing a number greater than 30 $\frac{1}{10}$
5. Choosing a number less than 2 0

Taking Math Deeper

Exercise 18

Students have not been formally introduced to the *complement* of an event, but they can still use this concept to solve the problem.

1 Find the probability of choosing a rubber duck that is *not* a winner.

The probability that you choose a winning rubber duck is 0.24, or 24%. So, 100% − 24% = 76% of the rubber ducks are *not* winners.

2 Find 76% of 25.

$0.76 \cdot 25 = 19$

a. There are 19 rubber ducks that are *not* winners.

3 Describe the likelihood of *not* choosing a winning duck.

b. There is a 76% chance of *not* choosing a winning rubber duck. So, it is *likely* that you will choose a rubber duck that is *not* a winner.

Project

Many games at carnivals can be modified so that it is harder for you to win a prize. Research a few carnival games and explain how they can be modified to decrease your probability of winning.

Reteaching and Enrichment Strategies

If students need help. . .	If students got it. . .
Resources by Chapter • Practice A and Practice B • Puzzle Time Record and Practice Journal Practice Differentiating the Lesson Lesson Tutorials Skills Review Handbook	Resources by Chapter • Enrichment and Extension • Technology Connection Start the next section

17. **CONTEST** The rules of a contest say that there is a 5% chance of winning a prize. Four hundred people enter the contest. Predict how many people will win a prize.

18. **RUBBER DUCKS** At a carnival, the probability that you choose a winning rubber duck from 25 ducks is 0.24.

 a. How many are *not* winning ducks?

 b. Describe the likelihood of *not* choosing a winning duck.

19. **DODECAHEDRON** A dodecahedron has twelve sides numbered 1 through 12. Find the probability and describe the likelihood of each event.

 a. Rolling a number less than 9

 b. Rolling a multiple of 3

 c. Rolling a number greater than 6

A Punnett square is a grid used to show possible gene combinations for the offspring of two parents. In the Punnett square shown, a boy is represented by *XY*. A girl is represented by *XX*.

Father's Genes \ Mother's Genes	*X*	*X*
X	*XX*	
Y		

20. Complete the Punnett square.

21. Explain why the probability of two parents having a boy or having a girl is equally likely.

22. **Critical Thinking** Two parents each have the gene combination *Cs*. The gene *C* is for curly hair. The gene *s* is for straight hair.

 a. Make a Punnett square for the two parents. When all outcomes are equally likely, what is the probability of a child having the gene combination *CC*?

 b. Any gene combination that includes a *C* results in curly hair. When all outcomes are equally likely, what is the probability of a child having curly hair?

Fair Game Review What you learned in previous grades & lessons

Solve the inequality. Graph the solution. *(Section 11.2 and Section 11.3)*

23. $x + 5 < 9$ 24. $b - 2 \geq -7$ 25. $1 > -\frac{w}{3}$ 26. $6 \leq -2g$

27. **MULTIPLE CHOICE** Find the value of x. *(Section 12.4)*

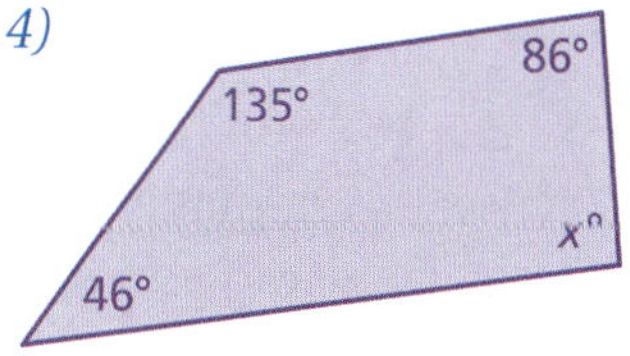

Ⓐ 85 Ⓑ 90

Ⓒ 93 Ⓓ 102

15.3 Experimental and Theoretical Probability

Essential Question How can you use relative frequencies to find probabilities?

When you conduct an experiment, the **relative frequency** of an event is the fraction or percent of the time that the event occurs.

$$\text{relative frequency} = \frac{\text{number of times the event occurs}}{\text{total number of times you conduct the experiment}}$$

1 ACTIVITY: Finding Relative Frequencies

Work with a partner.

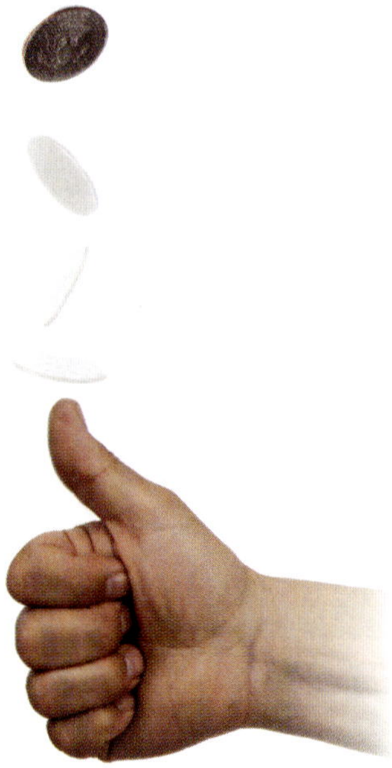

a. Flip a quarter 20 times and record your results. Then complete the table. Are the relative frequencies the same as the probability of flipping heads or tails? Explain.

	Flipping Heads	Flipping Tails
Relative Frequency		

b. Compare your results with those of other students in your class. Are the relative frequencies the same? If not, why do you think they differ?

c. Combine all of the results in your class. Then complete the table again. Did the relative frequencies change? What do you notice? Explain.

d. Suppose everyone in your school conducts this experiment and you combine the results. How do you think the relative frequencies will change?

2 ACTIVITY: Using Relative Frequencies

Probability and Statistics

In this lesson, you will

- find relative frequencies.
- use experimental probabilities to make predictions.
- use theoretical probabilities to find quantities.
- compare experimental and theoretical probabilities.

Work with a partner. You have a bag of colored chips. You randomly select a chip from the bag and replace it. The table shows the number of times you select each color.

Red	Blue	Green	Yellow
24	12	15	9

a. There are 20 chips in the bag. Can you use the table to find the exact number of each color in the bag? Explain.

b. You randomly select a chip from the bag and replace it. You do this 50 times, then 100 times, and you calculate the relative frequencies after each experiment. Which experiment do you think gives a better approximation of the exact number of each color in the bag? Explain.

Laurie's Notes

Introduction

Applying Mathematical Practices

- **Attend to Precision:** Mathematically proficient students try to communicate precisely to others. In these activities, students should pay close attention to the equally likely aspect of the outcomes based upon the relative frequencies of the experiments.

Motivate

- Ask ten students to stand at the front of the room. Hand each one a penny.
- ? "If [student 1] flips his or her penny, what is the probability it will land on heads?" $\frac{1}{2}$ "If [student 4] flips his or her penny, what is the probability it will land on heads?" $\frac{1}{2}$
- ? "If all of the students toss their pennies, what is the probability they will all land on heads?" Students may say $\frac{1}{2}$.
- ? Have all 10 students toss their coins, then ask, "Did everyone's penny land on heads?" It is unlikely that all 10 students' pennies will land on heads.

Discuss

- Discuss the definition of *relative frequency* and relate it to the Motivate.
- Make sure that students understand that tossing a coin 10 times and having it land on heads about half the time is a different experiment than tossing 10 coins and asking how likely it is that all 10 coins land on heads.

Activity Notes

Activity 1

- ? "What is the probability of flipping heads?" $\frac{1}{2}$ "What is the probability of flipping tails?" $\frac{1}{2}$ "Is each outcome equally likely?" yes
- **Use Appropriate Tools Strategically:** As data are gathered and recorded, several students with calculators can be summarizing the results.
- ? "Did the relative frequencies change as the results were combined?" Answers will vary. The additional data should smooth the results so that the relative frequencies approach 50% for each outcome.
- Online simulators are available to model this experiment. Be sure students understand as the number of trials increases, the experimental probability gets closer to the theoretical probability.

Activity 2

- ? "Why is it important to replace the chip each time?" The probability changes for each trial if the chip is not replaced.
- **Construct Viable Arguments and Critique the Reasoning of Others** and **Attend to Precision:** Listen to students' explanations of how the increase in the number of trials improves the accuracy of the approximation.

What Your Students Will Learn

- Find relative frequencies and experimental probabilities of events.
- Use experimental probabilities to make predictions of future outcomes.
- Find theoretical probabilities and use them to find quantities.
- Compare experimental and theoretical probabilities.

Previous Learning

Students should know how to find the probability of an event.

Technology for the Teacher

Dynamic Classroom

Lesson Plans
Complete Materials List

15.3 Record and Practice Journal

Essential Question How can you use relative frequencies to find probabilities?

When you conduct an experiment, the **relative frequency** of an event is the fraction or percent of the time that the event occurs.

$$\text{relative frequency} = \frac{\text{number of times the event occurs}}{\text{total number of times you conduct the experiment}}$$

1 ACTIVITY: Finding Relative Frequencies

Work with a partner. Check students' work.

a. Flip a quarter 20 times and record your results. Then complete the table. Are the relative frequencies the same as the probability of flipping heads or tails? Explain.

	Flipping Heads	Flipping Tails
Relative Frequency		

b. Compare your results with those of other students in your class. Are the relative frequencies the same? If not, why do you think they differ?

c. Combine all of the results in your class. Then complete the table again. Did the relative frequencies change? What do you notice? Explain.

d. Suppose everyone in your school conducts this experiment and you combine the results. How do you think the relative frequencies will change?
The relative frequencies should be close to the probabilities of $\frac{1}{2}$.

Differentiated Instruction

Kinesthetic

Set up 6 groups and assign each group a number from 1 to 6. Each group will predict whether a number less than, greater than, or equal to their assigned number will occur when rolling a number cube 20 times. Have each group track their results in a frequency table. Each group then determines what fraction of the results is less than, equal to, or greater than their assigned number and presents their findings to the class.

15.3 Record and Practice Journal

2 ACTIVITY: Using Relative Frequencies

Work with a partner. You have a bag of colored chips. You randomly select a chip from the bag and replace it. The table shows the number of times you select each color.

Red	Blue	Green	Yellow
24	12	15	9

a. There are 20 chips in the bag. Can you use the table to find the exact number of each color in the bag? Explain.
no

b. You randomly select a chip from the bag and replace it. You do this 50 times, then 100 times, and you calculate the relative frequencies after each experiment. Which experiment do you think gives a better approximation of the exact number of each color in the bag? Explain.
100 times

3 ACTIVITY: Conducting an Experiment

Work with a partner. You toss a thumbtack onto a table. There are two ways the thumbtack can land.

a. Your friend says that because there are two outcomes, the probability of the thumbtack landing point up must be $\frac{1}{2}$. Do you think this conclusion is true? Explain.
no

b. Toss a thumbtack onto a table 50 times and record your results. In a *uniform probability model*, each outcome is equally likely to occur. Do you think this experiment represents a uniform probability model? Explain.
no; Each outcome is not equally likely.

Use the relative frequencies to complete the following.

P(point up) = ______ P(on its side) = ______

Check students' work.

What Is Your Answer?

4. **IN YOUR OWN WORDS** How can you use relative frequencies to find probabilities? Give an example.
You can find the probability of an event occurring based on a relative frequency.

5. Your friend rolls a number cube 500 times. How many times do you think your friend will roll an odd number? Explain your reasoning.
250 times

6. In Activity 2, your friend says, "There are no orange-colored chips in the bag." Do you think this conclusion is true? Explain.
See Additional Answers.

7. Give an example of an experiment that represents a uniform probability model.
Check students' work.

8. Tell whether you can use each spinner to represent a uniform probability model. Explain your reasoning.

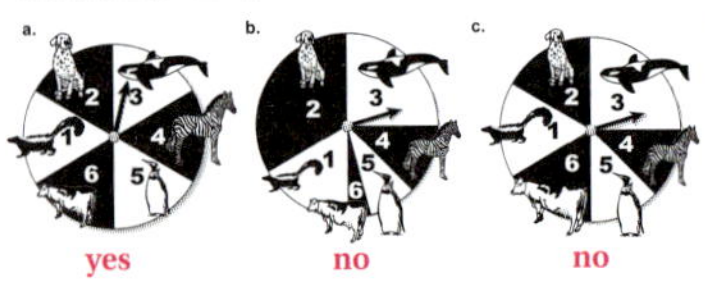

a. yes b. no c. no

Laurie's Notes

Activity 3

- Be aware of student safety for this activity. Instruct students to toss the thumbtacks carefully so that they stay on the table and no one is hurt.
- ? "When you toss a thumbtack, what are the possible outcomes?" It can land with the point up or land on its side. Some students may say it can land with the point down, but if it is tossed onto a hard surface, it is not possible.
- Similar to flipping a coin, tossing a tack also has two outcomes.
- Have students perform part (b) and discuss the results.
- ? "What is a *uniform probability model*?" An experiment where the outcomes are equally likely to occur.
- ? "Does this experiment represent a uniform probability model?" No; the two outcomes are not equally likely to occur.
- ? **Attend to Precision:** "How are the experiments in Activities 1 and 3 alike? How are they different?" Each activity has two possible outcomes. The two outcomes in Activity 1 are equally likely, and the two outcomes in Activity 3 are not equally likely.

What Is Your Answer?

- Ask volunteers to share their reasoning in Question 5. Are students able to distinguish between the relative frequency and the theoretical probability?
- Students often have difficulty thinking of experiments on their own that represent uniform or non-uniform probability models. If students are able to think of some, share these as they might trigger the thinking of other students in the class.
- In Question 8, these spinners were used in the previous section. Students should make a connection between the central angles that they measured and the relative frequencies that they tallied today. Each represents a method for understanding when outcomes of an experiment are equally likely.

Closure

- **Exit Ticket:**
  - **a.** You flip a coin 100 times. How many times do you expect to flip heads? about 50 times
  - **b.** A bag contains 3 red chips and 9 blue chips. You draw a chip and replace it 100 times. How many times do you expect to draw a blue chip? about 75 times

3 ACTIVITY: Conducting an Experiment

Work with a partner. You toss a thumbtack onto a table. There are two ways the thumbtack can land.

Point up

On its side

a. Your friend says that because there are two outcomes, the probability of the thumbtack landing point up must be $\frac{1}{2}$. Do you think this conclusion is true? Explain.

b. Toss a thumbtack onto a table 50 times and record your results. In a *uniform probability model,* each outcome is equally likely to occur. Do you think this experiment represents a uniform probability model? Explain. Use the relative frequencies to complete the following.

$P(\text{point up}) = \square \qquad P(\text{on its side}) = \square$

Math Practice

Analyze Relationships

How can you use the results of your experiment to determine whether this is a uniform probability model?

What Is Your Answer?

4. **IN YOUR OWN WORDS** How can you use relative frequencies to find probabilities? Give an example.
5. Your friend rolls a number cube 500 times. How many times do you think your friend will roll an odd number? Explain your reasoning.
6. In Activity 2, your friend says, "There are no orange-colored chips in the bag." Do you think this conclusion is truc? Explain.
7. Give an example of an experiment that represents a uniform probability model.
8. Tell whether you can use each spinner to represent a uniform probability model. Explain your reasoning.

a.

b.

c.

Practice Use what you learned about relative frequencies to complete Exercises 6 and 7 on page 649.

Check It Out
Lesson Tutorials
BigIdeasMath.com

Key Vocabulary
relative frequency, *p. 644*
experimental probability, *p. 646*
theoretical probability, *p. 647*

Key Idea

Experimental Probability

Probability that is based on repeated trials of an experiment is called **experimental probability**.

$$P(\text{event}) = \frac{\text{number of times the event occurs}}{\text{total number of trials}}$$

EXAMPLE 1 Finding an Experimental Probability

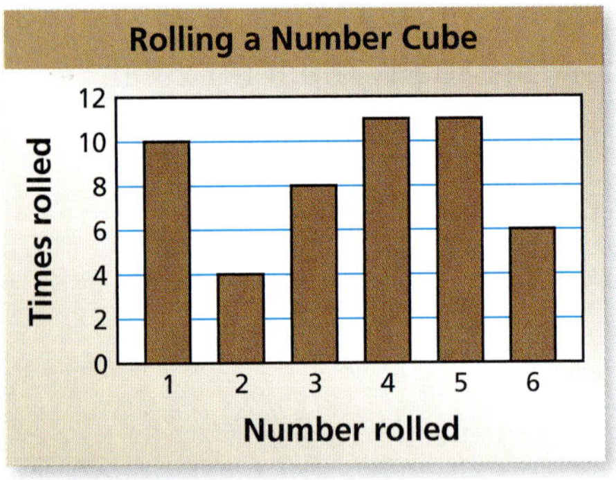

The bar graph shows the results of rolling a number cube 50 times. What is the experimental probability of rolling an odd number?

The bar graph shows 10 ones, 8 threes, and 11 fives. So, an odd number was rolled $10 + 8 + 11 = 29$ times in a total of 50 rolls.

$$P(\text{event}) = \frac{\text{number of times the event occurs}}{\text{total number of trials}}$$

$$P(\text{odd}) = \frac{29}{50}$$

An odd number was rolled 29 times.

There was a total of 50 rolls.

∴ The experimental probability is $\frac{29}{50}$, 0.58, or 58%.

EXAMPLE 2 Making a Prediction

"April showers bring May flowers." Old Proverb, 1557

It rains 2 out of the last 12 days in March. If this trend continues, how many rainy days would you expect in April?

Find the experimental probability of a rainy day.

$$P(\text{event}) = \frac{\text{number of times the event occurs}}{\text{total number of trials}}$$

$$P(\text{rain}) = \frac{2}{12} = \frac{1}{6}$$

It rains 2 days.

There is a total of 12 days.

To make a prediction, multiply the probability of a rainy day by the number of days in April.

$$\frac{1}{6} \cdot 30 = 5$$

∴ So, you can predict that there will be 5 rainy days in April.

Multi-Language Glossary at BigIdeasMath.com

Laurie's Notes

Introduction

Connect

- **Yesterday:** Students used relative frequencies to find probabilities.
- **Today:** Students will compute the experimental probability of an event and the theoretical probability of an event.

Motivate

- Play *Mystery Bag*. Before students arrive, place 10 cubes of the same shape and size in a paper bag; five of one color and five of a second color.
- Ask a volunteer to be the detective.
- ? "There are 10 cubes in my bag. Can you guess what color they are?" not likely
- Let the student remove a cube and look at its color.
- ? "Can you guess what color my cubes are?" not likely
- *Replace the cube.* Let the student pick again and see the color. Repeat your question.
- Try this 5–8 times until the student is ready to guess. The number of trials will depend upon the results and the student. You want students to see that they are collecting data and making a prediction.

Lesson Notes

Key Idea

- Discuss experimental probability and make the connection to the activities and to relative frequencies that students found.

Example 1

- ? "What information is given in the bar graph that will help answer the question?" The total number of times an odd number was rolled.
- ? "How do you write a fraction as a percent?" *Sample answer:* Write an equivalent fraction with a denominator of 100. Then write the numerator with the percent symbol.

Example 2

- Note the important phrase, *if this trend continues.* Knowing the weather for the last 12 days in March, you make a prediction about the weather in April.
- ? "Does it seem reasonable to use information from late March to predict weather in April?" Some students may say the weather in April is different, which is why the problem was phrased *if this trend continues.*
- **Big Idea:** The experimental probability is used to make a prediction when you expect a trend to continue, or you believe the experiment reflects what might be true about a larger population.

Goal Today's lesson is finding the **experimental probability** of an event and the **theoretical probability** of an event.

Technology for the Teacher

Dynamic Classroom

Lesson Tutorials
Lesson Plans
Answer Presentation Tool

Extra Example 1

Using the bar graph and results from Example 1, what is the experimental probability of rolling a prime number? $\frac{23}{50}$, 0.46, or 46%

Extra Example 2

It rains 3 out of the last 15 days in May. If this trend continues, how many rainy days would you expect in June? 6 days

JUNE

SUN	MON	TUE	WED	THU	FRI	SAT
			1	2	3	4
5	6	7	8	9	10	11
12	13	14	15	16	17	18
19	20	21	22	23	24	25
26	27	28	29	30		

On Your Own

1. $\frac{21}{50}$, 0.42, or 42%
2. 125

Extra Example 3

The letters in the word JACKSON are placed in a hat. You randomly choose a letter from the hat. What is the theoretical probability of choosing a vowel? $\frac{2}{7}$

Extra Example 4

The theoretical probability that you randomly choose a red marble from a bag is $\frac{5}{8}$. There are 40 marbles in the bag. How many are red? 25

Laurie's Notes

On Your Own

- **Neighbor Check:** Have students work independently and then have their neighbors check their work. Have students discuss any discrepancies.

? "What do you notice about P(odd) and P(even)?" sum to 1

Key Idea

- Write the Key Idea.
- Discuss theoretical probability and give several examples with which students would be familiar, such as cards, dice, and marbles in a bag. Stress that the outcomes must be equally likely.
- Explain to students that this is the type of probability we found in Section 15.2. Now we are calling it theoretical probability.

Example 3

- Work through the example.

? "What is the probability of *not* choosing a vowel?" $\frac{4}{7}$

? "How do you write $\frac{3}{7}$ as a percent?" *Sample answer:* Divide 3 by 7 and write as a decimal. Then write the decimal as a percent.

Example 4

? "What is the probability of winning a bobblehead?" $\frac{1}{6}$

? "If the spinner has 6 sections, then how many sections are bobblehead sections?" 1

? "Because there are 3 bobblehead sections on the prize wheel, what do you know about the total number of sections?" There must be more than 6.

- Write the proportion and solve.

On Your Own

Now You're Ready
Exercises 8–14

1. In Example 1, what is the experimental probability of rolling an even number?
2. At a clothing company, an inspector finds 5 defective pairs of jeans in a shipment of 200. If this trend continues, about how many pairs of jeans would you expect to be defective in a shipment of 5000?

Key Idea

Theoretical Probability

When all possible outcomes are equally likely, the **theoretical probability** of an event is the ratio of the number of favorable outcomes to the number of possible outcomes.

$$P(\text{event}) = \frac{\text{number of favorable outcomes}}{\text{number of possible outcomes}}$$

EXAMPLE 3 Finding a Theoretical Probability

You randomly choose one of the letters shown. What is the theoretical probability of choosing a vowel?

$$P(\text{event}) = \frac{\text{number of favorable outcomes}}{\text{number of possible outcomes}}$$

$$P(\text{vowel}) = \frac{3}{7}$$

There are 3 vowels.

There is a total of 7 letters.

The probability of choosing a vowel is $\frac{3}{7}$, or about 43%.

EXAMPLE 4 Using a Theoretical Probability

The theoretical probability of winning a bobblehead when spinning a prize wheel is $\frac{1}{6}$. The wheel has 3 bobblehead sections. How many sections are on the wheel?

$$P(\text{bobblehead}) = \frac{\text{number of bobblehead sections}}{\text{total number of sections}}$$

$\frac{1}{6} = \frac{3}{s}$ — Substitute. Let s be the total number of sections.

$s = 18$ — Cross Products Property

So, there are 18 sections on the wheel.

On Your Own

Now You're Ready
Exercises 15–23

3. In Example 3, what is the theoretical probability of choosing an X?

4. The theoretical probability of spinning an odd number on a spinner is 0.6. The spinner has 10 sections. How many sections have odd numbers?

5. The prize wheel in Example 4 was spun 540 times at a baseball game. About how many bobbleheads would you expect were won?

EXAMPLE 5 Comparing Experimental and Theoretical Probability

The bar graph shows the results of rolling a number cube 300 times.

a. What is the experimental probability of rolling an odd number?

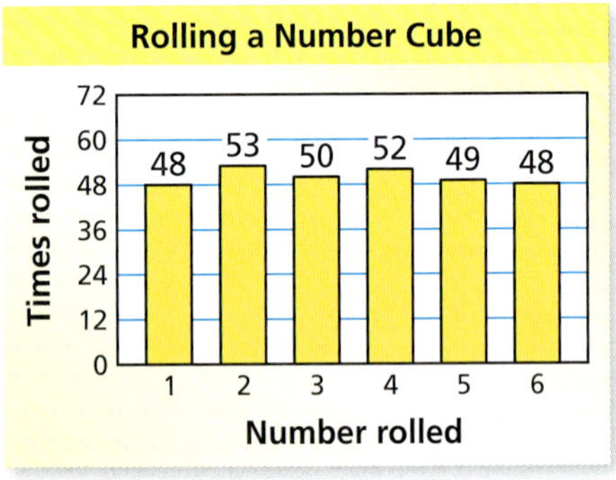

The bar graph shows 48 ones, 50 threes, and 49 fives. So, an odd number was rolled $48 + 50 + 49 = 147$ times in a total of 300 rolls.

$$P(\text{event}) = \frac{\text{number of times the event occurs}}{\text{total number of trials}}$$

$$P(\text{odd}) = \frac{147}{300}$$

An odd number was rolled 147 times.

There was a total of 300 rolls.

$$= \frac{49}{100}, \text{ or } 49\%$$

b. How does the experimental probability compare with the theoretical probability of rolling an odd number?

In Section 15.2, Example 2, you found that the theoretical probability of rolling an odd number is 50%. The experimental probability, 49%, is close to the theoretical probability.

c. Compare the experimental probability in part (a) to the experimental probability in Example 1.

As the number of trials increased from 50 to 300, the experimental probability decreased from 58% to 49%. So, it became closer to the theoretical probability of 50%.

On Your Own

Now You're Ready
Exercises 25–27

6. Use the bar graph in Example 5 to find the experimental probability of rolling a number greater than 1. Compare the experimental probability to the theoretical probability of rolling a number greater than 1.

Laurie's Notes

On Your Own

- **Think-Pair-Share:** Students should read each question independently and then work in pairs to answer the questions. When they have answered the questions, the pair should compare their answers with another group and discuss any discrepancies.

Example 5

? "What is different about the bar graph in Example 5 compared to the bar graph in Example 1?" The bar graph in Example 5 shows a greater number of trials; 300 instead of 50.

- Work through each part of the example as shown.
- **Reason Abstractly and Quantitatively:** Revisit the activities and discuss the fact that when the relative frequencies increase, the experimental probability gets closer and closer to the theoretical probability.

On Your Own

- **Neighbor Check:** Have students work independently and then have their neighbors check their work. Have students discuss any discrepancies.

Closure

- Use the colored cubes in the Mystery Bag. Reveal the contents, or use a different color ratio if you wish, and ask the following:
  - "If the contents of the Mystery Bag came from a bag of 1000 colored cubes, how many of each color would you predict in the bag of 1000?" Answers will vary depending upon materials.

On Your Own

3. $\frac{1}{7}$, or about 14.3%
4. 6 sections
5. 90 bobbleheads

Extra Example 5

Use the bar graph and results from Example 5.

a. What is the experimental probability of rolling an even number? $\frac{51}{100}$, 0.51, or 51%

b. How does the experimental probability compare with the theoretical probability of rolling an even number? The experimental probability, 51%, is close to the theoretical probability, 50%.

On Your Own

6. 84%; It is close to the theoretical probability of $83\frac{1}{3}$%.

English Language Learners

Vocabulary

This chapter contains many new terms that may cause English learners to struggle. Students should write the key vocabulary in their notebooks along with definitions and diagrams so that they become familiar and comfortable with the vocabulary.

Vocabulary and Concept Check

1. Perform an experiment several times. Count how often the event occurs and divide by the number of trials.
2. yes; You could flip tails 7 out of 10 times, but with more trials the probability of flipping tails should get closer to 0.5.
3. There is a 50% chance you will get a favorable outcome.
4. *Sample answer:* picking a 1 out of 1, 2, 3, 4
5. experimental probability; The population is too large to survey every person, so a sample will be used to predict the outcome.

Practice and Problem Solving

6. $\frac{7}{50}$, or 14%
7. $\frac{12}{25}$, or 48%
8. $\frac{7}{25}$, or 28%
9. $\frac{21}{25}$, or 84%
10. $\frac{17}{50}$, or 34%
11. 0, or 0%
12. $\frac{3}{20}$, or 15%
13. 45 tiles
14. 5 cards

Assignment Guide and Homework Check

Level	Assignment	Homework Check
Advanced	1–7, 14–34 even, 35–37	18, 22, 30, 32

Common Errors

- **Exercises 8–11** Students may forget to total all of the trials before writing the experimental probability. They may have an incorrect number of trials in the denominator. Remind them that they need to know the total number of trials when finding the probability.

15.3 Record and Practice Journal

Use the bar graph to find the experimental probability of the event.

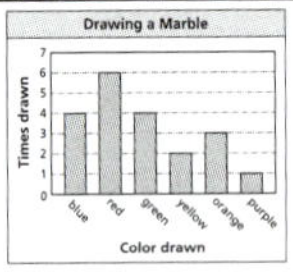

1. Drawing red $\frac{3}{10}$
2. Drawing orange $\frac{3}{20}$
3. Drawing *not* yellow $\frac{9}{10}$
4. Drawing a color with more than 4 letters in its name $\frac{1}{2}$
5. There are 25 students' names in a hat. You choose 5 names. Three are boys' names and two are girls' names. How many of the 25 names would you expect to be boys' names?
   15 boys' names

Use a number cube to determine the theoretical probability of the event.

6. Rolling a 2 $\frac{1}{6}$
7. Rolling a 5 $\frac{1}{6}$
8. Rolling an even number $\frac{1}{2}$
9. Rolling a number greater than 1 $\frac{5}{6}$

15.3 Exercises

Check It Out
Help with Homework
BigIdeasMath.com

Vocabulary and Concept Check

1. **VOCABULARY** Describe how to find the experimental probability of an event.
2. **REASONING** You flip a coin 10 times and find the experimental probability of flipping tails to be 0.7. Does this seem reasonable? Explain.
3. **VOCABULARY** An event has a theoretical probability of 0.5. What does this mean?
4. **OPEN-ENDED** Describe an event that has a theoretical probability of $\frac{1}{4}$.
5. **LOGIC** A pollster surveys randomly selected individuals about an upcoming election. Do you think the pollster will use experimental probability or theoretical probability to make predictions? Explain.

Practice and Problem Solving

Use the bar graph to find the relative frequency of the event.

6. Spinning a 6
7. Spinning an even number

Use the bar graph to find the experimental probability of the event.

1 8. Spinning a number less than 3
9. *Not* spinning a 1
10. Spinning a 1 or a 3
11. Spinning a 7

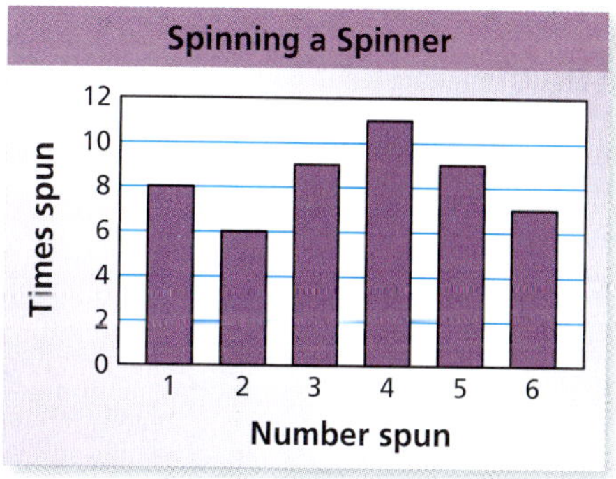

12. **EGGS** You check 20 cartons of eggs. Three of the cartons have at least one cracked egg. What is the experimental probability that a carton of eggs has at least one cracked egg?

2 13. **BOARD GAME** There are 105 lettered tiles in a board game. You choose the tiles shown. How many of the 105 tiles would you expect to be vowels?

14. **CARDS** You have a package of 20 assorted thank-you cards. You pick the four cards shown. How many of the 20 cards would you expect to have flowers on them?

Use the spinner to find the theoretical probability of the event.

3 **15.** Spinning red

16. Spinning a 1

17. Spinning an odd number

18. Spinning a multiple of 2

19. Spinning a number less than 7

20. Spinning a 9

21. **LETTERS** Each letter of the alphabet is printed on an index card. What is the theoretical probability of randomly choosing any letter except Z?

4 **22.** **GAME SHOW** On a game show, a contestant randomly chooses a chip from a bag that contains numbers and strikes. The theoretical probability of choosing a strike is $\frac{3}{10}$. The bag contains 9 strikes. How many chips are in the bag?

23. **MUSIC** The theoretical probability that a pop song plays on your MP3 player is 0.45. There are 80 songs on your MP3 player. How many of the songs are pop songs?

24. **MODELING** There are 16 females and 20 males in a class.

a. What is the theoretical probability that a randomly chosen student is female?

b. One week later, there are 45 students in the class. The theoretical probability that a randomly chosen student is a female is the same as last week. How many males joined the class?

The bar graph shows the results of spinning the spinner 200 times. Compare the theoretical and experimental probabilities of the event.

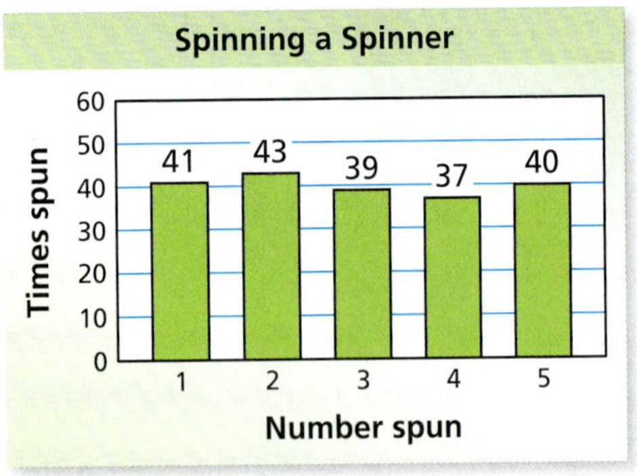

5 **25.** Spinning a 4

26. Spinning a 3

27. Spinning a number greater than 4

28. Should you use *theoretical* or *experimental* probability to predict the number of times you will spin a 3 in 10,000 spins?

29. **NUMBER SENSE** The table at the right shows the results of flipping two coins 12 times each.

HH	HT	TH	TT
2	6	3	1

a. What is the experimental probability of flipping two tails? Using this probability, how many times can you expect to flip two tails in 600 trials?

HH	HT	TH	TT
23	29	26	22

b. The table at the left shows the results of flipping the same two coins 100 times each. What is the experimental probability of flipping two tails? Using this probability, how many times can you expect to flip two tails in 600 trials?

c. Why is it important to use a large number of trials when using experimental probability to predict results?

Common Errors

- **Exercises 15–20** Students may write a different probability than what is asked, or forget to include a favorable outcome. For example, in Exercise 15 a student may not realize that there are two red sections and will write the probability as $\frac{1}{6}$ instead of $\frac{1}{3}$. Remind them to read the event carefully and to write the favorable outcomes before finding the probability.
- **Exercise 22** Students may write an incorrect proportion when finding how many strikes are in the bag. Encourage them to write the proportion in words before substituting and solving.

Practice and Problem Solving

15. $\frac{1}{3}$, or about 33.3%

16. $\frac{1}{6}$, or about 16.7%

17. $\frac{1}{2}$, or 50% **18.** $\frac{1}{2}$, or 50%

19. 1, or 100% **20.** 0, or 0%

21. $\frac{25}{26}$, or about 96.2%

22. 30 chips **23.** 36 songs

24. a. $\frac{4}{9}$, or about 44.4%

b. 5 males

25. theoretical: $\frac{1}{5}$, or 20%;

experimental: $\frac{37}{200}$, or 18.5%;

The experimental probability is close to the theoretical probability.

26. theoretical: $\frac{1}{5}$, or 20%;

experimental: $\frac{39}{200}$, or 19.5%;

The experimental probability is close to the theoretical probability.

27. theoretical: $\frac{1}{5}$, or 20%;

experimental: $\frac{1}{5}$, or 20%;

The probabilities are equal.

28–29. See Additional Answers.

English Language Learners

Group Activity

Set up groups of English learners and English speakers. Have each group predict the number of times a coin will land on *heads* when flipped 30 times. Each group should track their results in a frequency table. Ask groups to present their predictions and results to the class. Then combine the results of all the groups and discuss how the combined results compare to the individual group results.

Practice and Problem Solving

30–32. See *Taking Math Deeper*.

33. a. As the number of trials increases, the most likely sum will change from 6 to 7.

b. As an experiment is repeated over and over, the experimental probability of an event approaches the theoretical probability of the event.

34. See Additional Answers.

Fair Game Review

35. 4% **36.** 3.5%

37. D

Mini-Assessment

You have three sticks. Each stick has one red side and one blue side. You throw the sticks 10 times and record the results. Use the table to find the experimental probability of the event.

Outcome	Frequency
3 red	3
3 blue	2
2 blue, 1 red	4
2 red, 1 blue	1

1. Tossing 3 blue $\frac{1}{5}$

2. Tossing 2 blue, 1 red $\frac{2}{5}$

3. *Not* tossing all blue $\frac{4}{5}$

Use the spinner to determine the theoretical probability of the event.

4. P(purple) $\frac{1}{6}$ **5.** P(3) $\frac{1}{6}$

6. P(even) $\frac{1}{2}$ **7.** P(multiple of 3) $\frac{1}{3}$

Taking Math Deeper

Exercises 30–32

In this problem, students compare *experimental probabilities* (results of trials), shown in the bar graph, with the *theoretical probabilities* (all possible outcomes), shown in the table.

1 Make a list of the different sums and the number of times rolled from the bar graph. Check that the total is 60. Find the experimental probability of each sum.

2: 2/60 3: 4/60 4: 5/60 5: 6/60 6: 13/60 7: 10/60
8: 6/60 9: 8/60 10: 2/60 11: 3/60 12: 1/60

30. No, each sum is not equally likely. The experimental probabilities vary. The most likely sum is 6.

2 Make a list of all possible ways to get each sum.

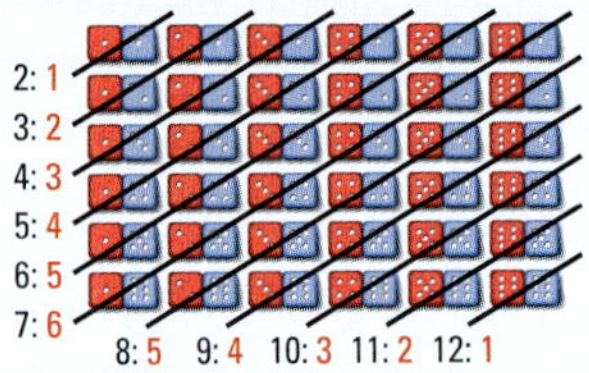

Find the theoretical probability of each sum.

2: 1/36 3: 2/36 4: 3/36 5: 4/36 6: 5/36 7: 6/36
8: 5/36 9: 4/36 10: 3/36 11: 2/36 12: 1/36

31. No, because there is not an equal number of possible ways to get each sum, each sum is not equally likely. So, the theoretical probabilities vary. The most likely sum is 7.

3 **32.** Compare the two types of probabilities with a double bar graph.

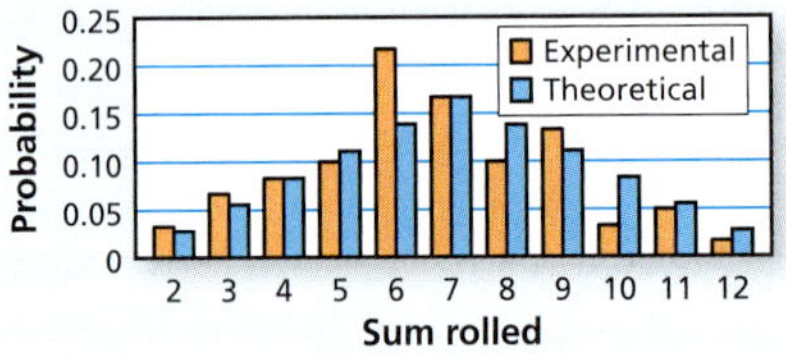

Reteaching and Enrichment Strategies

If students need help. . .	If students got it. . .
Resources by Chapter • Practice A and Practice B • Puzzle Time Record and Practice Journal Practice Differentiating the Lesson Lesson Tutorials Skills Review Handbook	Resources by Chapter • Enrichment and Extension • Technology Connection Start the next section

You roll a pair of number cubes 60 times. You record your results in the bar graph shown.

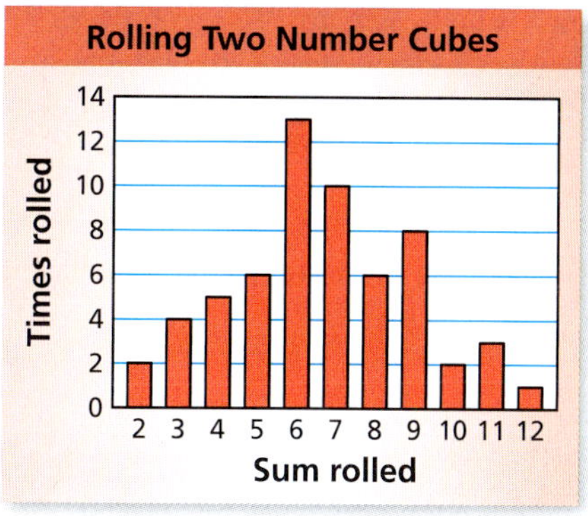

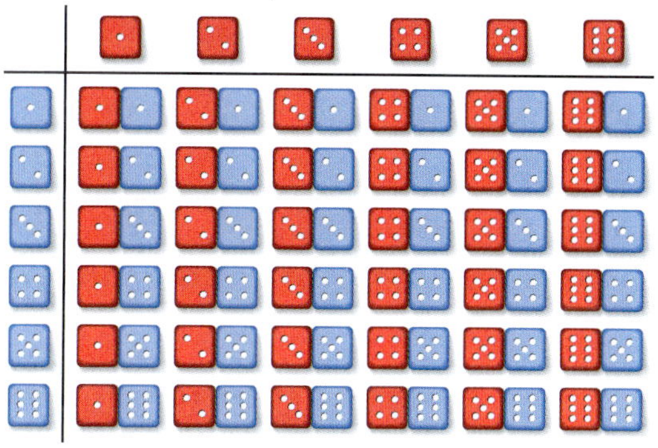

30. Use the bar graph to find the experimental probability of rolling each sum. Is each sum equally likely? Explain. If not, which is most likely?

31. Use the table to find the theoretical probability of rolling each sum. Is each sum equally likely? Explain. If not, which is most likely?

32. PROBABILITIES Compare the probabilities you found in Exercises 30 and 31.

33. REASONING Consider the results of Exercises 30 and 31.

a. Which sum would you expect to be most likely after 500 trials? 1000 trials? 10,000 trials?

b. Explain how experimental probability is related to theoretical probability as the number of trials increases.

34. **Project** When you toss a paper cup into the air, there are three ways for the cup to land: *open-end up, open-end down,* or *on its side.*

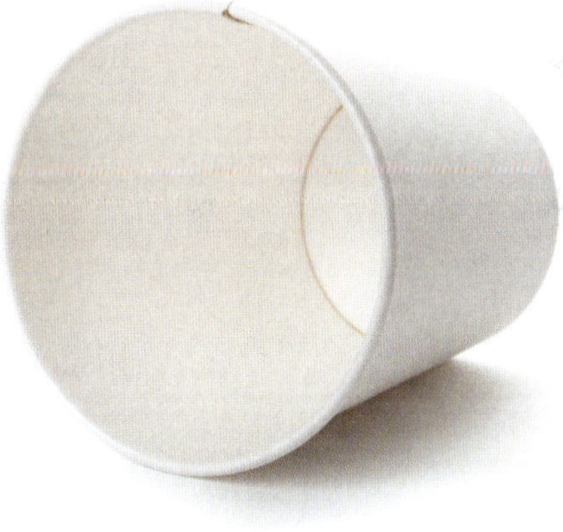

a. Toss a paper cup 100 times and record your results. Do the outcomes for tossing the cup appear to be equally likely? Explain.

b. What is the probability of the cup landing open-end up? open-end down? on its side?

c. Use your results to predict the number of times the cup lands on its side in 1000 tosses.

d. Suppose you tape a quarter to the bottom of the cup. Do you think the cup will be *more likely* or *less likely* to land open-end up? Justify your answer.

Fair Game Review What you learned in previous grades & lessons

Find the annual interest rate. *(Skills Review Handbook)*

35. $I = \$16$, $P = \$200$, $t = 2$ years

36. $I = \$26.25$, $P = \$500$, $t = 18$ months

37. MULTIPLE CHOICE The volume of a prism is 9 cubic yards. What is its volume in cubic feet? *(Section 14.4)*

Ⓐ 3 ft^3 Ⓑ 27 ft^3 Ⓒ 81 ft^3 Ⓓ 243 ft^3

15.4 Compound Events

Essential Question How can you find the number of possible outcomes of one or more events?

1 ACTIVITY: Comparing Combination Locks

Work with a partner. You are buying a combination lock. You have three choices.

a. This lock has 3 wheels. Each wheel is numbered from 0 to 9.

The least three-digit combination possible is .

The greatest three-digit combination possible is .

How many possible combinations are there?

b. Use the lock in part (a).

There are possible outcomes for the first wheel.

There are possible outcomes for the second wheel.

There are possible outcomes for the third wheel.

How can you use multiplication to determine the number of possible combinations?

c. This lock is numbered from 0 to 39. Each combination uses three numbers in a right, left, right pattern. How many possible combinations are there?

d. This lock has 4 wheels.

Wheel 1: 0–9

Wheel 2: A–J

Wheel 3: K–T

Wheel 4: 0–9

How many possible combinations are there?

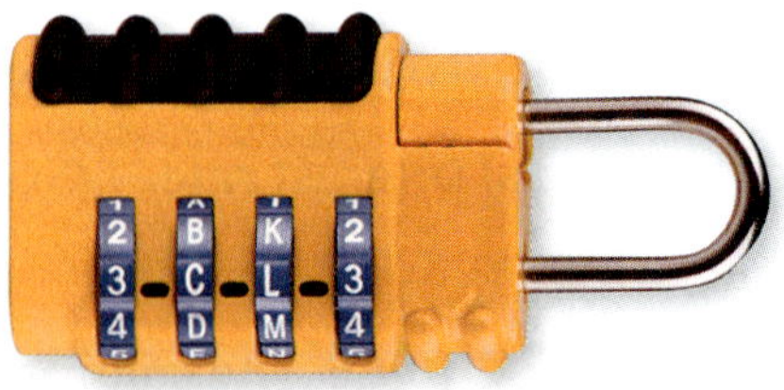

e. For which lock is it most difficult to guess the combination? Why?

Probability and Statistics

In this lesson, you will

- use tree diagrams, tables, or a formula to find the number of possible outcomes.
- find probabilities of compound events.

Laurie's Notes

Introduction

Applying Mathematical Practices

- **Model with Mathematics:** Mathematically proficient students use visual models to represent problems. Tree diagrams can be used to visualize the possible outcomes of events.

Motivate

- **Acting Time:** If possible, dress as Sherlock Holmes with overcoat, hat, eye monocle, and pipe. Probe students about your identity by sharing a little information:
  - You live on 221b Baker Street in London.
  - You owe your fame to Sir Arthur Conan Doyle (your creator).
  - You were first written about in 1887.
  - You have a friend named Watson and a foe named Professor Moriarty.
- Tell students you have a case that you're working on and they can help you with it. The principal put the teachers' paychecks in a locked box and forgot the combination.
- ? Hold up a combination lock and tell the students that you're trying to figure out the combination. "How many different combinations do you think I would need to try in order to open this lock?"
- Tell students that today's activity may help them answer this question.

Activity Notes

Activity 1

- If your school has lockers with combination locks, discuss their operation.
- ? "Do any of you have combination locks for your bike, skis, or other possessions?"
- **FYI:** Many local post offices have boxes with combinations. Some post offices now have digital locks where a code must be entered.
- Students are guided with questions in parts (a) and (b) to help them reason about the number of possible outcomes.
- ? "If the digits 0–9 are used, how many outcomes are there?" 10
- When students finish part (b), make sure they understand the connection to multiplication. This prepares them for parts (c) and (d), laying the foundation for the Fundamental Counting Principle.
- ? "In part (c), how many possible outcomes are there when the dial is turned to the right?" 40
- If students have made sense of parts (a) and (b), they will understand that the number of combinations in part (c) is 40 • 40 • 40.
- It may be helpful for some students if sample combinations are listed. For instance in part (d), possible combinations are 4AK0, 4AK1, 4AK2, etc.
- **Attend to Precision:** Ask a volunteer to explain his or her reasoning for parts (d) and (e).

What Your Students Will Learn

- Use tree diagrams, tables, or a formula to find the number of possible outcomes of events.
- Find probabilities of compound events.

Previous Learning

Students have counted the number of possible outcomes of an event.

Technology for the Teacher

Dynamic Classroom

Lesson Plans
Complete Materials List

15.4 Record and Practice Journal

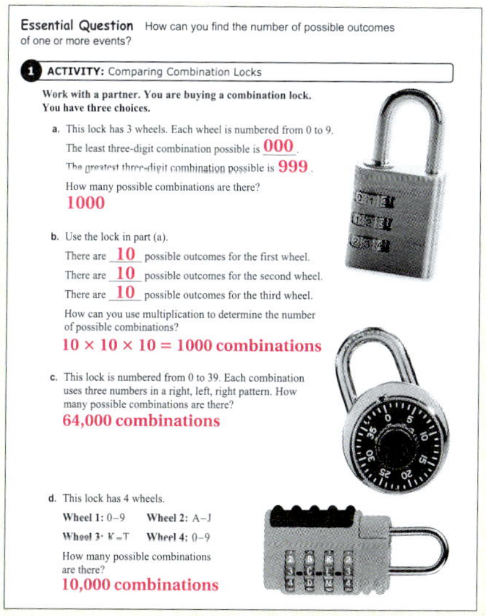

Essential Question How can you find the number of possible outcomes of one or more events?

1 ACTIVITY: Comparing Combination Locks

Work with a partner. You are buying a combination lock. You have three choices.

a. This lock has 3 wheels. Each wheel is numbered from 0 to 9.
The least three-digit combination possible is 000.
The greatest three-digit combination possible is 999.
How many possible combinations are there?
1000

b. Use the lock in part (a).
There are 10 possible outcomes for the first wheel.
There are 10 possible outcomes for the second wheel.
There are 10 possible outcomes for the third wheel.
How can you use multiplication to determine the number of possible combinations?
$10 \times 10 \times 10 = 1000$ combinations

c. This lock is numbered from 0 to 39. Each combination uses three numbers in a right, left, right pattern. How many possible combinations are there?
64,000 combinations

d. This lock has 4 wheels.
Wheel 1: 0–9 **Wheel 2:** A–J
Wheel 3: K–T **Wheel 4:** 0–9
How many possible combinations are there?
10,000 combinations

Differentiated Instruction

Visual

When solving counting problems like part (a) in Activity 2, have students write blanks to represent each outcome.

Step 1: Write 4 blanks to show that the password has 4 digits.

____ ____ ____ ____

Step 2: Because there are 10 choices for each of the digits, write:

10 10 10 10

Step 3: Multiply all of the values together to get 10,000 different possible passwords.

15.4 Record and Practice Journal

e. For which lock is it most difficult to guess the combination? Why?
lock in part (c)

2 ACTIVITY: Comparing Password Security

Work with a partner. Which password requirement is most secure? Explain your reasoning. Include the number of different passwords that are possible for each requirement.

a. The password must have four digits.
10,000 passwords

Username: funnydog
Password: 2335
Sign in

b. The password must have five digits.
100,000 passwords

Username: rascal1007
Password: 06772
Sign in

c. The password must have six letters.
308,915,776 passwords

Username: supergrowl
Password: AFYYWP
Sign in

d. The password must have eight digits or letters.
2,821,109,907,456 passwords; The password requirement in part (d) is most secure.

Username: jupitermars
Password: 7TT3PX4W
Sign in

What Is Your Answer?

3. **IN YOUR OWN WORDS** How can you find the number of possible outcomes of one or more events?
Sample answer: You can use an organized list, a table, a tree diagram, or multiplication.

4. **SECURITY** A hacker uses a software program to guess the passwords in Activity 2. The program checks 600 passwords per minute. What is the greatest amount of time it will take the program to guess each of the four types of passwords?

a. four digits
16 minutes 40 seconds

b. five digits
166 minutes 40 seconds

c. six letters
514,859 minutes 38 seconds, or about 358 days

d. eight digits or letters
4,701,849,846 minutes, or about 8946 years

Laurie's Notes

Activity 2

- If students have usernames and passwords for computers or any other aspect of school, be sure to discuss this.
- In this activity, the students are not considering the username, only the password.
- Point out that the passwords are *not* case sensitive, meaning the letters could be either upper- or lower-case.
- ? If students have difficulty getting started, ask a simpler question such as, "How many passwords are possible using just 2 digits, 0 to 9?" 100: 00, 01, 02, 03, . . . , 99
- Remind students that the digit "0" is included so there are 10 possible digits when numbers are considered.
- As with Activity 1, the numbers or letters can be repeated.
- For parts (c) and (d), you might have students describe how to find the answer instead of having them actually compute the answer.

What Is Your Answer?

- If time permits, have students convert their answer of minutes to a larger unit of time such as hours, days, and years.

Closure

- **Exit Ticket:** Help the teachers get their paychecks by determining how many different combinations there are for a lock that has 3 dials, each one containing the digits 0 through 4. $5 \cdot 5 \cdot 5 = 125$

2 ACTIVITY: Comparing Password Security

Work with a partner. Which password requirement is most secure? Explain your reasoning. Include the number of different passwords that are possible for each requirement.

a. The password must have four digits.

b. The password must have five digits.

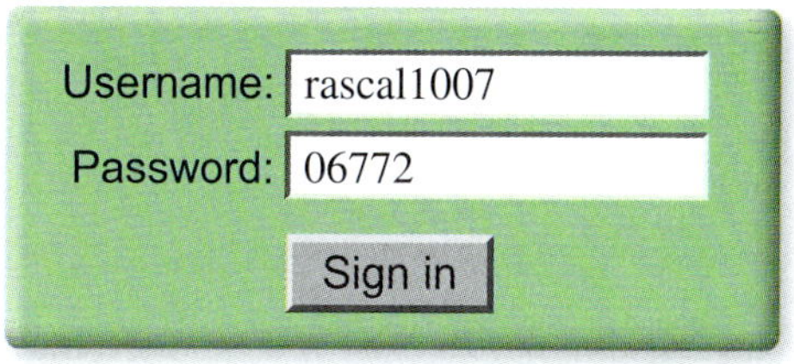

c. The password must have six letters.

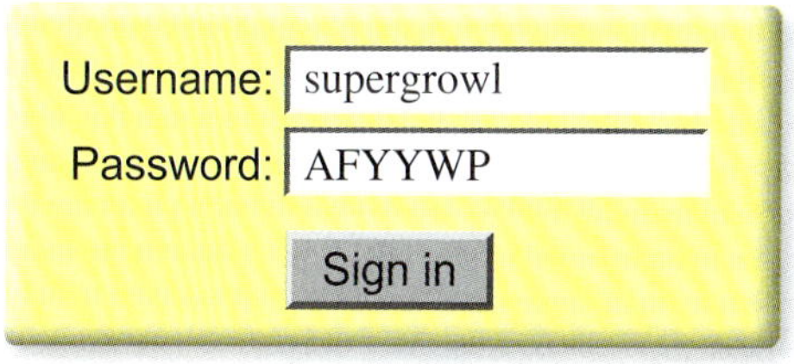

d. The password must have eight digits or letters.

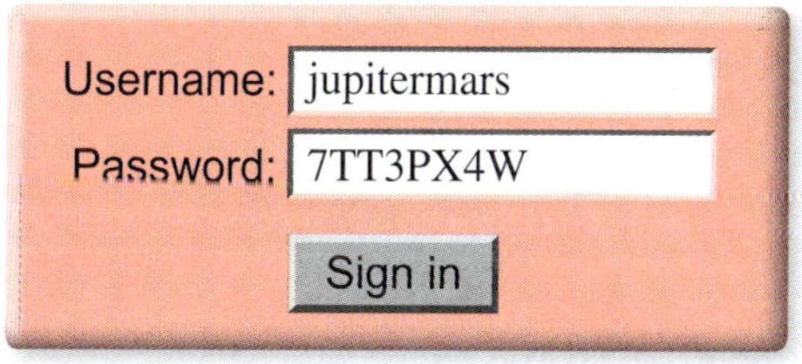

Math Practice

View as Components

What is the number of possible outcomes for each character of the password? Explain.

What Is Your Answer?

3. **IN YOUR OWN WORDS** How can you find the number of possible outcomes of one or more events?

4. **SECURITY** A hacker uses a software program to guess the passwords in Activity 2. The program checks 600 passwords per minute. What is the greatest amount of time it will take the program to guess each of the four types of passwords?

Practice Use what you learned about the total number of possible outcomes of one or more events to complete Exercise 5 on page 657.

15.4 Lesson

Check It Out
Lesson Tutorials
BigIdeasMath.com

Key Vocabulary
sample space, *p. 654*
Fundamental Counting Principle, *p. 654*
compound event, *p. 656*

The set of all possible outcomes of one or more events is called the **sample space**.

You can use tables and tree diagrams to find the sample space of two or more events.

EXAMPLE 1 Finding a Sample Space

Crust

- Thin Crust
- Stuffed Crust

Style

- Hawaiian
- Mexican
- Pepperoni
- Veggie

You randomly choose a crust and style of pizza. Find the sample space. How many different pizzas are possible?

Use a tree diagram to find the sample space.

Crust	*Style*	*Outcome*
Thin	Hawaiian	Thin Crust Hawaiian
	Mexican	Thin Crust Mexican
	Pepperoni	Thin Crust Pepperoni
	Veggie	Thin Crust Veggie
Stuffed	Hawaiian	Stuffed Crust Hawaiian
	Mexican	Stuffed Crust Mexican
	Pepperoni	Stuffed Crust Pepperoni
	Veggie	Stuffed Crust Veggie

There are 8 different outcomes in the sample space. So, there are 8 different pizzas possible.

On Your Own

Now You're Ready
Exercises 6 and 7

1. **WHAT IF?** The pizza shop adds a deep dish crust. Find the sample space. How many pizzas are possible?

Another way to find the total number of possible outcomes is to use the **Fundamental Counting Principle**.

Study Tip
The Fundamental Counting Principle can be extended to more than two events.

Key Idea

Fundamental Counting Principle

An event M has m possible outcomes. An event N has n possible outcomes. The total number of outcomes of event M followed by event N is $m \times n$.

Multi-Language Glossary at BigIdeasMath.com

Laurie's Notes

Introduction

Connect

- **Yesterday:** Students explored how the number of choices on locks and in passwords affected the total number of combinations possible.
- **Today:** Students will find the number of outcomes of compound events.

Motivate

- Display 3 different cups on your desk. I like to use a ceramic cup, a travel mug, and a foam cup. In addition, have a tea bag and hot cocoa mix.
- Pose to students: You are thirsty. You have three different cups to select from and two different beverages.
- ? "How many different ways can I select a cup and a beverage?" The answer of 6 may or may not be obvious. Hold up the travel mug and tea bag, and then the travel mug and cocoa. Repeat for the other two cups.
- ? "How many different ways if I add another cup, say a heavy plastic cup?" 8
- ? "How many different ways if I add another beverage, say coffee?" 9

Lesson Notes

Discuss

- Today's lesson is about outcomes of one event followed by one or more other events.
- Define sample space. Give examples.

Example 1

- ? Refer to the pizza shop menu and ask, "How many types of crust are available?" two Write the possible crusts with space between them as shown.
- ? "For either crust, how many different styles of pizza can you order?" four List the four styles with each crust using the tree diagram as shown.
- **Model with Mathematics:** The tree diagram helps students visualize the 8 outcomes in the sample space.
- **Extension:** Add a size to each (10" and 14"). Have students determine the number of possible outcomes. 16

On Your Own

- Ask a volunteer to share his or her thinking about the question.

Key Idea

- Write the Fundamental Counting Principle.
- Revisit previous problems to see if the Fundamental Counting Principle gives the same answer.

Goal Today's lesson is using **sample spaces** and the total number of possible outcomes to find probabilities of **compound events**.

Technology for the Teacher

Dynamic Classroom

Lesson Tutorials
Lesson Plans
Answer Presentation Tool

Extra Example 1

At a sub shop, you can choose ham, turkey, or roast beef on either white or wheat bread. You randomly choose a meat and bread. Find the sample space. How many subs are possible?

ham on white bread,
ham on wheat bread,
turkey on white bread,
turkey on wheat bread,
roast beef on white bread,
roast beef on wheat bread; 6

On Your Own

1. thin crust Hawaiian, thin crust Mexican, thin crust pepperoni, thin crust veggie, stuffed crust Hawaiian, stuffed crust Mexican, stuffed crust pepperoni, stuffed crust veggie, deep dish Hawaiian, deep dish Mexican, deep dish pepperoni, deep dish veggie; 12

English Language Learners

Writing

Pair or group English speakers with English learners. Have each group of students write a problem involving the Fundamental Counting Principle. On a separate sheet of paper, have the groups show a detailed solution to their problem. Next, have groups exchange their problems with other groups. Students in each group should work together to solve the problem they receive. Both groups should meet to discuss the problems and solutions.

Extra Example 2

Find the total number of possible outcomes of rolling two number cubes. $6 \times 6 = 36$

Extra Example 3

How many different outfits can you make from 5 T-shirts, 3 pairs of jeans, and 2 pairs of shoes? $5 \times 3 \times 2 = 30$

On Your Own

2. 20

3. 100

Laurie's Notes

Example 2

- Use a number cube and coin to model this example.
- Work through the example using each method.
- Discuss the efficiency of using the Fundamental Counting Principle instead of using a table. If you only need to know the number of outcomes, the Fundamental Counting Principle should be used. The table, however, shows the sample space instead of just the number of outcomes.

? "Is the answer the same if the coin is flipped first followed by rolling the number cube?" yes; The order changes but the number of outcomes is the same.

- **FYI:** Lists generated from tables and tree diagrams like those in Examples 1 and 2 are examples of organized lists.

Example 3

- Ask a volunteer to tell how many of each type of clothing is shown in the picture.
- In working through this question there will be students who suggest that certain shoes would not be worn with certain jeans, and so on. Remind students that the question is, "How many different outfits can be made, not would you wear the outfit."

? "Would you want to make a tree diagram or table to answer this question? Explain." no; It would take a long time to list the possible outcomes. The Fundamental Counting Principle is more efficient in finding the number of different outfits possible.

On Your Own

- Question 2 could be modeled using a spinner and five pieces of paper.

EXAMPLE 2 Finding the Total Number of Possible Outcomes

Find the total number of possible outcomes of rolling a number cube and flipping a coin.

Method 1: Use a table to find the sample space. Let H = heads and T = tails.

	1	2	3	4	5	6
	1H	2H	3H	4H	5H	6H
	1T	2T	3T	4T	5T	6T

∴ There are 12 possible outcomes.

Method 2: Use the Fundamental Counting Principle. Identify the number of possible outcomes of each event.

Event 1: Rolling a number cube has 6 possible outcomes.

Event 2: Flipping a coin has 2 possible outcomes.

$6 \times 2 = 12$ Fundamental Counting Principle

∴ There are 12 possible outcomes.

EXAMPLE 3 Finding the Total Number of Possible Outcomes

How many different outfits can you make from the T-shirts, jeans, and shoes in the closet?

Use the Fundamental Counting Principle. Identify the number of possible outcomes for each event.

Event 1: Choosing a T-shirt has 7 possible outcomes.

Event 2: Choosing jeans has 4 possible outcomes.

Event 3: Choosing shoes has 3 possible outcomes.

$7 \times 4 \times 3 = 84$ Fundamental Counting Principle

∴ So, you can make 84 different outfits.

On Your Own

Now You're Ready
Exercises 8–11

2. Find the total number of possible outcomes of spinning the spinner and choosing a number from 1 to 5.

3. How many different outfits can you make from 4 T-shirts, 5 pairs of jeans, and 5 pairs of shoes?

A **compound event** consists of two or more events. As with a single event, the probability of a compound event is the ratio of the number of favorable outcomes to the number of possible outcomes.

EXAMPLE 4 Finding the Probability of a Compound Event

In Example 2, what is the probability of rolling a number greater than 4 and flipping tails?

There are two favorable outcomes in the sample space for rolling a number greater than 4 and flipping tails: 5T and 6T.

$$P(\text{event}) = \frac{\text{number of favorable outcomes}}{\text{number of possible outcomes}}$$

$$P(\text{greater than 4 and tails}) = \frac{2}{12} \qquad \text{Substitute.}$$

$$= \frac{1}{6} \qquad \text{Simplify.}$$

∴ The probability is $\frac{1}{6}$, or $16\frac{2}{3}\%$.

EXAMPLE 5 Finding the Probability of a Compound Event

You flip three nickels. What is the probability of flipping two heads and one tails?

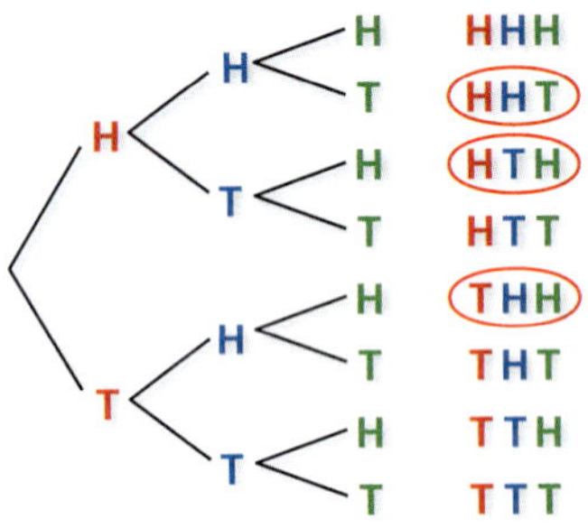

Use a tree diagram to find the sample space. Let H = heads and T = tails.

There are three favorable outcomes in the sample space for flipping two heads and one tails: HHT, HTH, and THH.

$$P(\text{event}) = \frac{\text{number of favorable outcomes}}{\text{number of possible outcomes}}$$

$$P(\text{2 heads and 1 tails}) = \frac{3}{8} \qquad \text{Substitute.}$$

∴ The probability is $\frac{3}{8}$, or 37.5%.

On Your Own

Now You're Ready
Exercises 15–24

4. In Example 2, what is the probability of rolling at most 4 and flipping heads?
5. In Example 5, what is the probability of flipping at least two tails?
6. You roll two number cubes. What is the probability of rolling double threes?
7. In Example 1, what is the probability of choosing a stuffed crust Hawaiian pizza?

Laurie's Notes

Discuss

- Define a *compound event.* Refer to first three examples and explain that they all involved compound events. Tell students that the next step is to determine probabilities of compound events.
- Define the probability of a compound event: the ratio of the number of favorable outcomes to the number of possible outcomes.

Example 4

? "How many possible outcomes are there when you roll a number cube and flip a coin as in Example 2?" 12

? "How many favorable outcomes are there for rolling a number greater than 4 and flipping tails? Explain." There are 2 favorable outcomes, 5T and 6T.

- So, $P(\text{rolling greater than 4 and flipping tails}) = \frac{2}{12} = \frac{1}{6}$.
- Students should now see the value in creating tree diagrams and tables to find sample spaces. When finding probabilities, it can help easily identify the number of favorable outcomes and number of possible outcomes.

Example 5

- **Model with Mathematics:** Read the problem and ask students to draw a tree diagram to find the sample space. The tree diagram makes the number of favorable outcomes and the number of possible outcomes easy to find.

? "If you use the Fundamental Counting Principle, how many possible outcomes are there?" $2 \cdot 2 \cdot 2 = 8$

- Continue to work the problem as shown.

On Your Own

- **Common Error:** In Exercise 6, students often say that there are 12 possible outcomes (6 + 6) instead of 36 possible outcomes (6 × 6).
- **Extension:** Ask students to find the probability of randomly selecting the green T-shirt, green jeans, and green shoes in Example 3. Rather than list the sample space, they should reason that there is only 1 favorable outcome, so the probability is $\frac{1}{84}$.

Closure

- **Writing Prompt:** The Fundamental Counting Principle is used to . . .
  Tree diagrams and tables can be used to . . .

Extra Example 4

In Example 2, what is the probability of rolling an even number and flipping heads? 25%

Extra Example 5

You flip three dimes. What is the probability of flipping three heads? 12.5%

On Your Own

4. $\frac{1}{3}$, or $33\frac{1}{3}\%$

5. $\frac{1}{2}$, or 50%

6. $\frac{1}{36}$, or about $2\frac{7}{9}\%$

7. $\frac{1}{8}$, or 12.5%

Vocabulary and Concept Check

1. A sample space is the set of all possible outcomes of an event. Use a table or tree diagram to list all the possible outcomes.
2. An event M has m possible outcomes and event N has n possible outcomes. The total number of outcomes of event M followed by event N is $m \times n$.
3. You could use a tree diagram or the Fundamental Counting Principle. Either way, the total number of possible outcomes is 30.
4. *Sample answer:* choosing two marbles from a bag

Practice and Problem Solving

5. 125,000
6. *Sample space:* Miniature golf 1 P.M.–3 P.M., Miniature golf 6 P.M.–8 P.M., Laser tag 1 P.M.–3 P.M., Laser tag 6 P.M.–8 P.M., Roller skating 1 P.M.–3 P.M., Roller skating 6 P.M.–8 P.M.; 6 possible outcomes
7. *Sample space:* Realistic Lion, Realistic Bear, Realistic Hawk, Realistic Dragon, Cartoon Lion, Cartoon Bear, Cartoon Hawk, Cartoon Dragon; 8 possible outcomes
8. 21
9. 20
10. 24
11. 60
12. See Additional Answers.
13. The possible outcomes of each question should be multiplied, not added. The correct answer is $2 \times 2 \times 2 \times 2 \times 2 = 32$.

Assignment Guide and Homework Check

Level	Assignment	Homework Check
Advanced	1–5, 6–12 even, 13, 14–30 even, 31–33	10, 18, 24, 26, 28

Common Errors

- **Exercises 8–11** Students may add the number of possible outcomes for each event rather than multiply them. Remind them that the total number of outcomes of two or more events is found by multiplying the number of possible outcomes of each event.

15.4 Record and Practice Journal

1. Use a tree diagram to find the total number of possible outcomes.

Bed Sheets	
Size	Twin, Twin XL, Full, Queen, King
Style	Solid, Patterned

Twin: S P; Twin XL: S P; Full: S P; Queen: S P; King: S P

Use the Fundamental Counting Principle to find the total number of possible outcomes.

2.

Photos	
Size	Wallet, 4 by 6, 5 by 7, 8 by 10, 11 by 14, 16 by 20
Finish	Matte, Glossy
Edits	Red eye, Black and white, Crop

36

3.

Laptops	
Hard Drive	250 GB, 320 GB, 500 GB
Style	HD, LCD
Color	Black, White, Red, Blue, Pink, Green, Purple

42

You spin the spinner and flip a coin. Find the probability of the events.

4. Spinning a 2 and flipping tails

$\frac{1}{10}$, or 10%

5. Spinning a 7 and flipping heads

0

6. *Not* spinning a 4 and flipping tails

$\frac{2}{5}$, or 40%

15.4 Exercises

Check It Out
Help with Homework
BigIdeasMath.com

Vocabulary and Concept Check

1. **VOCABULARY** What is the sample space of an event? How can you find the sample space of two or more events?

2. **WRITING** Explain how to use the Fundamental Counting Principle.

3. **WRITING** Describe two ways to find the total number of possible outcomes of spinning the spinner and rolling the number cube.

4. **OPEN-ENDED** Give a real-life example of a compound event.

Practice and Problem Solving

5. **COMBINATIONS** The lock is numbered from 0 to 49. Each combination uses three numbers in a right, left, right pattern. Find the total number of possible combinations for the lock.

Use a tree diagram to find the sample space and the total number of possible outcomes.

① 6.

Birthday Party	
Event	Miniature golf, Laser tag, Roller skating
Time	1:00 P.M.–3:00 P.M., 6:00 P.M.–8:00 P.M.

7.

New School Mascot	
Type	Lion, Bear, Hawk, Dragon
Style	Realistic, Cartoon

Use the Fundamental Counting Principle to find the total number of possible outcomes.

② 8.

Beverage	
Size	Small, Medium, Large
Flavor	Root beer, Cola, Diet cola, Iced tea, Lemonade, Water, Coffee

9.

MP3 Player	
Memory	2 GB, 4 GB, 8 GB, 16 GB
Color	Silver, Green, Blue, Pink, Black

③ 10.

Clown	
Suit	Dots, Stripes, Checkers board
Wig	One color, Multicolor
Talent	Balloon animals, Juggling, Unicycle, Magic

11.

Meal	
Appetizer	Nachos, Soup, Spinach dip, Salad, Fruit
Entrée	Chicken, Beef, Spaghetti, Fish
Dessert	Cake, Cookies, Ice cream

12. **NOTE CARDS** A store sells three types of note cards. There are three sizes of each type. Show two ways to find the total number of note cards the store sells.

13. **ERROR ANALYSIS** A true-false quiz has five questions. Describe and correct the error in using the Fundamental Counting Principle to find the total number of ways that you can answer the quiz.

> ✗ $2 + 2 + 2 + 2 + 2 = 10$
>
> You can answer the quiz in 10 different ways.

14. **CHOOSE TOOLS** You randomly choose one of the marbles. Without replacing the first marble, you choose a second marble.

    a. Name two ways you can find the total number of possible outcomes.

    b. Find the total number of possible outcomes.

You spin the spinner and flip a coin. Find the probability of the compound event.

15. Spinning a 1 and flipping heads

16. Spinning an even number and flipping heads

17. Spinning a number less than 3 and flipping tails

18. Spinning a 6 and flipping tails

19. *Not* spinning a 5 and flipping heads

20. Spinning a prime number and *not* flipping heads

You spin the spinner, flip a coin, then spin the spinner again. Find the probability of the compound event.

21. Spinning blue, flipping heads, then spinning a 1

22. Spinning an odd number, flipping heads, then spinning yellow

23. Spinning an even number, flipping tails, then spinning an odd number

24. *Not* spinning red, flipping tails, then *not* spinning an even number

25. **TAKING A TEST** You randomly guess the answers to two questions on a multiple-choice test. Each question has three choices: A, B, and C.

    a. What is the probability that you guess the correct answers to both questions?

    b. Suppose you can eliminate one of the choices for each question. How does this change the probability that your guesses are correct?

Common Errors

- **Exercises 15–24** Students may find the probability of each individual event. Remind them that they are finding the probability of a *compound event*, so they need to find the ratio of the number of favorable outcomes to the number of possible outcomes using tables, tree diagrams, or the Fundamental Counting Principle.

Practice and Problem Solving

14. **a.** tree diagram or the Fundamental Counting Principle

b. 12 possible outcomes

15. $\frac{1}{10}$, or 10% **16.** $\frac{1}{5}$, or 20%

17. $\frac{1}{5}$, or 20% **18.** 0, or 0%

19. $\frac{2}{5}$, or 40% **20.** $\frac{3}{10}$, or 30%

21. $\frac{1}{18}$, or $5\frac{5}{9}$%

22. $\frac{1}{9}$, or $11\frac{1}{9}$%

23. $\frac{1}{9}$, or $11\frac{1}{9}$%

24. $\frac{2}{9}$, or $22\frac{2}{9}$%

25. **a.** $\frac{1}{9}$, or about 11.1%

b. It increases the probability that your guesses are correct to $\frac{1}{4}$, or 25%, because you are only choosing between 2 choices for each question.

26. **a.** $\frac{1}{100}$, or 1%

b. It increases the probability that your choice is correct to $\frac{1}{25}$, or 4%, because each digit could be 0, 2, 4, 6, or 8.

English Language Learners

Pair Activity

Give pairs of students one or two problems similar to Example 4. Each pair is responsible for reaching a consensus on the answers. This will allow English learners to ask questions or explain concepts to another student.

Practice and Problem Solving

27. See Additional Answers.

28. See *Taking Math Deeper*.

29. See Additional Answers.

30. 10 ways

Fair Game Review

31. *Sample answer:* adjacent: $\angle XWY$ and $\angle ZWY$, $\angle XWY$ and $\angle XWV$; vertical: $\angle VWX$ and $\angle YWZ$, $\angle YWX$ and $\angle VWZ$

32. *Sample answer:* adjacent: $\angle LJM$ and $\angle LJK$, $\angle LJM$ and $\angle NJM$; vertical: $\angle KJL$ and $\angle PJN$, $\angle PJQ$ and $\angle MJL$

33. B

Mini-Assessment

1. Use a tree diagram to find the sample space and the total number of possible outcomes.

Snack	
Fruit	Apple, Banana, Pear
Drink	Water, Iced tea, Milk

AW, AI, AM, BW, BI, BM, PW, PI, PM; 9

2. Use the Fundamental Counting Principle to find the total number of possible outcomes.

Shirt	
Size	S, M, L, XL
Color	White, Blue, Red, Black, Gray
Style	T-shirt, Dress shirt

40

3. What is the probability of randomly choosing a banana and milk in Question 1? $11\frac{1}{9}\%$

Taking Math Deeper

Exercise 28

This problem can be used to introduce students to a new mathematical term: *factorial*. If n is any whole number, then ***n* factorial**, written as $n!$, is defined to be

$$n \cdot (n-1) \cdot (n-2) \cdot \cdots \cdot 3 \cdot 2 \cdot 1.$$

1 Use the Fundamental Counting Principle.

Number of choices for 1st car: 8
Number of choices for 2nd car: 7
Number of choices for 3rd car: 6
Number of choices for 4th car: 5
Number of choices for 5th car: 4
Number of choices for 6th car: 3
Number of choices for 7th car: 2
Number of choices for 8th car: 1

So, the total number of ways to arrange the cars in the train is

$$8! = 8 \cdot 7 \cdot 6 \cdot 5 \cdot 4 \cdot 3 \cdot 2 \cdot 1 = 40{,}320.$$

2 Factorials are used to count the number of **permutations** of n objects. Later in the study of probability, students will learn that there are $n!$ ways to order n objects.

3 *Solve A Simpler Problem:* Let's suppose there are only 3 train cars. We can list the $3! = 6$ ways as follows.

Reteaching and Enrichment Strategies

If students need help. . .	If students got it. . .
Resources by Chapter • Practice A and Practice B • Puzzle Time Record and Practice Journal Practice Differentiating the Lesson Lesson Tutorials Skills Review Handbook	Resources by Chapter • Enrichment and Extension • Technology Connection Start the next section

26. **PASSWORD** You forget the last two digits of your password for a website.

a. What is the probability that you randomly choose the correct digits?

b. Suppose you remember that both digits are even. How does this change the probability that your choices are correct?

27. **COMBINATION LOCK** The combination lock has 3 wheels, each numbered from 0 to 9.

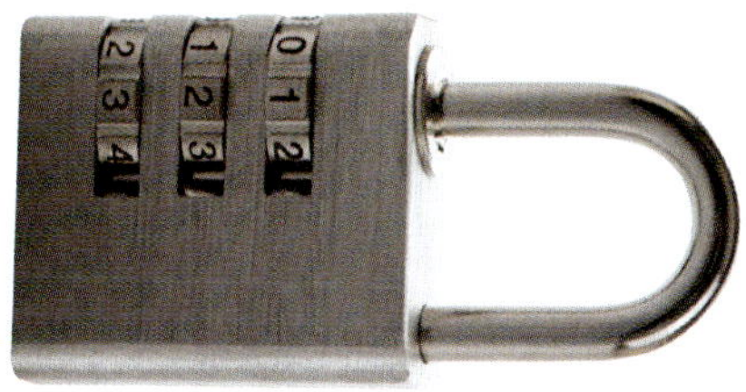

a. What is the probability that someone randomly guesses the correct combination in one attempt?

b. You try to guess the combination by writing five different numbers from 0 to 999 on a piece of paper. Explain how to find the probability that the correct combination is written on the paper.

28. **TRAINS** Your model train has one engine and eight train cars. Find the total number of ways you can arrange the train. (The engine must be first.)

29. **REPEATED REASONING** You have been assigned a 9-digit identification number.

a. Why should you use the Fundamental Counting Principle instead of a tree diagram to find the total number of possible identification numbers?

b. How many identification numbers are possible?

c. **RESEARCH** Use the Internet to find out why the possible number of Social Security numbers is not the same as your answer to part (b).

30. **Problem Solving** From a group of 5 candidates, a committee of 3 people is selected. In how many different ways can the committee be selected?

Fair Game Review What you learned in previous grades & lessons

Name two pairs of adjacent angles and two pairs of vertical angles in the figure. *(Section 12.1)*

31.

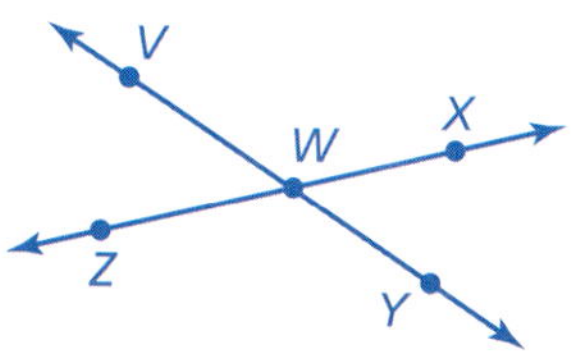

32.

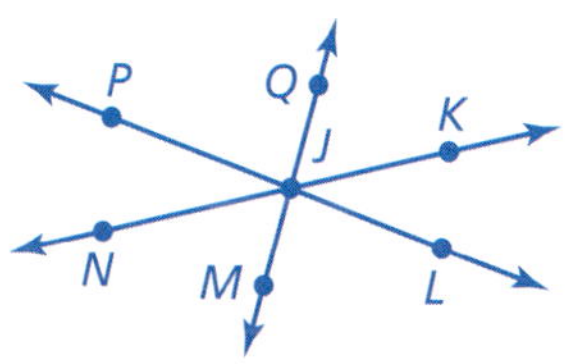

33. **MULTIPLE CHOICE** A drawing has a scale of 1 cm : 1 m. What is the scale factor of the drawing? *(Section 12.5)*

Ⓐ 1 : 1 Ⓑ 1 : 100 Ⓒ 10 : 1 Ⓓ 100 : 1

15.5 Independent and Dependent Events

Essential Question What is the difference between dependent and independent events?

1 ACTIVITY: Drawing Marbles from a Bag (With Replacement)

Work with a partner. You have three marbles in a bag. There are two green marbles and one purple marble. Randomly draw a marble from the bag. Then put the marble back in the bag and draw a second marble.

a. Complete the tree diagram. Let G = green and P = purple. Find the probability that both marbles are green.

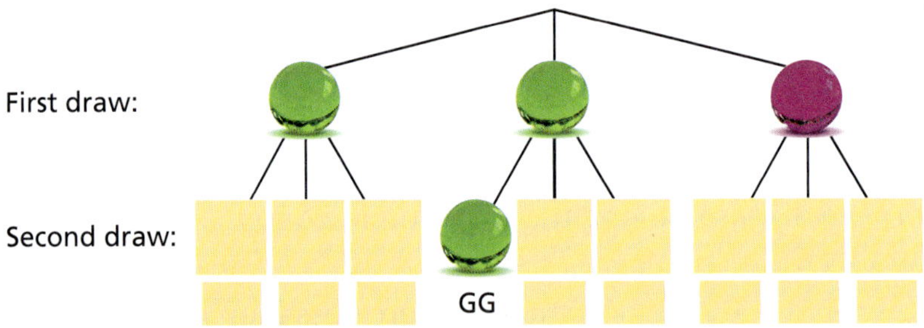

b. Does the probability of getting a green marble on the second draw *depend* on the color of the first marble? Explain.

2 ACTIVITY: Drawing Marbles from a Bag (Without Replacement)

Work with a partner. Using the same marbles from Activity 1, randomly draw two marbles from the bag.

a. Complete the tree diagram. Let G = green and P = purple. Find the probability that both marbles are green.

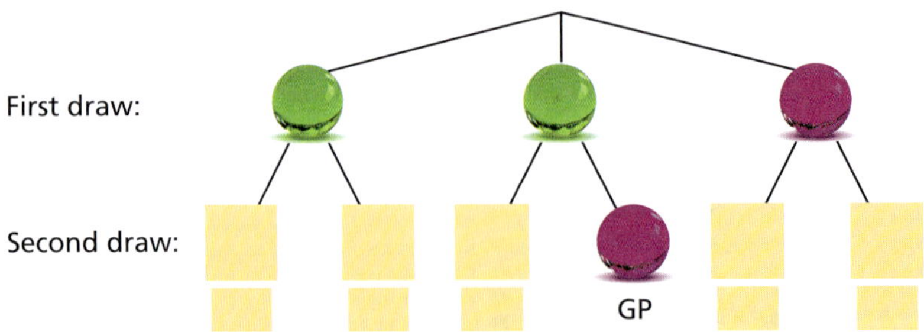

Is this event more likely than the event in Activity 1? Explain.

b. Does the probability of getting a green marble on the second draw *depend* on the color of the first marble? Explain.

Probability and Statistics

In this lesson, you will

- identify independent and dependent events.
- use formulas to find probabilities of independent and dependent events.

Laurie's Notes

Introduction

Applying Mathematical Practices

- **Attend to Precision:** Mathematically proficient students try to communicate precisely to others. In the activities today, students should pay attention to whether there is *replacement* after an event. This will be important in defining independent and dependent events in the lesson.

Motivate

- Hand two volunteers a bag containing 2 red and 2 blue cubes (any similar objects will work). The goal is to draw twice and to end up with two of the same color. The only difference is that person A *replaces* the cube after the first draw and person B *does not replace* the cube after the first draw. So, Person A is always drawing 1 out of 4 items. Person B is drawing 1 out of 4 items, and then 1 out of 3 items.
- Have each person do 10 trials and gather data. This should help students start to see a pattern.
- **Big Idea:** The probability on the second draw *depends upon* what is drawn on the first draw when you do *not* replace the cube.
- Draw a tree diagram to show the difference between Person A and Person B. This will help students read the tree diagrams in the activity.

Activity Notes

Activity 1 and Activity 2

- **Model with Mathematics:** The tree diagrams are used in the activities to help compute the theoretical probabilities.
- **Common Error:** Students read the tree diagram incorrectly and think a person is drawing nine times in Activity 1 and six times in Activity 2. Remind students that the first horizontal row of marbles displays the possible outcomes of the first draw and the second horizontal row displays the possible outcomes of the second draw.

? "On the first draw, you will either draw a green marble or a purple marble. Why does the tree diagram show more than just one of each?" There are two different green marbles that could be drawn and only one purple.

- **Teaching Tip:** Use the labels green marble 1 and green marble 2 to help students recognize why you need two greens and one purple in the tree diagram for the first draw.

? "When you *do* replace the marble, are the probabilities on the second draw the same as on the first draw? Explain." yes; There are the same marbles in the bag each time.

? "When you *do not* replace the marble, are the probabilities on the second draw the same as on the first draw? Explain." no; There are only two marbles instead of three to draw from. The probability of the second draw depends upon the first marble drawn.

? "For Activity 1, what is the probability of drawing 2 purple marbles? for Activity 2?" $\frac{1}{9}$, 0

What Your Students Will Learn

- Identify independent and dependent events.
- Use formulas to find probabilities of independent, dependent, and compound events.

Previous Learning

Students have used samples spaces to find probabilities of compound events.

Technology for the Teacher

Dynamic Classroom

Lesson Plans
Complete Materials List

15.5 Record and Practice Journal

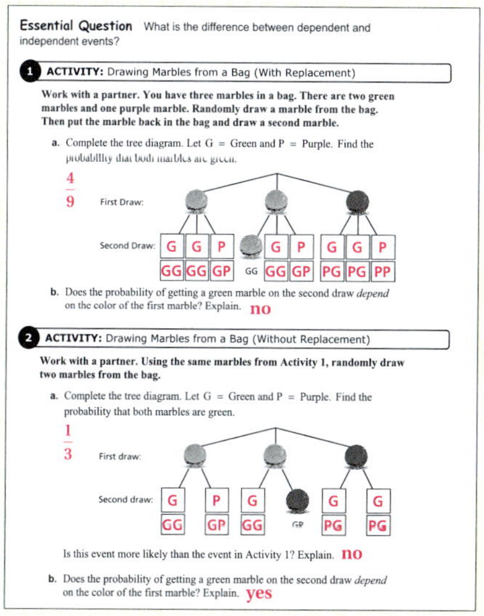

Essential Question What is the difference between dependent and independent events?

1 ACTIVITY: Drawing Marbles from a Bag (With Replacement)

Work with a partner. You have three marbles in a bag. There are two green marbles and one purple marble. Randomly draw a marble from the bag. Then put the marble back in the bag and draw a second marble.

a. Complete the tree diagram. Let G = Green and P = Purple. Find the probability that both marbles are green.

$\frac{4}{9}$

First Draw:
Second Draw: G G P | G P | G G P
GG GG GP GG GG GP PG PG PP

b. Does the probability of getting a green marble on the second draw *depend* on the color of the first marble? Explain. no

2 ACTIVITY: Drawing Marbles from a Bag (Without Replacement)

Work with a partner. Using the same marbles from Activity 1, randomly draw two marbles from the bag.

a. Complete the tree diagram. Let G = Green and P = Purple. Find the probability that both marbles are green.

$\frac{1}{3}$

First draw:
Second draw: G P G G G
GG GP GG GP PG PG

Is this event more likely than the event in Activity 1? Explain. no

b. Does the probability of getting a green marble on the second draw *depend* on the color of the first marble? Explain. yes

Differentiated Instruction

Kinesthetic

Before class, choose three books to put on your desk. Ask 3 students to each choose a book by writing the title on a piece of paper. Then ask another 3 students to select a book, one at a time, and take them back to their seats. Discuss when the selection of the books involved replacement (The students wrote their choice.) and when the selection of books did not involve replacement (The students took the books to their desks.).

15.5 Record and Practice Journal

3 ACTIVITY: Conducting an Experiment

Work with a partner. Conduct two experiments using two green marbles (G) and one purple marble (P).

a. In the first experiment, randomly draw one marble from the bag. Put it back. Draw a second marble. Repeat this 36 times. Record each result. Make a bar graph of your results.

GG	
GP	
PP	

Check students' work.

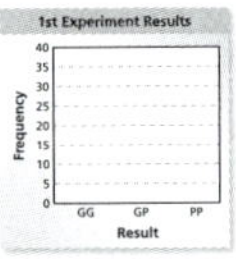

b. In the second experiment, randomly draw two marbles from the bag 36 times. Record each result. Make a bar graph of your results.

GG	
GP	
PP	

Check students' work.

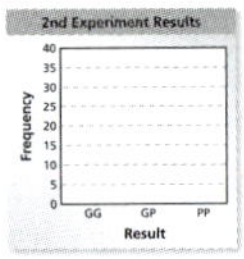

c. For each experiment, estimate the probability of drawing two green marbles.
Check students' work.

d. Which experiment do you think represents *dependent events*? Which represents *independent events*? Explain your reasoning.
The second experiment represents dependent events. The first experiment represents independent events.

What Is Your Answer?

4. **IN YOUR OWN WORDS** What is the difference between *dependent* and *independent* events? Describe a real-life example of each.
dependent events: occurrence of one event will affect the occurrence of another event
independent events: occurrence of one event will not affect the occurrence of another event

In Questions 5–7, tell whether the events are *independent* or *dependent*. Explain your reasoning.

5. You roll a 5 on a number cube and spin blue on a spinner.
independent

6. Your teacher chooses one student to lead a group, and then chooses another student to lead another group.
dependent

7. You spin red on one spinner and green on another spinner.
independent

8. In Activities 1 and 2, what is the probability of drawing a green marble on the first draw? on the second draw? How do you think you can use these two probabilities to find the probability of drawing two green marbles?
Activity 1: $\frac{2}{3}$; $\frac{2}{3}$; find the product of the probabilities
Activity 2: $\frac{2}{3}$; $\frac{1}{2}$; find the product of the probabilities

Laurie's Notes

Activity 3

- Students will now collect the data for the same events described in Activities 1 and 2. The experimental probabilities should be close to the theoretical probabilities.
- Explain that in part (a), you cannot draw two marbles at once. You draw one, replace it, and draw again. There are three possibilities: two green marbles (GG), one of each color (GP), or two purple marbles (PP).
- In part (b), it does not matter if you draw one marble followed by another, or if you draw two at one time. The result is the same. This drawing procedure eliminates the possibility of drawing two purple marbles (PP).
- Have pairs of students hold up the results of each experiment so that bar graphs can be analyzed visually.

? **Attend to Precision:** "For the first experiment, compare the lengths of the three bars. What do you notice? Interpret your answer." The GG bar and the GP bar are about the same length. The PP bar is not very tall, maybe about a fourth as tall as either of the other bars. So, you are as likely to draw two greens as you are to draw one of each color, and you are 4 times more likely to draw two greens or one of each color as you are to draw two purples.

? **Attend to Precision:** "For the second experiment, compare the lengths of the bars. What do you notice? Interpret your answer." The PP bar length is 0 because it is not a possible outcome. The GP bar should be about twice as long as the GG bar, so you are twice as likely to draw one of each color as you are to draw two greens.

What Is Your Answer?

- In Question 4, student language may not be precise at this point. Students may only be comfortable describing the difference in dependent and independent events in terms of purple and green marbles.
- Questions 5–7 will help extend student knowledge of independent and dependent events to other contexts.
- Question 8 is important in helping students discover the multiplication rule for independent and dependent events, which will be formalized in the lesson.

Closure

- You have a deck of playing cards (4 suits, 13 cards in each suit). Give an example of dependent events and independent events using the deck of cards.

  Sample answer:
  Dependent events—draw one card, set it aside, then draw another
  Independent events—draw one card, replace it, then draw another

3 ACTIVITY: Conducting an Experiment

Work with a partner. Conduct two experiments.

a. In the first experiment, randomly draw one marble from the bag. Put it back. Draw a second marble. Repeat this 36 times. Record each result. Make a bar graph of your results.

b. In the second experiment, randomly draw two marbles from the bag 36 times. Record each result. Make a bar graph of your results.

> **Math Practice**
>
> **Use Definitions**
>
> In what other mathematical context have you seen the terms *independent* and *dependent*? How does knowing these definitions help you answer the questions in part (d)?

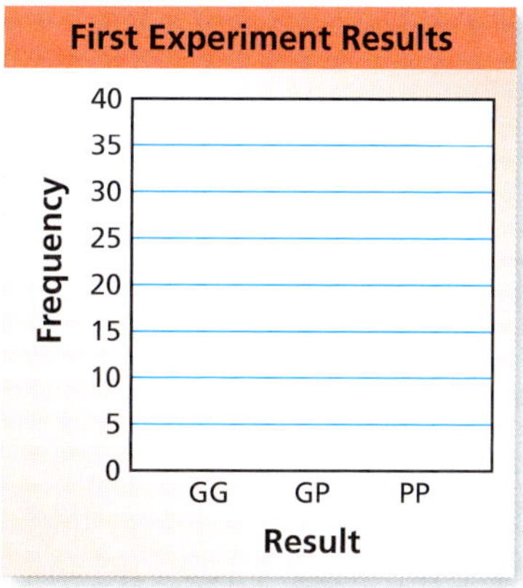

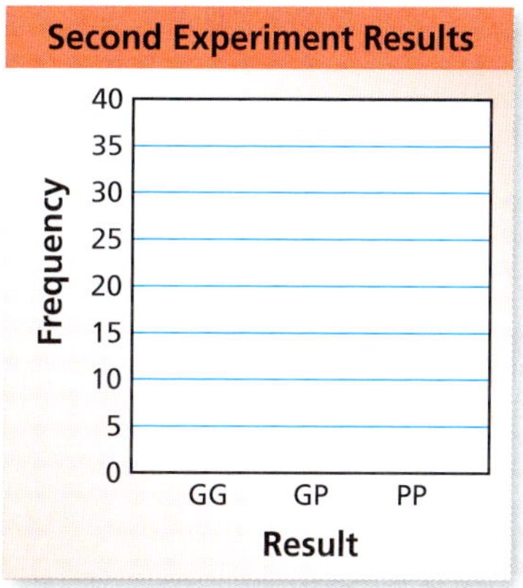

c. For each experiment, estimate the probability of drawing two green marbles.

d. Which experiment do you think represents *dependent events*? Which represents *independent events*? Explain your reasoning.

What Is Your Answer?

4. **IN YOUR OWN WORDS** What is the difference between dependent and independent events? Describe a real-life example of each.

In Questions 5–7, tell whether the events are *independent* or *dependent*. Explain your reasoning.

5. You roll a 5 on a number cube and spin blue on a spinner.
6. Your teacher chooses one student to lead a group, and then chooses another student to lead another group.
7. You spin red on one spinner and green on another spinner.
8. In Activities 1 and 2, what is the probability of drawing a green marble on the first draw? on the second draw? How do you think you can use these two probabilities to find the probability of drawing two green marbles?

Practice Use what you learned about independent and dependent events to complete Exercises 3 and 4 on page 665.

15.5 Lesson

Check It Out
Lesson Tutorials
BigIdeasMath.com

Key Vocabulary
independent events, *p. 662*
dependent events, *p. 663*

Compound events may be *independent events* or *dependent events*. Events are **independent events** if the occurrence of one event *does not* affect the likelihood that the other event(s) will occur.

Key Idea

Probability of Independent Events

Words The probability of two or more independent events is the product of the probabilities of the events.

Symbols $P(A \text{ and } B) = P(A) \cdot P(B)$

$P(A \text{ and } B \text{ and } C) = P(A) \cdot P(B) \cdot P(C)$

EXAMPLE 1 Finding the Probability of Independent Events

You spin the spinner and flip the coin. What is the probability of spinning a prime number and flipping tails?

The outcome of spinning the spinner does not affect the outcome of flipping the coin. So, the events are independent.

$P(\text{prime}) = \frac{3}{5}$ ← There are 3 prime numbers (2, 3, and 5). ← There is a total of 5 numbers.

$P(\text{tails}) = \frac{1}{2}$ ← There is 1 tails side. ← There is a total of 2 sides.

Use the formula for the probability of independent events.

$$P(A \text{ and } B) = P(A) \cdot P(B)$$

$$P(\text{prime and tails}) = P(\text{prime}) \cdot P(\text{tails})$$

$$= \frac{3}{5} \cdot \frac{1}{2} \quad \text{Substitute.}$$

$$= \frac{3}{10} \quad \text{Multiply.}$$

∴ The probability of spinning a prime number and flipping tails is $\frac{3}{10}$, or 30%.

On Your Own

Now You're Ready Exercises 5–8

1. What is the probability of spinning a multiple of 2 and flipping heads?

Multi-Language Glossary at BigIdeasMath.com

Laurie's Notes

Introduction

Connect

- **Yesterday:** Students developed an understanding of the difference between independent events and dependent events.
- **Today:** Students will use a formal definition to compute theoretical probabilities of independent events and dependent events.

Motivate

- Tell students that the bag contains slips of paper with each of their names. You will draw two names. The first name you draw will not have homework tonight (or wins a pencil). The second name you draw will get 5 bonus points on the test (or wins a certificate for an ice cream in the cafeteria).
- Draw and reveal the first name. Now slowly put the piece of paper back in the bag and wait for student reaction. They should recognize that they have greater probability of winning if the piece of paper is not replaced.

Lesson Notes

Discuss

- Write the definition of independent events on the board.
- Relate the definition to drawing cards from a deck. If you replace a drawn card before drawing again, the events are independent.

Key Idea

- Write the Key Idea on the board. Note that the probability of independent events is not defined in terms of two events only, it can be extended to three or more events by continuing to multiply.
- The formulas may seem daunting to students because of all the symbols. Be sure to use the words along with the notation.

Example 1

- **Model with Mathematics:** Use a spinner and a coin to model the example. If you do not have a spinner, sketch one on a transparency and use a paper clip, anchored at the center with a pen, as your spinner. You could also use an online simulator.

? "Are the two events independent? Explain." yes; The outcome of the spinner does not affect the outcome of flipping the coin.

- Students should know that a *prime number* is a number greater than 1 that is divisible by only two numbers, itself and 1, and that a *composite number* is a number having two or more prime factors.

? "What numbers on the spinner are prime?" 2, 3, and 5

- Write the formula for finding the probability of independent events. You are finding the probability of spinning a prime *and* flipping tails. This is not the same as spinning a prime *or* flipping tails.

On Your Own

- **Connection:** Drawing a tree diagram will help students see the probability of spinning a multiple of 2 and flipping heads.

Goal Today's lesson is identifying and finding probabilities of **independent events** and **dependent events**.

Technology for the Teacher

Dynamic Classroom

Lesson Tutorials
Lesson Plans
Answer Presentation Tool

Extra Example 1

In Example 1, what is the probability of spinning a composite number and flipping tails? $\frac{1}{10}$

On Your Own

1. $\frac{1}{5}$

English Language Learners

Group Activity

To assist English learners in understanding the concepts and ideas presented in this chapter, provide them with the opportunity to interact, discuss, and listen in a group situation.

Extra Example 2

In Example 2, what is the probability that one of your friends is chosen first, and then you are chosen second? $\frac{1}{1650}$, or about 0.061%

On Your Own

2. $\frac{58}{75}$, or about 77.3%

Laurie's Notes

Discuss

- Write the definition of dependent events on the board.
- Relate the definition to drawing cards from a deck. If you do *not* replace a drawn card before drawing again, the events are dependent.
- Explain that you can find probabilities of dependent events using a formula that is similar to the formula for the probability of independent events, except that the probability of the second event has been affected by the occurrence of the first event.

Key Idea

- Write the Key Idea on the board. Again, be sure to use the words along with the notation.
- Demonstrate the formula using four cards, two with even numbers and two with odd numbers. Ask what the probability of randomly drawing two even-numbered cards would be without replacement. The size of the sample space is 4 for the first draw but only 3 for the second draw. The occurrence of the first event (drawing an even-numbered card) affects the likelihood of the second event (drawing another even-numbered card). So, the probability of drawing both even-numbered cards would be $\frac{2}{4} \cdot \frac{1}{3} = \frac{1}{6}$.

Example 2

- **Representation:** Select five students to be the relatives and six students to be the friends. Give each relative a card marked with an R, and give each friend a card marked with an F. The whole class is the audience. Find the probability of choosing a relative. Then find the probability of choosing a friend after a relative is chosen. Finally, multiply the two probabilities.

On Your Own

? "Of the 100 audience members, how many are *not* you, your relatives, or your friends?" $100 - 12 = 88$

Events are **dependent events** if the occurrence of one event *does* affect the likelihood that the other event(s) will occur.

Key Idea

Probability of Dependent Events

Words The probability of two dependent events A and B is the probability of A times the probability of B after A occurs.

Symbols $P(A \text{ and } B) = P(A) \cdot P(B \text{ after } A)$

EXAMPLE 2 Finding the Probability of Dependent Events

People are randomly chosen to be game show contestants from an audience of 100 people. You are with 5 of your relatives and 6 other friends. What is the probability that one of your relatives is chosen first, and then one of your friends is chosen second?

Choosing an audience member changes the number of audience members left. So, the events are dependent.

$$P(\text{relative}) = \frac{5}{100} = \frac{1}{20}$$

There are 5 relatives.

There is a total of 100 audience members.

$$P(\text{friend}) = \frac{6}{99} = \frac{2}{33}$$

There are 6 friends.

There is a total of 99 audience members left.

Use the formula for the probability of dependent events.

$$P(A \text{ and } B) = P(A) \cdot P(B \text{ after } A)$$

$$P(\text{relative and friend}) = P(\text{relative}) \cdot P(\text{friend after relative})$$

$$= \frac{1}{20} \cdot \frac{2}{33} \qquad \text{Substitute.}$$

$$= \frac{1}{330} \qquad \text{Simplify.}$$

The probability is $\frac{1}{330}$, or about 0.3%.

On Your Own

Now You're Ready
Exercises 9–12

2. What is the probability that you, your relatives, and your friends are *not* chosen to be either of the first two contestants?

EXAMPLE 3 Finding the Probability of a Compound Event

A student randomly guesses the answer for each of the multiple-choice questions. What is the probability of answering all three questions correctly?

1. In what year did the United States gain independence from Britain?
A. 1492 **B.** 1776 **C.** 1788 **D.** 1795 **E.** 2000

2. Which amendment to the Constitution grants citizenship to all persons born in the United States and guarantees them equal protection under the law?
A. 1st **B.** 5th **C.** 12th **D.** 13th **E.** 14th

3. In what year did the Boston Tea Party occur?
A. 1607 **B.** 1773 **C.** 1776 **D.** 1780 **E.** 1812

Choosing the answer for one question does not affect the choice for the other questions. So, the events are independent.

Method 1: Use the formula for the probability of independent events.

$$P(\#1 \text{ and } \#2 \text{ and } \#3 \text{ correct}) = P(\#1 \text{ correct}) \cdot P(\#2 \text{ correct}) \cdot P(\#3 \text{ correct})$$

$$= \frac{1}{5} \cdot \frac{1}{5} \cdot \frac{1}{5} \quad \text{Substitute.}$$

$$= \frac{1}{125} \quad \text{Multiply.}$$

The probability of answering all three questions correctly is $\frac{1}{125}$, or 0.8%.

Method 2: Use the Fundamental Counting Principle.

There are 5 choices for each question, so there are $5 \cdot 5 \cdot 5 = 125$ possible outcomes. There is only 1 way to answer all three questions correctly.

$$P(\#1 \text{ and } \#2 \text{ and } \#3 \text{ correct}) = \frac{1}{125}$$

The probability of answering all three questions correctly is $\frac{1}{125}$, or 0.8%.

On Your Own

Now You're Ready
Exercises 18–22

3. The student can eliminate Choice A for all three questions. What is the probability of answering all three questions correctly? Compare this probability with the probability in Example 3. What do you notice?

Laurie's Notes

Example 3

- Note that this example involves more than two events.
- ? "Do you think it is wise to randomly guess on a true-false test or a multiple-choice test?" Answers will vary.
- Explain that this question is an example of guessing on a 3-question multiple-choice test with each question having 5 possible responses.
- You might have students try to answer the questions before finding the probability. B; E; B
- Work through the problem using both methods.
- **Construct Viable Arguments:** To make sure that students understand probabilities of independent and dependent events, ask them why they would or would not use a tree diagram or table to find the probabilities in Examples 1–3.

On Your Own

- The goal of this question is for students to recognize how the probability changes when a choice can be eliminated from each example.
- **Extension:** Ask students which choices they could eliminate from the questions and why.

Closure

- **Exit Ticket:** You and your friend are among 5 volunteers to help distribute workbooks. What is the probability that your teacher randomly selects you and your friend to distribute the workbooks? $\frac{1}{10}$, or 10%

Extra Example 3

A true-false quiz has 4 questions. What is the probability of randomly guessing and answering all of the questions correctly? $\frac{1}{16}$, or 6.25%

On Your Own

3. $\frac{1}{64}$; The probability of getting the correct answer increases when there are only four choices.

Vocabulary and Concept Check

1. What is the probability of choosing a 1 and then a blue chip?; $\frac{1}{15}$; $\frac{1}{10}$
2. For independent events, find the probability of the first event, find the probability of the second event, and then multiply. For dependent events, find the probability of the first event, find the probability of the second event after the first event occurs, and then multiply.

Practice and Problem Solving

3. independent; The outcome of the first roll does not affect the outcome of the second roll.
4. dependent; Your friend's lane number cannot be the same as your lane number. So, your friend's lane number depends on your lane number.
5. $\frac{1}{8}$
6. $\frac{1}{4}$
7. $\frac{3}{8}$
8. $\frac{3}{8}$
9. $\frac{1}{42}$
10. $\frac{1}{14}$
11. $\frac{2}{21}$
12. $\frac{2}{7}$
13. The two events are dependent, so the probability of the second event is $\frac{1}{3}$.

    $P(\text{red and green}) = \frac{1}{4} \cdot \frac{1}{3} = \frac{1}{12}$
14. dependent; The second draw is affected by the first draw.

Assignment Guide and Homework Check

Level	Assignment	Homework Check
Advanced	1–4, 6–12 even, 13, 14–26 even, 27–30	10, 16, 22, 24

Common Errors

- **Exercises 3 and 4** Students may mix up independent and dependent events or may have difficulty determining which type of event it is. Remind them that independent events are where you do two different things or events where you start over before the next trial. Dependent events have at least one less possible outcome after the first event.
- **Exercises 5–12** Students may find the probabilities of each individual event. Remind them that they are finding the probability of a compound event.

15.5 Record and Practice Journal

You roll a number cube twice. Find the probability of the events.

1. Rolling a 3 twice $\frac{1}{36}$
2. Rolling an even number and a 5 $\frac{1}{12}$
3. Rolling an odd number and a 2 or a 4 $\frac{1}{6}$
4. Rolling a number less than 6 and a 3 or a 1 $\frac{5}{18}$

You randomly choose a letter from a hat with the letters A through J. Without replacing the first letter, you choose a second letter. Find the probability of the events.

5. Choosing an H and then a D $\frac{1}{90}$
6. Choosing a consonant and then an E or an I $\frac{7}{45}$
7. Choosing a vowel and then an F $\frac{1}{30}$
8. Choosing a vowel and then a consonant $\frac{7}{30}$
9. You have 3 clasp bracelets, 4 watches, and 5 stretch bracelets. You randomly choose two from your jewelry box. What is the probability that you will choose 2 watches?

   $\frac{1}{11}$

You flip a coin, and then roll a number cube twice. Find the probability of the event.

10. Flipping heads, rolling a 5, and rolling a 2 $\frac{1}{72}$, or about 1.39%
11. Flipping tails, rolling an odd number, and rolling a 4 $\frac{1}{24}$, or about 4.17%
12. Flipping tails, rolling a 6 or a 1, and rolling a 3 $\frac{1}{36}$, or about 2.8%
13. Flipping heads, *not* rolling a 2, and rolling an even number $\frac{5}{24}$, or about 20.8%

15.5 Exercises

Check It Out
Help with Homework
BigIdeasMath.com

Vocabulary and Concept Check

1. **DIFFERENT WORDS, SAME QUESTION** You randomly choose one of the chips. Without replacing the first chip, you choose a second chip. Which question is different? Find "both" answers.

   What is the probability of choosing a 1 and then a blue chip?

   What is the probability of choosing a 1 and then an even number?

   What is the probability of choosing a green chip and then a chip that is *not* red?

   What is the probability of choosing a number less than 2 and then an even number?

2. **WRITING** How do you find the probability of two events *A* and *B* when *A* and *B* are independent? dependent?

Practice and Problem Solving

Tell whether the events are *independent* or *dependent*. Explain.

3. You roll a 4 on a number cube. Then you roll an even number on a different number cube.
4. You randomly draw a lane number for a 100-meter race. Then your friend randomly draws a lane number for the same race.

You spin the spinner and flip a coin. Find the probability of the compound event.

(1) 5. Spinning a 3 and flipping heads

6. Spinning an even number and flipping tails
7. Spinning a number greater than 1 and flipping tails
8. *Not* spinning a 2 and flipping heads

You randomly choose one of the tiles. Without replacing the first tile, you choose a second tile. Find the probability of the compound event.

(2) 9. Choosing a 5 and then a 6

10. Choosing an odd number and then a 20
11. Choosing a number less than 7 and then a multiple of 4
12. Choosing two even numbers

13. **ERROR ANALYSIS** Describe and correct the error in finding the probability.

✗ You randomly choose one of the marbles. Without replacing the first marble, you choose a second marble. What is the probability of choosing red and then green?

$P(\text{red and green}) = \frac{1}{4} \cdot \frac{1}{4} = \frac{1}{16}$

14. **LOGIC** A bag contains three marbles. Does the tree diagram show the outcomes for *independent* or *dependent* events? Explain.

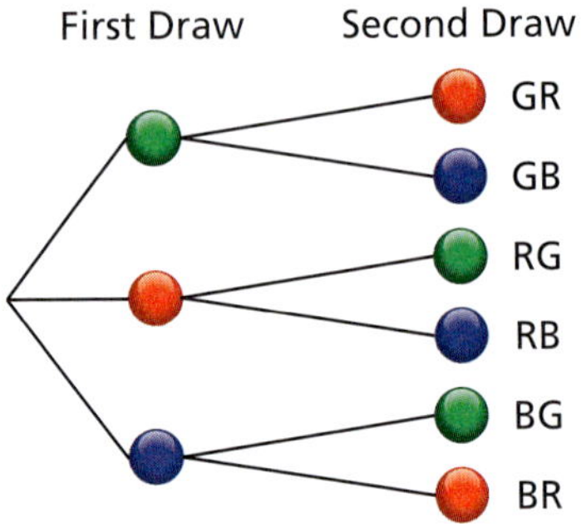

15. **EARRINGS** A jewelry box contains two gold hoop earrings and two silver hoop earrings. You randomly choose two earrings. What is the probability that both are silver hoop earrings?

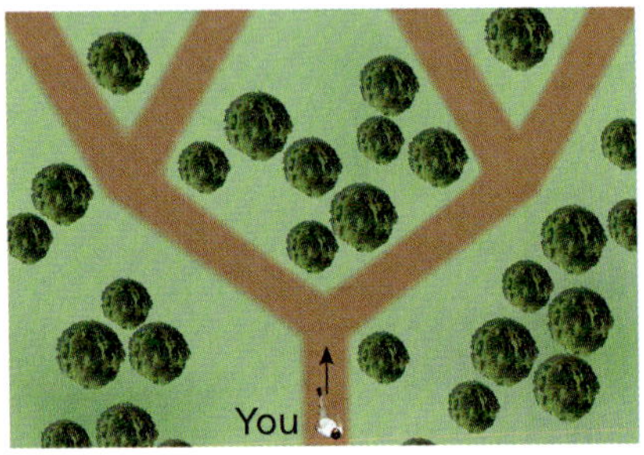

16. **HIKING** You are hiking to a ranger station. There is one correct path. You come to a fork and randomly take the path on the left. You come to another fork and randomly take the path on the right. What is the probability that you are still on the correct path?

17. **CARNIVAL** At a carnival game, you randomly throw two darts at the board and break two balloons. What is the probability that both of the balloons you break are purple?

You spin the spinner, flip a coin, then spin the spinner again. Find the probability of the compound event.

3 18. Spinning a 4, flipping heads, then spinning a 7

19. Spinning an odd number, flipping heads, then spinning a 3

20. Spinning an even number, flipping tails, then spinning an odd number

21. *Not* spinning a 5, flipping heads, then spinning a 1

22. Spinning an odd number, *not* flipping heads, then *not* spinning a 6

Common Errors

- **Exercise 17** Students may forget to decrease both the number of favorable outcomes and the number of possible outcomes by 1 when finding the probability of the second dart hitting a purple balloon. Remind them that the second throw has 1 less favorable outcome and 1 less possible outcome.
- **Exercises 18–22** Students may find the probabilities of each individual event. Remind them that they are finding the probability of a compound event.

Practice and Problem Solving

15. $\frac{1}{6}$, or about 16.7%

16. $\frac{1}{4}$, or 25%

17. $\frac{2}{35}$

18. $\frac{1}{162}$, or about 0.62%

19. $\frac{5}{162}$, or about 3.1%

20. $\frac{10}{81}$, or about 12.3%

21. $\frac{4}{81}$, or about 4.9%

22. $\frac{20}{81}$, or about 24.7%

23. $\frac{3}{4}$

24. 51.2%

25. **a.** If you and your best friend were in the same group, then the probability that you both are chosen would be 0 because only one leader is chosen from each group. Because the probability that both you and your best friend are chosen is $\frac{1}{132}$, you and your best friend are not in the same group.

b. $\frac{1}{11}$ **c.** 23

Differentiated Instruction

Vocabulary

Some students may have difficulty understanding how the independence or dependence of events affects probability. Allow students to experiment with bags containing similar small objects, as in the marble activity. Have them draw three objects from the bag with and without replacement and then compare probabilities. Ask volunteers to present their results to the class.

Practice and Problem Solving

26. See *Taking Math Deeper.*

Fair Game Review

27. See Additional Answers.

28.

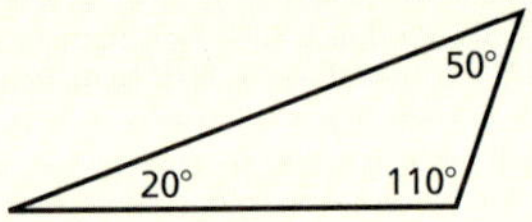

obtuse scalene

29.

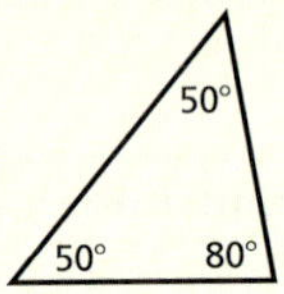

acute isosceles

30. C

Mini-Assessment

You spin the spinner and flip a coin. Find the probability of the compound event.

1. Spinning a five and flipping tails $\frac{1}{18}$
2. Spinning an odd number and flipping heads $\frac{5}{18}$

A bag holds six chips numbered 1–6. Without replacing the first chip, you choose a second chip. Find the probability of the compound event.

3. Choosing an even number and then choosing an odd number $\frac{3}{10}$
4. Choosing a number less than 3 and then choosing a number greater than 3 $\frac{1}{5}$

Taking Math Deeper

Exercise 26

This is an example of the strategy *Working Backward.*

1. Who was the oldest?
A. Ned **B.** Yvonne **C.** Sun Li **D.** Angel **E.** Dusty

2. What city was Stacey from?
A. Raleigh **B.** New York **C.** Roanoke **D.** Dallas **E.** San Diego

1 **a.** How can the probability of getting both answers correct be 25%?

Because $25\% = \frac{1}{4}$, you need two probabilities whose product is $\frac{1}{4}$. By eliminating all but 2 of the choices in each question, the probability of randomly guessing the correct answer to both questions is

$$P = \frac{1}{2} \cdot \frac{1}{2} = \frac{1}{4}.$$

2 **b.** How can the probability of getting both answers correct be $8\frac{1}{3}\%$?

Because $8\frac{1}{3}\% = \frac{1}{12}$, you have to find the different ways that you can multiply two probabilities to get $\frac{1}{12}$. As it turns out, there is only one way. You eliminate 1 of the choices to one question and eliminate 2 of the choices to another question. Then, the probability is

$$P = \frac{1}{4} \cdot \frac{1}{3} = \frac{1}{12}.$$

3 You might consider using this as an opportunity to talk about test-taking strategies. It should seem clear to students that the more choices you can eliminate, the greater your chance for a higher score.

Reteaching and Enrichment Strategies

If students need help. . .	If students got it. . .
Resources by Chapter • Practice A and Practice B • Puzzle Time Record and Practice Journal Practice Differentiating the Lesson Lesson Tutorials Skills Review Handbook	Resources by Chapter • Enrichment and Extension • Technology Connection Start the next section

23. **LANGUAGES** There are 16 students in your Spanish class. Your teacher randomly chooses one student at a time to take a verbal exam. What is the probability that you are *not* one of the first four students chosen?

24. **SHOES** Twenty percent of the shoes in a factory are black. One shoe is chosen and replaced. A second shoe is chosen and replaced. Then a third shoe is chosen. What is the probability that *none* of the shoes are black?

25. **PROBLEM SOLVING** Your teacher divides your class into two groups, and then randomly chooses a leader for each group. The probability that you are chosen to be a leader is $\frac{1}{12}$. The probability that both you and your best friend are chosen is $\frac{1}{132}$.

a. Is your best friend in your group? Explain.

b. What is the probability that your best friend is chosen as a group leader?

c. How many students are in the class?

26. **Structure** After ruling out some of the answer choices, you randomly guess the answer for each of the story questions below.

1. Who was the oldest?
A. Ned **B.** Yvonne **C.** Sun Li **D.** Angel **E.** Dusty

2. What city was Stacey from?
A. Raleigh **B.** New York **C.** Roanoke **D.** Dallas **E.** San Diego

a. How can the probability of getting both answers correct be 25%?

b. How can the probability of getting both answers correct be $8\frac{1}{3}\%$?

Fair Game Review What you learned in previous grades & lessons

Draw a triangle with the given angle measures. Then classify the triangle. *(Section 12.3)*

27. 30°, 60°, 90° **28.** 20°, 50°, 110° **29.** 50°, 50°, 80°

30. **MULTIPLE CHOICE** Which set of numbers is in order from least to greatest? *(Skills Review Handbook)*

Ⓐ $\frac{2}{3}$, 0.6, 67% Ⓑ 44.5%, $\frac{4}{9}$, $0.4\overline{6}$

Ⓒ 0.269, 27%, $\frac{3}{11}$ Ⓓ $2\frac{1}{7}$, 214%, $2.\overline{14}$

Extension 15.5 Simulations

Check It Out
Lesson Tutorials
BigIdeasMath.com

Key Vocabulary
simulation, *p. 668*

A **simulation** is an experiment that is designed to reproduce the conditions of a situation or process. Simulations allow you to study situations that are impractical to create in real life.

EXAMPLE 1 Simulating Outcomes That Are Equally Likely

HTH	HTT
HTT	HTH
HTT	TTT
HHH	HTT
HTT	TTT
HTT	HTH
HTH	HHH
HTT	HTT
TTT	HTH
HTH	HTT

A couple plans on having three children. The gender of each child is equally likely. (a) Design a simulation involving 20 trials that you can use to model the genders of the children. (b) Use your simulation to find the experimental probability that all three children are boys.

a. Choose an experiment that has two equally likely outcomes for each event (gender), such as tossing three coins. Let heads (H) represent a boy and tails (T) represent a girl.

b. To find the experimental probability, you need repeated trials of the simulation. The table shows 20 trials.

$$P(\text{three boys}) = \frac{2}{20} = \frac{1}{10}$$

HHH occurred 2 times.

There is a total of 20 trials.

The experimental probability is $\frac{1}{10}$, 0.1, or 10%.

EXAMPLE 2 Simulating Outcomes That Are Not Equally Likely

Study Tip

In Example 2, the digits 1 through 6 represent 60% of the possible digits (0 through 9) in the tens place. Likewise, the digits 1 and 2 represent 20% of the possible digits in the ones place.

There is a 60% chance of rain on Monday and a 20% chance of rain on Tuesday. Design and use a simulation involving 50 randomly generated numbers to find the experimental probability that it will rain on both days.

Use the random number generator on a graphing calculator. Randomly generate 50 numbers from 0 to 99. The table below shows the results.

Let the digits 1 through 6 in the tens place represent rain on Monday. Let digits 1 and 2 in the ones place represent rain on Tuesday. Any number that meets these criteria represents rain on both days.

```
randInt(0,99,50)
{52 66 73 68 75...
```

52	66	73	68	75	28	35	47	48	2
16	68	49	3	77	35	92	78	6	6
58	18	89	39	24	80	32	41	77	21
32	40	96	59	86	1	12	0	94	73
40	71	28	61	1	24	37	25	3	25

$$P(\text{rain both days}) = \frac{7}{50}$$

7 numbers meet the criteria.

There is a total of 50 trials.

The experimental probability is $\frac{7}{50}$, 0.14, or 14%.

Multi-Language Glossary at BigIdeasMath.com

Laurie's Notes

Introduction

Connect

- **Yesterday:** Students found probabilities of compound events.
- **Today:** Students will perform simulations to find probabilities of compound events.

Motivate

- It is near the end of the year and final exams are coming. Tell students that you heard the history exam will include 10 true-false questions.

? "Assume you don't read the questions. How do you think you could find the probability of randomly guessing the correct answer to *at least* 7 questions?" Students will have a range of ideas about this.

- Explain that this is an event that you probably don't want to perform. Studying is a much better option. However, you can *simulate* the results.

Lesson Notes

Discuss

- Define simulations and give an example. Explain that simulations should be kept simple, allowing you to repeatedly duplicate the results of an event.
- To perform a simulation: (a) state the situation, (b) describe a model that randomly generates appropriate outcomes, (c) use the model repeatedly and record the results, and (d) use the results to make a conclusion.
- Students know that experimental probability approximates theoretical probability as the number of trials increases. So, simulations can be designed to accurately approximate theoretical probabilities.

Example 1

? "How can you simulate an event that has two equally likely outcomes?" Answers will vary, but students will likely mention flipping coins.

- If time permits, students could use coins or an online tool to generate data.

? "If heads (H) represents a boy and tails (T) represents a girl, what outcome represents 3 boys?" HHH

- Students should view the data and interpret the first few results.
- The table shows that 2 out of 20, or 10% of the results represent 3 boys.
- **Extension:** Ask students to use the table to find the experimental probability that all three children are (a) girls and (b) of the same gender.

Example 2

? "Is it possible to simulate a situation with outcomes that are not equally likely?" Students may say no at first. If you tell them to consider a number cube or different colors of marbles in a bag, they may change their minds.

- **Use Appropriate Tools Strategically:** In this simulation, a graphing calculator generates random whole numbers between 0 and 99 using the *randInt* function. The syntax is *randInt(a,b,c)* where *a* and *b* represent the range of numbers and *c* represents how many are generated. You can use the right arrow key to scroll through the results. Commands and syntax may vary for different graphing calculators.

What Your Students Will Learn

- Use simulations to find experimental probabilities.

Goal Today's lesson is performing **simulations** to find probabilities of compound events.

Technology for the Teacher

Dynamic Classroom

Lesson Tutorials
Lesson Plans
Answer Presentation Tool

Extra Example 1

A couple plans on having four children. The gender of each child is equally likely.

a. Design a simulation involving 25 trials that you can use to model the genders of the children. Toss four coins, letting H = boy and T = girl.

b. Use your simulation to find the experimental probability that all four children have the same gender. Answers will vary, but should be close to 12.5%.

Extra Example 2

There is a 70% chance of snow on Thursday and a 40% chance of snow on Friday. Design and use a simulation involving 50 randomly generated numbers to find the experimental probability that it will snow both days. Randomly generate 50 numbers from 0 to 99 on a graphing calculator, let the digits 1–7 in the tens place represent snow on Thursday and let the digits 1–4 in the ones place represent snow in Friday; Answers will vary, but should be close to 28%.

Record and Practice Journal Extension 15.5 Practice

1–2. See Additional Answers.

Laurie's Notes

Example 2 (continued)

- Discuss the design of the simulation.
  - Randomly generating a digit from 0–9 has 10 possible outcomes, each digit having a 10% probability of being generated.
  - You can simulate rain on Monday (60% chance) using digits from 1–6 (60% probability of being generated).
  - You can simulate rain on Tuesday (20% chance) using digits of 1 or 2 (20% probability of being generated).
  - One way to simulate the weather on both days is to generate random numbers between 0 and 99, letting the tens digit represent Monday and the ones digit represent Tuesday. Results with 1, 2, 3, 4, 5, or 6 in the tens place and 1 or 2 in the ones place represent rain on both days.
- The table shows that 14% of the results represent rain on both days.

Example 3

? "What is the chance that weather causes a school cancellation?" 50%

- **Use Appropriate Tools Strategically:** The Study Tip shows how to randomly generate four-digit numbers in a spreadsheet. This could also be done using a graphing calculator.

? "If we randomly generate 50 numbers from 0 to 9999, what numbers represent school being canceled in at least three of the next four school years?" numbers with at least three of the four digits being 1, 2, 3, 4, or 5

- From the table, 17 out of 50, or 34% of the results represent a cancellation in at least three of the next four school years.
- Make sure students understand that assigning digits 1–5 to represent a school cancellation was arbitrary. They can use other digits, such as 0–4, or odd numbers. Have students repeat Examples 2 and 3 using different digits.
- **Teaching Tip:** Because the two outcomes are equally likely, you could also simulate by flipping coins. Have each student in a group of 4 flip a coin. If heads represents a cancellation and at least 3 heads result, school is cancelled three times in that four-year period. Have groups run several trials. Discuss the results.

Practice

- Have students share their simulation design for Questions 1 and 2.
- **Big Idea:** Students need to understand that as you increase the number of trials in a simulation, the experimental probability better approximates the theoretical probability.
- Question 4 helps students see that simulations can help them approximate theoretical probabilities that are difficult to compute directly. Finding the theoretical probability in Example 3 is difficult, but manageable. However, it would be unmanageable for 5 or more school years.

Closure

- **Exit Ticket:** Design and use a simulation to find the experimental probability that you correctly guess at least 7 of the 10 true-false questions in the Motivate.
- How does a probability of 15% in Example 2 or 25% in Example 3 change the simulation you design? How does it change the probabilities?

Extra Example 3

Each school year, there is a 20% chance that weather causes one or more days of school to be canceled. Design and use a simulation involving 50 randomly generated numbers to find the experimental probability that weather will cause school to be canceled in exactly two of the next five school years. Randomly generate 50 five-digit whole numbers in a spreadsheet, let the digits 1 and 2 represent a school year with a weather cancellation; Answers will vary, but should be close to 20% (the theoretical probability is 20.48%).

Practice

1–4. See Additional Answers.

Mini-Assessment

1. You randomly guess on five true-false questions. Design and use a simulation to find the experimental probability that you correctly answer exactly 3 questions. *Sample answer:* Toss five coins 50 times, letting H = correct and T = incorrect, probability should be close to 31.25%.
2. Suppose 40% of the people in a shopping mall are willing to take a survey. Design and use a simulation to find the experimental probability that out of the next 5 random shoppers, there will be 3 or more in a row that are willing to take the survey. Randomly generate 50 five-digit whole numbers in a spreadsheet, let the digits 1–4 represent a customer willing to take the survey; Answers will vary, but should be close to 11% (the theoretical probability is 11.008%).

EXAMPLE 3 Using a Spreadsheet to Simulate Outcomes

Probability and Statistics

In this extension, you will

- use simulations to find experimental probabilities.

Each school year, there is a 50% chance that weather causes one or more days of school to be canceled. Design and use a simulation involving 50 randomly generated numbers to find the experimental probability that weather will cause school to be canceled in at least three of the next four school years.

Use a random number table in a spreadsheet. Randomly generate 50 four-digit whole numbers. The spreadsheet below shows the results.

Let the digits 1 through 5 represent school years with a cancellation. The numbers in the spreadsheet that contain at least three digits from 1 through 5 represent four school years in which at least three of the years have a cancellation.

Study Tip

To create a four-digit random number table in a spreadsheet, follow these steps.

1. Highlight the group of cells to use for your table.
2. Format the cells to display four-digit whole numbers.
3. Enter the formula RAND()*10000 into each cell.

	A	B	C	D	E	F
1	7584	3974	8614	2500	4629	
2	3762	3805	2725	7320	6487	
3	3024	1554	2708	1126	9395	
4	4547	6220	9497	7530	3036	
5	1719	0662	1814	6218	2766	
6	7938	9551	8552	4321	8043	
7	6951	0578	5560	0740	4479	
8	4714	4511	5115	6952	5609	
9	0797	3022	9067	2193	6553	
10	3300	5454	5351	6319	0387	
11						

$$P\left(\begin{matrix}\text{cancellation in at least three} \\ \text{of the next four school years}\end{matrix}\right) = \frac{17}{50}$$

17 numbers contain at least three digits from 1 to 5.

There is a total of 50 trials.

The experimental probability is $\frac{17}{50}$, 0.34, or 34%.

Practice

1. **QUIZ** You randomly guess the answers to four true-false questions. (a) Design a simulation that you can use to model the answers. (b) Use your simulation to find the experimental probability that you answer all four questions correctly.
2. **BASEBALL** A baseball team wins 70% of its games. Assuming this trend continues, design and use a simulation to find the experimental probability that the team wins the next three games.
3. **WHAT IF?** In Example 3, there is a 40% chance that weather causes one or more days of school to be canceled each school year. Find the experimental probability that weather will cause school to be canceled in at least three of the next four school years.
4. **REASONING** In Examples 1–3 and Exercises 1–3, try to find the theoretical probability of the event. What do you think happens to the experimental probability when you increase the number of trials in the simulation?

15 Study Help

Check It Out
Graphic Organizer
BigIdeasMath.com

You can use a **notetaking organizer** to write notes, vocabulary, and questions about a topic. Here is an example of a notetaking organizer for probability.

Write important vocabulary or formulas in this space. →

	Probability
If *P*(event) = 0, the event is *impossible*. If *P*(event) = 0.25, the event is *unlikely*. If *P*(event) = 0.5, the event is *equally likely* to happen or not happen. If *P*(event) = 0.75, the event is *likely*. If *P*(event) = 1, the event is *certain*.	A number that measures the likelihood that an event will occur Can be written as a fraction, decimal, or percent Always between 0 and 1, inclusive
How do you find the probability of two or more events?	

← Write your notes about the topic in this space.

← Write your questions about the topic in this space.

On Your Own

Make notetaking organizers to help you study these topics.

1. experimental probability
2. theoretical probability
3. Fundamental Counting Principle
4. independent events
5. dependent events

After you complete this chapter, make notetaking organizers for the following topics.

6. sample
7. population

"I am using a notetaking organizer to plan my autobiography."

Sample Answers

1.

$P(\text{event}) = \dfrac{\text{number of times the event occurs}}{\text{total number of trials}}$ Experiment: an investigation or a procedure that has varying results Outcomes: the possible results of an experiment Event: a collection of one or more outcomes	Experimental Probability Probability that is based on repeated trials of an experiment Example: You flip a coin 100 times. You flip heads 52 times and tails 48 times. The experimental probabilites are $P(\text{heads}) = \frac{52}{100} = 0.52 = 52\%$, and $P(\text{tails}) = \frac{48}{100} = 0.48 = 48\%$.
How can I find the probability of an event without doing an experiment?	

2.

$P(\text{event}) = \dfrac{\text{number of favorable outcomes}}{\text{number of possible outcomes}}$ Outcomes: the possible results of an experiment Event: a collection of one or more outcomes Favorable outcomes: the outcomes of a specific event	Theoretical Probability The ratio of the number of favorable outcomes to the number of possible outcomes, when all possible outcomes are equally likely Example: You flip a coin. The theoretical probability of flipping heads and the theoretical probability of flipping tails is $P(\text{heads}) = \frac{1}{2}$, and $P(\text{tails}) = \frac{1}{2}$.
What if the possible outcomes are not equally likely?	

3–5. Available at *BigIdeasMath.com.*

List of Organizers

Available at *BigIdeasMath.com*

Comparison Chart
Concept Circle
Example and Non-Example Chart
Formula Triangle
Four Square
Idea (Definition) and Examples Chart
Information Frame
Information Wheel
Notetaking Organizer
Process Diagram
Summary Triangle
Word Magnet
Y Chart

About this Organizer

A **Notetaking Organizer** can be used to write notes, vocabulary, and questions about a topic. In the space on the left, students write important vocabulary or formulas. In the space on the right, students write their notes about the topic. In the space at the bottom, students write their questions about the topic. A notetaking organizer can also be used as an assessment tool, in which blanks are left for students to complete.

Technology for the Teacher

Editable Graphic Organizer

Answers

1. 2
2. 0
3. 4
4. $\frac{3}{10}$, or 30%
5. $\frac{1}{4}$, or 25%
6. $\frac{3}{4}$, or 75%
7. 0, or 0%
8. $\frac{2}{15}$, or about 13.3%
9. $\frac{11}{30}$, or about 36.7%
10. $\frac{43}{120}$, or about 35.8%
11. 1, or 100%
12. 12
13. 8
14. $\frac{2}{5}$, or 40%
15. $\frac{1}{10}$, or 10%

Technology for the Teacher

Online Assessment
Assessment Book
ExamView® Assessment Suite

Alternative Quiz Ideas

100% Quiz	Math Log
Error Notebook	Notebook Quiz
Group Quiz	Partner Quiz
Homework Quiz	Pass the Paper

Group Quiz

Students work in groups. Give each group a large index card. Each group writes five questions that they feel evaluate the material they have been studying. On a separate piece of paper, students solve the problems. When they are finished, they exchange cards with another group. The new groups work through the questions on the card.

Reteaching and Enrichment Strategies

If students need help. . .	If students got it. . .
Resources by Chapter • Practice A and Practice B • Puzzle Time Lesson Tutorials *BigIdeasMath.com*	Resources by Chapter • Enrichment and Extension • Technology Connection Game Closet at *BigIdeasMath.com* Start the next section

15.1–15.5 Quiz

Check It Out
Progress Check
BigIdeasMath.com

You randomly choose one butterfly. Find the number of ways the event can occur. *(Section 15.1)*

1. Choosing red
2. Choosing brown
3. Choosing *not* blue

6 Green
3 White
4 Red
2 Blue
5 Yellow

You randomly choose one paper clip from the jar. Find the probability of the event. *(Section 15.2)*

4. Choosing a green paper clip
5. Choosing a yellow paper clip
6. *Not* choosing a yellow paper clip
7. Choosing a purple paper clip

Use the bar graph to find the experimental probability of the event. *(Section 15.3)*

8. Rolling a 4
9. Rolling a multiple of 3
10. Rolling a 2 or a 3
11. Rolling a number less than 7

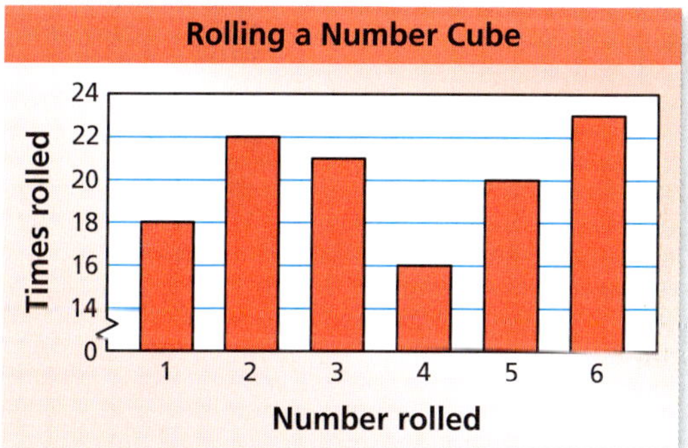

Use the Fundamental Counting Principle to find the total number of possible outcomes. *(Section 15.4)*

12.

Calculator	
Type	Basic display, Scientific, Graphing, Financial
Color	Black, White, Silver

13.

Vacation	
Destination	Florida, Italy, Mexico, England
Length	1 week, 2 weeks

14. **BLACK PENS** You randomly choose one of the pens shown. What is the theoretical probability of choosing a black pen? *(Section 15.3)*

15. **BLUE PENS** You randomly choose one of the five pens shown. Your friend randomly chooses one of the remaining pens. What is the probability that you and your friend both choose a blue pen? *(Section 15.5)*

15.6 Samples and Populations

Essential Question How can you determine whether a sample accurately represents a population?

A **population** is an entire group of people or objects. A **sample** is a part of the population. You can use a sample to make an *inference*, or conclusion, about a population.

Identify a population.		Select a sample.		Interpret the data in the sample.		Make an inference about the population.
Population	→	Sample	→	Interpretation	→	Inference

1 ACTIVITY: Identifying Populations and Samples

Work with a partner. Identify the population and the sample.

a.

The students in a school

The students in a math class

b.

The grizzly bears with GPS collars in a park

The grizzly bears in a park

c.

150 quarters

All quarters in circulation

d.

All books in a library

10 fiction books in a library

Probability and Statistics

In this lesson, you will

- determine when samples are representative of populations.
- use data from random samples to make predictions about populations.

2 ACTIVITY: Identifying Random Samples

Work with a partner. When a sample is selected at random, each member of the population is equally likely to be selected. You want to know the favorite extracurricular activity of students at your school. Determine whether each method will result in a random sample. Explain your reasoning.

a. You ask members of the school band.

b. You publish a survey in the school newspaper.

c. You ask every eighth student who enters the school in the morning.

d. You ask students in your class.

Laurie's Notes

Introduction

Applying Mathematical Practices

- **Construct Viable Arguments and Critique the Reasoning of Others:** Mathematically proficient students understand and use stated assumptions and previously established results in constructing arguments. They reason about data and make plausible arguments that take into account the context from which the data arose.

Motivate

- Conduct a quick survey of your class. Ask a couple of fun questions and then ask a math related question.
  - How many of you can roll your tongue?
  - Who likes spicy brown mustard better than yellow mustard?
  - Can you simplify a fraction?
- Discuss each of these questions, who would ask the question, and why they might be asked. Point out the following:
  - Tongue rolling is probably the most commonly used classroom example of a simple genetic trait in humans.
  - This question doesn't allow a person who doesn't like mustard at all to answer.
  - Teachers survey students all the time to help guide instruction.

Discuss

- **Teacher Note:** In a previous course, students answered simple statistical questions about a data set. The flowchart shows one of the basic principles of statistical reasoning, that a result from a sample can be used to make a generalization about the population from which it was selected.
- Define population and sample. In the Motivate, the students in the school can be the *population* and the students in the math class are the *sample.*
- **Construct Viable Arguments and Critique the Reasoning of Others:** Students should be able to give examples of cases where a sample would be representative (eye color) and cases where it would not be representative of a population (favorite musical group/singer).

Activity Notes

Activity 1

- Discuss student responses.

Activity 2

- Remind students that a survey is not the only way you can collect data from a sample, but it is likely the most common method.

? "What are the characteristics of a random sample?" Answers will vary.

? "Why do think a sample is taken instead of trying to survey an entire population?" You may not have the time, energy, or resources to do so.

- **Construct Viable Arguments and Critique the Reasoning of Others:** When students are finished, ask for volunteers to offer their reasoning about whether the method will result in a random sample.

What Your Students Will Learn

- Determine when samples are representative of populations.
- Use data from random samples to make predictions about populations.

Previous Learning

Students should know how to write and solve a proportion and how to interpret a circle graph.

Technology for the Teacher

Dynamic Classroom

Lesson Plans
Complete Materials List

15.6 Record and Practice Journal

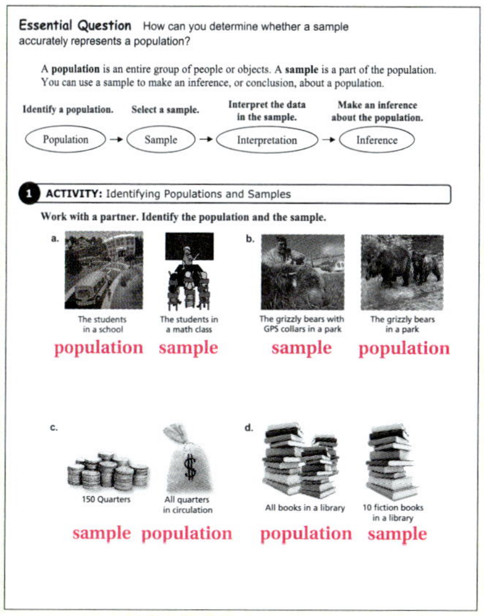

Essential Question How can you determine whether a sample accurately represents a population?

A **population** is an entire group of people or objects. A **sample** is a part of the population. You can use a sample to make an inference, or conclusion, about a population.

Identify a population. → **Select a sample.** → **Interpret the data in the sample.** → **Make an inference about the population.**

Population → Sample → Interpretation → Inference

1 ACTIVITY: Identifying Populations and Samples

Work with a partner. Identify the population and the sample.

a. The students in a school — population; The students in a math class — sample

b. The grizzly bears with GPS collars in a park — sample; The grizzly bears in a park — population

c. 150 Quarters — sample; All quarters in circulation — population

d. All books in a library — population; 10 fiction books in a library — sample

Differentiated Instruction

Visual

Ask students to discuss the pros and cons of showing survey results using visual representation. In Question 5, have students predict the types of visuals they would use to display and interpret their survey results.

15.6 Record and Practice Journal

2 ACTIVITY: Identifying Random Samples

Work with a partner. When a sample is selected at random, each member of the population is equally likely to be selected. You want to know the favorite extracurricular activity of students at your school. Determine whether each method will result in a random sample. Explain your reasoning.

a. You ask members of the school band.
no

b. You publish a survey in the school newspaper.
no

c. You ask every eighth student who enters the school in the morning.
yes

d. You ask students in your class.
no

3 ACTIVITY: Identifying Representative Samples

Work with a partner. A new power plant is being built outside a town. In each situation below and on the next page, residents of the town are asked how they feel about the new power plant. Determine whether each conclusion is valid. Explain your reasoning.

a. A local radio show takes calls from 500 residents. The table shows the results. The radio station concludes that most of the residents of the town oppose the new power plant.
not valid

New Power Plant	
For	70
Against	425
Don't know	5

b. A news reporter randomly surveys 2 residents outside a supermarket. The graph shows the results. The reporter concludes that the residents of the town are evenly divided on the new power plant.
not valid

c. You randomly survey 250 residents at a shopping mall. The table shows the results. You conclude that there are about twice as many residents of the town against the new power plant than for the new power plant.
valid

New Power Plant	
For	32%
Against	62%
Don't know	6%

What Is Your Answer?

4. **IN YOUR OWN WORDS** How can you determine whether a sample accurately represents a population?
The sample is large enough, it is random, and it accurately represents the entire population.

5. **RESEARCH** Choose a topic that you would like to ask people's opinions about, and then write a survey question. How would you choose people to survey so that your sample is random? How many people would you survey? Conduct your survey and display your results. Would you change any part of your survey to make it more accurate? Explain.
Check students' work.

6. Does increasing the size of a sample necessarily make the sample representative of a population? Give an example to support your explanation.
no

Laurie's Notes

Activity 3

- Give students time to discuss each scenario with their partners.
- ? "Can the results in part (a) be used to draw a valid conclusion? Explain." No; not everyone in the town may listen to the radio, only those who are really opposed to the new power plant may call the radio show, or it is possible that most of the people who call live close to the power plant.
- ? "Can the results in part (b) be used to draw a valid conclusion? Explain." No; the sample size is too small.
- ? "Can the results in part (c) be used to draw a valid conclusion? Explain." Yes; the sample size is large enough and randomly selected, and your conclusion is reasonable based on the results.
- **Reason Abstractly and Quantitatively** and **Construct Viable Arguments and Critique the Reasoning of Others:** Students should recognize that you can have a large sample or a random sample, and still not have one that accurately represents a population. Be sure students recognize the significance of the size of the sample. When the sample size is large enough, random sampling tends to produce representative samples that can be used to draw valid inferences.

What Is Your Answer?

- Ask volunteers to share their ideas for Question 5.

Closure

- **Exit Ticket:** What survey question would you ask to find out what vegetable should be served more often for the hot lunch program at your school? Answers will vary.

There are many different ways to select a sample from a population. To make valid inferences about a population, you must choose a random sample very carefully so that it accurately represents the population.

3 ACTIVITY: Identifying Representative Samples

Work with a partner. A new power plant is being built outside a town. In each situation below, residents of the town are asked how they feel about the new power plant. Determine whether each conclusion is valid. Explain your reasoning.

a. A local radio show takes calls from 500 residents. The table shows the results. The radio station concludes that most of the residents of the town oppose the new power plant.

New Power Plant	
For	70
Against	425
Don't know	5

Math Practice

Understand Quantities

Can the size of a sample affect the validity of a conclusion about a population?

New Power Plant

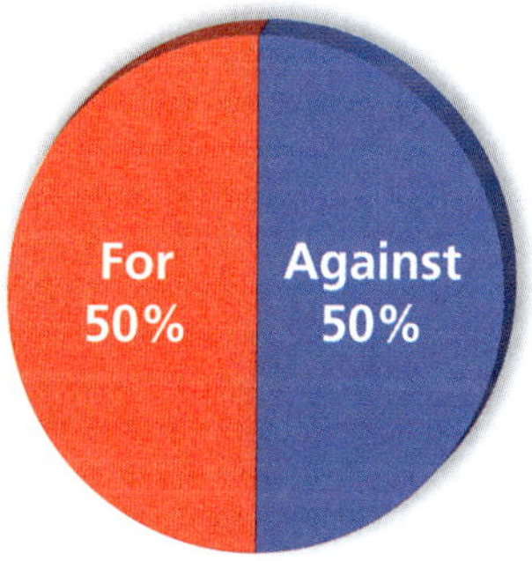

b. A news reporter randomly surveys 2 residents outside a supermarket. The graph shows the results. The reporter concludes that the residents of the town are evenly divided on the new power plant.

c. You randomly survey 250 residents at a shopping mall. The table shows the results. You conclude that there are about twice as many residents of the town against the new power plant than for the new power plant.

New Power Plant	
For	32%
Against	62%
Don't know	6%

What Is Your Answer?

4. **IN YOUR OWN WORDS** How can you determine whether a sample accurately represents a population?

5. **RESEARCH** Choose a topic that you would like to ask people's opinions about, and then write a survey question. How would you choose people to survey so that your sample is random? How many people would you survey? Conduct your survey and display your results. Would you change any part of your survey to make it more accurate? Explain.

6. Does increasing the size of a sample necessarily make the sample representative of a population? Give an example to support your explanation.

Practice Use what you learned about populations and samples to complete Exercises 3 and 4 on page 676.

15.6 Lesson

Check It Out
Lesson Tutorials
BigIdeasMath.com

Key Vocabulary
population, *p. 672*
sample, *p. 672*
unbiased sample, *p. 674*
biased sample, *p. 674*

An **unbiased sample** is representative of a population. It is selected at random and is large enough to provide accurate data.

A **biased sample** is not representative of a population. One or more parts of the population are favored over others.

EXAMPLE 1 Identifying an Unbiased Sample

You want to estimate the number of students in a high school who ride the school bus. Which sample is unbiased?

Ⓐ 4 students in the hallway

Ⓑ all students in the marching band

Ⓒ 50 seniors at random

Ⓓ 100 students at random during lunch

Choice A is not large enough to provide accurate data.

Choice B is not selected at random.

Choice C is not representative of the population because seniors are more likely to drive to school than other students.

Choice D is representative of the population, selected at random, and large enough to provide accurate data.

So, the correct answer is Ⓓ.

On Your Own

Now You're Ready
Exercises 5–7

1. **WHAT IF?** You want to estimate the number of seniors in a high school who ride the school bus. Which sample is unbiased? Explain.

2. You want to estimate the number of eighth-grade students in your school who consider it relaxing to listen to music. You randomly survey 15 members of the band. Your friend surveys every fifth student whose name appears on an alphabetical list of eighth graders. Which sample is unbiased? Explain.

The results of an unbiased sample are proportional to the results of the population. So, you can use unbiased samples to make predictions about the population.

Biased samples are not representative of the population. So, you should not use them to make predictions about the population because the predictions may not be valid.

Multi-Language Glossary at BigIdeasMath.com

Laurie's Notes

Introduction

Connect

- **Yesterday:** Students identified samples and populations.
- **Today:** Students will identify biased and unbiased samples and determine whether a sample can be used to draw conclusions and make predictions about a population.

Motivate

- Tell students about something you read recently that reported, "Four out of five students who responded to the survey said they should have more homework!"
- After students quiet down, restate what you read, omitting a few key words. "Four out of five students said they should have more homework!"

? "How are the two claims different?" Students should recognize that you could survey 100 people, only 5 respond to the survey and 4 of the 5 answered one way. 95 people did not respond. In the other scenario, 80 of the 100 students answered one way.

Discuss

- Describe unbiased and biased samples. Give a few examples of each type.

Lesson Notes

Example 1

- Work through the problem and discuss why the first three samples are not reasonable.

? "What other samples might not be reasonable?" Answers will vary.

On Your Own

- **Neighbor Check:** Have students work independently and then have their neighbors check their work. Have students discuss any discrepancies.

Discuss

? "How do you think the results from an unbiased sample can be used to make predictions about a population?" Listen for students to suggest you can use a proportion.

- State that the results of an unbiased sample are proportional to the results of the population.

Goal Today's lesson is identifying **biased** and **unbiased samples** and determining whether a sample can be used to draw conclusions and make predictions about a population.

Technology for the Teacher

Dynamic Classroom

Lesson Tutorials
Lesson Plans
Answer Presentation Tool

Extra Example 1

You want to know the number of students in your classroom who do their homework right after school. You survey every third student who arrives in the classroom.

a. What is the population of your survey? your class

b. What is the sample of your survey? every third student who arrives

c. Is the sample unbiased? Explain. Yes. The sample is selected at random, representative of the population, and large enough to provide accurate data.

On Your Own

1. C
2. Your friend's sample is unbiased.

Extra Example 2

You want to know how the residents of your town feel about a ban on texting while driving. Determine whether each conclusion is valid.

a. You survey 200 residents at random. One hundred sixty-four residents support the ban and thirty-six do not. So, you conclude that 82% of the residents of your town support the ban. The sample is unbiased and the conclusion is valid.

b. You survey the first 15 residents who drive into your neighborhood. Five support the ban and ten do not. So, you conclude that $33\frac{1}{3}\%$ of the residents of your town support the ban. The sample is biased and the conclusion is not valid.

Extra Example 3

You ask 50 randomly chosen students to name their favorite sport. There are 600 students in the school. Predict the number *n* of students in the school who would name soccer as their favorite sport. 144 students

Favorite Sport	Number of Students
Football	8
Soccer	12
Basketball	20
Baseball	10

On Your Own

3. No, firefighters are more likely to support the new sign.

4. about 384 students

English Language Learners

Vocabulary

English learners may easily grasp the concept of a sample, because it is similar to receiving a small amount of a product. However, the word *population* may be confusing. Relate both words to a sample of food. The sample is what the student tastes, and the population is all of the food.

Laurie's Notes

Example 2

- **Construct Viable Arguments and Critique the Reasoning of Others:** Ask a volunteer to read part (a) and ask whether the conclusion is valid. Students should recognize that the sample is biased because the survey was not random—you only survey nearby residents.
- **Construct Viable Arguments and Critique the Reasoning of Others:** Ask a volunteer to read part (b) and ask whether the conclusion is valid. Students should recognize that the sample is random and large enough so it is an unbiased sample.

? "If there are 2000 residents in the town, then how many would you expect to be in favor of the new sign?" 40% of 2000, or 800 residents

Example 3

- Read and discuss the information given.

? "Why is this sample unbiased?" The sample is representative of the population, the students were selected at random, and the sample is large enough to provide accurate data.

? "How many students were surveyed and how many of them watch one movie each week?" 75 were surveyed; 21 of them watch one movie each week

- Write a proportion to predict the number of students in the school who watch one movie each week.

On Your Own

- **Think-Pair-Share:** Students should read each question independently and then work in pairs to answer the questions. When they have answered the questions, the pair should compare their answers with another group and discuss any discrepancies.

Closure

- **Exit Ticket:** Describe an unbiased sample. Listen for students to mention the three qualities of an unbiased sample stated at the beginning of the lesson.

EXAMPLE 2 Determining Whether Conclusions Are Valid

You want to know how the residents of your town feel about adding a new stop sign. Determine whether each conclusion is valid.

a. You survey the 20 residents who live closest to the new sign. Fifteen support the sign, and five do not. So, you conclude that 75% of the residents of your town support the new sign.

The sample is not representative of the population because residents who live close to the sign are more likely to support it.

So, the sample is biased, and the conclusion is not valid.

b. You survey 100 residents at random. Forty support the new sign, and sixty do not. So, you conclude that 40% of the residents of your town support the new sign.

The sample is representative of the population, selected at random, and large enough to provide accurate data.

So, the sample is unbiased, and the conclusion is valid.

EXAMPLE 3 Making Predictions

Movies per Week

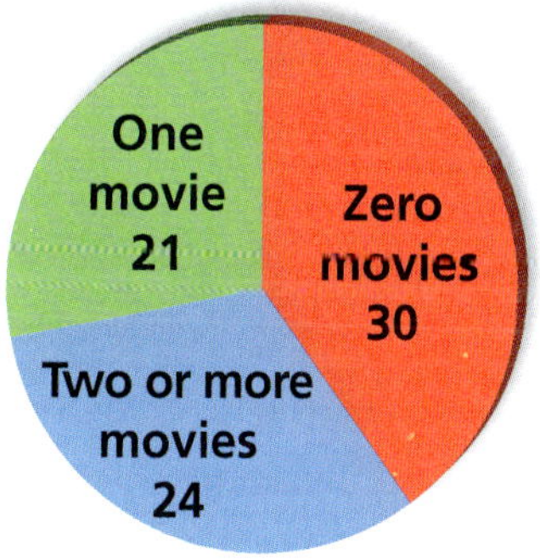

You ask 75 randomly chosen students how many movies they watch each week. There are 1200 students in the school. Predict the number *n* of students in the school who watch one movie each week.

The sample is representative of the population, selected at random, and large enough to provide accurate data. So, the sample is unbiased, and you can use it to make a prediction about the population.

Write and solve a proportion to find n.

$$\overset{\textbf{Sample}}{\frac{\text{students in survey (one movie)}}{\text{number of students in survey}}} = \overset{\textbf{Population}}{\frac{\text{students in school (one movie)}}{\text{number of students in school}}}$$

$$\frac{21}{75} = \frac{n}{1200} \qquad \text{Substitute.}$$

$$336 = n \qquad \text{Solve for } n.$$

So, about 336 students in the school watch one movie each week.

On Your Own

Now You're Ready
Exercises 8, 9, and 12

3. In Example 2, each of 25 randomly chosen firefighters supports the new sign. So, you conclude that 100% of the residents of your town support the new sign. Is the conclusion valid? Explain.

4. In Example 3, predict the number of students in the school who watch two or more movies each week.

15.6 Exercises

Check It Out
Help with Homework
BigIdeasMath.com

Vocabulary and Concept Check

1. **VOCABULARY** Why would you survey a sample instead of a population?
2. **CRITICAL THINKING** What should you consider when conducting a survey?

Practice and Problem Solving

Identify the population and the sample.

3. Residents of New Jersey  Residents of Ocean County 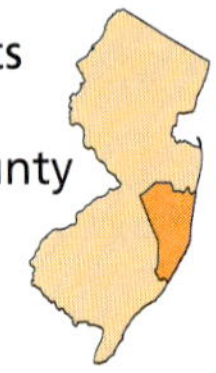

4. 
4 cards

All cards in a deck

Determine whether the sample is *biased* or *unbiased*. Explain.

1 5. You want to estimate the number of students in your school who play a musical instrument. You survey the first 15 students who arrive at a band class.

6. You want to estimate the number of books students in your school read over the summer. You survey every fourth student who enters the school.

7. You want to estimate the number of people in a town who think that a park needs to be remodeled. You survey every 10th person who enters the park.

Determine whether the conclusion is valid. Explain.

2 8. You want to determine the number of students in your school who have visited a science museum. You survey 50 students at random. Twenty have visited a science museum, and thirty have not. So, you conclude that 40% of the students in your school have visited a science museum.

9. You want to know how the residents of your town feel about building a new baseball stadium. You randomly survey 100 people who enter the current stadium. Eighty support building a new stadium, and twenty do not. So, you conclude that 80% of the residents of your town support building a new baseball stadium.

Which sample is better for making a prediction? Explain.

10.

Predict the number of students in a school who like gym class.	
Sample A	A random sample of 8 students from the yearbook
Sample B	A random sample of 80 students from the yearbook

11.

Predict the number of defective pencils produced per day.	
Sample A	A random sample of 500 pencils from 20 machines
Sample B	A random sample of 500 pencils from 1 machine

Assignment Guide and Homework Check

Level	Assignment	Homework Check
Advanced	1–4, 6–18 even, 19–24	8, 10, 12, 14, 18

Common Errors

- **Exercises 3 and 4** Students may get confused with the words population and sample. Encourage them to think about what it means to eat a sample of something, and then compare the whole to the population.
- **Exercise 5** Students may ignore the fact that the survey is conducted among students arriving at a band class. They may only see that it is a class. Ask them what kind of answers they would expect to get from different classes in the school (such as math), including band class. They should recognize that everyone going to band class will play an instrument, but not everyone in a math class will.
- **Exercises 13–15** Students may not understand why you would want to question the entire population for a survey. Ask them to estimate the population size for each survey, and then ask if it would be reasonable to ask everyone in that population for a response.

15.6 Record and Practice Journal

Determine whether the sample is *biased* or *unbiased*. Explain.

1. You want to estimate the number of students in your school who want a football stadium to be built. You survey the first 20 students who attend a Friday night football game.
   biased

2. You want to estimate the number of students in your school who drive their own cars to school. You survey every 8th person who enters the cafeteria for lunch.
   unbiased

Determine whether the conclusion is valid. Explain.

3. You want to determine the number of city residents who want to have 38th Street repaved. You randomly survey 15 residents who live on 38th Street. Twelve want the street to be repaved and three do not. So, you conclude that 80% of city residents want the street to be repaved.
   no

4. You want to determine how many students consider math to be their favorite school subject. You randomly survey 75 students. Thirty-three students consider math to be their favorite subject and forty-two do not. So, you conclude that 40% of students at your school consider math to be their favorite subject.
   yes

Vocabulary and Concept Check

1. Samples are easier to obtain.
2. You should make sure the people surveyed are selected at random and are representative of the population, as well as making sure your sample is large enough.

Practice and Problem Solving

3. Population: Residents of New Jersey
   Sample: Residents of Ocean County
4. Population: All cards in a deck
   Sample: 4 cards
5. biased; The sample is not selected at random and is not representative of the population because students in a band class play a musical instrument.
6. unbiased; The sample is representative of the population, selected at random, and large enough to provide accurate data.
7. biased; The sample is not representative of the population because people who go to a park are more likely to think that the park needs to be remodeled.
8. yes; The sample is representative of the population, selected at random, and large enough to provide accurate data. So, the sample is unbiased and the conclusion is valid.

9–10. See Additional Answers.

11. Sample A; it is representative of the population.

Practice and Problem Solving

12. 696 students

13. a sample; It is much easier to collect sample data in this situation.

14. a population; Because there are few enough students in your homeroom to not make the surveying difficult.

15–18. See Additional Answers.

19. See *Taking Math Deeper*.

Fair Game Review

20. 30% **21.** 140

22. 200 **23.** 3

24. A

Mini-Assessment

1. You want to know the number of students in your school who play a sport. You survey the first 10 students who arrive for lunch.
   a. What is the population of your survey? All students in your school the sample? First 10 students who arrive for lunch
   b. Is the sample *biased* or *unbiased*? Explain. biased; The sample is not large enough to provide accurate data or may not be representative of the population.
2. You ask 120 randomly chosen people at a stadium to name their favorite stadium food. There are about 50,000 people in the stadium. Predict the number of people in the stadium whose favorite stadium food is nachos. about 10,000 people

Favorite Stadium Food	
Nachos	24
Hot Dog	55
Peanuts	16
Popcorn	25

Taking Math Deeper

Exercise 19

This problem may look straightforward. In reality, it is filled with questions that encompass the nature of statistical sampling.

1 *Straightforward Approach:* 75% of the students in the sample said that they plan to go to college. Because 75% of 900 is 675, you can predict that 675 students in the high school plan to go to college.

Here, the guidance counselor assumes that the students who answered "Maybe" plan to attend college. So, because 75% + 5% = 80% and 80% of 900 is 720, the counselor predicts that 720 students in the high school plan to go to college.

Perhaps a better way to predict the number of students is to use the range and say, "675 to 720 students in the school plan to go to college."

2 Is the sample large enough to make an accurate prediction?

One of the most surprising results in statistics is that relatively small sample sizes can produce accurate results. *If* the sample of 60 students was random, then it is a large enough sample to provide results that are accurate.

3 Once you address the issues of sample size and randomness, there are still many other things to consider. Here are some of them:

- How were the students asked the question? Were they asked in such a way that they felt free to tell the truth?
- Do students in 9th grade really know whether they plan to go to college or not?
- What time of year was the survey taken? If it was in the spring, then the juniors and seniors should have a reasonable idea of whether they are going to college or not.

Project

Select a circle graph from the newspaper, a magazine, or online. Explain the graph. Explain another way that the data could be displayed.

Reteaching and Enrichment Strategies

If students need help. . .	If students got it. . .
Resources by Chapter • Practice A and Practice B • Puzzle Time Record and Practice Journal Practice Differentiating the Lesson Lesson Tutorials Skills Review Handbook	Resources by Chapter • Enrichment and Extension • Technology Connection Start the next section

③ **12.** **FOOD** You ask 125 randomly chosen students to name their favorite food. There are 1500 students in the school. Predict the number of students in the school whose favorite food is pizza.

Favorite Food	
Pizza	58
Hamburger	36
Pasta	14
Other	17

Determine whether you would survey the population or a sample. Explain.

13. You want to know the average height of seventh graders in the United States.

14. You want to know the favorite types of music of students in your homeroom.

15. You want to know the number of students in your state who have summer jobs.

Theater Ticket Sales	
Adults	Students
522	210

16. **THEATER** You su0.rvey 72 randomly chosen students about whether they are going to attend the school play. Twelve say yes. Predict the number of students who attend the school.

17. **CRITICAL THINKING** Explain why 200 people with email addresses may not be a random sample. When might it be a random sample?

18. **LOGIC** A person surveys residents of a town to determine whether a skateboarding ban should be overturned.

a. Describe how the person could conduct the survey so that the sample is biased toward overturning the ban.

b. Describe how the person could conduct the survey so that the sample is biased toward keeping the ban.

19. Reasoning A guidance counselor surveys a random sample of 60 out of 900 high school students. Using the survey results, the counselor predicts that approximately 720 students plan to attend college. Do you agree with her prediction? Explain.

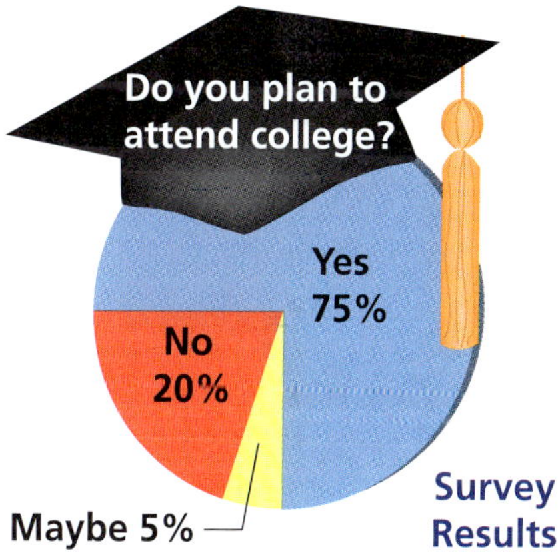

Fair Game Review What you learned in previous grades & lessons

Write and solve a proportion to answer the question. *(Skills Review Handbook)*

20. What percent of 60 is 18?

21. 70% of what number is 98?

22. 30 is 15% of what number?

23. What number is 0.6% of 500?

24. **MULTIPLE CHOICE** What is the volume of the pyramid? *(Section 14.5)*

Ⓐ 40 cm^3 Ⓑ 50 cm^3

Ⓒ 100 cm^3 Ⓓ 120 cm^3

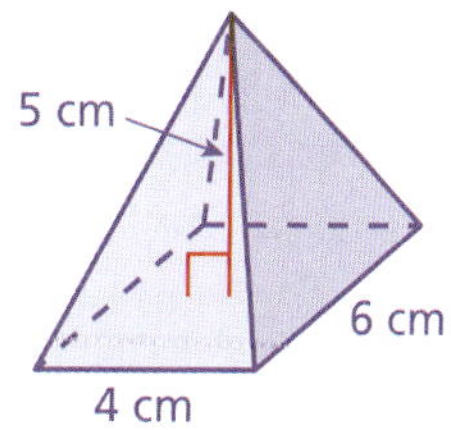

Extension 15.6 Generating Multiple Samples

Check It Out
Lesson Tutorials
BigIdeasMath.com

You have already used unbiased samples to make inferences about a population. In some cases, making an inference about a population from only one sample is not as precise as using multiple samples.

1 ACTIVITY: Using Multiple Random Samples

Work with a partner. You and a group of friends want to know how many students in your school listen to pop music. There are 840 students in your school. Each person in the group randomly surveys 20 students.

Step 1: The table shows your results. Make an inference about the number of students in your school who prefer pop music.

Favorite Type of Music			
Country	Pop	Rock	Rap
4	10	5	1

Step 2: The table shows Kevin's results. Use these results to make another inference about the number of students in your school who prefer pop music.

Favorite Type of Music			
Country	Pop	Rock	Rap
2	13	4	1

Compare the results of Steps 1 and 2.

Probability and Statistics
In this extension, you will
- use multiple samples to make predictions about populations.

Step 3: The table shows the results of three other friends. Use these results to make three more inferences about the number of students in your school who prefer pop music.

	Favorite Type of Music			
	Country	Pop	Rock	Rap
Steve	3	8	7	2
Laura	5	10	4	1
Ming	5	9	3	3

Step 4: Describe the variation of the five inferences. Which one would you use to describe the number of students in your school who prefer pop music? Explain your reasoning.

Step 5: Show how you can use all five samples to make an inference.

Practice

1. **PACKING PEANUTS** Work with a partner. Mark 24 packing peanuts with either a red or a black marker. Put the peanuts into a paper bag. Trade bags with other students in the class.

   a. Generate a sample by choosing a peanut from your bag six times, replacing the peanut each time. Record the number of times you choose each color. Repeat this process to generate four more samples. Organize the results in a table.

   b. Use each sample to make an inference about the number of red peanuts in the bag. Then describe the variation of the five inferences. Make inferences about the numbers of red and black peanuts in the bag based on all the samples.

   c. Take the peanuts out of the bag. How do your inferences compare to the population? Do you think you can make a more accurate prediction? If so, explain how.

Laurie's Notes

Introduction

Connect

- **Yesterday:** Students identified biased and unbiased samples and determined whether a sample can be used to draw conclusions and make predictions about a population.
- **Today:** Students will generate multiple samples of data and draw inferences about a population.

Motivate

- ? "Are you familiar with different groups that poll large groups of people?" Students may have heard of the Gallup Poll or Rasmussen Reports.
- ? "Have any of you or has someone you know ever been asked to participate in a survey, perhaps at the mall or on the telephone?" Answers will vary.
- Explain that in these activities, multiple samples are compared in order to make an inference about a population.

Activity Notes

Activity 1

- Read the introduction and context for the activity. Make sure students understand that more than one random sample is being taken of the 840 students in a school.
- ? "How can an unbiased sample be used to make an inference about a population?" Students should describe how a ratio table or proportion can be used to make predictions about the population based upon the unbiased sample.
- Give sufficient time for students to work through the steps.
- There are results from five different random samples. When students have finished Step 3, you could gather the results before students move on to Steps 4 and 5.
- **Construct Viable Arguments and Critique the Reasoning of Others:** Students may have different ways in which they arrive at their conclusions when describing the music preferences of the school. Ask volunteers to explain their reasoning.

Practice

- Tell students to vary the amount of peanuts they mark with red and black markers. It does not have to be 50-50. They can mark as many as they want but you can put a limit on the least amount for each color. So, a student cannot mark only 1 peanut black or all the peanuts red.

What Your Students Will Learn

- Use measures from multiple random samples and simulations to make predictions about populations.

Goal Today's lesson is generating multiple samples of data and drawing inferences about a population.

Technology for the Teacher

Dynamic Classroom

Lesson Plans

Complete Materials List

Extension 15.6 Record and Practice Journal

You have already used unbiased samples to make inferences about a population. In some cases, making an inference about a population from only one sample is not as precise as using multiple samples.

1 ACTIVITY: Using Multiple Random Samples

Work with a partner. You and a group of friends want to know how many students in your school listen to pop music. There are 840 students in your school. Each person in the group randomly surveys 20 students.

Step 1: The table shows your results. Make an inference about the number of students in your school who prefer pop music.

Favorite Type of Music			
Country	Pop	Rock	Rap
4	10	5	1

420 students

Step 2: The table shows Kevin's results. Use these results to make another inference about the number of students in your school who prefer pop music.

Favorite Type of Music			
Country	Pop	Rock	Rap
2	13	4	1

546 students

Compare the results of Steps 1 and 2.

Using Kevin's results, this inference is more than the inference in Step 1.

Step 3: The table shows the results of three other friends. Use these results to make three more inferences about the number of students in your school who prefer pop music.

	Favorite Type of Music			
	Country	Pop	Rock	Rap
Steve	3	8	7	2
Laura	5	10	4	1
Ming	5	9	3	3

Steve: 336 students
Laura: 420 students
Ming: 378 students

Extension 15.6
Record and Practice Journal

Step 4: Describe the variation of the five inferences. Which one would you use to describe the number of students in your school who prefer pop music? Explain your reasoning.

Sample answer: **The greatest is 546 students. The least is 336 students. 420 is the median and the mode of the data. So, use the inference of 420 students.**

Step 5: Show how you can use all five samples to make an inference.

Combine all the samples. 50 students out of the 100 students chose pop music.

Use the proportion $\frac{50}{100} = \frac{x}{840}$ to predict 420 students prefer pop music.

Practice

1. Work with a partner. Mark 24 packing peanuts with either a red or a black marker. Put the peanuts into a paper bag. Trade bags with other students in the class.

   a. Generate a sample by choosing a peanut from your bag six times, replacing the peanut each time. Record the number of times you choose each color. Repeat this process to generate four more samples. Organize the results in a table.

   Check students' work.

   b. Use each sample to make an inference about the number of red peanuts in the bag. Then describe the variation of the five inferences. Make inferences about the numbers of red and black peanuts in the bag based on all the samples.

   Check students' work.

   c. Take the peanuts out of the bag. How do your inferences compare to the population? Do you think you can make a more accurate prediction? If so, explain how.

   Check students' work; Yes, increase the number of random samples.

2 ACTIVITY: Using Measures from Multiple Random Samples

Work with a partner. You want to know the mean number of hours students with part-time jobs work each week. You go to 8 different schools. At each school, you randomly survey 10 students with part-time jobs. Your results are shown at the right.

Hours Worked Each Week
1: 6, 8, 6, 6, 7, 4, 10, 8, 7, 8
2: 10, 4, 4, 6, 8, 6, 7, 12, 8, 8
3: 10, 9, 8, 6, 5, 8, 6, 6, 9, 10
4: 4, 8, 4, 4, 5, 4, 4, 6, 5, 6
5: 6, 8, 8, 6, 12, 4, 10, 8, 6, 12
6: 10, 10, 8, 9, 16, 8, 7, 12, 16, 14
7: 4, 5, 6, 6, 4, 5, 6, 6, 4, 4
8: 16, 20, 8, 12, 10, 8, 8, 14, 16, 8

Step 1: Find the mean of each sample.

7, 7.3, 7.7, 5, 8, 11, 5, 12

Step 2: Make a box-and-whisker plot of the sample means.

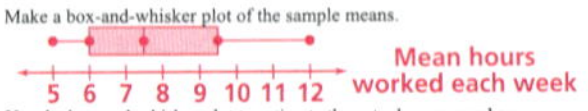

Step 3: Use the box-and-whisker plot to estimate the actual mean number of hours students with part-time jobs work each week. How does your estimate compare to the mean of the entire data set?

Sample answer: **7.5 hours**

3 ACTIVITY: Using a Simulation

Work with a partner. Another way to generate multiple samples of data is to use a simulation. Suppose 70% of all seventh graders watch reality shows on television.

Step 1: Design a simulation involving 50 packing peanuts by marking 70% of the peanuts with a certain color. Put the peanuts into a paper bag.

Check students' work.

Step 2: Simulate choosing a sample of 30 students by choosing peanuts from the bag, replacing the peanut each time. Record the results. Repeat this process to generate eight more samples. How much variation do you expect among the samples? Explain.

Sample answer: **Because the actual percent of students is 70%, expect the number of red peanuts in their sample to be between 60%–80%, or 55%–85%, or 65%–75%, etc.**

Step 3: Display your results.

Check students' work.

Practice

2. You want to know whether student-athletes prefer water or sports drinks during games. You go to 10 different schools. At each school, you randomly survey 10 student-athletes. The percents of student-athletes who prefer water are shown.

   60% 70% 60% 50% 80% 70% 30% 70% 80% 40%

   a. Make a box-and-whisker plot of the data.

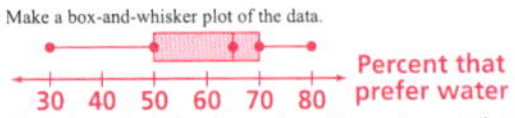

   b. Use the box-and-whisker plot to estimate the actual percent of student-athletes who prefer water. How does your estimate compare to the mean of the data?

   Sample answer: **60%**

3. Repeat Activity 2 using the medians of the samples. **7, 7.5, 8, 4.5, 8, 10, 5, 11**

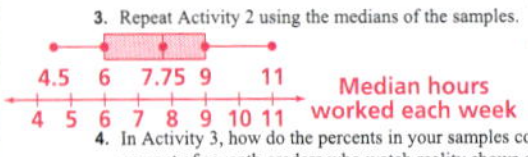

   Sample answer: **7.5**

4. In Activity 3, how do the percents in your samples compare to the actual percent of seventh graders who watch reality shows on television?

   Check students' work.

5. **REASONING** Why is it better to make inferences about a population based on multiple samples instead of only one sample? What additional information do you gain by taking multiple random samples? Explain.

   The more samples you have, the more accurate your inferences will be.

Laurie's Notes

Activity 2

- Discuss with students the description of the samples. The students all have part-time jobs and are from 8 different schools.
- **Use Appropriate Tools Strategically:** Students can use calculators to quickly find the means.
- The box-and-whisker plot is NOT using all of the data. Only the mean of each sample (from 8 different schools) is used to construct the box-and-whisker plot in Step 2.

? "Can you use your box-and-whisker plot to estimate the actual mean number of hours students work each week? Explain." yes; *Sample answer:* It probably lies somewhere within the "box."

? "How do you think your estimate compares to the mean of the entire data set?" Students should realize that it is a good estimate.

Activity 3

- This is a fun activity for students, if time permits. Make sure you have enough time to complete this simulation.
- The materials do not have to be peanuts. Any congruent shapes that differ in color will work, such as colored tiles.
- In Step 2, there is no wrong answer here. Students are just recognizing variability in their samples and determining a range where they think their percents will fall. They will compare in Question 4.
- Students can use percents or the actual numbers of red peanuts.
- Allow time for different groups to be able to share their method, and their results.

Practice

- Have students discuss Question 5 with their partners and then share their thoughts with the class.

Closure

- Describe how you can generate multiple samples using students in your school to determine their preference for an end-of-the-year field trip from four possible locations.

2 ACTIVITY: Using Measures from Multiple Random Samples

Hours Worked Each Week
1: 6, 8, 6, 6, 7, 4, 10, 8, 7, 8
2: 10, 4, 4, 6, 8, 6, 7, 12, 8, 8
3: 10, 9, 8, 6, 5, 8, 6, 6, 9, 10
4: 4, 8, 4, 4, 5, 4, 4, 6, 5, 6
5: 6, 8, 8, 6, 12, 4, 10, 8, 6, 12
6: 10, 10, 8, 9, 16, 8, 7, 12, 16, 14
7: 4, 5, 6, 6, 4, 5, 6, 6, 4, 4
8: 16, 20, 8, 12, 10, 8, 8, 14, 16, 8

Work with a partner. You want to know the mean number of hours students with part-time jobs work each week. You go to 8 different schools. At each school, you randomly survey 10 students with part-time jobs. Your results are shown at the left.

Step 1: Find the mean of each sample.

Step 2: Make a box-and-whisker plot of the sample means.

Step 3: Use the box-and-whisker plot to estimate the actual mean number of hours students with part-time jobs work each week.

How does your estimate compare to the mean of the entire data set?

3 ACTIVITY: Using a Simulation

Work with a partner. Another way to generate multiple samples of data is to use a simulation. Suppose 70% of all seventh graders watch reality shows on television.

Step 1: Design a simulation involving 50 packing peanuts by marking 70% of the peanuts with a certain color. Put the peanuts into a paper bag.

Step 2: Simulate choosing a sample of 30 students by choosing peanuts from the bag, replacing the peanut each time. Record the results. Repeat this process to generate eight more samples. How much variation do you expect among the samples? Explain.

Step 3: Display your results.

Practice

2. **SPORTS DRINKS** You want to know whether student-athletes prefer water or sports drinks during games. You go to 10 different schools. At each school, you randomly survey 10 student-athletes. The percents of student-athletes who prefer water are shown.

   60% 70% 60% 50% 80% 70% 30% 70% 80% 40%

   a. Make a box-and-whisker plot of the data.

   b. Use the box-and-whisker plot to estimate the actual percent of student-athletes who prefer water. How does your estimate compare to the mean of the data?

3. **PART-TIME JOBS** Repeat Activity 2 using the medians of the samples.

4. **TELEVISION** In Activity 3, how do the percents in your samples compare to the given percent of seventh graders who watch reality shows on television?

5. **REASONING** Why is it better to make inferences about a population based on multiple samples instead of only one sample? What additional information do you gain by taking multiple random samples? Explain.

15.7 Comparing Populations

Essential Question How can you compare data sets that represent two populations?

1 ACTIVITY: Comparing Two Data Distributions

Work with a partner. You want to compare the shoe sizes of male students in two classes. You collect the data shown in the table.

Male Students in Eighth-Grade Class														
7	9	8	$7\frac{1}{2}$	$8\frac{1}{2}$	10	6	$6\frac{1}{2}$	8	8	$8\frac{1}{2}$	9	11	$7\frac{1}{2}$	$8\frac{1}{2}$
Male Students in Sixth-Grade Class														
6	$5\frac{1}{2}$	6	$6\frac{1}{2}$	$7\frac{1}{2}$	$8\frac{1}{2}$	7	$5\frac{1}{2}$	5	$5\frac{1}{2}$	$6\frac{1}{2}$	7	$4\frac{1}{2}$	6	6

a. How can you display both data sets so that you can visually compare the measures of center and variation? Make the data display you chose.

b. Describe the shape of each distribution.

c. Complete the table.

	Mean	Median	Mode	Range	Interquartile Range (IQR)	Mean Absolute Deviation (MAD)
Male Students in Eighth-Grade Class						
Male Students in Sixth-Grade Class						

d. Compare the measures of center for the data sets.

e. Compare the measures of variation for the data sets. Does one data set show more variation than the other? Explain.

f. Do the distributions overlap? How can you tell using the data display you chose in part (a)?

g. The double box-and-whisker plot below shows the shoe sizes of the members of two girls basketball teams. Can you conclude that at least one girl from each team has the same shoe size? Can you conclude that at least one girl from the Bobcats has a larger shoe size than one of the girls from the Tigers? Explain your reasoning.

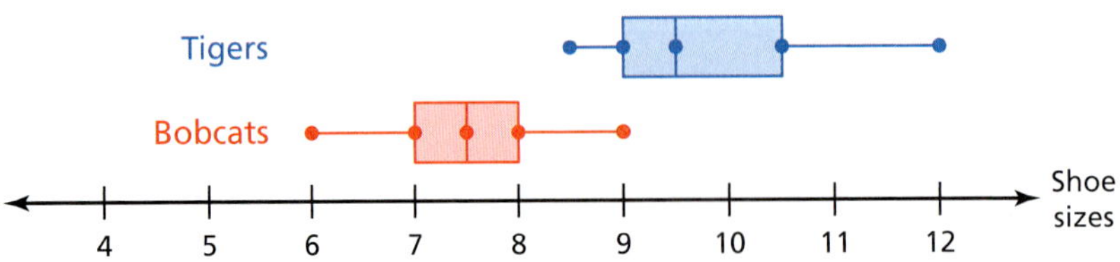

Probability and Statistics

In this lesson, you will

- use measures of center and variation to compare populations.
- use random samples to compare populations.

Laurie's Notes

Introduction

Applying Mathematical Practices

- **Reason Abstractly and Quantitatively:** Mathematically proficient students make sense of quantities and their relationships in problem situations. The overlap between two data sets can be compared visually using various data displays.

Motivate

- Display the double box-and-whisker plot showing the electricity produced (in kilowatt-hours, kWh) each day in July from solar panels on two houses.

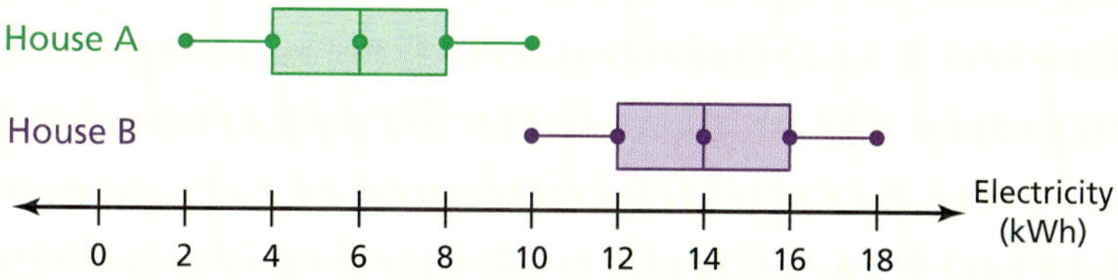

- Ask questions comparing the box-and-whisker plots.

? "Were there days in which House A generated more electricity than House B?" no

? "Could there have been days in which both houses generated the same amount of electricity?" yes

Activity Notes

Activity 1

- **FYI:** This first activity reviews statistical measures that students learned in a previous course as well as shapes of distributions. Students may need more guidance or probing questions for this activity than usual.
- Students should work with their partners to complete the activity.

? "If we were to collect shoe size data from the boys in this class, what would the data look like?" Answers vary, but students should mention whole and half sizes.

? "What data displays can you use to visually compare two data sets?" Students may say stem-and-leaf plots, box-and-whisker plots, or dot plots.

- Students may need to be reminded about measures of center, measures of variation, and shapes of distributions. You may need to review how to find the IQR and the MAD. This was all taught in a previous course.
- Discuss the measures of center and variation for the data sets.

? "Do the data sets represent samples or populations?" Because you are comparing shoe sizes of male students in two classes, not your grade or your school, these represent populations.

- If desired, have students draw dotted lines vertically through the measures of center as one way to visualize overlap.
- **Reason Abstractly and Quantitatively:** In Activity 1(g), students should conclude that just because two data sets overlap slightly doesn't necessarily mean they contain one or more of the same data values.

What Your Students Will Learn

- Use measures of center, variations, and random samples to compare populations.

Previous Learning

Students have found measures of center and measures of variation.

Technology for the Teacher

Dynamic Classroom

Lesson Plans
Complete Materials List

15.7 Record and Practice Journal

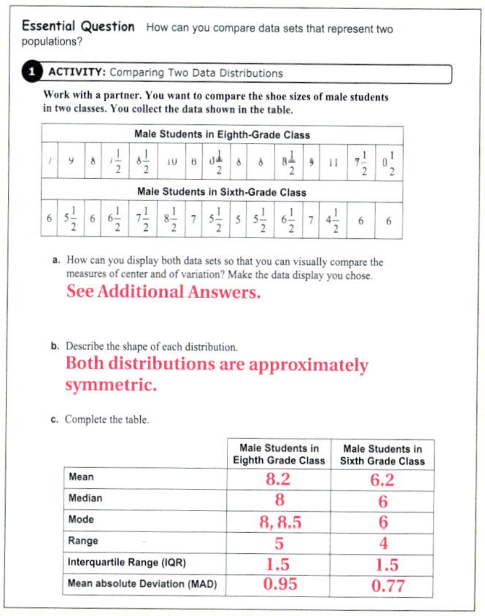

Essential Question How can you compare data sets that represent two populations?

1 ACTIVITY: Comparing Two Data Distributions

Work with a partner. You want to compare the shoe sizes of male students in two classes. You collect the data shown in the table.

Male Students in Eighth-Grade Class														
7	9	8	$7\frac{1}{2}$	$8\frac{1}{2}$	10	8	$6\frac{1}{2}$	8	8	$8\frac{1}{2}$	9	11	$7\frac{1}{2}$	$6\frac{1}{2}$

Male Students in Sixth-Grade Class														
6	$5\frac{1}{2}$	6	$6\frac{1}{2}$	$7\frac{1}{2}$	$8\frac{1}{2}$	7	$5\frac{1}{2}$	5	$5\frac{1}{2}$	$6\frac{1}{2}$	7	$4\frac{1}{2}$	6	6

a. How can you display both data sets so that you can visually compare the measures of center and of variation? Make the data display you chose.
See Additional Answers.

b. Describe the shape of each distribution.
Both distributions are approximately symmetric.

c. Complete the table.

	Male Students in Eighth Grade Class	Male Students in Sixth Grade Class
Mean	8.2	6.2
Median	8	6
Mode	8, 8.5	6
Range	5	4
Interquartile Range (IQR)	1.5	1.5
Mean absolute Deviation (MAD)	0.95	0.77

English Language Learners

Visual

Use a diagram of a generic box-and-whisker plot on an overhead as a visual aid for English learners. Have students identify the parts of the box-and-whisker plot: *median, first quartile, third quartile, least value,* and *greatest value.* Make sure students understand that they can interpret a box-and-whisker plot that does not have a scale.

15.7 Record and Practice Journal

d. Compare the measures of center for the data sets.
The mean and median for the eighth-grade class is 2 more than the mean and median for the sixth-grade class.

e. Compare the measures of variation for the data sets. Does one data set show more variation than the other? Explain.
yes

f. Do the distributions overlap? How can you tell using the data display you chose in part (a)?
yes

g. The double box-and-whisker plot below shows the shoe sizes of the members of two girls basketball teams. Can you conclude that at least one girl from each team has the same shoe size? Can you conclude that at least one girl from the Bobcats has a larger shoe size than one of the girls from the Tigers? Explain your reasoning. no; yes

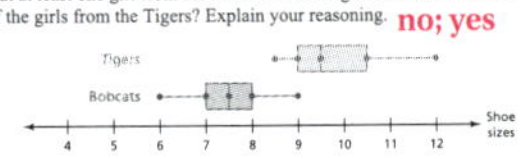

2 ACTIVITY: Comparing Two Data Distributions

Work with a partner. Compare the shapes of the distributions. Do the two data sets overlap? Explain. If so, use measures of center and the least and the greatest values to describe the overlap between the two data sets.

a.

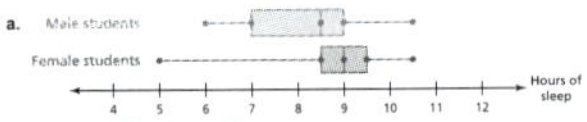

See Additional Answers.

b.

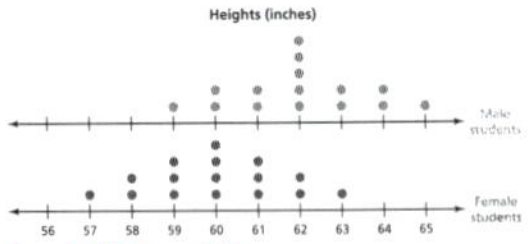

See Additional Answers.

c. Ages of People in Two Exercise Classes

10:00 A.M. Class		8:00 P.M. Class
	1	8 9
	2	1 2 2 7 9 9
	3	0 3 4 5 7
9 7 3 2 2 2	4	0
7 5 4 3 1	5	
7 0 0	6	
0	7	

Key: 1 | 8 = 18

8:00 P.M. class: symmetric; 10:00 A.M. class: skewed right; no; The oldest person in the 8:00 P.M. class is 40. The youngest person in the 10:00 A.M. class is 42.

What Is Your Answer?

3. **IN YOUR OWN WORDS** How can you compare data sets that represent two populations?
You can compare the measures of center, the measures of variation, the shapes of the distributions, and the overlap of the two distributions.

Laurie's Notes

Activity 2

- Once students have discussed Activity 1 thoroughly, Activity 2 should take much less time.
- This activity is a good review of the three data displays shown.
- **Reason Abstractly and Quantitatively** and **Use Appropriate Tools Strategically:** The analysis of each plot in this activity will give good insights into student understanding of each type of data display.
- If time permits, ask students to summarize what is known about each data set in each of the three parts.

? "Do you know how many students are represented in the box-and-whisker plots?" no

? "Do you know how many students are represented in the dot plots?" yes

? "Do you know how many people are represented in the stem-and-leaf plots?" yes

What Is Your Answer?

- Ask volunteers to share their answers.

Closure

- Refer back to the box-and-whisker plots in the Motivate. Write a summary that compares the data sets.

2 ACTIVITY: Comparing Two Data Distributions

Work with a partner. Compare the shapes of the distributions. Do the two data sets overlap? Explain. If so, use measures of center and the least and the greatest values to describe the overlap between the two data sets.

a.

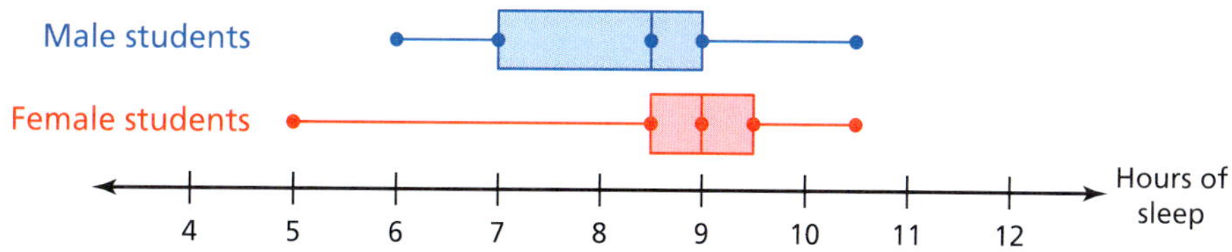

b.

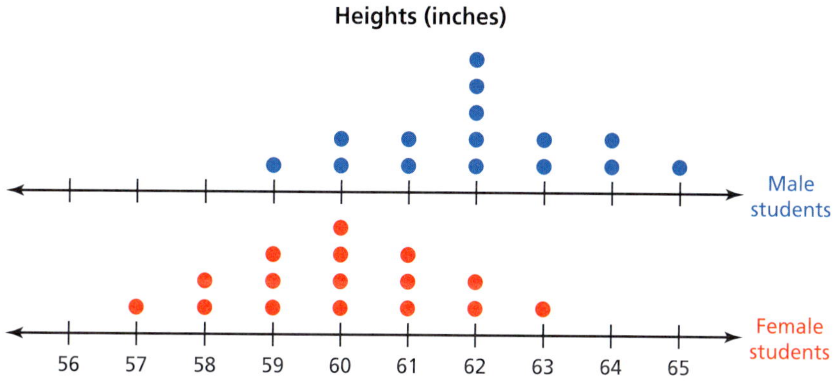

c. **Ages of People in Two Exercise Classes**

10:00 A.M. Class		8:00 P.M. Class
	1	8 9
	2	1 2 2 7 9 9
	3	0 3 4 5 7
9 7 3 2 2 2	4	0
7 5 4 3 1	5	
7 0 0	6	
0	7	

Key: 1 | 8 = 18

Math Practice

Recognize Usefulness of Tools

How is each type of data display useful? Which do you prefer? Explain.

What Is Your Answer?

3. **IN YOUR OWN WORDS** How can you compare data sets that represent two populations?

Practice Use what you learned about comparing data sets to complete Exercise 3 on page 684.

15.7 Lesson

Check It Out
Lesson Tutorials
BigIdeasMath.com

Recall that you use the mean and the mean absolute deviation (MAD) to describe symmetric distributions of data. You use the median and the interquartile range (IQR) to describe skewed distributions of data.

To compare two populations, use the mean and the MAD when both distributions are symmetric. Use the median and the IQR when either one or both distributions are skewed.

EXAMPLE 1 Comparing Populations

Study Tip
You can more easily see the visual overlap of dot plots that are aligned vertically.

The double dot plot shows the time that each candidate in a debate spent answering each of 15 questions.

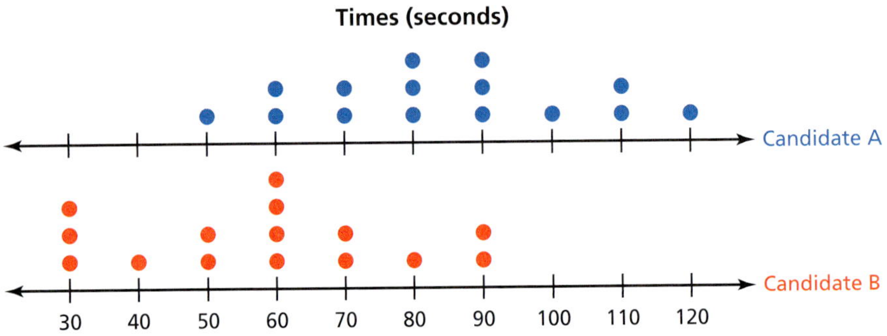

a. Compare the populations using measures of center and variation.

Both distributions are approximately symmetric, so use the mean and the MAD.

Candidate A

$$\text{Mean} = \frac{1260}{15} = 84$$

$$\text{MAD} = \frac{244}{15} \approx 16$$

Candidate B

$$\text{Mean} = \frac{870}{15} = 58$$

$$\text{MAD} = \frac{236}{15} \approx 16$$

So, the variation in the times was about the same, but Candidate A had a greater mean time.

Study Tip
When two populations have similar variabilities, the value in part (b) describes the visual overlap between the data. In general, the greater the value, the less the overlap.

b. Express the difference in the measures of center as a multiple of the measure of variation.

$$\frac{\text{mean for Candidate A} - \text{mean for Candidate B}}{\text{MAD}} = \frac{26}{16} \approx 1.6$$

So, the difference in the means is about 1.6 times the MAD.

On Your Own

Now You're Ready
Exercises 4–6

1. **WHAT IF?** Each value in the dot plot for Candidate A increases by 30 seconds. How does this affect the answers in Example 1? Explain.

Laurie's Notes

Introduction

Connect

- **Yesterday:** Students explored overlap between data sets.
- **Today:** Students will compare two populations using measures of center, measures of variation, and overlap.

Motivate

- Recall that in a previous course you used the mean and the MAD to describe symmetric distributions of data. You used the median and the IQR to describe skewed distributions of data.
- Tell students that when comparing two populations, they will use the mean and MAD when *both* distributions are symmetric. If either one or both of the distributions are skewed, they will use the median and IQR.

Lesson Notes

Example 1

- Refer to the double dot plot and ask students to describe what the dots represent. They should explain that each dot represents the number of seconds a candidate spent answering a question in a debate.
- ? "How can you describe the distributions shown in the display?" Candidate A's responses were from 50 seconds to 120 seconds and the distribution is approximately symmetric. Candidate B's responses were from 30 seconds to 90 seconds and the distribution is approximately symmetric.
- ? "How do you describe the centers and variation of symmetric distributions?" using the mean and the MAD
- **Use Appropriate Tools Strategically:** Students should use a calculator to quickly compute the mean and the MAD. Split the class with each half computing one of the two means and the associated MAD.
- **Attend to Precision:** Ask students to interpret the results in part (a) and to reference the visual display. The dot plots show two symmetric distributions with similar variabilities but different centers. Candidate A's graph is shifted to the right, meaning longer responses.
- ? "Do the dot plots overlap? If so, do you think we can *measure* the overlap?" yes; Students may think it can be measured but are unsure what it means.
- **Make Sense of Problems and Persevere in Solving Them:** In part (b), students need to find the difference in the means as a multiple of the MAD, so they must divide by the MAD. Make sure students understand the meaning of this number, as described in the Study Tip.
- Work to finish the problem and interpret the result.

On Your Own

- **Big Idea:** Increasing each data value by 30 seconds affects the mean but not the variability. The difference in the measures of center will be greater and the visual overlap will be much less. The variabilities are still similar, so the value in part (b) will increase, indicating less overlap.

Goal Today's lesson is comparing populations.

Technology for the Teacher

Dynamic Classroom

Lesson Tutorials
Lesson Plans
Answer Presentation Tool

Differentiated Instruction

Vocabulary

Key words in this lesson were introduced in a previous grade. Have students add the words and acronyms, *mean, mean absolute deviation (MAD), symmetric distribution, median, interquartile range (IQR),* and *skewed distribution,* to their math notebook glossaries.

Extra Example 1

The data sets below give the times (in seconds) that the candidates in Example 1 spent answering each of 15 questions in a second debate.

Candidate A: 40, 50, 50, 50, 50, 50, 60, 60, 60, 70, 70, 70, 80, 80, 90

Candidate B: 40, 50, 60, 70, 70, 70, 80, 80, 80, 80, 90, 90, 90, 90, 90

a. Compare the populations using measures of center and variation. same variation; Candidate B had a greater median time.

b. Express the difference in the measures of center as a multiple of the measure of variation. difference in medians is 1 times the IQR

On Your Own

1. Part (a): Candidate A's mean increases by 30 to 114 and the MAD does not change, so Candidate A still has a greater mean time.

   Part (b): The difference in the means is now 3.5 times the MAD. The number is greater, indicating less overlap in the data.

Extra Example 2

You want to compare the costs of speeding tickets in State C to the costs in States A and B in Example 2.

a. The box-and-whisker plot shows a random sample of 10 speeding tickets in State C. Compare the samples using measures of center and variation. Can you use this to make a valid comparison about speeding tickets in the three states? Explain.

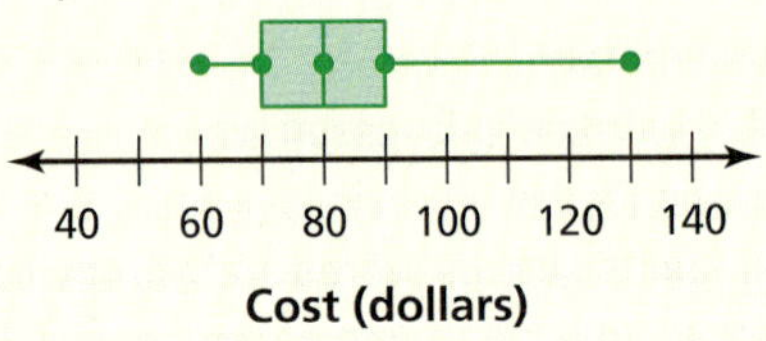

The median, 80, is the same as State B but greater than State A; The IQR, 20, is the same as State A but less than State B; No, the sample size is too small and the variability too great.

b. The box-and-whisker plot shows the medians of 100 random samples of 10 speeding tickets in State C. Compare the variability of the sample medians to the variability of the sample costs in part (a).

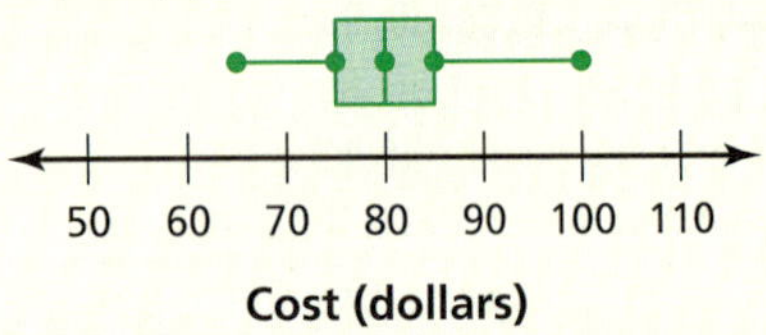

Sample medians vary much less than the sample costs.

c. Make a conclusion about the costs of speeding tickets in the three states. Speeding tickets generally cost less in State A than in States B and C, where the costs are about the same.

On Your Own

2. No, the sample size is too small to make a conclusion about the population.

Laurie's Notes

Discuss

- Explain that you do not need to have all of the data from two populations to make comparisons. You can use random samples to make comparisons.

Example 2

- Ask a volunteer to read part (a).
- ? "What do you notice about the distributions?" They are both skewed right.
- ? "How do you describe the centers and variation of skewed distributions?" using the median and the IQR
- Give students time to find the median and the IQR.
- **Attend to Precision:** Discuss part (a). The samples have different medians and different IQRs. The median and IQR for State A are each less than the median and IQR for State B. However, the sample size is too small and the variability is too great to make comparisons about the populations.
- In part (b), each box-and-whisker plot represents the medians of 100 random samples of 10 speeding tickets. The sample size is no longer small.
- Make sure students see the distinction between parts (a) and (b). In part (a), one random sample is taken (in each state), and the individual speeding ticket costs are used to make the box-and-whisker plot. In part (b), 100 random samples are taken, and the 100 *medians* of the samples are used to make the box-and-whisker plot.
- If students are confused about what this means, refer to part (a). The median, 70, for State A represents one of the 100 values for State A in part (b). A similar statement can be made for State B.
- ? "What can you say about the variation of the sample medians compared to the variation of the sample costs?" The sample medians vary much less than the sample costs.
- **Construct Viable Arguments** and **Attend to Precision:** Students should conclude from parts (a) and (b) that it is reasonable to assume that speeding tickets generally cost more in State B than in State A.

On Your Own

- **Neighbor Check:** Have students work independently and then have their neighbors check their work. Have students discuss any discrepancies.

Closure

- **Writing Prompt:** To compare two populations . . .

You do not need to have all the data from two populations to make comparisons. You can use random samples to make comparisons.

EXAMPLE 2 Using Random Samples to Compare Populations

You want to compare the costs of speeding tickets in two states.

a. The double box-and-whisker plot shows a random sample of 10 speeding tickets issued in two states. Compare the samples using measures of center and variation. Can you use this to make a valid comparison about speeding tickets in the two states? Explain.

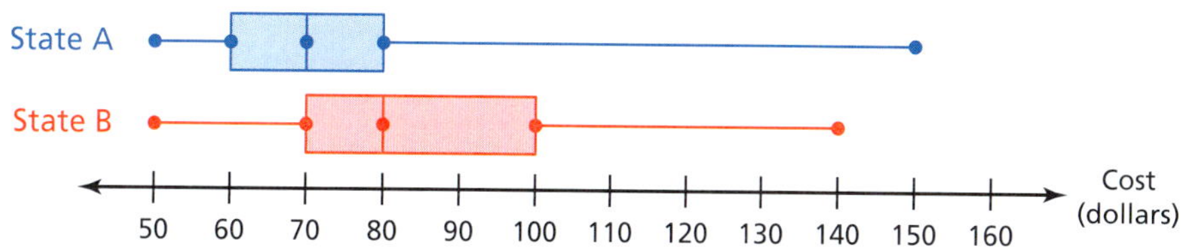

Both distributions are skewed right, so use the median and the IQR.

The median and the IQR for State A, 70 and 20, are less than the median and the IQR for State B, 80 and 30. However, the sample size is too small and the variability is too great to conclude that speeding tickets generally cost more in State B.

b. The double box-and-whisker plot shows the medians of 100 random samples of 10 speeding tickets for each state. Compare the variability of the sample medians to the variability of the sample costs in part (a).

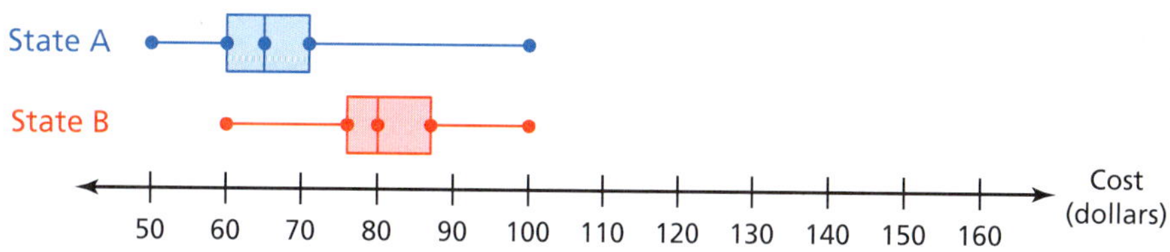

The IQR of the sample medians for each state is about 10.

So, the sample medians vary much less than the sample costs.

c. Make a conclusion about the costs of speeding tickets in the two states.

The sample medians show less variability. Most of the sample medians for State B are greater than the sample medians for State A.

So, speeding tickets generally cost more in State B than in State A.

On Your Own

Now You're Ready
Exercise 8

2. **WHAT IF?** A random sample of 8 speeding tickets issued in State C has a median of $120. Can you conclude that a speeding ticket in State C costs more than in States A and B? Explain.

15.7 Exercises

Check It Out
Help with Homework
BigIdeasMath.com

Vocabulary and Concept Check

1. **REASONING** When comparing two populations, when should you use the mean and the MAD? the median and the IQR?

2. **WRITING** Two data sets have similar variabilities. Suppose the measures of center of the data sets differ by 4 times the measure of variation. Describe the visual overlap of the data.

Practice and Problem Solving

3. **SNAKES** The tables show the lengths of two types of snakes at an animal store.

Garter Snake Lengths (inches)					
26	30	22	15	21	24
28	32	24	25	18	35

Water Snake Lengths (inches)					
34	25	24	35	40	32
41	27	37	32	21	30

   a. Find the mean, median, mode, range, interquartile range, and mean absolute deviation for each data set.

   b. Compare the data sets.

1 4. **HOCKEY** The double box-and-whisker plot shows the goals scored per game by two hockey teams during a 20-game season.

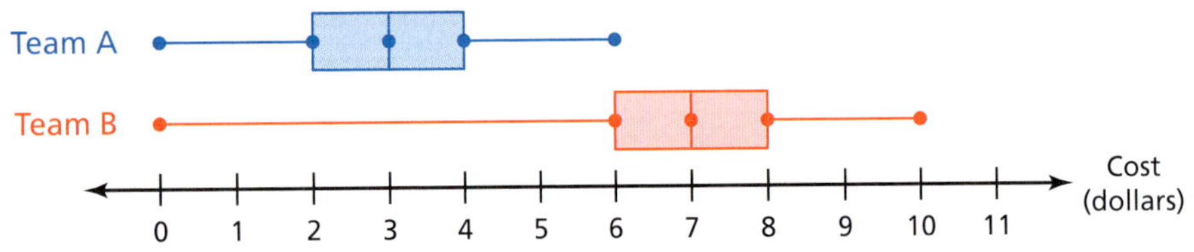

   a. Compare the populations using measures of center and variation.

   b. Express the difference in the measures of center as a multiple of the measure of variation.

5. **TEST SCORES** The dot plots show the test scores for two classes taught by the same teacher.

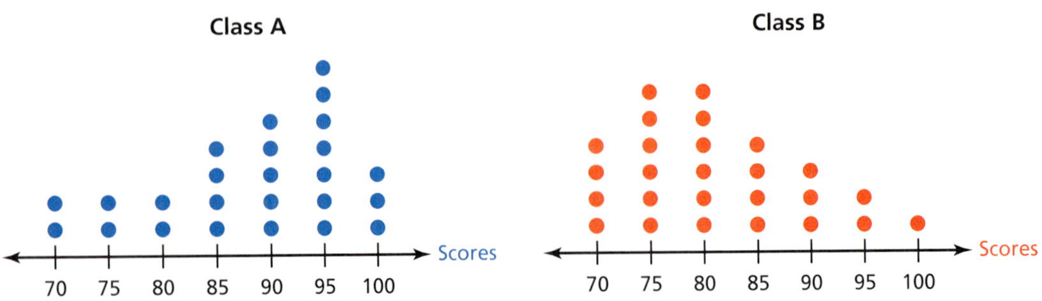

   a. Compare the populations using measures of center and variation.

   b. Express the difference in the measures of center as a multiple of each measure of variation.

Assignment Guide and Homework Check

Level	Assignment	Homework Check
Advanced	1 14	1, 5, 6, 7, 8

Common Errors

- **Exercises 4–6** Students may use the wrong measures of center and variation when comparing populations. Remind them to use the mean and the MAD when *both* distributions are symmetric. Otherwise, use the median and the IQR.

15.7 Record and Practice Journal

1. The dot plots show the quiz scores for two classes taught by the same teacher.

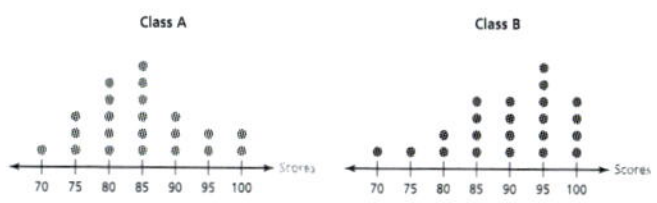

   a. Compare the populations using measures of center and variation.
   Class A: mean ≈ 85, MAD ≈ 19.3
   Class B: mean ≈ 90, MAD ≈ 20.7

   b. Express the difference in the measures of center as a multiple of the measure of variation.
   The difference in the means is about 0.2 to 0.3 times the MAD.

2. The double box-and-whisker plot shows the number of song downloads a month by two seventh grade classes.

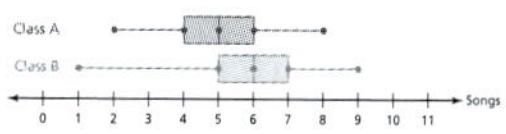

   a. Compare the populations using measures of center and variation.
   Class A: median = 5, IQR = 2
   Class B: median = 6, IQR = 2

   b. Express the difference in the measures of center as a multiple of the measure of variation.
   The difference in the medians is 0.5 times the IQR.

Vocabulary and Concept Check

1. When comparing two populations, use the mean and the MAD when each distribution is symmetric. Use the median and the IQR when either one or both distributions are skewed.
2. There will probably be little or no visual overlap of the data. The core (center) portions of the data are too far apart.

Practice and Problem Solving

3. a. garter snake: mean = 25, median = 24.5, mode = 24, range = 20, IQR = 7.5, MAD ≈ 4.33
   water snake: mean = 31.5, median = 32, mode = 32, range = 20, IQR = 10, MAD ≈ 5.08

   b. The water snakes have greater measures of center because the mean, median, and mode are greater. The water snakes also have greater measures of variation because the interquartile range and mean absolute deviation are greater.

4. a. Team A:
   median = 3, IQR = 2
   Team B:
   median = 7, IQR = 2
   The variation in the goals scored is the same, but Team B usually scores about 4 more goals per game.

   b. The difference in the medians is 2 times the IQR.

5. See Additional Answers.

Practice and Problem Solving

6. See Additional Answers.

7. See *Taking Math Deeper*.

8–9. See Additional Answers.

Fair Game Review

10. 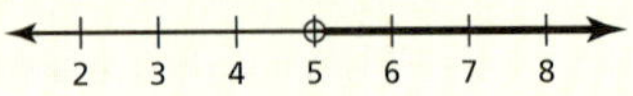

11. 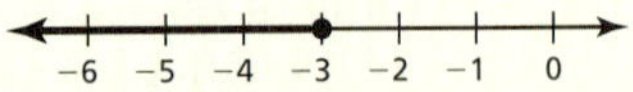

12. 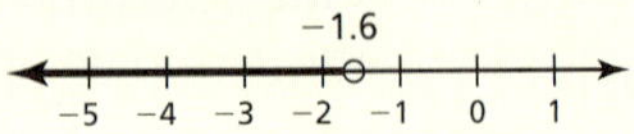

13. 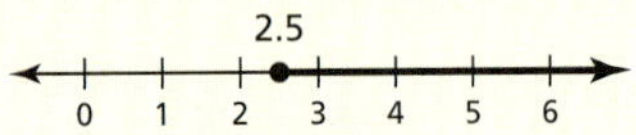

14. B

Mini-Assessment

The data sets below give the final grades of the females in a 7th grade math class and the females in an 8th grade math class.

Grade 7: 78, 82, 84, 87, 88, 89, 89, 90, 93, 100

Grade 8: 77, 80, 81, 84, 86, 87, 88, 90, 91, 96

1. Compare the populations using measures of center and variation. The variation was about the same, but Grade 7 females had a greater mean final grade.
2. Express the difference in the measures of center as a multiple of the measure of variation. difference in means is about 0.5 times the MAD
3. Now you want to compare the final grades of all females in 7th and 8th grade math classes. Can you conclude that the grades are better in Grade 7? Explain. No, sample size is too small.

Taking Math Deeper

Exercise 7

The values found in Exercises 4(b), 5(b), and 6(b) measure the overlap of the data sets. Begin by displaying the data so that you can visualize the overlap.

1 Describe the visual overlap.

Looking at the double box-and-whisker plot in Exercise 4, you can see that there is *some* overlap in the data.

Arrange the dot plots in Exercise 5 vertically. There is *a lot* of overlap.

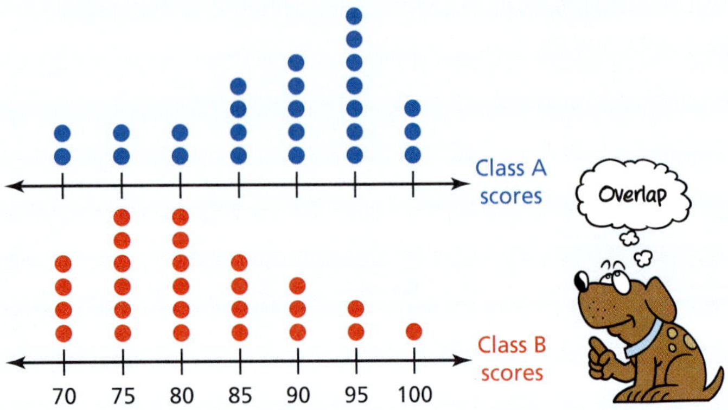

Make a double box-and-whisker plot in Exercise 6. There is *no* overlap.

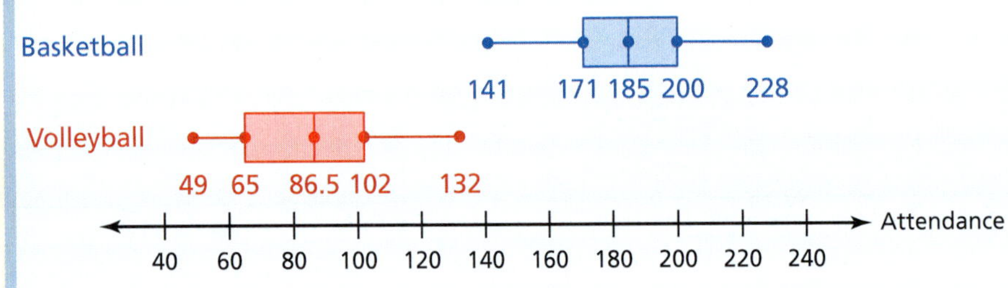

2 Compare the values in Exercises 4(b), 5(b), and 6(b): 2, 0.8 to 1, and 5.1 to 5.6. The value in Exercise 6(b) is the greatest, indicating that these data have less overlap than the data in Exercises 4 and 5.

3 Notice that greater numbers indicate less overlap, and lesser numbers indicate more overlap. The value in Exercise 5(b) is the least, indicating that these data have more overlap than the data in Exercises 4 and 6.

Reteaching and Enrichment Strategies

If students need help. . .	If students got it. . .
Resources by Chapter • Practice A and Practice B • Puzzle Time Record and Practice Journal Practice Differentiating the Lesson Lesson Tutorials Skills Review Handbook	Resources by Chapter • Enrichment and Extension • Technology Connection Start the next section

6. **ATTENDANCE** The tables show the attendances at volleyball games and basketball games at a school during the year.

Volleyball Game Attendance						
112	95	84	106	62	68	53
75	88	93	127	98	117	60
49	54	85	74	88	132	

Basketball Game Attendance						
202	190	173	155	169	188	195
176	141	152	181	198	214	179
163	186	184	207	219	228	

a. Compare the populations using measures of center and variation.

b. Express the difference in the measures of center as a multiple of each measure of variation.

7. **NUMBER SENSE** Compare the answers to Exercises 4(b), 5(b), and 6(b). Which value is the greatest? What does this mean?

2 8. **MAGAZINES** You want to compare the number of words per sentence in a sports magazine to the number of words per sentence in a political magazine.

a. The data represent random samples of 10 sentences in each magazine. Compare the samples using measures of center and variation. Can you use this to make a valid comparison about the magazines? Explain.

Sports magazine: 9, 21, 15, 14, 25, 26, 9, 19, 22, 30

Political magazine: 31, 22, 17, 5, 23, 15, 10, 20, 20, 17

b. The double box-and-whisker plot shows the means of 200 random samples of 20 sentences. Compare the variability of the sample means to the variability of the sample numbers of words in part (a).

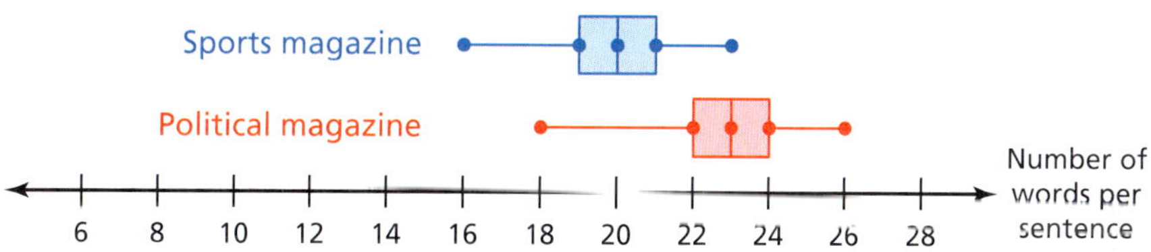

c. Make a conclusion about the numbers of words per sentence in the magazines.

9. **Project** You want to compare the average amounts of time students in sixth, seventh, and eighth grade spend on homework each week.

a. Design an experiment involving random sampling that can help you make a comparison.

b. Perform the experiment. Can you make a conclusion about which students spend the most time on homework? Explain your reasoning.

Fair Game Review What you learned in previous grades & lessons

Graph the inequality on a number line. *(Section 11.1)*

10. $x > 5$ 11. $b \leq -3$ 12. $n < -1.6$ 13. $p \geq 2.5$

14. **MULTIPLE CHOICE** The number of students in the marching band increased from 100 to 125. What is the percent of increase? *(Skills Review Handbook)*

(A) 20% (B) 25% (C) 80% (D) 500%

15.6–15.7 Quiz

Check It Out
Progress Check
BigIdeasMath.com

1. Which sample is better for making a prediction? Explain. *(Section 15.6)*

Predict the number of students in your school who play at least one sport.	
Sample A	A random sample of 10 students from the school student roster
Sample B	A random sample of 80 students from the school student roster

2. **GYMNASIUM** You want to estimate the number of students in your school who think the gymnasium should be remodeled. You survey 12 students on the basketball team. Determine whether the sample is *biased* or *unbiased*. Explain. *(Section 15.6)*

3. **TOWN COUNCIL** You want to know how the residents of your town feel about a recent town council decision. You survey 100 residents at random. Sixty-five support the decision, and thirty-five do not. So, you conclude that 65% of the residents of your town support the decision. Determine whether the conclusion is valid. Explain. *(Section 15.6)*

4. **FIELD TRIP** Of 60 randomly chosen students surveyed, 16 chose the aquarium as their favorite field trip. There are 720 students in the school. Predict the number of students in the school who would choose the aquarium as their favorite field trip. *(Section 15.6)*

5. **FOOTBALL** The double box-and-whisker plot shows the points scored per game by two football teams during the regular season. *(Section 15.7)*

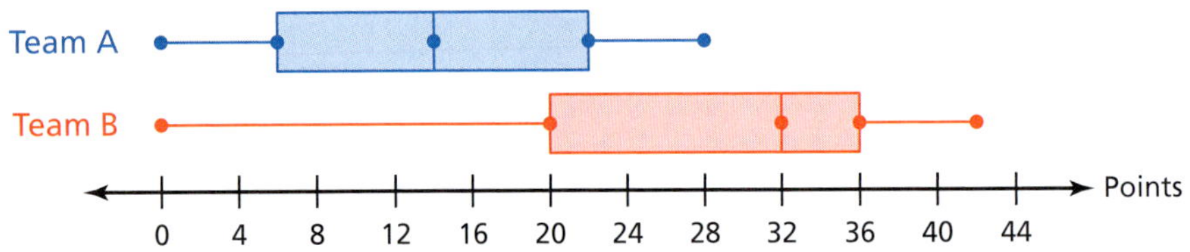

   a. Compare the populations using measures of center and variation.

   b. Express the difference in the measures of center as a multiple of the measure of variation.

6. **SUMMER CAMP** The dot plots show the ages of campers at two summer camps. *(Section 15.7)*

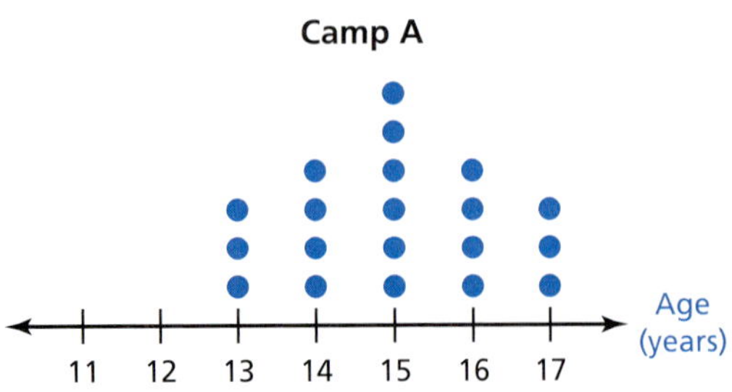

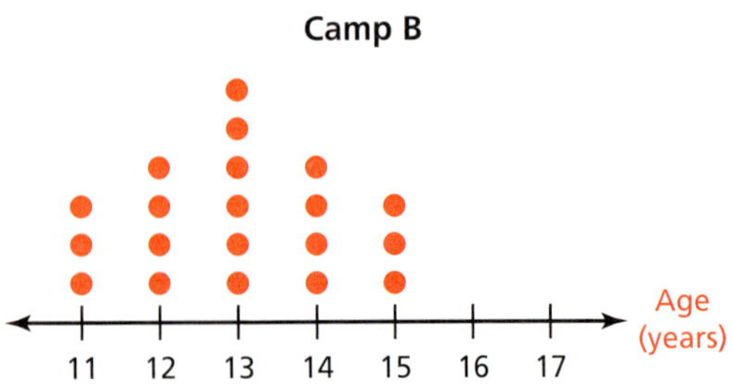

   a. Compare the populations using measures of center and variation.

   b. Express the difference in the measures of center as a multiple of the measure of variation.

Alternative Assessment Options

Math Chat | **Student Reflective Focus Question**
Structured Interview | Writing Prompt

Student Reflective Focus Question

Ask students to summarize the similarities and differences between samples and populations. Be sure that they include examples. Select students at random to present their summary to the class.

Study Help Sample Answers

Remind students to complete Graphic Organizers for the rest of the chapter.

6.

Population → Sample → Interpretation → Inference	Sample
An *unbiased sample* is representative of a population. It is selected at random and is large enough to provide accurate data. A *biased sample* is not representative of a population. One or more parts of the population are favored over others.	Part of a population The results of an *unbiased sample* are proportional to the results of the population. So, you can use unbiased samples to make predictions about the population. *Biased samples* are not representative of the population. So, you should not use them to make predictions about the population because the predictions may not be valid. Example: Population: All of the seventh-grade students in your school Unbiased sample: 100 seventh-grade students selected randomly during lunch Biased sample: The seventh-grade students at your lunch table
How do you select an unbiased sample from a large population?	

7. Available at *BigIdeasMath.com.*

Answers

1. Sample B because the sample size is larger.
2. biased; The sample is not selected at random and is not representative of the population because students on the basketball team use the gymnasium regularly when practicing.
3. yes; The sample is representative of the population, selected at random, and large enough to provide accurate data. So, the sample is unbiased and the conclusion is valid.
4. 192 students
5. **a.** Team A: median = 14, IQR = 16; Team B: median = 32, IQR = 16; The variation in the points scored is the same, but Team B generally has a greater score.

   b. The difference in the medians is 1.125 times the IQR.
6. **a.** Camp A: mean = 15, MAD = 1; Camp B: mean = 13, MAD = 1; The variation in the ages is the same, but Camp A has a greater age.

   b. The difference in the means is 2 times the MAD.

Reteaching and Enrichment Strategies

If students need help. . .	If students got it. . .
Resources by Chapter • Practice A and Practice B • Puzzle Time Lesson Tutorials *BigIdeasMath.com*	Resources by Chapter • Enrichment and Extension • Technology Connection Game Closet at *BigIdeasMath.com* Start the Chapter Review

Technology for the Teacher

Online Assessment
Assessment Book
ExamView® Assessment Suite

Review of Common Errors

Exercises 1–6

- Students may forget to include, or include too many, favorable outcomes. Encourage them to write out all of the possible outcomes and then circle the favorable outcomes for the given event.

For the Teacher
Additional Review Options

- *BigIdeasMath.com*
- Online Assessment
- Game Closet at *BigIdeasMath.com*
- Vocabulary Help
- Resources by Chapter

Answers

1. **a.** 2

   b. 1 green, 1 purple

2. **a.** 3

   b. 3 orange, 3 blue, 3 purple

3. **a.** 5

   b. 1 green, 1 purple, 3 orange, 3 blue, 3 purple

4. **a.** 3

   b. 2 blue, 2 orange, 2 green

5. **a.** 8

   b. 1 green, 1 purple, 2 blue, 2 orange, 2 green, 3 orange, 3 blue, 3 purple

6. **a.** 5

   b. 1 green, 1 purple, 2 blue, 2 orange, 2 green

15 Chapter Review

Check It Out
Vocabulary Help
BigIdeasMath.com

Review Key Vocabulary

experiment, *p. 634*
outcomes, *p. 634*
event, *p. 634*
favorable outcomes, *p. 634*
probability, *p. 640*
relative frequency, *p. 644*
experimental probability, *p. 646*
theoretical probability, *p. 647*
sample space, *p. 654*
Fundamental Counting Principle, *p. 654*
compound event, *p. 656*
independent events, *p. 662*
dependent events, *p. 663*
simulation, *p. 668*
population, *p. 672*
sample, *p. 672*
unbiased sample, *p. 674*
biased sample, *p. 674*

Review Examples and Exercises

15.1 Outcomes and Events *(pp. 632–637)*

You randomly choose one toy race car.

a. In how many ways can choosing a green car occur?

b. In how many ways can choosing a car that is *not* green occur? What are the favorable outcomes of choosing a car that is *not* green?

a. There are 5 green cars. So, choosing a green car can occur in 5 ways.

b. There are 2 cars that are *not* green. So, choosing a car that is *not* green can occur in 2 ways.

green	*not* green
green, green, green, green, green	blue, red

∴ The favorable outcomes of the event are blue and red.

Exercises

You spin the spinner. (a) Find the number of ways the event can occur. (b) Find the favorable outcomes of the event.

1. Spinning a 1

2. Spinning a 3

3. Spinning an odd number

4. Spinning an even number

5. Spinning a number greater than 0

6. Spinning a number less than 3

15.2 Probability (pp. 638–643)

You flip a coin. What is the probability of flipping tails?

$$P(\text{event}) = \frac{\text{number of favorable outcomes}}{\text{number of possible outcomes}}$$

$$P(\text{tails}) = \frac{1}{2}$$

There is 1 tails.

There is a total of 2 sides.

The probability of flipping tails is $\frac{1}{2}$, or 50%.

Exercises

7. You roll a number cube. Find the probability of rolling an even number.

15.3 Experimental and Theoretical Probability (pp. 644–651)

a. The bar graph shows the results of spinning the spinner 70 times. What is the experimental probability of spinning a 2?

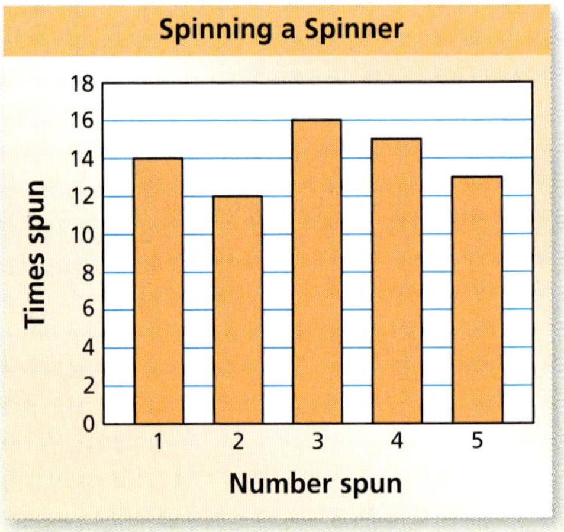

The bar graph shows 12 twos. So, the spinner landed on two 12 times in a total of 70 spins.

$$P(\text{event}) = \frac{\text{number of times the event occurs}}{\text{total number of trials}}$$

Two was landed on 12 times.

$$P(2) = \frac{12}{70} = \frac{6}{35}$$

There was a total of 70 spins.

The experimental probability is $\frac{6}{35}$, or about 17%.

b. The theoretical probability of choosing a purple grape from a bag is $\frac{2}{9}$. There are 8 purple grapes in the bag. How many grapes are in the bag?

$$P(\text{purple}) = \frac{\text{number of purple grapes}}{\text{total number of grapes}}$$

$\frac{2}{9} = \frac{8}{g}$ Substitute. Let g be the total number of grapes.

$g = 36$ Solve for g.

So, there are 36 grapes in the bag.

Review of Common Errors (continued)

Exercise 7

- Students may write the probability as the ratio of the number of favorable outcomes to the number of unfavorable outcomes. Remind them that the probability of an event is the ratio of the number of favorable outcomes to the number of possible outcomes.

Answers

7. $\frac{1}{2}$, or 50%

Answers

8. $\frac{8}{35}$, or about 22.9%

9. $\frac{43}{70}$, or about 61.4%

10. $\frac{57}{70}$, or about 81.4%

11. $\frac{2}{5}$, or 40%

12. $\frac{1}{4}$, or 25%

13. $\frac{3}{8}$, or 37.5%

14. $\frac{5}{8}$, or 62.5%

15. $\frac{1}{8}$, or 12.5%

16. 12

17. 90

18. $\frac{1}{8}$, or 12.5%

Review of Common Errors (continued)

Exercises 8–11

- Students may forget to total all of the trials before writing the experimental probability. They may have an incorrect number of trials in the denominator. Remind them that they need to know the total number of trials when finding the probability.
- Students may find the theoretical probability of the event instead of using the data to find the experimental probability. Remind them that they are using the experimental probability and assuming that this trend will continue to predict the outcome of an event.

Exercises 12–15

- Students may write a different probability than what is asked, or forget to include a favorable outcome. For example, in Exercise 13 a student may not realize that there are three "1" sections and will write the probability as $\frac{1}{8}$ or $\frac{1}{4}$ instead of $\frac{3}{8}$. Remind them to read the event carefully and to write the favorable outcomes before finding the probability.

Exercises 17 and 18

- Students may try to use a tree diagram to solve these problems. Although it is possible and not incorrect to do so, recommend that students use the Fundamental Counting Principle as a much less time consuming alternative.

Exercises 19–22

- Students may mix up independent and dependent events or may have difficulty determining which type of event it is. Remind them that independent events are where you do two different things or events where you start over before the next trial. Dependent events have at least one less possible outcome after the first draw, roll, or flip.

Exercise 23

- Students may ignore the fact that the survey is conducted among students arriving at a biology club meeting. They may only see that it is a meeting at school. Ask them what kind of answers they would expect to get from different club meetings (such as math club). They should recognize that everyone going to a biology club meeting likes biology, but not everyone in the math club will.

Exercise 24

- Students may use the wrong measures of center and variation when comparing populations. Remind them to use the mean and the MAD when *both* distributions are symmetric. Otherwise, use the median and the IQR.

Exercises

Use the bar graph on page 456 to find the experimental probability of the event.

8. Spinning a 3
9. Spinning an odd number
10. *Not* spinning a 5
11. Spinning a number greater than 3

Use the spinner to find the theoretical probability of the event.

12. Spinning blue
13. Spinning a 1
14. Spinning an even number
15. Spinning a 4
16. The theoretical probability of spinning an even number on a spinner is $\frac{2}{3}$. The spinner has 8 even-numbered sections. How many sections are on the spinner?

15.4 Compound Events *(pp. 652–659)*

a. How many different home theater systems can you make from 6 DVD players, 8 TVs, and 3 brands of speakers?

$6 \times 8 \times 3 = 144$ Fundamental Counting Principle

∴ So, you can make 144 different home theater systems.

b. You flip two pennies. What is the probability of flipping two heads?

Use a tree diagram to find the probability. Let H = heads and T = tails.

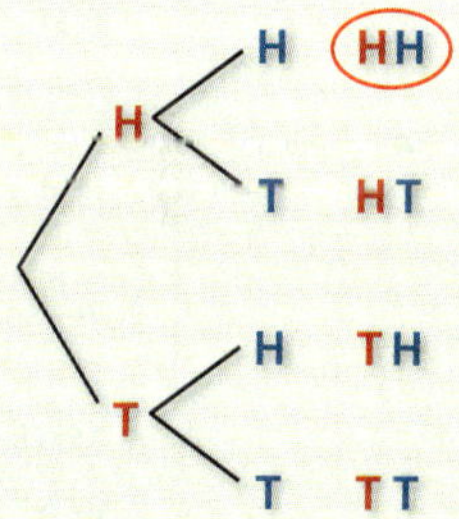

There is one favorable outcome in the sample space for flipping two heads: HH.

$$P(\text{event}) = \frac{\text{number of favorable outcomes}}{\text{number of possible outcomes}}$$

$$P(2 \text{ heads}) = \frac{1}{4} \quad \text{Substitute.}$$

∴ The probability is $\frac{1}{4}$, or 25%.

Exercises

17. You have 6 bracelets and 15 necklaces. Find the number of ways you can wear one bracelet and one necklace.

18. You flip two coins and roll a number cube. What is the probability of flipping two tails and rolling an even number?

15.5 Independent and Dependent Events *(pp. 660–669)*

You randomly choose one of the tiles and flip the coin. What is the probability of choosing a vowel and flipping heads?

Choosing one of the tiles does not affect the outcome of flipping the coin. So, the events are independent.

$P(\text{vowel}) = \frac{2}{7}$ ← There are 2 vowels (A and E). ← There is a total of 7 tiles.

$P(\text{tails}) = \frac{1}{2}$ ← There is 1 tails side. ← There is a total of 2 sides.

Use the formula for the probability of independent events.

$$P(A \text{ and } B) = P(A) \cdot P(B)$$

$$= \frac{2}{7} \cdot \frac{1}{2} = \frac{1}{7}$$

∴ The probability of choosing a vowel and flipping heads is $\frac{1}{7}$, or about 14%.

Exercises

You randomly choose one of the tiles above and flip the coin. Find the probability of the compound event.

19. Choosing a blue tile and flipping tails

20. Choosing the letter G and flipping tails

You randomly choose one of the tiles above. Without replacing the first tile, you randomly choose a second tile. Find the probability of the compound event.

21. Choosing a green tile and then a blue tile

22. Choosing a red tile and then a vowel

15.6 Samples and Populations *(pp. 672–679)*

You want to estimate the number of students in your school whose favorite subject is math. You survey every third student who leaves the school. Determine whether the sample is *biased* or *unbiased*.

The sample is representative of the population, selected at random, and large enough to provide accurate data.

∴ So, the sample is unbiased.

Review Game

Making Predictions

Materials per team:

- deck of cards
- paper
- pencil
- calculator

Directions:
Each team shuffles their deck of 52 cards. The first 39 cards are flipped over and placed in a discard pile. Teams are to keep track of what cards are discarded and what cards are left in the deck. The 13 cards remaining in the deck are shuffled by the teacher.

With the whole class, teams take turns predicting the next card to be flipped over from their deck. Predictions can include black card, red card, face card, single number card (if they are brave), etc. The team calculates the probability of the prediction on the board for the class to see.

Each team will do this for 5 cards. For each correct prediction, the team gets a point. The teacher holds onto each team's remaining cards until the game is over, in case of a tie.

Who Wins?
The team with the most points wins. If there is a tie, a one card draw will be the tie breaker. With the remaining cards in the team's deck, each team will make a prediction. The team whose prediction is correct wins.

For the Student Additional Practice

- Lesson Tutorials
- Multi-Language Glossary
- Self-Grading Progress Check
- *BigIdeasMath.com*
  Dynamic Student Edition
  Student Resources

Answers

19. $\frac{2}{7}$, or about 28.6%

20. $\frac{1}{14}$, or about 7.1%

21. $\frac{4}{21}$, or about 19.0%

22. $\frac{1}{21}$, or about 4.8%

23. biased; The sample is not selected at random and is not representative of the population because students in the biology club like biology.

24. **a.** Class A:
median = 88, IQR = 6;
Class B:
median = 91, IQR = 9;
In general, Class B has greater scores than Class A. Class A has less variation than Class B.

b. The difference in the medians is about 0.3 to 0.5 times the IQR.

My Thoughts on the Chapter

What worked. . .

What did not work. . .

What I would do differently. . .

> **Teacher Tip**
>
> Not allowed to write in your teaching edition? Use sticky notes to record your thoughts.

Exercises

23. You want to estimate the number of students in your school whose favorite subject is biology. You survey the first 10 students who arrive at biology club. Determine whether the sample is *biased* or *unbiased*. Explain.

15.7 Comparing Populations (pp. 680–685)

The double box-and-whisker plot shows the test scores for two French classes taught by the same teacher.

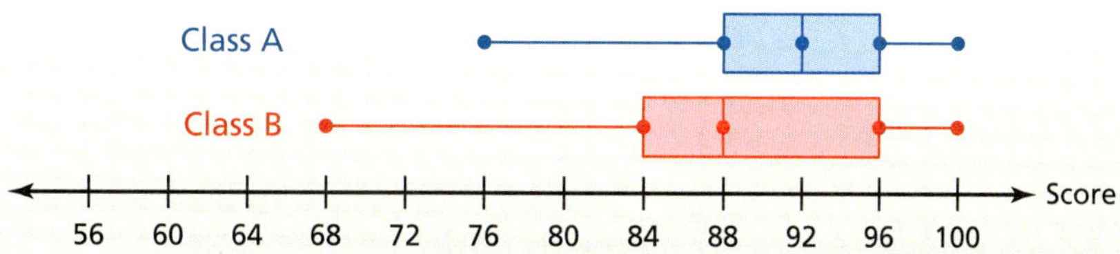

a. Compare the populations using measures of center and variation.

Both distributions are skewed left, so use the median and the IQR.

The median for Class A, 92, is greater than the median for Class B, 88. The IQR for Class B, 12, is greater than the IQR for Class A, 8. The scores in Class A are generally greater and have less variability than the scores in Class B.

b. Express the difference in the measures of center as a multiple of each measure of variation.

$$\frac{\text{median for Class A} - \text{median for Class B}}{\text{IQR for Class A}} = \frac{4}{8} = 0.5$$

$$\frac{\text{median for Class A} - \text{median for Class B}}{\text{IQR for Class B}} = \frac{4}{12} = 0.3$$

So, the difference in the medians is about 0.3 to 0.5 times the IQR.

Exercises

24. SPANISH TEST The double box-and-whisker plot shows the test scores of two Spanish classes taught by the same teacher.

a. Compare the populations using measures of center and variation.

b. Express the difference in the measures of center as a multiple of each measure of variation.

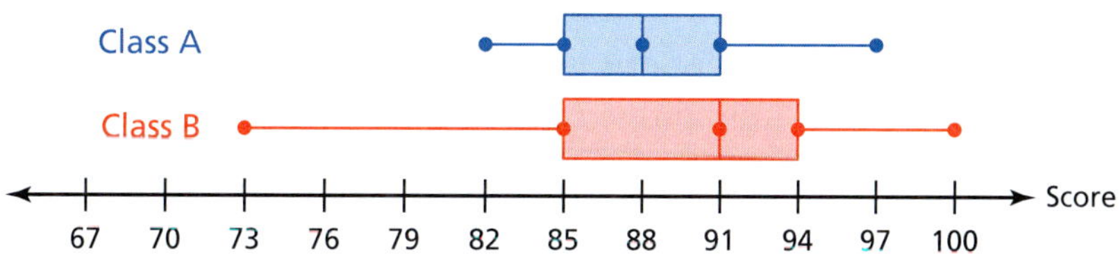

15 Chapter Test

Check It Out
Test Practice
BigIdeasMath.com

You randomly choose one game piece. (a) Find the number of ways the event can occur. (b) Find the favorable outcomes of the event.

1. Choosing green

2. Choosing *not* yellow

3. Use the Fundamental Counting Principle to find the total number of different sunscreens possible.

Sunscreen	
SPF	10, 15, 30, 45, 50
Type	Lotion, Spray, Gel

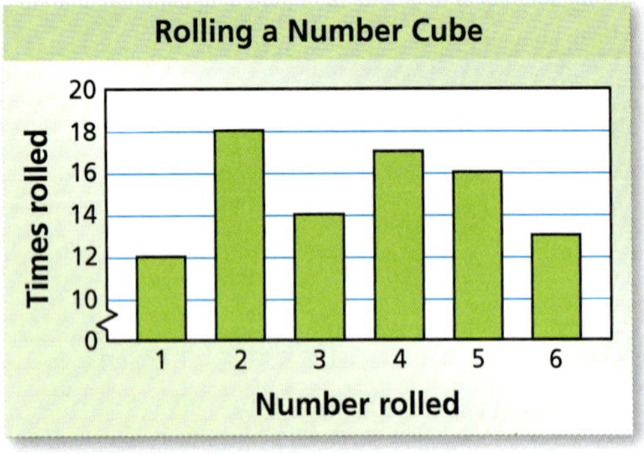

Use the bar graph to find the experimental probability of the event.

4. Rolling a 1 or a 2

5. Rolling an odd number

6. *Not* rolling a 5

Use the spinner to find the theoretical probability of the event(s).

7. Spinning an even number

8. Spinning a 1 and then a 2

You randomly choose one chess piece. Without replacing the first piece, you randomly choose a second piece. Find the probability of choosing the first piece, then the second piece.

9. Bishop and bishop

10. King and queen

11. **LUNCH** You want to estimate the number of students in your school who prefer to bring a lunch from home rather than buy one at school. You survey five students who are standing in the lunch line. Determine whether the sample is *biased* or *unbiased*. Explain.

12. **AGES** The double box-and-whisker plot shows the ages of the viewers of two television shows in a small town.

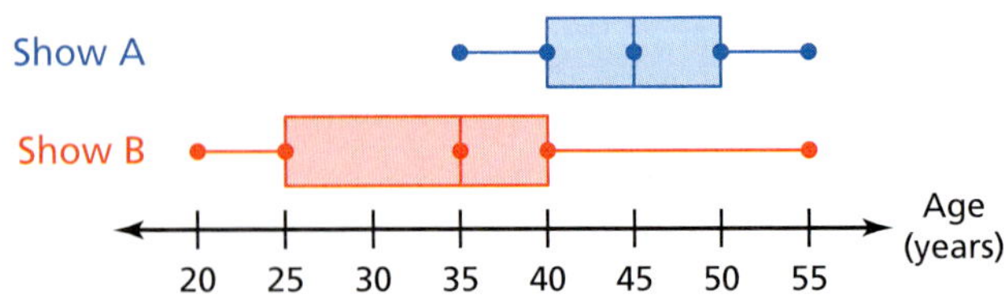

   a. Compare the populations using measures of center and variation.

   b. Express the difference in the measures of center as a multiple of each measure of variation.

Test Item References

Chapter Test Questions	Section to Review
1, 2	15.1
4–8	15.2
4–8	15.3
3	15.4
9, 10	15.5
11	15.6
12	15.7

Test-Taking Strategies

Remind students to quickly look over the entire test before they start so that they can budget their time. There is a lot of vocabulary in this chapter, so students should have been making flash cards as they worked through the chapter. Words that get mixed up should be jotted on the back of the test before they start. Students need to use the **Stop** and **Think** strategy before they answer a question.

Common Errors

- **Exercises 1 and 2** Students may forget to include, or include too many, favorable outcomes. Encourage them to write out all of the possible outcomes and then circle the favorable outcomes for the given event.
- **Exercise 3** Students may use a tree diagram to solve this problem. Although it is possible to do so, point out that the directions state to use the Fundamental Counting Principle. Also, point out that using the Fundamental Counting Principle is a much less time consuming method.
- **Exercises 4–6** Students may forget to total all of the trials before writing the experimental probability. Remind them that they need to know the total number of trials when finding the probability.
- **Exercises 9 and 10** Students may forget to subtract one from the total number of possible outcomes when finding the probability of choosing the second chess piece. Remind them that the second draw has one less possible outcome because they have removed one of the chess pieces.

Reteaching and Enrichment Strategies

If students need help. . .	If students got it. . .
Resources by Chapter • Practice A and Practice B • Puzzle Time Record and Practice Journal Practice Differentiating the Lesson Lesson Tutorials *BigIdeasMath.com* Skills Review Handbook	Resources by Chapter • Enrichment and Extension • Technology Connection Game Closet at *BigIdeasMath.com* Start Cumulative Assessment

Answers

1. **a.** 1 **b.** green
2. **a.** 5
   b. red, blue, red, green, blue
3. 15
4. $\frac{1}{3}$, or about 33.3%
5. $\frac{7}{15}$, or about 46.7%
6. $\frac{37}{45}$, or about 82.2%
7. $\frac{4}{9}$, or about 44.4%
8. $\frac{1}{81}$, or about 1.2%
9. $\frac{1}{120}$, or about 0.8%
10. $\frac{1}{240}$, or about 0.4%
11. biased; The sample size is too small and students standing in line are more likely to say they prefer to buy their lunches at school.
12. **a.** Show A: median = 45, IQR = 10; Show B: median = 35, IQR = 15; Show B generally has a younger audience and more variation in ages than Show A.
    b. The difference in the medians is about 0.7 to 1 times the IQR.

Technology for the Teacher

Online Assessment
Assessment Book
ExamView® Assessment Suite

Test-Taking Strategies

Available at *BigIdeasMath.com*

After Answering Easy Questions, Relax
Answer Easy Questions First
Estimate the Answer
Read All Choices before Answering
Read Question before Answering
Solve Directly or Eliminate Choices
Solve Problem before Looking at Choices
Use Intelligent Guessing
Work Backwards

About this Strategy

When taking a multiple choice test, be sure to read each question carefully and thoroughly. Sometimes you don't know the answer. So . . . guess intelligently! Look at the choices and choose the ones that are possible answers.

Answers

1. C
2. $\frac{1}{5}$, or 0.2

Technology for the Teacher

Performance Tasks
Online Assessment
Assessment Book
ExamView® Assessment Suite

Item Analysis

1. **A.** The student does not understand the concepts of certainty and likelihood.

   B. The student does not understand the difference between likely and unlikely.

   C. Correct answer

   D. The student does not understand that even a highly unlikely event is not impossible.

2. **Gridded Response:** Correct answer: $\frac{1}{5}$, or 0.2

   Common Error: The student only considers that Sunday is one day of the week and gets an answer of $\frac{1}{7}$.

15 Cumulative Assessment

Test-Taking Strategy
Use Intelligent Guessing

"You know it can't be -10% or 500%. So, you can intelligently guess between 40% and 60%."

1. A school athletic director asked each athletic team member to name his or her favorite professional sports team. The results are below:

- D.C. United: 3
- Florida Panthers: 8
- Jacksonville Jaguars: 26
- Jacksonville Sharks: 7
- Miami Dolphins: 22
- Miami Heat: 15
- Miami Marlins: 20
- Minnesota Lynx: 4
- New York Knicks: 5
- Orlando Magic: 18
- Tampa Bay Buccaneers: 17
- Tampa Bay Lightning: 12
- Tampa Bay Rays: 28
- Other: 6

One athletic team member is picked at random. What is the likelihood that this team member's favorite professional sports team is *not* located in Florida?

A. certain

B. likely, but not certain

C. unlikely, but not impossible

D. impossible

2. Each student in your class voted for his or her favorite day of the week. Their votes are shown below:

Favorite Day of the Week

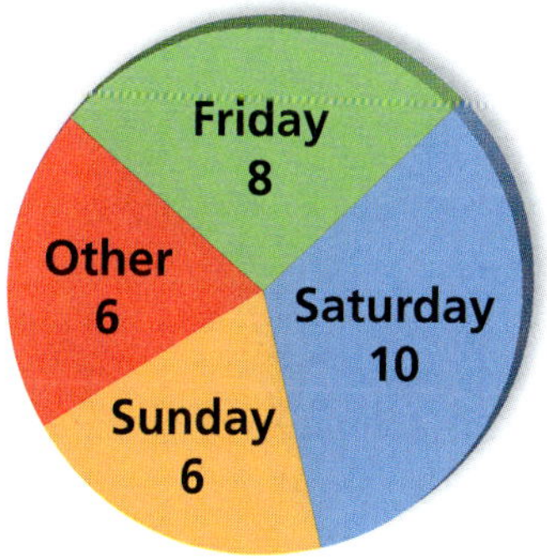

A student from your class is picked at random. What is the probability that this student's favorite day of the week is Sunday?

3. How far, in millimeters, will the tip of the hour hand of the clock travel in 2 hours? (Use $\frac{22}{7}$ for π.)

F. 44 mm

G. 88 mm

H. 264 mm

I. 528 mm

4. Nathaniel solved the proportion in the box below.

$$\frac{16}{40} = \frac{p}{27}$$

$$16 \cdot p = 40 \cdot 27$$

$$16p = 1080$$

$$\frac{16p}{16} = \frac{1080}{16}$$

$$p = 67.5$$

What should Nathaniel do to correct the error that he made?

A. Add 40 to 16 and 27 to p.

B. Subtract 16 from 40 and 27 from p.

C. Multiply 16 by 27 and p by 40.

D. Divide 16 by 27 and p by 40.

5. A North American hockey rink contains 5 face-off circles. Each of these circles has a radius of 15 feet. What is the total area, in square feet, of all the face-off circles? (Use 3.14 for π.)

F. 706.5 ft^2

G. 2826 ft^2

H. 3532.5 ft^2

I. 14,130 ft^2

Item Analysis (continued)

3. **F.** The student does not multiply the radius by two to find the circumference but then correctly divides by 6, or the student correctly finds the circumference but then divides by 12 to find how far the tip of the hour hand will travel in 1 hour.

G. Correct answer

H. The student does not multiply the radius by two to find the circumference and then uses the incorrect circumference to find how far the tip of the hour hand will travel in 12 hours.

I. The student finds how far the tip of the hour hand will travel in 12 hours.

4. **A.** The student incorrectly thinks that proportions involve addition.

B. The student incorrectly thinks that proportions involve subtraction.

C. Correct answer

D. The student switches the 40 and the 27 in the proportion, resulting in a proportion that is not equivalent to the original proportion.

5. **F.** The student finds the area of only 1 circle.

G. The student finds the area of only 1 circle and uses the diameter of the circle instead of the radius, finding $\pi \cdot 30^2$.

H. Correct answer

I. The student uses the diameters of the circles instead of the radii, finding $\pi \cdot 30^2$ for each of the circles.

Answers

3. G

4. C

5. H

Answers

6. $\frac{1}{16}$, or 0.0625

7. C

8. H

9. *Part A* independent

   Part B favorable outcomes: 3
   possible outcomes: 6

   Part C $\frac{1}{4}$, or 0.25

Item Analysis (continued)

6. **Gridded Response:** Correct answer: $\frac{1}{16}$, or 0.0625

   Common Error: The student does not realize the compound nature of the event and gets an answer of $\frac{1}{4}$ or 0.25 or equivalent.

7. **A.** The student finds the area of one lateral face.

   B. The student finds the area of the four lateral faces.

   C. Correct answer

   D. The student forgets to multiply the area of each triangular lateral face by $\frac{1}{2}$.

8. **F.** The student finds what percent $15.00 is of $6.00.

   G. The student subtracts $6.00 from $15.00 to get $9.00 and thinks that this means 90%.

   H. Correct answer

   I. The student finds what percent $6.00 is of $15.00.

9. **2 points** The student demonstrates a thorough understanding of determining probability. In Part A, the student determines the events are independent. In Part B, the student finds the number of possible outcomes is 6 and the number of favorable outcomes is 3. In Part C, the student gets an answer of $\frac{1}{4}$, or 0.25.

   1 point The student demonstrates a partial understanding of determining probability. The student gets a correct answer for Part A and Part B but the answer for Part C is incorrect.

   0 points The student provided no response, a completely incorrect or incomprehensible response, or a response that demonstrates insufficient understanding of probability.

6. A spinner is divided into eight congruent sections, as shown below.

You spin the spinner twice. What is the probability that the arrow will stop in a yellow section both times?

7. What is the surface area, in square inches, of the square pyramid?

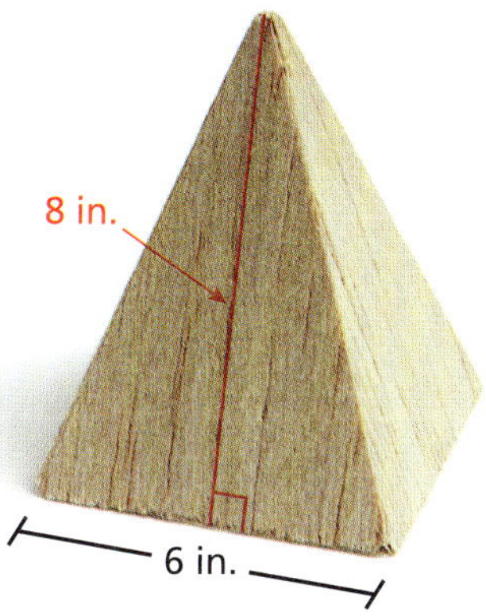

A. 24 in.^{2}

B. 96 in.^{2}

C. 132 in.^{2}

D. 228 in.^{2}

8. The value of one of Kevin's baseball cards was \$6.00 when he first got it. The value of this card is now \$15.00. What is the percent increase in the value of the card?

F. 40%

G. 90%

H. 150%

I. 250%

9. You roll a number cube twice. You want to roll two even numbers.

Think Solve Explain

Part A Determine whether the events are independent or dependent.

Part B Find the number of favorable outcomes and the number of possible outcomes of each roll.

Part C Find the probability of rolling two even numbers. Explain your reasoning.

Key Vocabulary Index

Mathematical terms are best understood when you see them used and defined *in context*. This index lists where you will find key vocabulary. A full glossary is available in your Record and Practice Journal and at *BigIdeasMath.com*.

adjacent angles, 504
angle of rotation, 62
base, 412
biased sample, 674
center, 550
center of dilation, 84
center of rotation, 62
circle, 550
circumference, 551
complementary angles, 510
composite figure, 558
compound event, 656
concave polygon, 119
congruent angles, 504
congruent figures, 44
congruent sides, 516
convex polygon, 119
corresponding angles, 44
corresponding sides, 44
cross section, 620
cube root, 296
dependent events, 663
diameter, 550
dilation, 84
distance formula, 320
event, 634
experiment, 634
experimental probability, 646
exponent, 412
exterior angles, 105
exterior angles of a polygon, 112
favorable outcomes, 634
function, 245
function rule, 250
Fundamental Counting Principle, 654
graph of an inequality, 467
hemisphere, 351
hypotenuse, 302
image, 50
independent events, 662
indirect measurement, 129
inequality, 466
input, 244
interior angles, 105
interior angles of a polygon, 112
irrational number, 310
joint frequency, 388
kite, 526
lateral surface area, 590
legs, 302
line of best fit, 381
line of fit, 380
line of reflection, 56
linear equation, 144
linear function, 258
literal equation, 28
mapping diagram, 244
marginal frequency, 388
nonlinear function, 268
outcomes, 634
output, 244
perfect cube, 296
perfect square, 290
pi, 551
point-slope form, 186
population, 672
power, 412
probability, 640
Pythagorean Theorem, 302
radical sign, 290
radicand, 290
radius, 550
real numbers, 310
reflection, 56
regular polygon, 121
regular pyramid, 596
relation, 244
relative frequency, 644
rise, 150
rotation, 62
run, 150
sample, 672
sample space, 654
scale, 532
scale drawing, 532
scale factor, 84, 533
scale model, 532
scatter plot, 374
scientific notation, 438
semicircle, 552
similar figures, 72
similar solids, 356
simulation, 668
slant height, 596
slope, 150
slope-intercept form, 168
solution of an inequality, 466
solution of a linear equation, 144
solution of a system of linear equations, 204
solution set, 466
sphere, 348
square root, 290
standard form, 174
supplementary angles, 510
system of linear equations, 204
theorem, 300
theoretical probability, 647
transformation, 50
translation, 50
transversal, 104
two-way table, 388
unbiased sample, 674
vertical angles, 504
x-intercept, 168
y-intercept, 168

Student Index

This student-friendly index will help you find vocabulary, key ideas, and concepts. It is easily accessible and designed to be a reference for you whether you are looking for a definition, real-life application, or help with avoiding common errors.

A

Addition
  Property
    of Equality, 4
    of Inequality, 472
  to solve inequalities, 470–475
Addition Property of Inequality, 472
Adjacent angles
  constructions, 502–507
  defined, 504
Algebra
  equations
    graphing linear, 142–147
    literal, 28
    multi-step, 10–15
    rewriting, 26–31
    simple, 2–9
    with variables on both sides, 18–25
  formulas, *See* Formulas
  functions
    linear, 256–263
    nonlinear, 266–271
    relations and, 242–247
    representing, 245–255
  linear equations
    graphing, 142–147
    lines of fit, 378–383
    slope of a line, 148–157
    slope-intercept form, 166–183
    standard form, 172–177
    systems of, 202–229
  properties, *See* Properties
Angle(s)
  adjacent
    constructions, 502–507
    defined, 504
  alternate exterior, 106
  alternate interior, 106
  classifying, 509–510
    triangles by, 516
  complementary
    constructions, 508–513
    defined, 510
  congruent
    defined, 504
    reading, 516
  corresponding, 104–105
    defined, 44
  exterior
    defined, 105
  interior, defined, 105
  measures
    of a quadrilateral, 527
    of a triangle, 520–521
  naming, 504
  of polygons, 118–125
    defined, 112
    reading, 120
    real-life application, 121
    similar, 126–131
  of rotation, 62
  sums
    for a quadrilateral, 527
    for a triangle, 520–521
  supplementary
    constructions, 508–513
    defined, 510
  of triangles, 110–115
    exterior, 112
    interior, 112
    real-life application, 113
    similar, 128
  vertical
    constructions, 502–507
    defined, 504
Angle of rotation, defined, 62
Area, *See also* Surface area
  of a circle, 564–569
    formula, 566
  of a composite figure, 570–575
  of similar figures, 76–81
    formula, 78
    writing, 80

B

Bar graphs, 394
Base, defined, 412
Biased sample(s), defined, 674
Box-and-whisker plots, 394

C

Center, defined, 550
Center of dilation, defined, 84
Center of rotation, defined, 62
Choose Tools, *Throughout.*
    For example, see:
  angles, 503
  circles and circumference, 553
  graphing linear equations, 143
  indirect measurement, 127
  probability, 638, 658
  scientific notation, 447
  slope, 154
  systems of linear equations, 203
Circle(s)
  area of, 564–569
    formula, 566
  center of, 550
  circumference and, 548–555
    defined, 551
    formula, 551
    research, 555
  defined, 550
  diameter of, 550
  pi, 548, 551
  semicircle, 552
Circle graphs, 394
Circumference
  and circles, 548–555
  defined, 551
  formula, 551
  research, 555
Common Error
  inequalities, 481
  linear functions, 259
  Pythagorean Theorem, 320
  Quotient of Powers Property, 424
  scientific notation, 445
  transformations
    rotations, 63
    similar figures, 72
Comparison chart, 264
Complementary angles
  constructing, 508–513
  defined, 510
Composite figure(s)
  area of, 570–575
  defined, 558
  perimeters of, 556–561
Compound event(s), *See also* Events, Probability
  defined, 656
  writing, 657
Concave polygon, defined, 119
Cone(s)
  volume of, 340–345
    formula, 342
    real-life application, 343
    writing, 344
Congruent angles
  defined, 504
  reading, 516
Congruent figures, 42–47
  corresponding angles, 44
  corresponding sides, 44
  defined, 44

  - identifying, 44
  - naming parts, 44
  - reading, 44
- **Congruent sides**
  - defined, 516
  - reading, 516
- Connections to math strands, *Throughout. For example, see:*
  - Algebra, 305
  - Geometry, 24, 25, 75, 81, 109, 125, 131, 147, 156, 157, 293, 299, 453, 598
- Constructions
  - angles
    - adjacent, 502–507
    - vertical, 502–507
  - quadrilaterals, 524–529
  - triangles, 514–521
- **Convex polygon,** defined, 119
- Coordinate plane(s)
  - transformations in the
    - dilations, 82–89
    - reflections, 55–59
    - rotations, 61–67
    - translations, 49–53
- **Corresponding angles**
  - defined, 44
  - naming, 44
  - symbol, 44
- **Corresponding sides**
  - defined, 44
  - naming, 44
  - symbol, 44
- Critical Thinking, *Throughout. For example, see:*
  - angle measures, 109
  - circles, area of, 569
  - composite figures
    - area of, 575
    - perimeter of, 561
  - cube roots, 299
  - cylinders, 604
  - equations
    - multi-step, 15
    - simple, 9
  - exponents, 415
    - Product of Powers Property, 420, 421
    - Quotient of Powers Property, 427
    - zero, 432
  - inequalities, 469
  - linear equations, 170
    - in slope-intercept form, 171, 183
    - solving systems of, 207, 223
    - in standard form, 177
    - writing, 183
  - probability, 643
  - proportional relationships, 163
  - samples, 676, 677
  - scale drawing, 535
  - scale factor, 536
  - scientific notation, 441, 453
  - similar triangles, 131
  - slant height, 598
  - slope, 153, 154, 155
  - slope-intercept form, 170
  - solids, 339
  - square roots, 292
  - surface area, 598, 604, 612
  - transformations
    - congruent figures, 47
    - dilations, 89
    - reflections, 59
    - rotations, 67
    - similar figures, 74, 75
  - triangles, 519
  - volume, 612, 613
    - of cones, 345
    - of cylinders, 339
- **Cross section(s)**
  - defined, 620
  - of three-dimensional figures, 620–621
- Cube(s), surface area of, 590
- **Cube root(s)**
  - defined, 296
  - finding, 294–299
    - real-life application, 297
  - perfect cube, 296
- Cylinder(s)
  - cross section of, 620–621
  - surface area of, 600–605
    - formula, 602
    - real-life application, 603
  - volume of, 334–339
    - formula, 336
    - modeling, 339
    - real-life application, 337

D

- Data, *See also* Equations; Graphs
  - analyzing
    - line of best fit, 381
    - writing, 382
  - displaying
    - bar graph, 394
    - box-and-whisker plot, 394
    - choosing a display, 392–399
    - circle graph, 394
    - dot plot, 394
    - histogram, 394
    - line graph, 394
    - pictograph, 394
    - project, 393
    - scatter plot, 372–377, 394
    - stem-and-leaf plot, 394
    - two-way table, 386–391
    - writing, 398
  - identifying relationships, 375
    - linear, 375
    - negative, 375
    - nonlinear, 375
    - positive, 375
  - joint frequencies, 388
  - marginal frequencies, 388
  - misleading displays, 396
- Decimal(s)
  - repeating, 316–317
- **Dependent events,** *See also* Events, Probability
  - defined, 663
  - formula, 663
  - writing, 665
- **Diameter,** defined, 550
- Different Words, Same Question, *Throughout. For example, see:*
  - angles of polygons, 123
  - area of a circle, 568
  - constructing triangles, 518
  - exponents, 432
  - functions, 253
  - inequalities, 468
  - probability, 665
  - rotations, 65
  - solving equations, 30
  - surface area of a prism, 591
  - triangles, 304
  - volume of cylinders, 338
- **Dilation(s),** 82–89
  - center of, 84
  - in the coordinate plane, 82–89
  - defined, 84
  - scale factor, 84
- Direct variation, *See also* Proportional relationships
- **Distance formula,** 319–323
  - defined, 320
  - real-life application, 321
- Distributive Property
  - equations with variables on both sides, 20
  - multi-step equations, 13
- Division
  - Property
    - of Equality, 5
    - of Inequality, 480–481
  - to solve inequalities, 478–485
- Division Property of Inequality, 480–481
- Dot plots, 394

Student Index

E

Equality
Addition Property of, 4
Division Property of, 5
Multiplication Property of, 5
Subtraction Property of, 4
Equation(s), *See also* Linear equations
function rules, 250
literal, 28
multi-step, 10–15
real-life application, 13
rewriting, 26–31
real-life application, 29
simple, 2–9
modeling, 8
real-life application, 6
solving
by addition, 4
by division, 5
by multiplication, 5
multi-step, 10–15
by rewriting, 26–31
simple, 2–9
by subtraction, 4
two-step, 12
with variables on both sides, 18–25
with variables on both sides, 18–25
real-life application, 22
writing, 23
Error Analysis, *Throughout. For example, see:*
angles
corresponding, 107
exterior, 115
of polygons, 123, 124
classifying triangles, 518
congruent figures, 47
corresponding sides, 47
distance formula, 322
equations
multi-step, 14
rewriting, 30
simple, 8
with variables on both sides, 23, 24
exponents
evaluating expressions, 414
negative, 432
functions
graphing, 254
relations and, 246
inequalities
solving, 474, 483, 484, 490
writing, 468
linear equations
graphing, 146
in slope-intercept form, 170, 182
solving systems of, 207, 213, 221, 222, 228
in standard form, 176
naming angles, 506
outcomes, 636
parallel lines, 107
perimeter of a composite figure, 560
powers
Product of Powers Property, 420
Quotient of Powers Property, 426
prisms
surface area, 592
volume of, 612
probability, 642
dependent events, 666
Fundamental Counting Principle, 658
outcomes, 636
Pythagorean Theorem, 304, 322
real numbers, 313
relations, 246
scale drawings, 535
scientific notation
operations in, 452
writing numbers in, 446
writing in standard form, 440
slope, 154
square roots, 313
finding, 292
surface area
of a cylinder, 604
of a prism, 592
systems of linear equations
solving by elimination, 221, 222
solving by graphing, 207
solving special, 228
solving by substitution, 213
transformations
dilations, 88
triangles, 518
exterior angles of, 115
Pythagorean Theorem, 304
volume
of a cone, 344
of a prism, 612
of similar solids, 360
Event(s), *See also* Probability
compound, 652–659
defined, 656
defined, 634
dependent, 660–667
defined, 663
writing, 665
independent, 660–667
defined, 662
writing, 665
outcomes of, 634
probability of, 638–643
defined, 640
Example and non-example chart, 116, 522
Experiment(s)
defined, 634
outcomes of, 632–637
project, 685
reading, 634
simulations, 668–669
defined, 668
Experimental probability, 644–651
defined, 646
formula, 646
Exponent(s)
defined, 412
evaluating expressions, 410–415
real-life application, 413
negative, 428–433
defined, 430
real-life application, 431
writing, 432
powers and, 410–421
real-life application, 425
writing, 426
properties of
Power of a Power Property, 418
Power of a Product Property, 418
Product of Powers Property, 416–421
Quotient of Powers Property, 422–427
quotients and, 422–427
scientific notation
defined, 438
operations in, 448–453
project, 453
reading numbers in, 436–441
real-life applications, 439, 445, 451
writing numbers in, 442–447
zero, 428–433
defined, 430
Expressions
evaluating exponential, 410–415
real-life application, 413
Exterior angle(s)
alternate, 106

angle sum of, 122
real-life application, 121
defined, 105, 112
of triangles, 110–115
Exterior angles of a polygon, defined, 112

F

Favorable outcome(s), defined, 634
Formula(s)
angles
sum for a quadrilateral, 527
sum for a triangle, 520
area
of a circle, 566
of a parallelogram, 573
of a rectangle, 573
of a semicircle, 573
of similar figures, 78
of a triangle, 573
circumference, 551
distance, 320
perimeter of similar figures, 78
pi, 548
probability
dependent events, 663
of an event, 640
experimental, 646
independent events, 662
relative frequency, 644
theoretical, 647
Pythagorean Theorem, 302
rewriting, 26–31
slope, 148, 150
surface area
of a cube, 590
of a cylinder, 602
of a prism, 589
of a pyramid, 596
of a rectangular prism, 588
of similar solids, 357
temperature conversion, 29
volume
of a cone, 342
of a cube, 610
of a cylinder, 336
of a hemisphere, 351
of a prism, 610
of a pyramid, 616
of similar solids, 358
of a sphere, 350
Formula triangle, 346
Four square, 306
Fraction(s)
repeating decimals written as, 316–317
Function(s)
defined, 245
function rules
defined, 250
real-life application, 252
writing, 250–255
linear, 256–263
compared to nonlinear, 266–271
defined, 258
modeling, 271
real-life applications, 259, 269
writing, 261
nonlinear
compared to linear, 266–271
defined, 268
real-life application, 269
relations and, 242–247
inputs, 244
mapping diagrams, 242–247
outputs, 244
research, 247
representing
with graphs, 248–255
with input-output tables, 248–255
with mapping diagrams, 245–247, 252
real-life application, 252
writing, 253
Function rule(s)
defined, 250
real-life application, 252
writing, 250–255
Fundamental Counting Principle
defined, 654
writing, 657

G

Geometry
angles, 102–115, 118–131
adjacent, 502–507
complementary, 508–513
congruent, 504
constructing, 502–513
corresponding, 44
exterior, 105, 112
interior, 105, 112
of polygons, 118–125
of rotation, 62
supplementary, 508–513
vertical, 502–507
area of similar figures, 76–81
circles
area of, 564–569
center of, 550
circumference, 548–555
diameter, 550
radius, 550
semicircles, 552
composite figures
area of, 570–575
perimeter of, 556–561
constructions and drawings, 502–521, 524–537
line of reflection, 56
parallel lines, 102–109
perimeter of similar figures, 76–81
pi, 548
polygons
angles of, 118–125
concave, 119
convex, 119
Pythagorean Theorem, 300–305
converse of, 320
defined, 302
using, 318–323
quadrilaterals, 526
constructing, 524–529
defined, 524
kite, 526
parallelogram, 526, 573
rectangle, 526, 573
rhombus, 526
square, 526
trapezoid, 526
sides, corresponding, 44
solids
cones, 340–345
cylinders, 334–339
similar, 354–361
spheres, 348–353
surface area of, 354–361
volume of, 334–345, 348–361, 354–361
surface area
of a cylinder, 600–605
of a prism, 586–593
of a pyramid, 594–599
of a rectangular prism, 588
tessellation, 48–49
transformations
congruent figures, 42–47
dilations, 82–89
reflections, 54–59
rotations, 60–67
similar figures, 70–81
translations, 48–53
transversals, 102–109
trapezoids, 526
triangles
angles of, 110–115
area of, 573
classifying, 516
congruent, 42–44
congruent sides, 516
constructing, 514–519

Student Index

hypotenuse, 302
legs, 302
right, 302
similar, 126–131
volume of a prism, 608–613
Graph of an inequality, defined, 467
Graphic Organizers
comparison chart, 264
example and non-example chart, 116, 522
formula triangle, 346
four square, 306
information frame, 384, 606
information wheel, 434
notetaking organizer, 214, 670
process diagram, 164
summary triangle, 68
word magnet, 562
Y chart, 16, 476
Graphing
inequalities, 464–469
defined, 467
Graphs
analyzing, 272–277
bar graphs, 394
box-and-whisker plots, 394
circle graphs, 394
dot plots, 394
of functions, 248–255
histograms, 394
line graphs, 394
linear, 142–147
defined, 144
of horizontal lines, 144
real-life application, 145
in slope-intercept form, 166–171
solution of, 144
in standard form, 172–177
of vertical lines, 144
misleading, 396
pictographs, 394
proportional relationships, 158–163
scatter plots, 372–377, 394
sketching, 272–277
slope, 148–157
defined, 148, 150
formula, 148, 150
reading, 150
stem-and-leaf plots, 394
used to solve linear equations, 230–231
real-life application, 231
used to solve systems of linear equations, 202–207
modeling, 207
real-life application, 205

H

Hemisphere(s)
defined, 351
volume formula, 351
Histograms, 394
Hypotenuse, defined, 302

I

Image(s)
defined, 50
reading, 50
Independent events, *See also* Events, Probability
defined, 662
formula, 662
writing, 665
Indirect measurement, 127–129
defined, 129
modeling, 127
project, 127
Inequality
Addition Property of, 472
defined, 466
Division Property of, 480–481
graphing, 464–469
defined, 467
Multiplication Property of, 480–481
solution of
defined, 466
reading, 473
solution set of
defined, 466
solving two-step, 486–491
real-life application, 489
writing, 490
solving using addition, 470–475
real-life application, 473
writing, 470
solving using division, 478–485
project, 485
writing, 479
solving using multiplication, 478–485
project, 485
writing, 479, 483
solving using subtraction, 470–475
writing, 470
Subtraction Property of, 472
symbols, 466
reading, 473
writing, 464–469
modeling, 469
Information frame, 384, 606
Information wheel, 434
Input(s), defined, 244
Input-output tables
using to represent functions, 248–255
Interior angle(s)
alternate, 106
defined, 105, 112
of triangles, 110–115
real-life application, 113
Interior angles of a polygon, defined, 112
Irrational number(s), defined, 310

J

Joint frequency, defined, 388

K

Kite, defined, 526

L

Lateral surface area, defined, 590
Leg(s), defined, 302
Like terms, combining to solve equations, 12
Line(s)
graphing
horizontal, 144
vertical, 144
parallel, 102–109
defined, 104
project, 108
slope of, 156
symbol, 104
perpendicular
defined, 104
slope of, 157
symbol, 104
of reflection, 56
slope of, 148–157
transversals, 102–109
x-intercept of, 168
y-intercept of, 168
Line of best fit, defined, 381
Line of fit, 378–383
defined, 380
line of best fit, 381
modeling, 378, 379, 383
writing, 382
Line graphs, 394
Line of reflection, defined, 56
Linear equation(s), *See also* Equations, Proportional relationships
defined, 144
graphing, 142–147
horizontal lines, 144

real-life applications, 145, 175, 231
in slope-intercept form, 166–171
to solve, 230–231
in standard form, 172–177
vertical lines, 144
lines of fit, 378–383
modeling, 378, 379, 383
point-slope form
defined, 186
real-life application, 187
writing, 188
writing in, 184–189
slope of a line, 148–157
defined, 148, 150
formula, 148, 150
reading, 150
slope-intercept form
defined, 168
real-life applications, 169, 181
writing in, 178–183
x-intercept, 168
y-intercept, 168
solution of, 144
standard form, 172–177
defined, 174
modeling, 177
real-life application, 175
writing, 176
systems of
defined, 202, 204
modeling, 207
reading, 204
real-life applications, 205, 211, 220
solution of a, 204
solving by elimination, 216–223
solving by graphing, 202–207
solving special, 224–229
solving by substitution, 208–213
writing, 206, 212, 221, 228
Linear function(s), 256–263
compared to nonlinear, 266–271
real-life application, 269
defined, 258
modeling, 271
real-life application, 259
writing, 261
Linear measures, 357
Literal equation(s), defined, 28
Logic, *Throughout. For example, see:*
angles, 513
interior, 110
measures, 108
circumference, 555
constructing a triangle, 519
cube roots, 299
equations
rewriting, 31
simple, 9
inequalities
solving, 485
writing, 469
linear equations
graphing, 142, 177
in slope-intercept form, 167
solving systems of, 217, 223, 229
prisms, 613
probability
dependent events, 666
experimental, 649
independent events, 666
theoretical, 649
samples, 677
scatter plots, 376
systems of linear equations, 217, 223, 229
transformations
similar figures, 75

M

Mapping diagram(s), 242–247
defined, 244
Marginal frequency, defined, 388
Meaning of a Word
adjacent, 502
dilate, 82
reflection, 54
rotate, 60
translate, 48
transverse, 102
Mental Math, *Throughout. For example, see:*
rotations, 65
Modeling, *Throughout. For example, see:*
equations, 8
indirect measurement, 127
inequalities, 469
linear equations
lines of fit, 378, 379, 383
solving systems of, 207
in standard form, 177
linear functions, 271
probability, 650
Pythagorean Theorem, 300
scale models, 537
volume of a cylinder, 339
Multiplication
inequalities, solving by, 478–485
Property
of Equality, 5
of Inequality, 480–481
Multiplication Property of Inequality, 480–481

N

Nonlinear function(s)
compared to linear, 266–271
real-life application, 269
defined, 268
Notetaking organizer, 214, 670
Number(s)
irrational, 310–315
defined, 310
rational, 310
real, 310–315
classifying, 310
defined, 310
Number Sense, *Throughout. For example, see:*
analyzing data, 382
angles
exterior, 114
of a polygon, 123
cube roots, 299
exponents, 414, 427, 432
functions, 271
inequalities, 475, 485
probability, 650
real numbers, 315
scientific notation, 441, 452
similar solids
surface area of, 359
volume of, 359
square roots, 292
statistics, 685
surface area of a prism, 593
systems of linear equations
solving by elimination, 221
solving by substitution, 212, 213
transformations
reflections, 59
similar figures, 80

O

Open-Ended, *Throughout. For example, see:*
angles, 507, 513
area of a composite figure, 574
composite figures, 560
data
histograms, 397
misleading displays, 397
scatter plots, 377
two-way tables, 390

Student Index

dilations, 89
equations
linear, 170
multi-step, 14
simple, 9
with variables on both sides, 23, 24
exponents, 433
inequalities, 483, 490
parallel lines, 107
probability, 642
compound events, 657
theoretical, 649
scale factor, 536
similar solids, 359
similar triangles, 131
slope, 153
square roots, 315
surface area of a prism, 592
volume
of a prism, 613
of a pyramid, 618, 619
Outcomes, *See also* Events, Probability
counting, 635
error analysis, 636
defined, 634
experiment, 634
favorable, 634
reading, 634
writing, 636
Output(s), defined, 244

P

Parallel line(s)
defined, 104
slope of, 156
symbol, 104
and transversals, 102–109
project, 108
Parallelogram(s)
area of, 573
defined, 526
Perfect cube, defined, 296
Perfect square, defined, 290
Perimeter
of composite figures, 556–561
of similar figures, 76–81
formula, 78
writing, 80
Perpendicular line(s)
defined, 104
slope of, 157
symbol, 104
Pi
defined, 551
formula, 548
Pictographs, 394
Point-slope form
defined, 186
writing equations in, 184–189
real-life application, 187
writing, 188
Polygon(s)
angles, 118–125
exterior, 112
interior, 112
measures of interior, 120
real-life application, 121
sum of exterior, 122
concave, 119
convex, 119
defined, 120
kite, 526
parallelogram, 526
quadrilateral, 524–529
reading, 120
rectangle, 526, 573
regular, 121
rhombus, 526
square, 526
trapezoid, 526
triangles, 110–115, 514–521
modeling, 127
project, 127
similar, 126–131
writing, 130
Population(s), 672–685
comparing, 680–685
project, 685
writing, 684
defined, 672
research, 673
samples, 672–679
biased, 674
defined, 672
unbiased, 674
Power(s), *See also* Exponents
base of, 412
defined, 412
exponent of, 412
of a power, 418
of a product, 418
product of, 416–421
Product of Powers Property, 418
quotient of, 422–427
Quotient of Powers Property, 424
real-life application, 425
writing, 426
scientific notation
defined, 438
operations in, 448–453
project, 453
reading numbers in, 436–441
real-life applications, 439, 445, 451
writing numbers in, 442–447
Power of a Power Property, 418
Power of a Product Property, 418
Precision, *Throughout.*
For example, see:
analyzing data, 391
angles of a triangle, 115
constructing angles, 507, 513
equations with variables on both sides, 24, 25
exponents, 433
functions, 246
indirect measurement, 127
inequalities
graphing, 468
solving, 483
linear equations
graphing, 142, 146
in slope-intercept form, 182
outcomes, 637
prisms, 613
Product of Powers Property, 420
Pythagorean Theorem, 305
relations, 246
similar solids, 361
square roots, 293
systems of linear equations, 229
transformations
rotations, 61
translations, 49
Prism(s)
cross section of, 620
surface area of, 586–593
formula, 589
real-life application, 590
rectangular, 588
writing, 591
volume of, 608–613
formula, 610
real-life application, 611
rectangular, 608
writing, 618
Probability, 638–651
defined, 640
events, 632–637
compound, 652–659
dependent, 660–667
independent, 660–667
writing, 665
of events
defined, 640
formula for, 640
experimental, 644–651
defined, 646

formula, 646
experiments
defined, 634
reading, 634
simulations, 668–669
Fundamental Counting Principle, 654
outcomes, 632–637
defined, 634
favorable, 634
reading, 634
writing, 636
project, 651
research, 659
sample space, 654
theoretical, 644–651
defined, 647
formula, 647
modeling, 650
writing, 657
Problem Solving, *Throughout. For example, see:*
angles, 513
angles of a polygon, 124
area and perimeter, 81
circumference, 555
data displays, 377
equations with variables on both sides, 25
inequalities, two-step, 491
linear equations
graphing, 147
in point-slope form, 189
solving systems of, 223
linear functions, 263
probability, 659, 667
proportional relationships, 163
Pythagorean Theorem, 301
scatter plots, 377
solids, 339
surface area, 599
transformations
dilations, 89
translations, 53
volume, 619
of a cylinder, 339
Process diagram, 164
Product of Powers Property, 416–421
defined, 418
Properties
Addition Property of Equality, 4
Addition Property of Inequality, 472
Division Property of Equality, 5
Division Property of Inequality, 480–481
Multiplication Property of Equality, 5
Multiplication Property of Inequality, 480–481
Power of a Power Property, 418
Power of a Product Property, 418
Product of Powers Property, 416–421
Quotient of Powers Property, 422–427
Subtraction Property of Equality, 4
Subtraction Property of Inequality, 472
real-life application, 473
Proportion(s), scale and, 532
Proportional relationships
direct variation, 160
graphing, 158–163
Proportions
similar figures, 70–81
Pyramid(s)
cross section of, 620
regular, 596
slant height, 596
surface area of, 594–599
formula, 596
real-life application, 597
volume of, 614–619
formula, 616
real-life application, 617
writing, 618
Pythagorean Theorem, 300–305
converse of, 320
defined, 302
modeling, 300
project, 305
real-life applications, 303, 321
using, 318–323
distance formula, 320
writing, 322

Q

Quadrilateral(s), 524–529
classifying, 526
constructions, 524–529
reading, 526
defined, 524
kite, 526
parallelogram, 526, 573
rectangle, 526, 573
rhombus, 526
square, 526
sum of angle measures, 527
trapezoid, 526
Quotient of Powers Property, 422–427
defined, 424
real-life application, 425
writing, 426

R

Radical sign, defined, 290
Radicand, defined, 290
Radius, defined, 550
Ratio(s), *See also* Proportions, Rates
scale and, 532
similar figures
areas of, 78
perimeters of, 78
Rational number(s), defined, 310
Reading
congruent angles, 516
congruent sides, 516
experiments, 634
images, 50
inequalities
solving, 473
symbols of, 466
outcomes, 634
polygons, 120
quadrilaterals, 526
slope, 150
symbol
congruent, 44
prime, 50
similar, 72
systems of linear equations, 204
Real number(s), 310–315
classifying, 310
defined, 310
Real-Life Applications, *Throughout. For example, see:*
angles of triangles, 113
cube roots, 297
distance formula, 321
equations
multi-step, 13
rewriting, 29
simple, 6
with variables on both sides, 22
exponents
evaluating expressions, 413
negative, 431
Quotient of Powers Property, 425
functions
graphing, 252
linear, 259, 269
nonlinear, 269
interior angles of a polygon, 121
linear equations
graphing, 145
in point-slope form, 187

in slope-intercept form, 169, 181
solving using graphs, 231
in standard form, 175
writing, 181
Pythagorean Theorem, 303, 321
scientific notation
operations in, 451
reading numbers in, 439
writing numbers in, 445
similar figures, 73
solving inequalities, 473, 489
square roots
approximating, 312
finding, 291
surface area
of a cylinder, 603
of a prism, 590
of a pyramid, 597
systems of linear equations, 205
solving by elimination, 220
solving by substitution, 211
volume
of cones, 343
of cylinders, 337
of a prism, 611
of a pyramid, 617
Reasoning, *Throughout. For example, see:*
analyzing graphs, 277
angle(s)
adjacent, 512
complementary, 512, 513
congruent, 507
constructing, 507
measures, 108, 115, 124
supplementary, 512, 513
circles, 554
circumference, 554
composite figures
area of, 574
perimeter of, 560
cones, 621
congruent figures, 47
cross sections
of a cone, 621
of a prism, 620
cube roots, 295, 298
data
analyzing, 387
displaying, 397, 399
scatter plots, 376, 377
two-way tables, 387, 391
distance formula, 323
equations
rewriting, 31
simple, 9
exponents, 433
exterior angles of polygons, 124
functions
graphing, 255
linear, 263
indirect measurement, 130
inequalities, 468
solving, 474, 475, 491
linear equations
in point-slope form, 189
in slope-intercept form, 171
lines of fit, 382, 383
perfect squares, 293
probability, 642
experimental, 649, 651
simulations, 669
theoretical, 651
Product of Powers Property, 420
proportional relationships, 162, 163
Pythagorean Theorem, 323
quadrilaterals, 525, 528
scale factor, 535
scale models, 536
scientific notation
operations in, 448
reading numbers in, 441
writing numbers in, 446, 447
slope, 153, 155
square roots, 292
statistics
comparing populations, 684
samples, 677, 679
surface area, 599
of a cylinder, 605
of a prism, 587
of a pyramid, 595
systems of linear equations
solving by elimination, 222, 223
solving by graphing, 207
solving special, 228, 229
solving by substitution, 212
transformations
congruent figures, 59
dilations, 87, 89
reflections, 58, 59
rotations, 67
similar figures, 75, 81
translations, 53
triangles
angle measures of, 521
constructions, 515
exterior angles of, 115
similar, 126, 130, 131
volume
of cones, 344, 345
of cylinders, 338, 353
of a prism, 619
of a pyramid, 618, 619
of spheres, 353
Rectangle(s)
area of, formula, 573
defined, 526
Reflection(s), 54–59
in the coordinate plane, 55–59
defined, 56
line of, 56
writing, 58
Regular polygon(s), defined, 121
Regular pyramid, defined, 596
Relation(s)
defined, 244
functions and, 242–247
research, 247
inputs, 244
mapping diagrams, 242–247
defined, 244
outputs, 244
Relative frequency, *See also* Probability
defined, 644
formula, 644
Repeated Reasoning, *Throughout. For example, see:*
circles
area, 569
circumference, 569
cube roots, 314
cylinders, 605
equations, 31
exponents, 410
negative, 429
zero, 428
inputs and outputs, 247
polygons
angles of, 111, 118, 119
probability, 659
repeating decimals, 316, 317
similar solids
surface area of, 355
volume of, 355, 361
slope, 149
surface area, 605
systems of linear equations, 213
Repeating decimals
writing as fractions, 316–317
Rhombus, defined, 526
Rise, defined, 150
Rotation(s), 60–67
angle of, 62
center of, 62
in the coordinate plane, 61–67
defined, 62
rotational symmetry, 66
Run, defined, 150

S

Sample(s), 672–679
biased, 674
defined, 672
research, 673
unbiased, 674
Sample space, defined, 654
Scale
defined, 532
research, 531
Scale drawing, 530–537
defined, 532
research, 531
scale, 532
scale factor, 533
Scale factor
defined, 84, 533
finding, 533
Scale model
defined, 532
finding distance in, 533
modeling, 537
Scatter plot(s), 372–379, 394
defined, 374
identifying relationships, 375
linear, 375
negative, 375
nonlinear, 375
positive, 375
interpreting, 374–375
line of best fit, 381
lines of fit, 378–383
defined, 380
modeling, 378, 379, 383
writing, 382
Scientific notation
defined, 438
operations in, 448–453
real-life application, 451
writing, 452
project, 453
reading numbers in, 436–441
real-life application, 439
writing, 440
writing numbers in, 442–447
real-life application, 445
standard form, 438–439
writing, 446
Semicircle(s)
area of, 573
defined, 552
Side(s)
classifying triangles by, 516
congruent, 516
defined, 516
reading, 516
corresponding, defined, 44
Similar figures, 70–81
areas of, 76–81
formula, 78
writing, 80
defined, 72
perimeters of, 76–81
formula, 78
writing, 80
reading, 72
real-life application, 73
Similar solids
defined, 356
surface area of, 354–361
linear measures, 357
volume of, 354–361
error analysis, 360
formula, 358
Simulation(s), 668–669
defined, 668
Slant height, defined, 596
Slope, 148–157
defined, 148, 150
formula, 148, 150
negative, 152
and parallel lines, 156
and perpendicular lines, 157
positive, 152
project, 154
reading, 150
rise, 150
run, 150
undefined, 152
zero, 152
Slope-intercept form, 166–171
defined, 168
graphing equations in, 166–171
real-life application, 169
writing equations in, 178–183
real-life application, 181
writing, 182
x-intercept, 168
y-intercept, 168
Solids, *See also:* Surface area, Three-dimensional figures, Volume
cones
real-life application, 343
volume of, 340–345
writing, 344
cross section, 620–621
cube, 590
cylinders, 600–605
modeling, 339
real-life application, 337
volume of, 334–339
hemispheres, 351
volume of, 351
prism, 586–593, 608–613
pyramid, 594–599, 614–619
similar
defined, 356
linear measures, 357
volume of, 354–361
spheres
volume of, 348–353
surface area of, 354–361
formula, 357
volume of, 334–345, 348–361
real-life applications, 337, 343
Solution of an inequality, defined, 466
Solution of a linear equation, defined, 144
Solution set, defined, 466
Solution of a system of linear equations, defined, 204
Sphere(s)
defined, 348
volume of, 348–353
formula, 350
Square, defined, 526
Square root(s)
approximating, 308–315
real-life application, 312
writing, 314
defined, 290
finding, 288–293
real-life application, 291
perfect square, 290
radical sign, 290
radicand, 290
Standard form of a linear equation
defined, 174
graphing equations in, 172–177
modeling, 177
real-life application, 175
writing, 176
Standard form of a number
scientific notation and, 438–439
Statistics, *See also* Probability
populations, 672–685
comparing, 680–685
defined, 672
project, 685
research, 673
writing, 684
samples, 672–679
biased, 674
defined, 672
research, 673
unbiased, 674
Stem-and-leaf plots, 394
Structure, *Throughout.*
For example, see:
angles, 507
angles of a polygon, 110, 119

Student Index

composite figures, 575
data displays, 399
distance formula, 323
equations, 3
exponents, 414, 429
linear equations
solving using graphs, 230
linear functions, 261
parallelograms, 529
probability, 667
Pythagorean Theorem, 323
real numbers, 315
repeating decimals, 316
scientific notation, 448
slope, 155
square roots, 308, 315
surface area of a prism, 593
systems of linear equations, 213, 217
transformations
dilations, 88, 89
similar figures, 81
volume
of a prism, 609
of a pyramid, 615
of solids, 345
Study Tip
analyzing graphs, 274
angles
alternate exterior, 106
alternate interior, 106
corresponding, 104
exterior, 113
circumference, 551
comparing populations, 682
constructing triangles, 517
direct variation, 160
equations, 13
exponents, 418
Quotient of Powers Property, 425
inequalities
graphing, 467
solving, 472
linear equations, 168, 231, 268
line of best fit, 381
line of fit, 380
in point-slope form, 187
in slope-intercept form, 168
in standard form, 174
system of, 205
writing, 180, 187
powers, 412
prisms, 610
probability, 640
Fundamental Counting Principle, 654
simulating outcomes, 668, 669
proportional relationships, 160
pyramids, 616
volume of, 616
Pythagorean triples, 320
quadrilaterals, 526
radius, 551
ratios, 534
real numbers, 310
right triangles, 302
scale, 532
scale drawings, 534
scientific notation, 438, 444, 450
in standard form, 450
slope, 150, 151, 160
solids, 351
cones, 342
similar solids, 358
squares, 526
square roots
approximating, 311
of zero, 290
systems of linear equations, 205, 211, 218, 219, 226
transformations
dilations, 85
rotations, 62, 63
transversals, 104, 106
volume
of cones, 342
of cylinders, 336
of a prism, 610
of a pyramid, 616
Substitution
to solve systems of linear equations, 208–213
Subtraction
Property
of Equality, 4
of Inequality, 472
to solve inequalities, 470–475
Subtraction Property of Inequality, 472
real-life application, 473
Summary triangle, 68
Supplementary angles
constructing, 508–513
defined, 510
Surface area
of a cylinder, 600–605
lateral, defined, 590
of a prism, 586–593
cube, 590
rectangular, 588
triangular, 589
of a pyramid, 594–599
of similar solids, 354–361
linear measures, 357
Symbols
congruent, 44
of inequality, 466
parallel lines, 104
perpendicular lines, 104
prime, 50
similar, 72
square root, 290
Symmetry, rotational, 66
System of linear equations
defined, 202, 204
reading, 204
solution of a
defined, 204
solving by elimination, 216–223
real-life application, 220
writing, 221
solving by graphing, 202–207
modeling, 207
real-life application, 205
writing, 206
solving special, 224–229
infinitely many solutions, 226
no solution, 226
one solution, 226
writing, 228
solving by substitution, 208–213
real-life application, 211
writing, 212

T

Theorem, defined, 300
Theoretical probability, *See also* Probability
defined, 647
formula, 647
modeling, 650
Three-dimensional figure(s), 620–621
cross sections of, 620
Transformation(s)
congruent figures, 42–47
corresponding angles of, 44
corresponding sides of, 44
defined, 44
identifying, 44
reading, 44
defined, 50
dilations, 82–89
center of, 84
in the coordinate plane, 82–89
defined, 84
scale factor, 84
image, 50
reflections, 54–59
in the coordinate plane, 55–57
defined, 56

line of, 56
writing, 58
rotations, 60–67
angle of, 62
center of, 62
in the coordinate plane, 61–67
defined, 62
rotational symmetry, 66
similar figures, 70–81
areas of, 76–81
defined, 72
perimeters of, 76–81
reading, 72
real-life application, 73
writing, 80
tessellations, 48–49
translations, 48–53
in the coordinate plane, 50
defined, 50
writing, 52
Translation(s), 48–53
in the coordinate plane, 49–53
defined, 50
tessellations, 48–49
writing, 52
Transversal(s), 102–109
alternate exterior angles and, 106
alternate interior angles and, 106
corresponding angles and, 104
defined, 104
exterior angles and, 105
interior angles and, 105
Trapezoid, defined, 526
Triangle(s)
acute, 516
angles of, 110 115
exterior, 112
interior, 112
real-life application, 113
area of, formula, 573
classifying
by angles, 516
by sides, 516
writing, 518
congruent, 42–44
congruent sides, 516
constructing, 514–519
equiangular, 516
equilateral, 516
isosceles, 516
obtuse, 516
Pythagorean Theorem, 300–305
defined, 302
project, 305
real-life applications, 303, 321
using, 318–323
right, 516
hypotenuse, 302
legs, 302
scalene, 516
sum of angle measures, 520–521
similar, 126–131
angles of, 128
modeling, 127
project, 127
writing, 130
Two-step inequalities, *See* Inequality
Two-way table(s), 386–391
defined, 388
joint frequencies, 388
marginal frequencies, 388

U

Unbiased sample(s), defined, 674

V

Vertical angles
constructions, 502–507
defined, 504
Volume
of composite solids, 351
of cones, 340–345
of cylinders, 334–339
of a prism, 608–613
of a pyramid, 614–619
of similar solids, 354–361
of spheres, 348–353

W

Which One Doesn't Belong?, *Throughout. For example, see:*
angle measures, 107
circles, 553
corresponding angles, 46
equations
linear, 146
simple, 7
exponents, 414
functions, 270
inequalities, 474
polygons, 123
powers, 426
pyramids, 598
Pythagorean Theorem, 322
quadrilaterals, 528
scientific notation, 440
similar triangles, 130
square roots, 313
systems of linear equations, 221
transformations
congruent figures, 46
reflections, 58
volume of solids, 352
Word magnet, 562
Writing, *Throughout. For example, see:*
comparing populations, 684
displaying data, 398
equations
multi-step, 14
with variables on both sides, 23
exponents, 432
functions
linear, 261
representing, 253
linear equations
in point-slope form, 188
in standard form, 176
lines of fit, 382
powers, 426
probability
events, 665
Fundamental Counting Principle, 657
outcomes, 636, 657
Pythagorean Theorem, 322
scientific notation, 440, 446, 452
similar triangles, 130
solving inequalities, 479, 483, 490
square roots, 313, 314
surface area of a prism, 591
systems of linear equations, 206
solving by elimination, 221
solving by graphing, 206
solving special, 228
solving by substitution, 212
transformations
reflections, 58
similar figures, 80
translations, 52
triangles, classifying, 518
volume
of a prism, 618
of a pyramid, 618
of solids, 344

X

***x*-intercept,** defined, 168

Y

Y chart, 16, 476
***y*-intercept,** defined, 168

Additional Answers

Chapter 11

Section 11.2

Practice and Problem Solving

12. $z \geq 3.1$;

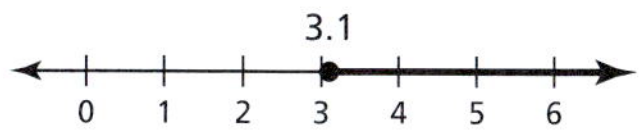

13. $-2.8 < d$;

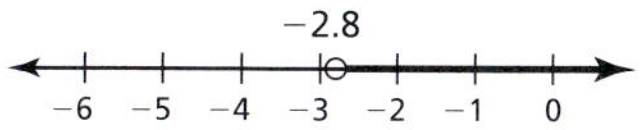

14. $-\frac{4}{5} > s$;

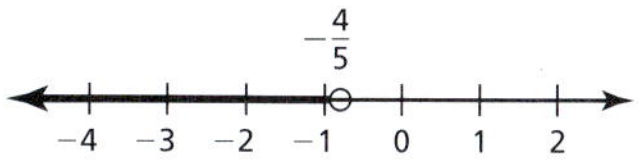

15. $\frac{3}{4} \geq m$;

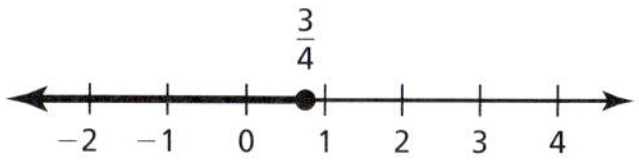

16. $r < -0.9$;

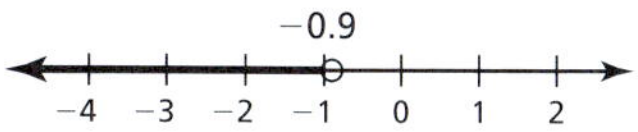

17. $h \leq -2.4$;

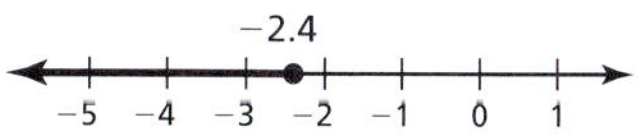

18. The 2 should have been subtracted rather than added.

$$\begin{aligned} x - 7 &> -2 \\ +7 \quad & \quad +7 \\ x &> 5 \end{aligned}$$

19. The wrong side of the number line is shaded.

20. **a.** $15 + p \leq 44$; $p \leq 29$ passengers

b. no; Only 29 more passengers can board the plane.

Section 11.3

Record and Practice Journal

5. If you multiply or divide each side of an inequality by the same positive number, the inequality remains true.

If you multiply or divide each side of an inequality by the same negative number, the direction of the inequality symbol must be reversed for the inequality to remain true.

Practice and Problem Solving

17. $y \leq -3$;

18. $b < 48.59$;

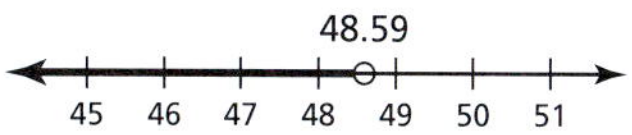

19. The inequality sign should not have been reversed.

$$\frac{x}{3} < -9$$

$$3 \cdot \frac{x}{3} < 3 \cdot (-9)$$

$$x < -27$$

20. $\frac{x}{4} \leq 5$; $x \leq 20$

21. $\frac{x}{7} < -3$; $x < -21$

22. $6x \geq -24$; $x \geq -4$

23. $-2x > 30$; $x < -15$

35. $b > 6$;

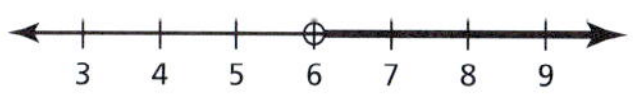

36. They forgot to reverse the inequality symbol.

$$-3m \geq 9$$

$$\frac{-3m}{-3} \leq \frac{9}{-3}$$

$$m \leq -3$$

37. $-2.5x < -20$; $x > 8$ h

38. **a.** $27x \leq 150$; $x \leq \frac{50}{9}$, or $5\frac{5}{9}$

b. no; The maximum height allowed is 150 inches, and 6 boxes has a height of 162 inches.

39. $10x \geq 120$; $x \geq 12$ cm

46. $x \geq 3$;

Additional Answers

47. $s < 14$;

Section 11.4

Practice and Problem Solving

12. $g > -1$;

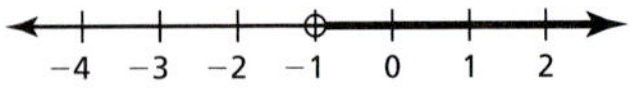

13. $w \leq 3$;

14. $k \geq -18$;

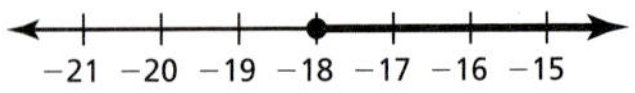

15. $d > -9$;

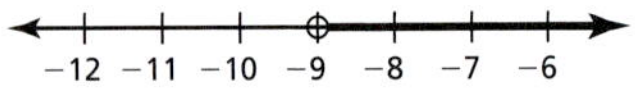

16. $n < -0.6$;

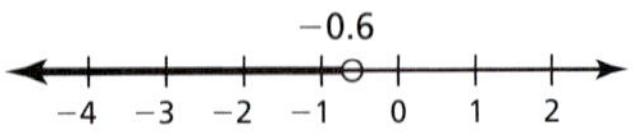

17. $c \geq -1.95$;

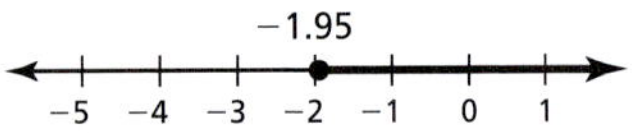

Chapter 12

Record and Practice Journal Fair Game Review

7.

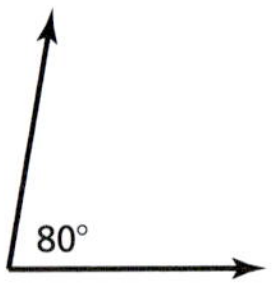

8.

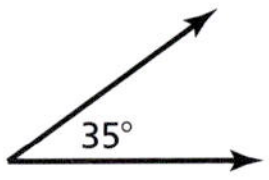

9.

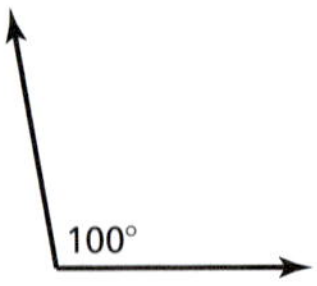

10.

Section 12.1

Practice and Problem Solving

15.

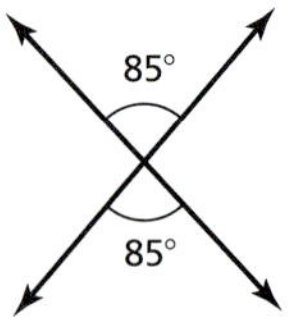

16.

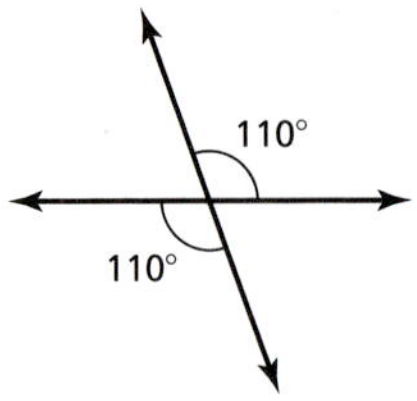

17.

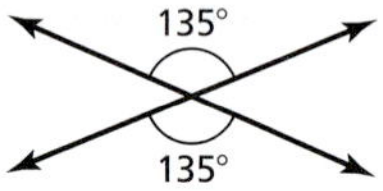

18. 43

19. **a.** *Sample answer:*

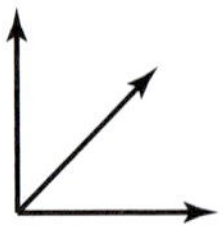

b. *Sample answer:*

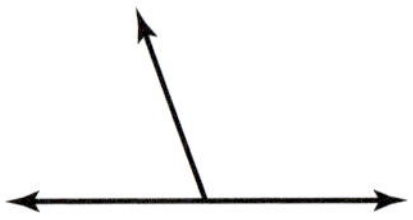

c. *Sample answer:*

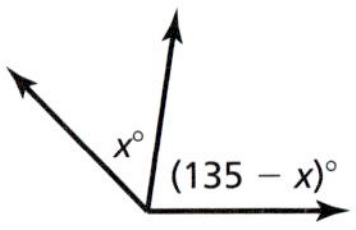

20. *Sample answer:*

1) Draw one angle, then draw the other angle using a side of the first angle.
2) Draw one angle that is the sum of the two angles, then draw the shared side.

21. never

22. always

23. sometimes

24. always

25.

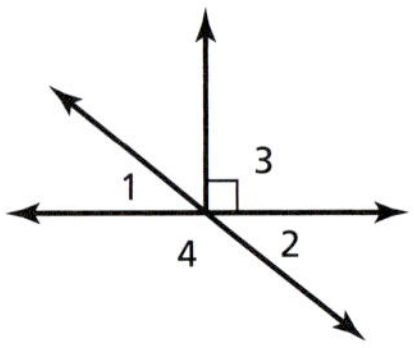

Section 12.2

Practice and Problem Solving

17.

18.

19.

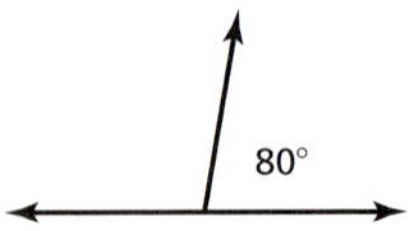

20.

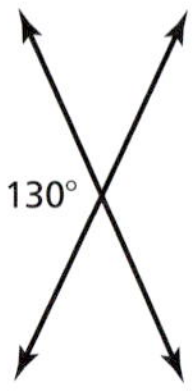

21. *Sample answer:* 1) Draw one angle, then draw the other using a side of the first angle; 2) Draw a right angle, then draw the shared side.

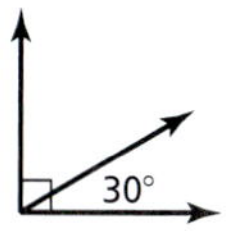

22. *Sample answer:* 120°; It is supplementary with a 60° angle, but it is greater than 90°, so it cannot be complementary with another angle.

23. **a.** 25° **b.** 65°

Section 12.3

On Your Own

4.

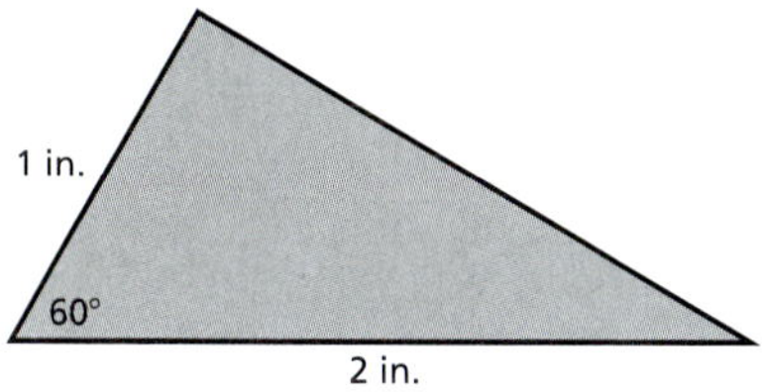

right scalene triangle

Practice and Problem Solving

3. *Sample answer:*

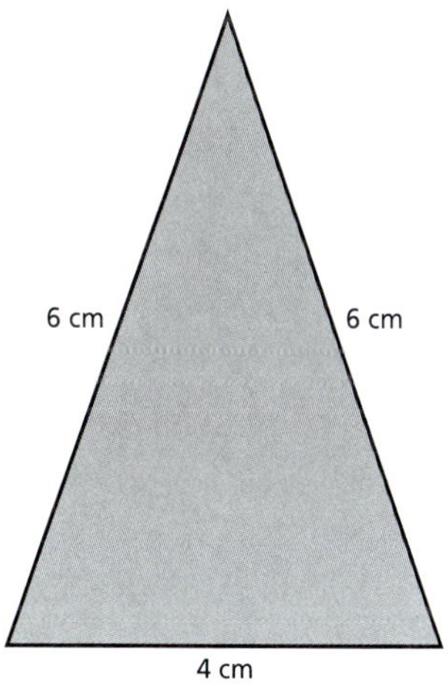

4. *Sample answer:*

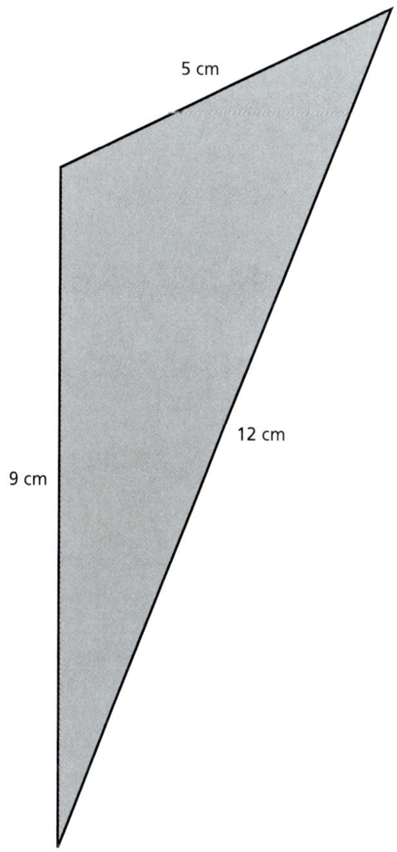

5. *Sample answer:*

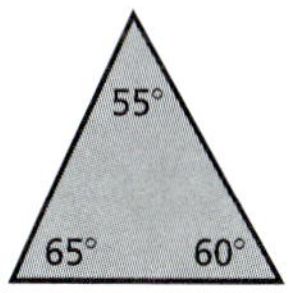

12. The triangle is not an acute triangle because acute triangles have 3 angles less than 90°. The triangle is an obtuse scalene triangle because it has one angle greater than 90° and no congruent sides.

13. acute isosceles

14.

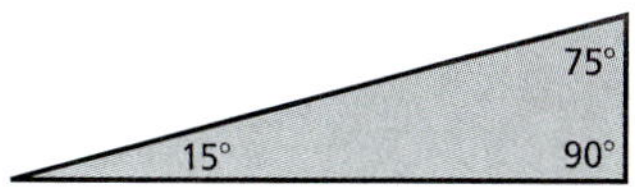

right scalene triangle

15.

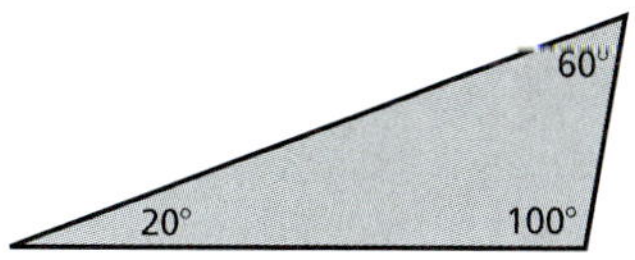

obtuse scalene triangle

16.

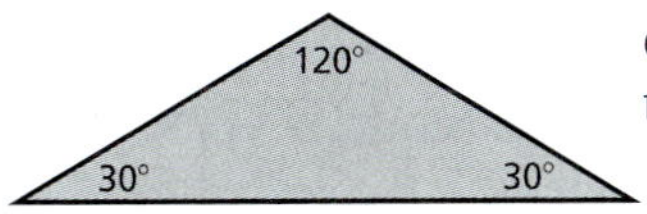

obtuse isosceles triangle

17.

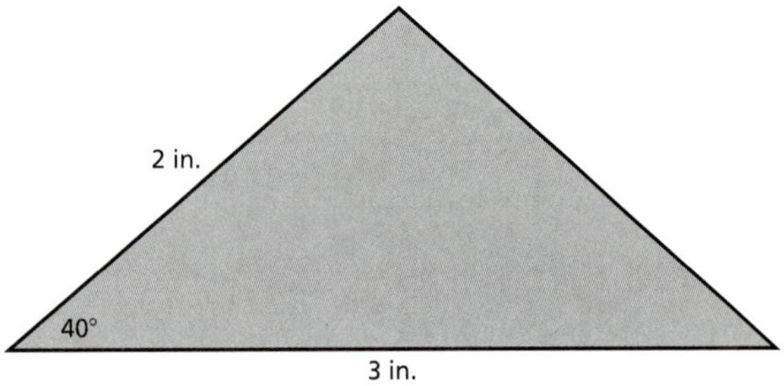

18.

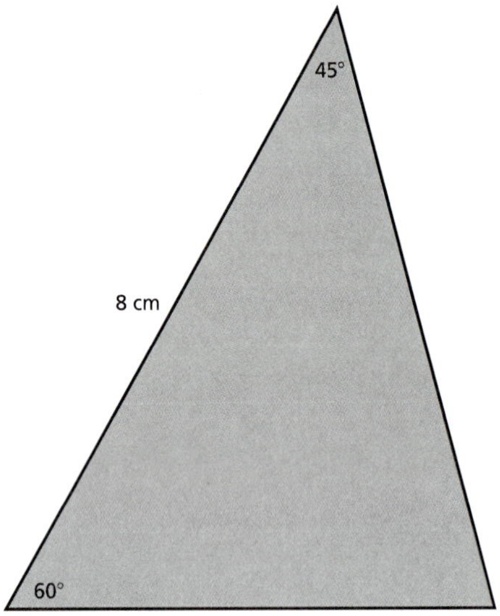

19.

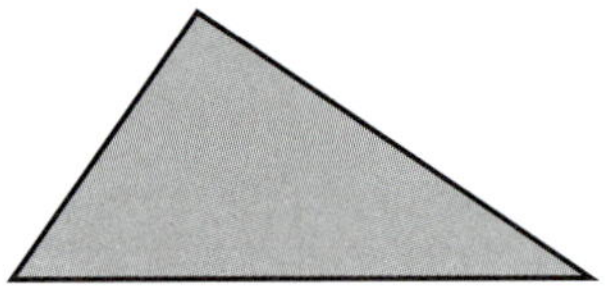

20. not necessarily; Just because one angle is acute doesn't mean it will be an acute triangle. The classification depends on the third side. It could form a right angle or an obtuse angle.

23. many; You can change the angle formed by the two given sides to create many triangles.

24. one; Only one line segment can be drawn between the end points of the two given sides.

25. no; The sum of any two side lengths must be greater than the remaining length.

26. no; An equilateral triangle cannot have a right angle.

27. **a.** green: 65; purple: 25; red: 45

b. The angles opposite the congruent sides are congruent.

c. An isosceles triangle has at least two angles that are congruent.

Extension 12.3

Practice

16. If two angle measures of a triangle were each greater than or equal to 90°, the sum of three angle measures would be greater than 180°, which is not possible.

17. **a.** 72

b. You can change the distance between the bottoms of the two upright cards; yes; x must be greater than 60 and less than 90; If x were less than or equal to 60, the two upright cards would have to be exactly on the edges of the base card or off the base card. It is not possible to stack cards at these angles. If x where equal to 90, then the two upright cards would be vertical, which is not possible. The card structure would not be stable. In practice, the limits on x are probably closer to $70 < x < 80$.

Record and Practice Journal Practice

1. $x = 60$; acute, isosceles

2. $x = 30$; obtuse, isosceles

3. $x = 31$; right, scalene

4. $x = 28$; acute, scalene

5. $x = 35$

6. yes

7. no; 54.5°

8. no; 86°

9. no; 28°

10. yes

11. yes

12.1–12.3 Quiz

9.

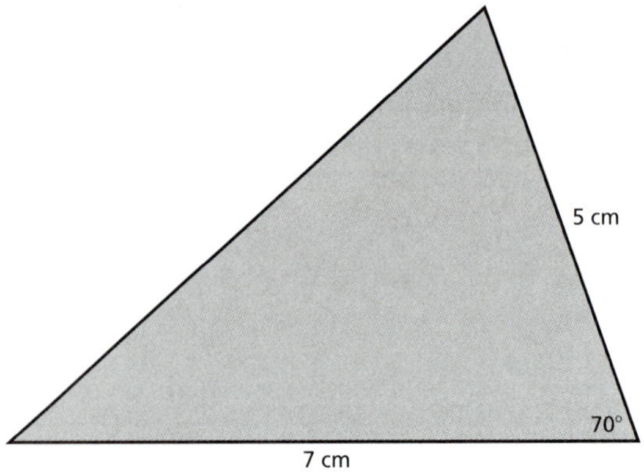

Section 12.4

Record and Practice Journal

1. Sample answers are given.

a. 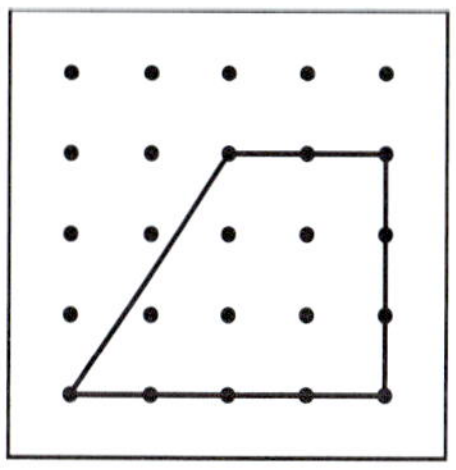

b. 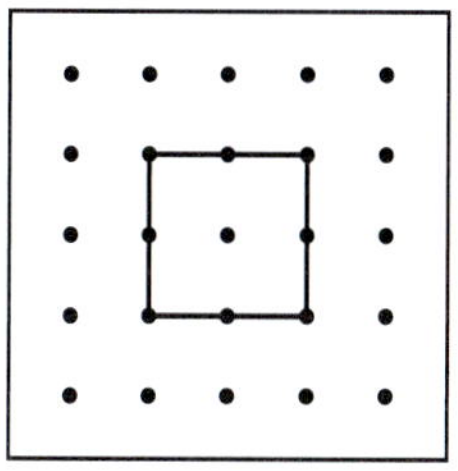

c. 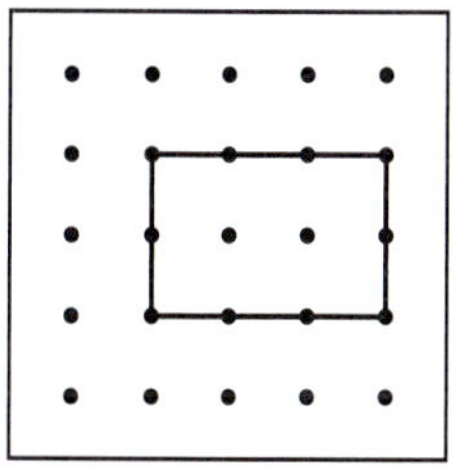

d. 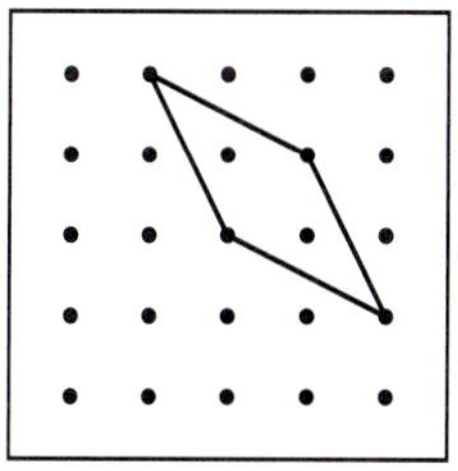

e. 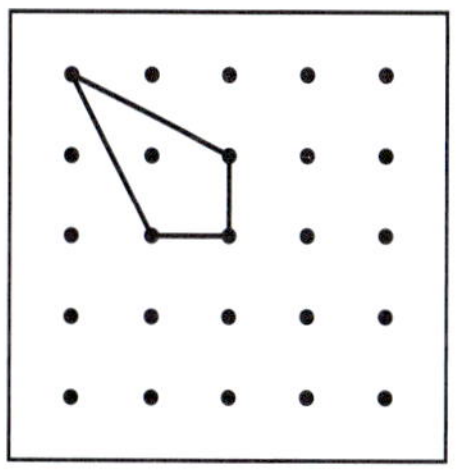

f. 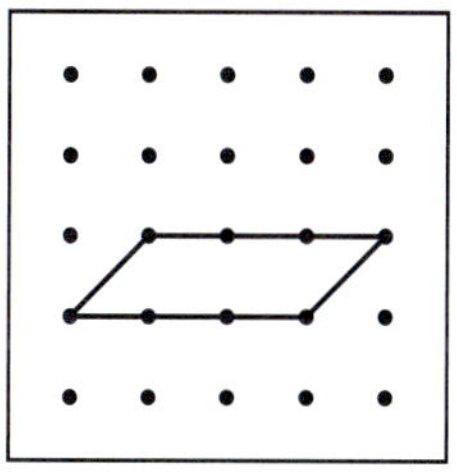

Practice and Problem Solving

25. 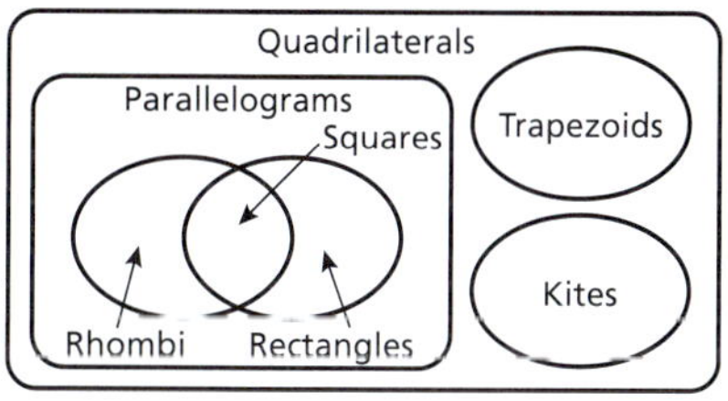

26. a. $x = 125$; $y = 55$

b. Opposite angles of a parallelogram are equal.

c. Consecutive interior angles of a parallelogram are supplementary.

Section 12.5

Record and Practice Journal

6. Painting:

	Actual Object	Original Drawing	Your Drawing
Perimeter	192 in.	24 units	8 units
Area	2304 in.^{2}	36 units^2	4 units^2

Check students' work. Students should conclude that the ratio of the perimeters for the original drawings to the actual objects is equal to the ratios they found in Activities 1(c) and 3(c). The ratio of the areas for the original drawings to the actual objects is equal to the square of those ratios. Students can find these ratios for their own drawings to the actual object and see the same relationships.

On Your Own

4. c. The scale factor and the ratio of the perimeters change to $\frac{10}{3}$, and the ratio of the areas changes to $\left(\frac{10}{3}\right)^2$; The change in scale results in a change to each of these three values, but the ratio of the perimeters is still the same as the scale factor, and the ratio of the areas is still the same as the square of the scale factor.

Practice and Problem Solving

25. 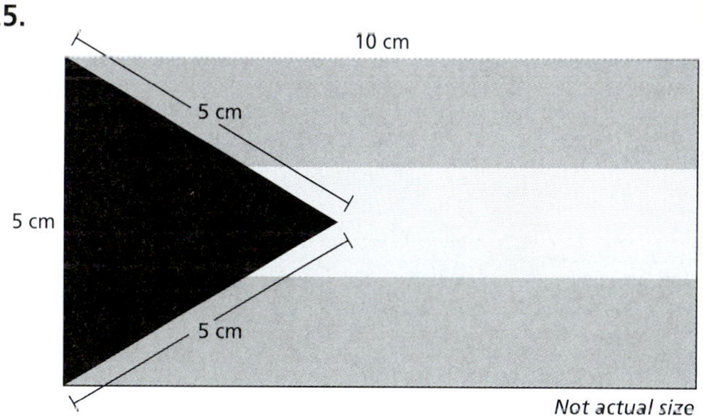

12.4–12.5 Quiz

6. 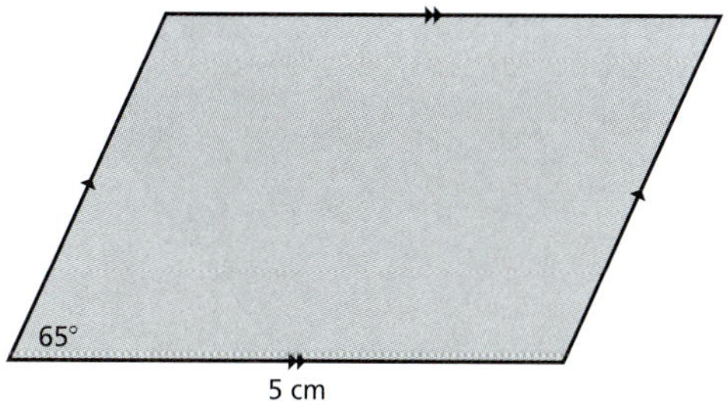

Chapter 12 Review

6.

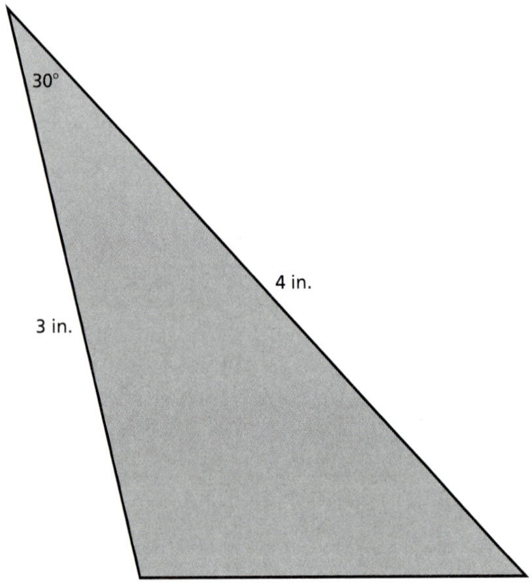

Chapter 12 Test

7.

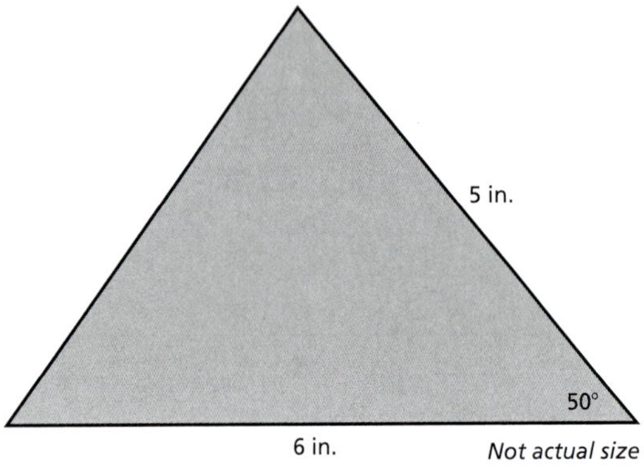

15.

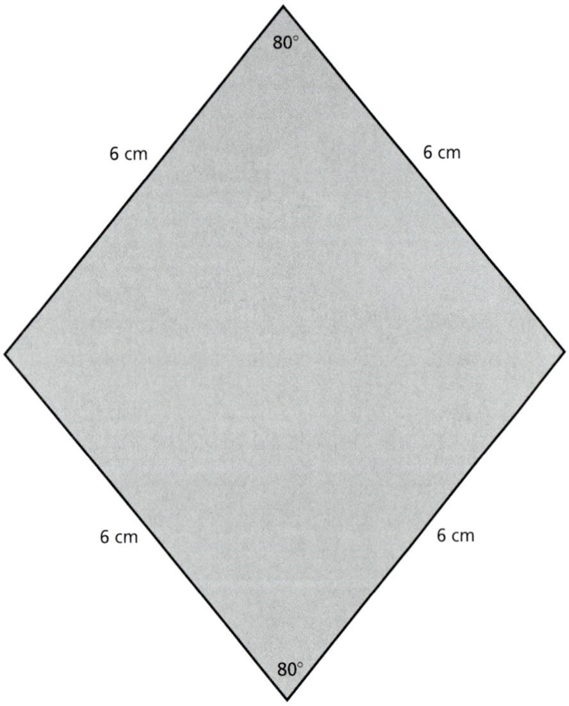

Chapter 13

Section 13.1

Practice and Problem Solving

25. b. about 320.28 mm; Subtract $\frac{1}{12}$ the circumference of a circle whose radius is the length of the hour hand from the circumference of a circle whose radius is the length of the minute hand.

Section 13.2

Practice and Problem Solving

18. The starting points are staggered so that each runner can run the same distance and use the same finish line. This is necessary because the circumference is different for each lane. The diagram shows this because the diameter is greater in the outer lanes.

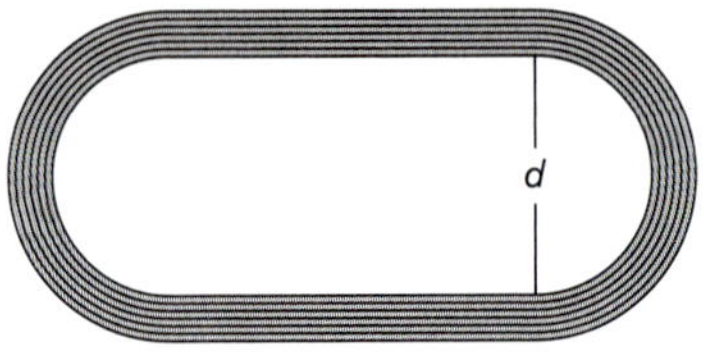

Section 13.3

Practice and Problem Solving

13. about 628 cm^2

14. about 226.08 $in.^2$

15. about 1.57 ft^2

16. a.

2	4π in.	4π $in.^2$
4	8π in.	16π $in.^2$
8	16π in.	64π $in.^2$
16	32π in.	256π $in.^2$

b. The circumference doubles; The area becomes four times as great.

c. The circumference triples; The area becomes 9 times as great.

19. about 9.8125 $in.^2$; The two regions are identical, so find one-half the area of the circle.

20. about 17.415 m^2; Subtract the area of a circle with a 9-meter diameter from the area of the square.

21. about 4.56 ft^2; Find the area of the shaded regions by subtracting the areas of both unshaded regions from the area of the quarter-circle containing them. The area of each unshaded region can be found by subtracting the area of the smaller shaded region from the semicircle. The area of the smaller shaded region can be found by drawing a square about the region.

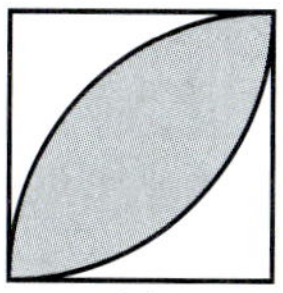

Subtract the area of a quarter-circle from the area of the square to find an unshaded area. Then subtract both unshaded areas from the square's area to find the shaded region's area.

Section 13.4

Practice and Problem Solving

12. 89 m^2

13. 23.5 $in.^2$

14. about 21.87 ft^2

15. 24 m^2

Chapter 14

Section 14.1

Practice and Problem Solving

26. The dimensions of the red prism are $\frac{3}{2}$ times the dimensions of the blue prism. The surface area of the red prism is $\frac{9}{4}$ times the surface area of the blue prism.

27. **a.** 0.125 pint

b. 1.125 pints

c. red and green: The ratio of the paint amounts (red to green) is 4 : 1 and the ratio of the side lengths is 2 : 1.

green and blue: The ratio of the paint amounts (blue to green) is 9 : 1 and the ratio of the side lengths is 3 : 1.

The ratio of the paint amounts is the square of the ratio of the side lengths.

28. **a.** 0.5 in.

b. 13.5 $in.^2$

Section 14.2

Record and Practice Journal

1. **a.** $S = 85{,}560\ m^2$

b. $S = 1404\ m^2$

c. $S = 1960\ m^2$

d. $S = 1276\ m^2$

2. **b.**

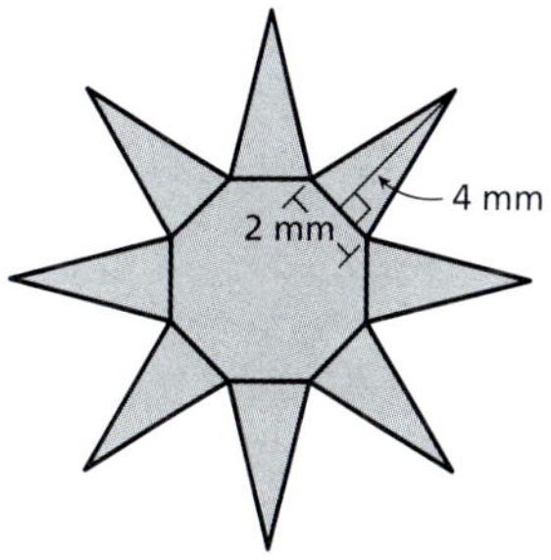

Practice and Problem Solving

18. The slant height is greater. The height is the distance between the top and the point on the base directly beneath it. The distance from the top to any other point on the base is greater than the height.

19. 124 cm^2

20. greater than; If it is less than or equal to, then the lateral face could not meet at a vertex to form a solid.

Section 14.3

Practice and Problem Solving

12. The area of only one base is added. The first term should have a factor of 2;

$$\begin{aligned} S &= 2\pi r^2 + 2\pi rh \\ &= 2\pi(5)^2 + 2\pi(5)(10.6) \\ &= 50\pi + 106\pi \\ &= 156\pi \approx 489.8\ yd^2 \end{aligned}$$

18. **a.** 4 times greater; 9 times greater; 25 times greater; 100 times greater

b. When both dimensions are multiplied by a factor of k, the surface area increases by a factor of k^2; 400 times greater

Section 14.4

Practice and Problem Solving

19. 1728 $in.^3$

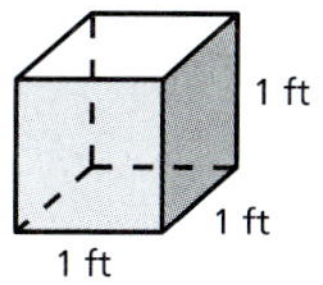

$1 \times 1 \times 1 = 1\ ft^3$

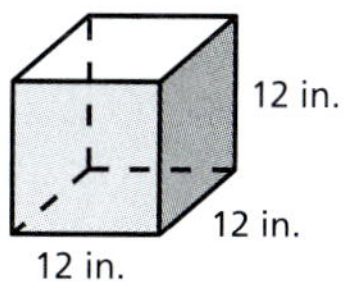

$12 \times 12 \times 12 - 1728\ in.^3$

Section 14.5

Practice and Problem Solving

17. 12,000 in.^{3}; The volume of one paperweight is 12 cubic inches. So, 12 cubic inches of glass is needed to make one paperweight. So, it takes $12 \times 1000 = 12{,}000$ cubic inches to make 1000 paperweights.

Chapter 15

Section 15.3

Record and Practice Journal

6. *Sample answer:* Most likely this is true because there are only 20 chips in the bag and you did not select an orange chip in 50 tries. However, you cannot say this for certain because there may be 1 orange chip in the bag and you never selected it.

Practice and Problem Solving

28. theoretical

29. **a.** $\frac{1}{12}$; 50 times

    b. $\frac{11}{50}$; 132 times

    c. A larger number of trials should result in a more accurate probability, which gives a more accurate prediction.

34. **a.** Check students' work. The cup should land on its side most of the time.

    b. Check students' work.

    c. Check students' work.

    d. more likely; Due to the added weight, the cup will be more likely to hit open-end up and thus more likely to land open-end up. Some students may justify by performing multiple trials with a quarter taped to the bottom of the cup.

Section 15.4

Practice and Problem Solving

12. Tree Diagram:

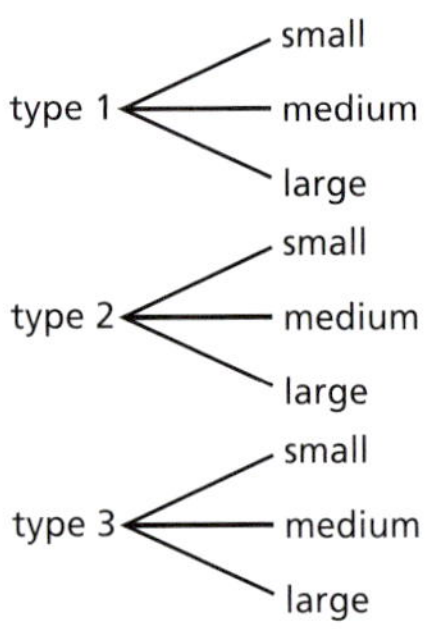

Fundamental Counting Principle: $3 \cdot 3 = 9$

27. **a.** $\frac{1}{1000}$, or 0.1%

    b. There are 1000 possible combinations. With 5 tries, someone would guess 5 out of the 1000 possibilities. So, the probability of getting the correct combination is $\frac{5}{1000}$, or 0.5%.

29. **a.** The Fundamental Counting Principle is more efficient. A tree diagram would be too large.

    b. 1,000,000,000 or one billion

    c. *Sample answer:* Not all possible number combinations are used for Social Security Numbers (SSN). SSNs are coded into geographical, group, and serial numbers. Some SSNs are reserved for commercial use and some are forbidden for various reasons.

Section 15.5

Fair Game Review

27. 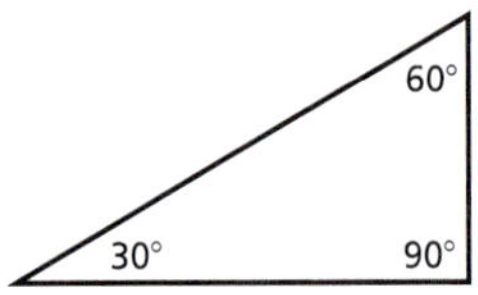

; right scalene

Extension 15.5

Practice

1. **a.** *Sample answer:* Roll four number cubes. Let an odd number represent a correct answer and an even number represent an incorrect answer. Run 40 trials.

   b. Check students' work. The probability should be "close" to 6.25% (depending on the number of trials, because that is the theoretical probability).

2. *Sample answer:* Place 7 green and 3 red marbles in a bag. Let the green marbles represent a win and the red marbles represent a loss. Randomly pick one marble to simulate the first game. Replace the marble and repeat two more times. This is one trial. Run 30 trials. Check students' work. The probability should be close to 34.3% (depending on the number of trials, because that is the theoretical probability).

3. *Sample answer:* Using the spreadsheet in Example 3 and using digits 1–4 as successes, the experimental probability is 16%.

4. Example 1: $\frac{1}{2} \cdot \frac{1}{2} \cdot \frac{1}{2} = \frac{1}{8}$

Example 2: $0.6 \times 0.2 = 0.12$, or 12%

Example 3: Students likely will not be able to find this theoretical probability. However, it can be found by examining the favorable outcomes and using logic:

$$P(1, 2, 3, 4) = \frac{1}{2} \cdot \frac{1}{2} \cdot \frac{1}{2} \cdot \frac{1}{2} = \frac{1}{16}$$

$$P(1, 2, 3, \text{not } 4) = \frac{1}{2} \cdot \frac{1}{2} \cdot \frac{1}{2} \cdot \frac{1}{2} = \frac{1}{16}$$

$$P(1, 2, \text{not } 3, 4) = \frac{1}{2} \cdot \frac{1}{2} \cdot \frac{1}{2} \cdot \frac{1}{2} = \frac{1}{16}$$

$$P(1, \text{not } 2, 3, 4) = \frac{1}{2} \cdot \frac{1}{2} \cdot \frac{1}{2} \cdot \frac{1}{2} = \frac{1}{16}$$

$$P(\text{not } 1, 2, 3, 4) = \frac{1}{2} \cdot \frac{1}{2} \cdot \frac{1}{2} \cdot \frac{1}{2} = \frac{1}{16}$$

So, the theoretical probability is $5\left(\frac{1}{16}\right) = \frac{5}{16} = 31.25\%$.

Or, they could realize that there are 5 favorable outcomes out of 16 (from the Fundamental Counting Principle), and all outcomes are equally likely, so the probability is $\frac{5}{16}$.

Exercise 1: $\frac{1}{2} \cdot \frac{1}{2} \cdot \frac{1}{2} \cdot \frac{1}{2} = \frac{1}{16}$

Exercise 2: $0.7^3 = \frac{343}{1000}$, or 34.3%

Exercise 3: Students likely will not be able to find this theoretical probability. However, it can be found by examining the favorable outcomes and using logic:

$$P(1, 2, 3, 4) = \frac{2}{5} \cdot \frac{2}{5} \cdot \frac{2}{5} \cdot \frac{2}{5} = \frac{16}{625}$$

$$P(1, 2, 3, \text{not } 4) = \frac{2}{5} \cdot \frac{2}{5} \cdot \frac{2}{5} \cdot \frac{3}{5} = \frac{24}{625}$$

$$P(1, 2, \text{not } 3, 4) = \frac{2}{5} \cdot \frac{2}{5} \cdot \frac{3}{5} \cdot \frac{2}{5} = \frac{24}{625}$$

$$P(1, \text{not } 2, 3, 4) = \frac{2}{5} \cdot \frac{3}{5} \cdot \frac{2}{5} \cdot \frac{2}{5} = \frac{24}{625}$$

$$P(\text{not } 1, 2, 3, 4) = \frac{3}{5} \cdot \frac{2}{5} \cdot \frac{2}{5} \cdot \frac{2}{5} = \frac{24}{625}$$

So, the theoretical probability is $\frac{16}{625} + 4\left(\frac{24}{625}\right) = \frac{112}{625} = 17.92\%$.

Notice that you still have 5 favorable outcomes and 16 possible outcomes but the outcomes are not equally likely, so the probability is not $\frac{5}{16}$ as before.

When you increase the number of trials in a simulation, the experimental probability approaches the theoretical probability of the event that you are simulating.

Record and Practice Journal Practice

1. a. *Sample answer:* Roll four number cubes. Let an odd number represent a *no* answer and an even number represent a *yes* answer. Run 40 trials.

b. Check students' work. The probability should be close to 6.25% (depending on the number of trials, because that is the theoretical probability).

2. a. Place 7 green and 3 red marbles in bag. Let the green marbles represent a snowy day and the red marbles represent a snowless day. Randomly pick one marble to simulate the today. Replace the marble and repeat one more time. This is one trial. Run 50 trials.

b. Check students' work. The probability should be close to 42% (depending on the number of trials, because that is the theoretical probability).

Section 15.6

Practice and Problem Solving

9. no; The sample is not representative of the population because people going to the baseball stadium are more likely to support building a new baseball stadium. So, the sample is biased and the conclusion is not valid.

10. Sample B because it is a larger sample.

15. sample; It is much easier to collect sample data in this situation.

16. 1260 students

17. Not everyone has an email address, so the sample may not be representative of the entire population. *Sample answer:* When the survey question is about technology or which email service you use, the sample may be representative of the entire population.

18. a. *Sample answer:* The person could ask, "Do you agree with the town's unfair ban on skateboarding on public property?"

b. *Sample answer:* The person could ask, "Do you agree that the town's ban on skateboarding on public property has made the town quieter and safer?"

Section 15.7

Record and Practice Journal

1. a. *Sample answer:* double box-and-whisker plot or double dot plot

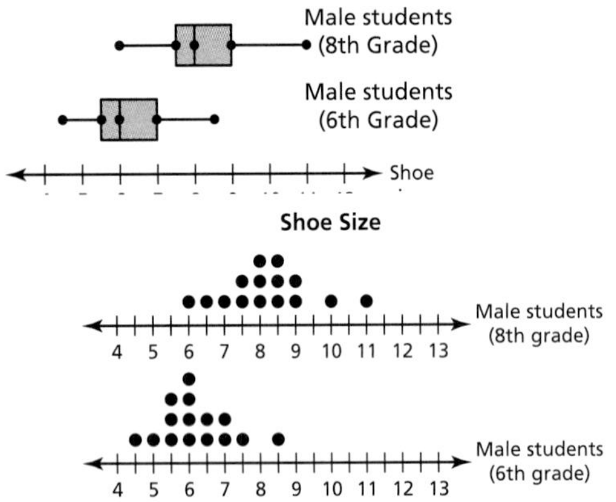

2. a. male students: symmetric; female students: skewed left; yes; *Sample answer:* The data set for the female students completely overlaps the data set for the male students. The overlaps between the centers and between the extreme values are shown.

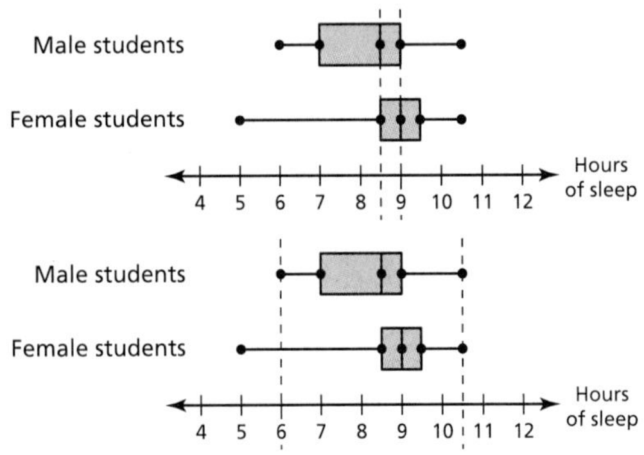

b. male students: symmetric; female students: symmetric; yes; *Sample answer:* The overlaps between the centers and between the extreme values are shown.

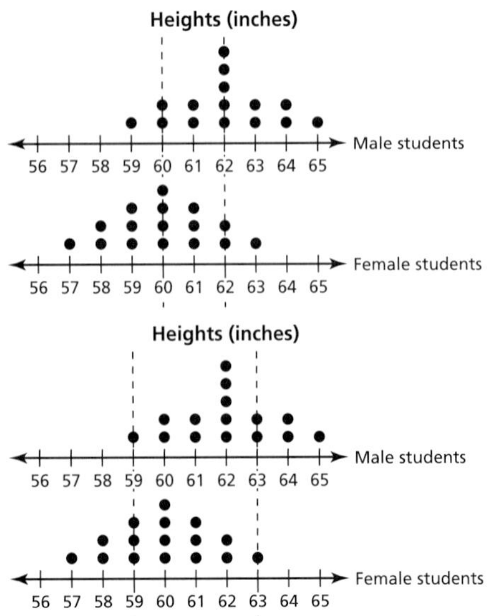

Practice and Problem Solving

5. a. Class A: median = 90, IQR = 12.5
Class B: median = 80, IQR = 10
The variation in the test scores is about the same, but Class A has greater test scores.

b. The difference in the medians is 0.8 to 1 times the IQR.

6. a. volleyball: mean = 86, MAD = 19.6
basketball: mean = 185, MAD = 17.7
The variation in the attendances is about the same, but basketball has a greater attendance.

b. The difference in the means is about 5.1 to 5.6 times the MAD.

8. a. The mean and MAD for the sports magazine, 19 and 5.8, are close to the mean and MAD for the political magazine, 18 and 5.2. However, the sample size is small and the variability is too great to conclude that the number of words per sentence is about the same.

b. The sample means vary much less than the sample numbers of words per sentence.

c. The number of words per sentence is generally greater in the political magazine than in the sports magazine.

9. a. Check students' work. Experiments should include taking many samples of a manageable size from each grade level. This will be more doable if the work of sampling is divided among the whole class, and the results are pooled together.

b. Check students' work. The data may or may not support a conclusion.

Photo Credits

Front matter

iv Big Ideas Learning, LLC; **viii** *top* ©iStockhphoto.com/Jonathan Larsen; *bottom* Evok20/Shutterstock.com; **ix** *top* ©iStockphoto.com/Lisa Thornberg, ©iStockphoto.com/Ann Marie Kurtz; *bottom* Sinisa Bobic/ Shutterstock.com; **x** *top* ©iStockphoto/Michael Flippo, ©iStockphoto.com/Ann Marie Kurtz; *bottom* Artpose Adam Borkowski/ Shutterstock.com; **xi** *top* stephan kerkhofs/Shutterstock.com, Cigdem Sean Cooper/Shutterstock.com, ©iStockphoto.com/Andreas Gradin; *bottom* ©iStockphoto.com/Aldo Murillo; **xii** *top* infografick/ Shutterstock.com, ©iStockphoto.com/Ann Marie Kurtz; *bottom* ©iStockphoto.com/kate_sept2004

Chapter 11

462 ©iStockphoto.com/Jonathan Larsen; **464** *Activity 1a* Filaphoto/ Shutterstock.com; *Activity 1b* Palmer Kane LLC/Shutterstock.com; *Activity 1c* bitt24/Shutterstock.com; *Activity 1d* kropic1/Shutterstock.com, ©iStockphoto.com/Chris Schmidt, ©iStockphoto.com/Jane norton; **469** Gregory James Van Raalte/Shutterstock.com; **470** *top left* Boy Scouts of America; *bottom right* William Seng/CC BY ND; **471** PILart/ Shutterstock.com; **473** eddtoro/Shutterstock.com; **474** Khafizov Ivan Harisovich/Shutterstock.com; **475** molchunya/Shutterstock.com; **483** Oleksiy Mark/Shutterstock.com; **484** LockStockBob/Shutterstock.com; **485** Jacek Chabraszewski/Shutterstock.com; **490** ©Keddie. Image from BigStockPhoto.com; **491** ©Lars Christenen/123RF; **492** *center left* Nattika/Shutterstock.com; *bottom right* Svetlana Mihailova/ Shutterstock.com; **496** *center right* ©iStockphoto.com/Jack Puccio; *bottom left* ©iStockphoto.com/itographer

Chapter 12

500 ©iStockphoto.com/Lisa Thornberg, ©iStockphoto.com/Ann Marie Kurtz; **507** mountainpix/Shutterstock.com; **509** Scott J. Carson/ Shutterstock.com; **513** ©iStockphoto.com/Jorgen Jacobsen; **521** *Exercise 11* ©iStockphoto.com/Chih-Feng Chen; *Exercise 12* ©iStockphoto.com/ Andreas Gradin; *Exercise 13* ©iStockphoto.com/Jim Lopes; *bottom right* ©iStockphoto.com/Zoran Kolundzija; **533** Andrew B Hall/ Shutterstock.com; **534** Serg64/Shutterstock.com; **535** ©iStockphoto.com/ Dan Moore; **536** *center right* ©iStockphoto.com/Aldo Murillo; *center left* MapMaster / CC-BY-SA-3.0; *bottom right* VectorWeb/Shutterstock.com

Chapter 13

546 ©iStockphoto.com/Michael Flippo, ©iStockphoto.com/Ann Marie Kurtz; **548** ©iStockphoto.com/HultonArchive; **549** Sir Cumference and the First Round Table; text copyright © 1997 by Cindy Neuschwander; illustrations copyright © 1997 by Wayne Geehan. Used with permission by Charlesbridge Publishing, Inc., 85 Main Street, Watertown, MA 02472; 617-926-0329; www.charlesbridge.com. All rights reserved.; **553** *Exercises 3 and 4* ©iStockphoto.com/zentilia; *Exercise 5* ©iStockphoto.com/Dave Hopkins; *Exercise 6* ©iStockphoto.com/ Boris Yankov; *Exercise 7* ©iStockphoto.com/ALEAIMAGE; *Exercise 8* ©iStockphoto.com/iLexx; *Exercise 9–11* ©iStockphoto.com/alexander mychko; **555** *center left* ©iStockphoto.com/HultonArchive; *center right* ©iStockphoto.com/winterling; **561** ©iStockphoto.com/Scott Slattery; **563** *Exercise 6* ©iStockphoto.com/DivaNir4A; *Exercise 7* ©iStockphoto.com/Stacey Walker; *Exercise 8* ©iStockphoto.com/ Creativeye99; *Exercise 11* Kalamazoo (Michigan) Public Library; **567** ©iStockphoto.com/Brian Sullivan; **568** *Exercise 3* ©iStockphoto.com/zentilia; *Exercises 4 and 5* ©iStockphoto.com/ PhotoHamster; *Exercise 6* ©iStockphoto.com/subjug; *Exercise 7* ©iStockphoto.com/Dave Hopkins; *Exercise 8* ©iStockphoto.com/7nuit; **574** Big Ideas Learning, LLC; **576** *Exercise 4* ©iStockphoto.com/ Mr_Vector; *Exercises 5 and 6* ©iStockphoto.com/sndr

Chapter 14

584 stephan kerkhofs/Shutterstock.com, Cigdem Sean Cooper/ Shutterstock.com, ©iStockphoto.com/Andreas Gradin; **587** ©iStockphoto.com/Remigiusz Załucki; **590** *center left* Bob the Wikipedian / CC-BY-SA-3.0; *center right* ©iStockphoto.com/Sherwin McGehee; **592** Yasonya/Shutterstock.com; **T-593** Yasonya/ Shutterstock.com; **593** Stankevich/Shutterstock.com; **594** *top left* ©iStockphoto.com/Luke Daniek; *top right* ©iStockphoto.com/Jeff Whyte; *bottom left* ©Michael Mattox. Image from BigStockPhoto.com; *bottom right* ©iStockphoto.com/Hedda Gjerpen; **595** ©iStockphoto.com/josh webb; **604** ©iStockphoto.com/Tomasz Pietryszek; **T-605** ©iStockphoto.com/scol22; **605** *center left* Newcastle Drum Centre; *center right* ©iStockphoto.com/scol22; **613** *top left* ©iStockphoto.com/david franklin; *top right* ©Ruslan Kokarev. Image from BigStockPhoto.com; *center right* ©iStockphoto.com/Lev Mel, ©iStockphoto.com/Ebru Baraz; **614** *bottom left* ©iStockphoto.com/ Jiri Vatka; *bottom right* Patryk Kosmider/Shutterstock.com; **T-616** Christophe Testi/Shutterstock.com; **618** ©iStockphoto.com/ ranplett, Image © Courtesy of Museum of Science, Boston; **T-619** ©iStockphoto.com/Yails; **619** *center right* ©iStockphoto.com/ James Kingman; *center left* ©iStockphoto.com/Yails; **621** *Exercise 12* ©iStockphoto.com/AlexStar; *Exercise 13* Knartz/Shutterstock.com; *Exercise 14* SOMMAI/Shutterstock.com

Chapter 15

630 infografick/Shutterstock.com, ©iStockphoto.com/Ann Marie Kurtz; **633** ryasick photography/Shutterstock.com; **634** *center* ©iStockphoto.com/sweetym; *bottom left* Big Ideas Learning, LLC; **635** ©iStockphoto.com/sweetym; **636** ©iStockphoto.com/Joe Potato Photo, ©iStockphoto.com/sweetym; **637** *top left* ©iStockphoto.com/ Jennifer Morgan; *top center* United States coin image from the United States Mint; **641** *top left* design56/Shutterstock.com; *center right* ©iStockphoto.com/spxChrome; **642** *center right* Daniel Skorodyelov/ Shutterstock.com; *bottom left* Mitrofanova/Shutterstock.com; **643** Jamie Wilson/Shutterstock.com; **644** *top right* James Steidl/ Shutterstock.com; *bottom right* traudl/Shutterstock.com; **645** Warren Goldswain/Shutterstock.com; **646** ©iStockphoto.com/Frank van de Bergh; **647** ©iStockphoto.com/Eric Ferguson; **651** Feng Yu/ Shutterstock.com; **652** *top right* John McLaird/Shutterstock.com; *center right* Robert Asento/Shutterstock.com; *bottom right* Mark Aplet/ Shutterstock.com; **657** ©iStockphoto.com/Justin Horrocks; ©iStockphoto.com/sweetym; **T-659** tele52/Shutterstock.com; **659** *top right* John McLaird/Shutterstock.com; *center* tele52/Shutterstock.com; **660 and 661** ©iStockphoto.com/Joe Potato Photo, ©iStockphoto.com/ sweetym; **663** Univega/Shutterstock.com; **664** James Steidl/ Shutterstock.com; **666** *top left* ©iStockphoto.com/sweetym; *center right* ©iStockphoto.com/Andy Cook; **667** -Albachiaraa-/Shutterstock.com; **671** ©iStockphoto.com/Doug Cannell; **672** *Activity 1a left* ©iStockphoto.com/Shannon Keegan; *Activity 1a right* ©iStockphoto.com/ Lorelyn Medina; *Activity 1b left* Joel Sartore/joelsartore.com; *Activity 1b right* Feng Yu/Shutterstock.com; *Activity 1c left* ©iStockphoto.com/ kledge; *Activity 1c right* ©iStockphoto.com/spxChrome; *Activity 1d* ©iStockphoto.com/Alex Slobodkin; **674** ©iStockphoto.com/Philip Lange; **676** ©iStockphoto.com/blackwaterimages, ©iStockphoto.com/Rodrigo Blanco, ©iStockphoto.com/7nuit; **679** Feng Yu/Shutterstock.com; **683** artis777/Shutterstock.com; **689** GoodMood Photo/Shutterstock.com; **690** Asaf Eliason/Shutterstock.com; **692** *top right* ©iStockphoto.com/ Frank Ramspott; *bottom left* ©iStockphoto.com/7nuit; **694** mylisa/ Shutterstock.com; **695** ra-design/Shutterstock.com

Cartoon illustrations Tyler Stout

Learning Progression

Kindergarten

Counting and Cardinality	– Count to 100 by Ones and Tens; Compare Numbers
Operations and Algebraic Thinking	– Understand and Model Addition and Subtraction
Number and Operations in Base Ten	– Work with Numbers 11–19 to Gain Foundations for Place Value
Measurement and Data	– Describe and Compare Measurable Attributes; Classify Objects into Categories
Geometry	– Identify and Describe Shapes

Grade 1

Operations and Algebraic Thinking	– Represent and Solve Addition and Subtraction Problems
Number and Operations in Base Ten	– Understand Place Value for Two-Digit Numbers; Use Place Value and Properties to Add and Subtract
Measurement and Data	– Measure Lengths Indirectly; Write and Tell Time; Represent and Interpret Data
Geometry	– Draw Shapes; Partition Circles and Rectangles into Two and Four Equal Shares

Grade 2

Operations and Algebraic Thinking	– Solve One- and Two-Step Problems Involving Addition and Subtraction; Build a Foundation for Multiplication
Number and Operations in Base Ten	– Understand Place Value for Three-Digit Numbers; Use Place Value and Properties to Add and Subtract
Measurement and Data	– Measure and Estimate Lengths in Standard Units; Work with Time and Money
Geometry	– Draw and Identify Shapes; Partition Circles and Rectangles into Two, Three, and Four Equal Shares

Grade 3

Operations and Algebraic Thinking	– Represent and Solve Problems Involving Multiplication and Division; Solve Two-Step Problems Involving Four Operations
Number and Operations in Base Ten	– Round Whole Numbers; Add, Subtract, and Multiply Multi-Digit Whole Numbers
Number and Operations—Fractions	– Understand Fractions as Numbers
Measurement and Data	– Solve Time, Liquid Volume, and Mass Problems; Understand Perimeter and Area
Geometry	– Reason with Shapes and Their Attributes

Grade 4

Operations and Algebraic Thinking	– Use the Four Operations with Whole Numbers to Solve Problems; Understand Factors and Multiples
Number and Operations in Base Ten	– Generalize Place Value Understanding; Perform Multi-Digit Arithmetic
Number and Operations—Fractions	– Build Fractions from Unit Fractions; Understand Decimal Notation for Fractions
Measurement and Data	– Convert Measurements; Understand and Measure Angles
Geometry	– Draw and Identify Lines and Angles; Classify Shapes

Grade 5

Operations and Algebraic Thinking	– Write and Interpret Numerical Expressions
Number and Operations in Base Ten	– Perform Operations with Multi-Digit Numbers and Decimals to Hundredths
Number and Operations—Fractions	– Add, Subtract, Multiply, and Divide Fractions
Measurement and Data	– Convert Measurements within a Measurement System; Understand Volume
Geometry	– Graph Points in the First Quadrant of the Coordinate Plane; Classify Two-Dimensional Figures

Mathematics Reference Sheet

Conversions

U.S. Customary
1 foot = 12 inches
1 yard = 3 feet
1 mile = 5280 feet
1 acre = 43,560 square feet
1 cup = 8 fluid ounces
1 pint = 2 cups
1 quart = 2 pints
1 gallon = 4 quarts
1 gallon = 231 cubic inches
1 pound = 16 ounces
1 ton = 2000 pounds
1 cubic foot ≈ 7.5 gallons

U.S. Customary to Metric
1 inch = 2.54 centimeters
1 foot ≈ 0.3 meter
1 mile ≈ 1.61 kilometers
1 quart ≈ 0.95 liter
1 gallon ≈ 3.79 liters
1 cup ≈ 237 milliliters
1 pound ≈ 0.45 kilogram
1 ounce ≈ 28.3 grams
1 gallon ≈ 3785 cubic centimeters

Time
1 minute = 60 seconds
1 hour = 60 minutes
1 hour = 3600 seconds
1 year = 52 weeks

Temperature
$$C = \frac{5}{9}(F - 32)$$

$$F = \frac{9}{5}C + 32$$

Metric
1 centimeter = 10 millimeters
1 meter = 100 centimeters
1 kilometer = 1000 meters
1 liter = 1000 milliliters
1 kiloliter = 1000 liters
1 milliliter = 1 cubic centimeter
1 liter = 1000 cubic centimeters
1 cubic millimeter = 0.001 milliliter
1 gram = 1000 milligrams
1 kilogram = 1000 grams

Metric to U.S. Customary
1 centimeter ≈ 0.39 inch
1 meter ≈ 3.28 feet
1 kilometer ≈ 0.62 mile
1 liter ≈ 1.06 quarts
1 liter ≈ 0.26 gallon
1 kilogram ≈ 2.2 pounds
1 gram ≈ 0.035 ounce
1 cubic meter ≈ 264 gallons

Number Properties

Commutative Properties of Addition and Multiplication

$a + b = b + a$

$a \cdot b = b \cdot a$

Associative Properties of Addition and Multiplication

$(a + b) + c = a + (b + c)$

$(a \cdot b) \cdot c = a \cdot (b \cdot c)$

Addition Property of Zero

$a + 0 = a$

Multiplication Properties of Zero and One

$a \cdot 0 = 0$

$a \cdot 1 = a$

Distributive Property:

$a(b + c) = ab + ac$

$a(b - c) = ab - ac$

Properties of Equality

Addition Property of Equality

If $a = b$, then $a + c = b + c$.

Subtraction Property of Equality

If $a = b$, then $a - c = b - c$.

Multiplication Property of Equality

If $a = b$, then $a \cdot c = b \cdot c$.

Multiplicative Inverse Property

$n \cdot \frac{1}{n} = \frac{1}{n} \cdot n = 1, n \neq 0$

Division Property of Equality

If $a = b$, then $a \div c = b \div c, c \neq 0$.

Squaring both sides of an equation

If $a = b$, then $a^2 = b^2$.

Cubing both sides of an equation

If $a = b$, then $a^3 = b^3$.

80235000021423

Properties of Inequality

Addition Property of Inequality
If $a > b$, then $a + c > b + c$.

Subtraction Property of Inequality
If $a > b$, then $a - c > b - c$.

Multiplication Property of Inequality
If $a > b$ and c is positive, then $a \cdot c > b \cdot c$.
If $a > b$ and c is negative, then $a \cdot c < b \cdot c$.

Division Property of Inequality
If $a > b$ and c is positive, then $a \div c > b \div c$.
If $a > b$ and c is negative, then $a \div c < b \div c$.

Properties of Exponents

Product of Powers Property: $a^m \cdot a^n = a^{m+n}$

Quotient of Powers Property: $\frac{a^m}{a^n} = a^{m-n}, a \neq 0$

Power of a Power Property: $(a^m)^n = a^{mn}$

Power of a Product Property: $(ab)^m = a^m b^m$

Zero Exponents: $a^0 = 1, a \neq 0$

Negative Exponents: $a^{-n} = \frac{1}{a^n}, a \neq 0$

Slope

$$m = \frac{\text{rise}}{\text{run}} = \frac{\text{change in } y}{\text{change in } x} = \frac{y_2 - y_1}{x_2 - x_1}$$

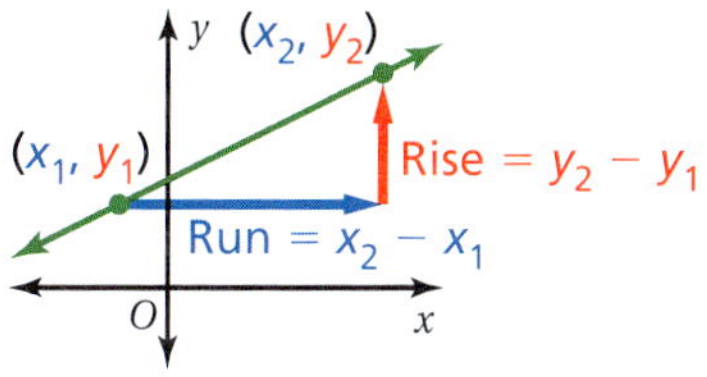

Equations of Lines

Slope-intercept form
$y = mx + b$

Standard form
$ax + by = c, a, b \neq 0$

Point-slope form
$y - y_1 = m(x - x_1)$

Pythagorean Theorem

$$a^2 + b^2 = c^2$$

Converse of the Pythagorean Theorem
If the equation $a^2 + b^2 = c^2$ is true for the side lengths of a triangle, then the triangle is a right triangle.

Distance Formula

$$d = \sqrt{(x_2 - x_1)^2 + (y_2 - y_1)^2}$$

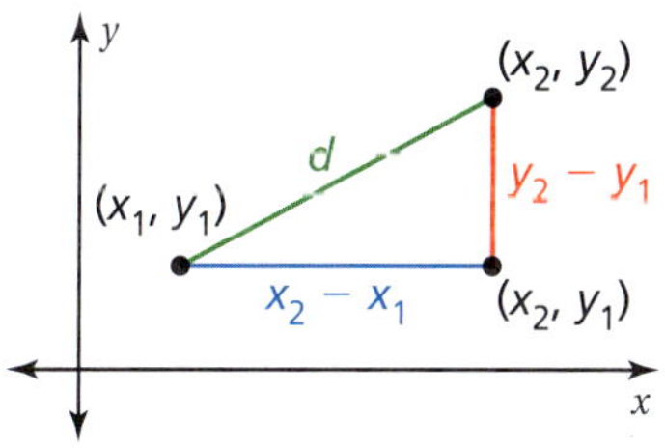

Formulas in Geometry

Prism

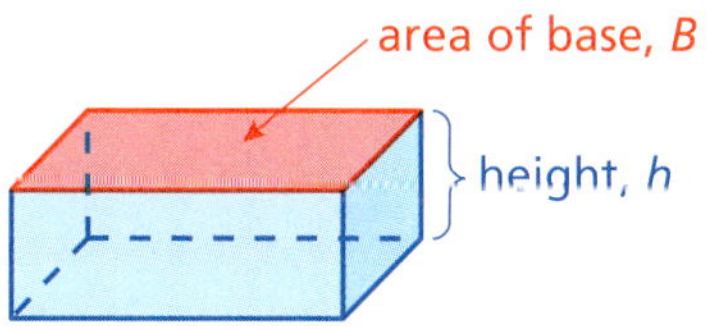

S = areas of bases + areas of lateral faces
$V = Bh$

Pyramid

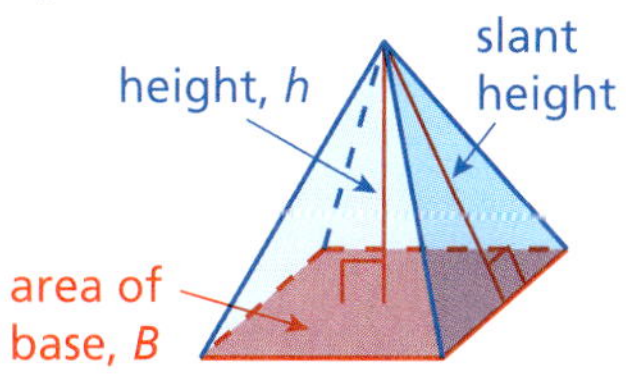

S = area of base + areas of lateral faces
$V = \frac{1}{3}Bh$

Circle

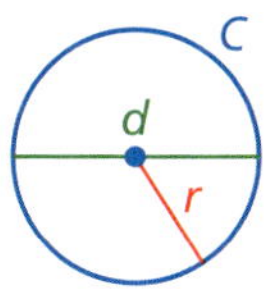

$C = \pi d$ or $C = 2\pi r$
$A = \pi r^2$
$\pi \approx \frac{22}{7}$, or 3.14

Cylinder

$V = Bh = \pi r^2 h$
$S = 2\pi r^2 + 2\pi rh$

Cone

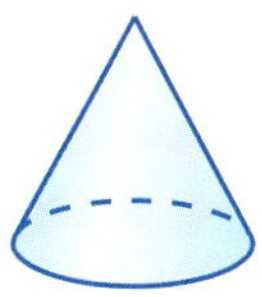

$V = \frac{1}{3}Bh = \frac{1}{3}\pi r^2 h$

Sphere

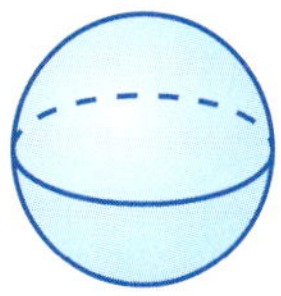

$V = \frac{4}{3}\pi r^3$